ABSTRACT ALGEBRA

Introduction to Groups, Rings and Fields with Applications

Second Edition

ABSTRACT ALGEBRA

Introduction to Groups, Rings and Fields with Applications

Second Edition

Clive Reis
Stuart A Rankin

University of Western Ontario, Canada

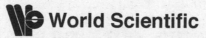 World Scientific

NEW JERSEY · LONDON · SINGAPORE · BEIJING · SHANGHAI · HONG KONG · TAIPEI · CHENNAI · TOKYO

Published by

World Scientific Publishing Co. Pte. Ltd.

5 Toh Tuck Link, Singapore 596224

USA office: 27 Warren Street, Suite 401-402, Hackensack, NJ 07601

UK office: 57 Shelton Street, Covent Garden, London WC2H 9HE

Library of Congress Cataloging-in-Publication Data
Names: Reis, Clive. | Rankin, Stuart A.
Title: Abstract algebra : introduction to groups, rings, and fields, with applications.
Description: Second edition / by Clive Reis (University of Western Ontario, Canada),
    Stuart A. Rankin (University of Western Ontario, Canada). |
    New Jersey : World Scientific, 2016. | Includes index.
Identifiers: LCCN 2016023596| ISBN 9789814730532 (hardcover : alk. paper) |
    ISBN 9789814730549 (pbk. : alk. paper)
Subjects: LCSH: Algebra, Abstract--Textbooks. | Group algebras--Textbooks. |
    Rings (Algebra)--Textbooks. | Algebraic fields--Textbooks.
Classification: LCC QA162 .R44 2016 | DDC 512/.02--dc23
LC record available at https://lccn.loc.gov/2016023596

British Library Cataloguing-in-Publication Data
A catalogue record for this book is available from the British Library.

Printed in Singapore

Preface to the Second Edition

Other than the inclusion of two new chapters, one dealing with the Fundamental Theorem of Finitely Generated Abelian Groups and the other providing an introduction to semigroups and the theory of automata and formal languages, the second edition covers essentially the same material as the first edition, although in some cases with expanded coverage. We have taken this opportunity to correct typographical errors that were identified in the first edition, and we have now provided the proofs of many results that had formerly been left as exercises for the reader.

Additionally, some of the topics are now covered with greater generality. For example, in the first edition, we introduced the simplest way of "stitching" groups together, namely the direct sum or product. In this edition, we define the semi-direct product and show how many of the more familiar groups can be viewed as semi-direct products of simpler groups, thus affording us a better understanding of the algebraic structure of these groups.

The section on semigroups (the topic of the first half of Chapter 14) introduces the reader to the fundamental ideas of the theory, including Green's relations, the egg-box picture of a semigroup and regular and inverse semigroups. We show how an inverse semigroup can be embedded as a subsemigroup of the semigroup of relations on a set, a result analogous to the Cayley representation theorem of group theory. Throughout, we point out the similarities of the semigroup-theoretic results with those of group theory while alerting the reader to the differences. In particular, for want of an analogue of normal subgroups, we are led to the notion of congruences which occur in the theory of groups disguised as relations defined in terms of normal subgroups.

Clive Reis and Stuart Rankin

Preface to the First Edition

This book has evolved over the years from notes I made for courses, ranging from second-year to fourth-year honors, I taught at the University of Western Ontario, Camosun College and the University of Victoria. As a result, it still bears some of the features of its humble origins: many side remarks, colloquial language and an informal style. I have also tried to convey to the reader how to translate into everyday English the somewhat stodgy language of formal definitions so that the ideas take on a more vivid and dynamic aspect.

The first nine chapters (together with the applications in Chapters 11 or 17) cover the basic material traditionally taught at the second-year level, although I have included topics which are often avoided at this stage: cosets, Lagrange's Theorem, the isomorphism theorems for groups and rings and an introduction to finite fields.

I have several reasons for doing this. First, a course which avoids the concept of a coset and consequently, all the ideas which stem from it, conveys very little of the flavor of abstract algebra and, moreover, makes for a very boring course. Second, in my experience, second-year students with some background in linear algebra are quite capable of dealing with the level of abstraction required to tackle these concepts provided they first encounter cosets in the guise of congruence classes in the integers. Time spent computing in the algebraic system consisting of congruence classes under the usual operations of addition and multiplication makes the transition to factor groups much smoother. Moreover, the natural map from the integers to the integers modulo n which identifies an integer with its congruence class prepares the student for the notion of a homomorphism. Lastly, by introducing these topics, the student is able to tackle some interesting applications to Number Theory, Combinatorics and Error-Correcting

Codes.

Chapters 10, 12, 15, 16, and 18 are more suitable for students at the third- or fourth-year honors level. In Chapter 12, the reader will find such topics as group actions, the class equation of a group, the Sylow theorems and applications to determining the structure of some of the simpler groups. The chapter ends with a discussion of simple groups and a proof that the alternating groups on five or more symbols form one class of simple groups.

Chapter 15 deals with isometries in Euclidean n-space with emphasis on finite Euclidean groups of dimensions 2 and 3. The rotation groups of the Platonic Solids are obtained and are shown to be the only finite Euclidean groups in Euclidean 3-space. A good background in linear algebra is needed.

Pólya-Burnside enumeration with applications to counting is the topic of Chapter 16. A theorem due to Pólya is proved and applied to counting the number of inequivalent switching circuits.

Applications of groups and polynomials to the construction of group and polynomial codes will be found in Chapters 17 and 18. Chapter 17 can be covered after Chapter 6 provided the students have acquired enough mathematical maturity to be comfortable with the notions of subgroup and homomorphism. Chapter 18, however, requires the material from Chapter 10 and a certain degree of mathematical maturity which comes only after solving a large number of the problems found at the end of each chapter.

To make the book as self-contained as possible, I have included two appendices, one on the more familiar fields of rational, real and complex numbers, the other on the elementary theorems of linear algebra with emphasis on linear transformations. These appendices can either be used in the classroom as abbreviated reviews of the material or as a resource for the student to be used as a reference.

Clive Reis

List of Figures

List of Tables

Contents

Chapter 1

Logic and Proofs

1 Introduction

Mathematics in the ancient world was born of practical necessity. In ancient Egypt, the yearly flooding of the Nile obliterated boundaries between properties which meant that these boundaries had to be redrawn once the waters receded, thus giving rise to a system of linear and areal measurement. In other ancient civilizations of the Middle East, increasing trade brought about the evolution of the concept of number to keep a tally of goods – by volume or by weight – and of earnings.

Out of these practical needs there grew a considerable body of "facts" and "formulae", some valid and some merely poor approximations. It was not, however, until the classical Greek period (ca 600-300 BC) that mathematics emerged as a deductive discipline wherein the validity of statements and formulae required justification. It was no longer acceptable to assert that a 3-4-5-triangle is right angled (a fact that was known to the ancient Egyptians); a "proof" was required.

What precisely is a mathematical proof of an assertion? It is a finite sequence of propositions, each deduced from previously established truths and ending with the required assertion. A moment's reflection will lead us to the conclusion that we must ultimately base our sequence of propositions on some fundamental assumptions which we agree to accept without proof; otherwise, we would be involved in an infinite regress by proving proposition A from B, B from C, C from D, etc.

These fundamental assumptions are called axioms or postulates and serve as a foundation upon which to build the various branches of mathematics. Later on we shall be studying a number of mathematical systems, among them groups and rings, and we shall see that each has its own set of axioms from which we shall deduce the various properties of the respective systems.

Since proofs are at the very heart of modern mathematics, it is essential that we familiarize ourselves with the fundamentals of deductive reasoning. This chapter is therefore devoted to making the reader aware of some of the simple rules of logic.

1.2   Statements, Connectives and Truth Tables

1.2.1 Definition. *A statement is an assertion which is either true or false. "True" and "False" are the truth values of the statement, and these are abbreviated as T and F, respectively.*

1.2.2 Example.
   (i)   The sea is salty. (Truth value T)
   (ii) The moon is made of blue cheese. (Truth value F)
   (iii) 4 is an even number. (Truth value T)
   (iv) This is a beautiful painting. (Since beauty is subjective and differ-ent people may differ in their assessment of the truth value of this assertion, we deem this assertion not to be a statement.)

We denote statements by lower case letters. For example we might let p stand for the statement "The sea is salty". We would indicate this by writing: p: the sea is salty.

We string two or more statements together to form a compound state-ment (or statement expression using logical connectives. The truth value of the compound statement naturally depends on the truth values of the com-ponents comprising the compound statement. We exhibit this dependence by means of a truth table as shown below.

The connectives we shall be using are: $\neg$, $\wedge$, $\vee$, $\rightarrow$, and $\leftrightarrow$.

The Connective $\neg$ (negation).

This connective is read "not".

1.2.3 Example. *$\neg$ is read "not ". Hence if p stands for "the sea is salty", then $\neg p$ stands for "not (the sea is salty)", which in English we would express as "the sea is not salty". If this connective is to reflect the meaning of "not" in ordinary English, then we must assign the truth value F to the negation of a true statement and T to the negation of a false statement. We obtain the truth table shown in Table 1.1.*

Table 1.1   The truth table for ¬

p	$\neg p$
T	F
F	T

Connective ∧ (conjunction).

This corresponds to the English "and". Bearing in mind that in ordinary English, if we assert that it is a sunny day and the temperature is below freezing, then we have told the truth if it is indeed both sunny and freezing. In all other cases we would be guilty of a falsehood.

Table 1.2   The truth table for ∧

p	q	$p \wedge q$
T	T	T
T	F	F
F	T	F
F	F	F

1.2.4 Example. *Suppose that we have statements p and q given by p: cobras are poisonous; and q: 5 is an odd number. Then p∧q is the statement "cobras are poisonous and 5 is an odd number".*

The Connective ∨ (disjunction).

We translate this by "or" in the inclusive sense. It has the meaning "either or both". It is thus clear that only when both are false will the compound statement be false. Observe that in colloquial English, "or" is often used in the exclusive sense. For example, if I say "For my graduate work, I shall go to Harvard or to Yale", clearly, only one of these events will occur. There is a symbol for exclusive "or" but we shall not have occasion to use it since the compound statement $(p \vee q) \wedge \neg (p \wedge q)$ has the same truth table as that for the exclusive "or" of p and q.

The Connective → (implication, conditional).

This is translated in English by "if ..., then ...". We remark that in English, the following are alternatives for "if p then q"; q only if p; p is a sufficient condition for q; q is a necessary condition for p; p implies q.

Here is an example which might help the reader understand how the truth values are arrived at: Let p: you do your homework; q: I shall take

Table 1.3   The truth table for ∨

p	q	$p \lor q$
T	T	T
T	F	T
F	T	T
F	F	F

Table 1.4   The truth table for →

p	q	$p \to q$
T	T	T
T	F	F
F	T	T
F	F	T

you to the movies, and consider under what circumstances you would deem me to have lied if I declare that "p implies q" is true. The first two rows of the truth table are clear. In the third row, you do not do your homework but I still take you to the movies. At worst I can be accused of being lenient, but certainly not of lying. Since we must assign a truth value, by default it must be T. As for the last line, neither you nor I have kept the bargain and so I cannot be accused of lying. Therefore, again, by default, the truth value is T.

The Connective ↔ (bi-implication, biconditional).

This corresponds in ordinary English to "... if, and only if, ...".

Table 1.5   The truth table for ↔

p	q	$p \leftrightarrow q$
T	T	T
T	F	F
F	T	F
F	F	T

1.2.5 Example. *Let p: I go to the movies; q: I finish my homework. In the statement $p \leftrightarrow q$, "I go to the movies" is synonymous with "I finish my homework" in the sense that either action can be deduced from the other: my presence at the movies means that I have finished my homework and, conversely, my completed homework means that I go to the movies. The truth values assigned in the table then become self-evident.*

3 Relations between Statements

1.3.1 Definition. *A statement p logically implies a statement q if, whenever p is true, so is q; that is, if $p \to q$ is true. In such a case, we write $p \Rightarrow q$.*

1.3.2 Example. $(p \leftrightarrow q) \Rightarrow (p \to q)$.

Let $P(p,q)$ denote the statement expression $p \leftrightarrow q$ and let $Q(p,q)$ denote the statement expression $p \to q$. Let p and q be statements. If $P(p,q)$ is false, then $P(p,q) \to Q(p,q)$ is true. Suppose that $P(p,q)$ is true. Then either both p and q are true or both are false. In either case, $p \to q$ is true; that is, $Q(p,q)$ is true. Thus in this case as well, we find that $P(p,q) \to Q(p,q)$ is true. It follows now that for any statements p and q, $P(p,q) \to Q(p,q)$ is true; that is, $P(p,q) \Rightarrow Q(p,q)$.
Note. For statement expressions P and Q on the same statement variables, to check that $P \Rightarrow Q$, one can draw up the truth table of $P \to Q$. Every truth value of $P \to Q$ should be T. Any statement expression for which all the calculated values in the truth table are T's is called a tautology. Thus the statement $(p \leftrightarrow q) \to (p \to q)$ is a tautology. Intuitively speaking, if $P \Rightarrow Q$, then P conveys more information than Q.

1.3.3 Definition. *A statement expression P is logically equivalent to a statement expression Q if, whenever P is true, so is Q and whenever Q is true, so is P; that is, $P \leftrightarrow Q$ is true for all values of the statement variables. We write $P \Leftrightarrow Q$ in this case.*

The reader should verify that $P \Leftrightarrow Q$ precisely when $P \leftrightarrow Q$ is a tautology.
Below is a list of logical equivalences which the reader is urged to prove.

1.3.4 Example.

(i) $\neg(\neg p) \Leftrightarrow p$;

(ii) $(p \to q) \Leftrightarrow (\neg p \vee q) \Leftrightarrow (\neg q \to \neg p)$ *($\neg q \to \neg p$ is called the contrapositive of $p \to q$)*;

(iii) $\neg(p \wedge q) \Leftrightarrow (\neg p \vee \neg q)$ *(one of two De Morgan's Laws)*;

(iv) $\neg(p \vee q) \Leftrightarrow (\neg p \wedge \neg q)$ *(the other De Morgan's Law)*;

(v) $p \vee (q \wedge r) \Leftrightarrow (p \vee q) \wedge (p \vee r)$ *($\vee$ is said to distribute over $\wedge$)*;

(vi) $p \wedge (q \vee r) \Leftrightarrow (p \wedge q) \vee (p \wedge r)$ *($\wedge$ is said to distribute over $\vee$)*;

(vii) $p \leftrightarrow q \Leftrightarrow (p \to q) \wedge (q \to p)$.

In later chapters, you will often be asked to prove statements of the form $p \leftrightarrow q$. This can be done by proving that $p \rightarrow q$ and then proving the converse of $p \rightarrow q$, namely, $q \rightarrow p$ (see (vii) above).

1.4   Quantifiers

An assertion such as $x^2 - 3x + 2 = 0$ has no truth value as it stands since x is a variable symbol. If we were to consider x as taking values from the set of real numbers, say, then the truth value of the assertion depends on the value of x. Such an assertion is called an open statement. By prefacing it by a quantifier, the open statement is converted to a statement, whose truth value is determined as we describe below.

We shall use two quantifiers, namely, "for all", called the universal quantifier, and "there exists", called the existential quantifier. In English, there are many ways of expressing each of these. Here are some examples: for "for all" we have: "for each" , "for every", "for any"; for "there exists", we have: "for some", "for at least one".

The symbols used for the universal and existential quantifiers are "$\forall$" and "$\exists$", respectively.

Let us preface the assertion $x^2 - 3x + 2 = 0$ by $\forall x \in \mathbb{R}$, where $\mathbb{R}$ is the set of real numbers and $\in$ is the symbol for membership of an element in a set (see Chapter 2). We thereby obtain the sentence

$$(\forall x \in \mathbb{R})(x^2 - 3x + 2 = 0).$$

This is a statement, determined to be true (have truth value T) if when x is replaced by any specific real number r, the resulting equation $r^2 - 3r + 2 = 0$ is valid, otherwise it is deemed to be false. Let us see if we can determine whether it is true. Consider the choice of $x = 0$. For $x = 0$, the equation becomes $2 = 0$, which is not valid. Thus the statement $(\forall x \in \mathbb{R})(x^2 - 3x + 2 = 0)$ is false.

Consider now the statement:

$$(\exists x \in \mathbb{R})(x^2 - 3x + 2 = 0).$$

In words, this sentence asserts that there is a real number r for which the equation $r^2 - 3r + 2 = 0$ is valid. That this is true can be verified by checking that for $r = 1$, we obtain the valid equation $1^2 - 3(1) + 2 = 0$. Thus the statement $(\exists x \in \mathbb{R})(x^2 - 3x + 2 = 0)$ is true.

As another example, consider the open statement "$y = x^2$" where the variables x and y are to take their values from $\mathbb{R}$. This open statement involves two variables, namely x and y. We may convert this open statement to a statement by quantifying both variables. As it turns out, there are in total six different ways to do this, and we list them all below.

1.4.1 Example.

(i) $(\forall x \in \mathbb{R} \, \forall y \in \mathbb{R})(y = x^2)$;

(ii) $(\forall x \in \mathbb{R} \, \exists y \in \mathbb{R})(y = x^2)$;

(iii) $(\exists x \in \mathbb{R} \, \forall y \in \mathbb{R})(y = x^2)$;

(iv) $(\exists x \in \mathbb{R} \, \exists y \in \mathbb{R})(y = x^2)$;

(v) $(\forall y \in \mathbb{R} \, \exists x \in \mathbb{R})(y = x^2)$;

(vi) $(\exists y \in \mathbb{R} \, \forall x \in \mathbb{R})(y = x^2)$.

We translate each of these six statements into words and identify the truth value of each.

(i) For all real numbers x and y, $y = x^2$; (F). Take, for example, $x = 1$ and $y = 0$.

(ii) For each real number x, there is a real number y such that $y = x^2$; (T).

(iii) There exists a real number x such that $y = x^2$ for every real number y. (F), since if it were true, then we would have $1 = x^2 = 0$, so $1 = 0$.

(iv) There exist real numbers x and y such that the equality holds; (T), for $x = 1 = y$ are such values.

(v) For every given value y, there exists x such that $y = x^2$ (F), since for no $x \in \mathbb{R}$ does $-1 = x^2$ hold.

(vi) There exists a value y such that for all values of x we have $y = x^2$ (F), since if it were true, then $1 = 1^2 = y = 0^2 = 0$, so $1 = 0$.

Suppose that P is some open statement involving finitely many variables $x, y, z, \ldots$ with each taking values from a set U. If $a, b, c, \ldots \in U$, then $P(a, b, c, \ldots)$ denotes the statement that is obtained from $P(x, y, z, \ldots)$ by replacing the variable x by a, the variable y by b, and so on. Thus for example, to say that $(\forall x, y, z, \ldots \in U) P(x, y, z, \ldots)$ is true is to say that for all possible choices of $a, b, c, \ldots, \in U$, $P(a, b, c, \ldots)$ is true.

We now establish rules for the negation of statements that are obtained from open statements by the application of quantifiers. For the sake of simplicity, we shall restrict ourselves to open statements of at most two variables, though the reader should be able to extend the rules to any number of variables.

We begin with the statement $(\exists x)(P(x))$, where $P(x)$ is a one-variable open statement. This asserts that there is a value x in the set from which

x is permitted to take its values such that the statement $P(x)$ is true. The negation of $(\exists x)(P(x))$ therefore asserts that no such x exists; that is, for every value x may take on, the statement $P(x)$ is false, or equivalently, the statement $\neg(P(x))$ is true. Thus

$$\neg(\exists x)(P(x)) \leftrightarrow (\forall x)(\neg P(x)).$$

Next, for a two-variable open statement $P(x, y)$, consider the statement $(\exists x \forall y)(P(x, y))$. This asserts that there is a value x in the set from which x is permitted to take its values such that the statement $(\forall y)(P(x, y))$ is true for each permitted value of the variable y. The negation would therefore assert that no such x exists. That is, whatever x we may choose, we can find a y such that $P(x, y)$ is false; that is, $\neg P(x, y)$ is true. Therefore, for each value of x, we can find a value of y so that $\neg P(x, y)$ is true. Thus

$$\neg[(\exists x \forall y)(P(x, y))] \Leftrightarrow (\forall x \exists y)(\neg P(x, y)).$$

In a similar manner, we can establish the following:

$$\neg[(\forall x \exists y)(P(x, y))] \Leftrightarrow (\exists x \forall y)(\neg P(x, y)).$$
$$\neg[(\forall x \forall y)(P(x, y))] \Leftrightarrow (\exists x \exists y)(\neg P(x, y)).$$
$$\neg[(\exists x \exists y)(P(x, y))] \Leftrightarrow (\forall x \forall y)(\neg P(x, y)).$$

The pattern is now apparent: in negating a statement with quantifiers, each $\forall$ turns into $\exists$ and conversely. In addition, the open statement whose quantification we are considering must be negated.

1.4.2 Example. *Suppose that P and Q are open statements, each with variable x, and that x is permitted to take its values from some set U. Negate $(\forall x \in U)(P(x)) \to Q(x)$.*

Recall that $p \to q \Leftrightarrow \neg p \vee q$. Hence,

$$\neg[(\forall x \in U)(P(x) \to Q(x))] \Leftrightarrow (\exists x \in U)(\neg[P(x) \to Q(x)])$$
$$\Leftrightarrow (\exists x \in U)(\neg[\neg P(x) \vee Q(x)])$$
$$\Leftrightarrow (\exists x \in U)(P(x) \wedge \neg Q(x)).$$

In the last equivalence, we have used one of De Morgan's Laws.

1.4.3 Example. *Negate $(\forall x \in \mathbb{R})(x^2 > 0 \to x > 0)$.*

The negation is, as we have seen above:

$$(\exists x \in \mathbb{R})[\neg[\neg(x^2 > 0) \vee (x > 0)]] \Leftrightarrow (\exists x \in \mathbb{R})[(x^2 > 0) \wedge (x \le 0)].$$

This last statement is true, as one can see by taking $x = -1$, and so the original statement is false.

5   Methods of Proof

As a student of mathematics, you will often be asked to prove (or disprove) a given statement. If the statement is universal, that is, one which asserts that all members of a given set have a certain property, then providing an example of a member of the set having the property is clearly not sufficient. On the other hand, if you are able to find a member of the set without the given property, then you have proved that the universal statement is false; that is, you have disproved the statement.

To firm up our ideas, let us consider some examples.

1.5.1 Example. *If n is an odd integer, then $n^2 - 1$ is divisible by 8.*

The assertion "$n^2 - 1$ is divisible by 8" is an open statement, where the variable n is permitted to take its values from the set $\mathbb{Z}$ of all whole numbers, positive, negative, or zero. The example asks us to investigate the statement

$$(\forall m \in \mathbb{Z})((n = 2m - 1) \to (n^2 - 1 \text{ is divisible by 8})).$$

We try a few values of m to test the validity of the statement, just to gain some familiarity with what is being asked. First, we try $m = 2$, so $n = 3$. We obtain $n^2 - 1 = 3^2 - 1$, which is a multiple of 8, $(1)(8)$ to be precise. We try another couple of values to convince ourselves that the validity of the statement is plausibles, say $m = 3$ and perhaps $m = 5$, which yield $n = 5$ and $n = 9$ respectively. We calculate $n^2 - 1 = 24 = (3)(8)$ in the first case, and $n^2 - 1 = 80 = (10)(8)$ in the second case. We could try many other values of m to convince ourselves of the validity of the statement. However, since there are infinitely many integers, no matter how many values of m we check, there will always be the possibility that for one of the integers we have not checked, our assertion is false. Therefore, we must construct an argument which, in a finite amount of time, will cover all cases.

Let $m \in \mathbb{Z}$, and set $n = 2m - 1$. Then $n^2 - 1 = (2m - 1)^2 - 1 = 4m^2 - 4m + 1 - 1 = 4(m^2 - m) = 4m(m - 1)$. Since $m - 1$ and m are two consecutive integers, one of them is a multiple of 2 and thus $4m(m - 1)$ is a multiple of 8, as claimed.

1.5.2 Example. *If n is a positive integer, then $2^n > 2n$.*

It is convenient to let $\mathbb{Z}^+$ denote the set of all positive integers, so $\mathbb{Z}^+ = \{1, 2, 3, \ldots\}$. With this notation, our assertion can be written symbolically as

$$(\forall n \in \mathbb{Z}^+)(2^n > 2n).$$

We test the validity of the statement. For $n = 3$ or 4, the statement is true. In fact, it is possible to prove that for each $n > 1$, the statement is valid. However, when $n = 1$, we get $2^1 > 2(1)$, which is false. This one instance is sufficient to establish that the statement is false. We say that $n = 1$ yields a counterexample to the assertion.

The proof we used to establish the validity of the statement of Example 1.5.2 is called a direct proof. An indirect proof would be one where we are to prove that statement P, but we do not see a direct approach that will work, so we assume that $\neg P$ is true and attempt to obtain a known contradiction as a result. This approach is called a proof by contradiction.

A related problem can arise when we try to prove some conditional statement $p \to q$ is true. It suffices to prove that if p is true, then q is also true, but it may be difficult to see how we may use the validity of p to establish the validity of q. In such a case, we can try to prove the logically equivalent statement $\neg q \to \neg p$ (the so-called contrapositive of $p \to q$). Alternatively, we may assume that both p and $\neg q$ are true and try to arrive at a contradiction, thus proving that p true and q false cannot co-exist; that is, the validity of p implies the validity of q.

1.5.3 Example. *Prove that if n^2 is an odd integer, then n is odd.*

Proof. The contrapositive of "n^2 odd$\to n$ odd" is "n even$\to n^2$ even". Therefore, let n be even. Then for some $m \in \mathbb{Z}$, $n = 2m$ and we have $n^2 = (2m)^2 = 4m$, which is even. Thus for every integer n, the statement "n even$\to n^2$ even" is true, and thus for every integer n, the statement "n^2 odd$\to n$ odd" is true. $\qquad\square$

Finally, we give an example of a proof by contradiction.

1.5.4 Example. *Prove that $\sqrt{2}$ is irrational (an irrational number is a real number which cannot be expressed as a quotient of two whole numbers – see Appendix A).*

Proof. Suppose the statement is false. Then $\sqrt{2}$ is rational, so there exist integers m and n such that $\sqrt{2} = \frac{m}{n}$. Since we may cancel common factors if necessary, we may assume that m and n have no factors in common and in particular, we may ensure that not both m and n are even. Square both sides of this equation to obtain $2 = \frac{m^2}{n^2}$; equivalently, $2n^2 = m^2$. We see that m^2 is even. Since the square of an odd integer is odd, this means that m itself is even, so there exists an integer k such that $m = 2k$. Thus $2n^2 = m^2 = (2k)^2 = 4k^2$. Cancel 2 from both sides to obtain that $n^2 = 2k^2$, so n^2 is even and thus n is even. But we have now established that both m and n are even, which we have ensured is not the case. This contradiction has arisen from the assumption that $\sqrt{2}$ is rational, so we conclude that $\sqrt{2}$ is not rational. $\qquad\qquad\qquad\qquad\qquad\qquad\qquad\qquad\qquad\qquad\qquad\quad$ □

Exercises

1. Let p: It is raining; q: It is cloudy. Use p and q to translate each of the following into symbolic form.

   a) If it rains, then it is cloudy.

   b) If it is cloudy, then it rains.

   c) It is cloudy if and only if it is raining.

   d) If it is not raining, then it is not cloudy.

   e) It is cloudy only if it is raining.

   f) It is not cloudy only if it is not raining.

2. For each part below, construct the truth table for the given statement formula. In each case, determine whether or not the statement formula is a tautology.

   a) $p \to (q \lor r)$.

   b) $(p \lor r) \land (p \to q)$.

   c) $[p \to (q \to r)] \to [(p \to q) \to (p \to r)]$.

   d) $(p \lor q) \leftrightarrow (\neg r \land \neg s)$.

   e) $(p \lor q) \to \neg(\neg p \lor (r \lor s))$.

   f) $\neg[(\neg p \land \neg q) \land (p \lor r)]$.

   g) $p \to [(r \lor q) \leftrightarrow \neg(r \land s)]$.

3. Find a simpler statement formula having the same truth table as the one found in Exercise 2 (f).

4. Show that the truth table in Exercise 2 (g) can be constructed much more easily by identifying the cases in which the statement is false.

5. A compound statement with statement variables p and q must have one of 16 possible truth tables. Find all of these tables and for each table, find a statement formula having that table as its truth table.

6. Give an example of an implication which is valid, but whose converse is false.

7. Establish each of the logical equivalences given in Example 1.3.4.

8. In each part below, x and y are permitted to take values in the set $\mathbb{R}$ of all real numbers. In each case, determine whether the statement is true or false.

    a) $(\exists x \forall y)(xy = 1)$.

    b) $(\forall x \exists y)(xy = 1)$.

    c) $(\forall x \forall y))([xy > 1] \to [(x > 1) \vee (y > 1)]$.

    d) $(\exists x \forall y)([xy > 0] \to [x + y > 0])$.

    e) $(\forall x \forall y)([xy > 0] \to [x + y > 0])$.

9. Prove that if m is a positive integer which is not divisible by any positive integer less than or equal to $\sqrt{m}$, then m has no positive factors other than m and 1; that is, m is prime.

10. Determine the truth value of each of the following:

    a) If the moon is made of blue cheese, then $2 + 3 = 6$.

    b) If it snows at the North Pole, then dogs are bipeds.

    c) If there are 12 months in the year, then $4 + 5 = 9$.

11. Negate each of the following:

    a) $(\forall x \in \mathbb{R})([x(x + 1) < 0] \to [x < -1])$.

    b) $(\exists x, y \in \mathbb{Z})([(x^2 > y^2) \wedge (xy > 0)] \leftrightarrow [(x > y) \vee (-x > -y)])$.

    c) $(\forall x \in \mathbb{Z} \, \exists y \in \mathbb{Z} \, \forall z \in \mathbb{Z})([xyz \leq 0] \to [[(x \geq 0) \wedge (y \geq 0)] \vee [z \leq 0]])$.

Chapter 2

Set Theory

1 Initial Concepts

The concept of a set and the language of set theory are of fundamental importance in mathematics. In this chapter, we shall familiarize ourselves with the ideas and notation of set theory that will be used freely in subsequent chapters.

2.1.1 Definition. *A set is a designated collection of objects. The designation is effected either by listing the members of the collection or by stipulating some property possessed by each member of the collection and not possessed by any object outside the collection.*

2.1.2 Example. *Consider the collection consisting of the numbers* 1, 2, *and* 3. *We denote this collection by* $\{1, 2, 3\}$ *and think of the curly brackets as a "bag" that contains the members of the set, which we call the elements of the set. Thus, the set above contains three elements, namely,* 1, 2, *and* 3.

To indicate that a particular element belongs to a given set, we use the symbol "$\in$". *In this example,* $1 \in \{1, 2, 3\}$, *while* $4 \notin \{1, 2, 3\}$. *The reader will infer that* "$\notin$" *means "is not a member of". Note that the order in which the elements of a set are listed is immaterial.*

Often we have occasion to refer to a specific set several times. In such a situation, to avoid writing out a lengthy expression every time we refer to it, we shall find it convenient to give the set a name, usually an upper case letter. In the example above, we could let A stand for the set in question. We convey this by writing $A = \{1, 2, 3\}$. As well, abstract, unspecified sets are usually denoted by upper case letters. The set that is distinguished by the fact that it has no elements is called the empty set, and (rather than using an upper case letter from the Roman alphabet) it will be denoted by $\varnothing$. This set can be thought of as an empty bag (think of it as { }). For any set A, we denote the number of elements in A by $|A|$ so that, for example, $|\varnothing| = 0$, and $|\{1, 2, 3\}| = 3$.

In the example above, we designated the set by listing all the elements belonging to it. Sometimes the set in question is either very big or even infinite. In the former case, it would be time-consuming and tedious to list all the elements, while in the latter case, it would be impossible to do so. We therefore resort to using a property which singles out the elements of the set.

2.1.3 Example. *Suppose we happen to be interested in the set of all Canadians. The property "Canadian" is used together with the so-called "set builder notation" as follows:*

$$\{ x \mid x \text{ is Canadian} \}.$$

The vertical line is to be read "such that". Therefore, this expression is read as: "The set of all x such that x is Canadian."

At this point we establish notation for some of the well-known sets:

2.1.4 Notation.

(i) $\mathbb{Z}^+ = \{ n \mid n \text{ is a positive whole number} \} = \{ 1, 2, 3, \ldots \}$. *We remark that the notation $\ldots$ is often used when the elements of the set in question can be listed in a sequence having an obvious pattern. The idea is to list enough elements of the sequence to establish the pattern, then follow with $\ldots$, which means "and so on". The set $\mathbb{Z}^+$ is called the set of positive integers.*

(ii) $\mathbb{N} = \{ n \mid n \text{ is a nonnegative whole number} \} = \{ 0, 1, 2, \ldots \}$. $\mathbb{N}$ *is called the set of natural numbers.*

(iii) $\mathbb{Z} = \{ n \mid n \text{ is a whole number} \} = \{ 0, \pm 1, \pm 2, \ldots \}$ *and is called the set of integers. The choice of capital Z to represent the set of all whole numbers stems from the fact that the German word for "number" is "Zahl".*

(iv) $\mathbb{Q} = \{ x \mid x = p/q \text{ where } p \text{ and } q \text{ are integers with } q \neq 0 \}$. $\mathbb{Q}$ *is called the set of rational numbers. The "Q" stands for "quotients".*

(v) $\mathbb{R} = \{ x \mid x \text{ is a real number} \}$. *Thus $\mathbb{R}$ consists of all the numbers, rational and irrational, which appear on the number line.*

(vi) $\mathbb{C} = \{ z \mid z = x + iy \text{ where } x, y \in \mathbb{R} \text{ and } i = \sqrt{-1} \}$ *is called the set of complex numbers (see Appendix A).*

2   Relations between Sets

2.2.1 Definition. *Let A and B be sets. If every element of A is an element of B, we say that A is contained in B, or A is a subset of B, and we write $A \subseteq B$. If $A \subseteq B$ and $A \neq B$, we say that A is properly contained in B or that A is a proper subset of B and write $A \subsetneq B$. If A is not a subset of B, we write $A \nsubseteq B$.*

2.2.2 Remarks.

(i) For any set A, $A \subseteq A$.

(ii) For any sets A, B, and C, if $A \subseteq B$ and $B \subseteq C$, then $A \subseteq C$. For suppose that $x \in A$. Then since $A \subseteq B$, it follows that $x \in B$. But now, since $B \subseteq C$ and $x \in B$, we obtain that $x \in C$, as required.

(iii) For any sets A and B, to prove that $A \subsetneq B$, we must prove that each element of A is an element of B and additionally, that there is at least one element of B that is not an element of A.

(iv) For sets A and B, to prove that $A \nsubseteq B$, we must show that there exists an element of A that is not an element of B. The reader may wish to verify this by negating the statement

$$(\forall x)(x \in A \rightarrow x \in B).$$

(v) The empty set $\varnothing$ mentioned above, is a subset of every set. Indeed, if it were not the case, then there would be some set A for which $\varnothing \nsubseteq A$. Then by (iv), there exists $x \in \varnothing$ for which $x \notin A$. However, since $\varnothing$ contains no elements, the existential part of the statement above is false. Therefore, the assertion that there exists a set A for which $\varnothing$ is not a subset of A is false, and we conclude that $\varnothing \subseteq A$ for every set A.

(vi) Sets A and B are equal, written $A = B$, if they each contain exactly the same elements. Thus $A = B$ if

$$(\forall x)(x \in A \text{ if, and only if, } x \in B)$$

is true.

We summarize the properties of "$\subseteq$" in the following proposition.

2.2.3 Proposition.

(i) *For every set A, $A \subseteq A$ (reflexivity).*

(ii) *For all sets A, B, C, $A \subseteq B$ and $B \subseteq C$ imply $A \subseteq C$ (transitivity).*

(iii) *For all sets A and B, if $A \subseteq B$ and $B \subseteq A$, then $A = B$ (antisymmetry).*

2.2.4 Definition. *Given a set S, the power set of S, denoted by $\mathscr{P}(S)$, is the set of all subsets of S.*

2.2.5 Example.

$$\mathscr{P}(\{\,1,2,3\,\}) = \{\,\varnothing, \{\,1\,\}, \{\,2\,\}, \{\,3\,\}, \{\,1,2\,\}, \{\,1,3\,\}, \{\,2,3\,\}, \{\,1,2,3\,\}\,\}.$$

It is interesting to observe that $\varnothing \in \mathscr{P}(\{\,1,2,3\,\})$ and $\varnothing \subseteq \mathscr{P}(\{\,1,2,3\,\})$, while $\varnothing \subseteq \{\,1,2,3\,\}$ but $\varnothing \notin \{\,1,2,3\,\}$.

2.3 Operations Defined on Sets—or New Sets from Old

We can use the logical operations "and", "or", and "negation " to define new sets from old, as described in the following definition.

2.3.1 Definition.

(i) *The union of a set $\mathscr{A}$ of sets, denoted by $\cup \mathscr{A}$, is the set defined by*

$$\cup \mathscr{A} = \{\,x \mid \exists A \in \mathscr{A}\,(x \in A)\,\}.$$

As a special case, for sets A and B, we denote $\cup\{\,A,B\,\}$ by $A \cup B$, so

$$A \cup B = \{\,x \mid x \in A \text{ or } x \in B \text{ (and possibly both)}\,\}.$$

(ii) *The intersection of a nonempty set $\mathscr{A}$ of sets, denoted by $\cap \mathscr{A}$, is the set defined by*

$$\cap \mathscr{A} = \{\,x \mid \forall A \in \mathscr{A}\,(x \in A)\,\}.$$

As a special case, for sets A and B, $\cap\{\,A,B\,\}$ shall be denoted by $A \cap B$, so

$$A \cap B = \{\,x \mid x \in A \text{ and } x \in B\,\}.$$

(iii) *The difference of two sets A and B, denoted by $A - B$, is the set consisting of all the elements of A which are not in B; that is,*

$$A - B = \{\,x \mid x \in A \text{ and } x \notin B\,\}.$$

(iv) *If S is a set, then for each $A \in \mathscr{P}(S)$, the complement of A (in S), denoted by A^c, is defined by $A^c = S - A$.*

To help us visualize the relationships between sets and the operations defined above, we draw representations of sets. These pictorial representations are called Venn diagrams. If we are working with subsets of a given

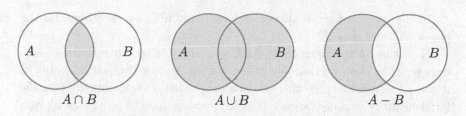

Fig. 2.1   Venn diagrams for $A \cap B$, $A \cup B$, and $A - B$

set U, we let a rectangle represent U, and the subsets are represented by circles drawn within the rectangle. If we are dealing with arbitrary sets, then we do not draw an enclosing rectangle. As shown in Figure 2.1, for sets A and B, we draw a circle whose interior is understood to represent the elements of A, and a circle whose interior is understood to represent the elements of B. The constructed sets $A \cap B$, $A \cup B$, and $A - B$ are shown as shaded regions in their respective Venn diagrams.

We collect some important facts about union, intersection and complementation in the following theorem.

2.3.2 Theorem. *Let A, B, and C be sets. Then the following hold:*

(i) $A \cup (B \cup C) = \cup \{A, B, C\} = (A \cup B) \cup C$ *($\cup$ is associative);*
(ii) $A \cap (B \cap C) = \cap \{A, B, C\} = (A \cap B) \cap C$ *($\cap$ is associative);*
(iii) $A \cup (B \cap C) = (A \cup B) \cap (A \cup C$ *($\cup$ distributes over $\cap$);*
(iv) $A \cap (B \cup C) = (A \cap B) \cup (A \cap C$ *($\cap$ distributes over $\cup$);*
(v) $C - (A \cup B) = (C - A) \cap (C - B)$ *(one of two De Morgan's laws);*
(vi) $C - (A \cap B) = (C - A) \cup (C - B)$ *(the other De Morgan's law);*
(vii) $A \cup B = B \cup A$ *and* $A \cap B = B \cap A$ *($\cup$ and $\cap$ are commutative).*

Proof. We prove only the first of these statements, leaving the rest to the reader. For any object x, we have

$$x \in A \cup (B \cup C) \text{ iff } x \in A \text{ or } x \in B \cup C$$
$$\text{iff } x \in A \text{ or } x \in B \text{ or } x \in \cup C$$
$$\text{iff } x \in \cup \{A, B, C\},$$

so $A \cup (B \cup C) = \cup \{A, B, C\}$. As well, for any object x, we have

$$x \in (A \cup B) \cup C \text{ iff } x \in A \cup B \text{ or } x \in C$$
$$\text{iff } x \in A \text{ or } x \in B \text{ or } x \in C$$
$$\text{iff } x \in \cup \{A, B, C\}.$$

Thus $(A \cup B) \cup C = \cup \{A, B, C\} = A \cup (B \cup C)$.   □

In the argument given above, we have used "iff" as a shorthand for the phrase "if, and only if".

Note that in the preceding proof, we were able to prove that two sets were equal by showing that any object was an element of one if, and only if, it was an element of the other. Such an approach is typically only possible for elementary demonstrations. It is far more common to prove that a given set A is equal to a given set B by proving that $A \subseteq B$ and $B \subseteq A$. In such a situation, it is not uncommon for one of these inclusions to be simple to demonstrate while the other inclusion may require quite a bit more work.

We observe that, in view of Theorem 2.3.2 (i) and (ii), we do not need to use brackets when forming the union or intersection of any finite number of sets. On the other hand, in a set expression that involves both union and intersection, brackets must be used. For example, $A \cup (B \cap C)$ is in general not equal to $(A \cup B) \cap C$. Similarly, brackets must be used in any expression involving set difference and another operation (even if the other operation is set difference itself). For example, the expression $A - B - C$ is ambiguous, since $A - (B - C)$ is not equal to $(A - B) - C$ in general. In Figure 2.2, we show the Venn diagrams for $A - (B - C)$ and for $(A - B) - C$. By viewing the Venn diagrams, the reader should find it easy to construct examples of sets A, B, and C such that $A - (B - C) \neq (A - B) - C$. It is evident from the Venn diagrams that for any sets A, B, and C, $(A - B) - C \subseteq A - (B - C)$.

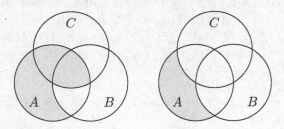

Fig. 2.2   Venn diagrams for $A - (B - C)$ and $(A - B) - C$

2.3.3 Notation. *We shall find it convenient to have the notion of an index set for a set $\mathscr{A}$ of sets. Given a set $\mathscr{A}$ of sets, if we are able to find a set I and construct a one-to-one correspondence between the elements of I and the elements of $\mathscr{A}$, then we can use each element of I as a label on an element of $\mathscr{A}$. We shall typically denote the element of $\mathscr{A}$ that corresponds to $i \in I$ by A_i. Then it is conventional to write $\mathscr{A} = \{ A_i \mid i \in I \}$. We say that $\mathscr{A}$ has been indexed by I, and we call I an index set for $\mathscr{A}$.*

For example, if we had a set $\mathscr{A}$ of sets, and $\mathscr{A}$ could be put in one-to-one correspondence with the elements of $\mathbb{N}$, where the element of $\mathscr{A}$ that corresponds to $i \in \mathbb{N}$ is denoted by A_i, then $\mathscr{A} = \{ A_i \mid i \in \mathbb{N} \}$, and we say that $\mathscr{A}$ is indexed by $\mathbb{N}$.

If we have a set $\mathscr{A}$ of sets that is indexed by I so that we can write $\mathscr{A} = \{ A_i \mid i \in I \}$, then we may use the notation $\underset{i \in I}{\cup} A_i$ to mean $\cup \mathscr{A}$, and $\underset{i \in I}{\cap} A_i$ to mean $\cap \mathscr{A}$. In the case of a set $\mathscr{A}$ that is indexed by $\mathbb{N}$, respectively $\mathbb{Z}^+$, we may write $\overset{\infty}{\underset{i=0}{\cup}} A_i$ in place of $\underset{i \in \mathbb{N}}{\cup} A_i$ and $\overset{\infty}{\underset{i=1}{\cup}} A_i$ in place of $\underset{i \in \mathbb{Z}^+}{\cup} A_i$.

2.3.4 Example. *For each $n \in \mathbb{Z}^+$, define $A_n = \{ x \in \mathbb{R} \mid \frac{1}{n} \leq x \leq 1 \}$. Then for $\mathscr{A} = \{ A_n \mid n \in \mathbb{Z}^+ \}$, we have $\overset{\infty}{\underset{n=1}{\cup}} A_n = \{ x \in \mathbb{R} \mid 0 < x \leq 1 \}$ and $\overset{\infty}{\underset{n=1}{\cap}} A_n = \{ 1 \}$.*

Exercises

1. Draw Venn diagrams to illustrate the statements of Theorem 2.3.2 (iii), (iv), (v), and (vi).

2. Let A, B, and C be sets. In each part below, draw a Venn diagram for both constructions, then prove the equality.
    a) $(A - B) - C = A - (B \cup C)$.
    b) $A - (B - C) = (A - B) \cup (A \cap C)$.
    c) $A \cup (B - C) = (A \cup B) - (C - A)$.
    d) $A \cap (B - C) = (A \cap B) - (A \cap C)$.

3. Prove that for any sets A and B, the following are equivalent:
    (i) $A \subseteq B$.
    (ii) $A \cap B = A$.
    (iii) $A \cup B = B$.

4. Determine $|\mathscr{P}(\{ 1, 2, 3, 4 \})|$. Can you determine $|\mathscr{P}(A)|$ for any finite set A?

For sets A and B, define their symmetric difference, denoted by $A \bigtriangleup B$, by $A \bigtriangleup B = (A - B) \cup (B - A)$.

---

5. For $S = \{ 1, 2, 3 \}$, compute $A \bigtriangleup B$ for all choices of $A, B \in \mathscr{P}(S)$.

6. Let A, B, and C be sets. In each case below, illustrate both set constructions with a Venn diagram, and prove the equality holds.

    a) $A \triangle (B \triangle C) = (A \triangle B) \triangle C$.

    b) $A \triangle A = \varnothing$.

    c) $A \triangle \varnothing = A$.

    d) $A \triangle B = B \triangle A$.

7. Let S denote the set of all strings of length 4 (that is; sequences of length 4) of 0's and 1's. For example, $0110 \in S$. Let A be a set of four elements. What can you say about $|\mathscr{P}(A)|$ and S? Generalize this to the set of all strings of length n of 0's and 1's and $\mathscr{P}(A)$ for any set A of size n. Can you prove the generalization?

8. a) If A and B are finite sets, prove that $|A \cup B| = |A| + |B| - |A \cap B|$.

    b) If A, B, C, and D are finite sets, find the corresponding expression for $|A \cup B \cup C|$ and $|A \cup B \cup C \cup D|$.

    c) If A, B, C, and D are finite sets such that $|A| = 9$, $|B| = 6$, $|C| = 11$, $|D| = 12$, $A \cap B| = 4$, $|A \cap C| = 7$, $|B \cap C| = 4$, $|A \cap B \cap C| = 2$, and $|(A \cup B \cup C) \cap D| = 6$, what is $|A \cup B \cup C \cup D|$?

9. Let $\mathscr{A}$ and $\mathscr{B}$ be sets of sets, indexed by sets I and J, respectively, so we may write $\mathscr{A} = \{ A_i \mid i \in I \}$ and $\mathscr{B} = \{ B_j \mid j \in J \}$. Prove each of the following:

    a) If X is a set such that $A_i \subseteq X$ for all $i \in I$, then $\bigcup_{i \in I} A_i \subseteq X$.

    b) If X is a set such that $X \subseteq A_i$ for all $i \in I$, then $X \subseteq \bigcap_{i \in I} A_i$.

    c) $\left(\bigcup_{i \in I} A_i \right) \cap \left(\bigcup_{j \in J} B_j \right) = \bigcup_{i \in I,\ j \in J} (A_i \cap B_j)$.

    d) $\left(\bigcap_{i \in I} A_i \right) \cup \left(\bigcap_{j \in J} B_j \right) = \bigcap_{i \in I,\ j \in J} (A_i \cup B_j)$.

    e) If S is a set such that $A_i \in \mathscr{P}(S)$ for all $i \in I$, then $\bigcup_{i \in I} A_i \in \mathscr{P}(S)$ and $\left(\bigcup_{i \in I} A_i \right)^c = \bigcap_{i \in I} A_i^c$.

    f) If S is a set such that $A_i \in \mathscr{P}(S)$ for all $i \in I$, then $\left(\bigcap_{i \in I} A_i \right)^c = \bigcup_{i \in I} A_i^c$.

10. In each part, either prove or disprove the statement.

    a) For all sets A and B, $\mathscr{P}(A) \cup \mathscr{P}(B) = \mathscr{P}(A \cup B)$.

    b) For all sets A and B, $\mathscr{P}(A) \cap \mathscr{P}(B) = \mathscr{P}(A \cap B)$.

11. Let S be a set, and let $A, B \in \mathscr{P}(S)$. Prove that the following are equivalent:

(i) $A \subseteq B$;

(ii) $A \cap B^c = \varnothing$;

(iii) $A^c \cup B = S$.

12. Let S be a set and let $A, B \in \mathscr{P}(S)$.

   a) Prove that $A \subseteq B$ if and only if $A^c \supseteq B^c$.

   b) Prove that if $A \cap B = A \cap B^c$, then $A = \varnothing$.

   c) Prove that if $A \cup B = A \cup B^c$, then $B \subseteq A$.

13. Let A, B, and C be sets. Prove that

$$[A \cup (B \cap C)] \cap [B \cup (A \cap C)] = (A \cap B) \cup (A \cap C) \cup (B \cap C)$$
$$= [B \cap (A \cup C)] \cup [C \cap (A \cup B)].$$

14. In each part, either prove or disprove the statement.

   a) For all sets A, B, and C, if $A \cap B = A \cap C$, then $B = C$.

   b) For all sets A, B, and C, if $A \cap B = A \cap C$ and $A \cup B = A \cup C$, then $B = C$.

   c) For all sets A, B, and C, if $A \bigtriangleup B = A \bigtriangleup C$, then $B = C$.

   d) For all sets A, B, and C, if $A - B = A - C$, then $B = C$.

   e) For all sets A, B, and C, if $A \cap C = B \cap C$ and $A - C = B - C$, then $A = B$.

   f) For all sets A, B, and C, if $A \cup C = B \cup C$ and $A - C = B - C$, then $A = B$.

15. Let S be a set, and let $A, B, C \in \mathscr{P}(S)$. Prove that $(A \cap B) \cup (B \cap C)^c = A \cup B^c \cup C^c$.

Chapter 3

Cartesian Products, Relations, Maps and Binary Operations

Introduction

Relations and maps occur in almost every branch of mathematics. We shall first study relations in general and then, as special cases of relations, equivalence relations and then maps (sometimes also called mappings or functions). Finally, we shall introduce binary operations as special kinds of maps. The framework we use is the Cartesian product of sets.

Cartesian Product

3.2.1 Definition. *Let A and B be sets. The Cartesian product of A and B, denoted by $A \times B$, is the set of all ordered pairs (a, b), where a is an element of A and b is an element of B. In symbols, we have*

$$A \times B = \{ (a, b) \mid (a \in A) \wedge (b \in B) \}.$$

3.2.2 Remark. When we say that a pair is ordered, we mean that the order in which the members of the pair are written is important. For example, the ordered pair $(1, 2)$ is different from the ordered pair $(2, 1)$. In general, the ordered pair (a, b) is defined to be $\{ \{ a \}, \{ a, b \} \}$. If $A \neq B$ and neither A nor B is the empty set, then it follows that $A \times B \neq B \times A$.

For example, the Cartesian product of $\mathbb{R}$ with itself, $\mathbb{R} \times \mathbb{R}$, can be thought of as the Cartesian plane of co-ordinate geometry. For another example, if $A = \{ 1, 2, 3 \}$ and $B = \{ 1, 2 \}$, then

$$A \times B = \{ (1, 1), (1, 2), (2, 1), (2, 2), (3, 1), (3, 2) \}.$$

Note that $A \times B \subseteq \mathbb{R} \times \mathbb{R}$.

Now that the notion of Cartesian product has been introduced, we may define the abstract notion of a relation from one set to another.

3.2.3 Definition. *Given sets A and B, we say that R is a relation from A to B if $R \subseteq A \times B$; that is, each subset of $A \times B$ is called a relation from A to B. For any set A, if $R \subseteq A \times A$, then rather than saying that R is a relation from A to A, we shall simply say that R is a relation on A.*

3.2.4 Example. *Let L be the relation on $\mathbb{R}$ given by*

$$L = \{\, (x,y) \in \mathbb{R} \times \mathbb{R} \mid x - y \leq 0 \,\}.$$

L is simply the usual order relation on the set of real numbers; that is, for any $x, y \in \mathbb{R}$, $(x,y) \in L$ if and only if $x \leq y$.

3.2.5 Example. *If A and B are (small) finite sets, then we may use a diagram to describe a relation from A to B. For example, if $A = \{\, 1, 2, 3, 4 \,\}$, $B = \{\, 1, 2, 3 \,\}$, and $R = \{\, (1,2), (2,1), (3,2) \,\}$, we may depict A, B, and R as shown in Figure 3.1. Note that to indicate the fact that $(a,b) \in R$, we have drawn an arrow from $a \in A$ to $b \in B$.*

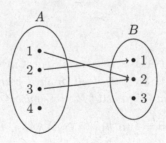

Fig. 3.1   A relation from $\{\, 1, 2, 3, 4 \,\}$ to $\{\, 1, 2, 3 \,\}$

3.2.6 Notation. *If A and B are sets and R is a relation from A to B, then we shall write $a\,R\,b$ to mean $(a,b) \in R$. Frequently, we shall use a symbol to represent R in this situation. For example, in Example 3.2.4, we usually write $a \leq b$ instead of $a\,L\,b$ or $(a,b) \in L$. As another example, for any set A, the identity (or diagonal) relation on A, denoted by $\mathbb{1}_A$, is defined by*

$$\mathbb{1}_A = \{\, (a,b) \in A \times A \mid a = b \,\},$$

and we shall write $a = b$ in place of $a\,\mathbb{1}_A\,b$ or $(a,b) \in \mathbb{1}_A$.

In the study of relations on a set A, there are certain properties that turn out to be of great importance.

3.2.7 Definition. *Let A be a set. A relation R on A is said to be:*

(i) *reflexive if $\mathbb{1}_A \subseteq R$;*

(ii) *symmetric if for all $a, b \in A$, $a\,R\,b$ implies that $b\,R\,a$;*

(iii) *antisymmetric if for all $a, b \in A$, $a\,R\,b$ and $b\,R\,a$ implies that $a = b$;*

(iv) *transitive if for all $a, b, c \in A$, if $a\,R\,b$ and $b\,R\,c$, then $a\,R\,c$.*

When we choose to illustrate an example of a relation R on a finite set A by a diagram, we may simplify the drawing by using only one copy of A, and simply draw an arrow from a to b in the diagram to indicate that $(a, b) \in R$.

3.2.8 Example. *Let $A = \{1, 2, 3, 4, 5\}$ and*

$$R = \{(1, 3), (1, 4), (2, 3), (2, 4), (2, 5), (3, 4), (5, 2), (5, 3), (5, 4)\}.$$

The diagram of R is shown in Figure 3.2.

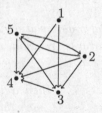

Fig. 3.2   A transitive relation on $\{1, 2, 3, 4, 5\}$

The reader can easily verify that for each i and j in $A = \{1, 2, 3, 4, 5\}$, if there is a "two-step" path from i to j (that is, if we may follow one arrow from i to some point, then follow another arrow from that point to j), then there is a "one-step" path from i to j; that is, $i\,R\,j$, and this means that R is a transitive relation on A. Since $(1, 1) \in \mathbb{1}_A$ but $(1, 1) \notin R$, R is not reflexive. Since $(1, 3) \in R$ but $(3, 1) \notin R$, R is not symmetric. Since $(2, 5) \in R$ and $(5, 2) \in R$, but $2 \neq 5$, R is not antisymmetric.

3.2.9 Example. *Consider the usual order relation $\leq$ on $\mathbb{R}$. It has the following properties:*

(i) *For all $a \in \mathbb{R}$, $a \leq a$ (reflexivity);*

(ii) *For all $a, b, c \in \mathbb{R}$, $a \leq b$ and $b \leq c$ imply $a \leq c$ (transitivity);*

(iii) For all $a, b \in \mathbb{R}$, if $a \le b$ and $b \le a$, then $a = b$ (antisymmetry).

3.2.10 Example. *Let S be any set, and consider the subset relation on $\mathscr{P}(S)$ given by $\{\, (A, B) \in \mathscr{P}(S) \mid A \subseteq B \,\}$. It has the following properties:*

(i) For all $A \in \mathscr{P}(S)$, $A \subseteq A$ (reflexivity);

(ii) For all $A, B, C \in \mathscr{P}(S)$, $A \subseteq B$ and $B \subseteq C$ imply $A \subseteq C$ (transitivity);

(iii) For all $A, B \in \mathscr{P}(S)$, if $A \subseteq B$ and $B \subseteq A$, then $A = B$ (antisymmetry).

3.2.11 Definition. *For any set A, a relation P on A is called a partial order if it is reflexive, transitive and antisymmetric. Often, the set together with the partial order is referred to as a poset.*

Thus Examples 3.2.9 and 3.2.10 are examples of partial order relations. Note that Example 3.2.9 also has the property that for all $a, b \in \mathbb{R}$, either $a \le b$ or else $b \le a$. Example 3.2.10 does not have this property (unless $|S| \le 1$).

3.2.12 Definition. *A partial order relation P on a set S is said to be a total order relation, or a linear order relation, on S if for all $a, b \in S$, either $a\,P\,b$ or $b\,P\,a$.*

3.2.13 Example. *Let $a, b \in \mathbb{Z}$. Then a is said to be a divisor of b, denoted by $a \mid b$, if there exists $k \in \mathbb{Z}$ such that $b = ka$. Let D denote the relation on $\mathbb{Z}^+$ for which $a\,D\,b$ if $a, b \in \mathbb{Z}^+$ and a is a divisor of b; that is, there exists an integer k such that $b = ka$. Then D is a partial order relation on $\mathbb{Z}^+$, but not a total order relation. Note that the antisymmetry of D follows from the fact that all of the integers involved are positive. D is called the divison relation on $\mathbb{Z}^+$, and we usually write $a\,D\,b$ as $a \mid b$.*

3.2.14 Example. *Let D_6 denote the set of all positive integer divisors of 6, so $D_6 = \{\, 1, 2, 3, 6 \,\}$, and let L be the relation on D_6 defined by $a\,L\,b$ if and only if $a, b \in D_6$ and a is a divisor of b; that is $L = D \cap (D_6 \times D_6)$, where D is the division relation of Example 3.2.13. Thus*

$$L = \{\, (1,1), (1,2), (1,3), (1,6), (2,2), (2,6), (3,3), (3,6), (6,6) \,\}.$$

Since D is a partial order relation on $\mathbb{Z}^+$ and $D_6 \subseteq \mathbb{Z}^+$, it follows that L is a partial order relation on D_6. Since D_6 is a finite set, we may represent L by a diagram as shown in Figure 3.3.

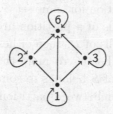

Fig. 3.3   A diagram of the division ordering on D_6, the set of positive divisors of 6

Note that a loop drawn at a node a indicates that $a\,L\,a$, so the presence of a loop at each node informs the reader that the relation is reflexive.

The next type of relation that we are about to introduce is of great importance in all areas of mathematics, and consequently will play an important role in later chapters of this book.

3.2.15 Definition. *Let S be a set. A relation E on S is called an equivalence relation if E is reflexive, symmetric, and transitive.*

3.2.16 Example. *Let*

$$E_4 = \{\,(a,b) \in \mathbb{Z} \times \mathbb{Z} \mid a - b = 4k \text{ for some } k \in \mathbb{Z}\,\},$$

so E_4 is a relation on $\mathbb{Z}$. For every $a \in \mathbb{Z}$, $a - a = 0 = (4)(0)$, so $a\,E_4\,a$. Thus E_4 is reflexive. If $a, b \in \mathbb{Z}$ are such that $a\,E_4\,b$, then there exists $k \in \mathbb{Z}$ such that $a - b = 4k$, and then $b - a = 4(-k)$ with $-k \in \mathbb{Z}$, so $b\,E_4\,a$. Thus E_4 is symmetric. Finally, if $a, b, c \in \mathbb{Z}$ are such that $a\,E_4\,b$ and $b\,E_4\,c$, then there exist $k, n \in \mathbb{Z}$ such that $b - a = 4k$ and $c - b = 4n$, so $c - a = (c - b) + (b - a) = 4(n - k)$, and since $n - k \in \mathbb{Z}$, we have $a\,E_4\,c$. Thus E_4 is transitive. Since E_4 is reflexive, symmetric, and transitive, it is an equivalence relation on $\mathbb{Z}$.

3.2.17 Definition. *Let S be a set, and E be an equivalence relation on S. For each $a \in S$,*

$$[a]_E = \{\, x \in S \mid a\,E\,x \,\}$$

is called the equivalence class of a mod E. If it is understood that we are talking about equivalence classes of the equivalence relation E, then it is customary to write $[a]$ instead of $[a]_E$.

3.2.18 Definition. *Let S be a set. A subset P of $\mathscr{P}(S) - \{\varnothing\}$ is called a partition of S if for each $a \in S$, there exists one and only one $C \in P$ with $a \in C$. Each $C \in P$ is called a cell of the partition P.*

Suppose that P is a partition for some set S. We observe that if $C \in P$, then $C \neq \varnothing$; that is, the cells of a partition are required to be nonempty. Moreover, if C and D are cells of P, then either $C = D$ or else $C \cap D = \varnothing$ (since no element of S can belong to more than one cell of P). Finally, $\cup P = S$ since each element of S belongs to some $C \in P$.

There is a close connection between partitions and equivalence relations, which we now establish.

3.2.19 Theorem. *Let S be a set. If E is an equivalence relation on S, then $\Pi_E = \{\, [a]_E \mid a \in S \,\}$ is a partition of S. Conversely, if Π is a partition of S, then the relation E_Π for which $a\, E_\Pi\, b$ if a and b belong to the same cell of Π is an equivalence relation on S. Moreover, $E = E_{\Pi_E}$, and $\Pi_{E_\Pi} = \Pi$; that is, the equivalence relation on S that is obtained from the partition of equivalence classes of E is E itself, and the equivalence classes of E_Π are the cells of Π.*

Proof. Suppose that E is an equivalence relation on S, and let $\Pi_E = \{\, [a]_E \mid a \in S \,\}$, so $\Pi_E \subseteq \mathscr{P}(S)$. Let $a \in S$. Then since $a\, E\, a$, $a \in [a]_E$. Thus no element of Π_E is empty, so $\Pi_E \subseteq \mathscr{P}(S) - \{\, \varnothing \,\}$, and each element of S belongs to at least one element of Π_E. It remains to prove that no element of S can belong to more than one element of Π_E. Suppose that $a, b \in S$ are such that $[a]_E \cap [b]_E \neq \varnothing$. Then we must prove that $[a]_E = [b]_E$. To prove this, we first show that for any $x \in S$, if $y \in [x]_E$, then $[y]_E = [x]_E$. Suppose that $x, y \in S$ are such that $y \in [x]_E$, and let $t \in [y]_E$. Then $x\, E\, y$ and $y\, E\, t$, so by transitivity, $x\, E\, t$ and therefore $t \in [x]_E$. Thus $[y]_E \subseteq [x]_E$. Then, since E is symmetric and $x\, E\, y$, we obtain $y\, E\, x$ and thus $x \in [y]_E$. But then $[x]_E \subseteq [y]_E$, and so $[x]_E = [y]_E$. Recall now that we have $[a]_E \cap [b]_E \neq \varnothing$, so there exists $x \in [a]_E \cap [b]_E$. But then $[a]_E = [x]_E = [b]_E$, as required. Thus Π_E is a partition of S.

Conversely, suppose that Π is a partition of S. We are to prove that E_Π is an equivalence relation on S. Let $a \in S$. Then there exists $C \in \Pi$ such that $a \in C$, so $a\, E_\Pi\, a$ and thus E_Π is reflexive. Next, suppose that $a, b \in S$, and $a\, E_\Pi\, b$. Then there exists $C \in \Pi$ with $a, b \in C$, and so $b\, E_\Pi\, a$. Thus E_Π is symmetric. Finally, suppose that $a, b, c \in S$ are such that $a\, E_\Pi\, b$ and $b\, E_\Pi\, c$. Then there exist $C, D \in \Pi$ with $a, b \in C$ and $b, c \in D$. Since $b \in C \cap D$, it follows that $C = D$, so $a, c \in C$ and thus $a\, E_\Pi\, c$, which proves that E_Π is transitive. Since E_Π is reflexive, symmetric, and transitive, E_Π is an equivalence relation on S. Next, we prove that the equivalence classes of E_Π are the cells of Π. Let $a \in S$. Then there exists a unique $C \in \Pi$

with $a \in C$. By definition of E_Π, for $x \in S$, $a \, E_\Pi \, x$ if and only if $x \in C$, so $[a]_{E_\Pi} = C$. Thus each equivalence class of E_Π is a cell of Π, so $\Pi_{E_\Pi} \subseteq \Pi$ and thus $\Pi_{E_\Pi} = \Pi$.

It remains to prove that $E_{\Pi_E} = E$. Suppose that $a, b \in S$ and $a \, E \, b$. Then $a, b \in [a]_E \in \Pi_E$, so $a \, E_{\Pi_E} \, b$. Thus $E \subseteq E_{\Pi_E}$. Now suppose that $a, b \in S$ and $a \, E_{\Pi_E} \, b$. Then there exists $[x]_E \in \Pi_E$ with $a, b \in [x]_E$, so $[a]_E = [x]_E = [b]_E$ and thus $b \in [a]_E$; that is, $a \, E \, b$. Thus $E_{\Pi_E} \subseteq E$ and so $E = E_{\Pi_E}$. $\qquad\square$

3.2.20 Example. *For the relation E_4 of Example 3.2.16, the equivalence classes are the four sets $[0]$, $[1]$, $[2]$, and $[3]$. To see that every equivalence class of E_4 is in this list, one need only observe that every integer a is either a multiple of 4, in which case, $a \in [0]$ and so $[a] = [0]$, or else there exists a multiple of 4, say $4k$, such that for $r = a - 4k$, $0 < r < 4$ and $a - r = 4k$; that is, $a \in [r]$ and so $[a] = [r]$ with $r = 1$, 2, or 3. It remains to determine that no two of these are the same. Suppose i and j are such that $0 \le i \le j \le 3$ and $[i] = [j]$. Then for some integer k, $4k = j - i \ge 0$, which means that $k \ge 0$, and as well, $4k = j - i \le 3 - i \le 3$, so $k < 1$. Thus $k = 0$ and so $j - i = 0$; that is, $j = i$. The reader is urged to verify that*

$$[0] = \{\, 4k \mid k \in \mathbb{Z} \,\}$$
$$[1] = \{\, 4k + 1 \mid k \in \mathbb{Z} \,\}$$
$$[2] = \{\, 4k + 2 \mid k \in \mathbb{Z} \,\}$$
$$[3] = \{\, 4k + 3 \mid k \in \mathbb{Z} \,\}.$$

Maps

The reader has undoubtedly already encountered the notion of a map, or function, in high school and in courses in calculus. In that setting, a function was likely represented by a formula in which one substituted one or more real numbers for the variables to obtain the value of this function at the substituted values. Our definition below is much more general than this although the idea is basically the same. From this point on, we shall refer to a function as a map, or a mapping.

3.3.1 Definition. *Let A and B be sets. A map f from A to B, denoted by $f : A \to B$, is a relation $f \subseteq A \times B$ with the property that for $a \in A$, there exists one and only one $b \in B$ with $(a, b) \in f$; that is, in f, each element*

*of A is related to exactly one element of B. The set A is called the domain
of f and the set B is called the codomain of f. Moreover, for each a ∈ A,
the unique b ∈ B for which (a, b) ∈ f is denoted by f(a); that is, b = f(a)
if and only if (a, b) ∈ f.*

Note that in the definition of a map $f : A \to B$, it is not required that
in f, each element of B be related to exactly one element of A.

This is the formal definition of a map, but in practice, we rarely think
of it this way. If $f : A \to B$, then rather than writing $(a, b) \in f$, we
shall instead write $b = f(a)$, and we think of f as somehow "sending"
or "transforming" $a \in A$ into $f(a) \in B$. $f(a)$ is called the image of a
under f.

3.3.2 Example. *Let $A = \{1, 2, 3, 4\}$, $B = \{1, 2, 3, 4, 5\}$, and*

$$f = \{(1, 2), (2, 2), (3, 1), (4, 4)\}.$$

*Then f is a map from A to B, and we have $f(1) = 2 = f(2)$, $f(3) = 1$,
and $f(4) = 4$. The diagram of the map is shown in Figure 3.4.*

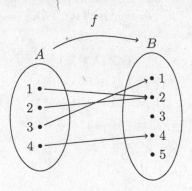

Fig. 3.4   The map $f : A \to B$

*In a situation such as this (where the domain of the map is finite), we
shall often use a two-row matrix to represent the map. For each domain
element x, the column that corresponds to x consists of x in the first row
and $f(x)$ in the second row. Thus we could represent the map f of this
example by $\left(\begin{smallmatrix} 1 & 2 & 3 & 4 \\ 2 & 2 & 1 & 4 \end{smallmatrix}\right)$.*

There will be a great many times when we will be called upon to prove
that two maps are equal, and we shall always make use of the following
result.

3.3.3 Proposition. *Let A and B be sets, and let f and g be maps from A to B. Then $f = g$ if and only if $f(a) = g(a)$ for all $a \in A$.*

Proof. First, suppose that $f = g$. Then for each $a \in A$, $(a, f(a)) \in f$, and $f = g$, so $(a, f(a)) \in g$. But then $f(a)$ is the unique element of B that g pairs with a, so $g(a) = f(a)$.

Conversely, suppose that for each $a \in A$, $f(a) = g(a)$. Then $f = \{(a, f(a)) \mid a \in A\} = \{(a, g(a)) \mid a \in A\} = g$, as required. $\square$

3.3.4 Definition. *Let A and B be sets, and let $X \subseteq A$, $Y \subseteq B$. Let f be a mapping from A to B.*

(i) *The image of X under f is the subset of B denoted by $f(X)$ and defined by*
$$f(X) = \{f(x) \mid x \in X\}.$$
$f(A)$ is called the image of f and is also denoted by $Im(f)$.

(ii) *The inverse image of Y under f is the subset of A denoted by $f^{-1}(Y)$ and defined by*
$$f^{-1}(Y) = \{a \in A \mid f(a) \in Y\}.$$
If $Y \subseteq B$ and $|Y| = 1$, say $Y = \{b\}$, then we usually write $f^{-1}(b)$ instead of $f^{-1}(\{b\})$, and if $X \subseteq A$ and $|X| = 1$, say $X = \{a\}$, then we usually write $f(a)$ instead of $f(\{a\})$. Admittedly, this has the potential to cause confusion, since $f(a) \in B$, while $f(\{a\}) \subseteq B$, but it is hoped that the context will make the intention clear.

Observe that if $f : A \to B$ and $X \subseteq A$, then $f(X) \subseteq Im(f) = f(A)$. Further, note that if $Im(f) \neq B$, then it is possible for there to exist $Y \subseteq B$ with $Y \neq \varnothing$, yet $f^{-1}(Y) = \varnothing$. More generally, the reader is urged to verify that for any subset Y of B, $f(f^{-1}(Y)) \subseteq Y$. In our next example, we illustrate that equality need not hold.

3.3.5 Example. *For the mapping f of Example 3.3.2, $Im(f) = \{1, 2, 4\}$, and $f^{-1}(1) = \{3\}$, $f^{-1}(2) = \{1, 2\}$, $f^{-1}(3) = \varnothing$, $f^{-1}(4) = \{4\}$, and $f^{-1}(5) = \varnothing$. Thus for example, $f^{-1}(\{4, 5\}) = \{4\}$, so*
$$f(f^{-1}(\{4, 5\})) = \{f(4)\} = \{4\} \subsetneq \{4, 5\}.$$

3.3.6 Definition. *Let A and B be sets.*

(i) *A map $f : A \to B$ is said to be injective, or one-to-one, if for all $x, y \in A$, if $x \neq y$, then $f(x) \neq f(y)$. Equivalently (using the logically equivalent contrapositive), f is injective if for all $x, y \in A$, $f(x) = f(y)$ implies that $x = y$.*

(ii) A map $f : A \to B$ is said to be surjective, or onto, if $Im(f) = B$.

(iii) A map $f : A \to B$ is said to be bijective if it is both injective and surjective.

3.3.7 Example. *We prove that the map $g : \mathbb{R} \to \mathbb{R}$ defined by $g(x) = 3x + 5$ for each $x \in \mathbb{R}$ is bijective. Note that when we say that g is defined by $g(x) = 3x + 5$ for each $x \in \mathbb{R}$, we mean that $g = \{ (x, 3x + 5) \mid x \in \mathbb{R} \}$. It is immediate that g is a map, since each $x \in \mathbb{R}$ is paired only with $y \in \mathbb{R}$ where $y = 3x + 5$. First, we prove that g is injective. Suppose that $x, y \in \mathbb{R}$ are such that $f(x) = f(y)$. Then $3x + 5 = 3y + 5$, so $3x = 3y$ and thus $x = y$, proving that g is injective. To prove that g is surjective, we must prove that $\mathbb{R} \subseteq Im(g)$; that is, for each $y \in \mathbb{R}$, we must prove that there exists $x \in \mathbb{R}$ with $y = g(x)$. Let $y \in \mathbb{R}$. Then we require the existence of $x \in \mathbb{R}$ such that $3x + 5 = y$. From this equation, we see that the only possible value of x is $\frac{y-5}{3}$, We compute $g(\frac{y-5}{3}) = 3(\frac{y-5}{3}) + 5 = (y - 5) + 5 = y$. Thus $y \in Im(g)$. This completes the proof that g is surjective. Since g is both injective and surjective, g is bijective.*

3.3.8 Proposition. *Let A and B be sets and let f be a map from A to B.*

(i) f is injective if, and only if, $f^{-1}(f(X)) = X$ for every subset X of A.

(ii) f is surjective if, and only if, $f(f^{-1}(Y)) = Y$ for every subset Y of B.

Proof. It is evident that for every subset X of A and every subset Y of B, $X \subseteq f^{-1}(f(X))$ and $f(f^{-1}(Y)) \subseteq Y$.

Suppose first that f is injective, and let X be a subset of A. Let $z \in f^{-1}(f(X))$. Then $f(z) \in f(X)$, and so there exists $x \in X$ with $f(z) = f(x)$. Since f is injective, it follows that $z = x \in X$ and thus $f^{-1}(f(X)) \subseteq X$. Conversely, suppose that $f^{-1}(f(X)) = X$ for every subset X of A. Let $x, y \in A$ be such that $f(x) = f(y)$, and let $X = \{ x \}$. Then $y \in f^{-1}(f(X)) = X$, and so $y = x$. Thus f is injective.

Next, suppose that f is surjective, and let Y be a subset of B. Let $y \in Y$. Then there exists $x \in X$ with $f(x) = y$, so $x \in f^{-1}(Y)$ and thus $y = f(x) \in f(f^{-1}(Y))$. Conversely, suppose that $f(f^{-1}(Y)) = Y$ for every subset Y of B. Let $y \in B$ and set $Y = \{ y \}$. Since $f(f^{-1}(Y)) = Y$, there exists $x \in f^{-1}(Y)$ such that $f(x) = y$. Thus f is surjective. $\square$

3.3.9 Definition. *Let A, B, and C be sets. If $f : A \to B$ and $g : B \to$*

C are maps, then the composite of f and g, denoted by $g \circ f$ and read g following f, is defined by

$$g \circ f = \{\, (a, g(b)) \mid a \in A, \ b \in B, \ \text{and} \ b = f(a) \,\}.$$

3.3.10 Proposition. *Let A, B, and C be sets, and $f : A \to B$ and $g : B \to C$ be maps. Then $g \circ f$ is a map from A to C, and for each $a \in A$, $g \circ f(a) = g(f(a))$.*

Proof. By definition, $g \circ f \subseteq A \times C$. For each $a \in A$, there is a unique $b \in B$ for which $(a, b) \in f$; namely $b = f(a)$, and since g is a map, there is a unique $c \in C$ that g pairs with $f(a)$; namely $g(f(a))$. Thus the unique element of C that $g \circ f$ pairs with $a \in A$ is $g(f(a))$. Thus $g \circ f$ is a map from A to C, and for each $a \in A$, $g \circ f(a) = g(f(a))$. $\qquad\square$

Note that for maps f and g, it is possible for $g \circ f$ to be defined but $f \circ g$ not be defined. If $f : A \to B$ and $g : B \to C$ and $C \neq A$, then $g \circ f$ is defined and $f \circ g$ is not defined. However, if $C = A$, then both $g \circ f$ and $f \circ g$ are defined, in fact, $g \circ f : A \to A$, and $f \circ g : B \to B$. If $A \neq B$, then we immediately see that $g \circ f \neq f \circ g$. However, even if $A = B = C$, so both $g \circ f$ and $f \circ g$ are maps from A to A, it is usually the case that $g \circ f \neq f \circ g$. For example, if $f, g : \mathbb{R} \to \mathbb{R}$ are defined by $f(x) = x^2$ and $g(x) = x - 1$ for each $x \in \mathbb{R}$, we have $g \circ f(x) = g(x^2) = x^2 - 1$, while $f \circ g(x) = f(x - 1) = (x - 1)^2 = x^2 - 2x + 1$. Thus, for example, $g \circ f(0) = -1$, while $f \circ g(0) = 1$. Since $g \circ f(0) \neq f \circ g(0)$, it follows from Proposition 3.3.3 that $g \circ f \neq f \circ g$.

3.3.11 Proposition. *Composition of maps is associative. More precisely, for any sets A, B, C, and D, and maps $f : A \to B$, $g : B \to C$, and $h : C \to D$, $h \circ (g \circ f) = (h \circ g) \circ f$.*

Proof. Let $a \in A$. Then $h \circ (g \circ f)(a) = h(g \circ f(a)) = h(g(f(a))) = h \circ g(f(a)) = (h \circ g) \circ f(a)$. By Proposition 3.3.3, $h \circ (g \circ f) = (h \circ g) \circ f$. $\quad\square$

3.3.12 Proposition. *Let A and B be sets, and let $f : A \to B$ be a map. Then $f \circ \mathbb{1}_A = f = \mathbb{1}_B \circ f$.*

Proof. Let $a \in A$. Then $f \circ \mathbb{1}_A(a) = f(\mathbb{1}_A(a)) = f(a) = \mathbb{1}_B(f(a)) = \mathbb{1}_B \circ f(a)$, so by Proposition 3.3.3, $f \circ \mathbb{1}_A = f = \mathbb{1}_B \circ f$. $\qquad\square$

Some mathematical computations can be reversed and some cannot. For example, the computation whereby 5 is added to a real number x can be reversed by subtracting 5 from the result, thus recovering the original number x. On the other hand, multiplication of a real number x by 0 is irreversible since $(0)x = 0$ for any $x \in \mathbb{R}$ and thus, given the outcome of 0, there is no way to determine what value of x was multiplied by 0 to obtain this outcome.

Analogously, for a given map $f : A \to B$, if we are given the image under f of a certain element $a \in A$, in general we cannot discern that it was a that led to this outcome. For example, in Example 3.3.2, if we know that the image of a certain $a \in \{1, 2, 3, 4\}$ under f is 2, we do not know whether the input a was 1 or 2, since both $f(1) = 2$ and $f(2) = 2$. It is necessary that the map f be injective in order that we be able to answer such a question. Let us continue to explore these ideas further.

3.3.13 Definition. *Let A and B be sets and $f : A \to B$ and $g : B \to A$ be maps. Then g is said to be a left inverse of f and f is said to be a right inverse of g if $g \circ f = \mathbb{1}_A$. Maps which possess left inverses are said to be left invertible and those with right inverses, right invertible.*

The following result tells us precisely how to identify which maps are left invertible and which are right invertible. We remark that one often finds that in attempting to prove that a bijective mapping is indeed bijective, the proof of injectivity is easier to obtain than the proof of surjectivity. The following results should help to explain why this might be expected.

3.3.14 Proposition. *Let A and B be sets, and $f : A \to B$ be a map. Then f is left invertible if, and only if, f is injective.*

Proof. Firstly, assume that f is left invertible, and that $g : B \to A$ is a left inverse of f, so $g \circ f = \mathbb{1}_A$. We wish to prove that f is injective, so suppose that $x, y \in A$ are such that $f(x) = f(y)$. Then $x = \mathbb{1}_A(x) = g \circ f(x) = g(f(x)) = g(f(y)) = g \circ f(y) = \mathbb{1}_A(y) = y$, so f is injective. Conversely, suppose that f is injective. We consider two cases: f surjective, and f not surjective. If f is surjective, then for each $b \in B$, there exists at least one $a \in A$ such that $f(a) = b$, and since f is injective, such a is unique. Thus if f is surjective, $g \subseteq B \times A$ defined by $g = \{(b, a) \in B \times A \mid f(a) = b\}$ is a map from B to A. Moreover, for each $a \in A$, $(f(a), a) \in g$, so $g(f(a)) = a$; that is, for each $a \in A$, $g \circ f(a) = a = \mathbb{1}_A(a)$, so $g \circ f = \mathbb{1}_A$ and thus g is a left inverse for f. On the other hand, if f is not surjective, then choose $a_0 \in A$ and define

$$g = \{(b, a) \in f(A) \times A \mid f(a) = b\} \cup ((B - f(A)) \times \{a_0\}).$$

Again, g is a map from B to A, and for each $a \in A$, $g(f(a)) = a$, so $g \circ f = \mathbb{1}_A$ and g is a left inverse for f. $\square$

One might expect an analogous result characterizing right invertible maps as the surjective maps, but here the story gets interesting.

3.3.15 Proposition. *Let A and B be sets and let $f : A \to B$ be a map. If f is right invertible, then f is surjective.*

Proof. Suppose that f is right invertible, with right inverse $g : B \to A$, so $f \circ g = \mathbb{1}_B$. Let $y \in B$. Then $y = \mathbb{1}_B(y) = f \circ g(y) = f(g(y))$, so $x = g(y) \in A$ has the property that $f(x) = y$. Thus f is surjective. $\square$

Now for the interesting part. Suppose that f is surjective. For each $b \in B$, let $X_b = f^{-1}(b)$. Then X_b is nonempty (since f is surjective) and $X_b \subseteq A$. For each $b \in B$, choose an element from X_b and label it as a_b, so for each $b \in B$, $a_b \in X_b$. Set $g = \{\, (b, a_b) \in B \times A \mid b \in B \,\}$. Then g is a map from B to A, and for each $b \in B$, $f \circ g(b) = f(g(b)) = f(a_b) = b$ since $a_b \in X_b = f^{-1}(b)$. Thus $f \circ g = \mathbb{1}_B$, and so g is a right inverse for f. The key to our success lies in our (assumed) ability to choose an element from X_b for each $b \in B$. The problem is that if B is infinite, it cannot be proven that such a choice is possible (for B finite, the construction works just fine).

Thus for B infinite, the construction requires an additional axiom of set theory, known as the Axiom of Choice.

3.3.16 Axiom (of Choice). *Every surjective map is right invertible.*

We shall assume the Axiom of Choice in this text.

3.3.17 Example. Let $A = \{\, 1, 2, 3, 4 \,\}$, $B = \{\, 1, 2, 3 \,\}$, and $f : A \to B$ be the map $f = \{\, (1, 2), (2, 3), (3, 1), (4, 2) \,\}$. Then $f(1) = 2$, $f(2) = 3$, $f(3) = 1$, and $f(4) = 2$, so $Im(f) = \{\, f(a) \mid a \in A \,\} = \{\, f(1), f(2), f(3), f(4) \,\} = \{\, 1, 2, 3 \,\}$, and so by definition, f is surjective. We have $X_1 = f^{-1}(1) = \{\, 3 \,\}$, $X_2 = f^{-1}(2) = \{\, 1, 4 \,\}$, and $X_3 = f^{-1}(3) = \{\, 2 \,\}$. There is only one way to choose one element from a one-element set, but there are two ways to make a choice of an element from X_2. Thus $g_1 = \{\, (1, 3), (2, 1), (3, 2) \,\}$ and $g_2 = \{\, (1, 3), (2, 4), (3, 2) \,\}$ are each right inverses for f (and they are the only right inverses for f).

3.3.18 Corollary. *Let A, B, and C be sets, and let $f : A \to B$ and $g : B \to C$ be maps. Then $g \circ f$ is injective if both f and g are injective, and $g \circ f$ is surjective if both f and g are surjective.*

Proof. Suppose that f and g are injective. Then by Proposition 3.3.14, f has a left inverse f_1, so $f_1 \circ f = \mathbb{1}_A$, and g has a left inverse g_1, so $g_1 \circ g = \mathbb{1}_B$. Then $f_1 \circ g_1 : C \to A$ satisfies $f_1 \circ g_1 \circ g \circ f = f_1 \circ \mathbb{1}_B \circ f = f_1 \circ f = \mathbb{1}_A$, and so $f_1 \circ g_1$ is a left inverse for $g \circ f$. By Proposition 3.3.14, $g \circ f$ is injective. A similar argument proves that if both f and g are surjective, then $g \circ f$ is surjective.   □

3.3.19 Corollary. *Let A and B be sets, and $f : A \to B$ be a map. Then f is both left and right invertible if, and only if, f is bijective. Moreover, if f is bijective, then f has a unique left inverse and a unique right inverse, and they are the same map.*

Proof. The first statement follows immediately from Propositions 3.3.14 and 3.3.15. As for the second, suppose that f is bijective. By Proposition 3.3.14, f has a left inverse g, and by the Axiom of Choice, f has a right inverse h. Then $g, h : B \to A$ and both $g \circ f = \mathbb{1}_A$ and $f \circ h = \mathbb{1}_B$. We have $g = g \circ \mathbb{1}_B = g \circ (f \circ h) = (g \circ f) \circ h = \mathbb{1}_A \circ h = h$.   □

3.3.20 Definition. *Let A and B be sets and $f : A \to B$ a bijective map. Then the unique left and right inverse $g : B \to A$ for f (that is, the unique map g for which $g \circ f = \mathbb{1}_A$ and $f \circ g = \mathbb{1}_B$) is called the inverse of f, denoted by f^{-1}. We also say that f is invertible with inverse f^{-1}.*

Note that if $f : A \to B$ is bijective, then $f^{-1} : B \to A$ is also a map.

3.3.21 Proposition. *Let A and B be sets and $f : A \to B$ be a bijective map. Then $f^{-1} = \{\, (b, a) \in B \times A \mid b = f(a) \,\} = \{\, (b, a) \in B \times A \mid (a, b) \in f \,\}$.*

Proof. Since $f^{-1} \circ f = \mathbb{1}_A$, we have for each $a \in A$, $a = f^{-1}(f(a))$, so $(f(a), a) \in f^{-1}$. Thus $\{\, (f(a), a) \mid a \in A \,\} \subseteq f^{-1}$. For $(y, x) \in f^{-1}$, $f^{-1}(y) = x$ and so $f(x) = f \circ f^{-1}(y) = \mathbb{1}_B(y) = y$, which means that (y, x) is in $\{\, (b, a) \in B \times A \mid b = f(a) \,\}$, and so $f^{-1} \subseteq \{\, (b, a) \in B \times A \mid b = f(a) \,\}$. This proves that $f^{-1} = \{\, (f(a), a) \mid a \in A \,\}$.   □

When dealing with maps between finite sets, the following result is often of great value.

3.3.22 Proposition. *Let A and B be finite sets with $|A| = |B|$. Then a map $f : A \to B$ is injective if, and only if, it is surjective.*

Proof. Since $f(A) \subseteq B$ and B is finite, f is surjective (that is, $f(A) = B$) if, and only if, $|f(A)| = |B|$. Since A is finite, f is injective if, and only if, $|A| = |f(A)|$. Since $|f(A)| \leq |B| = |A|$, it follows that $|A| = |f(A)|$ if, and only if, $|f(A)| = |B|$, and so f is injective if, and only if, f is surjective. $\square$

Let us recap what we have done. We started with a discussion of reversible mathematical computations. We can now see that a map f which has a left inverse g is "reversible" in the sense that, if we are told that b is in the image of f, then we can "undo" the effect of applying f and retrieve the pre-image of b under f. Indeed, since $g \circ f = \mathbb{1}_A$, the required element is $g(b)$. For suppose that $x \in A$ and $f(x) = b$. Then $g(f(x)) = g(b)$, so $g(b) = g \circ f(x) = \mathbb{1}_A(x) = x$.

Binary Operations

Our next and final topic in this chapter is binary operations. The notion of a binary operation on an arbitrary set A is motivated by the operations of ordinary addition and multiplication of real numbers.

Let us examine what is involved in these operations on the set of real numbers with a view to distilling the essence of the idea. For addition, we have a "rule" which allows us to associate with a given ordered pair (a, b) of real numbers (a first input and a second input) a real number c (their sum) denoted by $c = a + b$. We can think of addition as a map $+ : \mathbb{R} \times \mathbb{R} \to \mathbb{R}$ where instead of writing $+(a, b)$ for the image of (a, b) under $+$, we write instead $a + b$. A similar analysis shows that ordinary multiplication of real numbers may also be viewed as a map $\times : \mathbb{R} \times \mathbb{R} \to \mathbb{R}$, where again, instead of writing $\times(a, b)$, we usually simply write ab.

Bearing this discussion in mind, we arrive at the following definition.

3.4.1 Definition. *Let A be a set. A binary operation on A is a map $*$ from $A \times A$ to A.*

Instead of writing $*(a, b)$ for the image of (a, b) under $*$, we write $a * b$, or often simply ab. When the image of (a, b) under an operation $*$ is written as ab, we say that the operation is denoted by juxtaposition or that it is denoted multiplicatively, and we say that we are using multiplicative notation. We also say that ab is the product of a and b. Note that we are not saying that ab is the product of real numbers a and b, but rather we are simply saying that we have decided to denote $a * b$ by ab.

We offer a very simple example of a binary operation (note that the idea is applicable to any set A).

3.4.2 Example. *Let A be a set, and define $* : A \times A \to A$ by $a * b = a$ for all $a, b \in A$.*

Our next example is fundamental, and we shall study it extensively in subsequent sections.

3.4.3 Definition. *For any sets S and T, let $\mathscr{F}(S,T)$ denote the set of all maps from S to T, and for convenience, let $\mathscr{F}(S)$ denote $\mathscr{F}(S,S)$.*

3.4.4 Example. *Let S be a set, and let $A = \mathscr{F}(S)$. Then composition of maps provides a binary operation on A; that is, for $f, g \in A$, $\circ(f, g) = f \circ g \in A$. In this context, we will often write fg rather than $f \circ g$.*

Let us look at the specific situation of $S = \{1, 2, 3, 4\}$ in Example 3.4.4. Consider $f \in \mathscr{F}(\{1, 2, 3, 4\})$ defined by $f(1) = 3$, $f(2) = 4$, $f(3) = 4$, and $f(4) = 1$. As described in Example 3.3.2, we may use a two-row matrix to represent f as $\left(\begin{smallmatrix} 1 & 2 & 3 & 4 \\ 3 & 4 & 4 & 1 \end{smallmatrix}\right)$.

Sets of the form $\{1, 2, 3\}$ or $\{1, 2, 3, 4, 5\}$ have appeared often in this chapter, and they are used throughout mathematics in general, so it is convenient to introduce a more compact notation for them.

3.4.5 Notation. *For any positive integer n, let $J_n = \{i \in \mathbb{Z}^+ \mid i \leq n\}$.*

Thus for example, $J_2 = \{1, 2\}$, and $J_5 = \{1, 2, 3, 4, 5\}$.

3.4.6 Notation. *In the special case of maps from J_n to a set X, we shall use the two-row matrix notation, but we shall consider the first row to be understood to be $1, 2, \ldots, n$, and so we shall simply write the second row, surrounded by square brackets. For example, the map $\left(\begin{smallmatrix} 1 & 2 & 3 & 4 \\ 3 & 4 & 4 & 1 \end{smallmatrix}\right)$ would be written as*

$$[3\,4\,4\,1].$$

We continue our discussion of $\mathscr{F}(J_4)$. Let us compute the product (that is, the composition) $[3\,4\,4\,1][2\,3\,1\,3]$. Recall that in the composition of two maps $g \circ f$, f is the first map to be applied, so for each input a, $g \circ f(a) = g(f(a))$. We have

$$[3\,4\,4\,1][2\,3\,1\,3](1) = [3\,4\,4\,1]([2\,3\,1\,3](1)) = [3\,4\,4\,1](2) = 4,$$

and similarly,

$$[3\,4\,4\,1][2\,3\,1\,3](2) = 4,$$
$$[3\,4\,4\,1][2\,3\,1\,3](3) = 3, \text{ and}$$
$$[3\,4\,4\,1][2\,3\,1\,3](4) = 4,$$

so

$$[3\,4\,4\,1][2\,3\,1\,3] = [4\,4\,3\,4].$$

Note that $[2\,3\,1\,3][3\,4\,4\,1] = [1\,3\,3\,2]$, so $[3\,4\,4\,1][2\,3\,1\,3] \neq [2\,3\,1\,3][3\,4\,4\,1]$. Thus unlike, for example, addition of real numbers, or multiplication of real numbers, the result of a composition of maps does in general depend on the order of the arguments. However, just as for addition or multiplication of real numbers, map composition is associative (as proven in Proposition 3.3.11). Moreover, the identity mapping $\mathbb{1}_S$ plays a role for composition of maps that is analogous to that of 0 for addition of real numbers, or 1 for multiplication of real numbers, in that for any $f \in \mathscr{F}(A)$, $f\mathbb{1}_A = \mathbb{1}_A f = f$ (as proven in Proposition 3.3.12).

These observations suggest that such properties might be of interest in the study of binary operations, and so we introduce the following concepts.

3.4.7 Definition. *Let S be a set, and let $*$ be a binary operation on S.*

(i) *$*$ is said to be associative if for all $a, b, c \in S$, $(a*b)*c = a*(b*c)$.*
(ii) *$*$ is said to be commutative if for all $a, b \in S$, $a*b = b*a$.*
(iii) *An element $e \in S$ is said to be a left (respectively right) identity for $*$ if for all $a \in S$, $e*a = a$ (respectively, $a*e = a$).*
(iv) *$e \in S$ is said to be an identity for $*$ if e is both a left identity for $*$ and a right identity for $*$.*
(v) *An element $z \in S$ is said to be a left (respectively right) zero for $*$ if for all $a \in S$, $z*a = z$ (respectively $a*z = z$).*
(vi) *$z \in S$ is said to be a zero for $*$ if it is both a left and a right zero for $*$.*

Thus for example, the operation composition of maps on $\mathscr{F}(J_4)$ is associative, not commutative, and has identity $\mathbb{1}_{J_4}$. It has four left zeroes; namely $[1\,1\,1\,1]$, $[2\,2\,2\,2]$, $[3\,3\,3\,3]$, and $[4\,4\,4\,4]$, but no right zero at all. To see this, observe that if f is a right zero, then for every $g \in \mathscr{F}(J_4)$, $g \circ f = f$; that is, $g(f(i)) = f(i)$ for every $i = 1, 2, 3, 4$. In particular, set $j = f(1)$. Then for every $g \in \mathscr{F}(J_4)$, $g(j) = j$. Since this is not the case (there will be many g for which $g(j) \neq j$; for example, if $i \in J_4$, $i \neq j$, then for $g = [i\,i\,i\,i]$, we have $g \circ f(1) = g(j) = i \neq j$, so $g \circ f \neq f)$, we see that there can be no right zero. Since there is no right zero, then there is no zero either.

The set of all maps fron one set to another can be given a binary operation in a natural way if the target set has a binary operation defined on it.

3.4.8 Proposition. *Let X and Y be sets, and suppose that Y has a binary operation $*$ defined on it. Define a binary operation $\star$ on $\mathcal{F}(X,Y)$ as follows: for $f, g \in \mathcal{F}(X,Y)$, let $(f \star g)(x) = f(x) * g(x)$ for all $x \in X$. Then the following hold:*

- *(i) $\star$ is associative, respectively commutative, if $*$ is associative, respectively commutative.*
- *(ii) If $*$ has identity e, then $\star$ has identity $\hat{e}$, where $\hat{e} \in \mathcal{F}(X,Y)$ is the map for which $\hat{e}(x) = e$ for all $x \in X$.*

Proof. For (i), suppose that $*$ is associative, and let $f, g, h \in \mathcal{F}(X,Y)$. Then for any $x \in X$, we have $f \star (g \star h)(x) = f(x) * (g \star h)(x) = f(x) * (g(x) * h(x)) = (f(x) * g(x)) * h(x) = (f \star g)(x) * h(x) = (f \star g) \star h(x)$, and so $f \star (g \star h) = (f \star g) \star h$. Similarly, if $*$ is commutative, then $f \star g(x) = f(x) * g(x) = g(x) * f(x) = g \star f(x)$ and so $f \star g = g \star f$.

For (ii), suppose that $*$ has identity $e \in Y$. Define $\hat{e} : X \to Y$ by $\hat{e}(x) = e$ for all $x \in X$. Let $f \in \mathcal{F}(X,Y)$. Then for any $x \in X$, we have $f \star \hat{e}(x) = f(x) * \hat{e}(x) = f(x) * e = f(x)$, so $f \star \hat{e} = f$, and as well, $\hat{e} \star f(x) = \hat{e}(x) * f(x) = e * f(x) = f(x)$, so $\hat{e} \star f = f$. Thus $\hat{e}$ is an identity for $\star$. $\square$

We also observe that for any set S, there are some interesting subsets of $\mathcal{F}(S)$.

3.4.9 Example. *Let S be a set. Then $f \in \mathcal{F}(S)$ is called a constant map if for all $i, j \in S$, $f(i) = f(j)$. For instance, in $\mathcal{F}(J_4)$, there are four constant maps; $[1\,1\,1\,1]$, $[2\,2\,2\,2]$, $[3\,3\,3\,3]$, and $[4\,4\,4\,4]$. Note that these are the left zeros of the operation composition of maps on $\mathcal{F}(J_4)$. This is true in general, an element of $\mathcal{F}(S)$ is a left zero for composition if, and only if, it is a constant map. Note that in particular, the set of all constant maps has the property that the composition of constant maps f and g is again a constant map.*

3.4.10 Example. *Let A be a set. Then $f \in \mathcal{F}(A)$ is called a permutation of A if f is bijective. Note that by Corollary 3.3.18, the composition of* *permutations of A is again a permutation of A. The set of all permutations of A is usually denoted by S_A.*

In each of the preceding two examples, we encountered interesting subsets of a set with a binary operation, and in each case, the subsets exhibited the property that for any choice of two elements from the subset, their product was again an element of the subset in question.

3.4.11 Definition. *Let S be a set, and let $*$ be a binary operation on S. A subset X of S is said to be closed under $*$ if for all $x, y \in X$, $x * y \in X$.*

The importance of this notion is that if S is a set with a binary operation $*$, and X is a subset of S that is closed under $*$, then $*$ naturally induces a binary operation on X (and which we will continue to denote by $*$). Moreover, if $*$ is associative, then the induced operation on X is associative, and if $*$ is commutative, then the induced operation on X is commutative. For example, for any set A, composition of maps provides an associative (but in general, non-commutative) binary operation on S_A.

We shall find many opportunities to utilize this observation in the work to come. For now, we continue to introduce other examples of binary operations to help us understand the concepts introduced in Definition 3.4.7.

3.4.12 Example. *Let S be a set. On S, define a binary operation $*$ by $a * b = b$ for all $a, b \in S$. Then $*$ is associative, and every element of S is both a left identity and a right zero for $*$. Note that if $|S| \geq 2$, then $*$ is not commutative.*

3.4.13 Example. *On $\mathbb{Z}$, define a binary operation $*$ by $a * b = (2 + a)(2 + b) - 2$ for all $a, b \in \mathbb{Z}$, where we have used ordinary addition and multiplication in our expression of $a*b$. Then $*$ is associative, commutative, has identity -1, and zero -2.*

In the event that a set S is finite (well, small finite for practicality), then a binary operation on S can be represented by an operation table. An operation table has one row for each element of the set, and each row is labelled by the element to which it corresponds. Similarly, the table has one column for each element of the set, and each column is labelled by the element to which it corresponds. Then for $x, y \in S$, the value of $x * y$ is recorded at the intersection of the row labelled x and the column labelled y.

3.4.14 Example. *On the set* $\{0,1\}$, *define* AND *by the operation table*

AND	0	1
0	0	0
1	0	1

Then AND *is associative, commutative, has zero* 0 *and identity* 1.

As well, define OR by the operation table

OR	0	1
0	0	1
1	1	1

Then OR *is associative, commutative, has identity* 0, *and zero* 1.

Also, define NOT $: \{0,1\} \rightarrow \{0,1\}$ *by* NOT$(0) = 1$ *and* NOT$(1) = 0$. *Finally, define the binary operation* NOT-AND, *or* NAND, *by* NAND$(a,b) =$ NOT(AND(a,b)) *for any* $a,b \in \{0,1\}$. *Then* NAND *is commutative but not associative. Its operation table is*

NAND	0	1
0	1	1
1	1	0

Note that while NAND *is not associative, it is a very important logical operation in that each of* NOT, AND, *and* OR *can be expressed solely in terms of* NAND. *For example,* NOT$(a) =$ NAND(a,a).

In general, we shall be interested in sets with one or more binary operations defined on them.

3.4.15 Definition. *An algebraic system consists of a set* S, *and one or more binary operations on* S. *If we wish to denote the algebraic system with set* S *and operations* $*_1, *_2, \ldots$, *we shall write* $(S, *_1, *_2, \ldots)$. *We shall only allow finitely many operations to be specified, and by far the most common situations involve algebraic systems with one or two specified binary operations.*

For example, using · to denote multiplication on $\mathbb{Z}$, the set of integers with its binary operations of addition and multiplication would be denoted by $(\mathbb{Z}, +, \cdot)$ or $(\mathbb{Z}, \cdot, +)$.

The simplest algebraic systems are those with a single binary operation. Although there is not much that can be said about such systems without requiring some nice properties (such as those listed in Definition 3.4.7), we can obtain one very useful result.

3.4.16 Proposition. *Let $(S, *)$ be an algebraic system, and suppose that $*$ has a left identity e and a right identity f. Then $e = f$ and e is an identity for $*$. Moreover, e is the unique identity for $*$.*

Proof. We have $e = e * f = f$ since f is a right identity and e is a left identity. As a result, e is both a left and a right identity for $*$, hence an identity for $*$. If e' is an identity for $*$, then e' is a left identity for $*$, and thus $e' = f = e$. $\square$

The next concept is motivated by the characterizations of injective and surjective maps (see Proposition 3.3.14, Proposition 3.3.15, and the Axiom of Choice).

3.4.17 Definition. *Let $(S, *)$ be an algebraic system with identity e (by Proposition 3.4.16, e is the unique identity for $*$).*

   (i) For any $x \in S$, an element $y \in S$ is called a left inverse of x if $y * x = e$.
   (ii) For any $x \in S$, an element $y \in S$ is called a right inverse of x if $x * y = e$.
   (iii) For any $x \in S$, an element $y \in S$ is called an inverse of x if y is both a left and a right inverse of x. If x has an inverse in S, then x is said to be invertible.

3.4.18 Example. *Let $S = J_4$, and let $*$ be the binary operation on S whose operation table is shown below.*

$*$	1	2	3	4
1	1	2	3	4
2	2	4	1	1
3	3	1	4	2
4	4	1	2	3

Note that $*$ is commutative, has identity 1, and both 3 and 4 are inverses for 2. As it turns out, $*$ is not associative. For consider $2 * (2 * 4) = 2 * 1 = 2$, while $(2 * 2) * 4 = 4 * 4 = 3$, so $2 * (2 * 4) \neq (2 * 2) * 4$. As we see next, if $*$ had been associative, then any element that had an inverse would have had a unique inverse.

3.4.19 Proposition. *Let $(S, *)$ be an algebraic system such that $*$ is associative and has identity e. If $x \in S$ has a left inverse y and a right inverse z, then $y = z$, so y is an inverse for x, and in fact, y is the unique inverse for x.*

Proof. Suppose that $x, y, z \in S$ are such that $y * x = e = x * z$. Then $y = y * e = y * (x * z) = (y * x) * z = e * z = z$. If $t \in X$ is an inverse for x, then t is a right inverse for x and so $y = t$. Thus y is the unique inverse for x. $\square$

The preceding proposition, together with the fact that many of our familiar binary operations are associative, suggests that associative binary operations are worthy of further study. Accordingly, we shall introduce terminology to indicate that the binary operation in a single binary operation algebraic system is associative.

3.4.20 Definition. *An algebraic system $(S, *)$ is called a semigroup if $*$ is associative. Additionally, if $*$ has an identity, then $(S, *)$ is called a monoid.*

We should observe that even in monoids, an element may have several left inverses, or several right inverses (however, if an element has both left and right inverses, then it has exactly one left inverse and exactly one right inverse, and they are equal). The next example illustrates this phenomenon.

3.4.21 Example. *As we have seen, for any set S, $(\mathscr{F}(S), \circ)$ is a monoid since composition of maps is associative and $\mathbb{1}_S$ is the identity element for the binary operation of composition. Thus $(S, \circ)$ is a monoid. Furthermore, by Proposition 3.3.14, an element $f \in \mathscr{F}(S)$ has a left inverse if and only if f is injective, while by Proposition 3.3.15 and the Axiom of Choice, f has a right inverse if and only if f is surjective. If S is a finite set, then an element $f \in \mathscr{F}(S)$ is injective if and only if it is surjective, but the situation is radically changed when S is infinite. For example, consider $(\mathscr{F}(\mathbb{Z}), \circ)$. Let $f \in \mathscr{F}(\mathbb{Z})$ be the map defined by $f(n) = 2n$ for each $n \in \mathbb{Z}$. Let $h \in \mathscr{F}(\mathbb{Z})$ be chosen arbitrarily. Then define $g \in \mathscr{F}(\mathbb{Z})$ by*

$$g(n) = \begin{cases} n/2 & \text{if } n \text{ is even} \\ h(n) & \text{if } n \text{ is odd.} \end{cases}$$

Then $g \circ f(n) = g(2n) = n$ for each $n \in \mathbb{Z}$, so $g \circ f = \mathbb{1}_{\mathbb{Z}}$ and thus g is a left inverse for f. Since there are infinitely many choices for h, there are infinitely many left inverses for f (note that since $\circ$ is associative, if f had a right inverse, then it would have a unique left inverse, equal to the right inverse, so f does not have a right inverse – that is, f is injective but not surjective).

The fact that an invertible element in a monoid $(S, *)$ has a unique inverse allows us to intoduce notation for this unique inverse. The notation is modelled on that used in our familiar system of real numbers, but as there are two familiar operations on $\mathbb{R}$; namely addition and multiplication, there are two different conventions for denoting the inverse of an invertible element in a general monoid, depending on whether or not we are writing the result $a * b$ in multiplicative fashion as simply ab, or in additive fashion, by which we mean that we use the $+$ sign to denote the result of $a * b$, writing instead $a + b$. In the former case, we say that we are using multiplicative notation, while in the latter, we say that we are using additive notation. It is important to realize that these are just notational conventions, and one may decide to use one convention initially, then perhaps later on, decide that the other notation would have been preferable (for whatever reason), and so all results could be rewrittin in the other notation. In this book, we shall default to multiplicative notation, and we shall remind the reader when we arc using a result that was proven with multiplicative notation, but the usage occurs in a situation where additive notation was being used.

3.4.22 Notation. *Let $(S, *)$ be a monoid. If $x \in S$ is invertible, the unique inverse of x is denoted by x^{-1} in multiplicative notation, while if additive notation is being used, the unique inverse of x is denoted by $-x$.*

It is only reasonable to use additive notation for the operation of addition on $\mathbb{N}$, $\mathbb{Z}$, $\mathbb{Q}$, and $\mathbb{R}$, and multiplicative notation for the operation of multiplication on $\mathbb{N}$, $\mathbb{Z}$, $\mathbb{Q}$, and $\mathbb{R}$.

3.4.23 Proposition. *Let $(S, *)$ be a monoid with identity e. Then the following hold.*

(i) If $a \in S$ is invertible, then a^{-1} is invertible, and $(a^{-1})^{-1} = a$.
(ii) If $a, b \in S$ are invertible, then ab is invertible, and $(ab)^{-1} = b^{-1}a^{-1}$.

Proof. Let $a \in S$ be invertible. Then a^{-1} exists, and $aa^{-1} = e = a^{-1}a$. Thus by Proposition 3.4.19, a^{-1} has inverse a.

Next, let $a, b \in S$ be invertible. We must prove that $ab(b^{-1}a^{-1}) = e = b^{-1}a^{-1}ab$. We have $ab(b^{-1}a^{-1}) = a(bb^{-1})a^{-1} = aea^{-1} = aa^{-1} = e$, and $b^{-1}a^{-1}(ab) = b^{-1}(a^{-1}a)b = b^{-1}eb = b^{-1}b = e$ as well. Thus by Proposition 3.4.19, ab is invertible with inverse $b^{-1}a^{-1}$. $\qquad\square$

3.4.24 Example. *On the set $\mathbb{R} \times \mathbb{R}$, define a binary operation that we shall call addition (and therefore we shall use additive notation to denote it) by $(a, b) + (c, d) = (a + c, b + d)$ for all $(a, b), (c, d) \in \mathbb{R} \times \mathbb{R}$. Note that we have*

*used addition on $\mathbb{R}$ in each coordinate calculation. The reader is invited
to prove that this operation is associative, commutative, and has identity
$(0,0)$. We shall prove that each $(a,b) \in \mathbb{R} \times \mathbb{R}$ is invertible, and moreover,
$-(a,b) = (-a,-b)$. For let $(a,b) \in \mathbb{R} \times \mathbb{R}$. Then $(-a,-b) \in \mathbb{R} \times \mathbb{R}$ and we
have by definition of addition that $(a,b) + (-a,-b) = (a-a, b-b) = (0,0)$,
so $(-a,-b)$ is a right inverse for (a,b). Since the operation is commutative,
any right inverse for (a,b) is also a left inverse for (a,b), and thus $-(a,b)$,
which by definition is the unique element of $\mathbb{R} \times \mathbb{R}$ whose sum with (a,b) is
$(0,0)$, is $(-a,-b)$.*

Our next definition introduces a family of semigroups that are central
to much of mathematics, and will appear throughout this text in a variety
of contexts.

3.4.25 Definition. *Let $R = \mathbb{Z}$, $\mathbb{Q}$, $\mathbb{R}$, or $\mathbb{C}$. Then for any positive integers
m and n, an $m \times n$ matrix with entries from R is a rectangular array with
m rows and n columns whose entries come from R. The set of all such
matrices is denoted by $M_{m \times n}(R)$. For example, $\left[\begin{smallmatrix} 1 & 2 & 3 \\ 4 & 5 & 6 \end{smallmatrix}\right] \in M_{2\times 3}(\mathbb{Z})$. For any
$m \times n$ matrix A with entries from R, and any integers i and j with $1 \le i \le m$
and $1 \le j \le n$, the element of R that is in the ith row and j^{th} column of A
is denoted by A_{ij}. Define a binary operation on $M_{m \times n}(R)$ called addition
and denoted by $+$ as follows. For any $A, B \in M_{m \times n}(R)$, $A + B$ is the $m \times n$
matrix for which for any integers i and j with $1 \le i \le m$ and $1 \le j \le n$,
$(A+B)_{ij} = A_{ij} + B_{ij}$. Note that A_{ij} and B_{ij} are elements of R, and
we have used the addition operation in R to define the addition operation
in $M_{m \times n}(R)$. For example, if $A = \begin{bmatrix} -1 & 2 & 1 \\ 3 & \frac{1}{2} & 2 \end{bmatrix}$ and $B = \begin{bmatrix} -\frac{1}{2} & \frac{1}{2} & 2 \\ 1 & 0 & -1 \end{bmatrix}$, then
$A, B \in M_{2 \times 3}(\mathbb{Q})$ and*

$$A + B = \begin{bmatrix} -\frac{3}{2} & \frac{5}{2} & 3 \\ 4 & \frac{1}{2} & 1 \end{bmatrix}.$$

*Given positive integers m, n, and k, we also define a map from $M_{m \times n}(R) \times
M_{n \times k}(R)$ to $M_{m \times k}(R)$, called matrix multiplication, as follows. Let $A \in
M_{m \times n}(R)$ and $B \in M_{n \times k}(R)$. Then the matrix product AB is the $m \times k$
matrix with entries from R for which for each i and j, with $1 \le i \le m$ and
$1 \le j \le k$,*

$$(AB)_{ij} = \sum_{r=1}^{n} A_{ir} B_{rj}.$$

*If $m = n = k$, then matrix multiplication provides a binary operation on
$M_{n \times n}(R)$ (a matrix with the same number of rows as columns is called a
square matrix).*

3.4.26 Proposition. *Let $R = \mathbb{Z}$, $\mathbb{Q}$, or $\mathbb{R}$, and let m and n be positive integers. Then $(M_{m \times n}(R), +)$ is a commutative monoid with identity, and every element is invertible. Specifically, the following hold.*

(i) *For all $A, B \in M_{m \times n}(R)$, $A + B = B + A$.*

(ii) *For all $A, B, C \in M_{m \times n}(R)$, $A + (B + C) = (A + B) + C$.*

(iii) *The $m \times n$ matrix $\mathbb{0}_{m \times n}$ for which every entry is equal to 0 is the identity element for addition.*

(iv) *For each $A \in M_{m \times n}(R)$, $-A$ is the matrix defined by $(-A)_{ij} = -A_{ij}$ for each i and j with $1 \le i \le m$ and $1 \le j \le n$.*

Proof. Let $A, B, C \in M_{m \times n}(R)$.

For commutativity of addition, we observe that for any integers i and j with $1 \le i \le m$ and $1 \le j \le n$, $(A+B)_{ij} = A_{ij} + B_{ij} = B_{ij} + A_{ij} = (B+A)_{ij}$ since addition in R is commutative, and so $A + B = B + A$.

For associativity, we must prove that $A + (B + C) = (A + B) + C$. This holds if and only if for all integers i and j with $1 \le i \le m$ and $1 \le j \le n$, $(A + (B + C))_{ij} = ((A + B) + C)_{ij}$. Let i and j be integers for which $1 \le i \le m$ and $1 \le j \le n$. Then

$$
\begin{aligned}
(A + (B + C))_{ij} &= A_{ij} + (B + C)_{ij} = A_{ij} + (B_{ij} + C_{ij}) \\
&= (A_{ij} + B_{ij}) + C_{ij}) \quad \text{since addition in } R \text{ is associative} \\
&= (A + B)_{ij} + C_{ij} \\
&= (A + (B + C))_{ij}
\end{aligned}
$$

and so $A + (B + C) = (A + B) + C$.

Let $A \in M_{m \times n}(R)$. Then for any integers i and j for which $1 \le i \le m$ and $1 \le j \le n$, we have $(A + \mathbb{0}_{m \times n})_{ij} = A_{ij} + (\mathbb{0}_{m \times n})_{ij} = A_{ij} + 0 = A_{ij}$, so $A + \mathbb{0}_{m \times n} = A$. Since addition is commutative, this means that $\mathbb{0}_{m \times n} + A = A$ as well, and thus $\mathbb{0}_{m \times n}$ is the identity element for addition.

Finally, let $A \in M_{m \times n}$, and let $B \in M_{m \times n}(R)$ be the matrix defined by $B_{ij} = -A_{ij}$ for each i and j with $1 \le i \le m$ and $1 \le j \le n$.. Then for all integers i and j for which $1 \le i \le m$ and $1 \le j \le n$, we have $(A + B)_{ij} = A_{ij} + B_{ij} = A_{ij} + (-A_{ij}) = 0$ and so $A + B = \mathbb{0}_{m \times n}$. Since addition is commutative, we have $B + A = \mathbb{0}_{m \times n}$ as well, and thus B is the inverse for A with respect to the operation of addition. $\qquad \square$

Matrix multiplication is very interesting, and the reader is encouraged to read Appendix B to gain insight as to the origins of the concept.

3.4.27 Proposition. *Let $R = \mathbb{Z}$, $\mathbb{Q}$, or $\mathbb{R}$, and let m, n, k, p be positive integers. Then the following hold.*

(i) For any $A \in M_{m \times n}(R)$, $B \in M_{n \times k}(R)$, and $C \in M_{k \times p}(R)$,
     $A(BC) = (AB)C$.

(ii) For any $A, B \in M_{m \times n}(R)$, and any $C \in M_{n \times k}$, $(A + B)C = AC + BC$.

(iii) For any $A, B \in M_{n \times k}(R)$, and any $C \in M_{m \times n}$, $C(A + B) = CA + CB$.

(iv) For any $A \in M_{n \times k}(R)$, $A\mathbb{0}_{k \times p} = \mathbb{0}_{n \times p}$, and $\mathbb{0}_{m \times n}A = \mathbb{0}_{m \times k}$.

Proof. Let $A \in M_{m \times n}(R)$, $B \in M_{n \times k}(R)$, and $C \in M_{k \times p}(R)$. Then $AB \in M_{m \times k}(R)$, so $(AB)C \in M_{m \times p}$, and $BC \in M_{n \times p}$, so $A(BC) \in M_{m \times p}$. To prove that $A(BC) = (AB)C$, we must prove that for any integers i and j for which $1 \leq i \leq m$ and $1 \leq j \leq p$, $(A(BC))_{ij} = ((AB)C)_{ij}$. Let i and j be such integers. Then since in R, multiplication distributes over addition, and multiplication is associative, we have

$$(A(BC))_{ij} = \sum_{r=1}^{n} A_{ir}(BC)_{rj} = \sum_{r=1}^{n} A_{ir}\left(\sum_{t=1}^{k} B_{rt}C_{tj}\right)$$

$$= \sum_{r=1}^{n}\left(\sum_{t=1}^{k} A_{ir}(B_{rt}C_{tj})\right) = \sum_{r=1}^{n}\left(\sum_{t=1}^{k}(A_{ir}B_{rt})C_{tj}\right).$$

Then since addition in R is associative and commutative, we obtain

$$\sum_{r=1}^{n}\left(\sum_{t=1}^{k}(A_{ir}B_{rt})C_{tj}\right) = \sum_{t=1}^{k}\left(\sum_{r=1}^{n}(A_{ir}B_{rt})C_{tj}\right)$$

and so

$$(A(BC))_{ij} = \sum_{t=1}^{k}\left(\sum_{r=1}^{n}(A_{ir}B_{rt})C_{tj}\right) = \sum_{t=1}^{k}\left(\sum_{r=1}^{n}A_{ir}B_{rt}\right)C_{tj}$$

$$= \sum_{t=1}^{k}(AB)_{it}C_{tj} = ((AB)C)_{ij},$$

as required.

Next, let $A, B \in M_{m \times n}(R)$, and $C \in M_{n \times k}(R)$. Then $A + B \in M_{m \times n}(R)$, and so $(A + B)C$, AC, and $BC \in M_{m \times k}(R)$. Let i and j be integers for which $1 \leq i \leq m$ and $1 \leq j \leq k$. Then

$$((A + B)C)_{ij} = \sum_{r=1}^{n}(A + B)_{ir}C_{rj} = \sum_{r=1}^{n}(A_{ir} + B_{ir})C_{rj}$$

$$= \sum_{r=1}^{m}(A_{ir}C_{rj} + B_{ir}C_{rj}) = \sum_{r=1}^{m}A_{ir}C_{rj} + \sum_{r=1}^{n}B_{ir}C_{rj}$$

$$= (AC)_{ij} + (BC)_{ij} = (AC + BC)_{ij},$$

and thus $(A+B)C = AC + BC$. The proof of (iii) is similar and is omitted.

Finally, let $A \in M_{n \times k}(R)$. Then $A\mathbb{0}_{k \times p} \in M_{n \times p}(R)$. For any integers i and j for which $1 \le i \le n$ and $1 \le j \le p$, we have $(A\mathbb{0}_{k \times p})_{ij} = \sum_{r=1}^{k} A_{ir}(\mathbb{0}_{k \times p})_{rj} = \sum_{r=1}^{k} A_{ir}0 = \sum_{r=1}^{k} 0 = 0$, so $A\mathbb{0}_{k \times p} = \mathbb{0}_{n \times p}$. The proof that $\mathbb{0}_{m \times n}A = \mathbb{0}_{m \times k}$ is similar and is omitted. □

3.4.28 Definition. *Let $R = \mathbb{Z}$, $\mathbb{Q}$, or $\mathbb{R}$, and let n be a positive integer. Then the $n \times n$ matrix I_n defined by*

$$(I_n)_{ij} = \begin{cases} 0 & \text{if } i \ne j \\ 1 & \text{if } i = j, \end{cases}$$

where i and j are integers for which $1 \le i, j \le n$, is called the $(n \times n)$ identity matrix.

The reason for calling I_n the identity matrix is apparent from the following result.

3.4.29 Proposition. *Let $R = \mathbb{Z}$, $\mathbb{Q}$, or $\mathbb{R}$, and let m, n be positive integers. Then for any $A \in M_{m \times n}(R)$, $I_m A = A = A I_n$.*

Proof. Let $A \in M_{m \times n}(R)$. Then $I_m A, A I_n \in M_{m \times n}(R)$. Let i and j be integers with $1 \le i \le m$ and $1 \le j \le n$. Then $(I_m A)_{ij} = \sum_{r=1}^{m}(I_m)_{ir} A_{rj} = A_{rj}$ since $(I_m)_{ir} = 0$ for all $r \ne i$, while $(I_m)_{ir} = 1$ when $r = i$. Similarly, $(A I_n)_{ij} = \sum_{r=1}^{n} A_{ir}(I_n)_{rj} = A_{ij}$. Thus $I_m A = A = A I_n$. □

3.4.30 Corollary. *Let $R = \mathbb{Z}$, $\mathbb{Q}$, or $\mathbb{R}$, and let n be a positive integer. Then $M_{n \times n}(R)$, with the operation of matrix multiplication, is a monoid with identity I_n.*

Proof. By Proposition 3.4.27, matrix multiplication is an associative binary operation on $M_{n \times n}(R)$, and by Proposition 3.4.29, I_n is the identity element for matrix multiplication. □

Note that by Proposition 3.4.27 (iv), $\mathbb{0}_{n \times n}$ is not invertible with respect to multiplication, so not every element of $M_{n \times n}(R)$ has a multiplicative inverse.

We conclude this section with a general semigroup result.

3.4.31 Proposition. *Let S be a semigroup, and define a binary operation on $\mathscr{P}(S)$ by*

$$XY = \{\, xy \mid x \in X,\ y \in Y \,\}$$

for $X, Y \in \mathscr{P}(S)$. Then the following hold:

(i) The binary operation on $\mathscr{P}(S)$ is associative, so $\mathscr{P}(S)$ is a semigroup with this binary operation.

(ii) The binary operation on $\mathscr{P}(S)$ has an identity if, and only if, the binary operation on S has an identity. If so, then the identity element in $\mathscr{P}(S)$ is $\{\,e\,\}$, where e is the identity element in S.

(iii) Suppose that S has identity e, so $\{\,e\,\}$ is the identity of $\mathscr{P}(S)$. Then $X \in \mathscr{P}(S)$ is invertible if, and only if, $|X| = 1$ and $X = \{\,a\,\}$ where a is an invertible element of S.

Proof. (i) Let $X, Y, W \in \mathscr{P}(S)$; that is, $X, Y, W \subseteq S$. Then

$$(XY)W = \{\,uw \mid u \in XY,\ w \in W\,\} = \{\,(xy)w \mid x \in X,\ y \in Y,\ w \in W\,\}$$
$$= \{\,x(yw) \mid x \in X,\ y \in Y,\ w \in W\,\} = \{\,xv \mid x \in X,\ v \in YW\,\}$$
$$= X(YW)$$

amd so the binary operation on $\mathscr{P}(S)$ is associative.

(ii) Suppose first of all that $e \in S$ is an identity. Then for any $X \in \mathscr{P}(S)$, $\{\,e\,\}X = \{\,ex \mid x \in X\,\} = \{\,x \mid x \in X\,\} = X = \{\,xe \mid x \in X\,\} = X\{\,e\,\}$, and so $\{\,e\,\}$ is an identity for $\mathscr{P}(S)$. Conversely, suppose that E is an identity for $\mathscr{P}(S)$. Since $\varnothing X = \varnothing$ for every $X \in \mathscr{P}(S)$, $E \neq \varnothing$. Let $x \in E$. Then $\{\,x\,\} = \{\,x\,\}E = \{\,xy \mid y \in E\,\}$ and so for all $y \in E$, $xy = x$. Similarly, $\{\,x\,\} = E\{\,x\,\} = \{\,yx \mid y \in E\,\}$, and thus $yx = x$ for all $y \in E$. Thus for $a, b \in E$, we have $a = ab = b$ and so $|E| = 1$. Let e be the unique element of E. We prove that e is an identity for S. For any $x \in S$, we have $\{\,x\,\} = \{\,x\,\}E = \{\,x\,\}\{\,e\,\} = \{\,xe\,\}$, and thus $x = xe$ for every $x \in S$. Similarly, $x = ex$ for every $x \in S$, and thus e is an identity for S.

(iii) Suppose that S has identity e, so $\{\,e\,\}$ is the identity for $\mathscr{P}(S)$. If $a \in S$ is invertible, then $\{\,a\,\}\{\,a^{-1}\,\} = \{\,aa^{-1}\,\} = \{\,e\,\}$, and similarly, $\{\,a^{-1}\,\}\{\,a\,\} = \{\,e\,\}$, so $\{\,a\,\}$ is invertible in $\mathscr{P}(S)$ with inverse $\{\,a^{-1}\,\}$. Conversely, suppose that $X \in \mathscr{P}(S)$ is invertible, and let $Y = X^{-1}$. Then $XY = YX = \{\,e\,\}$. Let $x \in X$ and $y \in Y$. Then $xy = e = yx$, so x and Y are invertible elements of S. Let $x_1 \in X$. Then $x_1 y = e = y x_1$, and so x_1 is also an inverse of y. By Proposition 3.4.19, $x_1 = x$ and so $|X| = 1$, as required. $\qquad\square$

3.4.32 Notation. *If S is a semigroup, then in $\mathscr{P}(S)$ with the binary operation introduced in Proposition 3.4.31, it is conventional to represent $\{\,x\,\}Y$ by simply xY, and the set $Y\{\,x\,\}$ by Yx.*

Exercises

1. In each part, give an example of a (small) set S and a relation R on S such that:

   a) R is both reflexive and transitive, but not symmetric.

   b) R is both reflexive and symmetric, but not transitive.

   c) R is both transitive and symmetric, but not reflexive.

   d) R is both antisymmetric and transitive, but not reflexive.

2. If S is a finite set of size n, prove that the number of relations on S is finite, and determine the number of relations on S.

3. If A and B are finite sets, prove that $A \times B$ is finite, and determine $|A \times B|$ in terms of $|A|$ and $|B|$.

4. Let $F = \{ (a,b) \mid a,b \in \mathbb{Z}, b \neq 0 \}$. On F, define a relation, denoted by $\sim$, by: $(a,b) \sim (c,d)$ if and only if $ad - bc$. Prove that $\sim$ is an equivalence relation on F. What are the elements in the equivalence class of $(1,2)$?

5. Let $\mathscr{R}(A)$ denote the set of all relations on a set A. If R and S are elements of $\mathscr{R}(A)$, define the composition of S after R, denoted by $S \circ R$, by: $a \, S \circ R \, b$ if there exists $c \in A$ such that $(c,b) \in R$ and $(a,c) \in S$, where $a, b \in A$. Prove that composition is associative; that is, for any $R, S, T \in \mathscr{R}(A)$, $R \circ (S \circ T) = (R \circ S) \circ T$.

6. Let A and B be finite sets with $|A| \leq |B|$. How many injective maps are there from A to B?

7. Let $S(m,n)$ denote the number of surjective maps from a set of m elements to a set of n elements. Observe that $S(m,n) = 0$ if $n > m$. Let m be a positive integer, and prove each of the following.

   a) $S(m,m) = m!$, where $m!$ is the product of all positive integers from 1 to m.

   b) $S(m,1) = 1$.

   c) $S(m,n) = n[S(m-1,n) + S(m-1,n-1)]$ for every integer n with $m > n \geq 2$.

   d) Evaluate $S(5,3)$.

8. For any sets A and B, and any map $f : A \to B$, define the relation $\ker(f)$ on A by $x \ker(f) y$ if $f(x) = f(y)$ for $x, y \in A$.

   a) Prove that $\ker(f)$ is an equivalence relation on A.

b)   Let $x \in A$. Prove that $[x]_{\ker(f)} = f^{-1}(f(x))$.

c)   Denote the set of equivalence classes of $\ker(f)$ by $A/\ker(f)$. Define a relation $\hat{f}$ from $A/\ker(f)$ to B by $[x]_{\ker(f)} \, \hat{f} \, b$ if $f(x) = b$. Prove first that $\hat{f}$ is a map from $A/\ker(f)$ to B, then that $\hat{f}$ is injective.

9.   Let m be a fixed positive integer. Define a relation "congruence modulo m" on $\mathbb{Z}$, denoted by $a \equiv b \mod m$ if $m \mid (a-b)$ (see Example 3.2.13). Prove that congruence modulo m is an equivalence relation on $\mathbb{Z}$, and determine its equivalence classes.

10.   Let A and B be finite sets. Prove that $\mathscr{F}(A, B)$ is finite, and determine $|\mathscr{F}(A, B)|$ in terms of $|A|$ and $|B|$.

11.   Let S be a set, and let $f \in \mathscr{F}(S)$.

a)   Prove that f is injective if, and only if, for all $g, h \in \mathscr{F}(S)$, $f \circ g = f \circ h$ implies that $g = h$.

b)   Prove that f is surjective if, and only if, for all $g, h \in \mathscr{F}(S)$, $g \circ f = h \circ f$ implies that $g = h$.

12.   Let $f: \mathbb{N} \to \mathbb{N}$ be defined by $f(n) = n + 2$ for all $n \in \mathbb{N}$. Find at least three left inverses of f.

13.   Let $f: \mathbb{Z} \to \{0, 1, 2, 3, 4\}$ be defined by $f(n) = i \in \{0, 1, 2, 3, 4\}$ if $n \equiv i \mod 5$ (we know by the division theorem that for each $n \in \mathbb{Z}$, there exist unique $q, r \in \mathbb{Z}$ such that $n = 5q + r$ and $0 \le r < 5$, so f is a surjective mapping). Find at least three right inverses for f.

14.   Let A and B be sets, and let $f: A \to B$ be a map.

a)   Prove that for all $X, Y \subseteq A$, $f(X \cup Y) = f(X) \cup f(Y)$.

b)   Prove that for all $X, Y \subseteq A$, $f(X \cap Y) \subseteq f(X) \cap f(Y)$. Construct an example of a set A of size 2, a set B of size 1, a map $f: A \to B$ and sets $X, Y \subseteq A$ such that $f(X \cap Y) \neq f(X) \cap f(Y)$.

c)   Prove that for all $X, Y \subseteq B$, $f^{-1}(X \cup Y) = f^{-1}(X) \cup f^{-1}(Y)$.

d)   Prove that for all $X, Y \subseteq B$, $f^{-1}(X \cap Y) = f^{-1}(X) \cap f^{-1}(Y)$.

e)   Prove that for all $X \subseteq A$, $X \subseteq f^{-1}(f(X))$. Construct an example of a set A of size 2, a set B of size 1, a map $f: A \to B$ and $Y \subseteq A$ such that $X \neq f^{-1}(f(X))$.

f)   Prove that for all $Y \subseteq B$, $f(f^{-1}(Y)) \subseteq Y$. Construct an example of a set A of size 1, a set B of size 2, a map $f: A \to B$ and $Y \subseteq B$ such that $Y \neq f(f^{-1}(Y))$.

g)   Prove that f is injective if, and only if, for all $X \subseteq A$, $f^{-1}(f(X)) = X$.

h) Prove that f is surjective if, and only if, for all $Y \subseteq B$, $f(f^{-1}(Y)) = Y$.

15. Let A and B be sets, and let $f : A \to B$ be a map. Prove that if f is bijective, then $(f^{-1})^{-1} = f$.

16. Let A be a set of size $n \geq 1$.

a) Prove that the number of partitions of A that have exactly k cells, where $1 \leq k \leq n$, is $\frac{S(n,k)}{k!}$ (see Exercise 7 for the definition of $S(n,k)$). Thus the number of partitions of A is $\sum_{k=1}^{n} \frac{S(n,k)}{k!}$.

b) Determine the number of partitions of a set of size 3, respectively 4 or 5. For sets of size 3 or 4, list the partitions with one cell, two cells, three cells, and in the case of the 4 element set, four cells.

17. Define an operation $*$ on $\mathbb{Q}$ by $a * b = a + b - ab$ for all $a, b \in \mathbb{Q}$.

a) Prove that $*$ is associative and commutative.

b) Does $*$ have an identity? If so, which elements of $\mathbb{Q}$ have inverses (with respect to $*$)?

18. Let A be a nonempty set, and let $S = A \times A$. Define a binary operation $*$ on S by $(a, b) * (c, d) = (a, d)$ for all $(a, b), (c, d) \in S$.

a) Prove that $(S, *)$ is a semigroup.

b) Prove that for all $x, y \in S$, $x * y * x = x$.

c) Prove that in any semigroup $(T, \star)$ with the property that for all $x, y \in T$, $x \star y \star x = x$, then each element of T is an idempotent (an idempotent in a semigroup is an element x for which $x \star x = x$).

19. Let S be a nonempty set, and let $\{ X_i \mid i \in I \}$ be a partition of S. A subset X of S is said to be a cross-section of the partition if for each $i \in I$, we have $|X_i \cap X| = 1$. How many cross-sections are there for the partition

$$\{ \{1, 2, 3\}, \{4, 5\}, \{6, 7, 8, 9\} \}$$

of the set $\{1, 2, 3, 4, 5, 6, 7, 8, 9\}$?

20. Let S be a nonempty set. An element $f \in \mathscr{F}(S)$ is said to be an idempotent of $(\mathscr{F}(S), \circ)$ if $f^2 = f$.

a) Prove that $f \in \mathscr{F}(S)$ is an idempotent of $(\mathscr{F}(S), \circ)$ if and only if $f(X) \subseteq X$ for each $X \in S/\ker(f)$.

b) How many idempotents are there in $(\mathscr{F}(J_3), \circ)$?

21. Let $(S, *)$ and $(T, \cdot)$ be semigroups. On the set $S \times T$, define a binary operation, denoted by juxtaposition, by $(a, b)(c, d) = (a * c, b \cdot d)$ for all $(a, b), (c, d) \in S \times T$.

a) Prove that $S \times T$ is a semigroup under this operation.

b) Prove that each of $(S, *)$ and $(T, \cdot)$ are monoids if, and only if, $S \times T$ is a monoid under this operation.

c) If $S \times T$ is a monoid under this operation, determine necessary and sufficient conditions on an element $(a, b) \in S \times T$ in order that it be invertible.

22. Let A be a set, and let $\mathscr{R}(A)$ denote the set of all relations on S. Recall that for $R, S \in \mathscr{R}(A)$, we have defined

$$S \circ R = \{ (a, b) \in A \times A \mid \text{ there exists } c \in A \text{ with } a\,R\,c \text{ and } c\,S\,b \}.$$

a) Prove that $(\mathscr{R}(A), \circ)$ is a semigroup.

b) Prove that $R \in \mathscr{R}(A)$ is an idempotent if R is reflexive and transitive.

c) Prove that if $R \in \mathscr{R}(A)$ is an idempotent, then R is transitive. Give an example of a set A and a relation R on A such that R is an idempotent in $(\mathscr{R}(A), \circ)$, but R is not reflexive.

d) Prove that $\mathbb{1}_A$ is the identity element and $\varnothing$ is the zero element for the binary operation composition of relations.

23. Let $R = \mathbb{Z}$, $\mathbb{Q}$, or $\mathbb{R}$. For positive integers m and n, the transpose of $A \in M_{m \times n}(R)$ is the $n \times m$ matrix denoted by $A^t \in M_{n \times m}(R)$ that is defined by $(A^t)_{ij} = A_{ji}$ for all integers i and j for which $1 \le i \le n$ and $1 \le j \le m$.

a) Let m, n, and k be positive integers. Prove that for any $A \in M_{m \times n}(R)$ and any $B \in M_{n \times k}(R)$, $(AB)^t = B^t A^t$.

b) Let n be a positive integer. Prove that if $A \in M_{n \times n}(R)$ is invertible, then A^t is invertible and $(A^t)^{-1} = (A^{-1})^t$.

Chapter 4

The Integers

Introduction

The set of integers consists of all the whole numbers, positive, negative and zero. As indicated earlier, this set will be denoted by $\mathbb{Z}$. The subset $\mathbb{Z}^+ = \{\, n \in \mathbb{Z} \mid n \geq 1 \,\}$ is called the set of positive integers, and the subset $\mathbb{N} = \{\, n \in \mathbb{Z} \mid n \geq 0 \,\}$ is called the set of natural numbers, or the set of nonnegative integers.

In this chapter, we establish some basic properties of the integers. We assume as an axiom (called the Well Ordering Principle) that every nonempty set of positive integers contains a least integer and show how this gives rise to the principle of mathematical induction, a powerful tool we shall often use in this and subsequent chapters.

After developing properties of divisibility, we prove the Fundamental Theorem of Arithmetic which states that every integer greater than 1 is uniquely expressible as a product of primes and deduce that there are infinitely many primes. We end the chapter with a discussion of linear congruences and show how they can be viewed as equations in certain algebraic systems. It is hoped that a familiarization with these concrete algebraic systems will help the reader make the transition to the more abstract structures to be studied in subsequent chapters.

Elementary Properties

There are two familiar operations defined on $\mathbb{Z}$; namely, addition and multiplication. We shall assume that the reader is familiar with the following properties which we adopt as axioms.

4.2.1 Axioms for the integers.

(i) *For all $m, n, k \in \mathbb{Z}$, $m + (n + k) = (m + n) + k$; that is, addition is associative.*

(ii) *0 has the property that for all $m \in \mathbb{Z}$, $0 + m = m + 0 = m$; that is, addition has 0 as identity element.*

(iii) *For each $m \in \mathbb{Z}$, there exists a unique $n \in \mathbb{Z}$ such that $m + n = n + m = 0$; that is, each element of $\mathbb{Z}$ has an additive inverse. We denote this unique n by $-m$.*

(iv) *For each $m, n \in \mathbb{Z}$, $m + n = n + m$; that is, addition is commutative.*

(v) *For each $m, n, k \in \mathbb{Z}$, $m(nk) = (mn)k$; that is, multiplication is associative.*

(vi) *1 has the property that for all $m \in \mathbb{Z}$, $1m = m1 = m$; that is, multiplication has 1 as identity element.*

(vii) *For each $m, n \in \mathbb{Z}$, $mn = nm$; that is, multiplication is commutative.*

(viii) *For each $m \in \mathbb{Z}$, if $m \neq 0$ and $n, k \in \mathbb{Z}$ are such that $mn = mk$, then $n = k$. This property is referred to as the cancellation law for multiplication.*

(ix) *For each $m, n, k \in \mathbb{Z}$, $m(n + k) = mn + mk$. Note that by (vii), $(n + k)m = nm + km$. Multiplication is said to distribute over addition.*

We remark that there is no analogue of (iii) for multiplication; that is, not every element of $\mathbb{Z}$ has a multiplicative inverse. In fact, only 1 and -1 have multiplicative inverses, with $(1)^{-1} = 1$ and $(-1)^{-1} = -1$. One may view (viii) as a generalization for the requirement of the existence of multiplicative inverses, in the sense that if $m \in \mathbb{Z}$ is invertible (with respect to multiplication; that is, $m = \pm 1$), and $n, k \in \mathbb{Z}$ are such that $mn = mk$, then $n = k$.

There is a corresponding cancellation law for addition. It is in fact a consequence of property (iii). Indeed, if $m + n = m + k$, then adding $-m$ to both sides of this equation and using associativity yields $0 + n = 0 + k$ and so $n = k$.

4.2.2 Notation. *For $m, n \in \mathbb{Z}$, it is customary to denote $m + (-n)$ by $m - n$. We may think of this as a new binary operation on $\mathbb{Z}$, in which case it is called subtraction. Note that subtraction is not associative, as for example $(1 - 1) - 1 = -1$ while $1 - (1 - 1) = 1$.*

Note that $\mathbb{N}$ and $\mathbb{Z}^+$ are each closed under both addition and multiplication. We shall make use of these observations to establish some basic properties of the familiar order relation. For $m, n \in \mathbb{Z}$, $m \leq n$ if, and only if, $n - m \in \mathbb{N}$, and furthermore, $m < n$ if, and only if, $n - m \in \mathbb{Z}^+$.

4.2.3 Proposition. *Let $m, n \in \mathbb{Z}$ be such that $m \leq n$. Let $c \in \mathbb{Z}$.*

   (i) If $c > 0$, then $cm \leq cn$, and if $m < n$, then $cm < cn$.

   (ii) If $c < 0$, then $cm \geq cn$, and if $m < n$, then $cm > cn$.

   (iii) $m + c \leq n + c$, and if $m < n$, then $m + c < n + c$.

Proof. Suppose that $c > 0$, so $c \in \mathbb{Z}^+ \subseteq \mathbb{N}$. Since $n - m \in \mathbb{N}$ and $\mathbb{N}$ is closed under multiplication, we have $c(n - m) \in \mathbb{N}$, and so $cn - cm \in \mathbb{N}$, which means that $cm \leq cn$. If moreover, $m < n$, then $n - m \in \mathbb{Z}^+$ and then $c(n - m) \in \mathbb{Z}^+$, giving $cn < cm$. Now suppose that $c < 0$, so $0 - c = -c \in \mathbb{Z}^+ \subseteq \mathbb{N}$. Since $n - m \in \mathbb{N}$, we have $(-c)(n - m) = -cn + cm \in \mathbb{N}$ and so $cn \leq cm$. If $m < n$, then $n - m \in \mathbb{Z}^+$ and thus $-cn + cm \in \mathbb{Z}^+$, so $cn < cm$. Finally, since $(n + c) - (m + c) = n - m \in \mathbb{N}$, we have $m + c \leq n + c$. If in fact, $n - m \in \mathbb{Z}^+$, then $(n + c) - (m + c) \in \mathbb{Z}^+$ and thus $m + c < n + c$. $\square$

A further intuitively obvious property possessed by the integers is the so-called well ordering principle which states that *every nonempty subset of $\mathbb{N}$ contains a least member* (which is easily seen to be unique). In addition to the properties listed in 4.2.1, we shall assume as an axiom that the set of natural numbers is well ordered by the usual ordering, by which we mean that the well ordering principle holds.

4.2.4 Example. *The real numbers are not well ordered by the usual ordering. For example, suppose that the subset $\mathbb{R}^+$ of all positive real numbers has a least element, t say. Then for every $r \in \mathbb{R}^+$, $0 < t \leq r$. In particular, $t/2 > 0$ and so $t \leq t/2$; equivalently, $2 \leq 1$. Since this is not true, we conclude that $\mathbb{R}^+$ has no least element. It is true that 0 is less than every positive real number, but $0 \notin \mathbb{R}^+$.*

4.2.5 Definition. *If $S \subseteq \mathbb{Z}$ and $S \neq \varnothing$, then S is said to be bounded below if there exists an integer m such that $m \leq s$ for each $s \in S$ (we do not require that $m \in S$). Each such m is called a lower bound for S.*

In the following lemma, we extend the well ordering principle to subsets of the integers which are bounded below. We remark that the word lemma is used to denote a preliminary result which is needed in the proof of a subsequent proposition or theorem, but which is proved independently so as not to disrupt the flow of the proof of the main result.

4.2.6 Lemma. *Let $S \subseteq \mathbb{Z}$ be bounded below. Then S contains a least member.*

Proof. Let m be a lower bound for S and consider the set

$$T = \{ s - m \mid s \in S \}.$$

Since $s - m \geq 0$ for each $s \in S$, $T \subseteq \mathbb{N}$, and since $S \neq \varnothing$ (a requirement of bounded below), it follows that T is a nonempty subset of $\mathbb{N}$. By the well ordering principle, T has a least element, b say. Since $b \in T$, there exists $n \in S$ such that $b = n - m$. We prove that $n = b + m$ is the least element of S. We have $n \in S$, so it remains to prove that for each $s \in S$, $n \leq s$. Let $s \in S$. Then $s - n = s - (b + m) = (s - m) - b \in \mathbb{N}$ since $s \geq m$ and $b \in \mathbb{N}$. Thus $s \geq n$ for all $s \in S$, and so n is the least element of S. $\square$

An immediate consequence of this lemma is the following proposition.

4.2.7 Proposition. *Let $S \subseteq \mathbb{Z}$ be bounded below, and let $l \in S$ be the least element of S. Suppose that S has the property that for each $m \in \mathbb{Z}$, if $m \in S$, then $m + 1 \in S$. Then $S = \{ n \in \mathbb{Z} \mid n \geq l \}$.*

Proof. Observe first that l exists by virtue of Lemma 4.2.6. Suppose by way of contradiction that the assertion is false, and let

$$T = \{ m \in \mathbb{Z} \mid m \geq l, \; m \notin S \}.$$

By assumption, $T \neq \varnothing$, and l is a lower bound for T, so by Lemma 4.2.6, T has a least element, n say. Since $l \in S$, $n \geq l$, and $n \notin S$, and so it follows that $n \neq l$, which means that $n > l$ and thus $n - 1 \geq l$. Since n is the least element of T, $n - 1 \notin T$. Since $n - 1 \geq l$, this means that $n - 1 \in S$. But then $(n - 1) + 1 \in S$; that is, $n \in S$. Since $n \notin S$, our assumption that the assertion is false has led to a contradiction and so the assertion is true. $\square$

4.2.8 Corollary (First Principle of Mathematical Induction). *Let $P(n)$ be a predicate and $n_0 \in \mathbb{Z}$ be such that the following hold:*

  (i) $P(n_0)$ is true.
  (ii) For all $n \in \mathbb{Z}$ with $n \geq n_0$, $P(n)$ implies $P(n + 1)$.

Then $P(n)$ is true for all $n \in \mathbb{Z}$ with $n \geq n_0$.

Proof. Let $T = \{ n \in \mathbb{Z} \mid n \geq n_0$ and $P(n)$ is true $\}$. Then n_0 is a lower bound for T and by (i), $n_0 \in T$. Moreover, by (ii), $n \in T$ implies that $n + 1 \in T$. Thus by Proposition 4.2.7, $T = \{ n \in \mathbb{Z} \mid n \geq n_0 \}$ and the corollary is proved. $\square$

In the following, we give examples of how the First Principle of Mathematical Induction may be applied.

4.2.9 Example. *We prove that if $x \in \mathbb{R}$ and $x > -1$, then $(1 + x)^n \geq 1 + nx$ for all $n \in \mathbb{N}$. Let $x \in \mathbb{R}$ with $x > -1$. Then take as our predicate $P(n)$ the sentence "$(1 + x)^n \geq 1 + nx$", with base value $n_0 = 0$. $P(0)$ (sometimes referred to as the base case) is the assertion that $(1+x)^0 \geq 1+0$, which is true since $(1 + x)^0 = 1$. Suppose that $n \geq 0$ is an integer such that $P(n)$ is true. We are then to prove that $P(n+1)$ is true (this is sometimes referred to as the inductive step). We know that $(1 + x)^n \geq 1 + nx$, so $(1 + x)^{n+1} = (1 + x)^n (1 + x) \geq (1 + nx)(1 + x)$ since $1 + x > 0$. Thus $(1 + x)^{n+1} \geq 1 + nx + x + nx^2 \geq 1 + (n + 1)x$, as required. Thus for each $n \geq 0$, $P(n)$ implies $P(n + 1)$. It follows now by Corollary 4.2.8 that $P(n)$ is true for all $n \geq 0$.*

In the preceding example, we have written out the proof in full detail. Once the reader is more familiar with the process, we may be less formal. It is however very important to realize that the inductive hypothesis ($P(n)$) must be clearly identified, even if one does not formally refer to it as $P(n)$.

We shall write out the proof of the next example with somewhat less formality. Recall that for a positive integer n, $n!$ is the product of all integers between 1 and n inclusive, while $0! = 1$.

4.2.10 Example. *We shall apply the principle of mathematical induction (or more simply, we shall use induction) to prove that $n! > 2^{n+1}$ for all integers $n \geq 5$. In this example, the choice of predicate and base value are obvious (once one observes that the result is false for $n = 0, 1, 2, 3, 4$). We compute $5! = (5)(4)(3)(2)(1) = 120$ and $2^6 = 64$, so $5! > 2^{5+1}$. Suppose now that $n \geq 5$ is an integer such that $n! > 2^{n+1}$. Then $(n + 1)! = (n + 1)n! > (n + 1)2^{n+1} > (2)2^{n+1} = 2^{(n+1)+1}$, and the inductive step is proved. It follows now by the principle of mathematical induction (Corollary 4.2.8) that $n! > 2^{n+1}$ for all integers $n \geq 5$.*

It is interesting to note that the proof of the inductive step that we have presented in the preceding example is actually valid for all $n \geq 1$; that is, for all $n \geq 1$, $P(n)$ implies $P(n + 1)$ (where $P(n)$ would be the sentence "$n! > 2^{n+1}$"). This observation highlights the necessity of proving the base case.

Another consequence of well ordering is the Division Theorem.

4.2.11 Theorem (The Division Theorem). *Let a and b be integers*

*with $b \neq 0$. Then there exist unique integers q and r such that $a = qb + r$
and $0 \leq r < |b|$.*

Proof. First, we prove the existence of such integers q and r. Let $S = \{a - |b|t \mid t \in \mathbb{Z}\}$ and set $T = S \cap \mathbb{N}$. If $a \geq 0$, then $a = a - |b|(0) \in T$, while if $a < 0$, then $a(1 - |b|) \geq 0$ and so $a(1 - |b|) = a - |b|a \in T$. Thus $T \neq \varnothing$ and so T is a nonempty subset of $\mathbb{N}$. By the Well Ordering Principle, T has a least element, r say. Thus $r \geq 0$ and there exists $t \in \mathbb{Z}$ such that $r = a - |b|t$. We claim that $r < |b|$. Suppose by way of contradiction that $r \geq |b|$. Then $r - |b| \geq 0$ and $r - |b| = a - |b|t - |b| = a - |b|(t + 1)$, so $r - |b| \in T$. Since r is the least element of T, $r \leq r - |b|$, so $0 \leq -|b|$; that is, $|b| \leq 0$. As $b \neq 0$, this is not possible, and so we conclude that $r < |b|$. Thus $a = t|b| + r$ and $0 \leq r < |b|$. If $b > 0$, take $q = t$, while if $b < 0$, take $q = -t$ to obtain integers q and r such that $a = qb + r$ and $0 \leq r < |b|$.

Now we prove the uniqueness assertion. Suppose that $r_1, r_2, q_1, q_2 \in \mathbb{Z}$ are such that $a = q_1 b + r_1 = q_2 b + r_2$ and $0 \leq r_1, r_2 < |b|$. Then $(q_2 - q_1)b = r_1 - r_2$. Note that $|r_1 - r_2| < |b|$. If $q_2 - q_1 \neq 0$, then $|b| > |r_1 - r_2| = |b|(q_2 - q_1) \geq |b|$, which is not possible. Thus $q_2 - q_1 = 0$; that is, $q_1 = q_2$, and thus $r_1 - r_2 = 0$, giving $r_1 = r_2$, as required. $\qquad \square$

4.2.12 Definition. *For integers a and b with $b \neq 0$, if q and r are the unique integers such that $a = qb + r$ and $0 \leq r < |b|$, then we say that q is the quotient and r is the remainder when a is divided by b.*

We shall be using the Division Theorem in the very next theorem and again later, after we have discussed greatest common divisors.

4.2.13 Definition. *A nonempty set S of integers is said to be closed under subtraction if $a, b \in S$ implies $a - b \in S$.*

We observe that a nonempty set S of integers is closed under subtraction if, and only if, S is closed under addition and the formation of additive inverse; that is, $a, b \in S$ implies that $a + b \in S$ and $-a \in S$ (much of this observation is established in the proof of the next result).

As examples of nonempty subsets of $\mathbb{Z}$ that are closed under subtraction, we note that $\mathbb{Z}$ itself is closed under subtraction, and more generally (since $\mathbb{Z}$ is the set of all multiples of 1), the set of all multiples of a fixed integer is closed under subtraction. The interesting fact is that these are the only kinds of subsets of $\mathbb{Z}$ which are closed under subtraction, as we now prove.

4.2.14 Theorem. *If $S \subseteq \mathbb{Z}$ is nonempty and closed under subtraction,* *then there exists a unique nonnegative integer $n \in S$ such that $S = \mathbb{Z}n$, where by $\mathbb{Z}n$, we mean the set $\{ kn \mid k \in \mathbb{Z} \}$.*

Proof. Since $S \neq \varnothing$, there is $x \in S$. Since S is closed under subtraction, $0 = x - x \in S$. If $S = \{0\}$, then $S = \mathbb{Z}0$ (and 0 is the only integer n for which $\mathbb{Z}n = \{0\}$), so we may suppose that $S \neq \{0\}$. Let $x \in S$ be such that $x \neq 0$. Then since $0 \in S$ and S is closed under subtraction, we have $-x = 0 - x \in S$, so $x, -x \in S$. Thus S contains positive integers. Let $T = S \cap \mathbb{Z}^+$. Then T is a nonempty subset of $\mathbb{Z}$ bounded below by 1, and so by Lemma 4.2.6, T contains a least element n.

Our next step is to prove by induction that for all $k \in \mathbb{Z}^+$, $kn \in T$, and from this, to prove that $\mathbb{Z}n \subseteq S$. The base case is immediate, since $n \in T$. Suppose that $k \in \mathbb{Z}^+$ is such that $kn \in T$. Then $-kn = 0 - kn \in S$, and so $n - (-kn) = (k+1)n \in S$. Since $n \geq 1$ and $k + 1 \geq 1$, $(k+1)n \geq 1$ and thus $(k+1)n \in T$. It follows now by induction that $kn \in T$ for all $n \in \mathbb{Z}^+$, and thus $kn, -kn - 0 - kn \in S$ for all $k \in \mathbb{Z}^+$. Since $0 = n0 \in S$, we have $\mathbb{Z}n \subseteq S$.

It remains to prove that $S \subseteq \mathbb{Z}n$. Let $t \in S$. By the division theorem, there exist integers q and r such that $t = qn + r$ and $0 \leq r < n$. Since $qn \in S$ by the above, and S is closed under subtraction, we have $t - qn \in S$; that is, $r \in S$. Since n is the smallest positive element of S, and $0 \leq r < n$, it follows that $r = 0$ and thus $t = qn \in \mathbb{Z}n$.

Suppose now that m is a positive integer such that $S = \mathbb{Z}m$. Since n is the smallest positive integer in S, we have $n \leq m$. But since $n \in S = \mathbb{Z}m$, there is an integer k such that $n = km$, and since $m, n > 0$, we must have $k > 0$. Thus $n \geq m$, and so $m = n$, as required. $\qquad\square$

Divisibility

4.3.1 Definition. *Let a and b be integers. We say that a divides b, or that a is a factor of b, denoted by $a \mid b$, if $b = ka$ for some $k \in \mathbb{Z}$.*

The following lemma provides some elementary properties of division. They will be used freely from now on, usually without explicit mention.

4.3.2 Lemma. *Let a, b, and c be integers. Then the following hold:*

(i) *If $a \mid b$ and $b \mid a$, then $|a| = |b|$.*
(ii) *If $a \mid b$ and $b \mid c$, then $a \mid c$.*
(iii) *0 is the only integer that is divisible by every integer.*

(iv) If $a \mid b$, then $\pm a \mid \pm b$.

(v) If $a \mid b$ and $a \mid c$, then for any $r, s \in \mathbb{Z}$, $a \mid rb + sc$.

(vi) If $a \mid b$ and $a \mid qb + c$ for some $q \in \mathbb{Z}$, then $a \mid c$.

Proof. We shall only prove a couple of the parts, and the reader is invited to prove the rest. For (i), suppose that $a \mid b$ and $b \mid a$, so there exist integers m, n such that $b = ma$ and $a = nb$. Then $b = m(nb)$, so $b(1 - mn) = 0$. If $b = 0$, then $a = n(0) = 0$, so $a = b$ in this case. Otherwise, $b \neq 0$ and thus $a \neq 0$ since $b = ma$. By cancellation, we obtain that $1 - mn = 0$, and thus $m = n = \pm 1$. Thus $|a| = |nb| = |\pm 1||b| = |b|$.

For (v), suppose that $a \mid b$ and $a \mid c$, and let $r, s \in \mathbb{Z}$. There exist $m, n \in \mathbb{Z}$ such that $b = ma$ and $c = na$, so $rb + sc = r(ma) + s(na) = (rm + sn)a$. Thus $a \mid rb + sc$. Note that (vi) is an immediate consequence of (v). $\qquad \square$

4.3.3 Definition. *Let a and b be integers, not both zero. An integer c is said to be a common divisor of a and b if $c \mid a$ and $c \mid b$. A positive integer d is then said to be a greatest common divisor of a and b if d is a common divisor of a and b with the further property that every common divisor of a and b is a divisor of d.*

This definition makes considerable demands on d, and it is conceivable that our requirements are too stringent, with the result that d may not exist for some pairs of integers. The next theorem proves that greatest common divisors do exist and that, moreover, the positive integer d of Definition 4.3.3 is unique. We shall therefore be able to speak of the greatest common divisor of a and b, rather than a greatest common divisor of a and b, and we may introduce notation for the greatest common divisor of a and b.

4.3.4 Theorem. *Let a and b be integers, not both zero. Then there exists one and only one greatest common divisor of a and b. Moreover, there exist integers r and s such that $d = ra + sb$ is the greatest common divisor of a and b.*

Proof. Let $S = \{xa + yb \mid x, y \in \mathbb{Z}\}$. Since $a = (1)(a) + (0)(b)$ and $b = (0)(a) + (1)(b)$, we have $a, b \in S$, and since not both a and b can be zero, it follows that $S \neq \{0\}$. Furthermore, S is closed under subtraction. To see this, suppose that $m, n \in S$, so $m = x_1 a + y_1 b$ and $n = x_2 a + y_2 b$ for some integers x_1, y_1, x_2, y_2. But then $m - n = (x_1 - x_2)a + (y_1 - y_2)b \in S$. It follows now from Theorem 4.2.14, that there exists a unique nonnegative integer d such that $S = \mathbb{Z}d$. Since $S \neq \{0\}$, it follows that $d > 0$. As

$d \in S$, there exist $r, s \in \mathbb{Z}$ such that $d = ra + sb$, and thus by Lemma 4.3.2 (v), every common divisor of a and b is a divisor of d.

For the uniqueness assertion, suppose that m is a greatest common divisor of a and b. Then m is a divisor of d, and d is a divisor of m, so $|m| = |d|$. Since both m and d are positive, it follows that $m = d$, as required. $\qquad\square$

4.3.5 Notation. *For any integers a and b not both zero, the unique greatest common divisor of a and b shall be denoted by (a, b).*

Theorem 4.3.4 is an *existence theorem*, and it does not answer the question: how does one compute the greatest common divisor of two integers? As it turns out, there is an algorithm known to Euclid (and so called the Euclidean algorithm) which dates back more than two thousand years that yields the greatest common divisor of two integers very efficiently. Moreover, this algorithm provides an added bonus in that, from the computations involved in the derivation of the greatest common divisor of integers a and b, one is able to find integers r and s such that $(a, b) = ra + sb$. The Euclidean algorithm is based on the following lemma.

4.3.6 Lemma. *Let a, b, q, r be integers such that $a = qb + r$, with not both a and b equal to 0. Then $(a, b) = (b, r)$.*

Proof. Since $r = a - qb = a + (-q)b$, any common divisor of a and b is a divisor of r, so in particular, (a, b) is a divisor of r, and also of b, so (a, b) is a divisor of (b, r). Note that if both b and r equal 0, then $a = 0$, and not both a and b are 0, so not both b and r are 0. Since $r = (-q)b + a$, then as just shown, (b, r) is a divisor of (a, b). Thus $|(a, b)| = |(b, r)|$ and since both are positive, we obtain that $(a, b) = (b, r)$. $\qquad\square$

Note as well that for any nonzero integer a, $(a, 0) = |a|$, and furthermore, for any integers a and b, not both zero, $(a, b) = (|a|, |b|)$. Thus it is sufficient to have an algorithm for calculating the greatest common divisor of any two nonnegative integers not both of which are 0. Lemma 4.3.6 and these observations provide us with the Euclidean algorithm.

4.3.7 Algorithm (Euclidean Algorithm). *Let a and b be nonnegative integers, not both zero, with $a \geq b$. The following calculation is a fundamental step. Apply the division theorem to a and b to obtain integers q and r with $a = qb + r$ and $0 \leq r < b$ (note that $|b| = b$ since $b \geq 0$). Recall that r is called the remainder when b is divided by a. Then $(a, b) = (b, r)$ and*

$0 \leq r < b$. *Repeat the fundamental step until a step is reached for which the remainder is 0, in which case the previous step produced a remainder of $d > 0$ and we have $(a, b) = (d, 0) = d$.*

4.3.8 Example. *We use the Euclidean algorithm to find the greatest common divisor of 462 and 180. The data that result from the repetitions of the fundamental step is shown in the table below.*

$$(1)\ 462 = 2(180) + 102$$
$$(2)\ 180 = 1(102) + 78$$
$$(3)\ 102 = 1(78) + 24$$
$$(4)\ \ 78 = 3(24) + 6$$
$$(5)\ \ 24 = 4(6) + 0.$$

Thus $(462, 180) = 6$.

Let us now explore how the data presented in the preceding example can be used to produce integers r and s such that $6 = (462, 180) = r(462) + s(180)$. The penultimate line, line (4), of the computation gives

$$6 = 78 - 3(24).$$

From the next line up, line (3), we get $24 = 102 - 1(78)$. Substituting this expression for 24 in the displayed equation, we get

$$6 = 78 - 3(102 - 1(78))$$
$$= -3(102) + 4(78).$$

Then from the next line up, line (2), we find that $78 = 180 - 1(102)$, and so we substitute this expression for 78 in the preceding displayed equation to find that

$$6 = -3(102) + 4(180 - 1(102))$$
$$= 4(180) - 7(102).$$

Finally, line (1) allows us to write $102 = 462 - 2(180)$, and when this expression is substituted in place of 102 in the preceding displayed equation, we obtain

$$6 = 4(180) - 7(462 - 2(180))$$
$$= -7(462) + 18(180).$$

Thus $(462, 180) = 6 = (-7)(462) + (19)(180)$.

4.3.9 Definition. *For integers $a_1, a_2, \ldots, a_n$, not all zero, $(a_1, a_2, \ldots, a_n)$ is defined to be the unique positive integer d for which d is a divisor of a_i for $i = 1, 2, \ldots, n$, and if c is any divisor of a_i, $i = 1, 2, \ldots, n$, then c is a divisor of d.*

The proof of existence and uniqueness could either be presented by means of induction on n, or by mimicking the proof of the existence and uniqueness of (a, b). The latter approach is possible since the set $S = \{r_1 a_1 + \cdots + r_n a_n \mid r_1, r_2, \ldots, r_n \in \mathbb{Z}\}$ is closed under subtraction and contains $a_1, a_2, \ldots, a_n$, so by Theorem 4.2.14, there exists $d > 0$ such that $S = \mathbb{Z}d$. It then follows that d is a divisor of each a_i, $d = r_1 a_1 + \cdots + r_n a_n$ for some $r_1, r_2, \ldots, r_n \in \mathbb{Z}$, and thus any $c \in \mathbb{Z}$ for which c divides a_i for $i = 1, 2, \ldots, n$ is a divisor of d.

4.3.10 Definition. *Let a and b be nonzero integers. Then an integer n is said to be a common multiple of a and b if both a and b are divisors of n. An integer m is said to be a least common multiple of a and b if m is a positive common multiple of a and b with the property that m is a divisor of every common multiple of a and b.*

As was the case for the greatest common divisor, it is necessary to show existence of a least common multiple. As one might expect, even more is true.

4.3.11 Theorem. *Let a and b be nonzero integers, and let $d = (a, b)$. Then $\frac{|ab|}{d}$ is a least common multiple of a and b. Moreover, the least common multiple of a and b is unique.*

Proof. It is sufficient to prove the result for positive integers a and b, so assume that a and b are positive integers and let $m = \frac{ab}{d}$. Then

$$m = \left(\frac{a}{d}\right)b = a\left(\frac{b}{d}\right)$$

where we note that $\frac{a}{d}$ and $\frac{b}{d}$ are integers since d is a divisor of both a and b. Thus m is a common multiple of a and b. Suppose now that c is any common multiple of a and b. Then there exist integers k and l such that $c = ka = lb$. Furthermore, there exist integers r and s such that $d = ra + sb$. Divide through by d to obtain $1 = r\frac{a}{d} + s\frac{b}{d}$. Multiply by c to get $c = rc\frac{a}{d} + sc\frac{b}{d}$. Now, using the fact that $c = ka = lb$, we obtain

$$c = rlb\frac{a}{d} + ska\frac{b}{d} = rl\frac{ab}{d} + sk\frac{ab}{d} = (rl + sk)m,$$

and thus m is a divisor of c, as required. This completes the proof that $m = \frac{ab}{d}$ is a least common multiple of a and b. Finally, suppose that n is also a least common multiple of a and b. Then m is a divisor of n and n is a divisor of m, so $|m| = |n|$. Since both m and n are positive, we have $m = |m| = |n| = n$. $\square$

4.3.12 Notation. *For any nonzero integers a and b, the unique least common multiple of a and b shall be denoted by $[a, b]$.*

Our next major objective is to formulate and prove the Fundamental Theorem of Arithmetic. For its proof, we shall want a variant of the principle of mathematical induction. If we had proven the First Principle of Mathematical Induction (Corollary 4.2.8) by contradiction rather than directly, this variant would have been immediately apparent.

4.3.13 Theorem (Second Principle of Mathematical Induction).
Let $P(n)$ be a predicate and $n_0 \in \mathbb{Z}$ be such that the following hold:
  (i) $P(n_0)$ is true.
  (ii) For all $n \in \mathbb{Z}$ with $n \geq n_0$, $(P(n_0)$ and $P(n_0+1)$ and $\cdots$ and $P(n))$ implies $P(n+1)$; that is, for every integer $n \geq n_0$, if for each integer k with $n_0 \leq k \leq n$, the statement $P(k)$ is true, then $P(n+1)$ is true.
Then $P(n)$ is true for all $n \in \mathbb{Z}$ with $n \geq n_0$.

Proof. Let $Q(n)$: for all k with $n_0 \leq k \leq n$, $P(k)$. Then by (i), $Q(n_0)$ is true. Suppose that $n \geq n_0$ is an integer for which $Q(n)$ is true. Then by (ii), $P(n + 1)$ is true, and since $Q(n + 1)$ is logically equivalent to "$Q(n)$ and $P(n+1)$", it follows that $Q(n+1)$ is true. We may therefore apply the first principle of mathematical induction to obtain that $Q(n)$ is true for all integers $n \geq n_0$. Since for each integer $n \geq n_0$, $Q(n)$ implies $P(n)$, we have obtained that $P(n)$ is true for all integers $n \geq 0$. $\square$

From time to time, we shall come across situations where the seemingly natural choice of a predicate dictates that we must use the Second Principle of Mathematical Induction rather than the First Principle. Indeed, this is the case in the proof of the Fundamental Theorem of Arithmetic, which we shall tackle in the next section.

The Fundamental Theorem of Arithmetic

4.4.1 Definition. *An integer $p > 1$ is said to be a prime if the only positive factors of p are 1 and p. If an integer $n > 1$ is not prime, then n is said to be composite.*

The first eight primes are 2, 3, 5, 7, 11, 13, 17, 19. It is worth noting that 2 is the only even prime, since if $m > 2$ is even, then $m = 2k$ for some positive integer $k > 1$, so m has proper divisors 2 and k.

4.4.2 Definition. *Two distinct integers a and b are said to be relatively prime if $(a, b) = 1$.*

4.4.3 Proposition. *Let a, b, and c be integers such that not both a and b are zero, $a \mid bc$, and $(a, b) = 1$. Then $a \mid c$.*

Proof. By Theorem 4.3.4, since $(a, b) = 1$, there are integers r, s such that $1 = ra + sb$, and so $c = cra + sbc$. But then a is a divisor of cra and of sbc (since a is a divisor of bc), so a is a divisor of c. $\qquad\square$

Note that in the preceding result, if $a = 0$, then $bc = 0$ but $b \neq 0$, so $c = 0$.

4.4.4 Corollary. *Let $p \in \mathbb{Z}$ with $p > 1$. Then p is a prime if, and only if, for all $a, b \in \mathbb{Z}^+$, if $p \mid ab$, then $p \mid a$ or $p \mid b$.*

Proof. First, suppose that p is prime and $a, b \in \mathbb{Z}^+$ are such that p divides ab. If $p \mid a$, then the result holds. Suppose that p is not a divisor of a. Then since (p, a) is a divisor of both p and a and p does not divide a, it follows that $(p, a) \neq p$. Since p is a prime, the only possibility left is for $(p, a) = 1$, and then by Proposition 4.4.3, $p \mid b$.

Conversely, suppose that for all $a, b \in \mathbb{Z}$, if p divides ab, then p divides a or p divides b. Suppose that $a, b \in \mathbb{Z}^+$ are such that $p = ab$. Then p divides ab, and so by hypothesis, p divides a or p divides b. Suppose that p divides a, say $a = pk$ for some $k \in \mathbb{Z}^+$. Then $a = pk = abk$ and since $a \neq 0$, we obtain $1 = bk$. Since $b \in \mathbb{Z}^+$, this means that $b = k = 1$ and then $a = p$. Otherwise, p divides b and then $a = 1$, so $b = p$. Thus the only positive factors of p are 1 and p, and so p is prime. $\qquad\square$

4.4.5 Corollary. *Let p be a prime, $n \geq 2$ an integer, and $a_1, a_2, \ldots, a_n \in \mathbb{Z}$ such that $p \mid a_1 a_2 \cdots a_n$. Then for some i with $1 \leq i \leq n$, $p \mid a_i$.*

Proof. The proof is by induction on n, the number of factors. The base case of two factors is Corollary 4.4.4. Suppose that $n \geq 2$ is an integer for which the assertion holds, and let $a_1, a_2, \ldots, a_{n+1}$ be integers such that $p \mid a_1 a_2 \cdots a_{n+1} = (a_1 a_2 \cdots a_n) a_{n+1}$. If $p \mid a_{n+1}$, the assertion holds, so suppose that p does not divide a_{n+1}. Then by Corollary 4.4.4, $p \mid a_1 \cdots a_n$, and so for some i with $1 \leq i \leq n$, $p \mid a_i$. The result follows now by the first principle of mathematical induction. $\qquad\square$

The preceding corollary has a very nice consequence which we prove next.

4.4.6 Proposition. *Let p be a prime. Then for any integer $i = 1, 2, \ldots, p - 1$, p divides $\binom{p}{i}$.*

Proof. Let i be an integer with $1 \leq i \leq p - 1$. Since $\binom{p}{i} = \frac{p!}{i!(p-i)!}$ is an integer, and p is a divisor of $p!$, it suffices to prove that $(p, i!(p-1)!) = 1$. Since $(p, i!(p-i)!)$ is a positive divisor of the prime p, it must be either 1 or p. If it were p, then by Corollary 4.4.5, p would divide some positive integer less than p, which is not the case. Thus $(p, i!(p-i)!) = 1$. $\qquad\square$

We are now ready to prove the Fundamental Theorem of Arithmetic.

4.4.7 Theorem (Fundamental Theorem of Arithmetic). *Each integer greater than 1 is either a prime or a product of primes. This factorization is unique up to the order in which the factors occur.*

Proof. We first prove that every integer $n \geq 2$ is either prime or a product of primes by means of the second principle of mathematical induction. To begin with (the base case for the induction), 2 is prime. Suppose now that $n \geq 2$ is an integer such that every integer k for which $2 \leq k \leq n$ is either prime or a product of primes. We must now prove that $n+1$ is either prime or a product of primes. If $n + 1$ is prime, this assertion is true, so suppose that $n + 1$ is not prime. Then $n + 1 = rs$ for some integers r and s with $1 < r \leq n$ and $1 < s \leq n$. By hypothesis, each of r and s is either prime or a product of primes, and thus $n + 1 = rs$ is a product of primes. We may therefore apply the second principle of mathematical induction to establish that every integer greater than 1 is either prime or a product of primes.

It remains to prove the uniqueness assertion of the theorem. This we prove by using the first principle of mathematical induction on the number of prime factors. Our inductive hypothesis is the following assertion: for any positive integer l, if $p_1, p_2, \ldots, p_n$ and $q_1, q_2, \ldots, q_l$ are primes such that $p_1 p_2 \cdots p_n = q_1 q_2 \cdots q_l$, then $l = n$ and, relabelling if necessary, $p_i = q_i$ for $i = 1, 2, \ldots, n$. The assertion holds for $n = 1$, since that is just the assertion that a prime number has no proper factors. Suppose that $n \geq 1$ is an integer for which the assertion holds, and let $p_1, p_2, \ldots, p_{n+1}$ be primes (not necessarily distinct). Let l be a positive integer and suppose that $q_1, q_2, \ldots, q_l$ are primes (again, not necessarily distinct) such that $p_1 p_2 \cdots p_n p_{n+1} = q_1 q_2 \cdots q_l$. Since $l = 1$ would mean that $p_1 p_2 \cdots p_n p_{n+1}$ is prime, it follows that $l \geq 2$. Since p_{n+1} is a divisor of $q_1 q_2 \cdots q_l$, it follows from Corollary 4.4.4 that p_{n+1} is a divisor of q_i for some i with $1 \leq i \leq l$. By relabelling if necessary, we may assume that $i = l$ (and we have $l \geq 2$). Since q_l is a prime, it follows that $p_{n+1} = q_l$. Thus $p_1 p_2 \cdots p_n = q_1 \cdots q_{l-1}$, and so by hypothesis, $n = l - 1$, giving $n + 1 = l$, and, relabelling if necessary, $p_j = q_j$ for all $j = 1, 2, \ldots, n$. The result follows now by the first principle of mathematical induction. □

The following theorem was known to, and proved by, Euclid. Indeed, it seems possible that Euclid was the first to prove this result.

4.4.8 Theorem. *There are infinitely many primes.*

Proof. Suppose by way of contradiction that the assertion is false, and let $\{p_1, p_2, \ldots, p_n\}$ be the set of all primes. Consider the integer $m = p_1 p_2 \cdots p_n + 1 > 1$. By the fundamental theorem of arithmetic (Theorem 4.4.7), m has a prime divisor, so for some i, $p_i \mid m$. But p_i is a divisor of $p_1 p_2 \cdots p_n = m - 1$, so p_i is a divisor of $m - (m - 1) = 1$. Since no prime is a divisor of 1, we have obtained a contradiction. □

Congruence Modulo n and the Algebraic System $(\mathbb{Z}_n, +, \cdot)$

For a given positive integer n, we say that an integer a is related to an integer b modulo n, written $a \equiv b \mod n$, if n is a divisor of $b - a$. The integer n is said to be the modulus, and we show next that for each positive integer n, the relation "congruence modulo n" is an equivalence relation on $\mathbb{Z}$. The equivalence class of $a \in \mathbb{Z}$ modulo n will be denoted by $[a]_n$, or simply by $[a]$ if the modulus is understood.

4.5.1 Proposition. *For each positive integer n, congruence modulo n is an equivalence relation on $\mathbb{Z}$, and for each $a \in \mathbb{Z}$, $[a]_n = \{ a + kn \mid k \in \mathbb{Z} \}$. Moreover, the partition of $\mathbb{Z}$ whose elements are the equivalence classes of* congruence modulo n is the finite set of size n denoted by $\mathbb{Z}_n$, given by

$$\mathbb{Z}_n = \{\, [0]_n, [1]_n, \ldots, [n-1]_n \,\}.$$

Proof. Let n be a positive integer. Then for each $a \in \mathbb{Z}$, $a - a = 0 = (0)(n)$ and so $a \equiv a \mod n$; that is, congruence modulo n is reflexive. Next, if $a, b \in \mathbb{Z}$ and $a \equiv b \mod m$, then $n \mid (b - a)$ and thus $n \mid (a - b)$, so $b \equiv a \mod n$ and congruence modulo n is symmetric. Finally, if $a, b, c \in \mathbb{Z}$ are such that $a \equiv b \mod n$ and $b \equiv c \mod n$, then $n \mid (b - a)$ and $n \mid (c - b)$, so n is a divisor of $(b - a) + (c - b) = c - a$. Thus congruence modulo n is transitive. By definition, a reflexive, symmetric, transitive relation is an equivalence relation and so congruence modulo n is an equivalence relation on $\mathbb{Z}$.

For $a \in \mathbb{Z}$,

$$
\begin{aligned}
[a]_n &= \{\, b \in \mathbb{Z} \mid n \mid (b - a) \,\} \\
&= \{\, b \in \mathbb{Z} \mid \text{there exists } k \in \mathbb{Z} \text{ with } b - a = kn \,\} \\
&= \{\, a + kn \mid k \in \mathbb{Z} \,\}.
\end{aligned}
$$

Apply the division theorem to a and n to obtain integers q and r with $a = qn + r$ and $0 \le r < n$. Then $n \mid (a - r)$ and so $a \equiv r \mod n$. But then $[a]_n = [r]_n$, and $0 \le r \le n - 1$, so $\mathbb{Z}_n = \{\, [0]_n, [1]_n, \ldots, [n-1]_n \,\}$. It remains to prove that $|\mathbb{Z}_n| = n$; that is, for $0 \le i \le j \le n - 1$, $[i]_n = [j]_n$ implies that $i = j$. Suppose that $0 \le i \le j \le n - 1$ and $[i]_n = [j]_n$. Then $n \mid (j - i)$, and so for some integer q, $j - i = qn$, or $j = qn + i$. However, $j = (0)n + j$, and since $0 \le i < n$ and $0 \le j < n$, the uniqueness assertion of the division theorem tells us that $q = 0$ and $i = j$. $\qquad\square$

The next result establishes that congruence modulo n is compatible with both addition and multiplication on $\mathbb{Z}$. This fact will allow us to define binary operations on $\mathbb{Z}_n$ that we shall also call addition and multiplication.

4.5.2 Proposition. *Let n be a positive integer. Then for any $a, b, c, d \in \mathbb{Z}$, if $[a]_n = [b]_n$ and $[c]_n = [d]_n$, then $[a + c]_n = [b + d]_n$ and $[ac]_n = [bd]_n$.*

Proof. First, $[a]_n = [b]_n$ is equivalent to $a \equiv b \mod n$, and $[c]_n = [d]_n$ is equivalent to $c \equiv d \mod n$. We are to prove that, under these conditions, $a + c \equiv b + d \mod n$ and $ac \equiv bd \mod n$. We know that $n \mid (a - b)$ and $n \mid (c - d)$, so $n \mid (a - b + c - d) = ((a + c) - (b + d))$. Thus $a + c \equiv b + d \mod n$. As well, since $ac - bd = ac - bc + bc - bd = (a - b)c + b(c - d)$, $n \mid (ac - bd)$ and thus $ac \equiv bd \mod n$, as required. $\qquad\square$

4.5.3 Definition. *For* $[a]_n, [b]_n \in \mathbb{Z}_n$, *we define the sum* $[a]_n + [b]_n = [a+b]_n$, *and the product* $[a]_n [b]_n = [ab]_n$. *This defines two binary operations on* $\mathbb{Z}_n$ *that we shall call addition and multiplication, respectively. We shall use the symbol* $\times$ *to denote multiplication as an operation, but we shall simply use juxtaposition to denote the product of two elements of* $\mathbb{Z}_n$.

It follows from Proposition 4.5.2 that these operations of addition and multiplication on $\mathbb{Z}_n$ are well-defined; that is, $[a+b]_n$ and $[ab]_n$ do not depend on our choice of representative a for $[a]_n$ and b for $[b]_n$:

4.5.4 Example. *We give the addition and multiplication tables for* $\mathbb{Z}_4 = \{ [0]_4, [1]_4, [2]_4, [3]_4 \}$.

+	$[0]_4$	$[1]_4$	$[2]_4$	$[3]_4$
$[0]_4$	$[0]_4$	$[1]_4$	$[2]_4$	$[3]_4$
$[1]_4$	$[1]_4$	$[2]_4$	$[3]_4$	$[4]_4$
$[2]_4$	$[2]_4$	$[3]_4$	$[4]_4$	$[5]_4$
$[3]_4$	$[3]_4$	$[4]_4$	$[5]_4$	$[6]_4$

$\times$	$[0]_4$	$[1]_4$	$[2]_4$	$[3]_4$
$[0]_4$	$[0]_4$	$[0]_4$	$[0]_4$	$[0]_4$
$[1]_4$	$[0]_4$	$[1]_4$	$[2]_4$	$[3]_4$
$[2]_4$	$[0]_4$	$[2]_4$	$[4]_4$	$[6]_4$
$[3]_4$	$[0]_4$	$[3]_4$	$[6]_4$	$[9]_4$

However, we would like to be able to consult the addition and multiplication tables for $\mathbb{Z}_4$ *and see quickly which of* $[0]_4, [1]_4, [2]_4, [3]_4$ *is the result of any particular operation. Observe that since* $4 \equiv 0 \mod 4$, $[4]_4 = [0]_4$, *and since* $5 \equiv 1 \mod 4$, $[5]_4 = [1]_4$. *From* $6 \equiv 2 \mod 4$ *we obtain* $[6]_4 = [2]_4$, *and finally, from* $9 \equiv 1 \mod 4$, *we obtain* $[9]_4 = [1]$. *Thus the addition and multiplication tables for* $\mathbb{Z}_4$ *can be written as:*

+	$[0]_4$	$[1]_4$	$[2]_4$	$[3]_4$
$[0]_4$	$[0]_4$	$[1]_4$	$[2]_4$	$[3]_4$
$[1]_4$	$[1]_4$	$[2]_4$	$[3]_4$	$[0]_4$
$[2]_4$	$[2]_4$	$[3]_4$	$[0]_4$	$[1]_4$
$[3]_4$	$[3]_4$	$[0]_4$	$[1]_4$	$[2]_4$

$\times$	$[0]_4$	$[1]_4$	$[2]_4$	$[3]_4$
$[0]_4$	$[0]_4$	$[0]_4$	$[0]_4$	$[0]_4$
$[1]_4$	$[0]_4$	$[1]_4$	$[2]_4$	$[3]_4$
$[2]_4$	$[0]_4$	$[2]_4$	$[0]_4$	$[2]_4$
$[3]_4$	$[0]_4$	$[3]_4$	$[2]_4$	$[1]_4$

The reader is urged to verify that if the definition of addition and multiplication on $\mathbb{Z}$ is extended to the set of all subsets of $\mathbb{Z}$ by

$$X + Y = \{ x + y \mid x \in X, \, y \in Y \} \quad \text{and} \quad XY = \{ xy \mid x \in X, \, y \in Y \},$$

for any $X, Y \subseteq \mathbb{Z}$, then for any positive integer n, and any $a, b \in \mathbb{Z}$, calculated as a sum of subsets of $\mathbb{Z}$, we have $[a]_n + [b]_n = [a+b]_n$, and calculated as a product of subsets of $\mathbb{Z}$, we have $[a]_n [b]_n \subseteq [ab]_n$ with equality not holding in general.

We list the properties enjoyed by addition and multiplication on $\mathbb{Z}_n$ in the next result.

4.5.5 Proposition. *Let n be any positive integer. Then the following hold in $(\mathbb{Z}_n, +, \times)$.*

(i) *Addition and multiplication are associative; that is, for any $[a], [b], [c] \in \mathbb{Z}_n$, $[a] + ([b] + [c]) = ([a] + [b]) + [c]$, and $[a]([b][c]) = ([a][b])[c]$.*

(ii) *Addition and multiplication are commutative; that is, for any $[a], [b] \in \mathbb{Z}_n$, $[a] + [b] = [b] + [a]$, and $[a][b] = [b][a]$.*

(iii) *$[0]$ is an identity for addition, and $[1]$ is an identity for multiplication.*

(iv) *Each $[a] \in \mathbb{Z}_n$ has additive inverse $[-a]$; that is, $-[a] = [-a]$.*

(v) *Multiplication distributes over addition; that is, for any $[a], [b], [c] \in \mathbb{Z}_n$, $[a]([b] + [c]) = [a][b] + [a][c]$.*

Proof. We begin with the proof of (i). Let $[a], [b], [c] \in \mathbb{Z}_n$. Then $[a] + ([b] + [c]) = [a] + [b+c] = [a+(b+c)] = [(a+b)+c] = [a+b] + [c] = ([a] + [b]) + [c]$. Similarly, $[a]([b][c]) = [a][bc] = [a(bc)] = [(ab)c] = [ab][c] = ([a][b])[c]$.

As well, $[a] + [b] = [a + b] = [b + a] = [b] + [a]$, and $[a][b] = [ab] = [ba] = [b][a]$.

Furthermore, $[a] + [0] = [a + 0] = [a]$, and addition is commutative, so $[0] + [a] = [a]$. Similarly, $[a][1] = [a(1)] = [a]$, and multiplication is commutative, so $[1][a] = [a]$ as well. Thus $[0]$ is the additive identity element, and $[1]$ is the multiplicative identity element.

Since $[a] + [-a] = [a-a] = [0]$, and thus, since addition is commutative, $[-a] + [a] = [0]$, $[-a]$ is the additive inverse of $[a]$; that is, $-[a] = [-a]$.

Finally, $[a]([b] + [c]) = [a][b+c] = [a(b+c)] = [ab+ac] = [ab] + [ac] = [a][b] + [a][c]$. $\square$

The reader should compare these properties with those of the integers listed at the beginning of this chapter. It will be seen that every property of the integers except property (viii) has a matching property for $(\mathbb{Z}_n, +, \times)$ listed above. This means that any computation which is valid for the integers and which does not require the use of property (viii) will also be valid for $(\mathbb{Z}_n, +, \times)$.

We shall see shortly that if the modulus n is prime, then property (viii) of the integers does indeed hold for $(\mathbb{Z}_n, +, \times)$ but when n is not prime, then property (viii) does not hold. For if n is not prime, we can write $n = rs$ for some integers r, s with $1 < r < n$, $1 < s < n$. Then $[r] \neq [0]$, and $[r][s] = [rs] = [n] = [0] = [r][0]$, but $[s] \neq [0]$.

Before we begin to investigate the situation of a prime modulus, we must establish some preparatory results.

4.5.6 Proposition. *Let n be a positive integer. For any $a, b \in \mathbb{Z}$, if $[a]_n = [b]_n$, then $(a, n) = (b, n)$.*

Proof. Suppose that $a, b \in \mathbb{Z}$ are such that $[a]_n = [b]_n$. Then $n \mid a - b$, so there exists $q \in \mathbb{Z}$ such that $a - b = qn$. Thus $a = qn + b$. If $a = b = 0$, then $(a, n) = n = (b, n)$. Otherwise, not both a, b are zero, and then by Lemma 4.3.6, $(a, n) = (n, b)$, and $(n, b) = (b, n)$, so $(a, n) = (b, n)$. $\qquad\square$

4.5.7 Definition. *An element a of a commutative algebraic system $(S, \cdot)$ is said to be cancellable if, for all elements $x, y \in S$, $ax = ay$ implies that $x = y$.*

Note. When we say that an element of $\mathbb{Z}$ or $\mathbb{Z}_n$ is cancellable, we tacitly mean that it is cancellable with respect to multiplication, since every element of $\mathbb{Z}$ or $\mathbb{Z}_n$ is additively cancellable.

4.5.8 Theorem. *Let n be a positive integer. The following are equivalent for an element $[a]_n \in \mathbb{Z}_n$.*

    (i) $[a]_n$ has a multiplicative inverse in $\mathbb{Z}_n$;
    (ii) $(a, n) = 1$.
    (iii) For each $b \in [a]_n$, $(b, n) = 1$.
    (iv) $[a]_n$ is cancellable.

Proof. (i) implies (ii). Suppose that $[a]_n$ is invertible, with multiplicative inverse $[b]_n$. Then $[a]_n [b]_n = [1]_n$. Since $[a]_n [b]_n = [ab]_n$, it follows from Lemma 4.5.6 that $(ab, n) = (1, n) = 1$. Since (a, n) is a divisor of ab and of n, it follows that (a, n) is a divisor of $(ab, n) = 1$ and thus $(a, n) = 1$.

Next, (ii) implies (iii). If $(a, n) = 1$, then by Lemma 4.5.6, for each $b \in [a]_n$, $(b, n) = (a, n) = 1$.

(iii) implies (iv). Suppose that for each $b \in [a]_n$, $(b, n) = 1$. Then in particular, $(a, n) = 1$. Let $[x]_n, [y]_n \in \mathbb{Z}_n$ and suppose that $[a]_n [x]_n = [a]_n [y]_n$. Then $[ax]_n = [ay]_n$, and so n is a divisor of $ax - ay = a(x - y)$. Since $(a, n) = 1$, it follows from Proposition 4.4.3 that n is a divisor of $x - y$, and so $[x]_n = [y]_n$. Thus $[a]_n$ is cancellable.

Finally, (iv) implies (i). Suppose that $[a]_n$ is cancellable. Then the map called "left multiplication by $[a]_n$", denoted by $l_{[a]}$, from $\mathbb{Z}_n$ to $\mathbb{Z}_n$ is injective; that is, from $l_{[a]}([x]) = l_{[a]}([y])$ we obtain $[x] = [y]$. But $\mathbb{Z}_n$ is finite, and an injective map from a finite set to any finite set of the same size must also be surjective. Thus $l_{[a]}$ is surjective. In particular, there exists

$[b] \in \mathbb{Z}_n$ such that $l_{[a]}([b]) = [1]$; that is, $[a][b] = [1]$. Since multiplication is commutative, we have $[b][a] = [1]$ as well, and thus $[b] = [a]^{-1}$. □

Note. It is important for us to be able to find $[a]_n^{-1}$ when $[a]_n$ is invertible. We have seen how to use the Euclidean Algorithm for integers a, b, not both zero, to find integers r, s such that $(a, b) = ra + sb$. Suppose that $[a]_n$ is invertible. Then as shown above, $(a, n) = 1$ and thus there exist integers b, s such that $1 = ba + sn$. We conclude that n divides $1 - ba$, so $[ba]_n = [1]_n$; that is, $[b]_n [a]_n = [1]_n$, and thus $[b]_n = [a]_n^{-1}$.

4.5.9 Corollary. *Let n be a positive integer. Then every nonzero element of $\mathbb{Z}_n$ is invertible if, and only if, n is prime.*

Proof. If n were not prime, then n would have a factor r with $1 < r < n$, and then $[r]_n$ would be a nonzero, noninvertible element of $\mathbb{Z}_n$ since $(r, n) = r > 1$. Thus if every nonzero element of $\mathbb{Z}_n$ is invertible, then n is prime. Conversely, if n is prime, then for any i with $0 < i < n$, $(i, n) = 1$ and thus $[i]_n$ is invertible. □

4.5.10 Definition. *Let n be a positive integer. Then $[a] \in \mathbb{Z}_n$ is said to be a zero-divisor in $\mathbb{Z}_n$ if there exists a nonzero $[b] \in \mathbb{Z}_n$ with $[a][b] = [0]$ (note that in particular, $[0]$ is a zero-divisor), while $[a]$ is said to be a unit if $[a]$ is invertible.*

Let the set of all units of $\mathbb{Z}_n$ be denoted by $U(\mathbb{Z}_n)$, or more simply, by U_n, and the set of all zero-divisors of $\mathbb{Z}_n$ be denoted by $J(\mathbb{Z}_n)$.

4.5.11 Proposition. *Let $n \geq 2$ be an integer. Then $\mathbb{Z}_n = U_n \cup J(\mathbb{Z}_n)$ and $U_n \cap J(\mathbb{Z}_n) = \varnothing$ (if $n = 1$, then $[1]_1 = [0]_1$ and $U_1 = \mathbb{Z}_1 = J_1$).*

Proof. Let $n \geq 2$ be an integer. We first prove that if $[a]_n \notin U_n$, then $[a]_n \in J(\mathbb{Z}_n)$. Suppose $[a]_n \notin U_n$. Then by Theorem 4.5.8, $(a, n) = d > 1$, so $a = a_1 d$ and $n = n_1 d$ for some $a_1, n_1 \in \mathbb{Z}$. If $d = n$, then $[a]_n = [0]_n \in J(\mathbb{Z}_n)$. Suppose that $d < n$. Then $[d]_n \neq [0]_n$, and $1 < n_1 < n$, so $[n_1]_n \neq [0]_n$, but $[a]_n [n_1]_n = [a_1 d n_1]_n = [a_1 n]_n = [0]_n$, so $[a]_n \in J(\mathbb{Z}_n)$. Thus $\mathbb{Z}_n = U_n \cup J(\mathbb{Z}_n)$.

Suppose now that $U_n \cap J(\mathbb{Z}_n) \neq \varnothing$, and let $[a]_n \in U_n \cap J(\mathbb{Z}_n)$. Since $[a]_n \in J(\mathbb{Z}_n)$, there exists $[b]_n \in \mathbb{Z}_n$ with $[b]_n \neq [0]_n$ such that $[a]_n [b]_n = [0]_n$. However, since $[a]_n \in U_n$, there exists $[c]_n$ such that $[c]_n [a]_n = [1]_n$, and thus $[b]_n = ([c]_n [a]_n)[b]_n = [c]_n([a]_n [b]_n) = [c]_n [0]_n = [0]_n$, which is a contradiction. Thus $U_n \cap J(\mathbb{Z}_n) = \varnothing$. □

4.5.12 Example. $\mathbb{Z}_{12} = \{[0], [1], \ldots, [11]\}$, $U_{12} = \{[1], [5], [7], [11]\}$, *and* $J(\mathbb{Z}_{12}) = \mathbb{Z}_{12} - U_{12} = \{[0], [2], [3], [4], [6], [8], [9], [10]\}$. *Moreover,* $[1]^{-1} = [1]$, $[5]^{-1} = [5]$, $[7]^{-1} = [7]$, *and* $[11]^{-1} = [11]$.

4.5.13 Example. $\mathbb{Z}_{10} = \{[0], [1], \ldots, [9]\}$, $U_{10} = \{[1], [3], [7], [9]\}$, *and* $J(\mathbb{Z}_{10}) = \mathbb{Z}_{10} - U_{10} = \{[0], [2], [4], [5], [6], [8]\}$. *Moreover,* $[1]^{-1} = [1]$, $[3]^{-1} = [7]$, $[7]^{-1} = [3]$, *and* $[9]^{-1} = [9]$.

Linear Congruences in $\mathbb{Z}$ and Linear Equations in $\mathbb{Z}_n$

Equations of the form $ax + ny = b$, where a, n, and b are given integers and integral solutions x and y are sought, are called *linear Diophantine equations* after Diophantus of Alexandria (circa 250 CE) who first studied them. It is an easy matter to see that solving an equation $ax + ny = b$ for $x, y \in \mathbb{Z}$ is equivalent to solving the linear congruence $ax \equiv b \mod n$ (as $n \mid ax - b$ means that there exists $y \in \mathbb{Z}$ such that $ax - b = ny$, or $ax - ny = b$), and this in turn is equivalent to solving the linear equation $[a]_n X = [b]_n$ for $X \in \mathbb{Z}_n$. In this section, we develop an algorithm for the solution of a linear equation in $\mathbb{Z}_n$.

We should point out that if n is composite, then not every linear equation in $\mathbb{Z}_n$ has a solution. For example, in $\mathbb{Z}_{12}$, the linear equation $[8]X = [3]$ does not have a solution. To see this, observe that if $X \in \mathbb{Z}_n$, say $X = [x]$, were such that $[8]X = [3]$, then $x \in \mathbb{Z}$ would be an integer for which $8x - 3$ is a multiple of 12. But then $3 = 8x - 12k$ for some $k \in \mathbb{Z}$, which means that $2 \mid 3$. Since this is not the case, we conclude that $[8]X = [3]$ has no solution in $\mathbb{Z}_{12}$.

Observe that if we consider $[a]_n X = [b]_n$ in the case when $(a, n) = 1$, then $[a]_n$ has a multiplicative inverse in $\mathbb{Z}_n$, say $[a]_n^{-1} = [r]_n$, and then $X = [r]_n [a]_n X = [r]_n [b]_n$ is the unique solution in $\mathbb{Z}_n$. For example, consider the equation $[5]X + [8] = [3]$ in $\mathbb{Z}_{12}$. This is equivalent to the equation $[5]X = [3] - [8] = [-5] = [7]$. Since $(5, 12) = 1$, $[5]$ is invertible in $\mathbb{Z}_{12}$. In fact, since $[5]^2 = [25] = [1]$ in $\mathbb{Z}_{12}$, $[5]^{-1} = [5]$. Thus "to divide by $[5]$", we multiply by the inverse of $[5]$, which is $[5]$ itself, and so we find that $X = [5][7] = [35] = [-1] = [11]$.

Lest the reader believe that the reason no solution exists in the first example above is that $[8]_{12}$ is a zero divisor in $\mathbb{Z}_{12}$, consider the equation $[8]X = [4]$ in $\mathbb{Z}_{12}$. This equation has four solutions, namely $X = [2]$, $[5]$, $[8]$, and $[11]$, so the situation is more complicated than we might be led to believe. We shall give the full story shortly.

Note that solving the equation

$$[5]X + [8] = [3]$$

in $\mathbb{Z}_{12}$ is equivalent to solving

$$5x + 8 \equiv 3 \mod 12.$$

which in turn is equivalent to solving the linear Diophantine equation

$$5x + 12y = -5.$$

We have chosen to emphasize equations in $\mathbb{Z}_n$ rather than congruences because we feel that, conceptually, equations are easier to grasp.

Before we continue on with our study of linear equations in $\mathbb{Z}_n$, we briefly consider an example of a quadratic equation in $\mathbb{Z}_n$ in the case of a prime modulus.

4.6.1 Example. *Consider the quadratic equation* $[2]X^2 + [3]X + [5] = [0]$ *in* $\mathbb{Z}_7$. *Since* $[2]$ *has multiplicative inverse* $[4]$ *in* $\mathbb{Z}_7$, *the solutions to this equation are the same as those for the equation* $[4]([2]X^2 + [3]X + [5]) = [0]$; *that is,* $X^2 + [12]X + [20] = [0]$. *Since* $[12] = [5]$ *and* $[20] = [6]$ *in* $\mathbb{Z}_7$, *we see that we wish to solve the quadratic equation* $X^2 + [5]X + [6] = [0]$ *in* $\mathbb{Z}_7$. *Note that* $(X + [2])(X + [3]) = X^2 + ([2] + [3])X + [2][3] = X^2 + [5]X + [6]$, *so we are to solve* $(X + [2])(X + [3]) = [0]$ *in* $\mathbb{Z}_7$. *Since 7 is a prime, the only zero divisor of* $\mathbb{Z}_7$ *is* $[0]$, *so we must have either* $X + [2] = [0]$ *or else* $X + [3] = [0]$; *that is,* $X = -[2] = [5]$, *or* $X = -[3] = [4]$. *Thus there are exactly two solutions to the quadratic equation* $[2]X^2 + [3] + [5] = [0]$ *in* $\mathbb{Z}_7$, $X = [4]$ *or* $[5]$.

Observe that we could use the same completion of the square techniques that result in the formula for the roots of a quadratic polynomial with real coeficients to obtain a similar result in $\mathbb{Z}_n$ *when* n *is prime. For consider* $[a]X^2 + [b]X + [c] = [0]$, *where* $[a] \neq [0]$. *Then* $[a]^{-1}$ *exists in* $\mathbb{Z}_n$ *since* n *is prime, and so we may reformulate the equation as* $X^2 + [a]^{-1}[b]X + [a]^{-1}[c] = [0]$. *If* $[2]$ *is invertible in* $\mathbb{Z}_n$ *(which is the case if* n *any prime different from 2), then we may compute*

$$(X + [2a]^{-1}[b])^2 = X^2 + [a]^{-1}[b] + ([2a]^{-1}[b])^2,$$

and thus the equations

$$[a]X^2 + [b]X + [c] = [0]$$

and

$$(X + [2a]^{-1}[b])^2 - ([2a]^{-1}[b])^2 + [c] = [0]$$

have the same solutions. We may rewrite the last equation in the equivalent form

$$(X + [2a]^{-1}[b])^2 = ([2a]^{-1}[b])^2 - [c] = ([2a]^{-1})^2([b]^2 - [4a][c]).$$

If there exists (and there need not) $[d] \in \mathbb{Z}_n$ such that $[d]^2 = ([2a]^{-1})^2([b]^2 - [4a][c])$; that is, $([2][a][d])^2 = [b]^2 - [4a][c]$, then we write

$$\sqrt{[b]^2 - [4a][c]} = [2][a][d],$$

or

$$[d] = ([2][a])^{-1}\sqrt{[b]^2 - [4][a][c]},$$

and in such a case, we have $(X + [2a]^{-1}[b])^2 = [d]^2$. But then

$$(X + [2a]^{-1}[b])^2 - [d]^2 = [0],$$

and upon factoring the difference of squares, we obtain

$$(X + [2a]^{-1}[b] - [d])(X + [2a]^{-1}[b] + [d]) = [0]$$

and so, as before, we have $X + [2a]^{-1}[b] - [d] = [0]$ or $X + [2a]^{-1} + [d]) = [0]$; that is, $X - -[2a]^{-1}[b] + [d]$ or $X = -[2a]^{-1}[b] - [d]$. This would usually be written as $X = [2a]^{-1}(-[b] \pm [d])$, or

$$X = [2a]^{-1}(-[b] \pm \sqrt{[b]^2 - 4[a][c]}).$$

If however, there is no element $[d]$ for which $[d]^2 = [b]^2 - 4[a][c]$, then the quadratic equation has no solution in $\mathbb{Z}_n$. For example, consider the quadratic equation $X^2 + [5]X + [2] = [0]$ in $\mathbb{Z}_7$. In $\mathbb{Z}_7$, we have

X	$[0]$	$[1]$	$[2]$	$[3]$	$[4]$	$[5]$	$[6]$
$X^2 + [5]X + [2]$	$[2]$	$[1]$	$[2]$	$[5]$	$[3]$	$[3]$	$[5]$

so $X^2 + [5]X + [2] = [0]$ has no solution in $\mathbb{Z}_7$.

Equations of the form $[a]_n X = [b]_n$ in $\mathbb{Z}_n$.

We now focus our attention on the problem of finding all $X \in \mathbb{Z}_n$ for which $[a]_n X = [b]_n$ for given $[a]_n, [b]_n \in \mathbb{Z}_n$, and n is a fixed positive integer.

4.6.2 Theorem. *Let n be a positive integer. For any $a, b \in \mathbb{Z}$, the equation $[a]_n X = [b]_n$ has a solution $X \in \mathbb{Z}_n$ if and only if $(a, n) \mid b$. Moreover, if $(a, n) \mid b$, then for $d = (a, n)$, $a = a_0 d$, $n = n_0 d$, and $b = b_0 d$ for unique integers a_0, n_0, and b_0, and for any $x \in \mathbb{Z}$, $X = [x]_n \in \mathbb{Z}_n$ is a solution to $[a]_n X = [b]_n$ if and only if $Y = [x]_{n_0} \in \mathbb{Z}_{n_0}$ is a solution to $[a_0]_{n_0} Y = [b_0]_{n_0}$, and the complete set of solutions for the equation $[a]_n X = [b]_n$ is*

$$\{ [x_0 + kn_0]_n \mid k = 0, 1, 2, \ldots, d - 1 \},$$

a set of size d, where $[x_0]_n$ is any particular solution.

Proof. Let $d = (a, n)$, $a = a_0 d$, and $n = n_0 d$.

First, suppose that $X = [x]_n$ is a solution to the equation $[a]_n X = [b]_n$. Then $[ax]_n = [b]_n$, and so there exists an integer k such that $ax - b = kn$. Then $b = (a_0 x - l n_0) d$, so $d \mid b$.

Conversely, suppose that $d = (a, n)$ is a divisor of b, say $b = b_0 d$. By Theorem 4.3.4, there exist integers r and s such that $d = ra + sn$. Then $b = b_0 d = b_0 ra + b_0 sn$, so n is a divisor of $a(rb_0) - b$ and thus $[arb_0]_n = [b]_n$. If we let $x = rb_0$ and set $X = [x]_n$, then $[a]_n X = [b]_n$. This completes the proof that the equation $[a]_n X = [b]_n$ has a solution if and only if (a, n) divides b.

Suppose now that $d = (a, n)$ does divide b, and let b_0 be such that $b = b_0 d$. For any $x \in \mathbb{Z}$, $ax - b = d(a_0 x - b_0)$, and thus n is a divisor of $ax - b$ if and only if n_0 is a divisor of $a_0 x - b_0$. We have proven now that for $X = [x]_n$ and $Y = [x]_{n_0}$, X satisfies $[a]_n X = [b]_n$ if and only if Y satisfies $[a_0]_{n_0} Y = [b_0]_{n_0}$.

Finally, suppose that $X = [x_0]_n$ is a solution to the equation $[a]_n X = [b]_n$. For each $k = 0, 1, \ldots, d - 1$, $[a_0]_{n_0} [x_0 + k n_0]_{n_0} = [a_0]_{n_0} [x_0]_{n_0} + [k]_{n_0} [n_0]_{n_0} = [b_0]_{n_0} + [0]_{n_0} = [b_0]_{n_0}$, and thus $Y = [x_0 + k n_0]_{n_0}$ is a solution to $[a_0]_{n_0} Y = [b_0]_{n_0}$. As shown above, this means that $[x_0 + k n_0]_n$ is a solution to $[a]_n X = [b]_n$. Moreover, for $k_1, k_2 \in \{0, 1, \ldots, d - 1\}$, with $k_1 > k_2$, $(x_0 + k_1 n_0) - (x_0 + k_2 n_0) = n_0 (k_1 - k_2) < n_0 d = n$, and so $[x_0 + k_1 n_0]_n \neq [x + 0 + k_2 n_0]_n$. It follows that

$$\{ [x_0 + k n_0]_n \mid k = 0, 1, 2, \ldots, d - 1 \}$$

is a set of d distinct solutions to the equation $[a]_n X = [b]_n$. Suppose now that $[x]_n$ is a solution to the equation. Apply the division theorem to $x - x_0$ and n to obtain integers q and r for which $x - x_0 = qn + r$ and $0 \leq r < n$. Since $[a]_n [x]_n = [b]_n = [a]_n [x_0]_n$, it follows that

$$[a]_n ([x]_n - [x_0]_n) = [a]_n [x]_n - [a]_n [x_0]_n = [0]_n,$$

which means that n divides $a(x - x_0)$, and so n_0 divides $a_0(x - x_0)$. As above, there exist integers s, t such that $d = sa + tn$, so $1 = sa_0 + tn_0$. Thus $(a_0, n_0) = 1$, and so by Proposition 4.4.3, n_0 divides $x - x_0 = qn + r = qdn_0 + r$. It follows that n_0 divides r, say $r = kn_0$. Since $0 \leq r < n = dn_0$, we have $0 \leq kn_0 < dn_0$ and so $0 \leq k < d$, or $0 \leq k \leq d - 1$. Thus $x - x_0 = qn + kn_0$, and so $x = qn + x_0 + kn_0$, which means that $[x]_n = [qn + x_0 + kn]_n = [x_0 + kn]_n$, with $0 \leq k \leq d - 1$, as required. □

The preceding theorem tells us under what circumstances a general linear equation over $\mathbb{Z}_n$ has at least one solution. It also tells us how many solutions there are and what they are, provided we can find one solution. It also gave us a clue as to how that one solution could be found. For it was established that $X = [x]_n$ is a solution to $[a]_n X = [b]_n$ if and only if $Y = [x]_{n_0}$ is a solution to $[a_0]_{n_0} Y = [b_0]_{n_0}$, where $d = (a, n)$, $a = a_0 d$, $n = n_0 d$, and $b = b_0 d$ (since it was shown that b must be divisible by d if the equation has a solution). Moreover, in the proof, it was shown that $(a_0, n_0) = 1$, and so if we apply the theorem to the equation $[a_0]_{n_0} Y = [b_0]_{n_0}$, we find that this equation has a single solution. After a bit of reflection, it becomes evident that this is true because $[a_0]_{n_0}$ is an invertible element of $\mathbb{Z}_{n_0}$; that is, there exists $[c]_{n_0} \in \mathbb{Z}_{n_0}$ such that $[c]_{n_0} [a_0]_{n_0} = [1]_{n_0}$. Then $[a_0]_{n_0} Y = [b_0]_{n_0}$ if and only if $Y = [c]_{n_0} [b_0]_{n_0}$. These observations lead to the next result.

4.6.3 Proposition. *Let n be a positive integer, and let $a \in \mathbb{Z}$ be such that $(a, n) = 1$. Let $c, e \in \mathbb{Z}$ be such that $1 = ca + en$. Then $X = [cb]_n$ is the unique solution to the equation $[a]_n X = [b]_n$.*

Proof. We have $[a]_n ([c]_n [b]_n) = ([a]_n [c]_n)[b]_n = [1]_n ([b]_n = [b]_n$, so $X = [cb]_n$ is a solution to the equation $[a]_n X = [b]_n$. As observed above, this equation has a unique solution, and so this unique solution is $[cb]_n$. $\square$

Here is a summary of the steps to take to solve the equation $[a]_n X = [b]_n$.

(i) Use the Euclidean algorithm to calculate $d = (a, n)$ and determine whether or not d divides b. If not, the equation has no solution. Otherwise, continue on to the next step.

(ii) Calculate a_0, n_0, and b_0 for which $a = a_0 d$, $n = n_0 d$, and $b = b_0 d$. Necessarily, $(a_0, n_0) = 1$.

(iii) Use the data from the use of the Euclidean algorithm in step (i) to find integers r, s such that $d = ra + sn$, so $1 = ra_0 + sn_0$. Then $[a_0]_{n_0}^{-1} = [r]_{n_0}$. Set $x_0 = rb_0$. Then $Y = [x_0]_{n_0}$ is a solution to $[a_0]_{n_0} Y = [b_0]_{n_0}$, and so $X = [x_0]_n$ is a solution to $[a]_n X = [b]_n$.

(iv) The complete set of solutions (a set of size d) to the equation $[a]_n X = [b]_n$ is then

$$\{\, [x_0 + kn_0]_n \mid k = 0, 1, 2, \ldots, d-1 \,\}.$$

We illustrate with the following example.

4.6.4 Example. *Solve* $[84]_{198} X = [54]_{198}$.

(i) *We use the Euclidean algorithm to calculate* $(198, 84)$. *We find that* $198 = 2(84) + 30$, $84 = 2(30) + 24$, $30 = 1(24) + 6$, *and* $24 = 4(6)$, *so* $d = (198, 84) = 6$. *Since* $54 = 9(6)$, *the equation has solutions.*

(ii) $a_0 = 84/6 = 14$, $n_0 = 198/6 = 33$, *and* $b_0 = 54/6 = 9$.

(iii) *From step (i), we have* $6 = 30 - (84 - 2(30)) = -84 + 3(198 - 2(30)) = 3(198) - 7(84)$, *so for* $r = -7$ *and* $s = 3$, *we have* $6 = r(84) + s(198)$ *and thus* $1 = r(14) + s(33)$. *To verify that we have not made any computational errors, we compute* $(-7)(14) + 3(33) = -98 + 99 = 1$, *as expected. Set* $x_0 = rb_0 = (-7)(9) = -63$. *Then* $X = [-63]_{198} = [-63 + 198]_{198} = [135]_{198}$ *is a solution to the equation* $[84]_{198} X = [54]_{198}$.

(iv) *The complete set of solutions (a set of size 6) to the equation* $[84]_{198} X = [54]_{198}$ *is then*

$$\{ [135 + 33k]_{198} \mid k = 0, 1, 2, 3, 4, 5 \}.$$

Note that the problem in the preceding example could be formulated as a problem of solving a congruence in $\mathbb{Z}$. In this case, the equation would become the linear congruence $84x \equiv 54 \mod 198$, and we would be looking for the set of all integers x for which this congruence held. This set would be the union of the 6 congruence classes $[135 + 33k]_{198}$, $k = 0, 1, 2, 3, 4, 5$, or equivalently, as the single congruence class $[135]_{33} = [3]_{33}$.

4.7   Exercises

1. In each of the following, use induction to prove that the identity is valid for all positive integers.

    a) $\sum_{i=1}^{n} (2i - 1) = n^2$.

    b) $\sum_{i=1}^{n} i^2 = \frac{n(n+1)(2n+1)}{6}$.

    c) $\sum_{i=1}^{n} i^3 = \frac{n^2(n+1)^2}{4}$.

    d) $\sum_{i=1}^{n} \frac{1}{i(i+1)} = \frac{n(n+1)(n+2)}{3}$.

    e) $\sum_{i=1}^{n} \frac{1}{(i+1)^2 - 1} = \frac{3}{4} - \frac{1}{2(n+1)} - \frac{1}{2(n+2)}$.

2. Use induction to prove that if p is prime and $a_1, a_2, \ldots, a_n \in \mathbb{Z}$ are such that $p \mid a_1 a_2 \cdots a_n$, then $p \mid a_i$ for some i with $1 \leq i \leq n$.

3. Use induction to prove that for every positive integer n, if $a_1, a_2, \ldots, a_{2^n}$ are positive real numbers, then $(a_1 a_2 \cdots a_{2^n})^{\frac{1}{2^n}} \leq \frac{a_1, a_2 + \cdots + a_{2^n}}{2^n}$.

4. Use induction to prove that for every positive integer n, $6(7^n) - 2(3^n)$ is divisible by 4.

5. Prove that any postage of 4 cents or more can be achieved by using 2-cent and 5-cent stamps only.

6. Use induction to prove that any n distinct straight lines in the plane, no three of which have a common point, divide the plane into $\frac{(n^2+n+2)}{2}$ regions.

7. In each part below, find the greatest common divisor of the two given integers, and express the greatest common divisor as a linear combination of the two integers.

   a) 12 and 138.

   b) 130 and 210.

   c) 34 and 136.

   d) 17 and 164.

   e) 1211 and -6203.

8. For any integers a, b, let $a \vee b$ denote the maximum of a and b, and let $a \wedge b$ denote the minimum of a and b. For example, $(-5) \vee (-3) = -3$, and $2 \wedge (-5) = -5$. Note that $\vee$ and $\wedge$ are commutative; that is, $a \vee b = b \vee a$ and $a \wedge b = b \wedge a$ for all $a, b \in \mathbb{Z}$.

   a) Prove that $\wedge$ distributes over $\vee$; that is, for all $a, b, c \in \mathbb{Z}$, $a \wedge (b \vee c) = (a \wedge b) \vee (a \wedge c)$.

   b) Does $\vee$ distribute over $\wedge$? Either prove that it does, or provide a counterexample (that is, give an example of integers a, b, c such that $a \vee (b \wedge c) \neq (a \vee b) \wedge (a \vee c)$).

   c) Let k be a positive integer, and let $p_1, p_2, \ldots, p_k$ be distinct primes. Prove that for any nonnegative integers $m_1, m_2, \ldots, m_k$, $n_1, n_2, \ldots, n_k$,
   $$(p_1^{m_1} p_2^{m_2} \cdots p_k^{m_k}, p_1^{n_1} p_2^{n_2} \cdots p_k^{n_k}) = p_1^{m_1 \wedge n_1} p_2^{m_2 \wedge n_2} \cdots p_k^{m_k \wedge n_k}$$
   and
   $$[p_1^{m_1} p_2^{m_2} \cdots p_k^{m_k}, p_1^{n_1} p_2^{n_2} \cdots p_k^{n_k}] = p_1^{m_1 \vee n_1} p_2^{m_2 \vee n_2} \cdots p_k^{m_k \vee n_k}.$$

   d) Prove that for any positive integers a, b, c, $(a, [b, c]) = [(a, b), (a, c)]$.

9. Prove that both the set U_n of units of $\mathbb{Z}_n$ and the set $J(\mathbb{Z}_n)$ of zero divisors of $\mathbb{Z}_n$ are closed under multiplication. Are either of these sets closed under addition? Prove or give counterexamples.

10. Compute the multiplicative inverses of the following elements:

a) $[3]_{25}$.

b) $[12]_{31}$.

c) $[5]_{112}$.

11. Prove that if p is prime, then for any $X, Y \in \mathbb{Z}_p$, $(X+Y)^{p^n} = X^{p^n} + Y^{p^n}$ for every positive integer n.

12. a) Prove that for every positive integer n, $[10]_9^n = [1]_9$.

    b) Prove that a positive integer is divisible by 9 if, and only if, the sum of its digits is divisible by 9.

13. If the decimal expansion of an integer m is $a_n a_{n-1} \cdots a_0$, prove that m is divisible by 4 if, and only if, 4 divides $a_1 a_0$. Prove also that m is divisible by 8 if, and only if, 8 divides $a_2 a_1 a_0$ (for example, 8 divides $315,628,957,264$ since 8 divides 264, namely $264 = 33(8)$).

14. For given positive integers m and n, define addition (denoted by $+$) on $\mathbb{Z}_m \times \mathbb{Z}_n$ by

$$([a]_m, [b]_n) + ([c]_m, [d]_n) = ([a]_m + [c]_m, [b]_n + [d]_n)$$

and multiplication on $\mathbb{Z}_m \times \mathbb{Z}_n$ (denoted by juxtaposition) by

$$([a]_m, [b]_n)([c]_m, [d]_n) = ([a]_m[c]_m, [b]_n[d]_n).$$

Define a map $f : \mathbb{Z}_{mn} \to \mathbb{Z}_m \times \mathbb{Z}_n$ by $f(X) = ([a]_m, [a]_n)$ if $X = [a]_{mn}$.

    a) Prove that f is well-defined; that is, if $[a]_{mn} = [b]_{mn}$, then $[a]_m = [b]_m$ and $[a]_n = [b]_n$.

    b) Prove that $f(X + Y) = f(X) + f(Y)$ and $f(XY) = f(X)f(Y)$ for all $X, Y \in \mathbb{Z}_{mn}$.

    c) Prove that f is a bijection if m and n are relatively prime.

    d) Describe the set of elements of $\mathbb{Z}_m \times \mathbb{Z}_n$ that are invertible with respect to multiplication.

    e) The Euler φ-function is the map $\varphi : \mathbb{Z}^+ \to \mathbb{N}$ that is defined by $\varphi(m) = |\{ i \mid 1 \le i \le m, \ (i, n) = 1 \}|$. For example, $\varphi(12) = 4$, and $\varphi(1) = 1$. Prove that if m and n are relatively prime positive integers, then $\varphi(mn) = \varphi(m)\varphi(n)$. Hint: note that $\varphi(m) = |U_m|$, the number of units of $\mathbb{Z}_m$.

    f) Prove by induction on k that if $p_1, p_2, \ldots, p_k$ are distinct primes, then for any positive integers $m_1, m_2, \ldots, m_k$, $\varphi(p_1^{m_1} p_2^{m_2} \cdots p_k^{m_k}) = \varphi(p_1^{m_1})\varphi(p_2^{m_2}) \cdots \varphi(p_k^{m_k})$.

    g) Prove that if p is prime, then for every positive integer n, $\varphi(p^n) = p^n - p^{n-1}$.

h) Let n be an integer with $n > 1$. Let P denote the set of all prime divisors of n. Prove that $\varphi(n) = n \prod_{p \in P}(1 - \frac{1}{p})$. For example, we have observed that $\varphi(12) = 4$, and for $n = 12$, $P = \{2, 3\}$. We compute $12(1 - \frac{1}{2})(1 - \frac{1}{3}) = 6(\frac{2}{3}) = 4 = \varphi(12)$.

15. Let x, y, and z be pairwise relatively prime positive integers such that $x^2 + y^2 = z^2$. Prove that one of x and y is odd and the other even. Also prove that one of x and y is divisible by 3.

16. Solve the following linear equations:

   a) $[3]_7 X = [2]_7$.

   b) $[15]_{25} X = [10]_{25}$.

   c) $[980]_{1600} X = [1500]_{1600}$.

   d) $[5]_{35} X = [9]_{35}$.

17. Let p be a prime with $p > 2$. Prove that for any $[a]_p, [b]_p, [c]_p \in \mathbb{Z}_p$ with $[a]_p \neq [0]_p$, $[a]_p X^2 + [b]_p X + [c]_p = [0]_p$ has solutions $X = ([2]_p [a]_p)^{-1}(-[b]_p \pm \sqrt{[b]_p^2 - [4]_p [a]_p [c]_p})$ (note that just as in the case of quadratic equations over the real numbers, there is a solution if and only if $[b]_p^2 - [4]_p [a]_p [c]_p = [d]_p^2$ for some $[d]_p \in \mathbb{Z}_p$).

18. (Fermat's Little Theorem). If p is prime and $a \in \mathbb{Z}$ is such that $(a, p) = 1$, then $([a]_p)^{p-1} = [1]_p$ (equivalently, $a^{p-1} \equiv 1 \mod p$).

19. (Wilson's theorem). An integer p is prime if, and only if, $(p-1)! \equiv -1 \mod p$.

20. Let p be a prime. Prove that in $\mathbb{Z}_p$, every element is a square if, and only if, $p = 2$.

21. Let n be an integer with $n \geq 2$. For integers $a_1, a_2, \ldots, a_n$, not all zero, a positive integer d is said to be a greatest common divisor of $a_1, a_2, \ldots, a_n$ if and only if d divides each a_i, $i = 1, 2, \ldots, n$, and if c is any integer such that $c \mid a_i$ for each $i = 1, 2, \ldots, n$, then $c \mid d$.

   a) Use induction on $n \geq 2$ to prove that for any integers $a_1, a_2, \ldots, a_n$ not all zero, if the greatest common divisor exists, it is unique. Denote the greatest common divisor of $a_1, a_2, \ldots, a_n$ (when it exists) by $(a_1, a_2, \ldots, a_n)$. Prove that there exist integers $r_1, r_2, \ldots, r_n$ such that $(a_1, a_2, \ldots, a_n) = \sum_{i=1}^{n} r_i a_i$.

   b) Use induction on n to prove that for any $n \geq 2$, for any integers $a_1, a_2, \ldots, a_n$ not all zero, $(a_1, a_2, \ldots, a_n)$ exists and, if $n \geq 3$, $(a_1, a_2, \ldots, a_n) = (a_1, (a_2, a_3, \ldots, a_n))$.

c) Compute $(30, 45, 105, 210)$ and find integers a, b, c, d such that $(30, 45, 105, 210) = 30a + 45b + 105c + 210d$.

Chapter 5

Groups

Introduction

As early as 2000 BCE, the Babylonians knew how to solve quadratic equations and by the 16th century, formulae for the roots of cubics and quartics were known. In spite of the efforts of the finest mathematicians of the day, however, the solution of the general quintic proved to be elusive until, in the 1820's, the famous Norwegian mathematician, Niels Henrik Abel, using sets of permutations of the roots of equations (we would call them groups of permutations nowadays), proved the insolvability of the general quintic. More precisely, he proved that there does not exist a formula involving the usual arithmetic operations (addition, subtraction, multiplication, division and the extraction of roots) for the roots of a general quintic.

Although the abstract notion of a group was not formulated until the 1880's, Abel is given credit for showing the important role these sets of permutations play and we honour him by referring to a commutative group as an abelian group.

In the 1830's, Évariste Galois, an iconoclastic but brilliant French mathematician, developed what we now call Galois Theory in which groups play an important role. He showed why the general quintic cannot be solved in terms of the usual arithmetic operations listed above by associating a group with each equation and showing that in order for a polynomial equation to have a solution of the prescribed form, it is necessary and sufficient for the associated group to have a certain property which we now call solvability.

From the time that group theory emerged as an independent discipline, it has been found indispensable within mathematics and in applications in fields other than mathematics such as physics and computer science. The basic reason that it is such a powerful tool rests on the fact that it is eminently useful in describing symmetry in nature and mathematics.

5.2   Definitions and Elementary Properties

5.2.1 Definition. *A group is a set G together with a binary operation $*$ on G which satisfies the following three requirements:*

(i) *For all $a, b, c \in G$, $a * (b * c) = (a * b) * c$ (that is; the operation is associative).*

(ii) *There exists an element $e \in G$ such that for all $a \in G$, $e * a = a = a * e$ (that is; e is an identity for the operation).*

(iii) *For each $a \in G$, there exists $b \in G$ such that $a * b = e = b * a$ (every element is invertible).*

If in addition, for all $a, b \in G$, $a * b = b * a$, the group is said to be commutative, or abelian.

A few remarks are in order. In Proposition 3.4.16, it was established that a binary operation can have at most one identity element, so we may speak of "the" identity element of the group rather than "an" identity element.

Furthermore, it was established in Proposition 3.4.19 that in any monoid, an element can have at most one inverse, and so in any group G, each element has a unique inverse. This makes it possible to introduce notation to denote this unique inverse by means of a decoration added to the name of the element. As it turns out, our notation is influenced by our experience with the real number system, on which there are two important binary operations, multiplication and addition. Consequently, there are two different notations that may be used, one called multiplicative notation, and the other called additive notation (see the discussion on notation for binary operations, Notation 3.4.22). If, for a group G, we decide that we would like to use multiplicative notation (which will be our default choice of notation), this will be indicated by the use of juxtaposition to denote the result of the binary operation; that is, for $a * b$, we simply write ab. As well, the inverse of an element $a \in G$ shall be denoted by a^{-1}. On the other hand, if we have decided to use additive notation, we will indicate this by using the $+$ sign to denote the binary operation; that is; for $a * b$, we will write $a + b$, while the inverse of an element $a \in G$ will be denoted by $-a$. It is important for the reader to realize that these are just conventions. When a result is formulated and proven for an abstract group, we whall use multiplicative notation. When this result is to be applied to some particular group, if additive notation has been chosen for the particular group, then the statement of the result will have to be translated from multiplicative notation into additive notation before it can be applied.

Finally, since the operation is associative, the value of a product with several factors is independent of the choice of a computational recipe for its evaluation, it is conventional not to write brackets unless we wish to emphasize some particular factor in the product. Thus for example, $(a((bc)d))e$ would simple be written as $abcde$.

Our next result is a fundamental source of examples of groups.

5.2.2 Proposition. *Let M be a monoid, and let $U(M) = \{\, x \in M \mid x \text{ is invertible} \,\}$. Then $U(M)$ is a group, called the group of units of the monoid M.*

Proof. That $U(M)$ is closed under the binary operation of M follows from Proposition 3.4.23. Since the identity e of M is invertible (it is its own inverse), $e \in U(M)$. Thus the restriction of the associative binary operation on M to the subset $U(M)$ provides an associative binary operation on $U(M)$. Moreover, the identity of M is the identity of $U(M)$ and the inverse of each element of $U(M)$ is again an invertible element of M, hence belongs to $U(M)$, so each element of $U(M)$ has an inverse in $U(M)$. Thus $U(M)$ is a group under the restriction of the binary operation of M to $U(M)$. □

5.2.3 Example. `For any positive integer n, we determine the group of units $U(\mathbb{Z}_n)$ of the monoid $\mathbb{Z}_n$ under multiplication. By Theorem 4.5.8, $[i]_n \in U(\mathbb{Z}_n)$ if, and only if, $(n, i) = 1$. Thus $U(\mathbb{Z}_n) = \{\, [i]_n \mid (n, i) = 1 \,\}$.*

We will give other applications of this result as we proceed through our study of group theory. At this time, we turn our attention to the notion of exponentiation.

5.2.4 Definition. *Let G be a group with identity e. For $a \in G$, define $a^0 = e$, $a^1 = a$, and for $n \geq 1$, $a^{n+1} = (a^n)a$. Finally, for $n < 0$, define $a^n = (a^{-1})^{|n|}$.*

The preceding definition used multiplicative notation. For additive notation, for $n \geq 0$, we would write $(n+1)a = na + a$, and for $n < 0$, $na = |n|(-a)$.

We gather together in the next proposition some of the rules of computation in a group. We give the rules in multiplicative notation, leaving it up to the reader to formulate them in additive notation.

5.2.5 Proposition. *Let G be a group. Then the following hold.*

  (i) *For any $a, b, c \in G$, if either $ab = ac$ or $ba = ca$, then $b = c$*
       (cancellation laws). In particular, if $ab = e$ or $ba = e$, then $b = a^{-1}$.
  (ii) *For any $a, b \in G$, $(ab)^{-1} = b^{-1}a^{-1}$.*
  (iii) *For every $a \in G$, $(a^{-1})^{-1} = a$.*

Proof. For (i), let $a, b, c \in G$. If $ab = ac$, then $b = a^{-1}(ab) = a^{-1}(ac) = c$, while if $ba = ca$, then $b = (ba)a^{-1} = (ca)a^{-1} = c$. If $ab = e$, then $ab = aa^{-1}$ and so $b = a^{-1}$. If $ba = e$, then $ba = a^{-1}a$ and so $b = a^{-1}$.

We now use (i) to prove (ii) and (iii). Let $a, b \in G$. We have $(ab)(b^{-1}a^{-1}) = a(bb^{-1})a^{-1} = aea^{-1} = e$, and thus by (i), $(ab)^{-1} = b^{-1}a^{-1}$. As well, $a(a^{-1}) = e = (a^{-1})^{-1}(a^{-1})$, and thus by (i), $a = (a^{-1})^{-1}$. $\qquad\square$

Our next objective is to establish the fundamental laws of exponents; namely that for all $a \in G$, and for all $m, n \in \mathbb{Z}$, $a^{m+n} = a^m a^n$ and $a^{mn} = (a^m)^n$. We need one preliminary result.

5.2.6 Lemma. *Let G be a group, and let $a \in G$. Then for any $n \in \mathbb{Z}$, $(a^n)^{-1} = a^{-n}$.*

Proof. Let $a \in G$. By Proposition 5.2.5 (i), it suffices to prove that for every $n \in \mathbb{Z}$, $a^n a^{-n} = e$. We first prove the result for $n \geq 1$ by induction. The base case is immediate, so suppose that $n \geq 1$ is an integer such that $a^n a^{-n} = e$. Then $aea^{-1} = a(a^n a^{-n})a^{-1} = a^{n+1}(a^{-1})^n(a^{-1}) = a^{n+1}(a^{-1})^{n+1} = a^{n+1}a^{-(n+1)}$, and thus $a^{n+1}a^{-(n+1)} = aa^{-1} = e$. It follows now by induction that for all $n \geq 1$, $a^n a^{-n} = e$.

Since $a^0 a^{-0} = e(a^{-1})^0 = e$, it remains to prove that for all negative integers n, $a^n a^{-n} = e$; equivalently, for all positive integers n, $a^{-n}a^n = e$. Let n be a positive integer. Then by Proposition 5.2.5 (iii), $a^{-n}a^n = (a^{-1})^n((a^{-1})^{-1})^n = (a^{-1})^n(a^{-1})^{-n} = e$, where the last equality is obtained by applying the result of the first paragraph to the element a^{-1}. This concludes the proof of the lemma. $\qquad\square$

5.2.7 Proposition. *Let G be a group, and let $a \in G$. Then for any $m, n \in \mathbb{Z}$, $a^{m+n} = a^m a^n$ and $(a^m)^n = a^{mn}$.*

Proof. Let $a \in G$. We first prove that for every positive integer n, $a^{m+n} = a^m a^n$ for every positive integer m. The base case holds by definition of a^{m+1} for any positive integer m, so suppose that n is a positive integer such that for every positive integer m, $a^{m+n} = a^m a^n$. Let m be a positive integer. Then $a^{m+n+1} = a^{m+n}a = a^m a^n a = a^m a^{n+1}$, and so the claim follows by induction.

Next, let $a \in G$ and m, n be negative integers. Then $a^{m+n} = (a^{-1})^{(-m)+(-n)} = (a^{-1})^{-m}(a^{-1})^{-n} = a^m a^n$. As well, for any $m \in \mathbb{Z}$, $a^{m+0} = a^m = a^m e = a^m a^0$, and $a^{0+m} = a^m = e a^m = a^0 a^m$.

It remains to prove that for any $a \in G$, and any nonzero $m, n \in \mathbb{Z}$ with different signs, $a^{m+n} = a^m a^n$. Let $a \in G$ and $m, n \in \mathbb{Z}$ with $m > 0$ and $n < 0$. Let $k = m + n$. If $k = 0$, then $n = -m$, and by Lemma 5.2.6, $a^m a^n = a^m a^{-m} = e = a^0 = a^{m+n}$. Suppose $k > 0$. Then $m = k + (-n)$, so $a^m = a^{k+(-n)} = a^k a^{-n}$, and so by Lemma 5.2.6, $a^m a^n = a^k a^{-n} a^n = a^k e = a^k = a^{m+n}$. Finally, suppose that $k < 0$. We have $-k + m = -n$, so $a^{-n} = a^{(-k)+m} = a^{-k} a^m$, and thus $a^k a^{-n} a^n = a^k a^{-k} a^m a^n$. By Lemma 5.2.6 applied to a^k and a^n, we obtain $a^k = a^k e = a^k a^{-n} a^n = a^k a^{-k} a^m a^n = e a^m a^n = a^m a^n$; that is, $a^{m+n} = a^m a^n$, as required.

Finally, let $a \in G$ and $m, n \in \mathbb{Z}$ with $m < 0$ and $n > 0$. We apply the result established in the preceding paragraph to a^{-1}, $-m > 0$, and $-n < 0$ to obtain that $a^{m+n} = (a^{-1})^{(-m)+(-n)} = (a^{-1})^{-m}(a^{-1})^{-n} = a^m a^n$.

Now we prove that for every positive integer n, $(a^m)^n = a^{mn}$ for every integer m. This is true by definition when $n = 1$, so suppose that $n \geq 1$ is an integer for which $(a^m)^n = a^{mn}$ for every integer m. Let $m \in \mathbb{Z}$. Then $(a^m)^{n+1} = (a^m)^n(a^m) = a^{mn}a^m$, and as proven above, $a^{mn}a^m = a^{mn+m} = a^{m(n+1)}$. It now follows by induction that $(a^m)^n = a^{mn}$ for every integer m and every positive integer n. Let $m \in \mathbb{Z}$. Then $(a^m)^0 = e = a^{m(0)}$, and for any positive integer n, $(a^m)^{-n} = ((a^m)^n)^{-1} = (a^{mn})^{-1} = a^{-mn} = a^{m(-n)}$. Thus for all integers m and n, $(a^m)^n = a^{mn}$. $\qquad\square$

5.2.8 Corollary. *Let G be a group, and let $a \in G$. Then for any $m, n \in \mathbb{Z}$, $a^m a^n = a^n a^m$.*

Proof. By Proposition 5.2.7, $a^m a^n = a^{m+n} = a^{n+m} = a^n a^m$. $\qquad\square$

5.2.9 Example.

(i) $(\mathbb{Z}, +)$ *is an abelian group, with identity element* 0.

(ii) $(\mathbb{R}, \times)$ *is a monoid with identity* 1, *but is not a group since* 0 *has no multiplicative inverse. Let* $\mathbb{R}^* = U(\mathbb{R}, \times) = \mathbb{R} - \{0\}$, *the group of units of the monoid* $(\mathbb{R}, \times)$. *Then* $(\mathbb{R}^*, \times)$ *is an abelian group, with identity element* 1.

(iii) Let $R = \mathbb{Z}$, $\mathbb{Q}$, $\mathbb{R}$, or $\mathbb{C}$, and let m and n be positive integers. Then $(M_{m \times n}(R), +)$ is an abelian group (see Proposition 3.4.26 for the proof of this assertion).

(iv) Let $R = \mathbb{Z}$, $\mathbb{Q}$, $\mathbb{R}$, or $\mathbb{C}$. Then for any integer $m > 1$, $M_{m \times m}(R)$ with the operation of matrix multiplication is a nonabelian monoid, with identity I_m.

(v) Let $R = \mathbb{Z}$, $\mathbb{Q}$, $\mathbb{R}$, or $\mathbb{C}$. For any positive integer m, let $Gl_m(\mathbb{R})$ denote the group of units of the monoid $M_{m \times m}(R)$ under matrix multiplication. Then $(Gl_m(R), \times)$ is a group, nonabelian if $m > 1$. $Gl_n(R)$ is called the general linear group of degree n over R. For the proof of this assertion, see Corollary 3.4.30 and Proposition 3.4.23. To see that matrix multiplication is not commutative when $m > 1$, define for any positive integers r and s with $1 \leq r, s \leq n$ the matrix $E_{r,s} \in M_{n \times n}(R)$ for which $(E_{r,s})_{ij} = 0$ for all i and j with $1 \leq i, j \leq n$ and $i \neq r$ and $j \neq s$, while $(E_{r,s})_{rs} = 1$. Then for any integers r, s, k, p with $1 \leq r, s, k, p \leq n$, we have $E_{r,s} E_{k,p} = 0_{n \times n}$ if $s \neq k$, and equals $E_{r,p}$ if $s = k$. Thus if $n \geq 2$, we have $E_{1,2} E_{2,2} = E_{1,2}$, while $E_{2,2} E_{1,2} = 0_{n \times n} \neq E_{1,2}$.

(vi) Let n be a positive integer. Then $(\mathbb{Z}_n, +)$ is an abelian group of order n (the order $|G|$ of a group G is the number of elements in G, and (i)–(v) are groups of infinite order). The identity element is $[0]_n$, and for $[i]_n \in \mathbb{Z}_n$, $-[i]_n = [-i]_n$.

(vii) Let A be a nonempty set, and let $G = \mathscr{P}(A)$. Define an operation $\triangle$ on G by $X \triangle Y = (X - Y) \cup (Y - X)$ for all $X, Y \in G$. $(G, \triangle)$ is an abelian group with identity $\varnothing$, the empty set, and each element of G is its own inverse. The proof of the associativity of $\triangle$ is interesting, and not entirely straightforward, while the verification of the other properties is immediate. The reader is invited to show that for any $X, Y, Z \in G$, $(X \triangle Y) \triangle Z$ is equal to

$$(X - (Y \cup Z)) \cup (Y - (X \cup Z)) \cup (Z - (X \cup Y)) \cup (X \cap Y \cap Z).$$

Since $(X \triangle Y) \triangle Z = Z \triangle (Y \triangle X)$, if we simply switch X and Z in the preceding display, we find that $(X \triangle Y) \triangle Z$ is equal to

$$(Z - (Y \cup X)) \cup (Y - (Z \cup X)) \cup (X - (Z \cup Y)) \cup (Z \cap Y \cap X),$$

and thus $(X \triangle Y) \triangle Z = X \triangle (Y \triangle Z)$ for all $X, Y, Z \in G$. In Figure 5.1, we display the Venn diagram for $X \triangle (Y \triangle Z)$.

(viii) Let A be a nonempty set. In Example 3.4.21, it was shown that $(\mathscr{F}(A), \circ)$ is a monoid with identity element 1_A, the identity mapping on A, and the invertible elements are the bijective mappings

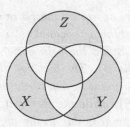

Fig. 5.1   A Venn diagram for $X \triangle (Y \triangle Z)$

from A to A. By Proposition 5.2.2, $U(\mathscr{F}(A))$ is a group and its elements are the bijective mappings from A to itself. It is conventional to refer to $U(\mathscr{F}(A))$ as the symmetric group on the set A, and denote it by S_A. If n is a positive integer and $A = J_n$, then it is customary to write S_n, and to refer to S_n as the symmetric group on n symbols. Since there are n! permutations of a set of n objects, the order of S_n is n!. For example, the order of S_4 is $4! = 24$, and the order of S_{10} is $10! = 3\,628\,800$.

(ix) Let $r = \begin{bmatrix} 0 & -1 \\ 1 & 0 \end{bmatrix} \in M_{2 \times 2}(\mathbb{Z})$. Then $r^2 = -I_2$, and thus $r^3 = -r$ and $r^4 = I_2$, so $r \in Gl_2(\mathbb{Z})$, and $r^{-1} = -r$. Let $h = \begin{bmatrix} 1 & 0 \\ 0 & -1 \end{bmatrix} \in M_{2 \times 2}(\mathbb{Z})$. Then $h^2 = I_2$, so $h \in Gl_2(\mathbb{Z})$ and $h^{-1} = h$. Let $d_+ = \begin{bmatrix} 0 & 1 \\ 1 & 0 \end{bmatrix} \in M_{2 \times 2}(\mathbb{Z})$. Then $(d_+)^2 = I_2$, so $d_+ \in Gl_2(\mathbb{Z})$ and $(d_+)^{-1} = d_+$. Then $-h$ and $-d_+ \in Gl_2(\mathbb{Z})$ and $(-h)^{-1} = -h$, $(-d_+)^{-1} = -d_+$. Let $D = \{\, I_2, r, -I_2, -r, h, -h, d_+, -d_+ \,\}$, so $D \subseteq Gl_2(\mathbb{Z})$. We compute $rh = d_+$, so $(-r)h = r(-h) = -d_+$, and $hr = (r^{-1}h^{-1})^{-1} = ((-r)h)^{-1} = (-d_+)^{-1} = -d_+$, and thus $(-h)r = h(-r) = d_+$. Furthermore, $rd_+ = r(rh) = -h$, so $(-r)d_+ = r(-d_+) = h$, and $hd_+ = h(h(-r)) = -r$, and thus $(-h)d_+ = h(-d_+) = r$. Finally, $d_+r = (-h)r^2 = h$, so $(-d_+)r = d_+(-r) = -h$, and $d_+h = (rh)h = -r$, so $(-d_+)h = d_+(-h) = r$. It follows that D is closed under matrix multiplication, so matrix multiplication provides an associative binary operation on D. Since $I_2 \in D$ and the inverse of every element of D also belongs to D, D, with operation matrix multiplication, is a group. We shall see (Example 5.4.6) that the elements of D represent certain transformations of the plane which leave any square with centre at the origin and sides parallel to the coordinate axes fixed.

(x) Let $R = \mathbb{Q}$, $\mathbb{R}$, or $\mathbb{C}$. For any positive integer n, the set

$$O(n)(R) = \{\, A \in Gl_n(R) \mid A^{-1} = A^t \,\}$$

is a group with binary operation matrix multiplication. The elements of $O(n)(R)$ are called the orthogonal $n \times n$ matrices with entries from R, and $O(n)(R)$ is called the orthogonal matrix group of degree n over R (see Appendix B). The fact that the product of two orthogonal matrices is orthogonal is due to the identity $(AB)^t = B^t A^t$ for any $A, B \in M_{n \times n}(R)$, while the fact that the inverse of an orthogonal matrix is orthogonal is due to the identity $(A^t)^{-1} = (A^{-1})^t$ for every $A \in M_{n \times n}(R)$. In the case of $O(n)(\mathbb{R})$, each $A \in O(n)(\mathbb{R})$ represents an isometry of $\mathbb{R}^n$ fixing the origin; that is, a map of $\mathbb{R}^n$ to itself which preserves distance (and thus also preserves angles).

To exhibit the algebraic structure of a group, we draw the so-called Cayley table of the group. We give a couple of examples.

5.2.10 Example. *Our first example will be the Cayley table for the group* $(\mathbb{Z}_6, +)$. *Recall that the elements of* $\mathbb{Z}_6$ *are* $[0]_6$, $[1]_6$, $[2]_6$, $[3]_6$, $[4]_6$, and $[5]_6$. *In order to reduce the clutter of the notation, we shall write these simply as* 0, 1, 2, 3, 4, *and* 5, *respectively.*

+	0	1	2	3	4	5
0	0	1	2	3	4	5
1	1	2	3	4	5	0
2	2	3	4	5	0	1
3	3	4	5	0	1	2
4	4	5	0	1	2	3
5	5	0	1	2	3	4

Fig. 5.2   The Cayley table for $(\mathbb{Z}_6, +)$

The elements in the first column and the first row are the first and second arguments, respectively, on which the operation of addition is to be performed. Specifically, the entry in the row with first element i and in the column with first entry j is the representative of $[i]_6 + [j]_6$ that belongs to $\{0, 1, 2, 3, 4, 5\}$. For example, $4 + 2 = 0$ since $[4]_6 + [2]_6 = [6]_6 = [0]_6$, and so in the row with first entry 4 and column with first entry 2, we have placed 0. Of course, since addition in $\mathbb{Z}_6$ is commutative, we also have $2 + 4 = 0$, so in the row with first entry 2 and column with first entry 4, a 0 also appears. In fact, that a binary operation is commutative is readily apparent from its Cayley table, since the table will be symmetric about its *main diagonal, by which we mean that for every i and j with $1 \leq i, j \leq n$, the entries in row i, column j and row j, column i are identical.*

The attentive reader will have noticed that in each column of the Cayley table show in Figure 5.2, all six elements of $\mathbb{Z}_6$ appear, each one appearing exactly once in the column. A similar observation applies to each row. That this is inevitable is a consequence of the cancellation laws in a group (see Proposition 5.2.5 (i)) (and the fact that $\mathbb{Z}_6$ is a finite set).

5.2.11 Example. *We construct the Cayley table for the group $D = \{ I_2, -I_2, r, -r, h, -h, d_+, -d_+ \}$ introduced in Example 5.2.9 (ix). The operation is matrix multiplication. Note that the Cayley table for D is not*

$\times$	I_2	$-I_2$	r	$-r$	h	$-h$	d_+	$-d_+$
I_2	I_2	$-I_2$	r	$-r$	h	$-h$	d_+	$-d_+$
$-I_2$	$-I_2$	I_2	$-r$	r	$-h$	h	$-d_+$	d_+
r	r	$-r$	$-I_2$	I_2	d_+	$-d_+$	$-h$	h
$-r$	$-r$	r	I_2	$-I_2$	$-d_+$	d_+	h	$-h$
h	h	$-h$	$-d_+$	d_+	I_2	$-I_2$	$-r$	r
$-h$	$-h$	h	d_+	$-d_+$	$-I_2$	I_2	r	$-r$
d_+	d_+	$-d_+$	h	$-h$	r	$-r$	I_2	$-I_2$
$-d_+$	$-d_+$	d_+	$-h$	h	$-r$	r	$-I_2$	I_2

Fig. 5.3   The Cayley table for D

symmetric about the main diagonal, which tells us that the operation is not commutative.

5.2.12 Example. *Our next example is the Cayley table of S_3. Recall that the elements of S_3 are the bijective mappings from J_3 to itself, and such a mapping $f: J_3 \to J_3$ is represented by the vector of its values, in the natural order; that is, f is represented by $[f(1)\, f(2)\, f(3)]$. The operation is composition of mappings, and for $f, g \in S_3$, $f \circ g$ is the mapping for which $f \circ g(i) = f(g(i))$ for each $i \in J_3$.*

Alternative Axioms for Groups

In this section we introduce several defining properties of groups which are sometimes easier to verify than those of Definition 5.2.1.

∘	[1 2 3]	[1 3 2]	[3 2 1]	[2 1 3]	[2 3 1]	[3 1 2]
[1 2 3]	[1 2 3]	[1 3 2]	[3 2 1]	[2 1 3]	[2 3 1]	[3 1 2]
[1 3 2]	[1 3 2]	[1 2 3]	[2 3 1]	[3 1 2]	[3 2 1]	[2 1 3]
[3 2 1]	[3 2 1]	[3 1 2]	[1 2 3]	[2 3 1]	[2 1 3]	[1 3 2]
[2 1 3]	[2 1 3]	[2 3 1]	[3 1 2]	[1 2 3]	[1 3 2]	[3 2 1]
[2 3 1]	[2 3 1]	[2 1 3]	[1 3 2]	[3 2 1]	[3 1 2]	[1 2 3]
[3 1 2]	[3 1 2]	[3 2 1]	[2 1 3]	[1 3 2]	[1 2 3]	[2 3 1]

Fig. 5.4 The Cayley table for S_3

In a group, we can solve equations of the form $ax = b$ and $ya = b$, where a and b are given elements of the group, and x and y are sought. For example, using Table 5.3, we find that the solutions to $hx = r$ and $yd^+ = -h$, respectively, are $x = -d^+$ and $y = r$.

The following proposition states that if we can always solve such equations in a semigroup, then the semigroup is a group. For it, we shall make use of the next lemma.

5.3.1 Lemma. *Let $(S, *)$ be a semigroup and suppose that $*$ has a left identity e, and furthermore, for each element $x \in S$, there exists $y \in S$ such that $y * x = e$. Then S is a group with identity e.*

Proof. We first prove that for all $x \in S$, $x * e = x$. Let $x \in S$. Then there exists $y \in S$ such that $y * x = e$, and there exists $z \in S$ such that $z * y = e$. Then $x = e * x = (z * y) * x = z * (y * x) = z * e$ and thus $x * e = (z * e) * e = z * (e * e) = z * e = x$, as required.

Next, we must prove that each $x \in S$ has an inverse. Let $x \in S$. Then there exist $y, z \in S$ such that $y * x = e = z * y$. We have $z = z * e = z * (y * x) = (z * y) * x = e * x = x$, so $x * y = z * y = e$ and $y * x = e$. Thus y is an inverse for x. □

5.3.2 Proposition. *Let G be a semigroup, with binary operation denoted by juxtaposition. Then G is a group if, and only if, given any a and b in G, the equations $ax = b$ and $ya = b$ have solutions $x, y \in G$.*

Proof. First, suppose that G is a group. Then for any $a, b \in G$, the equation $ax = b$ has solution $x = a^{-1}b$, and the equation $ya = b$ has solution $y = ba^{-1}$. To verify these assertions, we compute $a(a^{-1}b) = (aa^{-1})b = eb = b$, where e is the identity of the binary operation, and $(ba^{-1})a = b(a^{-1}a) =$

$be = b$. Note that we have used the fact that the binary operation is associative.

Next, suppose that for any $a, b \in G$, the equations $ax = b$ and $ya = b$ have solutions $x, y \in G$. We first prove that the binary operation has a left identity. Let $a \in G$. Then there exists an element, which we shall denote by e, such that $ea = a$. Let $b \in G$. We prove that $eb = b$. There exists $x \in G$ such that $ax = b$. Left multiply by e and use associativity to obtain that $(ea)x = eb$. Since $ea = a$ and $ax = b$, we have $b = eb$, as required. Thus e is a left identity for the binary operation. As well, for each $s \in G$, there exists $y \in G$ such that $ys = e$. It follows now from Lemma 5.3.1 that G is a group. $\qquad\square$

Note that in a group G, for any $a, b \in G$, the equation $ax = b$ has a unique solution (it was shown in the proof of Proposition 5.3.2 that a solution exists). For suppose that $x \in G$ satisfies the equation $ax = b$. Left multiply by a^{-1} to obtain that $x = a^{-1}b$, which is, of course, the solution found in the proof of Proposition 5.3.2. Similarly, $y = ba^{-1}$ is the unique solution to the equation $ya = b$.

Recall that as a special case of Notation 3.4.32, for a semigroup S, with binary operation denoted by juxtaposition, for any $a \in S$, the set $\{\, sa \mid s \in S \,\}$ is denoted by Sa and similarly, the set $\{\, as \mid s \in S \,\}$ is denoted by aS.

5.3.3 Corollary. *Let S be a semigroup. Then S is a group if, and only if, for each $a \in S$, $aS = S = Sa$.*

Proof. First, suppose that S is a group, and let $a, s \in S$. Then by Proposition 5.3.2, there exist $x, y \in S$ such that $ax = s$ and $ya = s$. Thus $s \in aS$ and $s \in Sa$, so $S \subseteq aS$ and $S \subseteq Sa$, which establishes that $aS = S = Sa$ for any $a \in S$.

Conversely, suppose that for any $a \in S$, $aS = S = Sa$. Let $a, b \in S$. Since $S = aS$ and $b \in S$, there exists $x \in S$ such that $b = ax$. As well, $b \in S = Sa$, so there exists $y \in S$ such that $ya = b$. By Proposition 5.3.2, S is a group. $\qquad\square$

We saw in Proposition 5.2.5 (i) that the cancellation laws hold in a group (that is, if G is a group, then for any $a, b, c \in G$ such that either $ab = ac$ or $ba = ca$ holds, then $b = c$). The cancellation laws do not characterize groups among semigroups; that is, a semigroup that satisfies the cancellation laws is not necessarily a group. For example, the set $\mathbb{Z}^+$ of positive integers, under either the operation of addition or of multiplication is a cancellative semigroup (commutative even), but of course, not a group. However, it is the case that a finite semigroup that satifies the cancellation laws is a group.

5.3.4 Proposition. *Let S be a finite semigroup in which the cancellation laws hold. Then S is a group.*

Proof. By Corollary 5.3.3, it suffices to prove that for every $a \in S$, $aS = S = Sa$. Let $a \in S$. Let $n = |S|$, say $S = \{a_1, a_2, \ldots, a_n\}$. Then $aS = \{aa_1, aa_2, \ldots, aa_n\} \subseteq S$. Let i and j be integers with $1 \leq i < j \leq n$, and suppose that $aa_i = aa_j$. Since the cancellation laws hold in S, we obtain that $a_i = a_j$, which is not possible since $i \neq j$. Thus $aa_i \neq aa_j$ and so $|aS| = n$, which means that $aS = S$. Similarly, $Sa = S$, and so by Corollary 5.3.3, S is a group. $\qquad\square$

5.4   Subgroups

If G is a group, then it may happen that G has subsets that are closed with respect to the binary operation of G. In such a case, the restriction of the binary operation on G to such a subset provides a binary operation on that subset.

5.4.1 Definition. *Let G be a group. If $H \subseteq G$ is closed under the binary operation of G and if, under the restriction of the binary operation of G to H, H is a group, then H is said to be a subgroup of G. If H is a subgroup of G and $H \neq G$, H is said to be a proper subgroup of G. We write $H \leq G$ to indicate that H is a subgroup of G, and $H < G$ to indicate that H is a proper subgroup of G.*

5.4.2 Example. *For any group G, G is a subgroup of itself, and if e denotes the identity of G, the subset $\{e\}$ of G is a subgroup of G, called the trivial subgroup of G.*

The group introduced in Example 5.2.11 is a subgroup of $Gl_2(\mathbb{R})$ (with binary operation multiplication of matrices). Also, $\mathbb{Z}$ and $\mathbb{Q}$ are each subgroups of $\mathbb{R}$ under addition.

Symmetry Groups.

For any positive integer n, consider the set $O(\mathbb{R}^n)$ of all rigid motions of $\mathbb{R}^n$ that fix the origin (by a rigid motion of $\mathbb{R}^n$, we mean a map from $\mathbb{R}^n$ to itself that preserves distance–see Chapter 15 for an in-depth discussion of rigid motions of $\mathbb{R}^n$). Observe that $O(\mathbb{R}^n) \subseteq S_{\mathbb{R}^n}$, the group of all bijective mappings from $\mathbb{R}^n$ to itself with binary operation composition of mappings. Since the composition of two rigid motions, each of which preserves the origin, is a rigid motion preserving the origin, it follows that $O(\mathbb{R}^n)$ is closed under the binary operation composition of mappings. Furthermore, the identity map is a rigid motion that preserves the origin, and the inverse of any rigid motion that preserves the origin is a rigid motion that preserves the origin. Thus $O(\mathbb{R}^n)$ is a subgroup of $S_{\mathbb{R}^n}$, and so $O(\mathbb{R}^n)$, with binary operation composition of mappings, is a group, called the orthogonal group of degree n. We shall examine the cases $n = 1$ or 2. The rigid motions of the line (the case $n = 1$) that preserve the origin are the identity map, I, and the map M that sends each point to its mirror image in the origin (if we put the usual Cartesian coordinate system on the line, this rigid motion sends x to $-x$ for each x on the line), and thus $O(\mathbb{R}^1) = \{\, I, M \,\}$.

5.4.3 Example. *Consider a line segment centered at the origin with end-points A and B. The rigid motions of the line that preserve the line segment AB are the identity, and the map M that reflects the line in the origin; that is, every rigid motion of the line that preserves the origin preserves the line segment AB. We call the set of rigid motions that preserve the origin and preserve the line segment AB the group of symmetries of the line segment, namely $\{\, I, M \,\}$. Note that $I \circ I = M \circ M = I$, and $I \circ M = M \circ I = M$, so the Cayley table for the symmetry group of the line segment AB is*

$\circ$	I	M
I	I	M
M	M	I

For the case $n = 2$, we are considering the rigid motions of the plane that preserve the origin, and these are the rotations by an angle θ, denoted by R_θ, where θ is the radian measure of the angle, with $\theta > 0$ denoting rotation in the counter-clockwise direction, and reflection in a line through the origin (see Chapter 15 for a comprehensive discussion of rigid motions of the plane). Since a line through the origin is uniquely determined by the angle it makes with the positive x-axis, where this angle is also measured in radians, reflection in the line through the origin that makes angle α with the positive x-axis will be denoted by M_α. Note that $R_\theta = R_{\theta+2\pi}$, and $M_\alpha = M_{\alpha+\pi}$. The identity element is the identity map, denoted by I. For any $\alpha \in \mathbb{R}$, the effect of R_α on $(x, y) \in \mathbb{R}^2$ is computed as follows. Let $r = \sqrt{x^2 + y^2}$, so r is the distance from the origin $(0,0)$ to the point (x,y). Let θ denote the angle the line segment from $(0,0)$ to (x,y) makes with the positive x-axis. Then $x = r\cos(\theta)$ and $y = r\sin(\theta)$, and so $(x,y) = (r\cos(\theta), r\sin(\theta))$. When R_α is applied to (x,y), the result is a point whose distance from $(0,0)$ is still r, but the line segment from $(0,0)$ to this point now makes an angle of $\theta + \alpha$ with the positive x-axis, so $R_\alpha(x,y) = (r\cos(\theta+\alpha), r\sin(\theta+\alpha))$. Now, for any $A, B \in \mathbb{R}$, $\cos(A+B) = \cos(A)\cos(B) - \sin(A)\sin(B)$ and $\sin(A+B) = \sin(A)\cos(B) + \sin(B)\cos(A)$. Thus we have $R_\alpha(x,y) = (r\cos(\theta)\cos(\alpha) - r\sin(\theta)\sin(\alpha), r\sin(\theta)\cos(\alpha) + r\sin(\alpha)\cos(\theta))$. Since $x = r\cos(\theta)$ and $y = r\sin(\theta)$, we finally obtain

$$R_\alpha(x,y) = (x\cos(\alpha) - y\sin(\alpha), y\cos(\alpha) + x\sin(\alpha)).$$

To determine the effect of M_α on a point $(x,y) \in \mathbb{R}^2$ is only slightly more involved. The first step is to determine the orthogonal projection of (x,y) onto the line l through the origin with direction vector $(\cos(\alpha), \sin(\alpha))$. The points on l all have the form $(r\cos(\alpha), r\sin(\alpha))$, $r \in \mathbb{R}$, and so we are to determine the value of r such that the line segment from $(r\cos(\alpha), r\sin(\alpha))$ to (x,y) is perpendicular to l; that is, the dot product of the vector $(r\cos(\alpha), r\sin(\alpha)) - (x,y)$ and the vector $(\cos(\alpha), \sin(\alpha))$ is 0. Thus r is the value for which

$$
\begin{aligned}
0 &= (r\cos(\alpha) - x, r\sin(\alpha) - y) \cdot (\cos(\alpha), \sin(\alpha)) \\
&= (r\cos(\alpha) - x)\cos(\alpha) + (r\sin(\alpha) - y)\sin(\alpha) \\
&= r(\cos^2(\alpha) + \sin^2(\alpha)) - (x\cos(\alpha) + y\sin(\alpha)) \\
&= r - (x\cos(\alpha) + y\sin(\alpha))
\end{aligned}
$$

and so we find that $r = x\cos(\alpha) + y\sin(\alpha)$. Now $M_\alpha(x,y)$ can be computed as the vector sum of $(r\cos(\alpha), r\sin(\alpha))$ and $(r\cos(\alpha), r\sin(\alpha)) - (x,y)$, as shown in Figure 5.5.
We calculate

$$
\begin{aligned}
M_\alpha(x,y) &= (r\cos(\alpha), r\sin(\alpha)) + (r\cos(\alpha), r\sin(\alpha)) - (x,y) \\
&= (2r\cos(\alpha) - x, 2r\sin(\alpha) - y) \\
&= (2(x\cos(\alpha) + y\sin(\alpha))\cos(\alpha) - x, \\
&\qquad\qquad\qquad 2(x\cos(\alpha) + y\sin(\alpha))\sin(\alpha) - y) \\
&= (x(2\cos^2(\alpha) - 1) + y(2\sin(\alpha)\cos(\alpha)), \\
&\qquad\qquad\qquad 2x\sin(\alpha)\cos(\alpha) + y(2\sin^2(\alpha) - 1)) \\
&= (x\cos(2\alpha) + y\sin(2\alpha), x\sin(2\alpha) - y\cos(2\alpha)).
\end{aligned}
$$

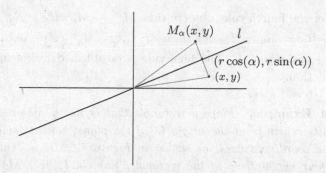

Fig. 5.5

Now that we have learned how to evaluate $R_\alpha(x,y)$ and $M_\alpha(x,y)$ for any $\alpha \in \mathbb{R}$ and any $(x,y) \in \mathbb{R}^2$, we are in a position to establish the following rules for computation in $O(\mathbb{R}^2)$.

(i) For any $\alpha, \beta \in \mathbb{R}$, $R_\alpha \circ R_\beta = R_{\alpha+\beta}$.
(ii) For any $\alpha \in \mathbb{R}$, $M_\alpha^2 = I$.
(iii) For any $\alpha, \beta \in \mathbb{R}$, $M_\alpha \circ M_\beta = R_{2(\alpha-\beta)}$.
(iv) For any $\alpha, \beta \in \mathbb{R}$, $M_\alpha \circ R_\beta = M_{\alpha-\frac{\beta}{2}}$.
(v) For any $\alpha, \beta \in \mathbb{R}$, $R_\alpha \circ M_\beta = M_{\beta+\frac{\alpha}{2}}$.

The first two rules are evident from the geometry. Note that the second rule shows that $M_\alpha^{-1} = M_\alpha$ for any $\alpha \in \mathbb{R}$. We prove the third rule. Let $\alpha, \beta \in \mathbb{R}$ and $(x,y) \in \mathbb{R}^2$. Then

$$M_\alpha \circ M_\beta(x,y) = M_\alpha(x\cos(2\beta) + y\sin(2\beta), x\sin(2\beta) - y\cos(2\beta)).$$

The first coordinate of $M_\alpha \circ M_\beta(x,y)$ is therefore

$$(x\cos(2\beta) + y\sin(2\beta))\cos(2\alpha) + (x\sin(2\beta) - y\cos(2\beta))\sin(2\alpha)$$
$$= x(\cos(2\alpha)\cos(2\beta) + \sin(2\alpha)\sin(2\beta))$$
$$+ y(\sin(2\beta)\cos(2\alpha) - \sin(2\alpha)\cos(2\beta))$$
$$= x(\cos(2\alpha)\cos(-2\beta) - \sin(2\alpha)\sin(-2\beta))$$
$$- y(\sin(-2\beta)\cos(2\alpha) + \sin(2\alpha)\cos(-2\beta))$$
$$= x\cos(2(\alpha-\beta)) - y\sin(2(\alpha-\beta)).$$

A similar calculation shows that the second coordinate of $M_\alpha \circ M_\beta(x,y)$ is

$$y\cos(2(\alpha-\beta)) + x\sin(2(\alpha-\beta))$$

which proves that $M_\alpha \circ M_\beta(x,y) = R_{2(\alpha-\beta)}(x,y)$.

For the fourth rule, observe that $M_{\alpha-\frac{\beta}{2}} \circ M_\alpha \circ R_\beta = R_{2(\alpha-\frac{\beta}{2}-\alpha)} \circ R_\beta = R_{-\beta} \circ R_\beta = R_0 = I$, and since $M_{\alpha-\frac{\beta}{2}}^{-1} = M_{\alpha-\frac{\beta}{2}}$, we obtain $M_\alpha \circ R_\beta = M_{\alpha-\frac{\beta}{2}}$, as required. The final rule is established by means of a similar calculation.

5.4.4 Example. *Place a rectangle that is not a square in such a way that its centre is at the origin O of the plane, and its sides are parallel to the coordinate axes, as shown in Figure 5.6. It is evident that there are four symmetries of the rectangle; namely I, R_π, M_0 (reflection in the x-axis), and $M_{\frac{\pi}{2}}$ (reflection in the y-axis), and the set of these four symmetries of the rectangle is a subgroup of $O(\mathbb{R}^2)$, called the group of symmetries of the rectangle. Observe that $M_0^2 = M_{\frac{\pi}{2}}^2 = R_\pi^2 = I$ (which*

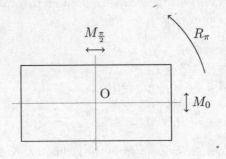

Fig. 5.6

tells us that $M_0^{-1} = M_0$, $M_{\frac{pi}{2}}^{-1} = M_{\frac{\pi}{2}}$, and $R_\pi^{-1} = R_\pi$). We calculate $M_0 \circ M_{\frac{\pi}{2}} = R_{2(0-\frac{\pi}{2})} = R_{-\pi} = R_\pi^{-1} = R_\pi = M_{\frac{\pi}{2}} \circ M_0$, $M_0 \circ R_\pi = M_{-\frac{\pi}{2}} = M_{\frac{\pi}{2}} = R_\pi \circ M_0$, $M_{\frac{\pi}{2}} \circ R_\pi = M_{\frac{\pi}{2}-\frac{\pi}{2}} = M_0$ *and* $R_\pi \circ M_{\frac{\pi}{2}} = M_\pi = M_0$. *Alternatively, due to the fact that the rectangle is not a square, the product of any two of these rigid motions can be determined by simply tracking the composite effect on a corner of the rectangle. For example, $M_0 M_{\frac{\pi}{2}}$ sends the top right corner of the rectangle first to the top left corner, then to the bottom left corner, and the only one of the symmetries that has this effect is R_π, so $M_0 M_{\frac{\pi}{2}} = R_\pi$.*

We have seen that the group of symmetries of the rectangle is a commutative group of order 4. It is convenient to use more suggestive labels for the elements of this group. Let $h = M_0$, $v = M_{\frac{\pi}{2}}$, and $r = R_\pi$. With this labelling, the group is known as the Klein 4-group, denoted by K_4, and its Cayley table is shown in Table 5.1.

Table 5.1 The Cayley table for K_4, the group of symmetries of a nonsquare rectangle

$\circ$	I	r	h	v
I	I	r	h	v
r	r	I	v	h
h	h	v	I	r
v	v	h	r	I

A more interesting example is obtained if the rectangle is replaced by a square, or more generally, by a regular n-gon centred at the origin (for $n \in \mathbb{Z}$ with $n \geq 3$. A regular n-gon is obtained by first marking n points on the circumference of a circle in such a way that the circumference is subdivided into n arcs of equal length, then, proceeding in the counter-clockwise direction, draw a line segment from each marked point to its successor.

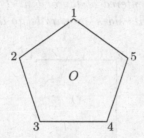

Fig. 5.7 A regular 5-gon centered at the origin O

5.4.5 Example. *For any integer $n \geq 3$, fix a regular n-gon centred at the origin, and consider the set D_n of all elements of $O(\mathbb{R}^2)$ that map the regular n-gon to itself. Evidently, $R_{\frac{2\pi}{n}} \in D_n$, and $R_{\frac{2\pi}{n}}^n = R_{2\pi} = I$, so for each vertex i of the regular n-gon, there is an element of D_n that maps vertex 1 to vertex i, namely $R_{\frac{2\pi}{n}}^{i-1}$. Moreover, if $f \in D_n$ maps vertex 1 to vertex i, then because of rigidity, f must map vertex 2 to one of the two vertices on the regular n-gon that are adjacent to i. If f maps vertex 2 to the vertex that follows vertex i in the counter-clockwise direction, then $f = R_{\frac{2\pi}{n}}^{i-1}$, while if f maps vertex 2 to the vertex that precedes vertex i in the counter-clockwise direction, then $f = M_\alpha \circ R_{\frac{2\pi}{n}}^{i-1}$, where α is the angle that the line segment from the origin to vertex i makes with the positive x-axis. Thus D_n is a subgroup of $O(\mathbb{R}^2)$ of order $2n$. Moreover, n of the elements of D_n are rotations; namely $R_{\frac{2\pi}{n}}^i$, $i = 1, 2, \ldots, n$, while by* *the rules for computation in $O(\mathbb{R}^2)$, a reflection following a rotation is a reflection, so the other n elements are reflections. An interesting distinction occurs between the cases of n even or odd. In the event that n is even, the vertices occur in diametrically opposed pairs. If $n = 2m$, then we have m pairs of diametrically opposed vertices, and reflection in the line through a diametrically opposed pair is a reflection in D_n. This accounts for m of the $n = 2m$ reflections. The remaining m reflections are found as follows. The $n = 2m$ sides of the n-gon can be partitioned into opposing pairs. For each of these m pairs of opposite sides, reflection in the line joining their midpoints belongs to D_n. When n is odd, the n reflections that belong to D_n are simply the reflections in the n lines that join the origin to a vertex of the n-gon. D_n is called the dihedral group of order $2n$.*

5.4.6 Example.   *Let us construct the Cayley table for D_4, the group of symmetries of a square centered at the origin. For convenience, we shall orient our square so that its sides are parallel to the coordinate axes.*

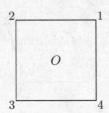

Fig. 5.8   Symmetries of the square

The four rotations in D_4 are $R_{\frac{\pi}{2}}$, $R_{\frac{\pi}{2}}^2$, $R_{\frac{\pi}{2}}^3$, and $R_{\frac{\pi}{2}}^4 = R_{2\pi} = R_0 = I$. The four reflections in D_4 are $M_{\frac{\pi}{4}}$ (reflection in the line through vertices 1 and 3), $M_{-\frac{\pi}{4}}$ (reflection in the line through vertices 2 and 4), M_0 (reflection across the x-axis), and $M_{\frac{\pi}{2}}$ (reflection across the y-axis). It is conventional to label these eight symmetries as follows: $r = R_{\frac{\pi}{2}}$ (which gives r, r^2, r^3 and $r^4 = I$), $h = M_0$, $v = M_{\frac{\pi}{2}}$, $d_+ = M_{\frac{\pi}{4}}$ (reflection across the positive slope diagonal of the square), and $d_- = M_{-\frac{\pi}{4}}$ (reflection across the negative slope diagonal).

The entries in the table were found using the rules for calculating in $O(\mathbb{R}^2)$. For example, $hv = M_0 \circ M_{\frac{\pi}{2}} = R_{2(0-\frac{\pi}{2})} = R_{-\pi} = (R_{\frac{\pi}{2}}^2)^{-1} = R_{\frac{\pi}{2}}^2 = r^2$. Alternatively, one could observe the effect of hv on vertex 1 in Figure 5.8. v sends vertex 1 to vertex 2, then h sends vertex 2 to vertex 3, so hv sends vertex 1 to vertex 3. There are only two symmetries of the square

∘	I	r	r^2	r^3	h	v	d_+	d_-
I	I	r	r^2	r^3	h	v	d_+	d_-
r	r	r^2	r^3	I	d_+	d_-	v	h
r^2	r^2	r^3	I	r	v	h	d_-	d_+
r^3	r^3	I	r	r^2	$-d_+$	d_+	h	v
h	h	d_-	v	d_+	I	r^2	r^3	r
v	v	d_+	h	d_-	r^2	I	r	r^3
d_+	d_+	h	d_-	v	r	r^3	I	r^2
d_-	d_-	v	d^+	h	r^3	r	r^2	I

Fig. 5.9   The Cayley table for D_4

that send vertex 1 to vertex 3, and they are r^2 and d_-. Where does hv send vertex 2? First v sends vertex 2 to vertex 1, then h sends vertex 1 to vertex 4, so hv sends vertex 2 to vertex 4. Since d_- fixes vertex 2, we see that hv must be equal to r^2.

5.4.7 Example.   *Consider the symmetries of a circle whose centre is at the origin. It is evident that for any $\alpha \in \mathbb{R}$, R_α is a symmetry, as is M_α. Thus the group of symmetries of a circle is $O(\mathbb{R})$ itself.*

In each of the examples given above, we have established that a subset H of a group G, where H was closed under the binary operation of G, was a subgroup by proving that the identity element of G was contained in H, which ensured that the induced binary operation on H had an identity element, and that the inverse of every element of H was again an element of H, which ensured that every element of H had an inverse (in H), and thus H, with the induced binary operation, was a group. We proceed now to establish that these two requirements are necessary and sufficient in order that H be a subgroup of G.

5.4.8 Lemma.   *Let G be a group, with identity e. If $x \in G$ satisfies $x^2 = x$, then $x = e$.*

Proof. Let $x \in G$ and suppose that $x^2 = x$. Then
$$x = ex = (x^{-1}x)x = x^{-1}x^2 = x^{-1}x = e,$$
as required.   □

5.4.9 Corollary.   *Let G be a group with identity e, and let H be a subgroup of G. Then $e \in H$.*

Proof. Since H is a group, it has an identity element, x say. Then $x^2 = x$, and so by Lemma 5.4.8, $x = e$. $\square$

5.4.10 Proposition. *A subset H of a group G is a subgroup of G if, and only if, the following hold:*

   (i) $H \neq \varnothing$.
   (ii) H is closed under the binary operation of G; that is, for all $a, b \in H$, $ab \in H$.
   (iii) H is closed under inversion; that is, for each $a \in H$, $a^{-1} \in H$ (where a^{-1} is the inverse of a in the group G).

Alternatively, H is a subgroup of G if, and only if,

   (i) $H \neq \varnothing$.
   (ii)' For all $a, b \in H$, $ab^{-1} \in H$.

Proof. Let e denote the identity of G.

Suppose firstly that $H \subseteq G$ is a subgroup of G. Then by Corollary 5.4.9, $e \in H$ and so (i) holds (and moreover, e is the identity of H). (ii) holds by definition of subgroup, so it remains to verify (iii). Let $a \in H$. Since H is a group, there exists $b \in H$ such that $ab = e = ba$. But $H \subseteq G$, so $b \in G$, and by Proposition 3.4.19, an element of G has a unique inverse. Thus $a^{-1} = b \in H$. Note that it is immediate that (ii) and (iii) imply (ii)'.

Conversely, suppose that $H \subseteq G$ satisfies (i), (ii), and (iii). Let $a \in H$ (by (i), $H \neq \varnothing$). By (iii), $a^{-1} \in H$, and then by (ii), $e = aa^{-1} \in H$. Thus the binary operation induced on H by that of G by (ii) has an identity element; namely e, the identity of G. As well, for each $a \in H$, $a^{-1} \in H$, so every element of H is invertible in H. By definition, H is a group, hence a subgroup of G. Note that if (i) and (ii)' hold, then in particular, for $a \in H$, we have $e = aa^{-1} \in H$, and then for any $b \in H$, $b = eb^{-1} \in H$, so (iii) holds, and for any $a, b \in H$, $ab = a(b^{-1})^{-1} \in H$, so (ii) holds as well. Thus the alernative criteria are equivalent to (i), (ii), and (iii).

Note that it was not necessary to prove that the operation induced on H by that of G is associative, since the operation on G is associative. $\square$

We remark that in order to prove that $H \neq \varnothing$ when trying to prove that H is a subgroup of G, one normally tries to prove that e, the identity of G, belongs to H (since by Corollary 5.4.9, if H is a subgroup of G, then $e \in H$). It might also seem strange to have to prove that $H \neq \varnothing$. The reason for this is that one often defines a subset H of a group G by selecting the elements . of G that have some property P. The property P might be such that one can prove that (ii) and (iii) hold, but if P is sufficiently complicated, it is in fact possible that there are no elements of G that satisfy P, yet this might not be obvious. It is in fact true that $\varnothing$ satisfies (ii) and (iii).

The formulation of (i), (ii), (iii), and (ii)' in additive notation are as follows:

(i) $H \neq \varnothing$.
(ii) H is closed under the binary operation of G; that is, for all $a, b \in H$, $a + b \in H$.
(iii) H is closed under inversion; that is, for each $a \in H$, $-a \in H$ (where $-a$ is the inverse of a in the group G).
(ii)' For all $a, b \in H$, $a - b \in H$ (recall that $a - b$ is short for $a + (-b)$).

As our first application of Proposition 5.4.10, we introduce the notion of the centre of a group.

5.4.11 Definition. *Let G be a group. The centre of G, denoted by $Z(G)$ is defined by $Z(G) = \{ g \in G \mid \text{ for all } x \in G, gx = xg \}$. An element of $Z(G)$ is called a central element of G.*

5.4.12 Proposition. *For any group G, $Z(G)$ is a subgroup of G.*

Proof. Certainly $e_G \in Z(G)$. Suppose that $g, h \in G$. Then for any $x \in G$, we have $gx = xg$ and $hx = xh$, so $(gh)x = g(hx) = g(xh) = (gx)h = (xg)h = x(gh)$ and thus $gh \in Z(G)$. As well, from $gx = xg$ we obtain $g^{-1}gxg^{-1} = g^{-1}xgg^{-1}$; that is, $xg^{-1} = g^{-1}x$ and thus $g^{-1} \in Z(G)$. By Proposition 5.4.10, $Z(G)$ is a subgroup of G. $\qquad\square$

As our next application of Proposition 5.4.10, we prove that the intersection of any set of subgroups of a group G is again a subgroup of G.

5.4.13 Proposition. *Let G be a group, and let $\mathscr{H}$ be a nonempty set of subgroups of G. Then $K = \cap \mathscr{H} = \{ h \in G \mid h \in H \text{ for all } H \in \mathscr{H} \}$ is a subgroup of G.*

Proof. Since $e \in H$ for each $H \in \mathscr{H}$, it follows that $e \in K$. Suppose now that $a, b \in K$. Then for each $H \in \mathscr{H}$, $a, b \in H$, and thus by Proposition 5.4.10, $ab^{-1} \in H$, so $ab^{-1} \in K$ for each $a, b \in K$. By (i) and (ii)' of Proposition 5.4.10, K is a subgroup of G. $\qquad\square$

We next introduce the concept of a generating set for a group.

5.4.14 Definition. *Let G be a group, and let T be a subset of G. The subgroup of G that is generated by T, denoted by $\langle T \rangle$, is defined by*

$$\langle T \rangle = \cap \{\, H \leq G \mid T \subseteq H \,\}.$$

If $T = \{\, g \,\}$, a singleton subset, we usually write $\langle g \rangle$ rather than $\langle \{\, g \,\} \rangle$.

We remark that for any subset T of a group G, there are subgroups of G that contain T. For example, G is a subgroup of itself. Furthermore, $\langle T \rangle$ is a subgroup of G by virtue of Proposition 5.4.13. Observe that $\langle \varnothing \rangle = \{\, e \,\}$ and if T is a subgroup of G, then $\langle T \rangle = T$. More generally, we may observe that for any subset T of a group G, $\langle T \rangle$ is the smallest subgroup of G that contains T, in the sense that if H is a subgroup of G and $T \subseteq H$, then $\langle T \rangle \subseteq H$ (this is immediate from the definition of $\langle T \rangle$).

The subgroups of a group G that are generated by singleton subsets are very important in the study of the group G, and fortunately, are easy to describe.

5.4.15 Proposition. *Let G be a group, and let $g \in G$. Then*

$$\langle g \rangle = \{\, g^i \mid i \in \mathbb{Z} \,\} = \langle g^{-1} \rangle.$$

Proof. Let $K = \{\, g^i \mid i \in \mathbb{Z} \,\}$. Suppose first that H is a subgroup of G and $g \in H$. We prove that $g^n \in H$ for all integers $n \geq 0$. To begin with, $g^0 = e \in H$. Suppose now that n is a nonnegative integer for which $g^n \in H$. Then $g^{n+1} = g^n g \in H$ since $g^n, g \in H$. We may therefore apply the principle of mathematical induction to conclude that $g^n \in H$ for all integers $n \geq 0$. Next, for any integer $n \geq 0$, $(g^n)^{-1} \in H$ since $g^n \in H$, and by Lemma 5.2.6, $(g^n)^{-1} = g^{-n}$, so for every nonnegative integer n, $g^{-n} \in H$. Thus $K \subseteq H$. In particular, $K \subseteq \langle g \rangle$. It remains to prove that $\langle g \rangle \subseteq K$. Since $g \in K$, this will follow if we prove that K is a subgroup of G. Let $x, y \in K$. Then there exist $m, n \in \mathbb{Z}$ such that $x = g^m$ and $y = g^n$. By Lemma 5.2.6, $y^{-1} = g^{-n}$ and then by Proposition 5.2.7, $xy^{-1} = g^m g^{-n} = g^{m-n} \in K$. Since $e = g^0 \in K$, it follows from Proposition 5.4.10 that K is a subgroup of G. Thus $\langle g \rangle \subseteq K$ and so $\langle g \rangle = K$, as required. Apply this result now to g^{-1} to obtain that $\langle g^{-1} \rangle = \{\, (g^{-1})^i \mid i \in \mathbb{Z} \,\} = \{\, g^{-i} \mid i \in \mathbb{Z} \,\} = \{\, g^i \mid i \in \mathbb{Z} \,\} = \langle g \rangle$. $\qquad\square$

The next concept is closely related to the result of Proposition 5.4.15, and is of fundamental importance in the theory of groups.

5.4.16 Definition. *Let G be a group with identity denoted by e and with the binary operation denoted multiplicatively. For any $g \in G$, if there exists* *a nonzero integer n such that $g^n = e$, then we say that g is an element of finite order. If $g \in G$ is an element for which no such nonzero integer exists, then g is said to be an element of infinite order.*

Note that if $g \in G$ and $n \in \mathbb{Z}$ are such that $g^n = e$, then $g^{-n} = (g^n)^{-1} = e^{-1} = e$. Thus if there is a nonzero integer n such that $g^n = e$, then there is a positive integer n such that $g^n = e$.

5.4.17 Notation. *Let G be a group and let $g \in G$. If g has finite order, then $m = \min\{ n \in \mathbb{Z}^+ \mid g^n = e \}$ is called the order of g (we note that by the preceding remark and the well-ordering principle, m exists), and we write $|g| = m$. Otherwise, we write $|g| = \infty$.*

5.4.18 Proposition. *Let G be a group, and let $g \in G$.*

(i) *If g has finite order m, then $\langle g \rangle = \{ g^k \mid k \in \mathbb{Z},\ 0 \le k < m \}$, and $|\langle g \rangle| = m$.*
(ii) *If g has infinite order, then the mapping $\mathbb{Z} \to \langle g \rangle$ for which $n \mapsto g^n$ is bijective.*

Proof. Suppose that g has finite order m, and let $H = \{ g^k \mid k \in \mathbb{Z},\ 0 \le k < m \}$. Since $H \subseteq \langle g \rangle$, we need only prove that $\langle g \rangle \subseteq H$ and that $|H| = m$. Let $n \in \mathbb{Z}$. Apply the division theorem to n and m to obtain the existence of $q, r \in \mathbb{Z}$ such that $n = qm + r$ and $0 \le r < m$. Then $g^n = g^{qm+r} = g^{qm} g^r = (g^m)^q g^r$. By definition, $g^m = e$, and as $e^q = e$, we have $g^n = g^r \in H$. Thus $\langle g \rangle \subseteq H$. Suppose now that $i, j \in \mathbb{Z}$ are such that $0 \le i \le j < m$ and $g^i = g^j$. Then $e = g^j (g^i)^{-1} = g^j g^{-i} = g^{j-i}$ and $0 \le j - i \le j < m$. By definition of m, $j - i$ must be 0, and so $j = i$. Thus $|H| = m$.

Now suppose that g has infinite order. By Proposition 5.4.15, the mapping $\mathbb{Z} \to \langle g \rangle$ determined by mapping $n \in \mathbb{Z}$ to g^n is surjective, so it suffices to prove that the mapping is injective. Suppose that $i, j \in \mathbb{Z}$ are such that $g^i = g^j$. As above, we then obtain that $e = g^{j-i}$. Since g has infinite order, this means that $j - i = 0$ and thus $j = i$, as required. $\qquad\square$

5.4.19 Corollary. *If G is a finite group, then every element of G has finite order.*

Proof. Let $g \in G$. Since G is finite, $\langle g \rangle$ is finite and so by Proposition 5.4.18, g does not have infinite order. $\qquad\square$

It is possible that an infinite group may consist entirely of elements of finite order. For example, if X is an infinite set, then the group presented in Example 5.2.9 (vii) is an infinite group. Moreover, for any set X, the identity element of $\mathscr{P}(X)$ with binary operation $\triangle$ is $\varnothing$, and for each $A \in \mathscr{P}(X)$, $A \triangle A = \varnothing$, which means that every nonidentity element has order 2.

Recall that in Example 5.2.9 (vi), we introduced the notation $|G|$ to mean the number of elements in the group G, and we called this number the order of G (and if G is infinite, then we said that G is a group of infinite order and write $|G| = \infty$). Now, for an element $g \in G$, we are using the notation $|g|$ to mean the order of the element g. It may be a little confusing for the beginner to have the same word (order) used in two seemingly different ways, but the context should make it clear as to whether we are thinking about the size of the subgroup generated by the element g or the smallest positive power to which we must raise g to obtain the identity (by Proposition 5.4.18, these are numerically the same).

Note that in additive notation, to say that $|g| = m$ is to say that m is the least positive integer for which $mg = 0$.

5.4.20 Proposition. *Let G be a group, and let $g \in G$ be an element of finite order m. Then for any integers i and j, $g^i = g^j$ if, and only if, $m \mid i - j$. In particular, for any integer n, $g^n = e$ if, and only if, $m \mid n$.*

Proof. Let $g \in G$, and suppose that g has finite order m. Let $i, j \in \mathbb{Z}$ be such that $i \leq j$ and $g^i = g^j$. Then $g^{j-i} = e$. Apply the division theorem to $j - i$ and m to obtain integers q and r with $j - i = qm + r$ and $0 \leq r < m$. We then have $e = g^{j-i} = g^{qm+r} = (g^m)^q g^r = e^q g^r = g^r$. By choice of m, we must have $r = 0$, so $j - i = qm$. Conversely, if $m \mid i - j$, then there exist $q \in \mathbb{Z}$ with $i - j = qm$, and thus $g^{i-j} = (g^m)^q = e^q = e$. From this, it follows that $g^i = g^j$. In particular, for any $n \in \mathbb{Z}$, we have $g^n = e = g^0$ if, and only if, $m \mid n - 0$. $\qquad\square$

We now give a number of examples which should help the reader to assimilate the preceding rather abstract results.

5.4.21 Example.

(i) *The orthogonal group $O(\mathbb{R}^2)$ has elements of all finite orders, as well as elements of infinite order. For example, for any positive integer n, $R_{\frac{2\pi}{n}}$, rotation of the plane about the origin by $\frac{1}{n}$ of a full rotation, satisfies $(R_{\frac{2\pi}{n}})^i = R_{\frac{2\pi i}{n}}$, so $(R_{\frac{2\pi}{n}})^n = I$, while for any integer i with*

$0 < i < n$, $(R_{\frac{2\pi i}{n}})$ is only $\frac{i}{n}^{th}$ of a full rotation. Thus $|R_{\frac{2\pi}{n}}| = n$. On the other hand, $R_{\sqrt{2}\pi}$ has infinite order. To see this, observe that $(R_{\sqrt{2}\pi})^n = I$ if, and only if $\sqrt{2}n\pi$ is an integral multiple of 2π, say $\sqrt{2}n\pi = 2k\pi$ for some $k \in \mathbb{Z}$. If $n \neq 0$, this would imply that $\sqrt{2} = \frac{2k}{n}$. Since $\sqrt{2}$ is not rational, we conclude that the only integer n for which $(R_{\sqrt{2}\pi})^n = I$ is $n = 0$. Thus $R_{\sqrt{2}\pi}$ has infinite order.

(ii) $[2\,3\,4\,1] \in S_4$ has order 4. For we have

$$[2\,3\,4\,1]^2 = [3\,4\,1\,2]$$
$$[2\,3\,4\,1]^3 = [3\,4\,1\,2]\circ[2\,3\,4\,1] = [4\,1\,2\,3] \text{ and}$$
$$[2\,3\,4\,1]^4 = [4\,1\,2\,3]\circ[2\,3\,4\,1] = [1\,2\,3\,4],$$

which is the identity element of S_4. Note that by Proposition 5.4.18 (i), we have

$$\langle[2\,3\,4\,1]\rangle = \{\,[1\,2\,3\,4],[2\,3\,4\,1],[3\,4\,1\,2],[4\,1\,2\,3]\,\}.$$

(iii) If G is a group with identity e, then for $g \in G$, $|g| = 1$ if, and only if, $g = e$.

(iv) The only element of finite order in $\mathbb{Z}$ under addition is the identity element, 0.

(v) In the group $\mathbb{R}^*$ of nonzero real numbers under multiplication, the only elements of finite order are the identity, 1, and -1, the latter having finite order 2. To prove this, it suffices to prove that for any $x \in \mathbb{R}^*$ with $x \neq 1, -1$, no positive power of x can equal 1. But we know that if x is a real number with either $x > 1$ or $x < -1$, the sequence x^{2n} is increasing to infinity as n tends to infinity. If $x^n = 1$ for some positive integer n, then this sequence would be periodic, so we see that $x^n \neq 1$ for every positive integer n. Similarly, if $-1 < x < 1$, $x \neq 0$, then x^{2n} tends to 0 as n tends to infinity. As before, if $x^n = 1$ for some positive integer n, then this sequence would be periodic, and thus $x^n \neq 1$ for every positive integer n.

Cyclic Groups

5.5.1 Definition. *A group G is said to be cyclic if there exists $g \in G$ such that $G = \langle g \rangle$. If so, g is said to be a generator of G.*

Cyclic groups have a particularly simple structure which we investigate in this section. The finite cyclic groups of prime power order form the building blocks (in a sense that will be made precise in Chapter 13) for all finite abelian groups.

5.5.2 Example.

(i) $\mathbb{Z}$, under addition, is an infinite cyclic group with generators 1 and -1 (and no others). To see this, observe that

$$\langle 1 \rangle = \{\, k(1) \mid k \in \mathbb{Z} \,\} = \mathbb{Z},$$

and by Proposition 5.4.15, $\langle 1 \rangle = \langle -1 \rangle$, so $\mathbb{Z}$ is cyclic with generators 1 and -1. For $n \in \mathbb{Z}$, if $\mathbb{Z} = \langle n \rangle = \{\, kn \mid k \in \mathbb{Z} \,\}$ then for some $k \in \mathbb{Z}$, we have $1 = kn$, and thus $k = n = \pm 1$.

(ii) $U(\mathbb{Z}_7)$, the group of units of $\mathbb{Z}_7$ under multiplication, is a cyclic group of order 6 with generators 3 and $3^{-1} = 5$ (for the sake of simplicity, we shall no longer write $[i]$, but simply i). We verify this: $3^2 = 9 = 2$, $3^3 = (3^2)3 = (2)(3) = 6$, $3^4 = (3^2)^2 = 4$, $3^5 = (3^4)3 = (4)(3) = 12 = 5$, $3^6 = (3^3)^2 = 6^2 = (-1)^2 = 1$, so $|3| = 6 = |U(\mathbb{Z}_7)|$. Thus $\langle 3 \rangle = U(\mathbb{Z}_7)$. As well, since $(3^5)3 = 1$, it follows that $3^{-1} = 3^5 = 5$, and so $\langle 5 \rangle = \langle 3^{-1} \rangle = U(\mathbb{Z}_7)$. Next, we demonstrate that $U(\mathbb{Z}_7)$ has no other generators. It suffices to prove that $|2| < 6$, $|4| < 6$, and $|6| < 6$. We have $2^2 = 4$, $2^3 = (2^2)2 = 8 = 1$, so $|2| = 2$. Since $4 = -2$, we have $4^2 = (-2)^2 = 1$, so $|4| = 2$ as well. Finally, $6 = -1$, so $6^2 = (-1)^2 = 1$ and thus $|6| = 2$. Since an element of $U(\mathbb{Z}_7)$ is a generator of $U(\mathbb{Z}_7)$ if, and only if, the order of the element is 6, it follows that the only generators of $U(\mathbb{Z}_7)$ are 3 and $3^{-1} = 5$.

Our first result establishes a very important property of cyclic groups.

5.5.3 Proposition. *If G is a cyclic group, then G is abelian.*

Proof. Let G be a cyclic group, say with generator g. For any $x, y \in G$, there exist $i, j \in \mathbb{Z}$ with $x = g^i$ and $y = g^j$, and so it follows by Corollary 5.2.8 that $xy = yx$. □

The converse of this result is not true. For example, $U(\mathbb{Z}_8) = \{1, 3, 5, 7\}$, under multiplication, is an abelian group but is not a cyclic group since each nonidentity element has order 2.

We now consider another interesting and important problem. Suppose that G is a group, and $g \in G$ is an element of finite order. Then $\langle g \rangle$ is a group of finite order, and so by Corollary 5.4.19, every element of $\langle g \rangle$ has finite order. The question we are interested in is the following: given $i \in \mathbb{Z}$, can we determine the order of $|g^i|$ if we know the order of g? The next result provides the complete (and very useful) answer.

5.5.4 Proposition. *Let G be a group, and let $g \in G$ be an element of finite order m. Then for any $i \in \mathbb{Z}$, $|g^i| = \frac{m}{(m,i)}$.*

Proof. Let $i \in \mathbb{Z}$, and suppose that $|g^i| = t$. Then $e = (g^i)^t = g^{it}$, and so by Proposition 5.4.20, m divides it, say $it = qm$. Let $d = (m, i)$. Then $\frac{i}{d}$ and $\frac{m}{d}$ are relatively prime integers, and we have $\frac{i}{d}t = q\frac{m}{d}$, so $\frac{i}{d}$ divides q, say $q = k\frac{i}{d}$. We therefore obtain $t = k\frac{m}{d}$; that is, $\frac{m}{d} \mid t$. On the other hand, $(g^i)^{\frac{m}{d}} = g^{m\frac{i}{d}} = (g^m)^{\frac{i}{d}} = e^{\frac{i}{d}} = e$, so by Proposition 5.4.20, $t \mid \frac{m}{d}$. Thus $t = \frac{m}{d}$, as required. $\square$

Our next result establishes a property of cyclic groups that has far-reaching consequences.

5.5.5 Proposition. *Every subgroup of a cyclic group is cyclic.*

Proof. Let G be a cyclic group, and let H be a subgroup of G. If $H = \{e\}$, then $H = \langle e \rangle$ and so H is cyclic. Suppose now that $H \neq \{e\}$. Let g be a generator of G. Let $h \in H$ with $h \neq e$. Then there exists $k \in \mathbb{Z}$ such that $h = g^k$. Since $h^{-1} = g^{-k}$ and $h^{-1} \in H$, we have established that there exists a positive integer n such that $g^n \in H$. By the Well Ordering Principle, $m = \min\{n \in \mathbb{Z}^+ \mid g^n \in H\}$ exists. Since $g^m \in H$, it follows that $\langle g^m \rangle$ is a subgroup of H. To prove that $H = \langle g^m \rangle$, it therefore suffices to prove that $H \subseteq \langle g^m \rangle$. Let $x \in H$. Then there exists $t \in \mathbb{Z}$ such that $x = g^t$. Apply the division theorem to t and m to obtain the existence of integers q and r such that $t = qm + r$ and $0 \leq r < m$. Thus $x = g^t = g^{qm+r} = (g^m)^q g^r$, and so $g^r = (g^m)^{-q}x$. Since $(g^m)^{-q} \in \langle g^m \rangle \subseteq H$, and $x \in H$, we have obtained that $g^r \in H$. If $r \neq 0$, then $0 < r < m$ and $g^r \in H$. Since this is not possible by definition of m, it follows that $r = 0$ and so $t = qm$ and $x = g^t \in \langle g^m \rangle$, as required. $\square$

The next two results describe how to determine whether one subgroup of a cyclic group is contained within another, thereby providing us with a clear picture of the subgroup structure of the group. The subgroup structure of an arbitrary group is generally quite complicated.

5.5.6 Proposition. *Let G be an infinite cyclic group with generator g. Then for any $n, m \in \mathbb{Z}$, $\langle g^n \rangle \subseteq \langle g^m \rangle$ if, and only if, $m \mid n$.*

Proof. Let $m, n \in \mathbb{Z}$. Suppose first that $\langle g^n \rangle \subseteq \langle g^m \rangle$. Then $g^n \in \langle g^m \rangle$, and so for some $k \in \mathbb{Z}$, we have $g^n = (g^m)^k = g^{km}$. By Proposition 5.4.18 (ii), $n = km$ and so $m \mid n$. Conversely, suppose that $m \mid n$, say $n = km$ for some $k \in \mathbb{Z}$. Then for any $i \in \mathbb{Z}$, $(g^n)^i = (g^k m)^i = (g^m)^{ki} \in \langle g^m \rangle$. Thus $\langle g^n \rangle \subseteq \langle g^m \rangle$. $\qquad\square$

5.5.7 Corollary. *An infinite cyclic group has exactly two generators.*

Proof. Let g be a generator of the infinite cyclic group G, and suppose that g^i is another generator. Then $\langle g^i \rangle = G = \langle g^1 \rangle$, and so i must be a divisor of 1; that is, $i = 1$ or -1. We remark that $g = g^{-1}$ if, and only if, $g^2 = e$, which means that g is an element of finite order (dividing 2). Thus $g \neq g^{-1}$. $\qquad\square$

5.5.8 Proposition. *Let G be a finite cyclic group of order m, and let g be any generator of G.*

   (i) *For each positive divisor k of m, there exists one, and only one, subgroup of G of order k; namely $\left\langle g^{\frac{m}{k}} \right\rangle$.*
   (ii) *The order of any subgroup of G is a divisor of m.*
   (iii) *For any $i, j \in \mathbb{Z}$, $\langle g^i \rangle \subseteq \langle g^j \rangle$ if, and only if, $(m, j) \mid i$.*
   (iv) *For any $i \in \mathbb{Z}$, g^i is a generator of G if, and only if, $(m, i) = 1$.*

Proof. For (i), suppose that k is a positive divisor of m, say $m = kt$ for some positive $t \in \mathbb{Z}$. By Proposition 5.5.4, $|g^t| = \frac{m}{(m,t)} = \frac{m}{t} = k$, and so $\langle g^t \rangle$ is a subgroup of G of order k. It remains to prove that $\langle g^t \rangle$ is the only subgroup of G of order k. To do so, suppose that H is a subgroup of G, and $|H| = k$. Then by Proposition 5.5.5, H is cyclic, say $H = \langle g^i \rangle$. Thus $|g^i| = k$, and Proposition 5.5.5, $|g^i| = \frac{m}{(m,i)}$, so $k = \frac{m}{(m,i)}$ and thus $t = (m, i)$. This means that $t \mid i$, and so $H = \langle g^i \rangle \subseteq \langle g^t \rangle$. Since $|H| = k = |\langle g^t \rangle|$, it follows that $H = \langle g^t \rangle = \left\langle g^{\frac{m}{k}} \right\rangle$.

For (ii), observe that by Proposition 5.5.5, every subgroup of G is cyclic. Let H be a subgroup of G. Then $H = \langle g^i \rangle$ for some integer i, and by Proposition 5.5.4, $|H| = |g^i| = \frac{m}{(m,i)} \mid m$.

For (iii), let $i, j \in \mathbb{Z}$. Suppose first that $\langle g^i \rangle \subseteq \langle g^j \rangle$. Then $\langle g^i \rangle$ is a subgroup of the cyclic group $\langle g^j \rangle$, and thus $|g^i|$ is a divisor of $|g^j|$; that is, $\frac{m}{(m,i)}$ divides $\frac{m}{(m,j)}$, say $\frac{m}{(m,j)} = q\frac{m}{(m,i)}$. Then $(m, i) = q(m, j)$ and so (m, j) divides (m, i), which in turn divides i. Thus $(m, j) \mid i$. Conversely, suppose that $(m, j) \mid i$. Then $(m, j) \mid (m, i)$ and thus $|g^i| = \frac{m}{(m,i)} \mid \frac{m}{(m,j)} = |g^j|$. The cyclic group $\langle g^j \rangle$ thus contains a unique subgroup of order $|g^i|$, and this must be the unique subgroup of G of order $|g^i|$, which is $\langle g^i \rangle$. Thus $\langle g^i \rangle \subseteq \langle g^j \rangle$.

Finally, for (iv), let $i \in \mathbb{Z}$. Then $\langle g^i \rangle = G$ if, and only if, $\frac{m}{(m,i)} = |g^i| = m$; that is, if, and only if, $(m,i) = 1$. $\qquad\square$

5.5.9 Example. *Let G be a cyclic group of order 12, and let g be a generator of G. Let us determine all of the subgroups of G. By Proposition 5.5.8 (i), for each $i = 1,2,3,4,6,12$, there is a unique subgroup of G of order i, and by Proposition 5.5.8 (ii), G has no other subgroups. Moreover, by Proposition 5.5.8 (i), the unique subgroup of G of order i is $\left\langle g^{\frac{6}{i}} \right\rangle$, so the subgroups of G are $\langle g^{12} \rangle = \langle e \rangle = \{\, e \,\}$, $\langle g^6 \rangle = \{\, e, g^6 \,\}$, $\langle g^4 \rangle = \{\, e, g^4, g^8 \,\}$, $\langle g^3 \rangle = \{\, e, g^3, g^6, g^9 \,\}$, $\langle g^2 \rangle = \{\, e, g^2, g^4, g^6, g^8, g^{10} \,\}$, and $\langle g^1 \rangle = G$.*

The generators of G are g, g^5, g^7, and g^{11}. What are the generators of the cyclic subgroup $\langle g^2 \rangle$? Since this is a (sub)group of order 6, we are looking for the elements of order 6 in $\langle g^2 \rangle$. These are $(g^2)^i$ where $(6,i) = 1$ and $1 \le i < 6$. Thus $(g^2)^1 = g^2$ and $(y^2)^5 = g^{10} = (g^2)^{-1}$ are the only generators of $\langle g^2 \rangle$. Another question we might ask would be whether or not $\langle g^8 \rangle \subseteq \langle g^{10} \rangle$. Since $(12,10) = 2 \,|\, 8$, the answer is yes.

Exercises

1. In each case below, find the order of the indicated element in the given group.

   a) $[2\,3\,4\,1\,6\,5]$ in S_6.

   b) $\frac{1}{\sqrt{2}} \left[\begin{smallmatrix} 1 & -1 \\ 1 & 1 \end{smallmatrix}\right]$ in $Gl_2(\mathbb{R})$.

   c) $\left[\begin{smallmatrix} 1 & -1 \\ 1 & 1 \end{smallmatrix}\right]$ in $Gl_2(\mathbb{R})$.

   d) $[8]$ in $\mathbb{Z}_{12}$ (the reader should infer that the operation is addition, since $\mathbb{Z}_n$ is a group under addition, but not under multiplication).

   e) $([2]_6, [3]_9)$ in the direct product $\mathbb{Z}_6 \times \mathbb{Z}_9$.

2. Let G be a group. Use induction on n to prove that for any $a_1, a_2, \ldots, a_n \in G$, $(a_1 a_2 \cdots a_n)^{-1} = a_n^{-1} a_{n-1}^{-1} \cdots a_2^{-1} a_1^{-1}$.

3. In each case below, construct a Cayley table for the indicated group.

   a) $\mathscr{P}(J_2)$, where the binary operation is $\triangle$.

   b) $\mathbb{Z}_5$ with binary operation addition. Find all generators of $\mathbb{Z}_5$.

   c) $U(\mathbb{Z}_7)$, the group of units of $\mathbb{Z}_7$, with binary operation multiplication. Prove that $U(\mathbb{Z}_7)$ is a cyclic group, and find all generators for it.

d) $U(\mathbb{Z}_{18})$, the group of units of $\mathbb{Z}_{18}$, with binary operation multiplication. Prove that $U(\mathbb{Z}_{18})$ is a cyclic group, and find all generators for it.

e) $U(\mathbb{Z}_{16})$, the group of units of $\mathbb{Z}_{16}$, with binary operation multiplication. Prove that $U(\mathbb{Z}_{16})$ is not a cyclic group.

4. Let G be a group. Prove that if H and K are subgroups of G, then $H \cup K$ is a subgroup of G if, and only if, either $H \subseteq K$ or $K \subseteq H$.

5. Let G be a group.

a) Let H be a finite subset of G. Prove that H is a subgroup of G if, and only if, $H \neq \varnothing$ and H is closed under the binary operation of G.

b) Give an example of a group G with an infinite subset H that is nonempty and closed under the binary operation of G, but which is not a subgroup of G.

6. Let G be a group. Then for any $g \in G$, the centralizer of g in G is the set $C_G(g) = \{\, x \in G \mid xg = gx \,\}$. Prove that for every $g \in G$, $C_G(g)$ is a subgroup of G.

7. Let G be an abelian group, and let H and K be subgroups of G. Prove that

$$HK = \{\, hk \mid h \in H,\ k \in K \,\}$$

is a subgroup of G. Give an example of a nonabelian group G with subgroups H and K for which HK is not a subgroup of G.

8. Let

$$Q_8 = \left\{\, \pm I_2, \pm \begin{bmatrix} 0 & -1 \\ 1 & 0 \end{bmatrix}, \pm \begin{bmatrix} 0 & i \\ i & 0 \end{bmatrix}, \pm \begin{bmatrix} -i & 0 \\ 0 & i \end{bmatrix} \,\right\} \subseteq Gl_2(\mathbb{C}).$$

For convenience, let $\mathbf{1} = I_2$, $\mathbf{i} = \begin{bmatrix} 0 & -1 \\ 1 & 0 \end{bmatrix}$, $\mathbf{j} = \begin{bmatrix} 0 & i \\ i & 0 \end{bmatrix}$, and $\mathbf{k} = \begin{bmatrix} -i & 0 \\ 0 & i \end{bmatrix}$. With this notation, $Q_8 = \{\, \pm\mathbf{1}, \pm\mathbf{i}, \pm\mathbf{j}, \pm\mathbf{k} \,\}$.

a) Prove that in $Gl_2(\mathbb{C})$, $(-\mathbf{1})^2 = \mathbf{1}$, $\mathbf{i}^2 = \mathbf{j}^2 = \mathbf{k}^2 = -\mathbf{1}$, and so deduce that $|-\mathbf{1}| = 2$, and $|\mathbf{i}| = |\mathbf{j}| = |\mathbf{k}| = 4$.

b) Prove that $(-\mathbf{1})^{-1} = -\mathbf{1}$, $\mathbf{i}^{-1} = \mathbf{i}^3 = -\mathbf{i}$, $\mathbf{j}^{-1} = \mathbf{j}^3 = -\mathbf{j}$, $\mathbf{k}^{-1} = \mathbf{k}^3 = -\mathbf{k}$, and $(-\mathbf{r})^{-1} = (-\mathbf{1})^{-1}\mathbf{r}^{-1} = (-\mathbf{1})\mathbf{r}^{-1} = -\mathbf{r}^{-1}$ for $\mathbf{r} = \mathbf{i}, \mathbf{j}, \mathbf{k}$. Thus Q_8 is closed under inversion in $Gl_2(\mathbb{C})$.

c) Prove that $\mathbf{ij} = \mathbf{k}$, $\mathbf{jk} = \mathbf{i}$, and $\mathbf{ki} = \mathbf{j}$. Then prove that $\mathbf{ji} = -\mathbf{k}$, $\mathbf{kj} = -\mathbf{i}$, and $\mathbf{ik} = -\mathbf{j}$. Deduce that Q_8 is closed under matrix multiplication.

Thus Q_8 is a subgroup of $Gl_2(\mathbb{C})$, called the quaternion group of order 8.

d) List all subgroups of Q_8.

9. Let G be a group with identity e. For each $g \in G$, define a map $c_g : G \to G$ by $c_g(x) = gxg^{-1}$ for all $x \in G$.

a) Prove that for each $g \in G$, $c_g \in S_G$, the group of all bijective mappings from G to itself under the binary operation composition of maps.

b) Prove that for each $g, h \in G$, $c_g \circ c_h = c_{gh}$, and $c_e = \mathbb{1}_G$.

c) Prove that $I(G) = \{ c_g \mid g \in G \}$ is a subgroup of S_G.

d) Prove that for each $g \in G$, $|g| = |c_g|$.

e) Prove that if H is a subgroup of G, then for each $g \in G$, $c_g(H)$ is also a subgroup of G.

10. Let G be a group, and let H and K be finite subgroups of G. On the Cartesian product $H \times K$, define a relation $\sim$ by $(h_1, k_1) \sim (h_2, k_2)$ if there exists $x \in H \cap K$ such that $h_2 = h_1 x$ and $k_1 = x k_2$.

a) Prove that $\sim$ is an equivalence relation on $H \times K$.

b) Prove that for each $(h, k) \in H \times K$, $|[(h, k)]| = |H \cap K|$.

c) Prove that the map from HK to $(H \times K)/\sim$ (where $(H \times K)/\sim$ denotes the set of all equvalence classes of $\sim$), for which $hk \mapsto [(h, k)]$ is a bijection.

d) Deduce that $|HK| = |H| \, |K| / |H \cap K|$.

11. This problem shows that in general, one cannot predict the order of a product of two elements if all we know is the order of each of the two. In each case below, compute the order of the listed elements in the given group.

a) $[2\,1\,4\,5\,3]$, $[2\,1\,3\,4\,5]$, and $[2\,1\,4\,5\,3] \circ [2\,1\,3\,4\,5]$ in S_5.

b) $\begin{bmatrix} 1 & 1 \\ 0 & 1 \end{bmatrix}$, $\begin{bmatrix} 0 & -1 \\ 1 & 1 \end{bmatrix}$, and $\begin{bmatrix} 1 & 1 \\ 0 & 1 \end{bmatrix}\begin{bmatrix} 0 & -1 \\ 1 & 0 \end{bmatrix}$ in $Gl_2(\mathbb{R})$.

c) $\begin{bmatrix} 1 & -1 \\ 1 & 0 \end{bmatrix}$, $\begin{bmatrix} 0 & 1 \\ -1 & 0 \end{bmatrix}$, and $\begin{bmatrix} 1 & -1 \\ 1 & 0 \end{bmatrix}\begin{bmatrix} 0 & 1 \\ -1 & 0 \end{bmatrix}$ in $Gl_2(\mathbb{R})$.

12. Let G be a group, and let $a, b \in G$ be such that $ab = ba$.

a) Prove that for every $n \in \mathbb{Z}$, $(ab)^n = a^n b^n$.

b) Suppose a and b have finite orders m and n, respectively. Prove that ab has finite order dividing $\frac{mn}{(m,n)}$.

13. Let n be a positive integer, and let $f : \mathbb{Z} \to \mathbb{Z}_n$ be the map for which $f(m) = [m]_n$ for all $m \in \mathbb{Z}$.

a)   Prove that for all $a, b \in \mathbb{Z}$, $f(a + b) = f(a) + f(b)$.

b)   Prove that for each subgroup H of $\mathbb{Z}$, $f(H)$ is a subgroup of $\mathbb{Z}_n$.

c)   Prove that if K is a subgroup of $\mathbb{Z}_n$, then $f^{-1}(K) = \{\, a \in \mathbb{Z} \mid f(a) \in K \,\}$ is a subgroup of $\mathbb{Z}$.

14. Let G and H be groups (denote the binary operation in each by juxta-position). Let $f : G \to H$ be a map for which

$$f(xy) = f(x)f(y) \qquad \text{for all } x, y \in G.$$

Note that the map in the preceding problem has this property (the binary operations in that problem being written additively).

a)   Prove that $f(e_G) = e_H$, where e_g, e_H are the identities of G and H, respectively.

b)   Prove that if K is a subgroup of G, then $f(K)$ is a subgroup of H.

c)   Prove that if L is a subgroup of H, then $f^{-1}(L) = \{\, g \in G \mid f(g) \in L \,\}$ is a subgroup of G.

d)   Prove that for each $g \in G$, if g has finite order, then $f(g)$ has finite order and $|f(g)|$ divides $|g|$.

15. Give an example of a semigroup S which is not a group, but which has a left identity e with the property that for each $x \in S$, there exists $y \in S$ such that $xy = e$ (compare with Lemma 5.3.1).

16. Let G be a group, and let H be a subgroup of G. Let $m, n \in \mathbb{Z}$ with $(m, n) = 1$. Prove that if $g \in G$ is such that both g^m and g^n are in H, then $g \in H$.

Chapter 6

Further Properties of Groups

Introduction

Given a subgroup H of a group G, we introduce two equivalence relations on G, each of which is a generalization of the notion of congruence modulo n on the integers (see Chapter 4), and we arrive quite naturally at the notion of a coset of H in G, the analogue of a congruence class modulo n. This then leads us to a fundamental theorem of group theory, known as Lagrange's theorem, without which not much can be said about finite groups. In the case of the integers, we were able to define a binary operation on the set $\mathbb{Z}_n$ of congruence classes modulo a positive integer n which made $\mathbb{Z}_n$ into a group, and we are thus led to ask whether it is possible to do the same for the set of all cosets of H in G. The situation is more complicated in an arbitrary group but under the right conditions, we can indeed turn the set of cosets of H in G into a group, which is then called a factor group, or a quotient group, of G. Closely related to factor groups are special maps from a group G to a group K called homomorphisms. It turns out that the image of G under a homomorphism is a subgroup of K which has, in general, a simpler structure than that of G but from which, nonetheless, we can infer properties of the original group G.

Cosets

Recall that in Chapter 4, for any positive integer n, we defined the relation congruence modulo n on $\mathbb{Z}$ by declaring that $a \equiv b \mod n$ if $n \,|\, (a - b)$. We can couch this in the language of group theory as follows: $a \equiv b \mod n$ if $a - b \in \langle n \rangle$, where $\langle n \rangle = \{\, kn \mid k \in \mathbb{Z} \,\} = n\mathbb{Z}$ is the subgroup generated by n (remember that since $\mathbb{Z}$ is cyclic, every subgroup of $\mathbb{Z}$ is cyclic). This definition is, of course, given in additive notation since the addition operation on $\mathbb{Z}$ is the very model for additive notation. We propose to generalize this definition to an arbitrary group G (where we shall use multiplicative notation). The role of $\langle n \rangle$ will be played by a subgroup H of G (not necessarily cyclic). Recall that for integers a, b, the expression $a - b$ is short for $a + (-b)$, the result of applying the binary operation (addition) to a and the inverse of b. Note that since addition in $\mathbb{Z}$ is commutative, $a + (-b) = (-b) + a$. In an arbitrary group G, we do not expect $ab^{-1} = b^{-1}a$ for all $a, b \in G$, and these observations lead us to the followig definition.

6.2.1 Definition. *Let G be a group and let H be a subgroup of G. Define relations $\sim_H$ and $\approx_H$ on G by*

$$a \sim_H b \quad \text{if } ab^{-1} \in H \qquad \text{and} \qquad a \approx_H b \quad \text{if } b^{-1}a \in H$$

for $a, b \in G$. The relations $\sim_H$ and $\approx_H$ are called the right congruence relation on G, respectively, the left congruence relation on G, determined by H.

6.2.2 Proposition. *Let G be a group, and let H be a subgroup of G. Then $\sim_H$, respectively $\approx_H$, is an equivalence relation on G, and for each $g \in G$, the equivalence class of g with respect to $\sim_H$, respectively $\approx_H$, is Hg, respectively gH, where $Hg = \{\, hg \mid h \in H \,\}$ and $gH = \{\, gh \mid h \in H \,\}$.*

Proof. For each of $\sim_H$ and $\approx_H$, the three requirements of equivalence relation (reflexivity, symmetry, and transitivity) match up exactly with the three properties of subgroup (nonempty–actually, contains the identity, closed under inversion, and closed under the binary operation, respectively). Let $g \in G$. Then $gg^{-1} = e \in H$, and thus $g \sim_H g$. As well, $g^{-1}g = e \in H$, and thus $g \approx_H g$. This proves that both $\sim_H$ and $\approx_H$ are reflexive. Next, suppose that $g, k \in G$ are such that $g \sim_H k$ (respectively, $g \approx_H k$). Then $gk^{-1} \in H$ (respectively, $k^{-1}g \in H$). Since H is closed under inversion, $kg^{-1} = (gk^{-1})^{-1} \in H$ (respectively, $g^{-1}k = (k^{-1}g)^{-1} \in H$), and so $k \sim_H g$ (respectively, $k \approx_H g$). Thus both $\sim_H$ and $\approx_H$ are symmetric. Finally, suppose that $g, k, m \in G$ are such that $g \sim_H k$ and $k \sim_H m$ (respectively, $g \approx_H k$ and $k \approx_H m$). Then $gk^{-1} \in H$ and $km^{-1} \in H$ (respectively, $k^{-1}g \in H$ and $m^{-1}k \in H$), and since H is closed under the binary operation of G, we have $gm^{-1} = (gk^{-1})(km^{-1}) \in H$ (respectively, $m^{-1}g = (m^{-1}k)(k^{-1}g) \in H$), and so $g \sim_H m$ (respectively, $g \approx_H m$). Thus $\sim_H$ and $\approx_H$ are transitive. It follows now that both $\sim_H$ and $\approx_H$ are equivalence relations on G.

Let $g \in G$, and let $x \in [g]_{\sim_H}$. Then $xg^{-1} \in H$, so $x \in Hg$. Thus $[g]_{\sim_H} \subseteq Hg$. Conversely, if $x \in Hg$, say $x = hg$ for some $h \in H$, then $xg^{-1} = h \in H$ and thus $x \in [g]_{\sim_H}$. This proves that $Hg \subseteq [g]_{\sim_H}$, and thus $[g]_{\sim_H} = Hg$. The proof that $[g]_{\approx_H} = gH$ is completely analogous to the above and is therefore omitted. $\square$

6.2.3 Corollary. *Let G be a group and let H be a subgroup of G. Then for any $a, b \in G$, $Ha = Hb$ if, and only if, $ab^{-1} \in H$, which in turn holds if, and only if, $a \in Hb$, and moreover, $aH = bH$ if, and only if, $a^{-1}b \in H$, which holds if, and only if, $b \in aH$.*

Proof. Let $a, b \in G$. By Proposition 6.2.2, $Ha = Hb$ if, and only if, $[a]_{\sim_H} = [b]_{\sim_H}$. By Proposition 3.2.19, this occurs if $a \in [b]_{\sim_H} = Hb$, equivalently, if $ab^{-1} \in H$. Since $a \in [a]_{\sim_H}$, it follows that if $[a]_{\sim_H} = [b]_{\sim_H}$, then $a \in [b]_{\sim_H} = Hb$, and thus $ab^{-1} \in H$. Thus $Ha = Hb$ if, and only if, $ab^{-1} \in H$, which in turn holds if, and only if, $a \in Hb$. The proof of the remaining assertion is similar and is therefore omitted. $\square$

6.2.4 Definition. *Let G be a group, and let H be a subgroup of G. For any $a \in G$, the set Ha is called a right coset of H in G. When the subgroup H is understood, we shall simply refer to the set Ha as the right coset of a. Similarly, the set aH is called a left coset of H in G, and if H is understood, we refer to aH as the left coset of a.*

We remark that by Proposition 6.2.2, the set of all equivalence classes of any equivalence relation is a partition of the set on which the relation is defined. Thus if H is a subgroup of a group G, the set of all right (respectively, all left) cosets of H in G is a partition of G. Further observe that $He = H = eH$, so the coset of e is the subgroup H. Since no other coset can contain the identity, it follows that the only coset (right or left) of H that is a subgroup of G is H itself.

6.2.5 Definition. *Let G be a group. For any $a \in G$, define $l_a : G \to G$, called left multiplication by a, by $l_a(g) = ag$ for all $g \in G$. Similarly, define $r_a : G \to G$, called right multiplication by a, by $r_a(g) = ga$ for all $g \in G$.*

6.2.6 Proposition. *Let G be a group. Then for each $a \in G$, $l_a, r_a \in S_G$; that is, both l_a and r_a are bijective maps from G to itself.*

Proof. Let $a \in G$. For any $x \in G$, $l_a \circ l_{a^{-1}}(x) = l_a(l_{a^{-1}}(x)) = l_a(a^{-1}x) = aa^{-1}x = x$, so $l_a \circ l_{a^{-1}} = \mathbb{1}_G$, the identity map on G. Replace a by a^{-1}

to obtain that $l_{a^{-1}} \circ l_a = \mathbb{1}_G$ as well, and so l_a is bijective, with inverse $(l_a)^{-1} = l_{a^{-1}}$. Similarly, r_a is bijective, with inverse $(r_a)^{-1} = r_{a^{-1}}$. □

6.2.7 Corollary. *Let G be a group, and let H be a subgroup of G. Then for any $a \in G$, $|Ha| = |H| = |aH|$.*

Proof. Let $a \in G$. Then $l_a(H) = \{ l_a(h) \mid h \in H \} = aH$, and thus $l_a|_H$ is a bijective map from H onto aH; hence $|H| = |aH|$. Similarly, $r_a|_H$ is a bijective map from H onto Ha and thus $|H| = |Ha|$. □

Our next step is to prove that the partition of the group G that consists of all right cosets of a subgroup H and the partition of G that consists of all left cosets of H contain the same number of cells (possibly an infinite number). In fact, we prove that there is a natural bijection between the two partitions of G.

6.2.8 Definition. *For any group G and subgroup H of G, let $\Lambda_H = \{ aH \mid a \in G \}$, the set of all left cosets of H in G, and let $P_H = \{ Ha \mid a \in G \}$, the set of all right cosets of H in G.*

The choice of notation is based on the Greek alphabet, where Λ is the upper-case verion of λ, and P is the upper-case version of ρ (representing left and right, respectively).

6.2.9 Proposition. *Let G be a group and let H be a subgroup of G. Then the mapping $\varphi : \Lambda_H \to P_H$ given by $\varphi(aH) = Ha^{-1}$ is a bijection, and thus $|\Lambda_H| = |P_H|$.*

Proof. The first task is to prove that φ is well-defined; that is, we must prove that if $a, b \in G$ are such that $aH = bH$, then $Ha^{-1} = Hb^{-1}$. By Proposition 6.2.3, $aH = bH$ implies that $a^{-1}b \in H$; that is, $(a^{-1})(b^{-1})^{-1} \in H$, and thus by the same proposition, $Ha^{-1} = Hb^{-1}$. Thus φ is a map from Λ_H into P_H. We now prove that it is a bijection.

Suppose first of all that $a, b \in G$ are such that $\varphi(aH) = \varphi(bH)$; that is, $Ha^{-1} = Hb^{-1}$. Then by Proposition 6.2.3, $(a^{-1})(b^{-1})^{-1} \in H$; that is, $a^{-1}b \in H$, and so by the same proposition, $aH = bH$. Thus φ is injective. Next, let $Ha \in P_H$. We have $\varphi(a^{-1}H) = H(a^{-1})^{-1} = Ha$, and so φ is surjective. Thus φ is a bijection from Λ_H onto P_H. □

6.2.10 Definition. *For any group G and subgroup H of G, define the index of H in G, denoted by $[G : H]$, by*

$$[G : H] = |\Lambda_H|.$$

Equivalently, $[G : H] = |P_H|$.

Note that if G is an infinite group, then it is entirely possible that $[G : H]$ is infinite as well, although this is not necessarily the case. For example, $\mathbb{Z}$ under addition is an infinite group, but for any positive integer n, $[\mathbb{Z} : \langle n \rangle] = n$.

We are now ready to present a fundamental group-theoretic result, due to the Italian mathematician Joseph-Louis Lagrange (born Giuseppe Lodovico Lagrangia in Turin, Italy, in 1736).

6.2.11 Theorem (Lagrange's theorem). *Let G be a finite group, and let H be a subgroup of G. Then $|G| = [G : H]\,|H|$.*

Proof. By Proposition 3.2.19 applied to $\sim_H$, we have $|G| = \sum_{A \in \Lambda_H} |A|$, and by Corollary 6.2.7, $|A| = |H|$ for every $A \in \Lambda_H$. Thus $|G| = \sum_{A \in \Lambda_H} |H| = |\Lambda_H|\,|H|$. Finally, by Definition 6.2.10, $|\Lambda_H| = [G : H]$. $\qquad\square$

Note that Lagrange's theorem tells us that the order of any subgroup of a finite group is a divisor of the order of the group.

We now begin to show just how fundamental Lagrange's theorem is, as we proceed to derive a number of important results directly from Lagrange's theorem.

6.2.12 Proposition. *Let G be a finite group. Then for any $g \in G$, $|g|$ divides $|G|$.*

Proof. By Proposition 5.4.18, $|g| = |\langle g \rangle|$, and by Lagrange's theorem, $|\langle g \rangle|$ is a divisor of $|G|$. $\qquad\square$

6.2.13 Corollary. *Let G be a finite group with identity e. Then for each $g \in G$, $g^{|G|} = e$.*

Proof. Let $g \in G$. By Proposition 6.2.12, there exists $k \in \mathbb{Z}$ such that $|G| = k|g|$, and thus $g^{|G|} = (g^{|g|})^k = e^k = e$. $\qquad\square$

6.2.14 Corollary. *Let G be a finite group of prime order. Then G is a cyclic group, and every nonidentity element of G is a generator.*

Proof. Let $g \in G$ with $g \neq e$, the identity of G. Then by Proposition 6.2.12, $|g|$ is a divisor of the prime $|G|$. Since $g \neq e$, $|g| > 1$ and thus $|g| = |G|$, which means that $G = \langle g \rangle$. Thus G is cyclic, generated by g. $\qquad\square$

Group theory has many applications in number theory, the study of the properties of the positive integers. This next result was first identified in a letter written by Pierre de Fermat in 1640. Fermat was a French lawyer and an amateur mathematician who communicated most of his mathematical results in letters to friends, often with proofs omitted, and this was the case with this one.

6.2.15 Proposition (Fermat's little theorem). *If p is a prime and $a \in \mathbb{Z}$ is not a multiple of p, then $a^{p-1} \equiv 1 \mod p$.*

Proof. Let p be a prime. Then $U(\mathbb{Z}_p) = \{\, [a]_p \mid a = 1, 2, \ldots, p-1 \,\}$, so $U(\mathbb{Z}_p)$ is a group of order $p-1$. Since $[a]_p \in U(\mathbb{Z}_p)$ if, and only if, $(a, p) = 1$, it follows from Corollary 6.2.13 that if $a \in \mathbb{Z}$ is not a multiple of p, then $[a]_p^{p-1} = [1]_p$; that is, $a^{p-1} \equiv 1 \mod p$. $\qquad\square$

6.2.16 Lemma. *If G is a group for which every nonidentity element has order 2, then G is abelian.*

Proof. Let G be such a group, and observe that for every $a \in G$, $a = a^{-1}$. Thus for any $a, b \in G$, we have $ab = (ab)^{-1} = b^{-1}a^{-1} = ba$, as required. $\quad\square$

As an interesting application of the elementary results on the order of an element that have been established above, we prove that every group of order at most 5 is abelian. Note that S_3 is a group of order 6 that is not abelian.

6.2.17 Proposition. *If G is a group of order at most 5, then G is abelian.*

Proof. The trival group is abelian. Suppose that G with $2 \leq |G| \leq 5$. If $|G| = 2, 3, 5$, then by Proposition 6.2.14, G is cyclic, and by Proposition 5.5.3, every cyclic group is abelian. Suppose that $|G| = 4$. Then by Proposition 6.2.12, every element of G has order a divisor of 4. If G has an element of order 4, then G is cyclic and thus abelian. Otherwise, every nonidentity element of G has order two, and then by Lemma 6.2.16, G is abelian. $\qquad\square$

In the preceding proof, the noncyclic groups of order 4 stood out. We prove now that there is essentially only one Cayley table for a noncyclic group of order 4. Let G be a noncyclic group of order 4 with identity e. The three nonidentity elements of G each have order 2, and we prove that the product of any two of the three elements of order 2 is the third element of order two. Let $x, y \in G - \{\, e \,\}$ with $x \neq y$. Since $x^{-1} = x$ and $y^{-1} = y$, it follows that $xy \neq e$, and so xy is an element of order two. If $xy = x$, then $xy = xe$ and so by left cancellation, $y = e$, which is not the case. Thus $xy \neq x$. If $xy = y$, then $xy = ey$ and so $x = e$, which is not the case. Thus $xy \neq y$, and so xy is the element of order two in $G - \{x, y\}$. Moreover, this quite graphically displays the fact that G is abelian, as we must have xy and yx each be the unique element of order 2 in $G - \{x, y\}$, so $xy = yx$. The interesting fact in all of this is that the binary operation in a noncyclic group of order 4 is uniquely determined. Now, we have already encountered a noncyclic group of order 4; namely Example 5.4.4, and by our observation above, this must be essentially the only noncyclic group of order 4. To make this notion precise, we require the concept of group isomorphism.

Isomorphisms and Homomorphisms

6.3.1 Definition. *Let G and H be groups. An isomorphism from G to H is a bijection $f : G \to H$ with the property*

$$f(xy) = f(x)f(y) \qquad \text{for all } x, y \in G.$$

If there is an isomorphism from the group G to the group H, then G is said to be isomorphic to H. An isomorphism from G to G is called an automorphism of G.

Note that in the preceding definition, both binary operations, that of G and that of H, are denoted by juxtaposition. For $x, y \in G$, the product xy is computed using the binary operation of G, while the product $f(x)f(y)$ is of course computed using the binary operation of H.

The origins of the word "isomorphism" are found in ancient Greek, where "isos" means "same" and "morphos" means "shape".

The algebraic implications of the existence of an isomorphism f from a group G to a group H is that by means of the isomorphism, computation in G can be carried out in H as follows: for $x, y \in G$, to compute xy, one may instead compute the elements $f(x), f(y) \in H$, then compute $f(x)f(y)$ in H. Since f is a bijection, the inverse mapping f^{-1} exists, and we have $f^{-1}(f(x)f(y)) = f^{-1}(f(xy)) = xy$. Note that f^{-1} is a map from H to G, and the question that arises naturally is whether or not f^{-1} must be an isomorphism from H to G. The answer is provided in the next result.

6.3.2 Proposition. *Let G and H be groups, and let f be an isomorphism from G to H. Then $f^{-1} : H \to G$ is an isomorphism.*

Proof. Let $x, y \in H$. Then $f(f^{-1}(x)f^{-1}(y)) = f(f^{-1}(x))f(f^{-1}(y)) = xy$, and thus $f^{-1}(x)f^{-1}(y) = f^{-1}(xy)$. Since the inverse of a bijection is a bijection, it follows that $f^{-1}: H \to G$ is an isomorphism. $\qquad \square$

Thus we may speak of isomorphic groups G and H without worrying about whether there is an isomorphism from G to H or vice-versa; there is an isomorphism from G to H if, and only if, there is an isomorphism from H to G. Since computation in one group of a pair of isomorphic groups can always be moved over to the other group by means of an isomorphism, it is apparent that the algebraic structure of G must be the same as the algebraic structure of H if G and H are isomorphic. We shall expand on this observation in the pages ahead.

Recall that the composition of two bijections is again a bijection. Is it the case that the composition of two isomorphisms is again an isomorphism?

6.3.3 Proposition. *Let G, H, and K be groups, and let $f: G \to H$ and $g: H \to K$ be isomorphisms. Then $g \circ f: G \to K$ is an isomorphism.*

Proof. As observed above, it suffices to prove that for all $x, y \in G$, $g \circ f(xy) = g \circ f(x)g \circ f(y)$. Since f is an isomorphism, $g \circ f(xy) = g(f(xy)) = g(f(x)f(y))$, and then since g is an isomorphism, $g(f(x)f(y)) = g(f(x))g(f(y)) = g \circ f(x)g \circ f(y)$. $\qquad \square$

6.3.4 Corollary. *Let G be a group. Then $\mathrm{Aut}(G)$, the set of all automorphisms of G, is a subgroup of S_G. $\mathrm{Aut}(G)$ is called the automorphism group of G.*

Proof. Since $\mathbb{1}_G$ is an automorphism of G, $\mathrm{Aut}(G) \neq \varnothing$. By Proposition 6.3.2, $\mathrm{Aut}(G)$ is closed under inversion, and by Proposition 6.3.3, $\mathrm{Aut}(G)$ is closed under the binary operation of S_G; that is, under the operation of composition of mappings. Thus $\mathrm{Aut}(G)$ is a subgroup of S_G. $\qquad \square$

6.3.5 Example.

    (i) The group $\mathbb{R}^+$ of positive real numbers under multiplication is isomorphic to the group $\mathbb{R}$ under addition. To support this claim, it is necessary to establish the existence of an isomorphism between the two groups. In this case, we can actually provide an isomorphism. Consider the natural logarithm mapping $\log: \mathbb{R}^+ \to \mathbb{R}$. Recall that the natural logarithm has as its inverse the exponential mapping, so the natural logarithm is a bijection. The familiar property

$\log(xy) = \log(x) + \log(y)$ *for all positive real numbers x and y tells us that the natural logarithm is a group isomorphism.*

(ii) *Let G be a group. Then for each $g \in G$, the mapping $c_g : G \to G$ given by $c_g(x) = gxg^{-1}$ for all $x \in G$ is an automorphism of G, called the inner automorphism of G induced by g (see Exercise 9 in Chapter 5). It is bijective since $(c_g)^{-1} = c_{g^{-1}}$, and for $x, y \in G$, $c_g(xy) = gxyg^{-1} = gxg^{-1}gyg^{-1} = c_g(x)c_g(y)$, so c_g is an isomorphism of G to itself; that is, an automorphism of G. Additionally, since c_e is the identity mapping on G and $c_{gh}(x) = ghx(gh)^{-1} = g(hxh^{-1})g^{-1} = c_g(c_h(x))$, so $c_g \circ c_h = c_{gh} \in Aut(G)$, we see that $I(G) = \{ c_g \mid g \in G \}$ is nonempty, closed under inversion, and closed under composition of mappings, so $I(G)$ is a subgroup of $Aut(G)$, called the group of inner automorphisms of G.*

The notion of conjugation in a group plays a very important role in the study of the structure of a group. We pause briefly in our study of isomorphisms and homomorphisms to present a few elementary conjugation results. These conjugacy results are actually a special case of the more general notion of group action (group actions will be studied in Chapter 12).

6.3.6 Definition. *Let G be a group. For $x, y \in G$, we say that x is conjugate to y in G if there exists $g \in G$ such that $y = gxg^{-1}$. We also say that y is the result of conjugating x by g.*

6.3.7 Proposition. *Let G be a group. The relation*

$$\{ (x,y) \in G \times G \mid x \text{ is conjugate to } y \text{ in } G \}$$

is an equivalence relation on G. Its equivalence classes are called the conjugacy classes of G.

Proof. For $x \in G$, $x = exe^{-1}$, so x is conjugate to itself in G. Thus the relation is reflexive. Let $x, y \in G$ be such that x is conjugate to y in G. Then there is $g \in G$ with $y = gxg^{-1}$, so $g^{-1}yg = x$. Since $g = (g^{-1})^{-1}$, for $h = g^{-1}$, we have $y = hxh^{-1}$ and thus y is conjugate to x in G. This proves that the relation is symmetric. Finally, suppose that $x, y, z \in G$ are such that x is conjugate to y in G and y is conjugate to z in G. Then there exist $g, h \in G$ such that $y = gxg^{-1}$ and $z = hyh^{-1}$. We have $z = h(gxg^{-1})h^{-1} = (hg)x(g^-h^{-1}) = (hg)x(hg)^{-1}$ and thus x is conjugate to z in G. This proves the relation is transitive, and so the relation is an equivalence relation on G. $\square$

Note that if g and x commute, then $gxg^{-1} = x$. In particular, we see that if x is a central element in the group G (that is, x commutes with every element of G), then the conjugacy class of x is just $\{x\}$, while if x is not central, then the conjugacy class of x has at least two elements in it. Thus the centre of the group G is the union of the singleton conjugacy classes. We remark that these observations make it clear that the notion of conjugation is of no value in the study of an abelian group.

We now return to our study of isomorphisms and homomorphisms. It turns out that removing the bijection requirement in the definition of isomorphism results in an indispensable tool in the study of groups.

6.3.8 Definition. *Let G and H be groups. A mapping $f : G \to H$ is called a homomorphism if for all $x, y \in G$, $f(xy) = f(x)f(y)$. If $f : G \to G$ is a homomorphism, then f is called an endomorphism of G.*

A homomorphism f from a group G to a group H need not preserve all of the algebraic structure of the group G. However, it is often the case that one can learn important information about G from knowledge about its homomorphic image.

6.3.9 Example.

(i) *If V and W are vector spaces, then V and W are abelian groups under their vector addition operations. If $T : V \to W$ is a linear transformation, then for all $u, v \in V$, $T(u + v) = T(u) + T(v)$ (and also for all $r \in \mathbb{R}$, $T(ru) = rT(u)$, but we do not make use of this fact here), and so T is a homomorphism from the group V under addition to the group W under addition.*

(ii) *For any positive integer n, the map $\pi_n : \mathbb{Z} \to \mathbb{Z}_n$ given by $\pi_n(m) = [m]_n$ for all $m \in \mathbb{Z}$ is a surjective homomorphism. It is not an isomorphism since it cannot possibly be injective: $\mathbb{Z}$ is infinite, while $|\mathbb{Z}_n| = n$.*

(iii) *For $R = \mathbb{Z}, \mathbb{Q}, \mathbb{R}$ or $\mathbb{C}$, the determinant map $\det : M_{2 \times 2}(R) \to R$, given by $\det(\left[\begin{smallmatrix} a & b \\ c & d \end{smallmatrix}\right]) = ad - bc$ for any $a, b, c, d \in R$, satisfies $\det(AB) = \det(A)\det(B)$ for all $A, B \in M_{2 \times 2}(R)$. The reader may directly verify that*

$$\det\left(\begin{bmatrix} a & b \\ c & d \end{bmatrix}\begin{bmatrix} r & s \\ t & u \end{bmatrix}\right) = adru + bcst - adts - bcru$$

$$= \det\left(\begin{bmatrix} a & b \\ c & d \end{bmatrix}\right)\det\left(\begin{bmatrix} r & s \\ t & u \end{bmatrix}\right)$$

for all $a, b, c, d, r, s, t, u \in R$. *Furthermore, as a result of this fact, the fact that* $\det(\left[\begin{smallmatrix} 1 & 0 \\ 0 & 1 \end{smallmatrix}\right]) = 1$, *and the fact that* $\left[\begin{smallmatrix} a & b \\ c & d \end{smallmatrix}\right] \left[\begin{smallmatrix} d & -b \\ -c & a \end{smallmatrix}\right] = (ad - bc)\left[\begin{smallmatrix} 1 & 0 \\ 0 & 1 \end{smallmatrix}\right]$, *it follows that* $A \in M_{2\times 2}(R)$ *is invertible if, and only if,* $\det(A)$ *is invertible in* R; *that is,* $A \in Gl_2(R)$ *if, and only if,* $\det(A) \in U(R)$, *the group of units (invertible elements of* R *with respect to multiplication) of* R. *The interesting observation here is that if* $A = \left[\begin{smallmatrix} a & b \\ c & d \end{smallmatrix}\right]$ *is such that* $\det(A) \in U(R)$, *then* $A^{-1} = (\det(A))^{-1}\left[\begin{smallmatrix} d & -b \\ -c & a \end{smallmatrix}\right]$. *Thus the restriction of the determinant map to* $Gl_2(R)$ *is a homomorphism from the group* $Gl_2(R)$ *(with binary operation matrix multiplication) to the group of units* $U(R)$ *(with binary operation multiplication). In fact, this homomorphism is surjective, since for any* $a \in U(R)$, $\det(\left[\begin{smallmatrix} a & 0 \\ 0 & 1 \end{smallmatrix}\right]) = a$.

We now present some of the elementary, but important, properties of homomorphisms.

6.3.10 Proposition. *For G and H groups and $f : G \to H$ a homomorphism, the following hold:*

(i) *If e_G denotes the identity of G, then $f(e_G) = e_H$, the identity of H.*

(ii) *For any $g \in G$, and any $n \in \mathbb{Z}$, $f(g^n) = (f(g))^n$. In particular, $(f(g))^{-1} = f(g^{-1})$.*

(iii) *If K is a subgroup of G, then $f(K)$ is a subgroup of H.*

(iv) *If L is a subgroup of H, then $f^{-1}(L)$ is a subgroup of G.*

(v) *If $g \in G$ has finite order, then $f(g)$ has finite order and $|f(g)|$ divides $|g|$.*

(vi) *For any $g \in G$, $f(\langle g \rangle) = \langle f(g) \rangle$. In particular, if G is cyclic, then $f(G)$ is cyclic.*

(vii) *If G is abelian, then $f(G)$ is abelian.*

(viii) *If L is a group and $g : H \to L$ is a homomorphism, then $g \circ f : G \to L$ is a homomorphism.*

Proof. (i) $f(e_G) = f(e_G e_G) = f(e_G)f(e_G)$, and so by Lemma 5.4.8, $f(e_G) = e_H$.

(ii) Let $g \in G$. We prove by induction that for every nonnegative integer n, $f(g^n) = (f(g))^n$. By (i), $f(g^0) = f(e_G) = e_H = (f(g))^0$. Suppose now that n is a nonnegative integer for which $f(g^n) = (f(g))^n$. Then $f(g^{n+1}) = f(g^n g) = f(g^n)f(g) = (f(g))^n f(g) = (f(g))^{n+1}$. It follows now by induction that for every nonnegative integer n, $f(g^n) = (f(g))^n$. Next, observe that by (i), $e_H = f(e_G) = f(gg^{-1}) = f(g)f(g^{-1})$, and thus

$(f(g))^{-1} = f(g^{-1})$. Let $n \in \mathbb{Z}^+$. Then $f(g^{-n}) = f((g^{-1})^n) = (f(g^{-1}))^n = ((f(g))^{-1})^n = (f(g))^{-n}$. Thus for every $n \in \mathbb{Z}$, $f(g^n) = (f(g))^n$.

(iii) Let K be a subgroup of G. Since $e_G \in K$, $e_H = f(e_G) \in f(K)$ and so $f(K) \neq \varnothing$. For $h \in f(K)$, $h = f(g)$ for some $g \in K$, and so $h^{-1} = (f(g))^{-1} = f(g^{-1}) \in f(K)$ since $g \in K$ implies that $g^{-1} \in K$. Finally, let $h, t \in f(K)$, so there exist $x, y \in K$ with $h = f(x)$ and $t = f(y)$. Then $ht = f(x)f(y) = f(xy) \in f(K)$ since $x, y \in K$ implies that $xy \in K$. Thus $f(K)$ is a subgroup of H.

(iv) Let L be a subgroup of H. Since $f(e_G) = e_H \in L$, $e_G \in f^{-1}(L)$ and thus $f^{-1}(L) \neq \varnothing$. Let $x, y \in f^{-1}(L)$. Then $f(x), f(y) \in L$, and so $f(x)(f(y))^{-1} = f(x)f(y^{-1}) = f(xy^{-1}) \in L$. Thus $xy^{-1} \in f^{-1}(L)$, and so by Proposition 5.4.10, $f^{-1}(L)$ is a subgroup of G.

(v) Suppose that $g \in G$ has finite order, say $|g| = n$. Then $e_H = f(e_G) = f(g^n) = (f(g))^n$, which means that $f(g)$ has finite order, and by Proposition 5.4.20, $|f(g)|$ divides $n = |g|$.

(vi) Let $g \in G$. Then $f(\langle g \rangle) = f(\{ g^k \mid k \in \mathbb{Z} \}) = \{ f(g^k) \mid k \in \mathbb{Z} \} = \{ (f(g))^k \mid k \in \mathbb{Z} \} = \langle f(g) \rangle$.

(vii) Suppose that G is abelian, and let $x, y \in f(G)$. Then there exist $a, b \in G$ with $x = f(a)$ and $y = f(b)$, so $xy = f(a)f(b) = f(ab) = f(ba) = f(b)f(a) = yx$, where we have used the fact that G is abelian in the equality $f(ab) = f(ba)$ since $ab = ba$.

(viii) Let $x, y \in G$. Then $g \circ f(xy) = g(f(xy)) = g(f(x)f(y)) = g(f(x))g(f(y)) = g \circ f(x) g \circ f(y)$, and so $g \circ f$ is a homomorphism from G to L. $\square$

6.3.11 Corollary. *Let G and H be isomorphic groups. Then G is abelian if, and only if, H is abelian. Furthermore, if $f : G \to H$ is an isomorphism, then for any $g \in G$, $|f(g)| = |g|$.*

Proof. Since G and H are isomorphic groups, there exists an isomorphism $f : G \to H$, and by Proposition 6.3.2, $f^{-1} : H \to G$ is an isomorphism. Suppose that G is abelian. Then by Proposition 6.3.10 (vii), $H = f(G)$ is abelian. On the other hand, suppose that H is abelian. Then by Proposition 6.3.10 (vii) applied to f^{-1}, $G = f^{-1}(H)$ is abelian. Thus G is abelian if, and only, H is abelian.

Let $g \in G$. Then $f^{-1}(f(g)) = g$, so by Proposition 6.3.10 (v) applied to f and to f^{-1}, g has finite order if, and only, $f(g)$ has finite order. Moreover, if g and $f(g)$ have finite order, then $|g|$ and $|f(g)|$ each divides the other and so $|g| = |f(g)|$. $\square$

We remark that to prove two groups are isomorphic, we must demonstrate the existence of an isomorphism from one to the other, and this can often prove to be a daunting task. On the other hand, to prove that two groups are not isomorphic, it suffices to find an algebraic property possessed by one but not the other. For example, by Corollary 6.3.11, if one of the groups has m elements of a given order, while the other group has n elements of that order, and $m \neq n$, then the two groups are not isomorphic. For example, the quaternion group Q_8 (see Exercise 8 in Chapter 5) has six elements of order 4, one element of order 2, and of coure, one element of order 1, while the dihedral group D_4 (see Example 5.4.6) has two elements of order 4, five elements of order 2, and one element of order 1. Thus Q_8 and D_4 are nonisomorphic groups of order 8. Another interesting application of this observation establishes that $\mathbb{Z}_4$ is not isomorphic to K_4, whose Cayley table was shown in Table 5.1. Both are abelian groups of order 4, But K_4 has no elements of order 4, while $\mathbb{Z}_4$ has two elements of order 4. Thus $\mathbb{Z}_4$ and K_4 are nonisomorphic groups of order 4.

Similarly, if one of the groups is abelian, but the other group is not, then the two groups are not isomorphic. A very simple criterion is size: since an isomorphism is a bijective mapping, isomorphic groups must have the same order. Thus two groups of different orders are not isomorphic.

We have seen that the homomorphic image of a cyclic group is cyclic, and it is natural to wonder about the existence of isomorphisms, or more generally, homomorphisms, from one cyclic group to another.

We have already seen in Example 6.3.9 (ii) that the map from $\mathbb{Z}$ to $\mathbb{Z}_n$ given by $m \mapsto [m]_n$ is a surjective homomorphism. More generally, we shall prove that if $G = \langle g \rangle$ is an infinite cyclic group, and H is a group, then for any $h \in H$, there is a unique homomorphism $\varphi : G \to H$ with $\varphi(g) = h$, while if $G = \langle g \rangle$ is a finite cyclic group, and H is a group, then for any $h \in H$ for which $|h|$ is a divisor of $|g|$, there is a unique homomorphism $\varphi : G \to H$ such that $\varphi(g) = h$. In either case, by Proposition 6.3.10 (vi), $\varphi(G) = \langle h \rangle$. This result has several important consequences, not least of which is the fact that two cyclic groups are isomorphic if, and only, they have the same order.

6.3.12 Proposition. *Let G be a cyclic group, and let g be a generator for G. If g has infinite order, then for any group H, and any $h \in H$, there exists a unique homomorphism from G to H which maps g to h. If g has finite order n, then for any group H, and any $h \in H$ for which $|h|$ divides n, there is a unique homomorphism from G to H which maps g to h.*

Proof. Suppose first that g has infinite order. Let H be a group and let $h \in H$. Since for any $i, j \in \mathbb{Z}$, $g^i = g^j$ if, and only if, $i = j$, it follows that $\varphi : G \to H$ defined by $\varphi(g^i) = h^i$ is a well-defined mapping (each $x \in G$ is uniquely of the form $x = g^i$ for some $i \in \mathbb{Z}$). Let $x, y \in G$. Then there exist $i, j \in \mathbb{Z}$ such that $x = g^i$ and $y = g^j$, so $\varphi(xy) = \varphi(g^i g^j) = \varphi(g^{i+j}) = h^{i+j}$, while $\varphi(x)\varphi(y) = h^i h^j = h^{i+j}$, and so $\varphi(xy) = \varphi(x)\varphi(y)$. Thus $\varphi : G \to H$ is a homomorphism, and $\varphi(g) = h$. Suppose now that $f : G \to H$ is a homomorphism such that $f(g) = h$. Then for any $x \in G$, there exists $i \in \mathbb{Z}$ such that $x = g^i$, so $f(x) = f(g^i)$, and thus by Proposition 6.3.10 (ii), $f(x) = (f(g))^i = h^i = \varphi(g^i) = \varphi(x)$. This proves that $f = \varphi$.

Now suppose that g has finite order n. Let H be a group and let $h \in H$ have finite order dividing n. Let $x \in G$. Then there exists $i \in \mathbb{Z}$ such that $x = g^i$. By Proposition 5.4.20, for any $j \in \mathbb{Z}$, $g^i = g^j$ if, and only if, n divides $i - j$, and so for such a $j \in \mathbb{Z}$, $h^{i-j} = e_H$. Thus we may define $\varphi(x) = h^i$, where $x = g^i$. Note that since $h^i = h^j$ if $g^i = g^j$, φ is a well-defined mapping. Morever, just as in the preceding paragraph, φ is a homomorphism, $\varphi(g) = h$, and φ is the unique homomorphism from G to H that maps g to h. $\square$

6.3.13 Corollary. *Let G and H be cyclic groups. Then the following hold:*

(i)   *G and H are isomorphic if, and only if, $|G| = |H|$ (that is, both are infinite cyclic, or both are finite cyclic of the same order).*

(ii) *There exists a surjective homomorphism from G onto H if, and only if, either G is infinite or else both G and H are finite, with $|H|$ a divisor of $|G|$.*

(iii) *If L is a subgroup of H, then there exists a homomorphism $f : G \to H$ such that $f(G) = L$ if, and only if, either G is infinite or else both G and L are finite and $|L|$ is a divisor of $|G|$.*

(iv) *$Aut(G)$ is isomorphic to $U(\mathbb{Z}) = \{1, -1\}$ (under multiplication) if G is infinite, while if $|G| = n$, then $Aut(G)$ is isomorphic to the multiplicative group $U(\mathbb{Z}_n)$.*

Proof. (i) If G and H are isomorphic, then there is a bijective mapping from G to H, and thus $|G| = |H|$. For the converse, suppose that $|G| = |H|$. We are to prove the existence of an isomorphism from G to H. Let g and h be generators for G and H, respectively. By Proposition 6.3.12, there is a unique homomorphism $\varphi : G \to H$ for which $\varphi(g) = h$, and a unique homomorphism $\psi : H \to G$ for which $\psi(h) = g$. Then $\psi \circ \varphi : G \to G$ is a homomorphism for which $\psi \circ \varphi(g) = g$. But by Proposition 6.3.12, there exists a unique homomorphism from G to itself that maps g to itself. Since the identity map is a homomorphism from G to G that maps g to itself, it follows that $\psi \circ \varphi = 1_G$. Similarly, $\varphi \circ \psi : H \to H$ is the identity map on H, and thus $\psi = \varphi^{-1}$; that is, φ is a bijective homomorphism, hence an isomorphism from G to H.

(ii) Let $G = \langle g \rangle$ and $H = \langle h \rangle$. If G is infinite, then by Proposition 6.3.12, there exists a homomorphism $\varphi : G \to H$ with $\varphi(g) = h$, and then by Proposition 6.3.10, (vi), $\varphi(G) = \langle h \rangle = H$. Thus φ is a surjective homomorphism from G to H. Suppose now that G is finite, with $|G| = n$, so $|g| = n$. If H is finite, with $|h| = |H|$ a divisor of n, then by Proposition 6.3.12, there exists a homomorphism $\varphi : G \to H$ for which $\varphi(g) = h$, and then $\varphi(G) = \varphi(\langle g \rangle) = \langle \varphi(g) \rangle = \langle h \rangle = H$, so φ is a surjective homomorphism from G to H.

Conversely, suppose that there is a surjective homomorphism $\varphi : G \to H$. If G is infinite, there is nothing further to show, so suppose that G is finite, say $|G| = n$. Since $H = \varphi(G) = \varphi(\langle g \rangle) = \langle \varphi(g) \rangle$, we have $|H| = |\varphi(g)|$. By Proposition 6.3.10 (v), $|H|$ is finite and a divisor of $|g| = |G|$.

(iii) Let L be a subgroup of H. By Proposition 5.5.5, L is cyclic. Since a homomorphism $\varphi : G \to H$ such that $\varphi(G) = L$ can be regarded as a surjective homomorphism from G to L, the result now follows from (ii).

(iv) Suppose first of all that $G = \langle g \rangle$ is infinite. Then by Corollary 5.5.7, G has exactly two generators g and g^{-1}. By Proposition 6.3.12, the identity map on G, and the unique homomorphism from G to itself that sends g to g^{-1} are automorphisms of G. Suppose that $\varphi \in \text{Aut}(G)$. Then $G = \varphi G = \varphi(\langle g \rangle) = \langle \varphi(g) \rangle$, and so $\varphi(g)$ is a generator of G. If $\varphi(g) = g$, then $\varphi = 1_G$, otherwise $\varphi(g) = g^{-1}$. Thus $|\text{Aut}(G)| = 2$. Since any finite group of prime order is cyclic, and $\{\pm 1\}$ is a subgroup of $\mathbb{R}^*$, the group of all nonzero real numbers under multiplication, it follows that $\text{Aut}(G)$ and $\{\pm 1\}$ are cyclic groups of order 2, and so by (i), $\text{Aut}(G)$ is isomorphic to $\{\pm 1\}$.

Suppose now that $G = \langle g \rangle$ is finite, say $|G| = n$. Then $|g| = n$. If $\varphi \in \text{Aut}(G)$, then $G = \varphi(G) = \langle \varphi(g) \rangle$, and so $|\varphi(g)| = n$; that is, $\varphi(g)$ is a generator of G. Thus if $\varphi \in \text{Aut}(G)$, φ maps g to a generator of G. The set of all generators of G is $\{g^i \mid |g^i| = n\} = \{g^i \mid \frac{n}{(n,i)} = n\} = \{g^i \mid (n,i) = 1\}$, and by Proposition 6.3.12, for each generator g^i of G, there exists a unique homomorphism from G to itself mapping g to g^i. Such a homomorphism is surjective since g^i is a generator of G. Since G

is finite, any surjective map from G to G is injective, so φ is bijective. This completes the proof that $\varphi \in \operatorname{Aut}(G)$. Thus for each generator g^i of G, there is a unique automorphism of G that maps g to g^i. As we have observed above, if g^i is a generator of G, then $(n, i) = 1$. Moreover, for any $i, j \in \mathbb{Z}$, if $g^i = g^j$, then n divides $i - j$ and so $[i]_n = [j]_n$. Thus the mapping Φ from $\operatorname{Aut}(G)$ to $U(\mathbb{Z}_n) = \{\, [i]_n \mid (n, i) = 1 \,\}$ that maps $\varphi \in \operatorname{Aut}(G)$ to $[i]_n$, where $\varphi(g) = g^i$ is well-defined and bijective. It remains to prove that it is a homomorphism. Suppose that $\varphi, \psi \in \operatorname{Aut}(G)$. Then there exist $i, j \in \mathbb{Z}$ with $\varphi(g) = g^i$, $\psi(g) = g^j$, and $(n, i) = 1 = (n, j)$. We have $\Phi(\varphi) = [i]_n$ and $\Phi(\psi) = [j]_n$. We know that $\psi \circ \varphi$ is an automorphism of G, and $\psi \circ \varphi(g) = \psi(g^i) = (\psi(g))^i = (g^j)^i = g^{ji}$. Since $[i]_n, [j]_n \in U(\mathbb{Z}_n)$, it follows that $[ji]_n = [j]_n [i]_n \in U(\mathbb{Z}_n)$, and so $\Phi(\psi \circ \varphi) = [ji]_n = [j]_n [i]_n = \Phi(\psi)\Phi(\varphi)$, as required. $\qquad\square$

We return now to the study of homomorphisms in general. Our first objective is to investigate subgroups of the domain and the codomain that are associated in a natural way with a homomorphism. We have already seen that if G and H are groups and $f : G \to H$, the image of f, $f(G)$, is a subgroup of H. As we have observed before, a study of this subgroup of H can often help us to obtain a better understanding of the algebraic structure of G. There is another subgroup that we associate with f that is also important in the study of the structure of G. It is a subgroup of G, called the kernel of the homomorphism f. Recall that by Proposition 6.3.10 (iv), for any subgroup L of H, $f^{-1}(L)$ is a subgroup of G. In particular, since $\{\, e_H \,\}$ is a subgroup of H, it follows that $f^{-1}(e_H)$ is a subgroup of G.

6.3.14 Definition. *Let G and H be groups, and let $f : G \to H$ be a homomorphism. Then the subgroup $f^{-1}(e_H)$ of G is called the kernel of f, denoted by $\ker(f)$.*

6.3.15 Example.

(i) *The map $f : \mathbb{Z} \to \mathbb{Z}_n$ defined by $f(m) = [m]_n$ for all $m \in \mathbb{Z}$ is a homomorphism (see Example 6.3.9 (ii)). Its kernel is $f^{-1}([0]_n) = \{\, i \in \mathbb{Z} \mid [i]_n = [0]_n \,\} = \{\, kn \mid k \in \mathbb{Z} \,\} = \mathbb{Z}n$.*

(ii) *In Example 5.2.9 (ix), the subgroup*

$$D = \{\, \pm I_2, \pm \begin{bmatrix} 0 & -1 \\ 1 & 0 \end{bmatrix}, \pm \begin{bmatrix} 1 & 0 \\ 0 & -1 \end{bmatrix}, \pm \begin{bmatrix} 0 & 1 \\ 1 & 0 \end{bmatrix} \,\}$$

of $Gl_2(\mathbb{Z})$ was introduced (see Figure 5.3 for its Cayley table). It was shown in Example 6.3.9 (iii) that the determinant mapping $\det : Gl_2(\mathbb{Z}) \to U(\mathbb{Z})$ is a homomorphism, and so its restriction to D is *a homomorphism* det $: D \to U(\mathbb{Z})$. *Recall that* $U(\mathbb{Z})$, *the group of units of the monoid* $\mathbb{Z}$ *under multiplication, is equal to* $\{\pm 1\}$. *Since the identity of* $U(\mathbb{Z})$ *is 1, the kernel of* det $: D \to \{\pm 1\}$ *consists of those elements of D whose determinant is 1; namely* $\{\pm I_2, \pm \begin{bmatrix} 0 & -1 \\ 1 & 0 \end{bmatrix}\} = \{\pm I_2, \pm r\}$ *in the notation of Example 5.2.9 (ix). Since $r^2 = -I_2$ and $r^3 = -r$ with $r^4 = I_2$, the kernel is a cyclic group of order 4, $\langle r \rangle$.*

In Example 6.3.15 (ii), we were investigating the group D introduced in Example 5.2.9 (ix). An important result in the theory of finite dimensional vector spaces (see Appendix B) asserts that the mapping l from $M_{2\times2}(\mathbb{R})$ to the monoid of all linear transformations from $\mathbb{R}^2$ to itself has the properties that for any $A, B \in M_{2\times2}(\mathbb{R})$, $l_{AB} = l_A \circ l_B$ (where the image of $A \in M_{2\times2}(\mathbb{R})$ under l is denoted by l_A), and A is invertible in $M_{2\times2}(\mathbb{R})$ if, and only if, l_A is an invertible linear transformation. Thus the restriction of l to $GL_2(\mathbb{R})$ provides a homomorphism from $Gl_2(\mathbb{R})$ to the group of all linear isomorphisms of $\mathbb{R}^2$ to itself. Moreover, in Chapter 15, it is shown that $A \in Gl_2(\mathbb{R})$ satisfies $A^{-1} = A^t$ if, and only if, l_A is an isometry fixing the origin of $\mathbb{R}^2$. In Appendix B, it is shown that (using the notation of Example 5.2.9 (ix))

(i) l_{I_2} is the identity map on $\mathbb{R}^2$;
(ii) l_r is rotation by $\frac{\pi}{4}$ radians in the counterclockwise direction;
(iii) l_{-I_2} is rotation by π radians in the counterclockwise direction;
(iv) l_{-r} is rotation by $\frac{3\pi}{4}$ radians in the counterclockwise direction;
(v) l_h is reflection across the x-axis;
(vi) l_{-h} is reflection across the y-axis;
(vii) l_{d_+} is reflection across the line with equation $y = x$; and
(viii) l_{d_-} is relection across the line with equation $y = -x$.

Each of these is an isometry that fixes a square centred at the origin, and thus l, when restricted to D, provides an isomomorphism from D to D_4, the group of symmetries of a square. Observe that the isometries that correspond to the elements of D with determinant 1 are all rotations (considering the identity map as a rotation of 0 radians), while the reflections all correspond to the elements of D with determinant -1.

6.4 Normal Subgroups and Factor Groups

In the study of the right and left congruences determined by a subgroup H of a group G, it is evident that if G is abelian, then for any subgroup H of G, $\approx_H = \sim_H$ since for any $a \in G$, the left coset of a is the same as the right coset of a; that is, $aH = Ha$. It is this fact about the equality of the right and left cosets of the subgroup $\langle n \rangle$ of $\mathbb{Z}$ that allowed us to define the operations of addition and multiplication on the set of equivalence classes of $\mathbb{Z}$ modulo n. In this section, we investigate the problem of determining necessary and sufficient conditions on a subgroup H of a group G under which $\approx_H = \sim_H$ and the binary operation on G allows us to define a binary operation on the set of all (left) cosets of H in G.

For example, the quaternion group (see Exercise 8 in Chapter 5) $Q_8 = \{\pm 1, \pm i, \pm j, \pm k\}$ has subgroup $H = \langle -1 \rangle = \{1, -1\}$. The $[Q_8 : H] = 4$ left cosets of H in Q_8 are:

$$1H = H = (-1)H$$
$$iH = \{i, (-i)\} = -iH$$
$$jH = \{j, (-j)\} = -jH$$
$$kH = \{k, (-k)\} = -kH$$

and the four right cosets of H in Q_8 are:

$$H1 = H = H(-1)$$
$$Hi = \{i, -i\} = H(-i)$$
$$Hj = \{j, -j\} = H(-j)$$
$$Hk = \{k, -k\} = H(-k)$$

and so we see that for every $x \in Q_8$, $xH = Hx$, even though Q_8 is not abelian. To see that this need not always happen, consider the example of S_3, a nonabelian group of order 6. The subgroup $H = \langle [2\,1\,3] \rangle$ is equal to $\{[2\,1\,3], [1\,2\,3]\}$. We have $[1\,3\,2]H = \{[1\,3\,2]\circ[2\,1\,3], [1\,3\,2]\} = \{[3\,1\,2], [1\,3\,2]\}$, while $H[1\,3\,2] = \{[2\,1\,3]\circ[1\,3\,2], [1\,3\,2]\} = \{[2\,3\,1], [1\,3\,2]\}$. Thus $[1\,3\,2]H \neq H[1\,3\,2]$.

6.4.1 Definition. *Let G be a group. A subgroup H of G is said to be normal in G if, for all $g \in G$, the left coset of g is the same as the right coset of g; that is, $gH = Hg$. If H is normal in G, we write $H \triangleleft G$.*

The following result is simple, yet surprisingly useful observation.

6.4.2 Lemma. *Let G be a group, and let H be a subgroup of index 2 in G. Then $H \triangleleft G$.*

Proof. There are exactly 2 left cosets of H in G, H itself, and $G - H$. Thus for any $g \in G - H$, $G - H = gH$. As well, there are exactly two right cosets of H in G, H itself, and $G - H$, and so for any $g \in G - H$, $G - H = Hg$. Thus for any $g \in G - H$, $gH = G - H = Hg$, while for any $g \in H$, $gH = H = Hg$, and so $H \triangleleft G$. $\qquad\square$

6.4.3 Example.

(i) For any group G, the trivial subgroup $\{e_G\}$ and the group G itself are normal subgroups of G.

(ii) In an abelian group, every subgroup is normal.

(iii) In S_3, the cyclic subgroup $\langle [2\,3\,1] \rangle$ has order 3, and so has index 2 in S_3. By Lemma 6.4.2, $\langle [2\,3\,1] \rangle \triangleleft S_3$.

(iv) The quaternion group Q_8 (see Exercise 8 in Chapter 5) is not abelian, but yet, every subgroup of Q_8 is normal. Aside from Q_8 and the trivial subgroup, there are three subgroups of order 4, all cyclic (namely, $\langle \mathbf{i} \rangle$, $\langle \mathbf{j} \rangle$, and $\langle \mathbf{k} \rangle$, and one subgroup of order 2, $\langle -\mathbf{1} \rangle$. The three subgroups of order 4 are normal in Q_8 by Lemma 6.4.2, while the subgroup of order 2 is normal since $-\mathbf{1}$ commutes with every element of Q_8. Recall that $\pm\mathbf{i}$ are elements of order 4, with $-\mathbf{i} = \mathbf{i}^{-1}$. Similarly, $-\mathbf{j} = \mathbf{j}^{-1}$ and $-\mathbf{k} = \mathbf{k}^{-1}$. By Lagrange's theorem, a subgroup of Q_8 must have order 1, 2, 4, or 8, and Q_8 has only one element of order 2, so there is a unique subgroup of order 2. A subgroup of order 4 must therefore contain at least one element of order 4 and is therefore a cyclic subgroup. Thus we have indeed examined all subgroups of Q_8.

Given a group G, we may apply Proposition 3.4.31 to extend the binary operation on G to the set $\mathscr{P}(G)$. Then $\mathscr{P}(G)$ is a monoid with identity $\{e_G\}$, and an element $X \in \mathscr{P}(G)$ is invertible if, and only if, $|X| = 1$. In Notation 3.4.32, it was declared that for $x \in G$ and $Y \in \mathscr{P}(G)$, we will write xY in place of $\{x\}Y$. More generally, for any $g, h \in G$ and $Y \in \mathscr{P}(G)$, we write gYh for $\{g\}Y\{h\} = \{xyh \mid y \in Y\}$.

6.4.4 Proposition. *Let G be a group. A subgroup H of G is normal in G if, and only if, $g^{-1}Hg \subseteq H$ for all $g \in G$.*

Proof. First, suppose that H is normal in G and let $g \in G$. Then $\{g\}H = gH = Hg = H\{g\}$. Since g is an invertible element of G, $\{g\}$ is an invertible element of $\mathscr{P}(G)$, and $\{g\}^{-1} = \{g^{-1}\}$. Thus $H = \{e\}H = \{g\}^{-1}(\{g\}H) = \{g^{-1}\}H\{g\} = g^{-1}Hg$; that is, $H = g^{-1}Hg$.

Conversely, suppose that for all $g \in G$, $g^{-1}Hg \subseteq H$. Let $g \in G$. Then in $\mathscr{P}(G)$, we have $\{g\}\{g^{-1}\}H\{g\} \subseteq \{g\}H$, and thus $\{e\}H\{g\} \subseteq \{g\}H$; that is, $H\{g\} \subseteq \{g\}H$, or $Hg \subseteq gH$. As well, we have $(g^{-1})^{-1}H(g^{-1}) \subseteq H$, and thus $\{g\}H\{g^{-1}\}\{g\} \subseteq H\{g\}$; that is, $gH \subseteq Hg$. Thus $Hg = gH$ for all $g \in G$, and so $H \triangleleft G$. $\qquad\square$

6.4.5 Corollary. *Let G be a group. If $H \triangleleft G$, then for every $g \in G$, $g^{-1}Hg = H$.*

Proof. Let $H \triangleleft G$, and let $g \in G$. By Proposition 6.4.4 applied to g, we obtain $g^{-1}Hg \subseteq H$, while when applied to g^{-1}, we obtain $(g^{-1})^{-1}H(g^{-1}) \subseteq H$; that is, $gHg^{-1} \subseteq H$ and thus $H \subseteq g^{-1}Hg$. These two inclusions establish that $g^{-1}Hg = H$. $\qquad\square$

Proposition 6.4.4 allows us to make an interesting observation about conjugacy classes. In our discussion of the conjugacy classes of a group G, it was apparent that the conjugacy classes of G do not, in general, interact in any reasonable way with an arbitrary subgroup of G. However, something useful can be said when the subgroup is normal.

6.4.6 Proposition. *Let G be a group, and let $H \triangleleft G$. Then H is a union of conjugacy classes of G.*

Proof. It suffices to prove that for each $x \in H$, if x is conjugate to y in G, then $y \in H$. Let $y \in G$ be such that x is conjugate to y in G, and let $g \in G$ be such that $y = gxg^{-1}$. Since $H \triangleleft G$, we have $gHg^{-1} = H$ and thus $y = gxg^{-1} \in H$. $\qquad\square$

While it can be difficult in general to find normal subgroups of a given group, there is one particular situation in which a normal subgroup can be identified.

6.4.7 Proposition. *Let G and H be groups, and let $f : G \to H$ be a homomorphism. Then $\ker(f) \triangleleft G$. Moreover, for every $g \in G$, $f^{-1}(f(g)) = g\ker(f) = \ker(f)g$, and in particular, f is injective if, and only if, $|\ker(f)| = 1$.*

Proof. It was noted in Definition 6.3.14 that $\ker(f)$ is a subgroup of G, so it suffices to prove that for all $g \in G$, $g\ker(f) = \ker(f)g$, or equivalently (by Proposition 6.4.4), that for all $g \in G$, $g^{-1}\ker(f)g = \ker(f)$. Let $g \in G$. For $x \in \ker(f)$, we have $f(g^{-1}xg) = f(g^{-1})f(x)f(g) = f(g^{-1})e_H f(g) = (f(g))^{-1}f(g) = e_H$, and thus $g^{-1}xg \in f^{-1}(e_H) = \ker(f)$. This proves that

$g^{-1} \ker(f) g \subseteq \ker(f)$ for all $g \in G$. By Proposition 6.4.4, $\ker \triangleleft G$, and then by Corollary 6.4.5, $g^{-1} \ker(f) g = \ker(f)$ for all $g \in G$.

Let $g \in G$. If $x \in f^{-1}(f(g))$, then $f(x) = f(g)$ and so $f(xg^{-1}) = e_H$; that is, $xg^{-1} \in \ker(f)$, or equivalently, $x \in \ker(f)g$. Thus $f^{-1}(f(g)) \subseteq \ker(f)g$. Conversely, for $x \in \ker(f)g$, there is $y \in \ker(f)$ such that $x = yg$ and so $f(x) = f(yg) = f(y)f(g) = e_H f(g) = f(g)$, so $x \in f^{-1}(f(g))$. Thus $\ker(f)g \subseteq f^{-1}(f(g))$, and so $f^{-1}(f(g)) = \ker(f)g$ (which equals $g \ker(f)$ since $\ker(f) \triangleleft G$). If f is injective, then $|\ker(f)| = 1$, and conversely, if $|\ker(f)| = 1$, then for any $g \in G$, $|f^{-1}(f(g))| = |\ker(f)g| = |\ker(f)| = 1$ (since right multiplication by g is injective by Proposition 6.2.6). If $f(x) = f(y)$ for $x, y \in G$, then $x, y \in f^{-1}(f(y))$, so if $|\ker(f)| = 1$, then $f(x) = f(y)$ implies that $x = y$; that is, f is injective. $\qquad\square$

The diagram shown in Figure 6.1 is an attempt to illustrate the results presented in Proposition 6.4.4. The diagram suggests that the homomorphism f is not surjective, in that the image $f(G)$ is shown as a proper subset of H. The regions pointed to by the arrows from $f^{-1}(f(g))$ and from $f^{-1}(f(e_G))$ are to be understood as representing the set of all element of G that f maps to the same element as g is mapped to, respectively that e_G is mapped to (e_H in the latter case).

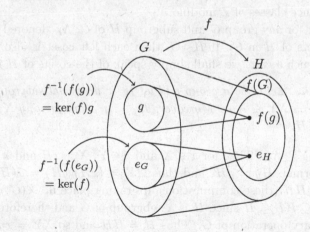

Fig. 6.1

We remark that the size of the kernel of a homomorphism can be thought of as a measure of the "shrinkage" which results upon the application of the homomorphism. For example, if the kernel of f has order 6, then the homomorphic image of f is one sixth the size of G (of course, this is really only meaningful when G is finite). This follows from Lagrange's theorem and Proposition 6.4.7, since $\ker(f)$ is a subgroup of G, and thus by Lagrange's theorem, $[G : \ker(f)] = |G|/|\ker(f)|$, while by Proposition 6.4.7, $|f(G)| = [G : \ker(f)]$. Thus $|f(G)| = |G|/|\ker(f)|$.

Also, note that every nontrivial homomorphism from a group whose only normal subgroups are the trivial subgroup and the whole group must be injective. This observation suggests that the class of groups whose only normal subgroups are either trivial or the entire group should be an interesting one to study.

6.4.8 Definition. *A nontrivial group whose normal subgroups are either the trivial subgroup or the whole group is called a simple group.*

For example, the only abelian simple groups are the abelian groups whose only nontrivial subgroup is the whole group, and these are the finite groups of prime order. See Section 5 of Chapter 7 for another interesting class of finite simple groups. Our next major objective is the so-called First Isomorphism Theorem for groups. For this, we will need the concept of factor group, or quotient group. We have already encountered this concept in one particular case; namely when we constructed the group $\mathbb{Z}_n$ from the set of congruence classes of $\mathbb{Z}$ modulo n.

Recall that for any group G and subgroup H of G, Λ_H denoted the set of all left cosets of H in G. If $H \lhd G$, then each left coset is also a right coset, and in such a case, we shall simply speak of the cosets of H in G.

6.4.9 Lemma. *Let G be a group and let H be a normal subgroup of G. For $X, Y \in \Lambda_H$, $XY \in \Lambda_H$. Moreover, $H \in \Lambda_H$, and for any $X \in \Lambda_H$, $HX = X = XH$.*

Proof. Let $X, Y \in \Lambda_H$. Then for $x \in X$ and $y \in Y$, $X = xH$ and $Y = yH$. Since H is normal, $Hy = yH$, and thus $XY = (xH)(yH) = x(Hy)H = x(yH)H = xyHH$ (these computations are taking place in $\mathscr{P}(G)$). Now, $H = \{e_G\}H \subseteq HH \subseteq H$ since H is a subgroup of G and therefore closed under the binary operation of G. Thus $H = HH$ and so $XY = xyHH = xyH \in \Lambda_H$. As well, $XH = xHH = xH = X$, and $HX = H(xH) = H(Hx) = HHx = Hx = xH = X$, as required. $\qquad\square$

Thus for a normal subgroup H of G, the subset Λ_H of $\mathscr{P}(G)$ is closed under the binary operation on $\mathscr{P}(G)$, and thus the binary operation on $\mathscr{P}(G)$ induces an associative binary operation on the set Λ_H. As established in the proof of Lemma 6.4.9, for any $x, y \in G$, the product of $xH, yH \in \Lambda_H$ is given by $(xH)(yH) = xyH$. Moreover, $H \in \Lambda_H$ is the identity element for this associative operation on Λ_H. Thus Λ_H is a monoid under the binary operation induced on it by that of G. Even more is true, as we establish shortly. But first, we remark that when H is normal in G, it is conventional to denote Λ_H by G/H.

6.4.10 Proposition. *Let G be a group and let $H \lhd G$. Then G/H is a group under the restriction of the binary operation on $\mathscr{P}(G)$ to G/H. For $x, y \in G$, $(xH)(yH) = xyH$, the identity element is H, and for $gH \in G/H$, $(gH)^{-1} = g^{-1}H$. More generally, the mapping $\pi_H : G \to G/H$ defined by $\pi_H(g) = gH$ is a surjective homomorphism with kernel H.*

Proof. It was established in Lemma 6.4.9 that the operation on $\mathscr{P}(G)$, when restricted to $G/H = \Lambda_H$, provides an associative binary operation on G/H, and that the coset of the identity, namely H, is the identity element for this binary operation on G/H. Let $X, Y \in G/H$, and let $x \in X$, $y \in Y$. Then $X = xH$ and $Y = yH$. Since $xy \in XY$, and $XY \in G/H$, it follows that $XY = xyH$; that is, $xHyH = xyH$. This also proves that $\pi_H(xy) = \pi_H(x)\pi_H(y)$, so π_H is a homomorphism. For $xH \in G/H$, $\pi_H(x) = xH$, so π_H is surjective, and $\pi_H(g) = H$ if, and only if, $g \in H$, so $\ker(\pi_H) = H$. It remains to prove that every element of G/H is invertible. Let $X \in G/H$, and let $x \in X$. Then $X = xH$, and $x^{-1}H \in G/H$ satisfies $xHx^{-1}H = xx^{-1}H = e_G H = H$ and $x^{-1}HxH = x^{-1}xH = H$, so $X = xH$ is invertible, and $(xH)^{-1} = x^{-1}H$. $\qquad\qquad\square$

In the particular case of $\mathbb{Z}$, with (normal) subgroup $\langle n \rangle$, we usually simplify the notation and write $\pi_n : \mathbb{Z} \to \mathbb{Z}_n$ rather than $\pi_{\langle n \rangle}$ (see Example 6.3.9 (ii) for earlier use of this notation).

We remark that if the operation in the group G is being denoted additively, then for $H \lhd G$, and $x, y \in G$, the binary operation on G/H is denoted by $(x + H) + (y + H) = (x + y) + H$, and $-(x + H) = (-x) + H$, where $x + H = \{ x + h \mid h \in H \}$. For example, in the group $\mathbb{Z}$ under addition, every subgroup is normal since addition is a commutative operation. Since $\mathbb{Z}$ is cyclic, every subgroup of $\mathbb{Z}$ is cyclic. Thus for any $n \in \mathbb{Z}$, $\langle n \rangle = n\mathbb{Z}$ is normal in $\mathbb{Z}$, and $\mathbb{Z}/\langle n \rangle = \{ a + \langle n \rangle \mid a \in \mathbb{Z} \}$. Note that $a + \langle n \rangle = b + \langle n \rangle$ if, and only if, $b - a \in \langle n \rangle$; that is, if, and only if, n divides $b - a$. Thus the cosets of $\langle n \rangle$ in $\mathbb{Z}$ are precisely the equivalence classes of the equivalence relation congruence modulo n. Addition of cosets is the same as addition of equivalence classes modulo n as defined in Definition 4.5.3.

Hence, the group $\mathbb{Z}/\langle n \rangle$ with its addition operation is none other than $\mathbb{Z}_n$ with addition as given in Definition 4.5.3.

The requirement that $H \triangleleft G$ is essential for the validity of Lemma 6.4.9. For example, the subgroup $H = \langle [2\,1\,3] \rangle = \{\,[1\,2\,3], [2\,1\,3]\,\}$ is not normal in S_3. To see this, it suffices to find one left coset of H that is not a right coset of H. Consider $[3\,1\,2]H = \{\,[3\,1\,2], [3\,1\,2][2\,1\,3]\,\} = \{\,[3\,1\,2], [1\,3\,2]\,\}$, while $H[3\,1\,2] = \{\,[3\,1\,2], [2\,1\,3][3\,1\,2]\,\} = \{\,[3\,1\,2], [3\,2\,1]\,\}$. Note that $[S_3 : H] = |S_3|/|H| = 6/2 = 3$. Since $H = \{\,[1\,2\,3], [2\,1\,3]\,\}$ is a left coset of itself, and we have computed the left coset $[3\,1\,2]H = \{\,[3\,1\,2], [1\,3\,2]\,\}$, it follows that the third left coset of H is

$$S_3 - (H \cup [3\,1\,2]H) = \{\,[1\,2\,3], [1\,3\,2], [2\,1\,3], [2\,3\,1], [3\,1\,2], [3\,2\,1]\,\}$$
$$- \{\,[1\,2\,3], [2\,1\,3], [3\,1\,2], [1\,3\,2]\,\}$$
$$= \{\,[2\,3\,1], [3\,2\,1]\,\},$$

and so $[2\,3\,1]H = [3\,2\,1]H = \{\,[2\,3\,1], [3\,2\,1]\,\}$. Let us compute the set product

$$([2\,3\,1]H)([3\,1\,2]H) = \{\,[2\,3\,1], [3\,2\,1]\,\}\{\,[3\,1\,2], [1\,3\,2]\,\}$$
$$= \{\,[2\,3\,1][3\,1\,2], [2\,3\,1][1\,3\,2], [3\,2\,1][3\,1\,2], [3\,2\,1][1\,3\,2]\,\}$$
$$= \{\,[1\,2\,3], [2\,1\,3], [1\,3\,2], [3\,1\,2]\,\}.$$

Since every $X \in S_3/H$ is a set of size $|H| = 2$, it follows that $([2\,3\,1]H)([3\,1\,2]H) \notin S_3/H$.

Our next result is of great importance in the theory of groups. Indeed, in many areas of abstract algebra, a result of essentially this nature is of fundamental importance in the study of the area (see for example the first isomorphism theorem for rings, Theorem 8.3.1).

6.4.11 Theorem (First Isomorphism Theorem for Groups). *Let G and H be groups, and let $f : G \to H$ be a surjective homomorphism. Then $G/\ker(f) \simeq H$. More precisely, the mapping $\overline{f} : G/\ker(f) \to H$ for which $\overline{f}(g\,\ker(f)) = f(g)$ is an isomorphism.*

Proof. By Proposition 6.4.7, $K = \ker(f)$ is normal in G, and so the quotient group $G/K = \{\,aK \mid a \in G\,\}$ exists, and for every $g \in G$, $f^{-1}(f(g)) = gK$. Thus for any $g \in G$, $f(gK) = f(g)$, and we define the mapping $\overline{f} : G/K \to H$ by $\overline{f}(X) = f(X)$ for $X \in G/K$. Note that $X \in G/K$ means that $X = xK$ for any $x \in X$, and so $f(X) = f(f^{-1}(f(x))) = f(x)$ for each $x \in X$. Thus for $X \in G/K$, $|f(X)| = 1$, and we are defining $\overline{f}(X)$ to be the value $h \in H$ such that $f(X) = y$; namely $y = f(x)$ for each and every $x \in X$. Thus $\overline{f}$ is well-defined. Let $X, Y \in G/K$, and let $x \in X$ and $y \in Y$. Then $X = xK$ and $Y = yK$, so $\overline{f}(XY) = \overline{f}(xKyK) = \overline{f}(xyK) = f(xy) = f(x)f(y) = \overline{f}(X)\overline{f}(Y)$. This proves that $\overline{f}$ is a homomorphism from G/K to H. It remains to prove that $\overline{f}$ is bijective. Suppose that $X, Y \in G/K$ are such that $\overline{f}(X) = \overline{f}(Y)$. Then for any $x \in X$ and any $y \in Y$, $f(x) = \overline{f}(X) = \overline{f}(Y) = f(y)$ and so $x^{-1}y \in \ker(f) = K$. Thus $y \in xK = X$, and so $Y = yK = xK = X$; that is, $\overline{f}$ is injective. Finally, suppose that $h \in H$. Since f is surjective, there exists $g \in G$ with $f(g) = h$. Then for $X = f^{-1}(h) = f^{-1}(f(g)) = gK$, we have $\overline{f}(X) = f(g) = h$ and thus $\overline{f}$ is surjective. □

Note that if $f : G \to H$ is a homomorphism with kernel K, then we may view f as a surjective homomorphism from G to $f(G) \leq H$, and thus by the First Isomorphism Theorem for groups, $G/K \simeq f(G)$, and $\overline{f}(G/\ker(f)) = f(G)$. In this case, $\overline{f} : G/\ker(f) \to H$ is an injective homomorphism.

We now apply the First Isomorphism Theorem to prove what are commonly known as the Second and Third Isomorphism theorems.

6.4.12 Lemma. *Let G be a group. If K is a normal subgroup of G, and H is any subgroup of G, then $HK = KH$ is a subgroup of G. Moreover, if H is also normal in G, then $HK \triangleleft G$ and additionally, if $H \cap K = \{e_G\}$, then for every $h \in H$ and $k \in K$, $hk = kh$.*

Proof. Since $e_G \in K \cap H$, it follows that $e_G \in HK$ and thus $HK \neq \varnothing$.

Note that since $K \triangleleft G$, $KH = \cup_{h \in H} Kh = \cup_{h \in H} hK = HK$, and thus $(HK)(HK) = H(KH)K = H(HK)K = (HH)(KK) = HK$, where the final equality is due to the fact that both H and K are closed under the binary operation of G. Thus HK is closed under the binary operation of G. Finally, for $x \in HK$, $x = hk$ for some $h \in H$ and $k \in K$, and so $x^{-1} = (hk)^{-1} = k^{-1}h^{-1} \in KH = HK$. Since HK is nonempty, closed under the binary operation of G, and closed with respect to inversion, HK is a subgroup of G.

Suppose now that $H \triangleleft G$, and let $g \in G$. Then $g^{-1}HKg = g^{-1}Hgg^{-1}Kg \subseteq HK$, and so by Proposition 6.4.4, $HK \triangleleft G$. Let $h \in H$ and $k \in K$. Then $h(kh^{-1}k^{-1}) \in H$, and $(hkh^{-1})k^{-1} \in K$, so $hkh^{-1}k^{-1} \in H \cap K$. Thus if $H \cap K = \{e_G\}$, then $hk = kh$ for all $h \in H$, $k \in K$. □

6.4.13 Definition. *Let G be a group, and let K be a subgroup of G. The normalizer of K in G, denoted by $N_G(K)$, is the set $N_G(K) = \{h \in G \mid$*

$hKh^{-1} \subseteq K$ }.

6.4.14 Proposition. *Let G be a group and let K be a subgroup of G. Then $K \subseteq N_G(K)$, $N_G(K)$ is a subgroup of G, and $K \triangleleft N_G(K)$.*

Proof. Since K is closed under inversion and multiplication, $K \subseteq N_G(K)$. If $h \in N_G(K)$, then $hKh^{-1} = K$ and so $h^{-1}Kh = h^{-1}(hKh^{-1})h = K$. Thus $h \in N_G(K)$ implies that $h^{-1} \in N_G(H)$. Now suppose that $g, h \in N_G(K)$. Then $(gh)K(gh)^{-1} = g(hKh^{-1})g^{-1} = gKg^{-1} = K$, and thus $gh \in N_G(K)$. We have now proven that $N_G(K)$ is a subgroup of G, and that K is a subgroup of $N_G(K)$. It is immediate now that $K \triangleleft N_G(K)$. $\square$

6.4.15 Theorem (The Second Isomorphism Theorem for Groups). *Let G be a group and let K be a subgroup of G. Then for any subgroup H of $N_G(K)$, HK is a subgroup of $N_G(K)$ containing K as a normal subgroup, $H \cap K \triangleleft H$, and*

$$HK/K \simeq H/(H \cap K).$$

Proof. Let K be a subgroup of G, and let H be a subgroup of $N_G(K)$. By Lemma 6.4.12, HK is a subgroup of $N_G(K)$, and $K \subseteq HK$, so K is a subgroup of HK. Then by definition of $N_G(K)$, K is normal in HK. Next, by Proposition 5.4.13, $H \cap K$ is a subgroup of $N_G(K)$. For any $h \in H$, $h^{-1}(H \cap K)h \subseteq h^{-1}Hh \cap h^{-1}Kh = H \cap K$, so $H \cap K$ is normal in H. Thus the quotient groups HK/K and $H/(H \cap K)$ exist. Consider the mapping $\theta: H \to HK/K$ defined by $\theta(h) = hK$ for each $h \in H$. We prove that θ is a surjective homomorphism with kernel $H \cap K$. Let $x, y \in H$. Then $\theta(xy) = xyK = (xK)(yK) = \theta(x)\theta(y)$, so θ is a homomorphism. Let $X \in HK/K$. Then there exist $h \in H$ and $k \in K$ such that $X = hkK = h(kK) = hK$, so $X = \theta(h)$. Thus θ is surjective. Finally, let $h \in \ker(\theta)$, so $\theta(h) = K \in HK/K$. Thus $hK = K$, which means that $h \in K$, so $\ker(\theta) \subseteq K$. Since $\theta: H \to HK/K$, $\ker(\theta) \subseteq H$ and so $\ker(\theta) \subseteq H \cap K$. Let $k \in H \cap K$. Then $\theta(k) = kK = K$, and so $H \cap K \subseteq \ker(\theta)$. Thus $\ker(\theta) = H \cap K$. By Theorem 6.4.11, $H/(H \cap K) \simeq HK/K$. $\square$

Note that if $K \triangleleft G$, then $N_G(K) = G$, and the preceding result states that for any subgroup H of G, HK is a subgroup of G, $K \triangleleft HK$, $H \cap K \triangleleft H$, and $HK/K \simeq H/(H \cap K)$.

6.4.16 Proposition. *Let G be a group, and let H and K be normal subgroups of G with $K \subseteq H$. Then the mapping $q_{H,K}: G/K \to G/H$ defined* by $q_{H,K}(gK) = gH$ *is a surjective homomorphism with kernel* $H/K = \{ hK \mid h \in H \}.$

Proof. The first task is to prove that $q_{H,K}$ is indeed a mapping. For $g, h \in G$, if $gK = hK$, then $g^{-1}h \in K \subseteq H$, and so $gH = hH$. Thus $q_{H,K}$ is well-defined. Now to prove that it is a surjective homomorphism. Let $xK, yK \in G/K$. Then $q_{H,K}(xKyK) = q_{H,K}(xyK) = xyH = xHyH = q_{H,K}(x)q_{H,K}(y)$, so $q_{H,K}$ is a homomorphism. For any $Y \in G/H$, there is $g \in G$ such that $Y = gH$, and since $q_{H,K}(gK) = gH = Y$, $q_{H,K}$ is surjective. Finally, $X = gK \in \ker(q_{H,K})$ if, and only if, $gH = q_{H,K}(X) = H$; that is, if, and only if, $g \in H$. Thus $\ker(q_{H,K}) = \{ gK \mid g \in H \}$, as required. $\qquad\square$

6.4.17 Proposition (The Third Isomorphism Theorem). *Let G be a group, and let H and K be normal subgroups of G for which $K \subseteq H$. Then H/K is a normal subgroup of G/K and $(G/K)/(H/K) \simeq G/H$.*

Proof. By Proposition 6.4.16, H/K is the kernel of the surjective homomorphism $q_{H,K} : G/K \to G/H$, and thus by the First Isomorphism Theorem, $(G/K)/(H/K) \simeq G/H$. $\qquad\square$

The following example is intended to provide some motivation for the final result of this section.

6.4.18 Example. *Consider the quaternion group $Q_8 = \{ \pm 1, \pm \mathbf{i}, \pm \mathbf{j}, \mathbf{k} \}$ (see Exercise 8 in Chapter 5, and Example 6.4.3 (iv)) and the Klein 4-group $K_4 = \{ I, r, h, v \}$ (see Example 5.4.4). Q_8 has three subgroups of order 4; namely $\langle \mathbf{i} \rangle$, $\langle \mathbf{j} \rangle$, and $\langle \mathbf{k} \rangle$, and one subgroup of order 2, $\langle -1 \rangle$, as well as the subgroups Q_8 itself and the trivial subgroup. K_4 has three subgroups of order 2; namely $\langle r \rangle$, $\langle h \rangle$, and $\langle v \rangle$, as well as the subgroups K_4 itself and the trivial subgroup. We define a surjective map $f : Q_8 \to K_4$ by $f(1) = f(-1) = I$, $f(\mathbf{i}) = f(-\mathbf{i}) = r$, $f(\mathbf{j}) = f(-\mathbf{j}) = h$ and $f(\mathbf{k}) = f(-\mathbf{k}) = v$. It may be readily verifed that f is a homomorphism. For example, $f(\mathbf{ij}) = f(\mathbf{k}) = v$, and $f(\mathbf{i})f(\mathbf{j}) = rh = v$ (see Table 5.1), so $f(\mathbf{ij}) = f(\mathbf{i})f(\mathbf{j})$ We leave the rest of the demonstration that f is a homomorphism to the reader. Note that $\ker(f) = \{ \pm 1 \}$, and that f induces a bijective correspondence between the set of subgroups of Q_8 that contain $\ker(f)$ and the set of subgroups of K_4 defined by $H \mapsto f^{-1}(H)$ for each subgroup H of K_4. Specifically, we have $f^{-1}\{ I \} = \{ \pm 1 \}$, $f^{-1}(\{ I, r \}) = \{ \pm 1, \pm \mathbf{i} \} = \langle \mathbf{i} \rangle$, $f^{-1}(\{ I, h \}) = \{ \pm 1, \pm \mathbf{j} \} = \langle \mathbf{j} \rangle$, $f^{-1}(\{ I, v \}) = \{ \pm 1, \pm \mathbf{k} \} = \langle \mathbf{k} \rangle$. Hasse diagrams for the set of all subgroups of Q_8, partially ordered by inclusion, and the set of all* *subgroups of K_4, also partially ordered by inclusion, are shown in Figure 6.2.*

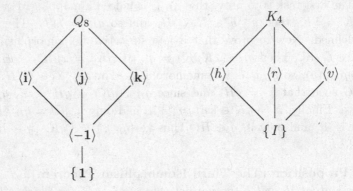

Fig. 6.2

6.4.19 Notation. *For any group G, the set of all subgroups of G shall be denoted by $\mathscr{S}_G$, and for any subgroup H of G, the set of all subgroups of G that contain H shall be denoted by $\mathscr{S}_G(H)$, so $\mathscr{S}_G(H) = \{ V \mid V \in \mathscr{S}_G,\ H \leq V \}$.*

6.4.20 Lemma. *Let G and H be groups and let $f : G \to H$ be a homomorphism.*

(i) For every $L \in \mathscr{S}_G(\ker(f))$, $f^{-1}(f(L)) = L$.

(ii) If f is surjective and $L \triangleleft G$, then $f(L) \triangleleft H$.

Proof. For (i), let $L \in \mathscr{S}_G(\ker(f))$. By definition, $L \subseteq f^{-1}(f(L))$. Let $g \in f^{-1}(f(L))$, so $f(g) \in f(L)$. Thus there is $z \in L$ such that $f(g) = f(z)$, and then we have $f(gz^{-1}) = e_H$; that is, $gz^{-1} \in \ker(f) \subseteq L$. But then $gz^{-1}, z \in L$ and so $g = (gz^{-1})z \in L$, which proves that $f^{-1}(f(L)) \subseteq L$ and thus $f^{-1}(f(L)) = L$, as required.

For (ii), suppose that f is surjective and $L \triangleleft G$. Let $h \in H$. Since f is surjective, there is $g \in G$ with $f(g) = h$ and thus $hf(L)h^{-1} = f(g)f(L)f(g)^{-1} = f(g)f(L)f(g^{-1}) = f(gLg^{-1}) = f(L)$. Thus $f(L) \triangleleft H$.   □

6.4.21 Proposition. *Let G and H be groups, and let $f : G \to H$ be a surjective homomorphism with kernel K. Then the assignment $L \mapsto f^{-1}(L)$*

for each subgroup L of H determines a bijective mapping $\varphi: \mathscr{S}_H \to \mathscr{S}_G(K)$. Moreover, for any $L_1, L_2 \in \mathscr{S}_H$, $\varphi(L_1) \subseteq \varphi(L_2)$ if, and only if, $L_1 \subseteq L_2$, and $L \lhd H$ if, and only if, $\varphi(L) \lhd G$.

Proof. We know by Proposition 6.3.10 (iv) that for each subgroup L of H, $f^{-1}(L)$ is a subgroup of G. Thus we may define a mapping $\varphi: \mathscr{S}_H \to \mathscr{S}_G$ by $\varphi(L) = f^{-1}(L)$ for each $L \in \mathscr{S}_H$. Note that for any $L \in \mathscr{S}_H$, $e_H \in L$ and thus $K = f^{-1}(e_H) \subseteq f^{-1}(L) = \varphi(L)$. We may therefore regard φ as a mapping from $\mathscr{S}_H$ to $\mathscr{S}_G(K)$. Furthermore, since f is surjective, it follows from Proposition 3.3.8 that for any $L \in \mathscr{S}_H$, $f(f^{-1}(L)) = L$. Suppose now that $L_1, L_2 \in \mathscr{S}_H$ are such that $\varphi(L_1) = \varphi(L_2)$. Then $f^{-1}(L_1) = f^{-1}(L_2)$ and so $L_1 = f(f^{-1}(L_1)) = f(f^{-1}(L_2)) = L_2$. Thus φ is injective. It remains to prove that φ is surjective.

Let $V \in \mathscr{S}_G(K)$. Then by Lemma 6.4.20, $f^{-1}(f(V)) = V$. Now, by Proposition 6.3.10 (iii), $f(V) \in \mathscr{S}_H$, and we have $\varphi(f(V)) = f^{-1}(f(V)) = V$. Thus φ is surjective.

Next, suppose that $L_1, L_2 \in \mathscr{S}_H$. It is immediate that if $L_1 \subseteq L_2$, then $\varphi(L_1) = f^{-1}(L_1) \subseteq f^{-1}(L_2) = \varphi(L_2)$. Suppose that $\varphi(L_1) \subseteq \varphi(L_2)$; that is, $f^{-1}(L_1) \subseteq f^{-1}(L_2)$. Then, again by Proposition 3.3.8, since f is surjective, we have $L_1 = f(f^{-1}(L_1)) \subseteq f(f^{-1}(L_2)) = L_2$.

It remains to prove that $L \in \mathscr{S}_H$ is normal in H if, and only if, $\varphi(L) \lhd G$. Let $L \lhd H$, and let $g \in G$. We wish to prove that $gf^{-1}(L)g^{-1} \subseteq f^{-1}(L)$. Let $x \in f^{-1}(L)$, so $f(x) \in L$. Then $f(g^{-1}xg) = (f(g))^{-1}f(x)f(g) \in L$ since $f(x) \in L$ and $L \lhd H$. Thus $g^{-1}xg \in f^{-1}(L)$, as required. Finally, suppose that $L \in \mathscr{S}_H$ and $\varphi(L) = f^{-1}(L) \lhd G$. Since f is surjective, $f(f^{-1}(L)) = L$, and so it follows from Lemma 6.4.20 that $L \lhd H$. $\square$

Direct Product of Groups

Given a set of groups, there are a number of ways of "stitching" them together to form a new group. In this section, we explore the simplest way of doing this by forming what is called the direct product of the groups. Recall that for any sets X and Y, $\mathscr{F}(X, Y)$ denotes the set of all maps from X to Y.

6.5.1 Definition. *Let I be a nonempty set, and for each $i \in I$, let G_i be* *a group (with binary operation denoted by juxtaposition). Let*

$$\prod_{i \in I} G_i = \{\, f \in \mathscr{F}(I, \cup_{i \in I} G_i) \mid \text{ for each } i \in I,\ f(i) \in G_i \,\}.$$

Define a binary operation on $\prod_{i \in I} G_i$ (denoted by juxtaposition as well) as follows: for $f, g \in \prod_{i \in I} G_i$, let $fg: I \to \cup_{i \in I} G_i$ be the mapping defined by $(fg)(i) = f(i)g(i)$ for each $i \in I$. Note that for each $i \in I$, $f(i), g(i) \in G_i$, and thus $f(i)g(i) \in G_i$ as well, so $fg \in \prod_{i \in I} G_i$.

6.5.2 Proposition. *Let I be a nonempty set, and for each $i \in I$, let G_i be a group (with binary operation denoted by juxtaposition and identity denoted by e_{G_i}). Then $\prod_{i \in I} G_i$, with the binary operation defined in Definition 6.5.1, is a group, called the direct product of the groups G_i, $i \in I$.*

Proof. First, we prove that the operation is associative. Let $f, g, h \in \prod_{i \in I} G_i$. Then for any $i \in I$, $(f(gh))(i) = f(i)((gh)(i)) = f(i)(g(i)h(i))$. Since $f(i), g(i), h(i) \in G_i$ and G_i is a group, the binary operation of G_i is associative and so we have $f(i)(g(i)h(i)) = (f(i)g(i))h(i) = (fg)(i)h(i) = ((fg)h)(i)$. Thus for each $i \in I$, $(f(gh))(i) = ((fg)h)(i)$ and so $f(gh) = (fg)h$.

Next, we prove that $e \in \prod_{i \in I} G_i$ defined by $e(i) = e_{G_i}$ for each $i \in I$ is the identity element for the binary operation on $\prod_{i \in I} G_i$. Let $f \in \prod_{i \in I} G_i$. To prove that $ef = fe = f$, we must prove that for each $i \in I$, $(ef)(i) = (fe)(i) = f(i)$. Let $i \in I$. Then $(ef)(i) = e(i)f(i) = e_{G_i}f(i)$. Since $f(i) \in G_i$, we have $e_{G_i}f(i) = f(i)$ and thus $(ef)(i) = f(i)$. Similarly, $(fe)(i) = f(i)e(i) = f(i)e_{G_i} = f(i)$. Thus for every $i \in I$, $(ef)(i) = f(i) = (fe)(i)$ and so $ef = f = fe$.

Finally, we must prove that each element of $\prod_{i \in I} G_i$ is invertible. Let $f \in \prod_{i \in I} G_i$. Then for each $i \in I$, $f(i) \in G_i$, and as each element of G_i is invertible, $(f(i))^{-1}$ exists in G_i. Define $g \in \prod_{i \in I} G_i$ by $g(i) = (f(i))^{-1}$ for each $i \in I$. We prove that $fg = e = gf$. Let $i \in I$. Then $(fg)(i) = f(i)g(i) = f(i)(f(i))^{-1} = e_{G_i} = e(i)$, and $(gf)(i) = g(i)f(i) = (f(i))^{-1}f(i) = e_{G_i} = e(i)$, so for each $i \in I$, we have $fg(i) = e(i) = gf(i)$ and thus $fg = e = gh$. This proves that g is an inverse for f, and so f is invertible.   □

In a great many situations, we will be considering the direct product of a finite set of groups. In such a case, it is often more convenient to use a simpler notation.

6.5.3 Notation. *Let n be a positive integer, and let $G_1, G_2, \ldots, G_n$ be groups. Then $\prod_{i \in J_n} G_i$ may be denoted by $G_1 \times G_2 \times \cdots \times G_n$, and an element $f \in \prod_{i \in J_n} G_i$ may be denoted by the n-tuple of image values $(f(1), f(2), \ldots, f(n))$. As a particularly simple case, if G and H are groups, then $G \times H = \{ (g, h) \mid g \in G, \ h \in H \}$.*

With the above notation, for groups $G_1, G_2, \ldots, G_n$, the binary operation in $G_1 \times G_2 \times \cdots \times G_n$ is computed by $(x_1, x_2, \ldots, x_n)(y_1, y_2, \ldots, y_n) = (x_1 y_1, x_2 y_2, \ldots, x_n y_n)$.

6.5.4 Proposition. *Let I be a nonempty set, and for each $i \in I$, let G_i be a group. Let $j \in I$. Then the mapping $\pi_j : \prod_{i \in I} G_i \to G_j$ defined by $\pi_j(f) = f(j)$ for all $f \in \prod_{i \in I} G_i$ is a surjective group homomorphism with $\ker(\pi_j) = \{ f \in \prod_{i \in I} G_i \mid f(j) = e_{G_j} \}$. Moreover, if H is a group such that for each $i \in I$, there is a homomorphism $\varphi_i : H \to G_i$, then there exists a unique homomorphism $\varphi : H \to \prod_{i \in I} G_i$ such that for each $i \in I$, $\pi_i \circ \varphi = \varphi_i$.*

Proof. Let $f, g \in \prod_{i \in I} G_i$. Then $\pi_j(fg) = (fg)(j) = f(j)g(j) = \pi_j(f)\pi_j(g)$, and so π_j is a homomorphism. Let $g \in G_j$. Define $f \in \prod_{i \in I} G_i$ by $f(i) = e_{G_i}$ for each $i \in I$ with $i \neq j$, and $f(j) = g$. Then $\pi_j(g) = g(j) = g$, so π_j is surjective. Finally, let $f \in \ker(\pi_j)$. Then $f(j) = \pi_j(f) = e_{G_j}$. Conversely, suppose that $f \in \prod_{i \in I} G_i$ and $f(j) = e_{G_j}$. Then $\pi_j(f) = f(j) = e_{G_j}$ and so $f \in \ker(\pi_j)$.

Now suppose that H is a group and for each $i \in I$, there is a homomorphism $\varphi_i : H \to G_i$. Let $\varphi : H \to \prod_{i \in I} G_i$ denote the map that sends $h \in H$ to $\varphi(h) \in \prod_{i \in I} G_i$ given by $\varphi(h)(i) = \varphi_i(h)$ for each $i \in I$. Since $(\pi_i \circ \varphi)(h) = \pi_i(\varphi(h)) = \varphi(h)(i) = \varphi_i(h)$ for each $h \in H$, it follows that $\pi_i \circ \varphi = \varphi_i$ for each $i \in I$. Next, we prove that φ is a homomorphism. Let $h_1, h_2 \in H$. Then for each $i \in I$, $\varphi(h_1 h_2)(i) = \varphi_i(h_1 h_2) = \varphi_i(h_1)\varphi_i(h_2) = \varphi(h_1)(i)\varphi(h_2)(i) = (\varphi(h_1)\varphi(h_2))(i)$ and thus $\varphi(h_1 h_2) = \varphi(h_1)\varphi(h_2)$, as required. The uniqueness assertion follows from the requirement that $\pi_i \circ \varphi = \varphi_i$ for each $i \in I$. For if $\alpha : H \to \prod_{i \in I} G_i$ is a homomorphism such that $\pi_i \circ \alpha = \varphi_i$ for each $i \in I$, then for each $h \in H$, $\alpha(h)(i) = \pi_i(\alpha(h)) = \pi_i \circ \alpha(h) = \varphi_i(h) = \varphi(h)(i)$ for each $i \in I$ and thus $\alpha(h) = \varphi(h)$ for all $h \in H$, which establishes that $\alpha = \varphi$. $\square$

6.5.5 Corollary. *Let I be a nonempty set, and for each $i \in I$, let G_i be a group. Then $\prod_{i \in I} G_i$ is abelian if, and only if, G_i is abelian for each $i \in I$.*

Proof. Suppose that $G = \prod_{i \in I} G_i$, and let $i \in I$. Then by Proposition 6.5.4, there is a surjective homomorphism from G onto G_i, and thus by Proposition 6.3.10 (vii), G_i is abelian. Conversely, suppose that for each $i \in I$, G_i is abelian, and let $f, g \in G$. Then for each $i \in I$, $(fg)(i) = f(i)g(i)$, and $f(i), g(i) \in G_i$, so $f(i)g(i) = g(i)f(i) = (gf)(i)$ and thus $(fg)(i) = (gf)(i)$. This proves that $fg = gf$, and so G is abelian. $\qquad\square$

6.5.6 Proposition. *Let I be a nonempty set, and for each $i \in I$, let G_i be a group. Let $j \in I$. Then the mapping $\iota_j \colon G_j \to \prod_{i \in I} G_i$, where for each $g \in G_j$, $\iota_j(g) \colon I \to \cup_{i \in I} G_i$ is the mapping given by*

$$(\iota_j(g))(i) = \begin{cases} e_{G_i} & \text{if } i \neq j \\ g & \text{if } i = j \end{cases},$$

is an injective homomorphism with image

$$\{ f \in \prod_{i \in I} G_i \mid f(i) = e_{G_i} \text{ for all } i \in I \text{ with } i \neq j \},$$

which is normal in $\prod_{i \in I} G_i$.

Proof. Let $x, y \in G_j$. Then $\iota_j(xy)(j) = xy = \iota_j(x)(j)\iota_j(y)(j) = (\iota_j(x)\iota_j(y))(j)$, while for $i \in I$ with $i \neq j$, $\iota_j(xy)(i) = e_{G_i} = e_{G_i}e_{G_i} = \iota_j(x)(i)\iota_j(y)(i) = (\iota_j(x)\iota_j(y))(i)$. This proves that for all $i \in I$, $\iota_j(xy)(i) = (\iota_j(x)\iota_j(y))(i)$ and so $\iota_j(xy) = \iota_j(x)\iota_j(y)$. Thus ι_j is a homomorphism. Suppose now that $x, y \in G_j$ are such that $\iota_j(x) = \iota_j(y)$. Then in particular, $x = (\iota_j(x))(j) = (\iota_j(y))(j) = y$, and so ι_j is injective. The definition of ι_j makes it clear that

$$\iota_j(G_j) \subseteq \{ f \in \prod_{i \in I} G_i \mid f(i) = e_{G_i} \text{ for all } i \in I \text{ with } i \neq j \},$$

so let $f \in \prod_{i \in I} G_i$ be such that $f(i) = e_{G_i}$ for all $i \in I$ with $i \neq j$. Let $g = f(j) \in G_j$. Then $\iota_j(g)(i) = e_{G_i} = f(i)$ for all $i \in I$ with $i \neq j$, while $\iota_j(g)(j) = g = f(j)$ and thus for all $i \in I$, $\iota_j(g)(i) = f(i)$. This means that $f = \iota_j(g)$ and thus $f \in \iota_j(G_j)$. Let $g \in \prod_{i \in I} G_i$ and $f \in \iota_j G_j$. We prove that $g^{-1}fg \in \iota_j(G_j)$. Let $i \in I$ with $i \neq j$. Then $(g^{-1}fg)(i) = g^{-1}(i)f(i)g(i) = g^{-1}(i)e_{G_i}g(i) = g^{-1}(i)g(i) = (g^{-1}g)(i) = e(i) = e_{G_i}$, and so $g^{-1}fg \in \iota_j(G_j)$. This proves that $g^{-1}\iota_j(G_j)g \subseteq \iota_j(G_j)$ and so $\iota_j(G_j)$ is normal in $\prod_{i \in I} G_i$. $\qquad\square$

Note that it is possible that for a nonempty set I, when we assign to each $i \in I$ a group G_i, $i \neq j$ need not imply that $G_i \neq G_j$. It may even be that there exists a group G such that for all $i \in I$, $G_i = G$, in which case we might write $\prod_{i \in I} G$. In the special case that $I = J_n$ and there exists a group G such that for all $i \in J_n$, $G_i = G$, we may write G^n in place of $\prod_{i \in J_n} G_i$.

6.5.7 Proposition. *Let n be a positive integer. For each $i \in J_n$, let G_i be a group with identity e_i. Then the following hold:*

(i) *For each $j \in I$, let $H_j = \iota_j(G_j)$. Then for each $j \in I$, $H_j \triangleleft \prod_{i \in J_n} G_i$, and $G_1 \times G_2 \times \cdots \times G_n = H_1 H_2 \cdots H_n$. Furthermore, for each $g \in G_1 \times G_2 \times \cdots \times G_n$, there exist unique $h_i \in H_i$, $i \in J_n$, for which $g = h_1 h_2 \cdots h_n$.*

(ii) *For each $i \in J_n$, $H_i \cap H_1 H_2 \cdots H_{i-1} H_{i+1} \cdots H_n = \{e\}$ (where if $i = 1$, we mean that $H_1 \cap H_2 \cdots H_n = \{e\}$, and if $i = n$, we mean that $H_n \cap H_1 H_2 \cdots H_{n-1} = \{e\}$).*

Proof. (i) Let $G = G_1 \times G_2 \times \cdots \times G_n$. By Proposition 6.5.7, for each $j \in I$, $H_j = \iota_j(G_j)$ is normal in G. Let $g \in G$. Then for some $g_i \in G_i$, $i \in J_n$, $g = (g_1, g_2, \ldots, g_n)$. For each $j \in J_n$, let $h_j = \iota_j \circ \pi_j(g) = \iota_j(\pi_j(g)) = \iota_j(g_j) \in H_j$. We have

$$h_1 h_2 \cdots h_n = (g_1, e_2, \ldots, e_n)(e_1, g_2, \ldots, e_n) \cdots (e_1, e_2, e_{n-1}, g_n)$$
$$= (g_1, g_2, \ldots, g_n) = g.$$

It follows that $G = H_1 H_2 \cdots H_n$. The proof of the uniqueness assertion will be deferred until the proof of (ii) has been completed.

(ii) Let $i \in J_n$, and let $g \in H_i \cap H_1 H_2 \cdots H_{i-1} H_{i+1} \cdots H_n$. There exist $g_j \in H_j$ for $j \in J_n$ with $g = (g_1, g_2, \ldots, g_n)$. We are to prove that for every $j \in J_n$, $g_j = e_j$. Since $g \in H_i = \iota_i(G_i)$, $g_j = e_j$ for each $j \in J_n$ with $j \neq i$. It remains to prove that $g_i = e_i$. We have $\pi_i(H_1 H_2 \cdots H_{i-1} H_{i+1} \cdots H_n) \subseteq \pi_i(H_1)\pi_i(H_2) \cdots \pi_i(H_{i-1})\pi_i(H_{i+1}) \cdots \pi_i(H_n)$. For each $j \in J_n$ with $j \neq i$, $\pi_i(H_j) = \pi_i(\iota_j(G_j)) = \{e_i\}$, and thus $g_i = \pi_i(g) \in \pi_i(H_1 H_2 \cdots H_{i-1} H_{i+1} \cdots H_n) \subseteq \{e_i\}$. It follows that $g_i = e_i$, as required.

We return now to the proof of the uniqueness assertion in (i). Observe that by Lemma 6.4.12, for $i, j \in J_n$ with $i \neq j$, $gh = hg$ for all $g \in H_i$ and $h \in H_j$ (since $H_i \cap H_j = \{e_G\}$). Suppose now that for each $i \in J_n$ we have $g_i, h_i \in H_i$ such that $g_1 g_2 \cdots g_n = h_1 h_2 \cdots h_n$. By the preceding commutativity observation, $(h_1 h_2 \cdots h_n)^{-1} = h_1^{-1} h_2^{-1} \cdots h_n^{-1}$ and, also by the commutativity observation, $e_g = g_1 g_2 \cdots g_n h_1^{-1} h_2^{-1} \cdots h_n^{-1} = g_1 h_1^{-1} g_2 h_2^{-1} \cdots g_n h_n^{-1}$. Since $g_i h_i^{-1} \in H_i$ for each $i \in J_n$, it therefore suffices to prove that if $r_i \in H_i$ for $i \in J_n$ satisfy $r_1 r_2 \cdots r_n = e_G$, then for each $i \in J_n$, $r_i = e_G$. Suppose that for each $i \in J_n$ we have $r_i \in H_i$

that satisfy $r_1 r_2 \cdots r_n = e_G$. By the commutativity result, for each $i \in J_n$, $r_i^{-1} = r_1 r_2 \cdots r_{i-1} r_{i+1} \cdots r_n \in H_i \cap H_1 H_2 \cdots H_{i-1} H_{i+1} \cdots H_n = \{e_G\}$, and so $r_i = e_G$ for each $i \in J_n$. $\qquad\square$

Proposition 6.5.7 tells us that if a group G is the direct product of finitely many groups $G_1, G_2, \ldots, G_n$, then G contains normal subgroups $H_1, H_2, \ldots, H_n$ such that $G = H_1 H_2 \cdots H_n$ and for each $i = 1, 2, \ldots, n$, $H_i \cap H_1 H_2 \cdots H_{i-1} H_{i+1} \cdots H_n = \{e\}$, and moreover, for each $g \in G$, there exist unique elements $g_i \in H_i$, $i = 1, 2, \ldots, n$ such that $g = g_1 g_2 \cdots g_n$. Our next result provides the converse to this proposition.

6.5.8 Proposition. *Let G be a group with normal subgroups* $H_1, H_2, \ldots, H_n$ *such that:*

  (i) $G = H_1 H_2 \cdots H_n$;
  (ii) For each $i = 1, 2, \ldots, n$, $H_i \cap H_1 H_2 \cdots H_{i-1} H_{i+1} \cdots H_n = \{e\}$ *(where if $i = 1$, we mean that $H_1 \cap H_2 \cdots H_n = \{e\}$, and if $i = n$, we mean that $H_n \cap H_1 H_2 \cdots H_{n-1} = \{e\}$).*

Then $G \simeq H_1 \times H_2 \times \cdots \times H_n$, and for each $g \in G$, there exist unique $h_i \in H_i$, $i \in J_n$ with $g = h_1 h_2 \cdots h_n$.

Proof. First, let i, j satisfy $1 \leq i < j \leq n$, and let $x \in H_i$ and $y \in H_j$. We prove that $xy = yx$. By (ii), $H_i \cap H_j \subseteq H_i \cap H_1 H_2 \cdots H_{i-1} H_{i+1} \cdots H_j \cdots H_n = \{e\}$. As well, $y^{-1} xy \in H_i$ since $H_i \triangleleft G$, and so $y^{-1} xyx^{-1} \in H_i$, while $yx^{-1} y^{-1} \in H_j$ since $H_j \triangleleft G$, and thus $y^{-1} xyx^{-1} \in H_j$. It follows that $y^{-1} xyx^{-1} = e$ and so $xy = yx$, as claimed. Next, let $g \in G$. We prove that there exist unique $h_i \in H_i$, $i \in J_n$ with $g = h_1 h_2 \cdots h_n$. By (i), there exist $h_i \in H_i$, $i \in J_n$ with $g = h_1 h_2 \cdots h_n$. Suppose now that there exist $y_i \in H_i$, $i \in J_n$ with $g = y_1 y_2 \cdots y_n$. Then $h_1 h_2 \cdots h_{n-1} = y_1 y_2 \cdots y_n h_n^{-1}$ and so $(y_1 y_2 \cdots y_{n-1})^{-1} h_1 h_2 \cdots h_{n-1} = y_n h_n^{-1} \in H_n$. Since elements of H_i commute with those of H_j when $i \neq j$, we have

$$(y_1 y_2 \cdots y_{n-1})^{-1} h_1 h_2 \cdots h_{n-1} = y_{n-1}^{-1} \cdots y_1^{-1} h_1 h_2 \cdots h_{n-1}$$
$$= y_1^{-1} h_1 y_2^{-1} h_2 \cdots y_{n-1}^{-1} h_{n-1} \in H_1 H_2 \cdots H_{n-1}$$

and thus $y_1^{-1} h_1 y_2^{-1} h_n \cdots y_{n-1}^{-1} h_{n-1} = y_n^{-1} h_n \in H_n \cap H_1 H_2 \cdots H_{n-1} = \{e\}$. It follows that $y_n = h_n$ and $(y_1 y_2 \cdots y_{n-1})^{-1} h_1 h_2 \cdots h_{n-1} = e$, so $y_1 y_2 \cdots y_{n-1} = h_1 h_2 \cdots h_{n-1}$. A simple inductive argument then yields $y_i = x_i$ for all $i = 1, 2, \ldots, n$.

It follows now that the map $\varphi : H_1 \times H_2 \times \cdots \times H_n \to G$ given by $\varphi(h_1, h_2, \ldots, h_n) = h_1 h_2 \cdots h_n$ is bijective. We prove that φ is a homomorphism, which will complete our demonstration. Let $(h_1, h_2, \ldots, h_n), (y_1, y_2, \ldots, y_n) \in H_1 \times H_2 \times \cdots \times H_n$. Then (again appealing to the fact that for any $i \neq j$, elements of H_i commute with elements of H_j) we have

$$
\begin{aligned}
\varphi((h_1, h_2, &\ldots, h_n)(y_1, y_2, \ldots, y_n)) \\
&= \varphi(h_1 y_1, h_2 y_2, \ldots, h_n y_n) \\
&= h_1 y_1 h_2 y_2 \cdots h_n y_n \\
&= h_1 h_2 y_2 \cdots h_n y_n y_1 = h_1 h_2 h_3 y_3 \cdots h_n y_n y_1 y_2 \\
&= \cdots = h_1 h_2 \cdots h_n y_1 y_2 \cdots y_n \\
&= \varphi(h_1, h_2, \ldots, h_n) \varphi(y_1, y_2, \ldots, y_n).
\end{aligned}
$$

Thus φ is a bijective homomorphism; that is, φ is an isomorphism. $\qquad\square$

When G has normal subgroups with the properties set out in the preceding proposition, we may say that G is the *internal direct product* of those subgroups. On the other hand, the group G constructed in Definition 6.5.1 from groups $G_1, G_2, \ldots, G_n$ is sometimes referred to as the *external direct product* of the groups. However, by Propositions 6.5.7 and 6.5.8, the external direct product $\prod_{i=1}^n G_i$ is isomorphic to the internal direct product of the subgroups $H_i = \iota_i(G_i)$, $i = 1, 2, \ldots, n$. For this reason, we usually make no distinction between the two, and refer to either as the direct product of the groups G_i (or of the subgroups H_i).

6.5.9 Example. *Recall that K_4, the Klein 4-group (see Example 5.4.4), is the group of symmetries of a non-square rectangle, with elements labelled as h, v, r, and I. Thus $|K_4| = 4$, and the three non-identity elements each have order 2. Since K_4 is abelian, all subgroups are normal. K_4 has three proper nontrivial subgroups, the cyclic subgroups generated by the three elements of order 2. Since the product of any two of the elements of order 2 is equal to the third element of order 2, it follows that K_4 is the internal direct product of any two of its three subgroups of order 2. Thus*

$$
K_4 \simeq \langle h \rangle \times \langle v \rangle \simeq \langle h \rangle \times \langle r \rangle \simeq \langle v \rangle \times \langle r \rangle .
$$

Note that for any groups G_1 and G_2, the map from $G_1 \times G_2$ to $G_2 \times G_1$ which sends $(x, y) \in G_1 \times G_2$ to $(y, x) \in G_2 \times G_1$ is an isomorphism, so for example, $\langle h \rangle \times \langle v \rangle \simeq \langle v \rangle \times \langle h \rangle$. If one examines the proof of Proposition 6.5.8, one finds that the map from, for example, $\langle h \rangle \times \langle v \rangle$ to K_4 for which

$(x, y) \in \langle h \rangle \times \langle v \rangle$ *is mapped to* $xy \in K_4$ *is an isomorphism from* $\langle h \rangle \times \langle v \rangle$ *to* K_4.

We conclude this section with a simple application of the direct product. This application will prove useful when we begin to discuss the theory of fields.

6.5.10 Theorem. *Let p be a prime, and let G be a nontrivial finite abelian group in which every non-identity element has order p. Then there exists a positive integer n such that $|G| = p^n$, and $G \simeq \prod_{i=1}^{n} \mathbb{Z}_p$.*

Proof. Observe first that if H is a subgroup of G, and $x \in G$, then either $x \in H$ or else $\langle x \rangle \cap H = \{ e \}$. For if $e \neq y \in \langle x \rangle \cap H$, then $\langle y \rangle \subseteq \langle x \rangle$, and as $|\langle y \rangle| = p = |\langle x \rangle|$, it follows that $\langle y \rangle = \langle x \rangle$ and so $x \in \langle y \rangle \subseteq H$. Now, since $|G| \in \{ m \in \mathbb{Z}^+ \mid$ there exists $S \subseteq G$ such that $\langle S \rangle = G$ and $|S| = m \}$, it follows by the well-ordering principle that there exists a smallest positive integer n for which there is a subset S of G such that $\langle S \rangle = G$ and $|S| = n$. Let $S = \{ g_1, g_2, \ldots, g_n \}$, and for each $i = 1, 2, \ldots, n$, let $H_i = \langle g_i \rangle$. Since G is commutative, for each $g \in G$, there exist integers $r_1, r_2, \ldots, r_n$ such that $g = g_1^{r_1} \cdots g_n^{r_n}$, and thus $G = H_1 H_2 \cdots H_n$. We prove that for each i, $H_i \simeq \mathbb{Z}_p$, and $G \simeq H_1 \times H_2 \times \cdots \times H_n$ (and thus $|G| = |H_1 \times H_2 \times \cdots \times H_n| = |H_1||H_2| \cdots |H_n| = p^n$). Let $i \in J_n$, and set $H = H_1 H_2 \cdots H_{i-1} H_{i+1} \cdots H_n$. Suppose that $H_i \cap H \neq \{ e \}$. Then by our initial observation, $g_i \in H$ and so $\langle S - \{ g_i \} \rangle = G$. Since this is not possible by choice of S, it follows that for every $i \in J_n$, $H_i \cap H_1 H_2 \cdots H_{i-1} H_{i+1} \cdots H_n = \{ e \}$. Since $G = H_1 H_2 \cdots H_n$, we may apply Proposition 6.5.8 to obtain that $G \simeq H_1 \times H_2 \times \cdots H_n$. Moreover, for each i, H_i is a cyclic group of order p, and by Corollary 6.3.13 (i), $H_i \simeq \mathbb{Z}_p$. For each $i = 1, 2, \ldots, n$, let $\alpha_i : H_i \to \mathbb{Z}_p$ be an isomorphism. Then the map from $\prod_{i=1}^{n} H_i$ to $\prod_{i=1}^{n} \mathbb{Z}_p$ for which $(h_1, h_2, \ldots, h_n) \in \prod_{i=1}^{n} H_i$ maps to $(\alpha_1(h_1), \alpha_2(h_2), \ldots, \alpha_n(h_n)) \in \prod_{i=1}^{n} \mathbb{Z}_p$ is an isomorphism and thus $G \simeq \prod_{i=1}^{n} \mathbb{Z}_p$. $\square$

6.6   Exercises

1. a) Let G be a group, and let H and K be finite subgroups of G such that $(|H|, |K|) = 1$. Prove that $H \cap K = \{ e \}$.

    b) Let G be a group.

      (i)   Let $a, b \in G$ be commuting elements of finite order for which $(|a|, |b|) = 1$. Prove that $|ab| = |a||b|$.

      (ii) Let $n \geq 2$, and let $a_1, a_2, \ldots, a_n \in G$ be commuting elements (that is, $a_i a_j = a_j a_i$ for all i, j) of G for which $(|a_i|, |a_j|) = 1$ for all $i \neq j$. Prove that $|a_1 a_2 \cdots a_n| = |a_1||a_2| \cdots |a_n|$.

  c) Let m and n be positive integers, and let G be a cyclic group of order mn with generator g. Let H be the unique subgroup of G of order m, and let K be the unique subgroup of G of order n.

      (i)   Find generators for H, K, and $H \cap K$.

      (ii) Prove that if m and n are relatively prime, then $G \simeq H \times K$.

2. a) Let G be a group, and let H and K be subgroups of G. Prove that HK is a subgroup of G if, and only if, $HK = KH$.

    b) Show by means of an example that in general, if H and K are subgroups of a group G, then HK is not necessarily a subgroup of G.

3. We have observed that if G is an abelian group, then every subgroup of G is normal in G. Prove that the converse of this statement is not true; that is, prove that there exists a group G for which every subgroup is normal, but G is not abelian. Hint: examine the groups of order 8 that have appeared in the text up to this point.

4. S_3, a group of order 6, has exactly four nontrivial proper subgroups, of which exactly one is normal in S_3. Find all four of these subgroups, and determine which one is the normal subgroup.

5. a) Prove that if G is a finite abelian group and m is the maximum order of the elements in G, then for any $g \in G$, $|g|$ divides m.

    b) Show by means of an example that this need not be the case if G is not abelian. Hint: as in (a), let m denote the maximum order of the elements of G, and let $h \in G$ be an element of order m. Assume that there exists $g \in G$ such that $|g|$ does not divide m. Take inspiration from Exercise 1 (b) and construct an element of G whose order is greater than m.

6. Let G be a finite abelian group.

    a) Let $T = \{ g \in G \mid g^2 = e \}$. Prove that $\prod_{g \in G} g = \prod_{g \in T} g$.

    b) Prove that if $|G|$ is odd, then $\prod_{g \in G} g = e$.

7. a) Let p be a prime. Prove that $\mathbb{Z}_p^*$, the group of units of the monoid $\mathbb{Z}_p$ under multiplication, has exactly one element of order 2.

b)   Use Exercise 6 and part (a) of the current exercise to prove Wilson's theorem, which states that $p \in \mathbb{Z}$ is prime if, and only if, $(p-1)! \equiv -1 \mod p$.

8.   Let n be a positive integer. Prove that $\sum_{d \mid n} \varphi(d) = n$, where φ is the Euler φ-function defined in Exercise 14 (e) of Chapter 4. Hint: Consider a cyclic group of order n and count the number of generators of each subgroup.

9.   Prove that D_3 (see Example 5.4.5) is isomorphic to S_3.

10.   In D_4 (see Example 5.4.5), let $K = \langle r^2 \rangle$, where $r = R_{\frac{2\pi}{4}}$.

a)   Prove that $K \triangleleft D_4$.

b)   Prove that $D_3/K \simeq K_4$.

11.   Let G be a group (not necessarily finite) with subgroups H and K of finite index in G such that $K \subseteq H$. Prove that $[G:K] = [G:H][H:K]$.

12.   Let m and n be positive integers. Prove that the number of homomorphisms from $\mathbb{Z}_m$ to $\mathbb{Z}_n$ is (m, n).

13.   Let $f : \mathbb{Z}_{30} \to \mathbb{Z}_{30}$ be a homomorphism such that $\ker(f) = \{ [0]_{30}, [10]_{30}, [20]_{30} \}$. If $f([23]_{30}) = [6]_{30}$, find all $x \in \mathbb{Z}_{30}$ for which $f(x) = [6]_{30}$.

14.   Let G and H be groups. Prove that $G \times H$ is cyclic if, and only if, both G and H are finite cyclic groups for which $(|G|, |H|) = 1$.

15.   Exhibit the bijective mapping established in Proposition 6.4.21 for the surjective homomorphism $\pi_{12} : \mathbb{Z} \to \mathbb{Z}_{12}$ (see Example 6.3.9 (ii)).

16.   Suppose that G is a group for which there exists a surjective homomorphism from G onto $\mathbb{Z}_{10}$. Prove that G has a normal subgroup of index 2 and a normal subgroup of index 5.

17.   Let G be a group, and suppose that there exists a surjective homomorphism $f : G \to H$, where H is a cyclic group of order 12. Further, suppose that $|\ker(f)| = 5$. Prove that G has normal subgroups of orders 5, 10, 15, 20, 30, and 60.

18.   Let G and H be groups for which there exists a surjective homomorphism $f : G \to H$. Prove that for any positive integer n, if H has an element of order n, then G has an element of order n.

19.   Let G be a group with distinct subgroups H and K each of index 2 in G. Prove that $H \cap K \triangleleft G$, and $G/(H \cap K) \simeq K_4$.

20.   Let G be an abelian group and n a positive integer. Define the map $f : G \to G$ by $f(g) = g^n$ for each $g \in G$.

a) Prove that f is a homomorphism.

b) If G is finite, say $m = |G|$, find an expression for $\ker(f)$.

c) Prove that if G is finite with $(n, |G|) = 1$, then f is an isomorphism.

21. Let G be a group. Prove that if the map $f : G \to G$ defined by $f(g) = g^2$ for all $g \in G$ is a homomorphism, then G is abelian.

22. Let G and H be cyclic groups of orders 12 and 9 respectively. Let a be a generator for G, and let b be a generator for H. Let $K = \langle (a^4, b^{-3}) \rangle$. Since G and H are cyclic, hence abelian, $G \times H$ is abelian and thus $K \triangleleft G \times H$. Let $L = (G \times H)/K$.

a) What is $|L|$?

b) Is L cyclic? Either provide a proof that L must be cyclic, or give an example of G and H for which L is not cyclic.

23. Let m and n be positive integers. Prove that if n divides m, then the number of surjective homomorphisms from $\mathbb{Z}_m$ to $\mathbb{Z}_n$ is the Euler φ-function $\varphi(n)$, and otherwise there are no surjective homomorphisms from $\mathbb{Z}_m$ to $\mathbb{Z}_n$.

24. Prove that the only homomorphism from $\mathbb{Z}_8 \times \mathbb{Z}_2$ to $\mathbb{Z}_4 \times \mathbb{Z}_4$ is the trivial homomorphism.

25. For any $a, b, c, d \in \mathbb{Q}$ with $ad - bc \neq 0$, let $f_{a,b,c,d} : \mathbb{Q} \to \mathbb{Q}$ be the map defined by $f_{a,b,c,d}(x) = \frac{ax+b}{cx+d}$ for all $x \in \mathbb{Q}$. Let

$$F = \{ f_{a,b,c,d} \mid a, b, c, d \in \mathbb{Q} \text{ and } ad - bc \neq 0 \}.$$

a) Prove that F is a subgroup of the group of units of the monoid $\mathscr{F}(\mathbb{Q})$.

b) Let $\alpha : Gl_2(\mathbb{Q}) \to F$ be the map defined by $\alpha(\begin{bmatrix} a & b \\ c & d \end{bmatrix}) = f_{a,b,c,d}$ for each $\begin{bmatrix} a & b \\ c & d \end{bmatrix} \in Gl_2(\mathbb{Q})$ (note that $ad - bc$ is the determinant of $\begin{bmatrix} a & b \\ c & d \end{bmatrix}$ and so $ad - bc \neq 0$). Prove that α is a homomorphism, and find $\ker(\alpha)$.

26. For $a, b \in \mathbb{R}$ with $a \neq 0$, let $T_{a,b} : \mathbb{R} \to \mathbb{R}$ be the map defined by $T_{a,b}(x) = ax + b$ for all $x \in \mathbb{R}$.

a) Prove that $G = \{ T_{a,b} \mid a, b \in \mathbb{R}, \ a \neq 0 \}$ is a subgroup of the group of units of $\mathscr{F}(\mathbb{R})$.

b) Given $T_{a,b} \in G$, determine the centralizer of $T_{a,b}$ in G; that is, find all $X \in G$ such that $T_{a,b} \circ X = X \circ T_{a,b}$.

c) Find the elements in the centre of G; that is, determine necessary and sufficient conditions on $a, b \in \mathbb{R}$, $a \neq 0$, so that $T_{a,b} \in Z(G)$.

d) Prove that $H = \{T_{1,b} \mid b \in \mathbb{R}\}$ is a normal subgroup of G, and prove that $G/H \simeq \mathbb{R}^*$, the group of units of the monoid $\mathbb{R}$ under multiplication.

27. a) Let n be a positive integer, and let G be a cyclic group of order n with identity e. Let m be a positive integer.

(i) Prove that for any $a \in G$, the equation $x^m = a$ has a solution in G if, and only if, $a^{\frac{n}{(m,n)}} = e$.

(ii) Let $H = \{x^m \mid x \in G\}$. Since G is cyclic and thus abelian, H is a subgroup of G. Prove that $|H| = \frac{n}{(m,n)}$. Moreover, if $G = \langle g \rangle$, prove that $H = \langle g^{(m,n)} \rangle$.

b) Let $n_1, n_2, \ldots, n_k$ be positive integers, and for each i, let G_i be a cyclic group of order n_i with identity e_i. Let m be a positive integer. Prove that $(a_1, a_2, \ldots, a_k) \in G = G_1 \times \cdots \times G_k$ is an mth power in G if, and only if, $a_i^{\frac{n_i}{(n_i,m)}} = e_i$ for all $i = 1, 2, \ldots, k$. Moreover, prove that $H = \{x^n \mid x \in G\}$ has order $\prod_{i=1}^{k} \frac{n_i}{(n_i,m)}$.

28. Let G be a group for which there exist homomorphisms $f, g : G \to \mathbb{R}^*$, the group of units of the monoid $\mathbb{R}$ under multiplication. Let

$$L_f = \{x \in G \mid f(x) < 1\} \quad \text{and} \quad L_g = \{x \in G \mid g(x) < 1\}.$$

a) Prove that if $L_f \subseteq L_g$, then $L_f = L_g$ and $\ker(f) = \ker(g)$.

b) Prove that if $L_f = L_g$ and there exists $x \in G - \ker(f)$ such that $f(x) = g(x)$, then $f = g$.

29. Let G be a finite group, and let $g \in G$. Recall (see Exercise 6) that the centralizer of g in G is $C_G(g) = \{x \in G \mid xg = gx\}$. Prove that the number of elements in the conjugacy class of g in G is equal to $[G : C_G(g)]$. Hint: define a map f from the set of left cosets of $C_G(g)$ in G to the conjugacy class of g in G by setting $f(xC_G(g)) = xgx^{-1}$ for each left coset $xC_G(g)$. Since the definition of f appears to depend on the selected representative x of the left coset $xC_G(g)$, you must first prove that f is well-defined. Then prove that f is a bijection.

Chapter 7

The Symmetric Groups

Introduction

In the investigation of finite groups, the symmetric groups play an important role. Often we are able to achieve a better understanding of a group if we can think of it as a group of permutations on some set, often one with some geometrical structure. In this chapter, we apply much of the theory we have developed so far to study the structure of S_n, the group of permutations on the set J_n, where n can be any positive integer. We shall show that when $n \geq 2$, S_n has a subgroup of index 2 that is of considerable importance. The two cosets of this subgroup form a partition of S_n, and we shall see that this partition can be refined by the formation of the conjugacy classes in S_n. This finer partition provides us with an important tool in the study of the finite symmetric groups. Among other things, we shall use it to demonstrate that the converse of Lagrange's theorem is not valid.

The Cayley Representation Theorem

We begin with a well-known theorem which says that any abstract group is isomorphic to a subgroup of some symmetric group. Of course, knowing that a group is contained in some other group does not, in general, tell us much about the group itself, especially if the containing group is big compared with the contained group, as is the case in this representation. Nevertheless, what is important is the idea that an abstract group can be realized as a more concrete object, an idea that has led to a branch of group theory called Representation Theory.

7.2.1 Theorem (Cayley Representation Theorem). *Every group is isomorphic to a subgroup of some symmetric group.*

Proof. Let G be a group. For each $g \in G$, let $\lambda_g : G \to G$ denote the map for which $\lambda_g(x) = gx$ for all $x \in G$. Let $g \in G$. We prove that $\lambda_g \in S_G$, the group of permutations of the set G. Note that for any $a, b \in G$, $\lambda_a \circ \lambda_b = \lambda_{ab}$ (that is; for each $x \in G$, $\lambda_a \circ \lambda_b(x) = \lambda_a(\lambda_b(x)) = \lambda_a(bx) = abx = \lambda_{ab}(x)$), and $\lambda_{e_G} = \mathbb{1}_G$. Thus we see that $\lambda_g \circ \lambda_{g^{-1}} = \lambda_{gg^{-1}} = \lambda_{e_G} = \mathbb{1}_G = \lambda_{g^{-1}}\lambda_g$, and so λ_g is invertible, with inverse $\lambda_{g^{-1}}$.

Now define a map α from G to S_G by $\alpha(g) = \lambda_g$ for each $g \in G$. We prove that α is an injective homomorphism, which then establishes that G is isomorphic to the subgroup $\alpha(G)$ of S_G. First, we prove that α is a homomorphism. Let $g, h \in G$. We have $\alpha(g) \circ \alpha(h) = \lambda_g \circ \lambda_h = \lambda_{gh} = \alpha(gh)$, so α is a homomorphism.

Next, suppose that $g, h \in G$ are such that $\alpha(g) = \alpha(h)$. Then $\lambda_g = \lambda_h$, and so in particular, $g = \lambda_g(e_G) = \lambda_h(e_G) = h$. This proves that α is injective. $\qquad\qquad\square$

Observe that the size of G is, in general, very small compared to the size of S_G. For example, if the order of G is 6, then the order of S_G is $6! = 720$. The subgroup (of order 6) of S_G that is isomorphic to G is inside a group 120 times its size!

7.2.2 Example. *The Klein 4-group K_4 was defined as a subgroup of $S_{\mathbb{R}^2}$ (it was defined as the group of symmetries of a non-square rectangle). But if we use the proof of the Cayley representation theorem, we see that we can regard K_4 as a subgroup of a much smaller (finite even) permutation group. Recall that $K_4 = \{I, h, v, r\}$ (see Table 5.1 for the multiplication table of K_4). We know that $\mathbb{1}_{K_4} = \lambda_I$, and we compute $\lambda_h = \begin{pmatrix} I & h & v & r \\ h & I & r & v \end{pmatrix}$, $\lambda_v = \begin{pmatrix} I & h & v & r \\ v & r & I & h \end{pmatrix}$, and $\lambda_r = \begin{pmatrix} I & h & v & r \\ r & v & h & I \end{pmatrix}$. If we relabel the elements of K_4 as $1 = I$, $2 = h$, $3 = v$, and $4 = r$, then (see Notation 3.4.6) $\lambda_1 = [1\,2\,3\,4]$, $\lambda_2 = [2\,1\,4\,3]$, $\lambda_3 = [3\,2\,4\,1]$, and $\lambda_4 = [4\,3\,2\,1]$, so K_4 is isomorphic to the subgroup $\{[1\,2\,3\,4], [2\,1\,4\,3], [3\,2\,4\,1], [4\,3\,2\,1]\}$ of S_4.*

Permutations as Products of Disjoint Cycles

In order to study the finite symmetric groups in detail, we need to introduce notation for each element which conveys at a glance the nature of the element. The notation we have been using so far does not do this. For example, $[1\,2\,3\,4]$ is a rather "tame" element of S_4 since it moves none of 1, 2, 3, or 4. On the other hand, $[2\,3\,4\,1]$ is quite "dynamic", since very element of J_4 is moved by it. The eye, however, does not immediately distinguish between the two.

7.3.1 Definition. *Let n be a positive integer. For $\sigma \in S_n$, define a relation $\sim_\sigma$ on J_n by $i \sim_\sigma j$ if there exists $k \in \mathbb{Z}$ such that $j = \sigma^k(i)$, where $i, j \in J_n$.*

7.3.2 Proposition. *Let n be a positive integer. For $\sigma \in S_n$, the relation $\sim_\sigma$ is an equivalence relation on J_n.*

Proof. Since $\sigma^0 = \mathbb{1}_{J_n}$, for $i \in J_n$, we have $i = \sigma^0(i)$ and thus $i \sim_\sigma i$. Now suppose that $i, j \in J_n$ are such that $i \sim_\sigma j$, so there is $k \in \mathbb{Z}$ such that $j = \sigma^k(i)$. Then $\sigma^{-k}(j) = \sigma^{-k}(\sigma^k(i)) = \sigma^{k-k}(i) = \sigma^0(i) = i$ and so $j \sim_\sigma i$. Finally, suppose that $i, j, k \in J_n$ are such that $i \sim_\sigma j$ and $j \sim_\sigma k$. Then there exist $m, t \in \mathbb{Z}$ such that $j = \sigma^m(i)$ and $k = \sigma^t(j)$, and so $k = \sigma^t(\sigma^m(i)) = \sigma^{t+m}(i)$, proving that $i \sim_\sigma k$. $\qquad\square$

7.3.3 Definition. *Let n be a positive integer and $\sigma \in S_n$. For each $i \in J_n$, the equivalence class of i under $\sim_\sigma$ is denoted by $O_\sigma(i)$ and is called an orbit of σ.*

7.3.4 Proposition. *Let n be a positive integer and $\sigma \in S_n$. For each $i \in J_n$, $O_\sigma(i) = \{\, \sigma^k(i) \mid i \in \mathbb{Z} \,\}$.*

Proof. Let $i \in J_n$. Then

$$O_\sigma(i) = \{\, j \in J_n \mid i \sim_\sigma j \,\} = \{\, j \in J_n \mid \text{ there is } k \in \mathbb{Z} \text{ with } j = \sigma^k(i) \,\}$$
$$= \{\, \sigma^k(i) \mid k \in \mathbb{Z} \,\}.$$

$\qquad\square$

7.3.5 Proposition. *Let n be a positive integer and $\sigma \in S_n$. Then for each $i \in J_n$, there exists a positive integer m such that $|O_\sigma(i)| = m$, and $O_\sigma(i) = \{\, i, \sigma(i), \sigma^2(i), \ldots, \sigma^{m-1}(i) \,\}$ with $\sigma^m(i) = i$.*

Proof. Let $i \in J_n$, and consider the infinite sequence $i = \sigma^0(i), \sigma^1(i) = \sigma(i), \sigma^2(i), \ldots$ of elements of J_n. Since J_n is finite, this sequence must have repetitions. Let m be the smallest positive integer such that there exists an integer t with $0 \le t < m$ and $\sigma^t(i) = \sigma^m(i)$. If $t > 0$, then we have $\sigma^{-1}(\sigma^t(i)) = \sigma^{-1}(\sigma^m(i))$; that is, $\sigma^{t-1}(i) = \sigma^{m-1}(i)$ and $0 \le t-1 < m-1$, which contradicts our choice of m. Thus $t = 0$, and we have $\sigma^m(i) = \sigma^0(i) = i$. We prove now that $O_\sigma(i) = \{i, \sigma(i), \ldots, \sigma^{m-1}(i)\}$. Now, we have $\{i, \sigma(i), \ldots, \sigma^{m-1}(i)\} \subseteq O_\sigma(i)$, so suppose that $j \in O_\sigma(i)$. Then $j = \sigma^t(i)$ for some $t \in \mathbb{Z}$. Apply the division theorem to m and t to obtain the existence of integers q and r with $t = qm + r$ and $0 \le r < m$. Then $j = \sigma^t(i) = \sigma^{qm+r}(i) = \sigma^r(\sigma^{qm}(i))$. Since $\sigma^m(i) = i$, it follows that $\sigma^{-m}(i) = i$, and thus for any integer s, $\sigma^{ms}(i) = (\sigma^m)^s(i) = i$. In particular, $\sigma^{qm}(i) = i$. This yields $j = \sigma^r(\sigma^{qm}(i)) = \sigma^r(i) \in \{\sigma^t(i) \mid 0 \le t < m\}$, and so $O_\sigma(i) = \{i, \sigma(i), \ldots, \sigma^{m-1}(i)\}$. By the choice of m, it follows that for any r, s with $0 \le r < s < m$, $\sigma^r(i) \ne \sigma^s(i)$, and thus $|O_\sigma(i)| = m$. $\qquad\square$

Note that for $\sigma \in S_n$ and $i \in J_n$, $|O_\sigma(i)| = 1$ if, and only if, $O_\sigma(i) = \{i\}$, which in turn holds if, and only if, $\sigma(i) = i$. In particular, $\sigma = \mathbb{1}_{J_n}$ if, and only if, every orbit of σ is trivial (that is, has size 1). If $i \in J_n$ is such that $\sigma(i) = i$, then i is called a fixed point of σ.

7.3.6 Definition. *Let n be a positive integer. For $\sigma \in S_n$, let $M_\sigma = \{i \in J_n \mid \sigma(i) \ne i\}$. If $\sigma, \tau \in S_n$ are such that $M_\sigma \cap M_\tau = \varnothing$, then we say that σ and τ are disjoint.*

Note that M_σ is just the union of the nontrivial orbits of σ, and $M_\sigma = \varnothing$ if, and only if, $\sigma = \mathbb{1}_{J_n}$.

7.3.7 Definition. *Let n and k be integers with $1 < k \le n$. Then $\sigma \in S_n$ is said to be a cycle of length k, or a k-cycle, if σ has exactly one nontrivial orbit, and that one nontrivial orbit has size k. In such a case, if $i \in J_n$ is such that $|O_\sigma(i)| = k$, then $O_\sigma(i) = \{\sigma^j(i) \mid j = 0, 1, \ldots, k-1\}$ and we write $\sigma = (i\,\sigma(i)\,\sigma^2(i)\cdots\sigma^{k-1}(i))$.*

7.3.8 Proposition. *Let n and k be positive integers with $2 \le k \le n$, and let $\{i_0, i_1, \ldots, i_{k-1}\} \subseteq J_n$ be a subset of size k. Define the map $\sigma : J_n \to J_n$ by $\sigma(i_t) = i_{t+1}$ for $t = 0, 1, \ldots, k-2$, $\sigma(i_{k-1}) = i_0$, and for any $i \in J_n - \{i_0, i_1, \ldots, i_{k-1}\}$, $\sigma(i) = i$. Then $\sigma \in S_n$, and σ is a k-cycle with $\sigma = (i_0\,i_1\cdots i_{k-1})$.*

Proof. σ is a surjective map from J_n to J_n, and thus by Proposition 3.3.22, σ is bijective; that is, $\sigma \in S_n$. We prove by induction on $t \geq 0$ that if $t \leq k - 1$, then $\sigma^t(i_0) = i_t$. The base case is immediate since $\sigma^0 = \mathbb{1}_{J_n}$, so suppose that $t \geq 0$ is an integer for which the hypothesis holds. If $t + 1 > k - 1$, there is nothing to do, so suppose that $t + 1 \leq k - 1$. Then $0 \leq t \leq k - 2$ and so $\sigma^t(i_0) = i_t$, from which we obtain $\sigma^{t+1}(i_0) = \sigma(\sigma^t(i_0)) = \sigma(i_t) = i_{t+1}$, as required. Thus we have $\sigma^t(i_0) = i_t$ for all t with $0 \leq t \leq k - 1$. Moreover, $\sigma^k(i_0) = \sigma(\sigma^{k-1}(i_0)) = \sigma(i_{k-1}) = i_0$. Then for any integer m, we have $m = qk + r$ for integers q and r with $0 \leq r < k$, and thus $\sigma^m(i_0) = \sigma^r(\sigma^{qm}(i_0)) = \sigma^r(i_0) = i_r \in \{i_0, i_1, \ldots, i_{k-1}\}$. Thus $O_\sigma(i_0) = \{i_0, i_1, \ldots, i_{k-1}\}$, while all other orbits of σ are trivial. It follows that σ is a k-cycle, and $\sigma = (i_0 \, \sigma(i_0) \cdots \sigma^{k-1}(i_0)) = (i_0 \, i_1 \cdots i_{k-1})$. $\qquad\square$

Some remarks are in order. It is convenient to let the identity permutation be denoted by (i), where i can be any element of J_n (we can think of this as a 1-cycle). Furthermore, for any integer $k > 1$, each k-cycle has k cycle representations. For example, in S_5 (or indeed, in S_n where $n \geq 3$), $(1\,2\,3) = (2\,3\,1) = (3\,1\,2)$. Finally, we remark that the inverse of a k-cycle is a k-cycle. More precisely, $(i_1 \, i_2 \cdots i_k)^{-1} = (i_k \, i_{k-1} \cdots i_1)$. Note that for the special case of a 2-cycle, we have $(i_1 \, i_2)^{-1} = (i_2 \, i_1) = (i_1 \, i_2)$.

It is often informative to draw a diagram to represent a given permutation. For each $i \in J_n$, we mark a point to represent i. Then for each $i \in J_n$, draw an arrow from i to $\sigma(i)$ (if $\sigma(i) = i$, then draw a loop at the point which represents i with an arrowhead on the loop) to represent the fact that σ maps i to $\sigma(i)$. If the points that represent $i \in J_n$ are chosen carefully, then cycles really stand out in this diagram.

7.3.9 Example. *Consider the 4-cycle $\sigma = (2\,5\,3\,7) \in S_7$. Then $\sigma(1) = 1$, $\sigma(4) = 4$, and $\sigma(6) = 6$, while $\sigma(2) = 5$, $\sigma(5) = 3$, $\sigma(3) = 7$, and $\sigma(7) = 2$. A diagram that represents σ would be:*

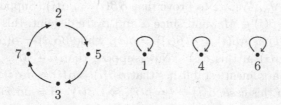

Fig. 7.1 A diagram to represent the cycle $(2\,5\,3\,7) \in S_7$

Our next example presents a diagram that represents a noncyclic permutation.

7.3.10 Example. *Let* $\sigma = [2\,1\,4\,5\,3\,6\,7] \in S_7$. *A diagram that represents* σ *would be:*

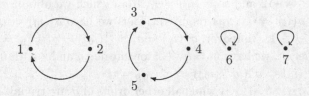

Fig. 7.2   A diagram to represent the permutation $[2\,1\,4\,5\,3\,6\,7] \in S_7$

It is conventional to leave out the fixed points of σ when one draws a diagram to represent σ, so in Figure 7.2, we would not draw points for 6 and 7. From now on, we shall adopt this convention.

Since cycles are permutations, they may be composed. One should note that the composition of two cycles is in general not a cycle. For example, in S_7, $(1\,2)(3\,4\,5)(6) = (1\,2)(3\,4\,5) = [2\,1\,4\,5\,3\,6\,7]$. The striking fact is that in S_n, every permutation (other than $\mathbb{1}_{J_n}$) can be written as a product (that is, with binary operation being composition of maps) of pairwise disjoint cycles. Moreover (as we are about to see), disjoint permutations commute, and aside from the order of the factors, each nonidentity permutation can be written uniquely as a product of disjoint cycles.

7.3.11 Proposition. *Let* $n > 1$ *be an integer, and let* $\alpha, \beta \in S_n$ *be disjoint. Then* $\alpha\beta = \beta\alpha$.

Proof. Let $i \in J_n$. We are to prove that $\alpha\beta(i) = \beta\alpha(i)$. Suppose first that $i \in M_\beta$. Then $\beta(i) \in M_\beta$, and since α and β are disjoint, this means that $i, \beta(i) \notin M_\alpha$. Thus $\alpha\beta(i) = \alpha(\beta(i)) = \beta(i)$, while $\beta\alpha(i) = \beta(\alpha(i)) = \beta(i)$, so $\alpha\beta(i) = \beta\alpha(i)$ in this case. Now suppose that $i \notin M_\beta$. If $i \in M_\alpha$, then a similar argument establishes that $\alpha\beta(i) = \alpha(i) = \beta\alpha(i)$. Otherwise, $i \notin M_\alpha$, and in this case, $\alpha(i) = i = \beta(i)$, so $\alpha\beta(i) = i = \beta\alpha(i)$. □

7.3.12 Proposition. *Let* n *be a positive integer, and let* $\sigma \in S_n$ *with* $\sigma \neq \mathbb{1}_{J_n}$. *Let* $i \in J_n$ *be such that* $m = |O_\sigma(i)| \geq 2$, *so* $O_\sigma(i) =$

$\{\,i, \sigma(i), \ldots, \sigma^{m-1}(i)\,\}$. *Let σ_i denote the m-cycle $(i\,\sigma(i)\,\cdots\,\sigma^{m-1}(i))$.*
Then $M_{\sigma_i} = O_\sigma(i)$, and for all $k \in \mathbb{Z}$, and for all $t \in M_{\sigma_i}$, $\sigma_i^k(t) = \sigma^k(t)$.

Proof. We have applied Proposition 7.3.8 to construct σ_i and to obtain
that it is an m-cycle. By definition of cycle, it then follows that for all
$k = 0, 1, \ldots, m-1$, $\sigma_i^k(i) = \sigma^k(i)$, and $\sigma_i^m(i) = i$, while by Proposition
7.3.5, $\sigma^m(i) = i$. For any integer k, there exist integers q and r with $qm + r$
and $0 \le r < m$, so $\sigma_i^k(i) = \sigma_i^r(i) = \sigma^r(i) = \sigma^k(i)$. Now let $t \in M_{\sigma_i}$
and $k \in \mathbb{Z}$. Then there exists $l \in \mathbb{Z}$ such that $t = \sigma_i^l(i) = \sigma^l(i)$, and so
$\sigma_i^k(t) = \sigma_i^k(\sigma_i^l(i)) = \sigma_i^{k+l}(i) = \sigma^{k+l}(i) = \sigma^k(\sigma^l(i)) = \sigma^k(t)$, as required. $\square$

7.3.13 Proposition. *For any positive integer n, each $\sigma \in S_n$ with $\sigma \ne$*
$\mathbb{1}_{J_n}$ is a product of pairwise disjoint cycles. Moreover, up to the order of
the factors, the representation of σ as a product of pairwise disjoint cycles
is unique.

Proof. Let $\sigma \in S_n$, with $\sigma \ne \mathbb{1}_{J_n}$. Let r be the number of nontrivial orbits
of σ, and choose $i_1, i_2, \ldots, i_r \in J_n$ such that the nontrivial orbits of σ are
$O_\sigma(i_j)$, $j = 1, 2, \ldots, r$. Apply Proposition 7.3.12 to each nontrivial orbit of
σ, thereby obtaining cycle σ_j from orbit $O_\sigma(i_j)$, $j = 1, 2, \ldots, r$. We claim
that $\sigma = \sigma_1 \sigma_2 \cdots \sigma_r$. Note that since the orbits of σ form a partition of J_n,
the cycles σ_j, $j = 1, 2, \ldots, r$ are pairwise disjoint and thus commute with
each other. Furthermore, for $i \in J_n - O_\sigma(i_j)$, $\sigma_j(i) = i$, while if $i \in O_\sigma(i_j)$,
then $\sigma_j^l(i) = \sigma^l(i)$ for every $l \in \mathbb{Z}$.

Let $i \in J_n$. If $i \notin M_\sigma = \cup_{j=1}^r O_\sigma(i_j)$, then i is fixed by σ, and
as observed above, i is fixed by each σ_j, $j = 1, 2, \ldots, r$, so $\sigma(i) = i = (\prod_{j=1}^r \sigma_j)(i)$. Suppose that $i \in M_\sigma$, and let t be such that $i \in O_{i_t}$. Then

$$(\prod_{j=1}^r \sigma_j)(i) = \sigma_t(\prod_{\substack{j=1 \\ j \ne t}}^r \sigma_j(i)) = \sigma_t(i) = \sigma(i),$$

where we have used the fact that the cycles σ_{i_j}, $j = 1, 2, \ldots, r$ are pairwise
disjoint and thus commute with each other. It follows that for every $i \in J_n$,
$(\prod_{j=1}^r \sigma_j)(i) = \sigma(i)$, and so $\prod_{j=1}^r \sigma_j = \sigma$.

It remains to prove the uniqueness assertion, and we shall prove this
by induction on the number r of nontrivial orbits of the permutation. Our
inductive hypothesis is the assertion that any $\sigma \in S_n$ that has r nontrivial
orbits has a unique representation (up to the order of factors) as a product
of pairwise disjoint cycles. The base case $r = 1$ is immediate, so suppose

that $r \geq 1$ is an integer for which the assertion is true. If there are no permutations in S_n with $r+1$ nontrivial orbits, then the assertion is vacuously true. Suppose then that $\sigma \in S_n$ has $r+1$ nontrivial orbits, and further, that $\sigma_1, \sigma_2, \ldots, \sigma_{r+1}$ are pairwise disjoint cycles, and $\tau_1, \tau_2, \ldots, \tau_{r+1}$ are pairwise disjoint cycles, and $\sigma_1 \sigma_2 \cdots \sigma_{r+1} = \sigma = \tau_1 \tau_2 \cdots \tau_{r+1}$. Let $O = M_{\sigma_{r+1}}$. Then O is an orbit of σ, and thus there is an index i such that $O = M_{\tau_i}$. Relabelling if necessary, we may assume that $i = r+1$ (since disjoint cycles commute). For any $j \in O$, $\sigma_{r+1}(j) = \sigma(j) = \tau_{r+1}(j)$, while for $j \in J_n - O$, $\sigma_{r+1}(j) = j = \tau_{r+1}(j)$, and thus $\sigma_{r+1} = \tau_{r+1}$. We may therefore cancel σ_{r+1} to obtain that $\sigma_1 \sigma_2 \cdots \sigma_r = \tau_1 \tau_2 \cdots \tau_r$. Since this is a permutation with r nontrivial orbits, we may apply the induction hypothesis to obtain that, after relabelling if necessary, $\sigma_i = \tau_i$ for $i = 1, 2, \ldots, r$. The result follows now by the first principle of mathematical induction. $\qquad \square$

7.3.14 Definition. *Let n be a positive integer. For $\sigma \in S_n - \{\mathbb{1}_{J_n}\}$, the unique representation of σ as a product of pairwise disjoint cycles is called the disjoint cycle decomposition of σ.*

In Figure 7.2, the cycle decomposition of $\sigma = [2\,1\,4\,5\,3\,6\,7]$, namely $(1\,2)(3\,4\,5)$, is completely obvious.

7.3.15 Example. *We determine the order of $\sigma = (1\,2)(3\,4\,5)(6\,7\,8\,9) \in S_9$; that is, we determine the value of the smallest positive integer k such that $\sigma^k = \mathbb{1}_{J_9}$. Recall that disjoint cycles commute, so for any positive integer r, $\sigma^r = ((1\,2)(3\,4\,5)(6\,7\,8\,9))^r = (1\,2)^r(3\,4\,5)^r(6\,7\,8\,9)^r$. We calculate the sequence $\sigma, \sigma^2, \sigma^3, \ldots$, until we arrive at a term $\sigma^k = \mathbb{1}_{J_9}$ for the first time. The value of this k is $|\sigma|$, the order of σ. For the computation of σ^2, we observe that $(1\,2)^2 = \mathbb{1}_{J_9}$, $(3\,4\,5)^2 = (3\,5\,4)$, and $(6\,7\,8\,9)^2 = (6\,8)(7\,9)$. Thus $\sigma^2 = (3\,5\,4)(6\,8)(7\,9)$. Since $\sigma^2 \neq \mathbb{1}_{J_9}$, we continue and calculate σ^3. Note that $(1\,2)^3 = (1\,2)^2(1\,2) = (1\,2)$, $(3\,4\,5)^3 = \mathbb{1}_{J_9}$, and $(6\,7\,8\,9)^3 = (6\,9\,8\,7)$, so $\sigma^3 = (1\,2)(6\,9\,8\,7) \neq \mathbb{1}_{J_9}$. To evaluate σ^4, we observe that $(1\,2)^4 = ((1\,2)^2)^2 = \mathbb{1}_{J_9}^2 = \mathbb{1}_{J_9}$, $(3\,4\,5)^4 = (3\,4\,5)^3(3\,4\,5) = (3\,4\,5)$, and $(6\,7\,8\,9)^4 = ((6\,7\,8\,9)^2)^2 = ((6\,8)^2(7\,9))^2 = \mathbb{1}_{J_9}$, so $\sigma^4 = (3\,4\,5)$. It follows that $\sigma^{12} = (\sigma^4)^3 = \mathbb{1}_{J_9}$, and so $|\sigma|$ is a divisor of 12, and is greater than 4. If we show that $\sigma^6 \neq \mathbb{1}_{J_9}$, then $|\sigma| = 12$. We have $\sigma^6 = (1\,2)^6(3\,4\,5)^6(6\,7\,8\,9)^6 = \mathbb{1}_{J_9}^3 \mathbb{1}_{J_9}^2(6\,8)^3(7\,9)^3 = (6\,8)(7\,9) \neq \mathbb{1}_{J_9}$, and so $|\sigma| = 12$.*

After reading through the preceding example, the following result will come as no surprise.

7.3.16 Proposition. *For any positive integers n and k, the order of any k-cycle in S_n is k.*

Proof. Let n be a positive integer, k an integer with $2 \leq k \leq n$, and let $\sigma \in S_n$ be a k-cycle. Then for some $i \in J_n$, $|O_\sigma(i)| = k$, and $O_\sigma(i) = \{\, i, \sigma(i), \sigma^2(i), \ldots, \sigma^{k-1}(i) \,\}$ with $\sigma^k(i) = i$. Since $i \neq \sigma^j(i)$ for $j = 1, 2, \ldots, k-1$, it follows that $\sigma^j \neq \mathbb{1}_{J_n}$ for $1 \leq j \leq k-1$. Moreover, we have $\sigma^k(i) = i$. For any $t \in O_\sigma(i)$, there exists $l \in \mathbb{Z}$ with $t = \sigma^l(i)$, and thus $\sigma^k(t) = \sigma^{k+l}(i) = \sigma^l(\sigma^k(i)) = \sigma^l(i) = t$. If $t \in J_n - O_\sigma(i)$, then $\sigma(t) = t$ and thus $\sigma^k(t) = t$, so $\sigma^k = \mathbb{1}_{J_n}$. It follows now that $|\sigma| = k$. $\quad\square$

Recall that the least common multiple of positive integers $m_1, m_2, \ldots, m_k$ is denoted by $[m_1, m_2, \ldots, m_k]$.

7.3.17 Proposition. *Let n be a positive integer. For any $\sigma \in S_n - \{\mathbb{1}_{J_n}\}$,*

$$|\sigma| = [m_1, m_2, \ldots, m_r],$$

where the disjoint cycle decomposition of σ is $\sigma = \sigma_1 \sigma_2 \cdots \sigma_r$, and for each i with $1 \leq i \leq r$, σ_i is a cycle of length m_i.

Proof. Let $\sigma \in S_n$ with $\sigma \neq \mathbb{1}_{J_n}$, and let $\sigma = \sigma_1 \sigma_2 \cdots \sigma_r$ be the disjoint cycle decomposition of σ. Let m_i denote the length of cycle σ_i for $i = 1, 2, \ldots, r$. By Proposition 7.3.16, $|\sigma_i| = m_i$ for each $i = 1, 2, \ldots, r$. Let $m = [m_1, m_2, \ldots, m_r]$. Then for each $i = 1, 2 \ldots, k$, m_i divides m and thus $\sigma_i^m = \mathbb{1}_{J_n}$. Consequently, $\sigma^m = (\sigma_1 \cdots \sigma_r)^m = \sigma_1^m \cdots \sigma_r^m = \mathbb{1}_{J_n}$ and so $|\sigma|$ is a divisor of m. On the other hand, for any $i = 1, 2, \ldots, r$, and any $j \in M_{\sigma_i}$, we have $\sigma_i^t(j) = \sigma^t(j)$ for every $t \in \mathbb{Z}$, while for $j \notin M_{\sigma_i}$, $\sigma_i(j) = j$ and thus $\sigma_i^t(j) = j$ for every $t \in \mathbb{Z}$. In particular, for $t = |\sigma|$, $\sigma_i^t(j) = \sigma^t(j) = j$ for every $j \in M_{\sigma_i}$ and so $\sigma_i^{|\sigma|}(j) = j$ for all $j \in J_n$; that is, $m_i = |\sigma_i|$ is a divisor of $|\sigma|$ for every $i = 1, 2, \ldots, r$. Thus $m = [m_1, m_2, \ldots, m_r]$ is a divisor of $|\sigma|$, and so $m = |\sigma|$. $\quad\square$

7.3.18 Example. *We determine the order of the permutation* $\sigma = (1\,3\,4)(9\,2\,6\,3\,1)(4\,7\,3\,5)(7\,8\,9) \in S_{10}$. *Note that these cycles are not disjoint. If we wish to use Proposition 7.3.17 to determine the order of σ (and we do), we must first determine the disjoint cycle decomposition of σ. This require that we determine the orbits of σ, and as we determine each non-trivial orbit of σ, we can write down the cycle that corresponds to the orbit. If $i \in J_{10}$ does not appear in any of the cycles whose product is σ, then i is a fixed point of σ. The only such i is $i = 10$, so we know immediately that*

$O_\sigma(10) = \{10\}$. *However, there may be other elements of J_{10} that are fixed by σ, and we will just have to discover such elements by computation. We proceed now to determine the orbit of each $i \in J_{10}$ that appears in at least one of the cycles whose product is σ. We shall (arbitrarily) start with 1. Just for this example, we shall denote the result of applying a permutation τ to $i \in J_n$ by $\tau[i]$, rather than the conventional $\tau(i)$, in order to avoid any possible confusion with cycle notation. We have*

$$\sigma[1] = (1\,3\,4)(9\,2\,6\,3\,1)(4\,7\,3\,5)(7\,8\,9)[1] = (1\,3\,4)[(9\,2\,6\,3\,1)[1]]$$
$$= (1\,3\,4)[9] = 9.$$

Similarly, we calculate $\sigma[9] = 3$, $\sigma[3] = 5$, and $\sigma[5] = 1$, so $\{1,3,5,9\}$ is an orbit of σ, and the corresponding cycle is $(1\,9\,3\,5)$. Since the orbits of σ form a partition of J_{10}, and $O_\sigma(i) = O_\sigma(1)$ for all $i \in O_\sigma(1)$, we next choose $i \in J_{10} - (O_\sigma(1) \cup O_\sigma(10))$, and in the absence of any other criteria, we shall choose the numerically least such element; that is, we choose 2. We have $\sigma[2] = 6$, $\sigma[6] = 4$, $\sigma[4] = 7$, $\sigma[7] = 8$, and $\sigma[8] = 2$, so $\{2,4,6,7,8\}$ is an orbit of σ, and the associated cycle is $(2\,6\,4\,7\,8)$. Since $\{1,3,5,9\} \cup \{2,4,6,7,8\} \cup \{10\} = \{1,2,3,4,5,6,7,8,9,10\} = J_{10}$, we have found all of the orbits of σ, two of which are nontrivial, so the disjoint cycle decomposition of σ is $\sigma = (1\,9\,3\,5)(2\,6\,4\,7\,8)$. Since σ is a product of two disjoint cycles, one a 4-cycle and one a 5-cycle, by Proposition 7.3.17, the order of σ is the least common multiple of 4 and 5; namely 20.

By the remarks following Definition 7.3.7, we may quickly obtain the disjoint cycle decomposition for σ^{-1} from that of σ. For example, if $\sigma = (1\,9\,3\,5)(2\,6\,4\,7\,8)$, then $\sigma^{-1} = ((1\,9\,3\,5)(2\,6\,4\,7\,8))^{-1} = (2\,6\,4\,7\,8)^{-1}(1\,9\,3\,5)^{-1} = (8\,7\,4\,6\,2)(5\,3\,9\,1) = (5\,3\,9\,1)(8\,7\,4\,6\,2)$. Note that we have used the fact that disjoint cycles commute.

7.4 Even and Odd Permutations

In this section, we establish that for any positive integer n, there is a natural way to partition S_n into two classes, with the elements of one class being called the even permutations in S_n, and the elements in the other class being called the odd permutations. This labelling is due to the fact (which we will establish) that the product of two permutations of the same type is even, while the product of two permutations of different types is odd.

There are numerous ways to prove this result, and our approach was chosen because it puts homomorphisms to use.

7.4.1 Definition. *For any positive integer n, each 2-cycle in S_n is called a transposition.*

7.4.2 Proposition. *Let $n > 1$ be a integer, and let k be an integer with $2 \le k \le n$. Then any k-cycle in S_n is a product of $k - 1$ transpositions.*

Proof. The proof is by induction on k. Since a 2-cycle is a transposition, the result holds when $k = 2$. Suppose that $k \ge 2$ is an integer for which the assertion holds (note that there need not be any k-cycles in S_n – this is the case if, and only if, $k > n$, and in such a case, the assertion is trivially true since there are no k-cycles). If $k + 1 > n$, then the assertion holds as just explained. Suppose that $k + 1 \le n$, and let $\tau = (j_1 \, j_2 \cdots j_{k+1})$ be any $(k + 1)$-cycle in S_n. Since $(j_1 \, j_{k+1})(j_1 \, j_2 \cdots j_{k+1}) = (j_1 \, j_2 \cdots j_k)$ and $(j_1 \, j_{k+1})^{-1} = (j_1 \, j_{k+1})$, we obtain that $(j_1 \, j_2 \cdots j_{k+1}) = (j_1 \, j_{k+1})(j_1 \, j_2 \cdots j_k)$. As $(j_1 \, j_2 \cdots j_k)$ is a k-cycle, it can be written as a product of $k - 1$ transpositions, and thus τ can be written as a product of $k = (k + 1) - 1$ transpositions. The result follows now by the first principle of mathematical induction. $\square$

Note that the inductive argument in the proof of Proposition 7.4.2 could be used to prove that a k-cycle $(j_1 \, j_2 \cdots j_k)$ can be written in the following way as a product of $k - 1$ transpositions:

$$(j_1 \, j_2 \cdots j_k) = (j_1 \, j_k)(j_1 \, j_{k-1}) \cdots (j_1 \, j_2).$$

7.4.3 Corollary. *For any integer $n > 1$, every element of S_n can be written as a product of transpositions.*

Proof. Let $\sigma \in S_n$ and suppose that $\sigma \ne \mathbb{1}_{J_n}$. By Proposition 7.3.13, σ can be written as a product of cycles, and by Proposition 7.4.2, every cycle can be written as a product of transpositions, so σ can be written as a product of transpositions. This leaves only $\mathbb{1}_{J_n}$ to consider, but since $n > 1$, we have $\mathbb{1}_{J_n} = (1 \, 2)^2$ and thus every permutation in S_n can be written as a product of transpositions. $\square$

For example, in Example 7.3.18, we investigated the permutation $(1 \, 3 \, 4)(9 \, 2 \, 6 \, 3 \, 1)(4 \, 7 \, 3 \, 5)(7 \, 8 \, 9)$. Even though this is not the disjoint cycle decomposition for the permutation, we still have

$(1 \, 3 \, 4)(9 \, 2 \, 6 \, 3 \, 1)(4 \, 7 \, 3 \, 5)(7 \, 8 \, 9)$

$\qquad = (1 \, 4)(1 \, 3)(9 \, 1)(9 \, 3)(9 \, 6)(9 \, 2)(4 \, 5)(4 \, 3)(4 \, 7)(7 \, 9)(7 \, 8)$.

Of course, from its disjoint cycle deposition, we also obtain

$$(1\,3\,4)(9\,2\,6\,3\,1)(4\,7\,3\,5)(7\,8\,9) = (1\,9\,3\,5)(2\,6\,4\,7\,8)$$
$$= (1\,5)(1\,3)(1\,9)(2\,8)(2\,7)(2\,4)(2\,6).$$

Given that the representation of a permutation as a product of transpositions is not unique, one might wonder if there is any value to the notion. It turns out (as we shall shortly see) that the parity of the number of transposition factors that appear in any representation of a permutation σ as a product of transpositions depends only on σ, and this is what leads to the notion of even and odd permutations.

Our approach will be to define a map from S_n, where $n \geq 2$, into $\mathbb{Q}^*$, the group of all nonzero rational numbers under multiplication, and then prove that this map is a homomorphism with image $\{1, -1\}$. Under this homomorphism, the preimage of 1 will be the set of all even permutatations, while the preimage of -1 will be the set of all odd permutations.

7.4.4 Definition. *Let $n \geq 2$ be an integer. For any $\sigma \in S_n$, let*

$$S(\sigma) = \prod_{1 \leq i < j \leq n} \frac{\sigma(i) - \sigma(j)}{i - j}.$$

Note that in this product, $i - j \neq 0$ for every i and j under consideration, and thus $S(\sigma) \in \mathbb{Q}$. As well, for every i and j under consideration, $i \neq j$ and σ is a permutation, so $\sigma(i) \neq \sigma(j)$. Thus $S(\sigma) \in \mathbb{Q}^*$ for every $\sigma \in S_n$.

For example, for $\sigma = (1\,2)(3\,4) \in S_4$, we have

$$S(\sigma) = \prod_{1 \leq i < j \leq 4} \frac{\sigma(i) - \sigma(j)}{i - j}$$

$$= \left(\frac{\sigma(1) - \sigma(2)}{1 - 2}\right)\left(\frac{\sigma(1) - \sigma(3)}{1 - 3}\right)\left(\frac{\sigma(1) - \sigma(4)}{1 - 4}\right)$$
$$\left(\frac{\sigma(2) - \sigma(3)}{2 - 3}\right)\left(\frac{\sigma(2) - \sigma(4)}{2 - 4}\right)\left(\frac{\sigma(3) - \sigma(4)}{3 - 4}\right)$$

$$= \left(\frac{2-1}{1-2}\right)\left(\frac{2-4}{1-3}\right)\left(\frac{2-3}{1-4}\right)\left(\frac{1-4}{2-3}\right)\left(\frac{1-3}{2-4}\right)\left(\frac{4-3}{3-4}\right)$$

$$= \left(\frac{2-1}{1-2}\right)\left(\frac{2-4}{2-4}\right)\left(\frac{1-4}{1-4}\right)\left(\frac{2-3}{2-3}\right)\left(\frac{1-3}{1-3}\right)\left(\frac{4-3}{3-4}\right)$$

$$= (-1)(-1) = 1.$$

7.4.5 Proposition. *For any integer $n > 1$, the map $S : S_n \to \mathbb{Q}^*$ is a homomorphism.*

Proof. First, observe that for any $\sigma \in S_n$, for any $i, j \in J_n$ with $i \neq j$, $\dfrac{\sigma(i) - \sigma(j)}{i - j} = \dfrac{\sigma(j) - \sigma(i)}{j - i}$. Let $P = \{\, \{i, j\} \mid i, j \in J_n,\ i \neq j \,\}$ denote the set of all subsets of size 2 with elements from J_n. Then for any $\sigma \in S_n$,

$$S(\sigma) = \prod_{\{i,j\} \in P} \frac{\sigma(i) - \sigma(j)}{i - j}.$$

Let $\sigma, \tau \in S_n$. Then

$$
\begin{aligned}
S(\sigma\tau) &= \prod_{\{i,j\} \in P} \frac{\sigma\tau(i) - \sigma\tau(j)}{i - j} \\
&= \prod_{\{i,j\} \in P} \left(\frac{\sigma(\tau(i)) - \sigma(\tau(j))}{\tau(i) - \tau(j)} \right) \left(\frac{\tau(i) - \tau(j)}{i - j} \right) \\
&= \prod_{\{i,j\} \subset P} \left(\frac{\sigma(\tau(i)) - \sigma(\tau(j))}{\tau(i) - \tau(j)} \right) \prod_{\{i,j\} \in P} \left(\frac{\tau(i) - \tau(j)}{i - j} \right).
\end{aligned}
$$

Since τ is a permutation, it is both injective and surjective, and so as $\{i, j\}$ ranges through P, so does $\{\tau(i), \tau(j)\}$; that is,

$$P = \{\, \{\tau(i), \tau(j)\} \mid i, j \in J_n,\ i \neq j \,\}.$$

Thus $\prod_{\{i,j\} \in P} \left(\frac{\sigma(\tau(i)) - \sigma(\tau(j))}{\tau(i) - \tau(j)} \right) = \prod_{\{i,j\} \in P} \left(\frac{\sigma(i) - \sigma(j)}{i - j} \right) = S(\sigma)$, and so $S(\sigma\tau) = S(\sigma)S(\tau)$. Thus S is a homomorphism from S_n into $\mathbb{Q}^*$. $\quad\square$

Our next major step is to prove that for any transposition τ, $S(\tau) = -1$. This will be facilitated by the following observation.

7.4.6 Corollary. *Let $n \geq 2$ be an integer. Then for any $\sigma, \tau \in S_n$, $S(\sigma\tau\sigma^{-1}) = S(\sigma)$.*

Proof. By Proposition 7.4.5 (and the fact that multiplication of rational numbers is commutative), $S(\sigma\tau\sigma^{-1}) = S(\sigma)S(\tau)S(\sigma^{-1}) = S(\sigma)S(\sigma^{-1})S(\tau) = S(\sigma\sigma^{-1})S(\tau) = S(\mathbb{1}_{J_n})S(\tau)$. Since S is a homomorphism, $S(\mathbb{1}_{J_n}) = 1$ and thus $S(\sigma\tau\sigma^{-1}) = S(\tau)$. $\quad\square$

We are almost ready to prove that the image under S of any transposition is -1. Our approach will be to prove that $S(1\,2) = -1$, and then use Corollary 7.4.6 together with the fact that any two 2-cycles are conjugate in S_n. More generally, we have the following result.

7.4.7 Proposition. *Let n and k be integers with $1 < k \leq n$. For any $\sigma \in S_n$, the conjugate of any k-cycle in S_n by σ is a k-cycle. Specifically, if $(j_1\, j_2 \cdots j_k)$ is a k-cycle in S_n, then*

$$\sigma(j_1\, j_2 \cdots j_k)\sigma^{-1} = (\sigma(j_1)\, \sigma(j_2) \cdots \sigma(j_k)).$$

Moreover, any two k-cycles in S_n are conjugate in S_n.

Proof. Let t be an integer for whihc $1 \leq t < k$. Then

$$\sigma(j_1\, j_2 \cdots j_k)\sigma^{-1}(\sigma(j_t)) = \sigma(j_1\, j_2 \cdots j_k)(j_t) = \sigma(j_{t+1}),$$

while $\sigma(j_1\, j_2 \cdots j_k)\sigma^{-1}(\sigma(j_k)) = \sigma(j_1)$. For any $i \in J_n$ for which $i \notin \{\sigma(j_1), \sigma(j_2), \ldots, \sigma(j_k)\}$, it follows that $\sigma^{-1}(i) \notin \{j_1, j_2, \ldots, j_k\}$ and thus $\sigma(j_1\, j_2 \cdots j_k)\sigma^{-1}(i) = \sigma(\sigma^{-1}(i)) = i$. This proves that $M_{\sigma(j_1\, j_2 \cdots j_k)\sigma^{-1}} = \{\sigma(j_1), \sigma(j_2), \ldots, \sigma(j_k)\}$, and that $\sigma(j_1\, j_2 \cdots j_k)\sigma^{-1} = (\sigma(j_1)\, \sigma(j_2) \cdots \sigma(j_k))$.

Now suppose that $(i_1\, i_2 \cdots i_k)$ and $(j_1\, j_2 \cdots j_k)$ are any two k-cycles in S_n. Let $\sigma \in S_n$ be any permutation for which $\sigma(i_t) = j_t$ for each $t = 1, 2, \ldots, k$ (as there are $(n-k)!$ bijective mappings between any two sets of size $n-k$, in particular, there are $(n-k)!$ bijective mappings between $J_n - \{i_1, i_2, \ldots, i_k\}$ and $J_n - \{j_1, j_2, \ldots, j_k\}$, each of which gives rise to a permutation σ with the desired property). Then $\sigma(i_1\, i_2 \cdots i_k)\sigma^{-1} = (\sigma(i_1)\, \sigma(i_2) \cdots \sigma(i_k)) = (j_1\, j_2 \cdots j_k)$. $\qquad\square$

7.4.8 Proposition. *Let $n > 1$ be an integer. Then for any transposition $\tau \in S_n$, $S(\tau) = -1$.*

Proof. If τ is any 2-cycle in S_n, then by Proposition 7.4.7, τ is conjugate to $(1\,2)$, and so by Corollary 7.4.6, it suffices to prove that $S(1\,2) = -1$. The proof will be by induction on $n \geq 2$. For $n = 2$, $S(1\,2) = \frac{\tau(1)-\tau(2)}{1-2} = -1$, and for $n = 3$, $S(1\,2) = \frac{\tau(1)-\tau(2)}{1-2}\frac{\tau(1)-\tau(3)}{1-3}\frac{\tau(2)-\tau(3)}{2-3} = \frac{(2-1)(2-3)(1-3)}{(1-2)(1-3)(2-3)} = -1$. Suppose now that $n \geq 3$ is such that in S_n, $S(1\,2) = -1$, and consider $S(\tau)$

for $\tau = (1\,2)$ in S_{n+1}. We have

$$S(1\,2) = \prod_{1 \le i < j \le n+1} \frac{\tau(i) - \tau(j)}{i - j}$$

$$= \prod_{1 \le i < j \le n} \frac{\tau(i) - \tau(j)}{i - j} \prod_{1 \le i \le n} \frac{\tau(i) - \tau(n+1)}{i - (n+1)}$$

$$= (-1) \prod_{1 \le i \le n} \frac{\tau(i) - (n+1)}{i - (n+1)}$$

$$= (-1) \frac{2 - (n+1)}{1 - (n+1)} \frac{1 - (n+1)}{2 - (n+1)} \prod_{3 \le i \le n} \frac{i - (n+1)}{i - (n+1)}$$

$$= -1.$$

$\square$

7.4.9 Corollary. *The image of S is $\{1, -1\}$.*

Proof. By Corollary 7.4.3, every permutation can be written as a product of transpositions, and by Proposition 7.4.8, if τ is a transposition, then $S(\tau) = -1$. Thus for every $\sigma \in S_n$, $S(\sigma) = 1$ or -1. Since $n > 1$, S_n contains transpositions and so $-1 \in S(S_n)$. Thus the image of S is the two-element set $\{1, -1\}$. $\square$

7.4.10 Definition. *Let $n > 1$ be an integer. Then $\sigma \in S_n$ is said to be even if $S(\sigma) = 1$, otherwise σ is said to be odd. We shall henceforth write $\mathrm{sgn}(\sigma) = S(\sigma)$, and refer to $\mathrm{sgn}(\sigma)$ as the sign of the permutation σ.*

A few observations are in order. Firstly, since S is a homomorphism, it follows that if σ is even, then every way of writing σ as a product of transpositions must have an even number of factors, and similarly, if σ is odd, then every way of writing σ as a product of transpositions must have an odd number of factors. Secondly, the set of even permutations is the kernel of S, and as S has image $\{1, -1\}$, there are two cosets of $\ker(S)$, the kernel itself, and the set of odd permutations. Since all cosets of any subgroup of a finite group have the same size as the subgroup itself, it follows that the number of even permutations is equal to the number of odd permutations, and since $|S_n| = n!$, we deduce that the number of even permutations is $n!/2$. Finally, note that when $n = 1$, the symmetric group S_1 contains only the identity map, and $S: S_1 \to \mathbb{Q}^*$ is the map which sends $\mathbb{1}_{J_1}$ to 1. Thus in this case as well, S is a homomorphism, but the image is only $\{1\}$.

Note also that by Proposition 7.4.2, a k-cycle can be written as the product of $k-1$ transpositions, and thus if σ is a k-cycle, $S(\sigma) = (-1)^{k-1}$. It follows that a k-cycle is even if, and only if, k is odd. Thus the parity of an arbitrary permutation can be determined immediately from its disjoint cycle representation, as the product of two even permutations or two odd permutations is even, while the product of an even and an odd permutation is odd.

7.4.11 Definition. *For any positive integer n, the alternating group of degree n, denoted by A_n, is defined by $A_n = \ker(S)$.*

As we have observed above, $|A_n| = |\ker(S)| = n!/2$.

The fact that a permutation has a unique representation as a product of disjoint cycles allow us to classify permutations according to their cycle structure.

7.4.12 Definition. *Let $n > 1$ be an integer. For $\sigma \in S_n - \{\mathbb{1}_{J_n}\}$, let $\sigma = \sigma_1 \sigma_2 \cdots \sigma_k$ be a disjoint cycle decomposition of σ. The cycle structure s_σ of σ is the sequence formed by sorting the list $l(\sigma_1), l(\sigma_2), \ldots, l(\sigma_k)$ in non-decreasing order, where for each integer i with $1 \le i \le k$, $l(\sigma_i)$ is the length of the cycle σ_i. The cycle structure of $\mathbb{1}_{J_n}$ might reasonably be defined as the sequence $(1, 1, 1, \ldots, 1)$ of length n, but for the sake of simplicity, we shall define it to be $()$, the empty sequence. For $\sigma, \tau \in S_n$, we say that σ and τ have the same cycle structure if $s_\sigma = s_\tau$.*

It is possible to count the number of elements of S_n with a given cycle structure. To begin with, let us determine the number of k-cycles in S_n. Suppose that $2 \le k \le n$. There are $n(n-1) \cdots (n-(k-1)) = n!/(n-k)!$ sequences of length k with entries chosen without repetition from J_n. If $j_1, j_2, \ldots, j_k$ is such a sequence, we form the k-cycle $(j_1\, j_2 \cdots j_k)$. However, due to the nature of the cycle notation,

$$(j_1\, j_2 \cdots j_k) = (j_2 \cdots j_k\, j_1) = \cdots = (j_k\, j_1 \cdots j_{k-1}),$$

so each cycle corresponds to k different sequences. Thus the actual number of cycles is $\frac{1}{k}\frac{n!}{(n-k)!}$. For example, in S_4, there are $\frac{1}{3}\frac{4!}{(4-3)!} = (4)(2) = 8$ 3-cycles; namely $(1\,2\,3), (1\,3\,2), (1\,2\,4), (1\,4\,2), (1\,3\,4), (1\,4\,3), (2\,3\,4), (2\,4\,3)$. In S_5, the number of 3-cycles is $\frac{1}{3}\frac{5!}{(5-3)!} = \frac{(5)(4)(3)}{3} = 15$. Still in S_5, the number of permutations with cycle structure $(2, 2)$; that is, the number of permutations that are a product of two disjoint 2-cycles, is $\binom{5}{2}\binom{3}{2}/2$, since there are $\binom{5}{2}$ 2-cycles in S_5, and $\binom{3}{2}$ ways to choose a second 2-cycle disjoint from the first (we must then divide by 2 in order to count the number of ways of choosing two dijoint 2-cycles, rather than the number of ways of choosing an ordered pair of disjoint 2-cycles–remember, disjoint cycles commute). Thus there are $(10)(3)/2 = 15$ permutations in S_5 with cycle structure $(2, 2)$.

7.4.13 Example. *The possible cycle structures for permutations in S_4 are (), (2), (2, 2), (3), and (4). In Figure 7.3, for each possible cycle structure, we list the number of permutations in S_4 with that cycle structure, the order of any permutation with that cycle structure, and the parity of each such permutation. The most interesting calculation is the number of permutations with cycle structure $(2, 2)$. By an analysis similar to that above for the number of permutations with cycle structure $(2, 2)$ in S_5, we find this to be $\binom{4}{2}\binom{2}{2}/2 = (6)(1)/2 = 3$.*

Cycle Structure	Number	Order	Parity
()	1	1	even
(2)	6	2	odd
(2,2)	3	2	even
(3)	8	3	even
(4)	6	4	odd

Fig. 7.3   Cycle structure of S_4

It turns out that conjugation (see Definition 6.3.6) is the key to understanding the classification of permutations according to their cycle structure. Accordingly, we shall now investigate conjugation in S_n. We have already seen in Proposition 7.4.7 that any two cycles of the same length are conjugate in S_n, and that any conjugate of a cycle is a cycle of the same length. Our next result is a very striking generalization of this observation.

7.4.14 Proposition. *Let n be a positive integer. For any $\sigma, \tau \in S_n$, σ is conjugate to τ in S_n if, and only if, $s_\sigma = s_\tau$; that is, two permutations in S_n are conjugate in S_n if, and only, they have the same cycle structure.*

Proof. Let $\sigma, \tau \in S_n$, and suppose that there exists $\lambda \in S_n$ such that $\lambda \sigma \lambda^{-1} = \tau$. If $\sigma = \mathbb{1}_{J_n}$, then $\tau = \mathbb{1}_{J_n} = \sigma$. Suppose then that $\sigma \neq \mathbb{1}_{J_n}$, and let $\sigma_1, \sigma_2, \ldots, \sigma_k$ be pairwise disjoint cycles such that $\sigma = \sigma_1 \sigma_2 \cdots \sigma_k$.

Then

$$\lambda \sigma \lambda^{-1} = \lambda \sigma_1 \sigma_2 \cdots \sigma_k \lambda^{-1}$$
$$= \lambda \sigma_1 \lambda^{-1} \lambda \sigma_2 \lambda^{-1} \lambda \cdots \lambda^{-1} \lambda \sigma_k \lambda^{-1}.$$

By Proposition 7.4.7, for each $i = 1, 2, \ldots, k$, $\lambda \sigma_i \lambda^{-1}$ is a cycle of the same length as σ_i. We prove that for any i and j with $1 \leq i, j \leq k$ and $i \neq j$, $\lambda \sigma_i \lambda^{-1}$ and $\lambda \sigma_j \lambda^{-1}$ are disjoint; that is, $M_{\lambda \sigma_i \lambda^{-1}} \cap M_{\lambda \sigma_j \lambda^{-1}} = \varnothing$. Suppose by way of contradiction that there are $i, j \in J_k$ with $i \neq j$ for which there is $t \in M_{\lambda \sigma_i \lambda^{-1}} \cap M_{\lambda \sigma_j \lambda^{-1}}$. Then $\lambda \sigma_i \lambda^{-1}(t) \neq t$ and $\lambda \sigma_j \lambda^{-1}(t) \neq t$, from which we deduce that for $r = \lambda^{-1}(t)$, $\sigma_i(r) \neq r$ and $\sigma_j(r) \neq r$; that is, $r \in M_{\sigma_i} \cap M_{\sigma_j} = \varnothing$. Since this is not possible, we conclude that for all $i, j \in J_k$, $\lambda \sigma_i \lambda^{-1}$ and $\lambda \sigma_j \lambda^{-1}$ are disjoint, and so σ and τ have the same cycle structure.

For the converse, suppose that $\sigma, \tau \in S_n$ are such that $s_\sigma = s_\tau = (l_1, l_2, \ldots, l_k)$, so that a disjoint cycle decomposition for σ and a disjoint cycle decomposition for τ will each have k factors. We are to prove that σ is conjugate to τ in S_n. Let $\sigma = \alpha_1 \alpha_2 \cdots \alpha_k$ and $\tau = \beta_1 \beta_2 \cdots \beta_k$ be disjoint cycle representations of σ and τ, respectively. Since disjoint cycles commute, we may assume without loss of generality that for each $i = 1, 2, \ldots, k$, $|M_{\alpha_i}|$, the length of α_i, is the same as $|M_{\beta_i}|$, the length of β_i. Note that $M_\sigma = \cup_{i=1}^k M_{\alpha_i}$ and $M_\tau = \cup_{i=1}^k M_{\beta_i}$ have the same size, and thus $J_n - M_\sigma$ and $J_n - M_\tau$ have the same size. Define $\pi \in S_n$ piecemeal as follows: first, choose any bijective mapping $\gamma : J_n - M_\sigma$ to $J_n - M_\tau$, and set $\pi|_{J_n - M_\sigma} = \gamma$, while for each $i = 1, 2, \ldots, k$, the restriction of π to M_{α_i} is to be a bijective mapping from M_{α_i} to M_{β_i} obtained by writing $\alpha_i = (a_1 a_2 \cdots a_{l_i})$, $\beta_i = (b_1 b_2 \cdots b_{l_i})$, and setting $\pi(a_t) = b_t$ for $t = 1, 2, \ldots, l_i$. Note that by Proposition 7.4.7, for each $i = 1, 2, \ldots, k$, $\pi \alpha_i \pi^{-1} = \beta_i$, and thus $\pi \sigma \pi^{-1} = \pi \alpha_1 \alpha_2 \cdots \alpha_k \pi^{-1} = \pi \alpha_1 \pi^{-1} \pi \alpha_2 \pi^{-1} \pi \cdots \pi^{-1} \pi \alpha_k \pi^{-1} = \beta_1 \beta_2 \cdots \beta_k = \tau$. $\qquad \square$

We remark that many choices were made in the construction of π in the preceding proof, and so in general, there are many permutations π for which $\pi \sigma \pi^{-1} = \tau$.

7.4.15 Example. *Let* $\sigma = (1\,2)(3\,5\,10)(12\,4\,6\,7)$ *and* $\tau = (10\,11)(2\,4\,3)(9\,8\,1\,5)$ *in* S_{12}. *Note that* $s_\sigma = (2, 3, 4) = s_\tau$, *so by Proposition 7.4.14, σ and τ are conjugate in* S_{12}. *We follow the steps in the proof of Proposition 7.4.14 to construct* $\pi \in S_{12}$ *such that* $\pi \sigma \pi^{-1} = \tau$. *We have* $M_\sigma = \{1, 2, 3, 4, 5, 6, 7, 10, 12\}$ *and* $M_\tau = \{1, 2, 3, 4, 5, 8, 9, 10, 11\}$, *so* $J_{12} - M_\sigma = \{8, 9, 11\}$ *and* $J_{12} - M_\tau = \{6, 7, 12\}$. *Choose* $\gamma = \begin{pmatrix} 8 & 9 & 11 \\ 6 & 7 & 12 \end{pmatrix}$, *a* bijection from $J_{12} - M_\sigma$ to $J_{12} - M_\tau$. The number of cycles in the disjoint cycle decomposition of σ and of τ is $k = 3$. Let $\alpha_1 = (1\,2)$, $\alpha_2 = (3\,5\,10)$, $\alpha_3 = (12\,4\,6\,7)$, $\beta_1 = (10\,11)$, $\beta_2 = (2\,4\,3)$, and $\beta_3 = (9\,8\,15)$. Then we construct $\delta_1 = \left(\begin{smallmatrix} 1 & 2 \\ 10 & 11 \end{smallmatrix}\right)$, $\delta_2 = \left(\begin{smallmatrix} 3 & 5 & 10 \\ 2 & 4 & 3 \end{smallmatrix}\right)$, and $\delta_3 = \left(\begin{smallmatrix} 12 & 4 & 6 & 7 \\ 9 & 8 & 1 & 5 \end{smallmatrix}\right)$ to obtain $\pi = \left(\begin{smallmatrix} 1 & 2 & 3 & 4 & 5 & 6 & 7 & 8 & 9 & 10 & 11 & 12 \\ 10 & 11 & 2 & 8 & 4 & 1 & 5 & 6 & 7 & 3 & 12 & 9 \end{smallmatrix}\right) = (1\,10\,3\,2\,11\,12\,9\,7\,5\,4\,8\,6)$. By contruction, $\pi\sigma\pi^{-1} = \tau$. Note that in this instance, π turned out to be cycle. This is not always the case.

As our last result of this section, we show that the converse of Lagrange's theorem does not hold. For a given finite group G, Lagrange's theorem can be stated in the following form: for every positive integer k, if there exists a subgroup H of G with $|H| = k$, then k divides $|G|$. In this form, the converse would be: for every positive integer k, if k divides $|G|$, then there exists a subgroup H of G with $|H| = k$.

7.4.16 Proposition. *The alternating group A_4 has order 12, but has no subgroup of order 6.*

Proof. Suppose by way of contradiction that A_4 has a subgroup H of order 6. Then $[A_4 : H] = 2$, and so by Lemma 6.4.2, $H \lhd A_4$. Now, A_4 consists of the identity, the eight 3-cycles, and the three products of two disjoint 2-cycles. Since $|H| = 6$, H must contain at least one 3-cycle. Suppose that $(i\,j\,k)$ is a 3-cycle in H. Since $H \lhd A_4$, any conjugate of $(i\,j\,k)$ by any element of A_4 will belong to H. In particular, every conjugate of $(i\,j\,k)$ by a 3-cycle will be in H. As well, the inverse of any element of H is in H. Let $J_4 - \{i, j, k\} = \{l\}$. We find that in addition to the identity and the 3-cycle $(i\,j\,k)$ and its inverse $(i\,k\,j)$, H must contain the following six elements:

$$(i\,j\,l)(i\,j\,k)(i\,j\,l)^{-1} = (j\,l\,k), \ (j\,l\,k)^{-1} = (j\,k\,l)$$
$$(j\,i\,l)(i\,j\,k)(j\,i\,l)^{-1} = (l\,i\,k) = (i\,k\,l), \ (i\,k\,l)^{-1} = (i\,l\,k)$$
$$(j\,l\,k)(i\,j\,k)(j\,l\,k)^{-1} = (i\,l\,j), \ (i\,l\,j)^{-1} = (i\,j\,l).$$

Thus $6 = |H| \geq 9$, which is impossible. It follows that A_4 has no subgroup of order 6. $\square$

7.5   The Simplicity of A_n

We shall prove that for every integer $n \geq 5$, A_n is simple, and we shall thoroughly explore A_n for integers n with $1 \leq n \leq 4$. We shall want to use Cauchy's theorem, which states that if a prime p divides the order of a finite group G, then G contains an element of order p. Cauchy's theorem is a general fact about finite groups, but it is easily proven with the facts about permutations that we have established in Section 7.3.

7.5.1 Proposition. *Let p be a prime, and let X be a finite set for which* $p \,|\, |X|$. *Let $\sigma \in S_X$ be an element of order p. If $|M_\sigma| < |X|$, then $|M_\sigma| \leq |X| - p$.*

Proof. By Proposition 7.3.17, the disjoint cycle decomposition of σ presents σ as a product of disjoint p-cycles, and so $|M_\sigma|$ is a multiple of p. Since p divides $|X|$, it follows that p divides $|X - M_\sigma| = |X| - |M_\sigma|$, say $|X - M_\sigma| = kp$. Thus if $X \neq M_\sigma$, then we have $|X - M_\sigma| = kp > 0$, which means that $k \geq 1$ and thus $|X - M_\sigma| = kp \geq p$.   □

7.5.2 Theorem (Cauchy's Theorem). *Let G be a finite group, and let p be a prime divisor of $|G|$. Then G contains an element of order p.*

Proof. Let $X = \{\,(g_1, g_2, \ldots, g_p) \in G^p \mid g_1 g_2 \cdots g_p = e\,\}$, where e is the identity of G. Note that for any $(g_1, g_2 \ldots, g_{p-1}) \in G^{p-1}$, if we let $g = g_1 g_2 \cdots g_{p-1}$, then $(g_1, g_2, \ldots, g_{p-1}, g^{-1}) \in X$, and every element of X is obtained this way. Thus $|X| = |G|^{p-1}$, and since p is a divisor of $|G|$, it follows that p is a divisor of X. Note that for any $x, y \in G$, if $xy = e$, then $y = x^{-1}$ and so $yx = e$ as well. Thus if $(g_1, g_2, \ldots, g_p) \in X$, then $(g_1 g_2 \cdots g_{p-1}) g_p = e$ and so $g_p (g_1 g_2 \cdots g_{p-1}) = e$, whereby we have that $(g_p, g_1, g_2, \ldots, g_{p-1}) \in X$. Thus we have a map $\sigma : X \to X$ defined by $\sigma(g_1, g_2, \ldots, g_p) = (g_p, g_1, g_2, \ldots, g_{p-1})$ for each $(g_1, g_2, \ldots, g_p) \in X$. It is immediate from the definition that σ is bijective, so $\sigma \in S_X$. Moreover, since the effect of σ on a p-tuple is to cyclically permute the p-tuple by one position, it follows that $\sigma^p = \mathbb{1}_X$. Thus $|\sigma|$ divides p, and since $\sigma \neq \mathbb{1}_X$, it follows that $|\sigma| = p$. Observe that $(g_1, g_2, \ldots, g_p) \in X - M_\sigma$ if, and only if, $g_1 g_2 \cdots g_p = e$ and $g_1 = g_2 = \cdots = g_p$. In particular, $(e, e, \ldots, e) \in X - M_\sigma$, so $X \neq M_\sigma$. By Proposition 7.5.1, $|M_\sigma| \leq |X| - p$. Since p is a prime, $p \geq 2$ and thus $|M_\sigma| \leq |X| - p \leq |X| - 2$; equivalently, $|X - M_\sigma| \geq 2$. Thus there is at least one $g \in G$ with $g \neq e$ such that $g^p = e$, and so we conclude that G has an element of order p.   □

As an interesting consequence of Cauchy's theorem, we observe that if G is a finite group for which there exists a prime p such that every element of G has order a power of p, then there exists a positive integer k such that $|G| = p^k$.

We are ready now to proceed with our investigation of the alternating groups. Note that $S_1 = A_1$ is the trivial group, and S_2 is cyclic of order 2, while A_2 is trivial, so we need only investigate the situation for $n \geq 3$. We have observed that for any integer $n > 1$, A_n consists of all even permutations, and since every permutation can be written as the product of transpositions, it follows that every permutation in A_n can be written as the product of an even number of transpositions. In particular, for $n \geq 3$, A_n is generated, as a subgroup of S_n, by the set of all products of two disjoint transpositions.

Let $n \geq 3$ be an integer. If α and β are two different transpositions, then $M_\alpha \cap M_\beta$ is either empty or a singleton. If i, j, k, m are distinct elements of J_n, then $(i\,j)(j\,k) = (ijk)$, while $(i\,j)(k\,m) = (i\,j)(j\,k)(j\,k)(k\,m) = (i\,j\,k)(j\,k\,m)$. Thus every element of A_n can be written as a product of 3-cycles; that is, A_n is generated by the set of all 3-cycles in S_n.

Recall that for any group G and subgroup H of G, the normalizer of H in G, denoted by $N_G(H)$, is the subgroup $\{\, g \in G \mid gHg^{-1} = H \,\}$. We shall show that if H is a nontrivial subgroup of S_n such that $A_n \subseteq N_{S_n}(H)$, then either H contains a 3-cycle, or else $n = 4$ and $H = \{\, \mathbb{1}_{J_4}, (1\,2)(3\,4), (1\,3)(2\,4), (1\,4)(2\,3) \,\}$.

We shall rely heavily on the fact that if H is a subgroup of S_n that is normalized by A_n (that is, $A_n \subseteq N_{S_n}(H)$), then for any $\sigma \in H$ and any 3-cycle $\alpha = (a_1\,a_2\,a_3) \in H$, we have $\sigma\alpha\sigma^{-1} = (\sigma(a_1)\,\sigma(a_2)\,\sigma(a_3))$ and thus $\sigma\alpha\sigma^{-1}\alpha^{-1} = (\sigma(a_1)\,\sigma(a_2)\,\sigma(a_3))(a_1\,a_3\,a_2) \in H$. Our investigations will make extensive use of this observation, where in each instance, our choice of a particular α will depend on the structure and entries of σ.

7.5.3 Proposition. *Let $n \geq 3$. If H is a nontrivial subgroup of S_n such that $\alpha H \alpha^{-1} = H$ for all $\alpha \in A_n$, then either H contains a 3-cycle or else $n = 4$ and $H = \{\, \mathbb{1}_{J_4}, (1\,2)(3\,4), (1\,3)(2\,4), (1\,4)(2\,3) \,\}$, which is a normal subgroup of S_4 contained in A_4.*

Proof. Suppose that H does not contain any 3-cycle. Then we are to prove that $n = 4$ and $H = \{\, \mathbb{1}_{J_4}, (1\,2)(3\,4), (1\,3)(2\,4), (1\,4)(2\,3) \,\}$.

Let p be the largest prime divisor of $|H|$. By Cauchy's theorem, H contains an element σ of order p. Then the disjoint cycle structure of σ

presents σ as a product of disjoint p-cycles. We consider three cases.

Case 1: $p > 3$. Let $(a_1 \, a_2 \cdots a_p)$ be one of the p-cycles in the disjoint cycle representation of σ, so that $\sigma = (a_1 \, a_2 \cdots a_p)\gamma$ for some $\gamma \in S_n$ disjoint from $(a_1 \, a_2 \cdots a_p)$. Let $\alpha = (a_1 \, a_2 \, a_3)$. Then $\alpha \in A_n$ and we have

$$\sigma\alpha\sigma^{-1}\alpha^{-1} = (\sigma(a_1)\,\sigma(a_2)\,\sigma(a_3))(a_1 \, a_3 \, a_2)$$

$$= (a_2 \, a_3 \, a_4)(a_1 \, a_3 \, a_2) = (a_1 \, a_4 \, a_2).$$

Thus $(a_1 \, a_4 \, a_2) = \sigma(\alpha\sigma^{-1}\alpha^{-1}) \in H$. Since H does not contain any 3-cycles, this is impossible, so it must be that $p \le 3$.

Case 2: $p = 3$. Then σ is a product of disjoint 3-cycles, and since H does not contain any 3-cycles, there are disjoint 3-cycles $(a_1 \, a_2 \, a_3)$ and $(a_4 \, a_5 \, a_6)$ such that $\sigma = (a_1 \, a_2 \, a_3)(a_4 \, a_5 \, a_6)\gamma$ for some $\gamma \in S_n$ disjoint from $(a_1 \, a_2 \, a_3)$ and $(a_4 \, a_5 \, a_6)$. Let $\alpha = (a_1 \, a_2 \, a_4)$. Then $\alpha \in A_n$ and we have

$$\sigma\alpha\sigma^{-1}\alpha^{-1} = (\sigma(a_1)\,\sigma(a_2)\,\sigma(a_4))(a_1 \, a_4 \, a_2)$$

$$= (a_2 \, a_3 \, a_5)(a_1 \, a_4 \, a_2) = (a_1 \, a_4 \, a_3 \, a_5 \, a_2).$$

Thus $(a_1 \, a_4 \, a_3 \, a_5 \, a_2) = \sigma(\alpha\sigma^{-1}\alpha^{-1}) \in H$. However, this means that H contains a 5-cycle, and so 5 divides the order of H. Since this contradicts the fact that 3 is the largest prime divisor of $|H|$, we see that this case does not occur.

Case 3: $p = 2$. Since p is the largest prime divisor of $|H|$, it follows that the order of H is some power of 2. We claim that H does not contain any transpositions. Suppose to the contrary that H contains a transposition $(a_1 \, a_2)$. Since $n \ge 3$, there exists $a_3 \in J_n - \{a_1, a_2\}$. Then for $\alpha = (a_1 \, a_2 \, a_3) \in A_n$, we have $(a_1 \, a_2)\alpha(a_1 \, a_2)^{-1}\alpha^{-1}) = (a_1 \, a_2)(a_1 \, a_2 \, a_3)(a_1 \, a_2)^{-1}(a_1 \, a_3 \, a_2) = (a_2 \, a_1 \, a_3)(a_1 \, a_3 \, a_2) = (a_1 \, a_3 \, a_2)^2 = (a_1 \, a_2 \, a_3)$. Since $(a_1 \, a_2) \in H$ and $\alpha \in A_n$, it follows that $(a_1 \, a_2 \, a_3) = (a_1 \, a_2)\alpha(a_1 \, a_2)^{-1}\alpha^{-1} \in H$, which is not possible since H does not contain any 3-cycle. Thus H does not contain any transposition. Now, since σ is an element of H of order 2, and H does not contain any transpositions, σ is a product of at least two disjoint transpositions (which implies in particular that $n \ge 4$), so there exist disjoint transpositions $(a_1 \, a_2)$ and $(a_3 \, a_4)$ such that $\sigma = (a_1 \, a_2)(a_3 \, a_4)\gamma$ where γ is disjoint from $(a_1 \, a_2)$ and $(a_3 \, a_4)$.

Suppose first that $n \ge 5$. Then there exists $a_5 \in J_n - \{a_1, a_2, a_3, a_4\}$. Let $\alpha = (a_1 \, a_3 \, a_5)$. Then $\alpha \in A_n$ and

$$\sigma\alpha\sigma^{-1}\alpha^{-1} = (\sigma(a_1)\,\sigma(a_3)\,\sigma(a_5))(a_1 \, a_5 \, a_3)$$

$$= (a_2 \, a_4 \, \gamma(a_5))(a_1 \, a_5 \, a_3).$$

Since γ is disjoint from $(a_1 \, a_2)$ and $(a_3 \, a_4)$, either $\gamma(a_5) = a_5$ or else $\gamma(a_5) = a_6 \notin \{a_1, a_2, a_3, a_4, a_5\}$. In the latter case, $(a_2 \, a_4 \, a_6)(a_1 \, a_5 \, a_3) =$

$\sigma \alpha \sigma^{-1} \alpha^{-1} \in H$, so H contains a product of two disjoint 3-cycles, which is an element of order 3, and this implies that 3 divides $|H|$, while in the former case, $(a_1\, a_2\, a_4\, a_5\, a_3) = (a_2\, a_4\, a_5)(a_1\, a_5\, a_3) = \sigma \alpha \sigma^{-1} \alpha^{-1} \in H$, so H contains an element of order 5 and thus 5 divides $|H|$. But H is a 2-group, so neither 3 nor 5 divides $|H|$. Thus $4 \le n < 5$, which means that $n = 4$. Since $\sigma = (a_1\, a_2)(a_3\, a_4)\gamma \in H$ and γ is disjoint from $(a_1\, a_2)$ and $(a_3\, a_4)$, it follow that γ is the identity. Thus $\sigma = (a_1\, a_2)(a_3\, a_4) \in H$. Recall that H does not contain any transpositions. Since the only elements of S_4 that have order a power of 2 are the transpositions and the three products of two disjoint transpositions, of which at least one belongs to H, we see that it suffices to prove that H contains a second product of two disjoint transpositions in order to conclude that $|H| \ge 3$ and so $H = \{\, e, (1\,2)(3\,4), (1\,3)(2\,4), (1\,4)(2\,3) \,\}$. Since $\alpha = (a_1\, a_2\, a_3) \in A_4$, we have

$$(a_1\, a_2\, a_3)(a_1\, a_2)(a_3\, a_4)(a_1\, a_2\, a_3)^{-1} = (a_2\, a_3)(a_1\, a_4) \in H,$$

as required. $\qquad\square$

7.5.4 Theorem. *For $n \ge 5$, A_n is simple, while*

$$\{\, \mathbb{1}_{J_4}, (1\,2)(3\,4), (1\,3)(2\,4), (1\,4)(2\,3) \,\}$$

is a normal subgroup of S_4 contained in A_4. A_3 is a cyclic group of order 3, hence simple, and A_2 is trivial.

Proof. There is nothing to do for $n = 2$, 3, or 4, so assume that $n \ge 5$. We prove that any two 3-cycles are conjugate in A_n (of course, they are conjugate in S_n, but it is not obvious that given two 3-cycles α, β, there is $\sigma \in A_n$ such that $\sigma \alpha \sigma^{-1} = \beta$). Choose a 3-cycle, say $(i\,j\,k)$, and let $(r\,s\,t)$ be any 3-cycle in S_n. We prove that there exists $\sigma \in A_n$ such that $\sigma(i\,j\,k)\sigma^{-1} = (r\,s\,t)$. Since $(i\,j\,k)$ and $(r\,s\,t)$ have the same cycle structure, there exists $\sigma \in S_n$ such that $\sigma(i\,j\,k)\sigma^{-1} = (r\,s\,t)$. If $\sigma \in A_n$, then we are done. Suppose that $\sigma \notin A_n$. Since $n \ge 5$, there exist $a, b \in J_n - \{\, r\,s\,t \,\}$ with $a \ne b$. Then $(a\,b)\sigma \in A_n$ and $(a\,b)\sigma(i\,j\,k)((a\,b)\sigma^{-1})^{-1} = (a\,b)(r\,s\,t)(a\,b) = (r\,s\,t)$, as required.

Now let H be a nontrivial normal subgroup of A_n. By Proposition 7.5.3, H contains a 3-cycle, and since we now know that any two 3-cycles are conjugate in A_n and H is normal in A_n, it follows that H contains all 3-cycles. Finally, since A_n is generated by the set of all 3-cycles, $A_n \subseteq H \subseteq A_n$ and so $H = A_n$. $\qquad\square$

7.5.5 Corollary. *For* $n = 1, 2$, $Z(S_n) = S_n$, *while for* $n \geq 3$, $Z(S_n) = \{1_{J_n}\}$.

Proof. S_1 is trivial, and S_2 is a cyclic group of order 2, hence abelian. Suppose now that $n \geq 3$. Suppose that $Z(S_n)$ is not trivial.

If $Z(S_n)$ contains a 3-cycle, then $Z(S_n) \cap A_n$ is a normal subgroup of S_n that contains a 3-cycle. But in S_n, the conjugacy class of any 3-cycle is the set of all 3-cycles, and A_n is generated by the set of all 3-cycles, so $A_n \subseteq Z(S_n)$. However, for $n \geq 3$, A_n is nonabelian, so this is not possible.

Suppose now that $Z(S_n)$ does not contain a 3-cycle. Since $Z(S_n)$ is normal in S_n, we may apply Proposition 7.5.3 to $Z(S_n)$ to conclude that $n = 4$ and $Z(S_4) = \{1_{J_4}, (12)(34), (13)(24), (14)(23)\} \subseteq A_4$. But then A_4 would have a cyclic quotient by a central subgroup, which would mean that A_4 is abelian. Since A_4 is not abelian, we have arrived at a contradiction to the assumption that $Z(S_n)$ is not trivial. $\qquad\square$

7.6   Exercises

1. Let $G = \langle g \rangle = \{g, g^2, g^3, g^4, g^5 = e\}$ be a cyclic group of order 5. The Cayley representation of G (see Proposition 7.2.1) provides a subgroup of S_G that is isomorphic to G. Use the proof of Proposition 7.2.1 to find this subgroup of S_G.

2. Determine the order and the parity of each of the following permutations. As well, write each one as a product of transpositions.

   a) $(3\,4\,7)(2\,3\,1\,5)(4\,7\,3\,5)(2\,1\,9\,8\,4\,7) \in S_9$.

   b) $(1\,2)(1\,5\,3)(2\,4\,3)(3\,6) \in S_6$.

   c) $(1\,3\,5\,7)(2\,4\,6\,3)(1\,2\,3\,4\,5\,6\,7) \in S_7$.

   d) $(1\,2\,3)(1\,3\,4\,2)(5\,4\,6\,1)(1\,3\,5) \in S_5$.

3. Construct a table similar to that shown in Figure 7.3 to provide the corresponding information for S_5.

4. Let m and n be relatively prime positive integers.

   a)   Let G be a group, and let $g \in G$ be an element of order mn. Prove that there exist $x, y \in G$ with $g = xy$, $xy = yx$, and $|x| = m$, $|y| = n$. Hint: Find integers r, s such that $1 = rm + sn$ and let $x = g^{sn}$ and $y = g^{rm}$.

   b)   Express $(1\,2\,3\,4\,5\,6) \in S_6$ as a product of two commuting elements x and y, with x of order 2 and y of order 3. Further, find the disjoint cycle decomposition of x and of y.

5. Find the disjoint cycle decomposition of any $\sigma \in S_7$ for which:

   a) $\sigma(1\,2\,3\,4) = (3\,7\,2\,6)(1\,4)$. Further prove that there is exactly one such σ.

   b) $\sigma(1\,2\,3\,4)(5\,6)\sigma^{-1} = (3\,7\,2\,6)(1\,4)$. Determine the exact number of $\sigma \in S_7$ for which this holds.

6. a) Find $\sigma \in S_{13}$ with $|\sigma| = 42$.

   b) Find $\sigma \in S_{13}$ for which $M_\sigma = J_{13}$ and $|\sigma| = 20$.

7. We know that for any integer $n \geq 2$, every permutation in S_n can be written as a product of transpositions, which means that the set $\{(i\,j) \mid 1 \leq i < j \leq n\}$ of all $\binom{n}{2} = \frac{n(n-1)}{2}$ transpositions in S_n generates S_n.

   a) Prove that for any integers i and j with $1 < i < j \leq n$, $(1\,i)(1\,j)(1\,i)^{-1} = (i\,j)$. Deduce that the set $\{(1\,i) \mid 1 < i \leq n\}$ of size $n-1$ generates S_n.

   b) Prove that for any integer i with $1 < i < n$, $(i\,i+1)(1\,i)(i\,i+1)^{-1} = (1\,i+1)$. Deduce that the set $\{(i\,i+1) \mid 1 \leq i < n\}$ of size $n-1$ generates S_n.

   c) Prove that for any integer i with $1 < i < n$,

   $$(1\,2\,3\cdots n)(1\,i)(1\,2\,3\cdots n)^{-1} = (1\,i+1).$$

   Deduce that if $n \geq 3$, then the set $\{(1\,2\,3\cdots n), (1\,2)\}$ of size 2 generates S_n.

   d) Prove that S_n has a one-element generating set if, and only if, $n = 2$.

8. Let $n > 1$ be an integer.

   a) Prove that if τ_1 and τ_2 are distinct transpositions, then $|\tau_1\tau_2| = 2$ or 3.

   b) Let $\tau_1, \tau_2, \tau_3 \in S_n$ be transpositions (not necessarily distinct). Prove that $\tau_1\tau_2\tau_3 \neq 1_{J_n}$.

9. We have seen in the proof of Theorem 7.5.4 that if $n \geq 5$, then any two 3-cycles are conjugate in A_n. Recall that all cycles of odd length are even, so in particular, $(1\,2\,3\,4\,5) \in A_5$. Prove that the conjugacy class of $(1\,2\,3\,4\,5)$ in A_5 is exactly one-half the size of its conjugacy class in S_5.

10. Recall that for an element g of a group G, the centralizer of g in G, denoted by $C_G(g)$, is the subgroup $\{\, h \in G \mid gh = hg \,\}$ of G. Apply the results in Exercise 29 of Chapter 6 in each of the following to determine the order of $C_{S_n}(\sigma)$.

    a) $n = 5,\ \sigma = (1\,2)(3\,4)$.

    b) $n \geq 2,\ \sigma = (1\,2\cdots n)$.

11. Determine the number of conjugacy classes in A_6, and determine the size of each.

12. Let n be a positive integer, and let $\sigma \in S_n$. Prove that if $|\sigma|$ is odd, then $\sigma \in A_n$.

13. If $\sigma \in S_{13}$ and $\sigma^2 = (1\,10\,11\,6\,3\,12\,2\,5\,13\,7\,4\,8)$, what is σ?

14. Let n be a positive integer. Prove that if H is a subgroup of S_n, then either $H \subseteq A_n$ or else $|H \cap A_n| = \frac{1}{2}|H|$.

15. If $\sigma = \left(\begin{smallmatrix} 1\,2\,3\,4\,5\,6\,7\,8\,9 \\ 3\,1\,2\,?\,?\,?\,7\,8\,9\,6 \end{smallmatrix}\right) \in S_9$ is even, what are the values of $\sigma(4)$ and $\sigma(5)$?

16. Give another proof of the fact that A_4 has no subgroup of order 6 as follows. Assume that there is a subgroup H of order 6 in A_4. You know that $[A_4 : H] = 2$ and thus $H \triangleleft A_4$.

    a)   Prove that for any $g \in A_4$, $g^2 \in H$.

    b)   Deduce that H must then contain all eight 3-cycles.

17. If $\sigma = (1\,3\,5\,7\,9)(2\,4\,6)(8\,10) \in S_{10}$, and m is a positive integer such that σ^m is a 5-cycle, what can be said about m?

18. Let n be a positive integer. Prove that for any $\sigma, \tau \in S_n$, $\sigma\tau$ and $\tau\sigma$ have the same cycle structure.

19. Consider the symmetric group S_7.

    a)   Does S_7 have an element of order 5?

    b)   Does S_7 have an element of order 10?

    c)   Does S_7 have an element of order 15?

    d)   What is the maximum of the orders of the elements of S_7? Find an element of maximum order in S_7.

20. For any positive integer n, let $\sigma = \left(\begin{smallmatrix} 1 & 2 & 3 & 4 & \dots & 2n \\ n+1 & 1 & n+2 & 2 & \dots & n \end{smallmatrix}\right) \in S_{2n}$. What is $|\sigma|$?

Chapter 8

Rings, Integral Domains and Fields

Rings

For any positive integer n, the set $M_{n \times n}(\mathbb{R})$ of all $n \times n$ matrices with real entries is a typical example of a ring. Let us distill the essence of this system.

We have a set of elements, namely, the matrices, with two operations which we call matrix addition and matrix multiplication, denoted by $+$ and juxtaposition, respectively. Under addition, the matrices form an abelian group and under multiplication, a semigroup (in fact, a monoid, but we shall not require the existence of a multiplicative identity as an essential property). Moreover, these two operations are linked by distributivity. These are the properties which we extract and call any system possessing them a ring. More formally, we have:

8.1.1 Definition. *A ring is a set R with two binary operations, addition and multiplication, denoted by $+$ and juxtaposition (or $\cdot$), respectively, and satisfying the following requirements:*

(i) *$(R, +)$ is an abelian group with identity denoted by 0_R or simply by 0, called the zero of the ring;*

(ii) *$(R, \cdot)$ is a semigroup;*

(iii) *Multiplication is both left and right distributive over addition; that is, for all $a, b, c \in R$, $a(b + c) = ab + ac$ and $(b + c)a = ba + ca$.*

If the multiplication operation is commutative, then R is said to be a commutative ring.

If there exists an identity for multiplication, and the multiplicative identity is different from 0_R, then the multiplicative identity element shall be denoted by 1_R, or simply by 1, and R shall be called a unital ring, or a

183

ring with unity, and 1_R is called the unity of the ring. The ring with one element is called the trivial ring, and is not called a unital ring.

Although all of the matrix rings $M_{n\times n}(\mathbb{R})$, $n \in \mathbb{Z}^+$, are rings with unity, the only commutative ring in this collection is $M_{1\times 1}(\mathbb{R})$. There do exist rings without a multiplicative identity. For example, any abelian group G can be made into a ring by letting the binary operation of G be the addition operation, while multiplication is defined by $ab = 0_G$ for all $a, b \in G$, where 0_G denotes the identity for the addition operation.

We now offer several other examples of rings.

8.1.2 Example.

(i) *$\mathbb{Z}$, $\mathbb{Q}$, $\mathbb{R}$, and $\mathbb{C}$ with the usual addition and multiplication operations are each commutative rings with unity.*

(ii) *For any $n \geq 2$, $\mathbb{Z}_n$, with the usual addition and multiplication operations, is a commutative ring with unity. $\mathbb{Z}_1$ is the trivial (one-element) ring, and so it is a commutative ring. However, it is not a unital ring, since to be a unital ring, it was specifically required that the multiplicative identity (when it exists) be different from the additive identity.*

(iii) *Let T be a nonempty set. Then by Example 5.2.9 (vii), $\mathscr{P}(T)$ is an abelian group with the binary operation $X \triangle Y = (X - Y) \cup (Y - X)$, the identity element being $\varnothing$. As well, by Proposition 2.3.2, $\cap$ is a commutative, associative binary operation on $\mathscr{P}(T)$. We consider $\triangle$ to be the addition operation and $\cap$ to be the multiplication operation. It also follows from Proposition 2.3.2 that for $X, Y, Z \in \mathscr{P}(T)$, $X \cap (Y \triangle Z) = X \cap ((Y - Z) \cup (Z - Y)) = (X \cap (Y - Z)) \cup (X \cap (Z - Y)) = ((X \cap Y) - (X \cap Z)) \cup ((X \cap Z) - (X \cap Y)) = (X \cap Y) \triangle (X \cap Y)$, and thus $\cap$ left distributes over $\triangle$. Since $\cap$ is commutative, it also holds that $\cap$ right distributes over $\triangle$ and thus $\mathscr{P}(T)$, with operations $\triangle$ and $\cap$ is a commutative ring. Note that it is a unital ring, since for all $X \in \mathscr{P}(T)$, $X \cap T = T \cap X = X$, which proves that T is the unity element.*

(iv) *The set $\mathscr{F}(\mathbb{R})$ of all maps from $\mathbb{R}$ to itself is a ring with operations defined as follows: for all $f, g \in \mathscr{F}(\mathbb{R})$, $f + g : \mathbb{R} \to \mathbb{R}$ is the map defined by $(f + g)(x) = f(x) + g(x)$, where the latter addition is the usual addition of real numbers, and $fg(x) = f(x)g(x)$, where the latter multiplication is the usual multiplication of real numbers. By Proposition 3.4.8, each of addition and multiplication in $\mathscr{F}(\mathbb{R})$ is as-*

sociative and commutative, the identity for addition is $\hat{0}$, where for all $x \in \mathbb{R}$, $\hat{0}(x) = 0$, and the identity for multiplication is $\hat{1}$, where for all $x \in \mathbb{R}$, $\hat{1}(x) = 1$. The reader is invited to verify that multiplication does (left, and then since multiplication is commutative, right) distribute over addition.

(v) *For any ring R, and any positive integer n, the set $M_n(R)$ of all $n \times n$ matrices with entries from R, with the usual definition of matrix addition and matrix multiplication, is a ring, in general non-commutative. The identity element is the $n \times n$ matrix with all entries equal to the identity of R. If R has a unity 1, then $M_n(R)$ has a unity; namely the $n \times n$ matrix with all off-diagonal entries equal to 0, while each diagonal entry is equal to 1.*

There is an important general construction that we can apply to any abelian group.

8.1.3 Definition. *Let A be an abelian group. The set of all endomorphisms of A (that is, all group homomorphisms from A to itself) shall be denoted by $End(A)$.*

8.1.4 Proposition. *Let A be an abelian group. Then $End(A)$ is a unital ring, where addition is defined by $(f+g)(a) = f(a)+g(a)$ for all $f, g \in \mathscr{F}(A)$ and $a \in A$, and multiplication is composition of maps. The unity element is $\mathbb{1}_A$.*

Proof. By Proposition 3.4.8, $\mathscr{F}(A)$ is a commutative monoid under the binary operation of addition, where for $f, g \in \mathscr{F}(A)$, $f + g \in \mathscr{F}(A)$ was defined by $(f + g)(a) = f(a) + g(a)$ for all $a \in A$, where the latter $+$ is the binary operation of A. The identity for addition is $\hat{0} \in \mathscr{F}(A)$, where for all $a \in A$, $\hat{0}(a) = 0$, where 0 is the identity for $+$ in A. We prove that for each $f \in \mathscr{F}(A)$, there exists $g \in \mathscr{F}(A)$ such that $f + g = \hat{0}$ (and by commutativity, $g + f = \hat{0}$ then as well). Let $f \in \mathscr{F}(A)$. Define $g \in \mathscr{F}(A)$ by $g(a) = -f(a)$ for all $a \in A$ (recall that A is a group, so every element has an inverse in A). Then for any $a \in A$, $(f + g)(a) = f(a) + g(a) = f(a) + (-f(a)) = 0 = \hat{0}(a)$, and thus $f + g = \hat{0}$. This proves that f has inverse g; that is, for each $a \in A$, $(-f)(a) = -f(a)$. Thus $\mathscr{F}(A)$ is an abelian group under addition. We prove that $End(A)$ is a subgroup of $\mathscr{F}(A)$. To prove that $End(A)$ is closed under addition, let $f, g \in End(A)$. We are to prove that $f + g \in End(A)$; that is, that $f + g$ is a homomorphism, given that f and g are homomorphisms. Let $a, b \in A$.

Then since f and g are homomorphisms, we have $(f+g)(a+b) = f(a+b) + g(a+b) = f(a) + f(b) + g(a) + g(b)$. As addition in A is commutative, we have $(f+g)(a+b) = f(a) + g(a) + f(b) + g(b) = (f+g)(a) + (f+g)(b)$, and so $f + g \in \text{End}(A)$. Next, we prove that $\text{End}(A)$ is closed under taking of inverse. Let $f \in \text{End}(A)$. Then for any $a, b \in A$, $(-f)(a+b) = -(f(a+b)) = -(f(a) + f(b)) = -f(b) + (-f(a)) = -f(a) + (-f(b)) = (-f)(a) + (-f)(b)$, and so $-f \in \text{End}(A)$. This completes the proof that $\text{End}(A)$ is a subgroup of $\mathscr{F}(A)$.

Next, we note that by Example 5.2.9 (viii), $\circ$ is an associative binary operation on $\mathscr{F}(A)$, with identity element $\mathbb{1}_A$. By Proposition 6.3.10 (viii), for any $f, g \in \text{End}(A)$, $f \circ g \in \text{End}(A)$, and so $\text{End}(A)$ is closed under $\circ$. Thus $\text{End}(A)$ is an abelian group under addition, and a monoid under composition. It remains to prove that composition distributes, both left and right (since composiiton is not commutative, we must prove each separately), over addition. Let $f, g, h \in \text{End}(A)$. Then for any $a \in A$, we have $f \circ (g+h)(a) = f((g+h)(a)) = f(g(a) + h(a)) = f(g(a)) + f(h(a)) = f \circ g(a) + f \circ h(a) = (f \circ g + f \circ h)(a)$, and so $f \circ (g+h) = f \circ g + f \circ h$. Note that the fact that f is a homomorphism was required. Next, consider $(g+h) \circ f(a) = (g+h)(f(a)) = g(f(a)) + h(f(a)) = g \circ f(a) + h \circ f(a) = (g \circ f + h \circ f)(a)$, and so $(g+h) \circ f = g \circ f + h \circ f$, as required. $\square$

Let us bear in mind that since a ring is an abelian group under addition, any property that we have already proved about abelian groups will hold for the additive structure of the ring. Since our choice of generic notation for a group operation was multiplicative, it will frequently be necessary to reformulate an earlier result into additive notation for application to the additive group of a ring. As well, since a ring under multiplication is a semigroup, any results we have established about semigroups are applicable to the multiplicative semigroup of the ring.

There are several very useful elementary properties of rings that we establish now.

8.1.5 Proposition. *Let R be a ring. Then the following hold:*

(i) *For any $x \in R$, $x0_R = 0_R x = 0_R$.*

(ii) *For all $x, y \in R$, $x(-y) = (-x)y = -(xy)$.*

(iii) *For all $x \in R$, $-(-x) = x$.*

(iv) *For all $x, y \in R$, $(-x)(-y) = xy$.*

Proof. For (i), let $x \in R$. We have $x0_R = x(0_R + 0_R) = x0_R + x0_R$, and thus $0_R = -x0_R + x0_R = -x0_R + x0_R + x0_R = 0_R + x0_R = x0_R$. Similarly,

$0_R = -0_R x + 0_R x = -0_R x + 0_R x + 0_R x = 0_R + 0_R x = 0_R x.$

Next, for (ii), let $x, y \in R$. Then $0_R = x 0_R = x(y + (-y)) = xy + x(-y)$, and so $-(xy) = x(-y)$. Similarly, $0_R = 0_R y = (x + (-x))y = xy + (-x)y$, and so $-(xy) = (-x)y$.

For (iii), let $x \in R$. Since $x + (-x) = 0_R = (-x) + x$, it follows that $-(-x) = x$.

For (iv), let $x, y \in R$. Then by (ii), $(-x)(-y) = -x(-y) = -(-(xy))$, and then by (iii), $-(-(xy)) = xy$, so $(-x)(-y) = xy$. $\qquad \square$

Observe that (i), (ii), (iii), and (iv) above are familiar to us as properties of multiplication of real numbers, but we now know that they are valid in any ring.

8.1.6 Notation. *For any ring R, for $a, b \in R$, we shall henceforth write $a - b$ for $a + (-b)$. We may think of this computation as defining a non-commutative, nonassociative binary operation called subtraction. It is the case that multiplication distributes over subtraction.*

From this point on in the chapter on rings, the reader will find many concepts and results that are reminiscent of concepts and results we have encountered in our study of groups. For example, instead of subgroups, we shall be talking about subrings; instead of group homomorphisms, we shall speak of ring homomorphisms; instead of normal subgroups, ideals.

8.1.7 Definition. *Let R be a ring. A subring S of R is a subgroup of the additive group of R that is also closed under the operation of multiplication.*

8.1.8 Proposition. *Let R be a ring. If S is a subring of R, then S is a ring under the induced operations of addition and multiplication (S is closed under both addition and multiplication), and the identity of R is the identity of S. If R is a unital ring, it is not necessary that S be unital, while on the other hand, S might be unital, but the unity element of S could be different from that of R.*

Proof. Let S be a subring of R. Since S is an abelian group under the operation of addition in R, and S is closed with respect to multiplication, we need only observe that for any $a, b, c \in S$, $a(b + c) = ab + ac$ and $(b + c)a = ba + ca$ since the computation is carried out in R and in R, multiplication distributes both left and right over addition.

Even if R has unity, the subset $\{0_R\}$ is a subring of R without unity element. To see that a ring R with unity can have a subring with unity, but the unity element of the subring may be different from the unity element of R, consider $R = M_{2\times 2}(\mathbb{Z})$, the ring of all 2×2 matrices with integer entries. The reader should verify that $S = \{ \begin{bmatrix} a & 0 \\ 0 & 0 \end{bmatrix} \mid a \in \mathbb{Z} \}$ is a subring of R. Note that the unity of R is $\begin{bmatrix} 1 & 0 \\ 0 & 1 \end{bmatrix} \notin S$. However, S does have as unity the element $\begin{bmatrix} 1 & 0 \\ 0 & 0 \end{bmatrix}$. To see this, let $\begin{bmatrix} a & 0 \\ 0 & 0 \end{bmatrix} \in S$. Then

$$\begin{bmatrix} a & 0 \\ 0 & 0 \end{bmatrix} \begin{bmatrix} 1 & 0 \\ 0 & 0 \end{bmatrix} = \begin{bmatrix} a(1) + 0(0) & a(0) + 0(0) \\ 0(1) + 0(0) & 0(0) + 0(0) \end{bmatrix} = \begin{bmatrix} a & 0 \\ 0 & 0 \end{bmatrix}$$

and

$$\begin{bmatrix} 1 & 0 \\ 0 & 0 \end{bmatrix} \begin{bmatrix} a & 0 \\ 0 & 0 \end{bmatrix} = \begin{bmatrix} 1(a) + 0(0) & 1(0) + 0(0) \\ 0(0) + 0(0) & 0(0) + 0(0) \end{bmatrix} = \begin{bmatrix} a & 0 \\ 0 & 0 \end{bmatrix}.$$

$\square$

8.1.9 Example.

(i) *For any integer n, the set $n\mathbb{Z}$ of all multiples of n is a subring of $\mathbb{Z}$. In particular, the set of all even integers is a subring of $\mathbb{Z}$. Note that while $\mathbb{Z}$ is a (commutative) unital ring, $2\mathbb{Z}$ has no unity element.*

(ii) *$\mathbb{Z}$ is a subring of $\mathbb{Q}$, $\mathbb{Q}$ is a subring of $\mathbb{R}$, and $\mathbb{R}$ is a subring of $\mathbb{C}$.*

(iii) *Let n be a positive integer, and let R be a ring. A matrix $A \in M_{n\times n}(R)$ is said to be upper-triangular if every entry of A below the main diagonal is 0_R; that is, if $A_{ij} = 0_R$ for all integers i and j with $1 \leq j < i \leq n$. If in addition, $A_{ii} = 0_R$ for all $i = 1, 2, \ldots, n$, then A is called strictly upper-triangular. The set of all upper-triangular $n \times n$ matrices with entries from R is a subring (with unity that of $M_{n\times n}(R)$) of $M_{n\times n}(R)$. The set of all strictly upper-triangular $n \times n$ matrices with entries from R is a subring of the ring of all upper-triangular matrices (and has no unity element).*

8.1.10 Proposition. *Let R be a ring. A subset S of R is a subring of R if, and only if, S is nonempty and closed under both subtraction and multiplication.*

Proof. By Proposition 5.4.10 (with the multiplicative notation recast into additive notation), a subset S is a subgroup of the additive group of R if, and only if, S is nonempty and closed under subtraction. By definition, an additive subgroup of R is a subring if, and only if the subgroup is also closed under multiplication. $\square$

8.1.11 Example. *We prove that the set $S = \{ m + n\sqrt{2} \mid m, n \in \mathbb{Z} \} \subseteq \mathbb{R}$ is a subring of $\mathbb{R}$. Since $1 = 1 + 0\sqrt{2} \in S$, $S \neq \varnothing$. Suppose now that*

$m, n, r, s \in \mathbb{Z}$. Then $(m + n\sqrt{2}) - (r + s\sqrt{2}) = (m - r) + n\sqrt{2} - s\sqrt{2} = (m - r) + (n - s)\sqrt{2} \in S$, and $(m + n\sqrt{2})(r + s\sqrt{2}) = mr + ms\sqrt{2} + nr\sqrt{2} + ns\sqrt{2}\sqrt{2} = (mr + 2ns) + (ms + nr)\sqrt{2} \in S$. *Since S is nonempty and closed under both subtraction and multiplication, it follows by Proposition 8.1.10 that S is a subring of $\mathbb{R}$.*

8.1.12 Proposition. *Let R be a ring, and let Λ be a nonempty set such that for each $\lambda \in \Lambda$, S_λ is a subring of R. Then $\cap_{\lambda \in \Lambda} S_\lambda$ is a subring of R.*

Proof. Let $S = \cap_{\lambda \in \Lambda} S_\lambda$. Since $0 \in S_\lambda$ for each $\lambda \in \Lambda$, $0 \in S$ and so $S \neq \varnothing$. Let $x, y \in S$. Then for each $\lambda \in \Lambda$, $x, y \in S_\lambda$, so $x - y, xy \in S_\lambda$ and thus $x - y, xy \in S$. It follows now by Proposition 8.1.10 that S is a subring of R. □

In Definition 4.5.10, we introduced the notion of unit element and zero divisor for $\mathbb{Z}_n$. As we now see, these notions are properly viewed in a more general ring-theoretic setting.

8.1.13 Definition. *If R is a unital ring, then R with the operation of multiplication is a monoid, and by Proposition 5.2.2, the set of all elements of R that are invertible with respect to multipication is a group under multiplication, called the group of units of R and denoted by $U(R)$. An element of $U(R)$ is called a unit of R (and the inverse of the unit a is denoted by a^{-1}).*

If R is a commutative ring, then an element a of R is called a zero divisor of R if there is a nonzero element $b \in R$ such that $ab = 0$.

We remark that while it is possible to define zero divisor in a noncommutative ring, it would then be necessary to distinguish between left and right zero divisors, which would complicate matters beyond the usefulness of the concept in an introductory algebra course.

Note also that if R is a commutative unital ring, then the subset of all units of R is disjoint from the subset of all zero divisors of R. For suppose that a is both a unit and a zero divisor of R. Then a^{-1} exists in R, and there exists nonzero $b \in R$ with $ab = 0$. Then $b = a^{-1}(ab) = a^{-1}(0) = 0$, which is not possible.

It is sometimes the case that the set of units of R (which always contains 1_R) and the set of zero divisors (which always contains 0_R) form a partition of R, but this need not always be so, as we have observed in $\mathbb{Z}$.

We now offer examples of the group of units for various familiar rings.

8.1.14 Example.

(i) *For any ring R with unity, the unity element is always a unit of R. Sometimes, the unity is the only unit. For example, in the powerset ring described in Example 8.1.2 (iii), for any nonempty set T, T is the unity element of $\mathscr{P}(T)$, and T is in fact the only unit of $\mathscr{P}(T)$; that is, $U(\mathscr{P}(T)) = \{T\}$. On the other hand, every element of $\mathscr{P}(T)$ except for the unit T is a zero divisor in $\mathscr{P}(T)$.*

(ii) *The only zero divisor of $\mathbb{Z}$ is 0, while $U(\mathbb{Z}) = \{\pm 1\}$.*

(iii) *For any positive integer n, the set of units of $M_{n \times n}(\mathbb{R})$ is by definition $Gl_n(\mathbb{R})$.*

(iv) *$U(\mathbb{Q}) = \mathbb{Q}^* = \mathbb{Q} - \{0\}$, and $U(\mathbb{R}) = \mathbb{R}^* = \mathbb{R} - \{0\}$, while in each, 0 is the only zero divisor.*

8.1.15 Example. *Let R be a commutative unital ring, and let S denote the ring of all 2×2 upper-triangular matrices with entries from R. Then*

$$U(S) = \left\{ \begin{bmatrix} a & b \\ 0 & c \end{bmatrix} \middle| a, b \in U(R) \right\}.$$

For suppose that $\begin{bmatrix} a & b \\ 0 & c \end{bmatrix} \in U(S)$. Then there exists $\begin{bmatrix} r & s \\ 0 & t \end{bmatrix} \in S$ such that $\begin{bmatrix} a & b \\ 0 & c \end{bmatrix}\begin{bmatrix} r & s \\ 0 & t \end{bmatrix} = \begin{bmatrix} 1 & 0 \\ 0 & 1 \end{bmatrix}$. In particular, we have $ar = 1 = ct$ and since R is commutative, this means that $ra = 1 = tc$ and both a and c are invertible in R. Conversely, suppose that $\begin{bmatrix} a & b \\ 0 & c \end{bmatrix} \in S$ and that $a, c \in U(R)$. Then $\begin{bmatrix} a & b \\ 0 & c \end{bmatrix}\begin{bmatrix} a^{-1} & -a^{-1}bc^{-1} \\ 0 & c^{-1} \end{bmatrix} = \begin{bmatrix} 1 & 0 \\ 0 & 1 \end{bmatrix} = \begin{bmatrix} a^{-1} & -a^{-1}bc^{-1} \\ 0 & c^{-1} \end{bmatrix}\begin{bmatrix} a & b \\ 0 & c \end{bmatrix}$ and so $\begin{bmatrix} a & b \\ 0 & c \end{bmatrix} \in U(S)$.

In our discussion of the properties of the integers in Chapter 4, we observed that nonzero integers can be cancelled; that is, if $a, b, c \in \mathbb{Z}$ with $a \neq 0$ and $ab = ac$, then $b = c$. As we may now see, this is a consequence of the fact that in $\mathbb{Z}$, a nonzero integer is not a zero divisor. For in $\mathbb{Z}$, $ab = ac$ if, and only if, $a(b - c) = 0$. If $a \neq 0$, then a is not a zero divisor, and so if $a(b - c) = 0$, it must be the case that $b - c = 0$; that is, $b = c$.

8.1.16 Definition. *A ring is said to be a cancellative ring if it is commutative and the only zero divisor is the zero of the ring.*

Thus in a cancellative ring R, for any $a, b, c \in R$ with $a \neq 0$ and $ab = ac$, it follows that $b = c$.

8.1.17 Definition. *A cancellative unital ring is said to be an integral domain, and if R is an integral domain for which $U(R) = R^* = R - \{0\}$, then R is called a field.*

Note that $\mathbb{Z}$ is an integral domain which is not a field, as 2 has no multiplicative inverse in $\mathbb{Z}$. Thus not every integral domain is a field. However, it is the case that every finite integral domain (for example, $\mathbb{Z}_p$ when p is a prime) is a field. In fact, we do not even need to require the existence of unity! Also observe that for a commutative ring R with unity, no unit of R is a zero divisor of R, and thus we could have defined a field to be a commutative ring R with unity for which $U(R) = R^*$.

8.1.18 Proposition. *Every finite cancellative ring is a field.*

Proof. Let R be a finite cancellative ring. Since R is cancellative, R^*, the set of nonzero elements of R, is a finite cancellative semigroup under multiplication, and so by Proposition 5.3.4, R^* is a group under multiplication. Thus R has unity, and $U(R) = R^*$, so R is a field. $\qquad\square$

8.1.19 Example.

(i) $\mathbb{Q}$, $\mathbb{R}$, and $\mathbb{C}$, *the set of complex numbers, are all fields with the usual operations of addition and multiplication.*

(ii) *For any positive integer n, $\mathbb{Z}_n$ is a field if, and only if, n is a prime. For $\mathbb{Z}_n$ is a commutative unital ring, and by Corollary 4.5.9, $U(\mathbb{Z}_n) = \mathbb{Z}_n^*$ if, and only if, n is prime.*

(iii) *In Example 8.1.11, it was observed that the set*

$$D = \{\, m + n\sqrt{2} \mid m, n \in \mathbb{Z} \,\}$$

is a subring of $\mathbb{R}$, and since the only zero divisor of $\mathbb{R}$ is 0, it follows that D is an integral domain (since $1 \in D$). We could have allowed $m, n \in \mathbb{Q}$ and obtained a similar result. More generally, for any positive integer n, $D = \{\, r + s\sqrt{n} \mid r, s \in \mathbb{Q} \,\}$ is a subring of $\mathbb{R}$ and is thus an integral domain (again, $1 = 1 + 0\sqrt{n} \in D$). We claim that D is in fact a field. If n is a perfect square, then $n = m^2$ for some positive integer m and $\{\, r + s\sqrt{n} \mid r, s \in \mathbb{Q} \,\} = \{\, r + sm \mid r, s \in \mathbb{Q} \,\} = \mathbb{Q}$, so we may suppose that n is not a perfect square. Then in the prime decomposition of n, at least one prime factor appears to an odd power, and there exist unique positive integers m and k such that $n = m^2 k$ and k is square-free, by which we mean that in the prime decomposition of k, every prime factor appears only to the first power. It is evident that

$$\{\, r + s\sqrt{n} \mid r, s \in \mathbb{Q} \,\} = \{\, r + sm\sqrt{k} \mid r, s \in \mathbb{Q} \,\}$$
$$= \{\, r + s\sqrt{k} \mid r, s \in \mathbb{Q} \,\}.$$

Now, if $\sqrt{k} \in \mathbb{Q}$, say $\sqrt{k} = \frac{m}{n}$ with $m, n \in \mathbb{Z}$, then we have $m^2 = kn^2$. But every prime factor in m^2 must appear to an even power, while every prime factor of k will appear to an odd power. By the uniqueness of the prime decomposition, this is not possible and so we conclude that $\sqrt{k} \notin \mathbb{Q}$. Suppose now that $a, b \in \mathbb{Q}$. If $a^2 - b^2k = 0$, then either $a = b = 0$, or else $b \neq 0$ and $\sqrt{k} = \frac{|a|}{|b|} \in \mathbb{Q}$. Since $\sqrt{k} \notin \mathbb{Q}$, we see that either $a = b = 0$, or else $a^2 - b^2k \neq 0$. Let $a, b \in \mathbb{Q}$ with $a \neq 0$ or $b \neq 0$, so $a^2 - b^2k \neq 0$. Then $\frac{a}{a^2 - b^2k} - \frac{b}{a^2 - b^2k}\sqrt{k} \in D$ and $(a + b\sqrt{k})(\frac{a}{a^2 - b^2k} - \frac{b}{a^2 - b^2k}\sqrt{k}) = 1$, so $a + b\sqrt{k} \in U(D)$. Thus D is an integral domain for which $U(D) = D^*$; that is, D is a field.

8.2 Ring Homomorphisms, Ring Isomorphisms, and Ideals

The next few definitions and results should give the reader a feeling of "déjà vu", as they should call to mind analogous definitions and results in our treatment of groups.

8.2.1 Definition. *Let R and S be rings. A mapping $f : R \to S$ is a ring homomorphism, or simply a homomorphism, if for all $x, y \in R$, $f(x + y) = f(x) + f(y)$ and $f(xy) = f(x)f(y)$. If both R and S are unital rings and $f(1_R) = 1_S$, then f is said to be a unital homomorphism. A bijective homomorphism is called an isomorphism, and if there exists an isomorphism from R to S, then we say that R and S are isomorphic, and write $R \simeq S$.*

A ring homomorphism f from a ring R to a ring S is in particular a group homomorphism of the additive group of R to the additive group of S, and consequently, all of the results that have been established for group homomorphisms apply. In particular, the identity of R is mapped to the identiy of S, and the kernel of f is an additive subgroup of R.

We note that if R and S are unital rings and $f : R \to S$ is a ring homomorphism, it is not necessarily the case that f is a unital homomorphism; that is, the requirement that a homomorphism preserve both addition and multiplication is not enough to ensure that the unity of R is mapped to the unity of S. For example, the map from $f : \mathbb{R} \to M_{2 \times 2}(\mathbb{R})$ defined by $f(a) = \left[\begin{smallmatrix} a & 0 \\ 0 & 0 \end{smallmatrix}\right]$ satisfies $f(a + b) = f(a) + f(b)$ and $f(ab) = f(a)f(b)$ for all $a, b \in \mathbb{R}$, but $f(1) = \left[\begin{smallmatrix} 1 & 0 \\ 0 & 0 \end{smallmatrix}\right]$ is not the unity of $M_{2 \times 2}(\mathbb{R})$, so f is a ring homomorphism but not a unital ring homomorphism. On the other hand, if we consider f to be a map from R to the subring $T = f(R)$ of S, then T is a unital ring, and f does map the unity of R to the unity of T, and then $f : R \to T$ is a unital ring homomorphism.

Note that if $f : R \to S$ is a homomorphism, it might very well be the case that R has no unity but S does. For example, let R denote the subring $2\mathbb{Z}$ of $\mathbb{Z}$, and let $f : R \to \mathbb{Z}$ be the inclusion map; that is, $f(n) = n$ for all $n \in R$. Then f is a ring homomorphism, R has no unity, while of course, $\mathbb{Z}$ has unity 1. There are also examples of rings R and S and a ring homomorphism $f : R \to S$, where R is unital while S is not unital (for a class of such examples, see Definition 8.4.4 and the discussion which follows it).

Our first result establishes that if R and S are rings and R is isomorphic to S, then S is isomorphic to R. As we shall soon come to realize, isomorphic rings are algebraically indistinguishable.

8.2.2 Proposition. *Let R and S be rings. If $f : R \to S$ is an isomorphism, then f is bijective and $f^{-1} : S \to R$ is an isomorphism.*

Proof. By Proposition 6.3.2, f^{-1} is a group isomorphism from the additive group of S to the additive group of R, so it remains to verify that for all $r, s \in S$, $f^{-1}(rs) = f^{-1}(r)f^{-1}(s)$. Let $r, s \in S$. Then $f(f^{-1}(rs)) = rs = f(f^{-1}(r))f(f^{-1}(s)) = f(f^{-1}(r)f^{-1}(s))$, and since f is injective, we conclude that $f^{-1}(rs) = f^{-1}(r)f^{-1}(s)$. $\qquad\square$

8.2.3 Proposition. *Let R and S be rings, and let $f : R \to S$ be a (ring) homomorphism. Then the following hold:*

(i) $f(0_R) = 0_S$.

(ii) for all $x \in R$ and all $n \in \mathbb{Z}$, $f(nx) = nf(x)$, while for all positive $n \in \mathbb{Z}$, $f(x^n) = (f(x))^n$. If both rings are unital and f is unital, then for any $x \in U(R)$, $f(x) \in U(S)$ and for all $n \in \mathbb{Z}$, $f(x^n) = (f(x))^n$.

(iii) If T is a subring of R, then $f(T)$ is a subring of S.

(iv) If T is a subring of S, then $f^{-1}(T)$ is a subring of R.

(v) If T is a ring and $g : S \to T$ is a homomorphism, then $g \circ f : R \to T$ is a homomorphism.

Proof. (i) is a consequence of Proposition 6.3.10 (i), and from Proposition 6.3.10 (ii), we obtain that $f(nx) = nf(x)$ for all $x \in R$ and $n \in \mathbb{Z}$. In fact, the proof of Proposition 6.3.10 (ii) also can be used to prove that $f(x^n) = (f(x))^n$ for all $x \in R$ and every positive integer n. Moreover, if both R and S are unital, and f is a unital homomorphism, and $x \in U(R)$, then $f(x)f(x^{-1}) = f(xx^{-1}) = f(1_R) = 1_S$, and similarly, $f(x^{-1})f(x) = 1_S$, so

$f(x) \in U(S)$ and $(f(x))^{-1} = f(x^{-1})$. Thus the restriction of f to $U(R)$ is a homomorphism from the multiplicative group $U(R)$ to the multiplicative group $U(S)$, and so the rest of (ii) follows from Proposition 6.3.10 (ii).

For (iii), let T be a subring of R. Then by Proposition 6.3.10 (iii), $f(T)$ is an additive subgroup of S. Let $r, s \in f(T)$. Then there exist $x, y \in T$ such that $r = f(x)$ and $s = f(y)$, and we have $rs = f(x)f(y) = f(xy) \in f(T)$. Thus $f(T)$ is a subring of S.

For (iv), let T be a subring of S. By Proposition 6.3.10 (iv), $f^{-1}(T)$ is an additive subgroup of R. Let $r, s \in f^{-1}(T)$. Then $f(r), f(s) \in T$, and so $f(rs) = f(r)f(s) \in T$. Thus $rs \in f^{-1}(T)$, and it follows that $f^{-1}(T)$ is a subring of R.

Finally, for (v), let $x, y \in R$. Then $g \circ f(x+y) = g(f(x+y)) = g(f(x) + f(y)) = g(f(x)) + g(f(y)) = g \circ f(x) + g \circ f(y)$. Similarly, $g \circ f(xy) = g(f(xy)) = g(f(x)f(y)) = g(f(x))g(f(y)) = g \circ f(x)g \circ f(y)$, and so $g \circ f$ is a homomorphism. $\square$

We proceed now to determine the correct analogue of the group-theoretic notion of normal subgroup. Recall that for a group G, the normal subgroups of G are distinguished among the subgroups of G as those for which the set of all left cosets of the subgroup forms a group under the binary operation of multiplication of subsets of G. Since a ring has two binary operations fundamental to its structure, we shall be examining the subrings of the ring. Of course, since a ring is an abelian group under addition, every subring is an additive subgroup and thus normal in the additive group of the ring, so we can form the set of all (left) cosets of the subring, and they form an additive group under the operation of addition of subsets of the ring. We determine appropriate requirements on the subring which, when met, allow us to define a multiplication operation on the set of cosets of the subring by using multiplication of subsets. This leads us to the notion of ideal.

8.2.4 Definition. *Let R be a ring. A subring I of R is an ideal of R, denoted by $I \lhd R$, if for all $r \in R$ and all $x \in I$, $rx \in I$ and $xr \in I$.*

Note that in a ring R, we may now define two associative binary operations on $\mathscr{P}(R)$. For $X, Y \in \mathscr{P}(R)$, $X + Y = \{\, x + y \mid x \in X, \ y \in Y \,\}$, and $XY = \{\, xy \mid x \in X, \ y \in Y \,\}$. A subring I of a ring R is thus an ideal of R if, and only if, $IR \subseteq I$ and $RI \subseteq I$.

8.2.5 Proposition. *Let R be a ring. A subset I of R is an ideal of R if,* *and only if, the following hold:*

(i) $I \neq \varnothing$;

(ii) *For all* $x, y \in I$, $x - y \in I$;

(iii) *For all* $x \in I$ *and all* $y \in R$, $xy \in I$ *and* $yx \in I$.

Proof. Of course, if $I \lhd R$, then (i), (ii), and (iii) hold. Conversely, if $I \subseteq R$ satisfies (i) and (ii), then by Proposition 5.4.10, I is an additive subgroup of R, and if in addition, (iii) holds, then I is a subring of R and moreover, $IR \subseteq I$ and $RI \subseteq I$, so $I \lhd R$. □

8.2.6 Example. *If R is a commutative ring, then for any $r \in R$, the subset $\{ a \} \cup aR = \{ a \} \cup \{ ax \mid x \in R \}$ is an ideal of R, called the principal ideal generated by a. This follows immediately from Proposition 8.2.5 since multiplication in R is commutative. Note that if R has unity, then $a \in aR$ and so aR is the principal ideal generated by a in that case. For example, in $\mathbb{Z}$, every additive subgroup of $\mathbb{Z}$ is also an ideal of $\mathbb{Z}$, since the additive group of $\mathbb{Z}$ is cyclic and so the additive subgroups of $\mathbb{Z}$ are the subsets of the form $n\mathbb{Z}$, $n \in \mathbb{Z}$. Since $\mathbb{Z}$ is commutative with unity, it follows that every additive subgroup of $\mathbb{Z}$ is a principal ideal of $\mathbb{Z}$.*

8.2.7 Proposition. *Let R be a ring, and let $\mathscr{I}$ be a nonempty set such that for each $\lambda \in \mathscr{I}$, $I_\lambda \lhd R$. Then $\cap_{\lambda \in \mathscr{I}} I_\lambda$ is an ideal of R.*

Proof. Let $J = \cap_{\lambda \in \mathscr{I}} I_\lambda$. Since $0 \in I_\lambda$ for each $\lambda \in \mathscr{I}$, $0 \in J$ and so $J \neq \varnothing$. Let $x, y \in J$ and $r \in R$. Then for each $\lambda \in \mathscr{I}$, $x, y \in I_\lambda$, so $x - y, xr, rx \in I_\lambda$ and thus $x - y, xr, rx \in J$. It follows now by Proposition 8.2.5 that $J \lhd R$. □

8.2.8 Corollary. *Let R be a ring and let $I, J \lhd R$. Then $I \cap J \lhd R$.*

Proof. This follows immediately from Proposition 8.2.7 since $I \cap J = \cap \{ I, J \}$. □

As one might expect, in view of Proposition 6.4.7 and the preceding discussion, ideals are associated with homomorphisms.

8.2.9 Proposition. *Let R and S be rings, and let $f : R \to S$ be a homomorphism. Then the following hold:*

(i) *For any ideal I of S, $f^{-1}(I) = \{ x \in R \mid f(x) \in I \}$ is an ideal of R. In particular, $\ker(f) = f^{-1}(0_S)$ is an ideal of R.*

(ii) *For any ideal I of R, $f(I)$ is an ideal of the subring $f(R)$ of S.*

(iii) *f is injective if, and only if, $\ker(f) = \{ 0_R \}$.*

(iv) For all $r \in R$, $f^{-1}(f(r)) = r + \ker(f)$.

Proof. For (i), let I be an ideal of S. By Proposition 6.3.10 (iv), $f^{-1}(I)$ is a subgroup of the additive group of R. Let $x \in f^{-1}(I)$ and $r \in R$. Then $f(xr) = f(x)f(r) \in I$ and $f(rx) = f(r)f(x) \in I$ since $f(x) \in I$ and $I \triangleleft S$. Thus $xr, rx \in f^{-1}(I)$ and so $f^{-1}(I) \triangleleft R$. In particular, since $\{0_S\} \triangleleft S$, $\ker(f) = f^{-1}(0_S) \triangleleft R$.

For (ii), let I be an ideal of R. Then by Proposition 6.3.10 (iii), $f(I)$ is a subgroup of the additive group of S, and since $f(I) \subseteq f(R)$, it follows that $f(I)$ is a subgroup of the additive group of $f(S)$. Let $x \in f(I)$ and $y \in f(S)$. Then there exists $a \in I$ and $b \in R$ such that $f(a) = x$ and $f(b) = y$. We have $xy = f(a)f(b) = f(ab) \in f(I)$ since $ab \in I$, and $yx = f(b)f(a) = f(ba) \in f(I)$ since $ba \in I$. Thus $f(I) \triangleleft f(R)$.

(iii) and (iv) both follow immediately from Propositon 6.4.7.   □

8.2.10 Lemma. *Let R be a ring, and let I be a subring of R. Then I is an ideal of R if, and only if, for every $X, Y \in R/I$ (the set of all left cosets of the subgroup I of the additive group of R), there exists $W \in R/I$ such that $XY \subseteq W$. In particular, for any $x, y \in R$, $(x + I)(y + I) \subseteq xy + I$.*

Proof. Suppose first of all that $I \triangleleft R$, and let $X, Y \in R/I$. Let $x \in X$ and $y \in Y$, so $X = x + I$ and $Y = y + I$. Then

$$XY = \{(x + r)(y + s) \mid r, s \in I\}$$
$$= \{xy + (xs + ry + rs) \mid r, s \in I\}$$
$$\subseteq \{xy + t \mid t \in I\} = xy + I \in R/I.$$

Conversely, suppose that for all $X, Y \in R/I$, there exists $W \in R/I$ such that $XY \subseteq W$. We wish to prove that for all $r \in I$ and $x \in R$, both rx and xr belong to I. Let $x \in R$. By assumption, there exists $W \in R/I$ such that $(x + I)(0_R + I) \subseteq W$. In particular, $0_R = x0_R = (x + 0_R)(0_R + 0_R) \in W$ and so $W = 0_R + I = I$. Thus $(x + I)I \subseteq I$, and so for any $r \in I$, we have $xr = (x + 0_R)r \in (x + I)I \subseteq I$. Similarly, from the observation that $(0_R + I)(x + I) \subseteq 0_R + I$, we obtain that for all $r \in I$, $rx \in I$. It follows that $I \triangleleft R$.   □

It is interesting to observe that if $I \triangleleft R$ and $X, Y \in R/I$, it is not necessarily the case that $XY \in R/I$. For example, consider the ideal $2\mathbb{Z}$ of the ring $\mathbb{Z}$. For $X = Y = 2\mathbb{Z}$, we have $XY = \{2n2m \mid m, n \in \mathbb{Z}\} = 4\mathbb{Z}$, but $4\mathbb{Z}$ is not a coset of $2\mathbb{Z}$; that is, if $4\mathbb{Z}$ were a coset of $2\mathbb{Z}$, then for any $i \in 4\mathbb{Z}$, $4\mathbb{Z} = i + 2\mathbb{Z}$. In particular, $0 \in 4\mathbb{Z}$, but $4\mathbb{Z} \neq 2\mathbb{Z}$.

The importance of Lemma 8.2.10 is that it allows us to define a binary operation on R/I, the set of all cosets of an ideal I in a ring R, by using the multiplication operation on $\mathscr{P}(R)$. Even though the product of two cosets of I need not be equal to a coset of I, there is a coset of I that contains the product of the two and that is sufficient.

8.2.11 Proposition. *Let R be a ring and let $I \triangleleft R$. Then I is a normal subgroup of the additive group of R, and R/I is an abelian group under addition of cosets. Define multiplication on R/I by the rule $XY = xy + I$ for any $x \in X$ and $y \in Y$; that is, for all $x, y \in R$, $(x + I)(y + I) = xy + I$ by definition. Then R/I with these operations is a ring, called the quotient ring R modulo I, and the mapping $\pi_I : R \to R/I$ is a surjective ring homomorphism with kernel I. π_I is called the natural ring homomorphism from R to R/I.*

Proof. By Proposition 6.4.10, all that we need to establish is that the multiplication operation is associative and distributes both left and right over addition, and that $\pi_I : R \to R/I$ satisfies $\pi_I(xy) = \pi_I(x)\pi_I(y)$ for all $x, y \in R$. Note that the multiplication operation is well defined by virtue of Lemma 8.2.10. For associativity, suppose that $X, Y, Z \in R/I$, and let $x \in X$, $y \in Y$, and $z \in Z$. Then $X(YZ) = (x + I)(yz + I) = x(yz) + I = (xy)z + I = (xy + I)(z + I) = (XY)Z$, and so multiplication is associative. As well, $X(Y + Z) = (x + I)(y + z + I) = x(y + z) + I = xy + xz + I = (xy + I) + (xz + I) = XY + XZ$, and $(Y + Z)X = (y + I + z + I)(x + I) = (y + z + I)(x + I) = (y + z)x + I = yx + zx + I = (yx + I) + (zx + I) = YX + ZX$, as required. Finally, let $x, y \in R$. Then $\pi_I(xy) = xy + I = (x + I)(y + I) = \pi_I(x)\pi_I(y)$, and so π_I is a ring homomorphism. $\square$

8.2.12 Example.

(i) *As observed in Example 8.2.6, every additive subgroup of $\mathbb{Z}$ is an ideal of $\mathbb{Z}$, and so the ideals of $\mathbb{Z}$ are the subsets of the form $n\mathbb{Z}$, $n \in \mathbb{Z}$. Since $n\mathbb{Z} = (-n)\mathbb{Z}$ for any $n \in \mathbb{Z}$, the ideals of $\mathbb{Z}$ are the subsets $n\mathbb{Z}$, n a nonnegative integer. Moreover, the trivial ideal $\{0\}$ is $0\mathbb{Z}$, while the nontrivial ideals are the subsets of the form $n\mathbb{Z}$, n a positive integer. For a positive integer n, the quotient ring $\mathbb{Z}/n\mathbb{Z}$ has as its elements $\{n\mathbb{Z}, 1 + n\mathbb{Z}, \ldots, n - 1 + n\mathbb{Z}\}$, with addition defined by $(i + n\mathbb{Z}) + (j + n\mathbb{Z}) = i + j + n\mathbb{Z}$ and multiplication defined by $(i + n\mathbb{Z})(j + n\mathbb{Z}) = ij + n\mathbb{Z}$. In Chapter 4 (see Proposition 4.5.5), the properities of addition and multiplication on the set $\mathbb{Z}_n$ were* *established, and in light of our current knowledge, we see that the ring $\mathbb{Z}_n$ is exactly the quotient ring $\mathbb{Z}/n\mathbb{Z}$.*

(ii) *Let S be a commutative ring and let $n \geq 2$ be a positive intgeger. In the ring R of all upper-triangular $n \times n$ matrices with entries from a ring S, the set I of all strictly upper-triangular matrices is an ideal of R and the quotient ring R/I is commutative, even though R is not commutative. The reader is urged to work through the specific case for $n = 2$ and $n = 3$.*

8.3   Isomorphism Theorems

Each of the three isomorphism theorems for groups has a direct analogue for rings, and these are the object of this section.

8.3.1 Theorem (First Isomorphism Theorem for Rings). *Let R and S be rings, and let $f : R \to S$ be a homomorphism with kernel I. Then $R/I \simeq f(R)$, a subring of S. In fact, the rule $\hat{f}(X) = f(x)$ for any $X \in R/I$ and any $x \in X$ determines an injective ring homomorphism $\hat{f} : R/I \to S$ such that $\hat{f} \circ \pi_I = f$, where π_I is the natural homomorphism from R to R/I.*

Proof. Apply Theorem 6.4.11 to obtain the injective additive homomorphism $\hat{f} : R/I \to S$ such that $\hat{f} \circ \pi_I = f$. Let $X, Y \in R/I$, and let $x \in X$, $y \in Y$. Then

$$\hat{f}(XY) = \hat{f}(xy + I) = f(xy) = f(x)f(y) = \hat{f}(x + I)\hat{f}(y + I) = \hat{f}(X)\hat{f}(Y),$$

and thus $\hat{f}$ preserves products.   □

We should emphasize that in Theorem 8.3.1, for any $x \in R$, $\hat{f}(x + I) = f(x)$. This is exactly the content of the assertion $\hat{f} \circ \pi_I = f$.

8.3.2 Lemma. *Let R be a ring, and let S be a subring of R and $I \lhd R$. Then $S + I$ is a subring of R and I is an ideal of $S + I$.*

Proof. As each of S and I are normal subgroups of the additive group of R, it follows from Lemma 6.4.12 that $S + I$ is a subgroup of the additive group of R. To complete the proof that $S + I$ is a subring of R, we must prove that $S + I$ is closed under multplication. But $(S + I)(S + I) = SS + SI + IS + II \subseteq S + I + I + I = S + I$, and so $S + I$ is closed under multiplication. Thus $S + I$ is a subring of R. Finally, $I = \{0\} + I \subseteq S + I$ and thus $I \lhd (S + I)$.   □

8.3.3 Theorem (Second Isomorphism Theorem for Rings). *Let R be a ring, and let S be a subring of R and $I \triangleleft R$. Then $(S+I)/I \simeq S/(S \cap I)$.*

Proof. By Lemma 8.3.2, $S + I$ is a subring of R and $I \triangleleft (S + I)$. Let $f : S \to (S + I)/I$ be the map defined by $f(s) = s + I$ for $s \in S$. Then for $s, t \in S$, $f(s+t) = s + t + I = (s+I) + (t+I) = f(s) + f(t)$, and $f(st) = st + I = (s+I)(t+I) = f(s)f(t)$, so f is a ring homomorphism. Moreover, f is surjective as for each $s \in S$ and $i \in I$, $s + i + I = s + I = f(s)$. By the First Isomorphism Theorem for rings, $(S + I)/\ker(f) \simeq f(S) = (S+I)/I$. But $\ker(f) = \{ s \in S \mid s + I = I \} = \{ s \in S \mid s \in I \}$, and so $\ker(f) = S \cap I$. The result follows now by the First Isomorphism theorem for rings. $\qquad \square$

8.3.4 Theorem (Third Isomorphism Theorem for Rings). *Let R be a ring and let I and J be ideals of R with $I \subseteq J$. The canonical homomorphism from R to R/I is surjective, so the image of J is an ideal of R/I. Denote the image of J under the canonical homomorphism from R to R/I by J/I. The assignment $a + I \mapsto a + J$ for all $a \in R$ defines a surjective homomorphism from R/I onto R/J whose kernel is J/I, and so $(R/I)/(J/I) \simeq R/J$.*

Proof. We prove first that $f : R/I \to R/J$ defined by $f(a + I) = a + J$ is well defined. Suppose that $a, b \in R$ are such that $a + I = b + I$. Then $a - b \in I \subseteq J$ and so $a + J = b + J$. Thus f is well defined, and is obviously a surjective ring homomorphism. Suppose that $a + I \in \ker(f)$. Then $a + J = 0_R + J$; that is, $a \in J$. On the other hand, for any $a \in J$, $f(a+I) = a+J = 0_R+J$, and so $a+I \in \ker(f)$. We have therefore proven that $\ker(f) = \{ a + I \mid a \in J \} = J/I$. By the First Isomorphism Theorem, $(R/I)/(J/I)$ is isomorphic to R/J. $\qquad \square$

Direct Product and Direct Sum of Rings

8.4.1 Definition. *Let $\mathscr{I}$ be a nonempty set, and for each $\lambda \in \mathscr{I}$, let R_λ be a ring. On the direct product of the additive groups of R_λ, $\lambda \in \mathscr{I}$, define multiplication as follows: for each $f, g \in \prod_{\lambda \in \mathscr{I}} R_\lambda$ (see Definition 6.5.1), let $fg \in \prod_{\lambda \in \mathscr{I}} R_\lambda$ be the map from $\mathscr{I}$ to $\cup_{\lambda \in \mathscr{I}} R_\lambda$ for which $(fg)(\lambda) = f(\lambda)g(\lambda)$ for each $\lambda \in \mathscr{I}$.*

8.4.2 Proposition. *Let $\mathscr{I}$ be a nonempty set, and for each $\lambda \in \mathscr{I}$, let R_λ be a ring. Then the additive group $\prod_{\lambda \in \mathscr{I}} R_\lambda$, with multiplication as defined in Definition 8.4.1, is a ring, called the direct product of the rings*

R_λ, $\lambda \in \mathscr{I}$. Moreover, $\prod_{\lambda \in \mathscr{I}} R_\lambda$ is unital if, and only if, for each $\lambda \in \mathscr{I}$, R_λ is unital, and if so, the map $\mathbb{1} : \mathscr{I} \to \cup_{\lambda \in \mathscr{I}} R_\lambda$ for which $\mathbb{1}(\lambda) = 1_{R_\lambda}$ for each $\lambda \in \mathscr{I}$ is the unity of $\prod_{\lambda \in \mathscr{I}} R_\lambda$.

Proof. Since the additive group of each R_λ, $\lambda \in \mathscr{I}$, is abelian, it follows from Corollary 6.5.5 that the additive group $\prod_{\lambda \in \mathscr{I}} R_\lambda$ is abelian. As well, for $f, g \in R = \prod_{\lambda \in \mathscr{I}} R_\lambda$, $fg \in R$ since for each $\lambda \in \mathscr{I}$, $(fg)(\lambda) = f(\lambda)g(\lambda) \in R_\lambda$ as both $f(\lambda)$ and $g(\lambda)$ are elements of R_λ. It remains to prove that multiplication is associative and distributes both left and right over addition. Let $f, g, h \in R$. Then for each $\lambda \in \mathscr{I}$, we have

$$(f(gh))(\lambda) = f(\lambda)((gh)(\lambda)) = f(\lambda)(g(\lambda)h(\lambda))$$
$$= (f(\lambda)g(\lambda))h(\lambda) = ((fg)(\lambda))h(\lambda) = ((fg)h)(\lambda)$$

and so $f(gh) = (fg)h$. As well, for each $\lambda \in \mathscr{I}$, we have

$$(f(g+h))(\lambda) = f(\lambda)(g+h)(\lambda) = f(\lambda)(g(\lambda) + h(\lambda))$$
$$= f(\lambda)g(\lambda) + f(\lambda)h(\lambda)) = (fg)(\lambda) + (fh)(\lambda)$$
$$= (fg + fh)(\lambda)$$

and thus $f(g+h) = fg + fh$. The argument that $(g+h)f = gf + hf$ is similar and is omitted.

Suppose now that for each $\lambda \in \mathscr{I}$, R_λ has unity, which we denote by 1_{R_λ}. Define $\mathbb{1} : \mathscr{I} \to \cup_{\lambda \in \mathscr{I}} R_\lambda$ by $\mathbb{1}(\lambda) = 1_{R_\lambda}$ for each $\lambda \in \mathscr{I}$. Since $\mathbb{1}(\lambda) \in R_\lambda$ for each $\lambda \in \mathscr{I}$, $\mathbb{1} \in R$. Let $f \in R$. Then for each $\lambda \in \mathscr{I}$, $\mathbb{1}f(\lambda) = \mathbb{1}(\lambda)f(\lambda) = 1_{R_\lambda}f(\lambda) = f(\lambda)$ since $f(\lambda) \in R_\lambda$. Thus $\mathbb{1}f = f$, and similarly, $f\mathbb{1} = f$, so $\mathbb{1}$ is the unity of R.

Conversely, suppose that R has unity, denoted by $\mathbb{1}$. Let $\gamma \in \mathscr{I}$, and let $x \in R_\gamma$. We prove that $x\mathbb{1}(\gamma) = x = \mathbb{1}(\gamma)x$. The additive group homomorphisms $\iota_\gamma : R_\gamma \to R$ were introduced in Proposition 6.5.6, and we have $\mathbb{1}\iota_\gamma(x) = \iota_\gamma(x) = \iota_\gamma(x)\mathbb{1}$. By definition of ι_γ, we have $\mathbb{1}\iota_\gamma(x)(\gamma) = \mathbb{1}(\gamma)x = \iota_\gamma(x) = x$, and $\iota_\gamma(x)\mathbb{1}(\gamma) = \iota_\gamma(x)(\gamma)\mathbb{1}(\gamma) = x\mathbb{1}(\gamma) = \iota_\gamma(x)(\gamma) = x$. Thus $\mathbb{1}(\gamma)$ is a unity for R_γ for each $\gamma \in \mathscr{I}$. $\square$

8.4.3 Proposition. *Let $\mathscr{I}$ be a nonempty set, and for each $\lambda \in \mathscr{I}$, let R_λ be a ring. Let $\gamma \in \mathscr{I}$. Then the mapping $\pi_\gamma : \prod_{\lambda \in \mathscr{I}} R_\lambda \to R_\gamma$ defined by $\pi_\gamma(f) = f(\gamma)$ for all $f \in \prod_{\lambda \in \mathscr{I}} R_\lambda$ is a surjective ring homomorphism with kernel $\{ f \in \prod_{\lambda \in \mathscr{I}} R_\lambda \mid f(\gamma) = 0_{R_\lambda} \}$. Moreover, if S is a ring such that for each $\lambda \in \mathscr{I}$, there is a ring homomorphism $\varphi_\lambda : S \to R_\lambda$, then there exists a unique ring homomorphism $\varphi : S \to \prod_{\lambda \in \mathscr{I}} R_\lambda$ such that for each $\lambda \in \mathscr{I}$, $\pi_\lambda \circ \varphi = \varphi_\lambda$.*

Proof. Let $R = \prod_{\lambda \in \mathscr{I}} R_\lambda$. Since each R_λ is an abelian group under addition, we may apply Proposition 6.5.4 to obtain the surjective additive group homomorphisms $\pi_\gamma : R \to R_\gamma$ with the specified kernel for each $\gamma \in \mathscr{I}$, so it remains to verify the multiplicative property. Let $f, g \in R$, and let $\gamma \in \mathscr{I}$. Then $\pi_\gamma(fg) = (fg)(\gamma) = f(\gamma)g(\gamma) = \pi_\gamma(f)\pi_\gamma(g)$, as required. Thus for each $\gamma \in \mathscr{I}$, π_γ is a ring homomorphism. $\square$

8.4.4 Definition. *Let $\mathscr{I}$ be a nonempty set, and for each $\lambda \in \mathscr{I}$, let R_λ be a ring. For each $f \in \prod_{\lambda \in \mathscr{I}} R_\lambda$, let $S_f = \{ \lambda \in \mathscr{I} \mid f(\lambda) \neq 0_{R_\lambda} \}$. S_f is called the support of f. The subset*

$$\{ f \in \prod_{\lambda \in \mathscr{I}} R_\lambda \mid S_f \text{ is finite} \}$$

is called the direct sum of the rings R_λ, $\lambda \in \mathscr{I}$, and denoted by $\oplus_{\lambda \in \mathscr{I}} R_\lambda$.

8.4.5 Proposition. *Let $\mathscr{I}$ be a nonempty set, and for each $\lambda \in \mathscr{I}$, let R_λ be a ring. Then $\oplus_{\lambda \in \mathscr{I}} R_\lambda$ is an ideal of $\prod_{\lambda \in \mathscr{I}} R_\lambda$.*

Proof. Let $R = \prod_{\lambda \in \mathscr{I}} R_\lambda$. Since the support of 0_R is $\varnothing$, which is finite, $0_R \in \oplus_{\lambda \in \mathscr{I}} R_\lambda$ and so $\oplus_{\lambda \in \mathscr{I}} R_\lambda \neq \varnothing$. Let $f, g \in \oplus_{\lambda \in \mathscr{I}} R_\lambda$ and $h \in R$. Since $S_{f-g} \subseteq S_f \cup S_g$, both of which are finite, it follows that S_{f-g} is finite and therefore $f - g \in \oplus_{\lambda \in \mathscr{I}} R_\lambda$. As well, if $\lambda \in \mathscr{I}$ is such that $(fg)(\lambda) \neq 0_{R_\lambda}$, then $f(\lambda)g(\lambda) \neq 0_{R_\lambda}$ and in particular, $f(\lambda) \neq 0_{R_\lambda}$. Thus $\lambda \in S_{fg}$ implies that $\lambda \in S_f$ and so $S_{fg} \subseteq S_f$. Since S_f is finite, so is S_{fg} and thus $fg \in \oplus_{\lambda \in \mathscr{I}} R_\lambda$. Similarly, $gf \in \oplus_{\lambda \in \mathscr{I}} R_\lambda$. By Proposition 8.2.5, $\oplus_{\lambda \in \mathscr{I}} R_\lambda$ is an ideal of $\prod_{\lambda \in \mathscr{I}} R_\lambda$. $\square$

As a consequence of Proposition 8.4.5, we see that the direct sum of rings R_λ, $\lambda \in \mathscr{I}$ is, in fact, a ring. In general, the direct sum is a proper ideal of the direct product of rings. By Proposition 8.4.2, $\prod_{\lambda \in \mathscr{I}} R_\lambda$ has unity if, and only if, for each $\lambda \in \mathscr{I}$, R_λ has unity. Suppose that for each $\lambda \in \mathscr{I}$, R_λ has unity 1_{R_λ}. Then by Proposition 8.4.2 and the definition of support, $S_{\mathbb{1}} = \mathscr{I}$. Thus if $\mathscr{I}$ is an infinite set, the unity of $\prod_{\lambda \in \mathscr{I}} R_\lambda$ is not an element of $\oplus_{\lambda \in \mathscr{I}} R_\lambda$. The argument used in the proof of Proposition 8.4.2 can be used to establish that if $\oplus_{\lambda \in \mathscr{I}} R_\lambda$ had unity, it would be $\mathbb{1}$, and thus if $\mathscr{I}$ is infinite, $\oplus_{\lambda \in \mathscr{I}}$ has no unity. It is apparent from the definition however that if $\mathscr{I}$ is a finite (nonempty) set, then $\oplus_{\lambda \in \mathscr{I}} R_\lambda = \prod_{\lambda \in \mathscr{I}} R_\lambda$, and it is this situation that most interests us.

8.4.6 Notation. *For any positive integer n, the direct sum of n rings will always use J_n as its index set. For rings $R_1, R_2, \ldots, R_n$, the direct* sum $\oplus_{i \in J_n} R_i$ will usually be denoted by $R_1 \oplus R_2 \oplus \cdots R_n$, and its elements will be written as sequences of length n whose i^{th} entry belongs to R_i for $i = 1, 2, \ldots, n$.

For example, $\mathbb{Z} \oplus \mathbb{Z} = \{(m, n) \mid m, n \in \mathbb{Z}\}$, and

$$\mathbb{Z} \oplus \mathbb{Q} \oplus \mathbb{R} = \{(m, l, r) \mid m \in \mathbb{Z}, \ l \in \mathbb{Q}, \ r \in \mathbb{R}\}.$$

With the above notation, the operations of addition and multiplication in $R_1 \oplus R_2 \oplus \cdots \oplus R_n$ are given by:

$$(r_1, r_2, \ldots, r_n) + (s_1, s_2, \ldots, s_n) = (r_1 + s_1, r_2 + s_2, \ldots, r_n + s_n)$$
$$(r_1, r_2, \ldots, r_n)(s_1, s_2, \ldots, s_n) = (r_1 s_1, r_2 s_2, \ldots, r_n s_n)$$

for each $(r_1, r_2, \ldots, r_n), (s_1, s_2, \ldots, s_n) \in R_1 \oplus R_2 \oplus \cdots \oplus R_n$.

We illustrate the above results with an interesting example.

8.4.7 Example. *Consider the ring* $\mathbb{Z}_2 = \mathbb{Z}/2\mathbb{Z} = \{0 + 2\mathbb{Z}, 1 + 2\mathbb{Z}\}$. *To simplify notation, it is conventional to denote* $0 + 2\mathbb{Z}$ *simply by* 0, *and similarly, denote* $1 + 2\mathbb{Z}$ *by* 1. *Recall that the addition and multiplication operations in* $\mathbb{Z}_2$ *are as shown in Figure 8.1.*

+	0	1
0	0	1
1	1	0

·	0	1
0	0	0
1	0	1

Fig. 8.1

Let n *be a positive integer, and consider the direct sum of* $R_1, R_2, \ldots, R_n$, *where for each* $i = 1, 2, \ldots, n$, $R_i = \mathbb{Z}_2$. *We could denote this either by* $\oplus_{i=1}^n \mathbb{Z}_2$ *or, since it is a finite direct sum and is therefore equal to the direct product, by* $\prod_{i=1}^n \mathbb{Z}_2$. *We shall use the latter notation, and as a further simplication of the notation, we shall write* $\mathbb{Z}_2^n$ *instead of* $\prod_{i=1}^n \mathbb{Z}_2$. *Thus for example,*

$$\mathbb{Z}_2^3 = \{(0,0,0), (0,0,1), (0,1,0), (0,1,1), (1,0,0), (1,0,1), (1,1,0), (1,1,1)\}.$$

By the definition of support (see Definition 8.4.4), for each $f \in \mathbb{Z}_2^n$, $S_f \subseteq J_n$, *or equivalently,* $S_f \in \mathscr{P}(J_n)$. *For example, the supports of the eight elements of* $\mathbb{Z}_2^3$ *are* $S_{(0,0,0)} = \varnothing$, $S_{(0,0,1)} = \{3\}$, $S_{(0,1,0)} = \{2\}$, $S_{(0,1,1)} = \{2,3\}$, $S_{(1,0,0)} = \{1\}$, $S_{(1,0,1)} = \{1,3\}$, $S_{(1,1,0)} = \{1,2\}$, *and* $S_{(1,1,1)} = \{1,2,3\}$. $(0,0,0)$ *and* $(1,1,1)$ *are the zero and the unity elements, respectively, of* $\mathbb{Z}_2^3$.

Note that for $f, g \in \mathbb{Z}_2^n$, $S_{f+g} \subseteq S_f \cup S_g$, and if $i \in S_f \cap S_g$, then $(f + g)(i) = f(i) + g(i) = 1 + 1 = 0$, so $i \notin S_{f+g}$. Thus

$$S_{f+g} \subseteq (S_f \cup S_g) - (S_f \cap S_g) = S_f \bigtriangleup S_g.$$

On the other hand, if $i \in S_f \bigtriangleup S_g$, then $i \in S_f$ or $i \in S_g$, but $i \notin S_f \cap S_g$, so either $f(i) = 0$ and $g(i) = 1$, or else $f(i) = 1$ and $g(i) = 0$, but in either case, $(f + g)(i) = f(i) + g(i) = 1 \neq 0$ and so $i \in S_{f+g}$. Thus $S_f \bigtriangleup S_g \subseteq S_{f+g}$ and so $S_{f+g} = S_f \bigtriangleup S_g$. A similar argument proves that $S_{fg} = S_f \cap S_g$. Now consider the mapping $\theta : \mathbb{Z}_2^n \to \mathscr{P}(J_n)$ given by $\theta(f) = S_f$ for each $f \in \mathbb{Z}_2^n$. For $f, g \in \mathbb{Z}_2^n$, we have $\theta(f + g) = S_{f+g} = S_f \bigtriangleup S_g = \theta(f) \bigtriangleup \theta(g)$ and $\theta(fg) = S_{fg} = S_f \cap S_g = \theta(f) \cap \theta(g)$. Recall (see Example 8.1.2 (iii)) that $\mathscr{P}(J_n)$ is a ring with addition given by symmetric difference and multiplication given by intersection. Moreover, $\theta(1_{\mathbb{Z}_2^n}) = J_n = \mathbb{1}_{\mathscr{P}(J_n)}$, so θ is a unital ring homomorphism from $\mathbb{Z}_2^n$ to $\mathscr{P}(J_n)$. Moreover, the only element of $\mathbb{Z}_2^n$ with empty support is the zero element, and so $\ker(\theta) = \{ 0_{\mathbb{Z}_2^n} \}$. Thus θ is injective. Finally, since $|\mathbb{Z}_2^n| = 2^n$ and $|\mathscr{P}(J_n)| = 2^n$, θ is bijective, and so θ is a ring isomorphism from $\mathbb{Z}_2^n$ to $\mathscr{P}(J_n)$. Thus abstractly, $\mathscr{P}(J_n)$ is simply the direct sum of n copies of $\mathbb{Z}_2$.

8.4.8 Example. *Let n be a positive integer. We determine the ideals of $\mathscr{P}(J_n)$ (and thereby determine the ideals of $\mathbb{Z}_2^n$). Suppose that I is an ideal of $\mathscr{P}(J_n)$, and let $A = \{ i \in X \mid X \in I \}$. Let $i \in A$. Then there exists $X \in I$ such that $i \in X$ and thus $\{i\} \cap X = \{i\}$ is an element of I. This proves that for each $i \in A$, $\{i\} \in I$. Since the symmetric difference of pairwise-disjoint sets is simply their union, it follows that every subset of A is in I, so $\mathscr{P}(A) \subseteq I$. On the other hand, let $X \in I$. Then by definition of A, $X \subseteq A$ and thus $X \in \mathscr{P}(A)$. This proves that $I = \mathscr{P}(A)$; that is, every ideal of $\mathscr{P}(J_n)$ is of the form $\mathscr{P}(A)$ for some $A \subseteq J_n$. Moreover, for any $A \subseteq J_n$, $\mathscr{P}(A)$ is a subring of $\mathscr{P}(J_n)$. Let $X \in \mathscr{P}(A)$ and $Y \in \mathscr{P}(J_n)$. Then $X \cap Y \subseteq X \subseteq A$ and so $X \cap Y \in \mathscr{P}(A)$, which proves that $\mathscr{P}(A)$ is an ideal of $\mathscr{P}(J_n)$. Note that for any $A \subseteq J_n$, $\mathscr{P}(A) = \{ A \cap X \mid X \in \mathscr{P}(J_n) \}$, so $\mathscr{P}(A)$ is the principal ideal of $\mathscr{P}(J_n)$ generated by A. Thus every ideal of $\mathscr{P}(J_n)$ is principal.*

Under the isomorphism between $\mathbb{Z}_2^n$ and $\mathscr{P}(J_n)$ that was constructed in Example 8.4.7, for $A \subseteq J_n$, the ideal $\mathscr{P}(A)$ of $\mathscr{P}(J_n)$ corresponds to the principal ideal of $\mathbb{Z}_n^n$ generated by $(\delta_{1,A}, \delta_{2,A}, \ldots, \delta_{n,A})$, where for each $i = 1, 2, \ldots, n$, $\delta_{i,A} = 0$ if $i \notin A$, otherwise $\delta_{i,A} = 1$.

Our next goal is to establish the analogues of the group-theoretic results contained in Propositions 6.5.6, 6.5.7, and 6.5.8.

8.4.9 Proposition. *Let $\mathscr{I}$ be a nonempty set, and for each $\lambda \in \mathscr{I}$, let R_λ be a ring. Let $\gamma \in \mathscr{I}$. Then the mapping $\iota_\gamma : R_\gamma \to R = \oplus_{\lambda \in \mathscr{I}} R_\lambda$, where for each $x \in R_\gamma$, $\iota_\gamma(x) \in \oplus_{\lambda \in \mathscr{I}} R_\lambda$, is the mapping given by*

$$(\iota_\gamma(x))(\lambda) = \begin{cases} 0_{R_\lambda} & \text{if } \lambda \neq \gamma \\ x & \text{if } \lambda = \gamma \end{cases}$$

for each $\lambda \in \mathscr{I}$, is an injective homomorphism with image

$$\{ f \in \oplus_{\lambda \in \mathscr{I}} R_\lambda \mid f(\lambda) = 0_{R_\lambda} \text{ for all } \lambda \in \mathscr{I} \text{ with } \lambda \neq \gamma \},$$

which is an ideal of $\oplus_{\lambda \in \mathscr{I}} R_\lambda$. Moreover, for each $\lambda \in \mathscr{I}$, $\pi_\lambda \circ \iota_\lambda$ is the identity map on R_λ, while for any $\delta \in \mathscr{I}$ with $\lambda \neq \delta$, $\pi_\delta \circ \iota_\lambda$ is the zero map from R_λ into R_δ.

Proof. Let $x, y \in R_\gamma$. Then $\iota_\gamma(xy)(\gamma) = xy = \iota_\gamma(x)(\gamma)\iota_\gamma(y)(\gamma) = (\iota_\gamma(x)\iota_\gamma(y))(\gamma)$, while for $\delta \in \mathscr{I}$ with $\delta \neq \gamma$, $\iota_\gamma(xy)(\delta) = 0_{R_\delta} = 0_{R_\delta}0_{R_\delta} = \iota_\gamma(x)(\delta)\iota_\gamma(y)(\delta) = (\iota_\gamma(x)\iota_\gamma(y))(\delta)$. This proves that for all $\delta \in \mathscr{I}$, $\iota_\gamma(xy)(\delta) = (\iota_\gamma(x)\iota_\gamma(y))(\delta)$ and so $\iota_\gamma(xy) = \iota_\gamma(x)\iota_\gamma(y)$. Thus ι_γ is a ring homomorphism. Suppose now that $x, y \in R_\gamma$ are such that $\iota_\gamma(x) = \iota_\gamma(y)$. Then in particular, $x = (\iota_\gamma(x))(\gamma) = (\iota_\gamma(y))(\gamma) = y$, and so ι_γ is injective. The definition of ι_γ makes it clear that

$$\iota_\gamma(R_\gamma) \subseteq \{ f \in \oplus_{\lambda \in \mathscr{I}} R_\lambda \mid f(\lambda) = 0_{R_\lambda} \text{ for all } \lambda \in \mathscr{I} \text{ with } \lambda \neq \gamma \},$$

so let $f \in \oplus_{\lambda \in \mathscr{I}} R_\lambda$ be such that $f(\lambda) = 0_{R_\lambda}$ for all $\lambda \in \mathscr{I}$ with $\lambda \neq \gamma$. Let $x = f(\gamma) \in R_\gamma$. Then $\iota_\gamma(x)(\lambda) = 0_{R_\lambda} = f(\lambda)$ for all $\lambda \in \mathscr{I}$ with $\lambda \neq \gamma$, while $\iota_\gamma(x)(\gamma) = x = f(\gamma)$ and thus for all $\lambda \in \mathscr{I}$, $\iota_\gamma(x)(\lambda) = f(\lambda)$. This means that $f = \iota_\gamma(x)$ and thus $f \in \iota_\gamma(R_\gamma)$. Let $g \in R$ and $f \in \iota_\gamma(R_\gamma)$. We prove that gf and fg belong to $\iota_\gamma(R_\gamma)$. Let $\lambda \in \mathscr{I}$ with $\lambda \neq \gamma$. Then $(gf)(\lambda) = g(\lambda)f(\lambda) = g(\lambda)0_{R_\lambda} = 0_{R_\lambda}$ while $(gf)(\gamma) = f(\gamma)g(\gamma) \in R_\gamma$. Let $x = (gf)(\gamma)$. Then $(gf)(\delta) = \iota_\gamma(x)(\delta)$ for all $\delta \in \mathscr{I}$, and so $gf = \iota_\gamma(x) \in \iota_\gamma(R_\gamma)$. A similar argument would prove that $fg \in \iota_\gamma(R_\gamma)$, and thus $\iota_\gamma(R_\gamma) \triangleleft R$.

Finally, let $\delta \in \mathscr{I}$. Then by definition of π_δ and of ι_γ, respectively, we have that for any $x \in R_\gamma$,

$$\pi_\delta \circ \iota_\gamma(x) = \iota_\gamma(x)(\delta) = \begin{cases} 0_{R_\gamma} & \text{if } \delta \neq \gamma \\ x & \text{if } \delta = \gamma, \end{cases}$$

and thus $\pi_\delta \circ \iota_\gamma$ is the zero map when $\delta \neq \gamma$, while $\pi_\gamma \circ \iota_\gamma$ is the identity map on R_γ. $\square$

8.4.10 Proposition. *Let n be a positive integer, and for each $i \in J_n$, let R_i be a ring. Then in the ring $R = \oplus_{i \in J_n} R_i$, the following hold:*

(i) *For each $i \in J_n$, let $S_i = \iota_i(R_i)$, so $S_i \lhd R$. Then $R = S_1 + S_2 + \cdots + S_n$, and in fact, for each $r \in R$, there exist unique $s_i \in S_i$, $i \in J_n$, for which $r = s_1 + s_2 + \cdots + s_n$.*

(ii) *For each $i \in J_n$, $S_i \cap (S_1 + S_2 + \cdots + S_{i-1} + S_{i+1} + \cdots + S_n) = \{0_R\}$ (where if $i = 1$, we mean that $S_1 \cap (S_2 + \cdots + S_n) = \{0_R\}$, and if $i = n$, we mean that $S_n \cap (S_1 + S_2 + \cdots + S_{n-1}) = \{0_R\}$).*

Proof. Let $r = (r_1, r_2, \ldots, r_n) \in R$. For each $i \in J_n$, let $s_i = \iota_i(r_i) \in S_i$. Then

$$r = (r_1, r_2, \ldots, r_n) = \iota_1(r_1) + \iota_2(r_2) + \cdots + \iota_n(r_n)$$

$$= s_1 + s_2 + \cdots + s_n,$$

and so $R = S_1 + S_2 + \cdots + S_n$. Suppose now that for each $i \in J_n$, $t_i \in S_i$ satisfy $r = t_1 + t_2 + \cdots + t_n$. Then $0_R = (s_1 - t_1) + \cdots + (s_n - t_n)$, and for each $i \in J_n$, $s_i - t_i \in S_i$. Let $i \in J_n$. Then $0_i = \pi_i(0_R) = \pi_i((s_1 - t_1) + \cdots + (s_n - t_n)) = \pi_i(s_1 - t_1) + \cdots + \pi_i(s_n - t_n) = \pi_i(s_i - t_i)$ since for all $j \in J_n$ with $j \neq i$, $s_j - t_j \in S_j = \iota_j(R_j)$, and by Proposition 8.4.9, $\pi_i \circ \iota_j$ is the zero map when $j \neq i$. As well, since $s_i - t_i \in S_i = \iota_i(R_i)$, there exists $x \in R_i$ such that $s_i - t_i = \iota_i(x)$. Then $0_i = \pi_i \circ \iota_i(x) = x$ since by Proposition 8.4.9, $\pi_i \circ \iota_i$ is the identity map on R_i. But then $s_i - t_i = \iota_i(0_i) = 0_R$, and thus $s_i = t_i$.

Finally, let $i \in J_n$, and let $x \in S_i \cap (S_1 + S_2 + \cdots + S_{i-1} + S_{i+1} + \cdots + S_n)$. Then $x \in S_i$, and for each $j \in J_n$ with $j \neq i$, there exists $x_j \in S_j$ such that

$$x = \sum_{\substack{j \in J_n \\ j \neq i}} x_j.$$

Let $x_i = -x$. Then $0_R = \sum_{j \in J_n} x_j$, and by the uniqueness established above, since $0_R \in S_j$ for every $j \in J_n$ we have $x_j = 0_R$ for every $j \in J_n$. In particular, $x = -x_i = 0_R$, as required. $\qquad\qquad\square$

For example, if we consider the direct sum (which is the same as the direct product since we only have a finite number of rings) $R = \mathbb{Z} \oplus \mathbb{Q} \oplus \mathbb{R}$, then upon indexing in the natural order, we have taken $R_1 = \mathbb{Z}$, $R_2 = \mathbb{Q}$, and $R_3 = \mathbb{R}$. Thus $\iota_1 : \mathbb{Z} \to R$ is the map $\iota_1(m) = (m, 0, 0)$, $\pi_1 : R \to \mathbb{Z}$ is the map $\pi_1(m, q, r) = m$, $\iota_2 : \mathbb{Q} \to R$ is the map $\iota_1(q) = (0, q, 0)$, $\pi_2 : R \to \mathbb{Q}$ is the map $\pi_2(m, q, r) = q$, and $\iota_3 : \mathbb{R} \to R$ is the map $\iota_3(r) = (0, 0, r)$, $\pi_3 : R \to \mathbb{R}$ is the map $\pi_3(m, q, r) = r$. $S_1 = \{(m, 0, 0) \mid m \in \mathbb{Z}\}$, $S_2 = \{(0, q, 0) \mid q \in \mathbb{Q}\}$, and $S_3 = \{(0, 0, r) \mid r \in \mathbb{R}\}$.

8.4.11 Proposition. *Let R be a ring, and suppose that for some positive integer n, R has ideals $S_1, S_2, \ldots, S_n$ for which the following hold:*

(i) $R = S_1 + S_2 + \cdots + S_n$.

(ii) For each $i \in J_n$, $S_i \cap (S_1 + S_2 + \cdots + S_{i-1} + S_{i+1} + \cdots + S_n) = \{0_R\}$ (where if $i = 1$, we mean that $S_1 \cap (S_2 + \cdots + S_n) = \{0_R\}$, and if $i = n$, we mean that $S_n \cap (S_1 + S_2 + \cdots + S_{n-1}) = \{0_R\}$).

Then for each element $r \in R$, for each $i \in J_n$, there exists a unique $s_i \in S_i$ such that $r = s_1 + s_2 + \cdots + s_n$, and the map $\theta : R \to \oplus_{i \in J_n} S_i$ for which $\theta(r) = (s_1, s_2, \ldots, s_n)$, where $s_i \in S_i$, $i \in J_n$, are the unique elements such that $r = s_1 + \cdots + s_n$, is a ring isomorphism such that for each $i \in J_n$, $\theta\big|_{S_i} = \iota_i$.

Proof. Let $r \in R$. Since $R = S_1 + S_2 + \cdots + S_n$, there exist $s_i \in S_i$, $i \in J_n$, such that $r = r_1 + r_2 + \cdots + r_n$. Suppose that $t_i \in S_i$, $i \in J_n$ are such that $r = t_1 + \cdots + t_n$. Then for each $i \in J_n$, $t_i - s_i \in S_i \cap (S_1 + S_2 + \cdots + S_{i-1} + S_{i+1} + \cdots + S_n) = \{0_R\}$, and so $s_i = t_i$. Define $\theta : R \to \oplus_{i \in J_n} S_i$ by $\theta(r) = (s_1, s_2, \ldots, s_n)$, where $r = s_1 + \cdots + s_n$ and $s_i \in S_i$ for each $i \in J_n$. For $x, y \in R$, there exist $s_i, t_i \in S_i$, $i \in J_n$ such that $x = s_1 + \cdots + s_n$ and $y = t_1 + \cdots + t_n$. Then $x + y = (s_1 + t_1) + \cdots + (s_n - t_n)$, and $xy = \sum_{i \in J_n} \sum_{j \in J_n} s_i t_j$. Now, for $i, j \in J_n$ with $i \neq j$, $s_i t_j \in S_i \cap S_j$ since each of S_i and S_j are ideals of R. But by (ii), $S_i \cap S_j = \{0_R\}$, and thus $s_i t_j = 0_R$. We conclude that $xy = \sum_{i \in J_n} s_i t_i$. By definition, $\theta(x + y) = (s_1 + t_1, \ldots, s_n + t_n) = (s_1, \ldots, s_n) + (t_1, \ldots, t_n) = \theta(x) + \theta(y)$, and $\theta(xy) = (s_1 t_1, \ldots, s_n t_n) = (s_1, \ldots, s_n)(t_1, \ldots, t_n) = \theta(x)\theta(y)$, so θ is a ring homomorphism. Furthermore, since $\theta(x) = (0, 0, \ldots, 0)$ means that $x = 0 + 0 + \cdots + 0 = 0_R$, θ is injective. Finally, for any $(s_1, s_2, \ldots, s_n) \in \oplus_{i \in J_n} S_i$, $r = s_1 + s_2 + \cdots + s_n \in R$ satisfies $\theta(r) = (s_1, s_2, \ldots, s_n)$ and so θ is surjective. Let $i \in J_n$, and let $x \in S_i$. Then $\iota_i(x) = (0, \cdots, x, 0, \cdots, 0)$, and $\theta(x) = (0, 0, \ldots, x, \ldots, 0)$, since $x = 0 + 0 + \cdots + x + \cdots + 0$. Thus $\iota_i = \theta\big|_{S_i}$. $\qquad\square$

As an interesting example of Proposition 8.4.11, consider the subring

$$R = \left\{ \begin{bmatrix} A & 0_{2\times 3} \\ 0_{3\times 2} & B \end{bmatrix} \,\middle|\, A \in M_{2\times 2}(\mathbb{R}),\ B \in M_{3\times 3}(\mathbb{R}) \right\}$$

of $M_{5\times 5}(\mathbb{R})$, where $0_{2\times 3}$ is the 2×3 matrix with all entries 0, and $0_{3\times 2}$ is the 3×2 matrix with all entries 0. The reader is urged to verify that $S_1 = \left\{ \begin{bmatrix} A & 0_{2\times 3} \\ 0_{3\times 2} & 0_{3\times 3} \end{bmatrix} \,\middle|\, A \in M_{2\times 2}(\mathbb{R}) \right\}$ and $S_2 = \left\{ \begin{bmatrix} 0_{2\times 2} & 0_{2\times 3} \\ 0_{3\times 2} & B \end{bmatrix} \,\middle|\, B \in M_{3\times 3}(\mathbb{R}) \right\}$ are ideals of R such that $R = S_1 + S_2$ and $S_1 \cap S_2 = \{0_{5\times 5}\}$. Thus $R \simeq S_1 \oplus S_2$.

Principal Ideal Domains and Unique Factorization Domains

Integral domains were introduced in Definition 8.1.17 as cancellative unital rings (recall that a cancellative ring R is a commutative ring whose only divisor of 0_R is 0_R itself), and an integral domain D for which the group of units is $D^* = D - \{0_D\}$ is called a field. Thus every field is an integral domain, while $\mathbb{Z}$ is the classical example of an integral domain that is not a field. However, $\mathbb{Z}$ is rather a special type of integral domain. It was shown in Example 8.2.6 that the ideals of $\mathbb{Z}$ are precisely the subsets of the form $n\mathbb{Z}$ for $n \in \mathbb{Z}$.

8.5.1 Lemma. *Let R be a commutative unital ring. For each $r \in R$, the set $rR = \{rx \mid x \in R\}$ is an ideal of R containing a.*

Proof. Let $a, b \in rR$. Then $a = rx$ and $b = ry$ for some $x, y \in R$, and thus $a + b = rx + ry = r(x + y) \in rR$, while for any $t \in R$, $at = ta = rxt \in rR$. Since $0_R = r(0_R) \in rR$, it follows that rR is an ideal of R. Finally, since R has unity 1_R, $r = r1_R \in rR$. $\qquad\qquad\square$

8.5.2 Definition. *In a commutative unital ring R, an ideal of the form aR, $a \in R$, is called a principal ideal of R with generator a, and is denoted by (a). An integral domain in which every ideal is principal is called a principal ideal domain, or PID for short.*

Note that any field F is a principal ideal domain, since if $I \triangleleft F$, then either $I = \{0_F\} = (0_F)$, or else there exists $a \in I$ with $a \neq 0_F$, in which case for any $x \in F$, $x = aa^{-1}x \in I$, so $I = F = (1_F)$. Thus our familiar examples of integral domains are PID's. In the following example, we provide an integral domain that is not a PID.

8.5.3 Example. *Let $D = \{m + n\sqrt{10} \mid m, n \in \mathbb{Z}\}$, so $D \subseteq \mathbb{R}$. Since $1 = 1 + 0\sqrt{10} \in D$, D is not empty. For $m, n, m_1, n_1 \in \mathbb{Z}$, $m + n\sqrt{10} + m_1 + n_1\sqrt{10} = (m + m_1) + (n + n_1)\sqrt{10} \in D$, and $(m + n\sqrt{10})(m_1 + n_1\sqrt{10}) = mm_1 + 10nn_1 + (nm_1 + mn_1)\sqrt{10} \in D$, so D is a subring of $\mathbb{R}$. Since $\mathbb{R}$ is a field and thus an integral domain, it follows that the only zero divisor of D is 0. Since D is a commutative unital ring with only zero divisor 0, D is an integral domain. We prove that D is not a PID. Note that for $m, n, r, s \in \mathbb{Z}$, $m + n\sqrt{10} = r + s\sqrt{10}$ if, and only if, $m - r = (s - n)\sqrt{10}$. By using the method used in Example 1.5.4, in which it was shown that $\sqrt{2}$ is not rational, one may show that $\sqrt{10}$ is not rational. Thus $m - r = (s - n)\sqrt{10}$ holds if, and only if, $m = r$ and $s = n$. Now, let I denote the sum of the*

principal ideals (3) *and* $(1+\sqrt{10})$, *so* $I = (3)+(1+\sqrt{10}) \triangleleft D$. *We claim that* $1 \notin D$. *For suppose that* $1 \in D$. *Then there exist* $m, n, r, s \in \mathbb{Z}$ *such that*
$$1 = 3(m+n\sqrt{10})+(1+\sqrt{10})(r+s\sqrt{10}) = (3m+r+10s)+(3n+s+r)\sqrt{10},$$
from which we obtain $1 = 3m + r + 10s$, *and* $3n + r + s = 0$. *But then* $1 = 3m+9s-3n$, *which means that* 3 *divides* 1. *Since this is not the case, we conclude that* $1 \notin I$, *and so* $I \neq D$. *We claim that* I *is not a principal ideal of* D. *For suppose to the contrary that there exists* $a + b\sqrt{10} \in I$ *such that* $I = (a+b\sqrt{10})$. *Then since* $3 \in I$ *and* $1+\sqrt{10} \in I$, *there exist* $m, n, r, s \in \mathbb{Z}$ *such that* $3 = (a+b\sqrt{10})(m+n\sqrt{10})$ *and* $1 + \sqrt{10} = (a+b\sqrt{10})(r+s\sqrt{10})$. *From the first equation, we obtain* $3 + 0\sqrt{2} = am + 10bn + (an + bm)\sqrt{10}$ *and thus*

$$am + 10bn = 3 \quad and \quad an + bm = 0.$$

Upon multiplication by a *and* $10b$, *respectively, we obtain* $a^2m+10abn = 3a$ *and* $10ban + 10b^2m = 0$, *so* $(a^2 - 10b^2)m = 3a$. *Again, since* $\sqrt{10}$ *is not rational,* $a^2 - 10b^2 \neq 0$, *and thus* $a^2 - 10b^2$ *is a divisor of* $3a$. *Now go back and multiply by* b *and* a, *respectively, to obtain* $bam + 10b^2n = 3b$ *and* $a^2n + abm = 0$, *so* $(a^2 - 10b^2)n = -3b$ *and so* $a^2 - 10b^2$ *is a divisor of* $3b$. *From the equation* $1 + \sqrt{10} = (a + b\sqrt{10})(r + s\sqrt{10})$, *we obtain* $1 = ar + 10bs + (as + rb)\sqrt{10}$ *and so* $1 = ar + 10bs$. *Multiply this by* 3 *to get* $3 = 3ar + (10s)(3b)$. *Since* $a^2 - 10b^2$ *is a divisor of both* $3a$ *and* $3b$, $a^2 - 10b^2$ *is a divisor of* 3. *Thus* $a^2 - 10b^2 = \pm 1, \pm 3$. *Since* $a^2 - 10b^2 = (a+b\sqrt{10})(a-b\sqrt{10}) \in (a+b\sqrt{10})D = I$, *and* $I \neq D$, *it follows that* $a^2 - 10b^2 \neq \pm 1$ *and thus* $a^2 - 10b^2 = \pm 3$. *In* $\mathbb{Z}_5$, $[a^2]_5 = [a^2 - 10b^2]_5 = \pm[3]_5 = [2]_5, [3]_5$. *However,* $[a]_5$ *is one of* $[0]_5, [1]_5, [2]_5, [3]_5, [4]_5$, *and the squares of these five elements in* $\mathbb{Z}_5$ *are* $[0]_5, [1]_5, [4]_5, [4]_5, [1]_5$, *respectively. Thus* $a^2 - 10b^2 = \pm 3$ *is not possible either, and so we conclude that* I *is not a principal ideal, from which it follows that* D *is not a PID.*

The importance of the notion of PID stems from the fact that we can prove the existence of greatest common divisors in a PID.

8.5.4 Definition. *Let* D *be an integral domain. For* $a, b \in D$, *we say that* a *divides* b *in* D *if* $(b) \subseteq (a)$; *that is, if* $b \in (a)$, *in which case there exists* $x \in D$ *such that* $b = ax$. *An element* a *of* D *is said to be prime in* D *if* $a \neq 0_D$, $a \notin U(D)$, *and if for all* $x, y \in D$, *if* a *divides* xy, *then either* a *divides* x *or else* a *divides* y, *while* a *is said to be irreducible in* D *if for all* $x, y \in D$, *if* $a = xy$, *then either* $x \in U(D)$ *or else* $y \in U(D)$. *Finally, for* $a, b \in D$, *not both zero, a greatest common divisor of* a *and* b *is an element* $d \in D$ *such that* d *is a divisor of both* a *and* b, *and if* $c \in D$ *is any common* divisor of a and b (that is, c is a divisor of both a and of b), then c is a divisor of d.

It is important to know the relationship between two elements a and b of an integral domain D in the situation when $(a) = (b)$.

8.5.5 Lemma. *Let D be an integral domain, and let $a, b \in D$. Then $(a) = (b)$; that is, a divides b and b divides a, if, and only if, there exists $u \in U(D)$ such that $a = bu$.*

Proof. Suppose first that $(a) = (b)$. Then $a \in (b)$ and so there exists $x \in D$ such that $a = bx$. As well, $b \in (a)$, and so there exists $y \in D$ such that $b = ay$. Thus $b = bxy$ and so $b(1_D - xy) = 0_D$. Either $b = 0_D$, in which case $a = bx = 0_D$ and so $a = bx$ where $x = 1_D \in U(D)$, or else $1_D - xy = 0_D$ and so $x, y \in U(D)$, giving $a = bx$ with $x \in U(D)$, as required.

Conversely, suppose that there exists $u \in U(D)$ such that $a = bu$. Then $a \in (b)$, so $(a) \subseteq (b)$. But as $u \in U(D)$, u^{-1} exists and we have $b = au^{-1}$, so $b \in (a)$ and thus $(b) \subseteq (a)$. This proves that $(a) = (b)$. $\square$

8.5.6 Proposition. *In an integral domain, every prime element is irreducible.*

Proof. Let D be an integral domain, and let a be prime in D. Suppose that $x, y \in D$ are such that $a = xy$. Then $xy = a(1_D)$ and so a divides xy. Since a is prime, either a divides x or else a divides y. Suppose first that a divides x. Then there exists $r \in D$ such that $x = ra$, so $a = ray = rya$ and thus $(1_D - ry)a = 0_D$. Since $a \neq 0_D$ and the only zero divisor of D is 0_D, it follows that $1_D = ry$ and thus $y \in U(D)$. Similarly, if a divides y, then $x \in U(D)$, and thus a is irreducible. $\square$

It is in general not the case that every irreducible element of an integral domain is prime.

8.5.7 Example. *In the integral domain D presented in Example 8.5.3, we claim that the element $1 + \sqrt{10}$ is irreducible but not prime. To prove that it is irreducible, suppose that $m + n\sqrt{10}, r + s\sqrt{10} \in D$ are such that $1 + \sqrt{10} = (m + n\sqrt{10})(r + s\sqrt{10}) = (mr + 10ns) + (ms + nr)\sqrt{10}$, and so*

$1 - \sqrt{10} = (mr + 10ns) - (ms + nr)\sqrt{10}$. *Then*

$$-9 = (1 + \sqrt{10})(1 - \sqrt{10})$$
$$= \big((mr + 10ns) + (ms + nr)\sqrt{10}\big)\big((mr + 10ns) - (ms + nr)\sqrt{10}\big)$$
$$= (mr + 10ns)^2 - 10(ms + nr)^2$$
$$\qquad + \big((ms + nr)(mr + 10ns) - (mr + 10ns)(ms + nr)\big)\sqrt{10}$$
$$= m^2r^2 + 100n^2s^2 + 20mrns - 10m^2s^2 - 10n^2r^2 - 20mnrs$$
$$= m^2r^2 + 100n^2s^2 - 10m^2s^2 - 10n^2r^2$$
$$= (m^2 - 10n^2)(r^2 - 10s^2).$$

Thus $m^2 - 10n^2 = \pm1, \pm3, \pm9$. *If* $m^2 - 10n^2 = \pm1$, *then* $m - n\sqrt{10}, m + n\sqrt{10} \in D$ *and* $(m - n\sqrt{10})(m + n\sqrt{10}) = m^2 - 10n^2 = \pm1$, *which means that* $m + n\sqrt{10} \in U(D)$. *If* $m^2 - 10n^2 = \pm9$, *then* $r^2 - 10s^2 = \pm1$ *and in this case,* $r + s\sqrt{10} \in U(D)$. *The last case to consider occurs when* $m^2 - 10s^2 = \pm3$. *But this is not possible, since it implies that* $\pm[3]_5$ *is a square in* $\mathbb{Z}_5$, *and it was observed in Example 8.5.3 that this is not the case. Thus* $1 + \sqrt{10} = (m + n\sqrt{10})(r + s\sqrt{10})$ *implies that either* $m + n\sqrt{10} \in U(D)$ *or else* $r + s\sqrt{10} \in U(D)$, *and so* $1 + \sqrt{10}$ *is irreducible in* D. *Since* $9 = -(1 - 10) = -(1 + \sqrt{10})(1 - \sqrt{10}) \in (1 + \sqrt{10})D$, *but* $3 \notin (1 + \sqrt{10})D$ *(otherwise* $I = (3) + (1 + \sqrt{10}) = (1 + \sqrt{10})$ *is principal, and it was shown in Example 8.5.3 that* I *was not principal). Thus* $1 + \sqrt{10}$ *divides* $(3)(3)$, *but* $1 + \sqrt{10}$ *does not divide* 3, *and so* $1 + \sqrt{10}$ *is not prime in* D.

However, in a principal ideal domain, every irreducible element is prime.

8.5.8 Proposition.   *In a PID, every irreducible element is prime.*

Proof. Let D be a PID, and suppose that $a \in D$ is irreducible. Let $x, y \in D$ be such that a divides xy; that is, $xy \in (a)$. We are to prove that either $x \in (a)$ or else $y \in (a)$. Since every ideal is principal, there exists $d \in D$ with $(a) + (x) = (d)$. But then d divides a, and so there exists $u \in D$ such that $a = du$. Since a is irreducible, either $d \in U(D)$, in which case $(a) + (x) = D$, or else $u \in U(D)$, in which case $(d) = (a)$. If $(d) = (a)$, then $(a) + (x) = (d) = (a)$, and so $(x) \subseteq (a)$, while if $(a) + (x) = D$, then $1 = au + xv$ for some $u, v \in D$, and then $y = a(uy) + v(xy) \in (a)$. $\square$

For example, in $\mathbb{Z}$, primes and irreducibles are the same (it is customary in fact to use the definition of an irreducible element as the definition of a prime in $\mathbb{Z}$).

8.5.9 Proposition. *Let D be a PID, and let $a, b \in D$ with not both a, b equal to 0_D. Then there exists $d \in D$ such that $(a) + (b) = (d)$, and d is a greatest common divisor of a and b. Moreover, if d' is any greatest common divisor of a and b, then $(d') = (d)$, and there exists $u \in U(D)$ such that $d' = du$.*

Proof. Let $I = (a) + (b)$, so $I \lhd D$. Since D is a PID, there exists $d \in I$ such that $I = (d)$. Since $a, b \in I = (d)$, $(a) \subseteq (d)$ and $(b) \subseteq (d)$; that is, d divides each of a and b in D. Let $c \in D$ be a common divisor of a and b, so $(a) \subseteq (c)$ and $(b) \subseteq (c)$. Then $(d) = (a) + (b) \subseteq (c)$, and so c divides d. Thus d is a greatest common divisor of a and b in D. $\qquad\qquad\square$

We note that in an arbitrary PID, there is no algorithm for the determination of a greatest common divisor of two elements, unlike the situation for $\mathbb{Z}$, where we have the Euclidean algorithm (Algorithm 4.3.7) for the calculation of the greatest common divisor of two integers.

Embedding an Integral Domain in a Field

The integral domain $\mathbb{Z}$ is a subring of the field $\mathbb{Q}$ of rational numbers. In point of fact, $\mathbb{Q}$ is constructed as an extension of $\mathbb{Z}$ by a process that allows us to think of each rational number as a ratio of two integers. We may regard each integer as a rational number by identifying the integer m with the rational number $\frac{m}{1}$. The precise way of stating this is to say that the map $\theta : \mathbb{Z} \to \mathbb{Q}$ given by mapping $m \in \mathbb{Z}$ to $\frac{m}{1} \in \mathbb{Q}$ is an injective ring homomorphism for which $\frac{m}{n} = \theta(m)(\theta(n))^{-1}$ for each $\frac{m}{n} \in \mathbb{Q}$. We show that for every PID D, there is a field Q_D and an injective ring homomorphism $\theta_D : D \to Q_D$ such that for each $x \in Q_D$, there exist $a, b \in D$ with $x = \theta_D(a)(\theta_D(b))^{-1}$. As a result, Q_D (which is then unique up to ring isomorphism) is called the field of quotients for D. It will then follow that for any field F and any injective ring homomorphism $\alpha : D \to F$, there exists a unique homomorphism $\hat{\alpha} : Q_D \to F$ such that $\hat{\alpha} \circ \theta_D = \alpha$.

Let D be an integral domain. On the set $D \times (D - \{0_D\})$, define operations called addition and multiplication, denoted respectively by $+$ and juxtaposition. For $(a, b), (c, d) \in D \times (D - \{0_D\})$, we define

$$(a, b) + (c, d) = (ad + bc, bd)$$
$$(a, b)(c, d) = (ac, bd).$$

Note that since D is an integral domain, $D - \{0_D\}$ is closed under multiplication, so these are indeed binary operations on $D \times (D - \{0_D\})$. We prove that each operation is associative, commutative, and has identity. Let $(a,b), (c,d), (r,s) \in D \times (D - \{0_D\})$. Then $((a,b) + (c,d)) + (r,s) = (ad + bc, bd) + (r,s) = (ads + bcs + bdr, bds)$, while $(a,b) + ((c,d) + (r,s)) = (a,b) + (cs + dr, ds) = (ads + bcs + bdr, bds)$, so $((a,b) + (c,d)) + (r,s) = (a,b) + ((c,d) + (r,s))$. As well, $((a,b)(c,d))(r,s) = (ac, bd)(r,s) = (acr, bds) = (a,b)(cr, ds) = (a,b)((c,d)(r,s))$. Thus addition and multiplication are each associative. We have $(a,b) + (c,d) = (ad + bc, bd) = cb + da, db) = (c,d) + (a,b)$ and $(a,b)(c,d) = (ac, bd) = (ca, db) = (c,d)(a,b)$ since both addition and multiplication in D are commutative. Finally, $(a,b) + (0_D, 1_D) = (a,b)$, so $(0_D, 1_D)$ is the additive identity in $D \times (D - \{0_D\})$, and $(a,b)(1_D, 1_D) = (a,b)$, so $(1_D, 1_D)$ is the multiplicative identity.

Next, on the set $D \times (D - \{0_D\})$, define a relation $\sim$ by $(a,b) \sim (c,d)$ if $ad = bc$. We prove that $\sim$ is an equivalence relation. For $(a,b) \in D \times (D - \{0_D\})$, $(a,b) \sim (a,b)$ since $ab = ba$. Thus $\sim$ is reflexive. Suppose that $(a,b), (c,d) \in D \times (D - \{0_D\})$ are such that $(a,b) \sim (c,d)$; that is, $ad = bc$. Then $cb = da$ and so $(c,d) \sim (a,b)$, which proves that $\sim$ is symmetric. Finally, suppose that $(a,b), (c,d), (r,s) \in D \times (D - \{0_D\})$ are such that $(a,b) \sim (c,d)$ and $(c,d) \sim (r,s)$. Then $ad = bc$ and $cs = dr$, so $ads = bcs = bdr$. Since $d \neq 0_D$ and D is an integral domain, it follows that $as = br$ and thus $(a,b) \sim (r,s)$, proving that $\sim$ is transitive.

Denote the equivalence class of $(a,b) \in D \times (D - \{0_D\})$ by $[(a,b)]$, and recall that any binary operation on a set X naturally extends to a binary operation on $\mathscr{P}(X)$.

8.6.1 Lemma. *For $(a,b), (c,d) \in D \times (D - \{0_D\})$, $[(a,b)][(c,d)] \subseteq [(a,b)(c,d)]$ and $[(a,b)] + [(c,d)] \subseteq [(a,b) + (c,d)]$.*

Proof. Let $(a',b') \in [(a,b)]$ and $(c',d') \in [(c,d)]$, so that $a'b = b'a$ and $c'd = d'c$, and note that $(a',b') + (c',d') = (a'd' + b'c', b'd')$. Since

$$(a'd' + b'c')(bd) = a'bdd' + b'bc'd = b'add' + b'bd'c = (ad + bc)(b'd'),$$

it follows that $(a',b') + (c',d') \sim (a,b) + (c,d)$ and thus $[(a,b)] + [(c,d)] \subseteq [(a,b) + (c,d)]$. As well, $(a',b')(c',d') = (a'c', b'd') \sim (a'c'bd, b'd'bd) = (b'ad'c, b'd'bd) \sim (ac, bd) = (a,b)(c,d)$, and thus $[(a,b)][(c,d)] \subseteq [(a,b)(c,d)]$. $\square$

Now define $Q_D = \{ [(a,b)] \mid (a,b) \in D \times (D - \{0_D\}) \}$. On Q_D define binary operations of addition, denoted by $+$, and multiplication, denoted by juxtaposition, as follows: for $X, Y \in Q_D$, there exist $(a, b), (c, d) \in D \times (D - \{0_D\})$ such that $X = [(a, b)]$ and $Y = [(c, d)]$. Then define

$$X + Y = [(a, b) + (c, d)]$$
$$XY = [(a, b)(c, d)].$$

By Lemma 8.6.1, both addition and multiplication are well-defined operations on Q_D. Furthermore, since addition and multiplication on $D \times (D - \{0_D\})$ are associative and commutative, it follows that addition and multiplication are associative and commutative on Q_D. As well, $[(0_D, 1_D)]$ is the additive identity on Q_D and $[(1_D, 1_D)]$ is the multiplicative identity on Q_D.

Our immediate objective is to prove that the operations of addition and multiplication make Q_D a field. It remains to prove that every element of Q_D has an additive inverse, every nonzero (that is, not $[(0_D, 1_D)]$) element of Q_D has a multiplicative inverse, and that multiplication distributes over addition. Let $[(a, b)] \in Q_D$. Then $[(-a, b)] \in Q_D$ and $[(a, b)] + [(-a, b)] = [(ab + b(-a), b^2)] = [(0_D, b^2)] = [(0_D, 1_D)]$ since $(0_D, b^2) \sim (0_D, 1_D)$. Since addition is commutative, it follows that the additive inverse of $[(a, b)]$ is $[(-a, b)]$; that is, $-[(a, b)] = [(-a, b)]$. Next, let $[(a, b)] \in Q_D$ with $[(a, b)] \neq [(0_D, 1_D)]$. Then $a \neq (b)(0_D) = 0_D$, and so $(b, a) \in D \times (D - \{0_D\})$. We have $[(a, b)][(b, a)] = [(ab, ba)] = [(1_D, 1_D)]$ since $ab1_D = ba1_D$. Thus the multiplicative inverse of $[(a, b)]$ is $[(b, a)]$; that is, $[(a, b)]^{-1} = [(b, a)]$. Finally, let $(a, b), (c, d), (r, s) \in D \times (D - \{0_D\})$. Then

$$([(a, b)] + [(c, d)])[(r, s)] = [(ad + bc, bd)][(r, s)] = [(adr + bcr, bds)]$$

and

$$[(a, b)][(r, s)] + [(c, d)][(r, s)] = [(ar, bs)] + [(cr, ds)]$$
$$= [(ards + bscr, bsds)] = [((ard + bcr)s, bsds)]$$
$$= [(ard + bcr, bsd)] = ([(a, b)] + [(c, d)])[(r, s)].$$

Thus multiplication in Q_D distributes over addition. This completes the proof that Q_D, with these addition and multiplication operations, is a field.

8.6.2 Proposition. *For any integral domain D, the map $\varphi_D : D \to Q_D$ for which $\varphi_D(a) = [(a, 1_D)]$ for all $a \in D$ is an injective unital ring homomorphism, and for any $[(a, b)] \in Q_D$, $[(a, b)] = [(a, 1_D)][(b, 1_D)]^{-1} = \varphi_D(a)(\varphi_D(b))^{-1}$. Moreover, if F is a field and $\alpha : D \to F$ is an injective ring homomorphism, then there exists a unique ring homomorphism $\hat{\alpha} : Q_D \to F$ such that $\hat{\alpha} \circ \varphi_D = \alpha$.*

Proof. Let $a, b \in D$. Then $\varphi_D(a+b) = [(a+b, 1_D)] = [(a, 1_D)] + [(b, 1_D)] = \varphi_D(a) + \varphi_D(b)$, and $\varphi_D(ab) = [(ab, 1_D)] = [(a, 1_D)][(b, 1_D)] = \varphi_D(a)\varphi_D(b)$, so φ_D is a ring homomorphism. Since $\varphi_D(1_D) = [(1_D, 1_D)]$, which is the unity of Q_D, φ_D is a unitary homomorphism. If $\varphi_D(a) = [(0_D, 1_D)]$, then $[(a, 1_D)] = [(0_D, 1_D)]$ and thus $(a)(1_D) = (1_D)(0_D) = 0_D$, so $a = 0_D$. Thus φ_D is injective. Suppose now that F is a field, and $\alpha \colon D \to F$ is an injective ring homomorphism. Note that $\alpha(1_D)$ is a nonzero multiplicative idempotent of F and thus $\alpha(1_D) = 1_F$; that is, α is necessarily unital. Define a map $\hat{\alpha}$ from Q_D to F by $\hat{\alpha}([(a, b)]) = \alpha(a)\alpha(b)^{-1}$. The first thing to establish is that $\hat{\alpha}$ is well-defined; that is, it does not depend on our choice of representative of an element of Q_D. Suppose that $[(a, b)] = [(c, d)]$. Then $ad = bc$, and so $\alpha(a)\alpha(d) = \alpha(ad) = \alpha(bc) = \alpha(b)\alpha(c)$. Since α is injective and $b, d \neq 0_D$, it follows that $\alpha(b)$ and $\alpha(d)$ are invertible elements of F. Thus $\alpha(a)\alpha(b)^{-1} = \alpha(c)\alpha(d)^{-1}$, and so $\hat{\alpha}$ is well-defined. Let $[(a, b)], [(c, d)] \in Q_D$. Then $\hat{\alpha}([(a, b)] + [(c, d)]) = \hat{\alpha}([(ad + bc, bd)]) = \alpha(ad + bc)\alpha(bd)^{-1} = (\alpha(a)\alpha(d) + \alpha(b)\alpha(c))\alpha(d)^{-1}\alpha(b)^{-1} = \alpha(a)\alpha(b)^{-1} + \alpha(c)\alpha(d)^{-1} = \hat{\alpha}([(a, b)]) + \hat{\alpha}([(c, d)])$, and $\hat{\alpha}([(a, b)][(c, d)]) = \hat{\alpha}([(ac, bd)]) = \alpha(ac)\alpha(bd)^{-1} = \alpha(a)\alpha(b)^{-1}\alpha(c)\alpha(d)^{-1} = \hat{\alpha}([(a, b)])\hat{\alpha}([(c, d)])$. Thus $\hat{\alpha}$ is a ring homomorphism from Q_D to F. Now, for $a \in D$, $\hat{\alpha}(\varphi_D(a)) = \hat{\alpha}([(a, 1_D)]) = \alpha(a)\alpha(1_D)^{-1} = \alpha(a)1_F = \alpha(a)$, and thus $\hat{\alpha} \circ \varphi_D = \alpha$.   □

In keeping with the representation of a rational number as a ratio of integers, we denote $[(a, b)]$ by $\frac{a}{b}$, so $\frac{a}{b} = \frac{c}{d}$ if, and only if, $[(a, b)] = [(c, d)]$; that is, if, and only if, $ad = bc$. Furthermore, we usually identity $a \in D$ with $\frac{a}{1_D} = \varphi_D(a)$ in Q_D, and in this way, we may think of D as a subring of Q_D.

Note that $\mathbb{Q}$, the field of rational numbers, is the field of quotients of the integral domain $\mathbb{Z}$.

8.7   The Characteristic of an Integral Domain

Since a ring is an abelian group under the operation of addition, we can speak of the additive order of an element. It turns out that for integral domains, this is a very useful concept. Of course, it is entirely possible that all nonzero elements of a given integral domain have infinite order, as is the case with $\mathbb{Z}$, but there are many interesting integral domains (even infinite ones) in which every element has finite order. As it turns out, it is not possible for an integral domain to have some elements of infinite order and other (nonzero) elements of finite order. Even more can be said. If an integral domain has nonzero elements of finite order, then there is a prime p such that every nonzero element of the integral domain has order p. The proof of these assertions is the content of this section. One simple example would be $\mathbb{Z}_p$, p a prime (actually, $\mathbb{Z}_p$ is not just an integral domain, but is in fact a field). Every nonzero element of $\mathbb{Z}_p$ has additive order p.

8.7.1 Definition. *Let D be an integral domain. If 1_D has infinite additive order, then we say that D is of characteristic 0, and we write $char(D) = 0$. Otherwise, 1_D has finite additive order, and we say that D has characteristic n, where n is the additive order of 1_D.*

8.7.2 Proposition. *Let D be an integral domain. The map $c \colon \mathbb{Z} \to D$ for which $n \in \mathbb{Z}$ maps to $n1_D \in D$ is a ring homomorphism, and so $\ker(c) \lhd \mathbb{Z}$. If $\ker(c) = (0)$, then $char(D) = 0$ and every nonzero element of D has infinite order, and otherwise $\ker(c) = (p)$, where p is a prime, and then every nonzero element of D has additive order p; that is, $pa = 0_D$ for every $a \in D$, and $char(D) = p$.*

Proof. By Proposition 5.2.7 (in its additive notation version), for any $m, n \in \mathbb{Z}$, we have $c(m+n) = (m+n)1_D = m1_D + n1_D = c(m) + c(n)$. We also need to establish that $c(mn) = c(m)c(n)$, and this we do by induction on n, with hypothesis the following: for any integer m, $c(mn) = c(m)c(n)$. The base case of $n = 1$ is immediate, since $c(1) = 1_D$ by definition of exponentiation. Suppose now that $n \geq 1$ is an integer for which the hypothesis holds, and let $m \in \mathbb{Z}$. Then $c(m(n+1)) = c(mn+m) = c(mn) + c(m) = c(m)c(n) + c(m) = c(m)(c(n) + 1_D) = c(m)c(n) + c(1)) = c(m)c(n+1)$, as required. It follows now by induction that for any integers m and n with $n \geq 1$, $c(mn) = c(m)c(n)$. Since c is an additive homomorphism, $c(0) = 0_D$, and so for any $m \in \mathbb{Z}$, $c(m(0)) = c(0) = 0_D = c(m)0_D = c(m)c(0)$. Finally, let $m, n \in \mathbb{Z}$ with $n > 0$. By Proposition 8.1.5, $c(m(-n)) = c(-(mn)) = (-(mn))1_D$. Since $(mn)1_D + (-(mn)1_D = (mn - mn)1_D = 01_D = 0_D$, it follows that $-((mn)1_D) = (-(mn))1_D$ and so $c(m(-n)) = -c(mn) = -c(m)c(n) = c(m)(-c(n)) = c(m)c(-n)$. Thus for all $m, n \in \mathbb{Z}$, $c(mn) = c(m)c(n)$, which completes the proof that $c \colon \mathbb{Z} \to D$ is a ring homomorphism. Since $c(1) = 1_D$, c is in fact a unitary ring homomorphism. If $\ker(c) = (0)$, then c is injective, and so for $n \in \mathbb{Z}$, if $n1_D = 0_D$, then $c(n) = 0_D$ and so $n \in (0)$, which means that $n = 0$. Moreover, in this case, for any nonzero $a \in D$, if $n \in \mathbb{Z}$ is such that $na = 0_D$, then $0_D = na = n(1_Da) = (n1_D)a$ since multiplication distributes over addition. Now, since D is an integral domain, we obtain $0_D = n1_D$; that is, $c(n) = 0_D$, and so $n \in (0)$, establishing that $n = 0$. Thus in this case, every nonzero element of D has infinite additive order. If $\ker(c) \neq (0)$, then there exists $p \in \mathbb{Z}^+$ such that $\ker(c) = (p)$. Suppose that p divides mn for some $m, n \in \mathbb{Z}^+$, say $mn = pk$. Then $c(mn) = c(pk) = c(p)c(k) = 0_D c(k) = 0_D$, and thus $0_D = c(mn) = c(m)c(n)$. Since D is an integral domain, it follows that either $c(m) = 0_D$ or else $c(n) = 0_D$; that is, either $m \in (p)$ or else $n \in (p)$. Thus whenever p divides a product of two positive integers, it must divide one of the two integers, and so by Corollary 4.4.4, p is a prime. Since $0_D = c(p) = p1_D$, it follows that the additive order of 1_D is a divisor of p, and thus it must be equal to p; that is, $char(D) = p$. As before, for any nonzero $a \in D$, $pa = p(1_D a) = (p1_D)a = 0_D a = 0_D$, and thus the additive order of a is a divisor of p that is greater than 1 (since $a \neq 0_D$). Thus the additive order of a is p. $\qquad\square$

If D is an integral domain, then a subdomain of D is a unital subring of D. Note that the unity element of a unital subring of D is necessarily the unity of D.

8.7.3 Corollary. *Let D be an integral domain. Then the intersection of all subdomains of D is equal to $c(\mathbb{Z})$, and if D has prime characteristic p, then $c(\mathbb{Z})$ is a field isomorphic to $\mathbb{Z}_p$, while if D has characteristic 0, then $c(\mathbb{Z})$ is isomorphic to $\mathbb{Z}$. In particular, if k is a field of prime characteristic p, then the intersection of all subfields of k is a field isomorphic to $\mathbb{Z}_p$, while if k is a field of characteristic 0, then the intersection of all subfields of k is isomorphic to $\mathbb{Q}$, the field of quotients of $\mathbb{Z}$.*

Proof. If D' is a subdomain of D, then $1_D \in D'$ and so for all $n \in \mathbb{Z}$, $c(n) = n1_D \in D'$. Thus $c(\mathbb{Z}) \subseteq D'$. Since $c(\mathbb{Z})$ is itself a unital subring of D, it follows that the intersection of all subdomains of D is equal to $c(\mathbb{Z})$. If $char(D) = p$, a prime, then by the first isomorphism theorem for rings, $c(\mathbb{Z}) \simeq \mathbb{Z}/(p) = \mathbb{Z}_p$, while if $char(D) = 0$, then $c(\mathbb{Z}) \simeq \mathbb{Z}$. If k is a field, then every subfield of k is a subdomain of k, and thus $c(\mathbb{Z})$ is contained in the intersection of all subfields of k. In the case $char(k) = p$, $c(\mathbb{Z})$ is a subfield of k and thus the intersection of all subfields of k is $c(\mathbb{Z})$, a field isomorphic to $\mathbb{Z}_p$. In the event that $char(k) = 0$, then the intersection of all subfields of k is a subfield k' containing $c(\mathbb{Z})$, and so $c : \mathbb{Z} \to k'$ is an injective unitary ring homomorphism, and thus by Proposition 8.6.2, c extends to an injective homomorphism $\hat{c}$ from $\mathbb{Q}$ into k'. But $\hat{c}(\mathbb{Q})$ is a subfield of k' and thus a subfield of k, which means that $\hat{c}(\mathbb{Q}) = k'$. Thus the intersection of all subfields of k is a subfield isomorphic to $\mathbb{Q}$. $\qquad\square$

8.7.4 Definition. *Let k be a field. Then the intersection of all subfields of k is a subfield of k, called the prime subfield of k.*

Thus if k is a field of prime characteristic p, then the prime subfield of k isomorphic to $\mathbb{Z}_p$, otherwise the prime subfield of k is isomorphic to $\mathbb{Q}$.

Exercises

1. Let R be a ring. Prove that if 0_R is also a multiplicative identity element for R, then $R = \{0_R\}$ (and this is why we require that $1_R \neq 0_R$ in the definition of a unital ring).

2. Let n be a positive integer, and let R be the ring of upper-triangular $n \times n$ matrices with entries from $\mathbb{R}$ (see Example 8.1.9). Let I denote the subring of R whose elements are the strictly upper-triangular $n \times n$ matrices with entries from $\mathbb{R}$. Prove that $I \lhd R$ and that $T/I \simeq \oplus_{i=1}^n \mathbb{R}$. Conclude that R is a noncommutative ring but R/I is commutative.

3. In Lemma 8.3.2, it was proven that in a ring R, the sum of a subring S of R and an ideal I of R is a subring of R. Give an example of a ring R with subrings S and T such that $S + T = \{s + t \mid s \in S,\ t \in T\}$ is not a subring of R.

4. Let R be a commutative unital ring, and let n be a positive integer. Prove that the ideals of $M_{n \times n}(R)$ are exactly the subrings of the form $M_{n \times n}(I)$, $I \lhd R$ as follows.

    a) Prove that if $I \lhd R$, then $M_{n \times n}(I) \lhd M_{n \times n}(R)$.

    b) For each i and j with $1 \le i, j \le n$, let $D_{ij} \in M_n(R)$ denote the matrix whose entries are zero except for the entry in row i, column j, which is 1, and let $E_{i,j}$ denote the elementary matrix obtained from I_n by exchanging rows i and j.

        (i) Prove that for each $A \in M_n(R)$,
        $$(D_{i,j}A)_{r,s} = \begin{cases} 0 & \text{if } r \neq i \\ A_{j,s} & \text{if } r = i \end{cases}.$$

        (ii) Prove that for each $A \in M_n(R)$,
        $$(AD_{i,j})_{r,s} = \begin{cases} 0 & \text{if } s \neq j \\ A_{r,i} & \text{if } s = j \end{cases}.$$

    (iii)   Prove that for each $A \in M_n(R)$, $D_{i,j}AD_{i,j} = A_{j,i}D_{i,j}$.

    (iv)   Prove that for any indices i, j, r, s, $E_{r,i}D_{i,j} = D_{r,j}$ and $D_{r,j}E_{j,s} = D_{r,s}$. Conclude that for any $a \in R$, $E_{r,i}(aD_{i,j})E_{j,s} = aD_{r,s}$.

  c) Let $J \lhd M_{n \times n}(R)$, and let $I(J) = \{\, A_{i,j} \mid A \in J,\ i, j = 1 \ldots, n \,\}$.

    (i)   Prove that $I(J)$ is an ideal of R.

    (ii) Prove that $J = M_n(I(J))$.

  d) What are the ideals of $M_n(R)$ if R is a field?

5. Let R be a ring. This exercise shows that there exists a unital ring that contains an isomorphic copy of R, so that R can be thought of as a subring (in fact, an ideal) of some unital ring.

On the direct product $S = \mathbb{Z} \times R$ of the additive groups of $\mathbb{Z}$ and R, define multiplication by

$$(m, r)(n, s) = (mn, ms + nr + rs).$$

  a) Prove that this multiplication operations makes the abelian additive group $\mathbb{Z} \times R$ into a ring with unity $(1, 0)$.

  b) Prove that the map $\beta : R \to \mathbb{Z} \times R$ for which $\beta(r) = (0, r)$ for $r \in R$ is an injective ring homomorphism.

  c) Prove that for any unital ring S, and any ring homomorphism $f : R \to S$, there is a unique unital ring homomorphism $g : \mathbb{Z} \times R \to S$ such that $g \circ \beta = f$. Hint: Try $g(m, r) = m1_S + f(r)$.

  d) Prove that the projection map $\alpha : \mathbb{Z} \times R \to \mathbb{Z}$ for which $\alpha(m, r) = m$ is a surjective unital ring homomorphism and that $\ker(\alpha) = \beta(R)$. Conclude that since $\beta(R) \lhd \mathbb{Z} \times R$, the first isomorphism theorem for rings implies that $(\mathbb{Z} \times R)/\beta(R) \simeq \mathbb{Z}$.

6. a) Prove that a commutative ring R is a field if, and only if, $\{0_R\}$ and R are the only ideals of R and $\{\, xy \mid x, y \in R \,\} \neq \{0_R\}$ (note that $\{\, xy \mid x, y \in R \,\}$ is an ideal of R).

  b) Let R be a commutative unital ring. Prove that R is a field if, and only if, the only ideals of R are (0_R) and R.

7. Let D be an integral domain. Prove that if $a, b \in D$ are such that $a^m = b^m$ and $a^n = b^n$ for some relatively prime integers m and n, then $a = b$.

8. Prove that if R is a ring such that $a^2 = a$ for every $a \in R$, then R is commutative.

9. Give an example of a ring R for which $R \oplus R \simeq R$.

10. Let R be a ring, and let $I_1, I_2, \ldots, I_n$ be ideals of R. Prove that $R/(\cap_{i=1}^n I_i)$ is isomorphic to a subring of $\oplus_{i=1}^n R/I_i$.

11. Let R be a nonempty set on which is given a binary operation, denoted by $+$ and referred to as addition, such that $(R, +)$ is a group (not postulated to be abelian). Suppose that there is a second binary operation, denoted by juxtaposition and referred to as multiplication, which is associative and has unity element 1_R, and is such that multiplication distributes both on the left and on the right over addition. Prove that addition is commutative (so R, with this addition and multiplication, is a unital ring). Hint: For $a, b \in R$, expand $(a + b)(1_R + 1_R)$ in two different ways.

12. An ideal I of a commutative unital ring R is said to be a prime ideal if $I \neq R$ and for all $a, b \in R$, if $ab \in I$, then $a \in I$ or $b \in I$, while I is said to be a maximal ideal of R if $I \neq R$ and for any ideal J of R, if $I \subseteq J \subsetneq R$, then $I = J$.

   a) Let R be a commutative unital ring. Prove that an ideal I of R is prime if, and only if, R/I is an integral domain.

   b) What are the prime ideals of $\mathbb{Z}$?

   c) Let R be a commutative unital ring. Prove that an ideal I of R is maximal if, and only if, R/I is a field.

   d) Prove that every maximal ideal of a commutative unital ring R is a prime ideal of R.

   e) Let p be a prime and let $I = \{ (ip, j) \mid i, j \in \mathbb{Z} \}$.

      (i) Prove that I is a maximal ideal of $\mathbb{Z} \oplus \mathbb{Z}$.

      (ii) By the preceding part, I is a maximal ideal of $\mathbb{Z} \oplus \mathbb{Z}$, so $(\mathbb{Z} \oplus \mathbb{Z})/I$ is a field. What is the characteristic of this field?

13. Let R be a commutative unital ring. For ideals I and J of R, let $IJ = \cap \{ U \triangleleft R \mid ij \in U \text{ for all } i \in I, j \in J \}$, so that $IJ \triangleleft R$.

   a) Prove that for ideals I, J, K of R, $I(JK) = (IJ)K$.

   Thus in the set of all ideals of R, the operation of multiplication of ideals is associative, which means that we may write the product of ideals $I_1, I_2, \ldots, I_N$ without the use of brackets.

   b) Prove that if $I_1, I_2, \ldots, I_n \triangleleft R$, then $\prod_{i=1}^n I_i \subseteq \cap_{i=1}^n I_i$.

c) Let n be a positive integer. Prove that for any ideals $I_1, I_2, \ldots, I_{n+1}$ of R such that for all $i, j = 1, 2, \ldots, n+1$ with $i \neq j$, $I_i + I_j = R$, $I_{n+1} + \prod_{i=1}^{n} I_i = R$.

Hint: Use induction on $n \geq 2$. For the base case, let I, J, and K be ideals of R such that $I + J = I + K = J + K = R$. Then there exist $u, v \in I$, $w \in J$ and $y \in K$ such that $1 = u + w = v + y$, and then $1 = (u+w)(v+y)$. Use this to prove that $1 \in I + JK$, and conclude that $I + JK = R$. In the proof of the inductive step, for ideals $I_1, I_2, \ldots, I_{n+1}$ for which $I_i + I_j = R$ for all $i, j = 1, 2, \ldots, n+1$ with $i \neq j$, consider $I_{n+1} + \prod_{i=1}^{n} I_i = I_{n+1} + (\prod_{i-1}^{n-1} I_i) I_n$.

d) Prove that if $I, J \triangleleft R$ are such that $I + J = R$, then $I \cap J = IJ$. Then use induction on $n \geq 2$ to prove that if $I_1, I_2, \ldots, I_n \triangleleft R$ for which $I_i + I_j = R$ for all $i, j = 1, 2, \ldots, n$ with $i \neq j$, then $\cap_{i=1}^{n} I_i = \prod_{i=1}^{n} I_i$.

Hint: In the proof of the inductive step (where in view of part (b), the inductive postulate need only stipulate that $\cap_{i=1}^{n} I_i \subseteq \prod_{i=1}^{n} I_i$), suppose that $I_1, I_2, \ldots, I_{n+1} \triangleleft R$ meet the condition, so that $I_{n+1} \cap (\cap_{i=1}^{n} I_i) \subseteq I_{n+1} \cap \prod_{i=1}^{n} I_i$. Now use part (c) and the base case for ideals I_{n+1}, I_n, and $\prod_{i=1}^{n-1} I_i$.

e) Suppose now that R is a PID, and let $I_1, I_2, \ldots, I_n$ be ideals of R such that $I_i + I_j = R$ for all $i, j = 1, 2, \ldots, n$ with $i \neq j$. For $x, y \in R$, write $x \equiv y \mod (I_k)$ if $x + I_k = y + I_k$. Prove that for any $a_i \in I_i$, $i = 1, 2, \ldots, n$, there exists $x \in R$ such that $x \equiv a_i \mod (I_i)$ for $i = 1, 2, \ldots, n$. This fact is known as the **Chinese Remainder Theorem**.

14. An element a of a ring R is called an **idempotent** of R if $a^2 = a$. Note that the zero of R is an idempotent of R.

a) Prove that for any prime p and any positive integer n, the only nonzero idempotent of $\mathbb{Z}_{p^n}$ is $[1]_{p^n}$.

b) For any integer $m > 1$, prove that the number of idempotents of $\mathbb{Z}_m$ is 2^k, where k is the number of distinct prime divisors of m.

Hint: If $m = p_1^{r_1} p_2^{r_2} \cdots p_n^{r_n}$ is the prime decomposition of n, then $\mathbb{Z}_m$ is isomorphic to $\oplus_{i=1}^{n} \mathbb{Z}_{p_i^{r_i}}$.

15. Let R be a ring, and define the **centre** of R, denoted by $Z(R)$, by

$$Z(R) = \{\, a \in R \mid ax = xa \text{ for all } x \in R \,\}.$$

a) Prove that $Z(R)$ is a subring of R.

b) Suppose that R is commutative and unital, and let n be a positive integer. Determine the centre of $M_{n \times n}(R)$ as follows.

    (i) For any integers $i, j = 1, 2, \ldots, n$, let D_{ij} denote the element of $M_{n \times n}(R)$ which has 1_R in row i, column j, and 0_R for every other entry. If $A \in Z(M_{n \times n}(R))$, then for each $i, j = 1, 2, \ldots, n$, $D_{in}A = AD_{in}$. Prove that if $A \in Z(M_{n \times n}(R))$, then A is a diagonal matrix.

    (ii) For any integers $i, j = 1, 2, \ldots, n$, let $E_{i,j}$ denote the matrix obtained from I_n by exchanging rows i and j. Use the fact that if $A \in Z(M_{n \times n}(R))$, then for each $i, j = 1, 2, \ldots, n$, $E_{ij}A = AE_{ij}$ to prove that for each $i, j = 1, 2, \ldots, n$, $A_{ii} = A_{jj}$.

    (iii) Deduce that $Z(M_{n \times n}(R)) = \{\, aI_n \mid a \in R \,\}$.

c) Suppose that R is unital, and that $e \in R$ is a central idempotent; that is, $e^2 = e \in Z(R)$. Prove that $1-e$ is also a central idempotent of R, and that Re and $R(1-e)$ are ideals of R with $Re \cap R(1-e) = \{\, 0_R \,\}$ and $R = Re \oplus R(1-e)$.

16. Let R be a commutative ring. Prove that the binomial theorem holds in R; that is, for any $a, b \in F$, and any positive integer n, $(a+b)^n = \sum_{i=0}^{n} \binom{n}{i} a^i b^{n-i}$.

17. An element a of a ring R is said to be nilpotent if for some positive integer n, $a^n = 0_R$.

a) Suppose that R is a commutative unital ring, and that b is a nilpotent element of R.

    (i) Prove that $1 + b \in U(R)$.

    (ii) Prove that if $x, y \in R$ are such that $xy \in U(R)$, then both $x \in U(R)$ and $y \in U(R)$.

    (iii) Prove that if $a \in U(R)$, then $a + b \in U(R)$.

b) Let R be a commutative ring.

    (i) Prove that the set $N(R) = \{\, a \in R \mid b \text{ is nilpotent} \,\}$ is an ideal of R.

    (ii) Prove that the only nilpotent element of $R/N(R)$ is its zero element.

    (iii) Prove that $\mathbb{Z}_{108}/N(\mathbb{Z}_{108})$ is isomorphic to $\mathbb{Z}_6$.

    (iv) For which positive integers n is $N(\mathbb{Z}_n)$ the trivial subring of $\mathbb{Z}_n$?

18. Let p be a prime.

   a) What is the size of the ring $M_{2\times 2}(\mathbb{Z}_p)$?

   b) What is the size of $U(M_{2\times 2}(\mathbb{Z}_p))$?

19. Let p be a prime. Up to ring isomorphism, what is the number of unital rings R of order p?

20. Give an example of a noncommutative ring of order 16.

21. Let $R = ([2]_{10}) \subseteq \mathbb{Z}_{10}$, so R is an ideal of $\mathbb{Z}_{10}$, and thus a subring of $\mathbb{Z}_{10}$. Prove that R is in fact a field.

22. Let R and S be rings, and let $\varphi : R \to S$ be a ring homomorphism. By Proposition 8.2.9 (i), for any ideal J of S, $\varphi^{-1}(J) \lhd R$. Prove that for any ideal J of S, $R/\varphi^{-1}(J) \simeq S/J$.

23. Let $R = \{\, m + ni \mid m, n \in \mathbb{Z} \,\} \subseteq \mathbb{C}$.

   a) Prove that R is a subring of $\mathbb{C}$.

   b) Let $I = (2 + 2i)$, the principal ideal of R generated by $2 + 2i$. Prove that I is not a prime ideal of R (see Problem 12 for the definition of a prime ideal). Determine the characteristic of R/I.

   c) Let $J = (1 - i)$, the principal ideal of R generated by $1 - i$. Prove that R/J is a field. What is the order of R/J?

24. Let A be an abelian group with binary operation denoted by $+$ and identity element denoted by 0. Let $End(A)$ denote the set of all group homomorphisms from A to itself. For $f, g \in End(A)$, define $f + g : A \to A$ by $(f + g)(a) = f(a) + g(a)$ for each $a \in A$.

   a) Prove that for all $f, g \in End(A)$, $f + g \in End(A)$.

   b) Prove that the binary operation on $End(A)$, denoted by $+$, that assigns $f + g$ to $f, g \in End(A)$, is an associative, commutative operation with identity element the constant function $\mathbb{0} : A \to A$ defined by $\mathbb{0}(a) = 0$ for all $a \in A$. Further, prove that for any $f \in End(A)$, the mapping denoted by $-f$ from A to A and defined by $(-f)(a) = -f(a)$ for each $a \in A$ is a ring homomorphism, and $f + (-f) = \mathbb{0}$.

   c) Since the composition of ring homomorphisms is a ring homomorphism, it follows that composition of maps determines a binary operation on $End(A)$, known to be associative. Prove that the identity map is the identity element for composition of maps, and composition left and right distributes over addition; that is, for ell $f, g, h \in End(A)$, $f \circ (g + h) = f \circ g + f \circ h$, and $(g + h) \circ f =$

$g \circ f + h \circ f$. Thus $End(A)$ is a unital ring, in general not commutative. $End(A)$ is called the endomorphism ring of A.

d) Determine, up to ring isomorphism, the endomorphism ring of a cyclic group of order 6.

25. Let R be a ring and n be a positive integer for which there are ring endomorphisms $\pi_1, \pi_2, \ldots, \pi_n$ of R such that the following hold:

(i) for any $i, j = 1, 2, \ldots, n$ with $i \neq j$, $\pi_i \circ \pi_j = 0$, the zero endomorphism of R;

(ii) for any $i = 1, 2, \ldots, n$, $\pi_i \circ \pi_i = \pi_i$;

(iii) $\pi_1 + \pi_2 + \cdots + \pi_n = \mathbb{1}_R$.

For each $i = 1, 2, \ldots, n$, let $R_i = \pi_i(R)$.
Prove that $R = R_1 \oplus R_2 \oplus \cdots \oplus R_n$.

26. For any prime p, let $E_p = \{ a + b\sqrt{p} \mid a, b \in \mathbb{Q} \}$, so that $E_p \subseteq \mathbb{R}$.

a) Prove that E_p is a subfield of $\mathbb{R}$.

b) Prove that E_2 and E_3 are not isomorphic.

27. Let F be a finite field, and let $\varphi : F \to F$ be the map for which $\varphi(a) = a^2$ for each $a \in F$.

a) Prove that $\varphi(F^*) \subseteq F^*$, and that the map $\varphi : F^* \to F^*$ is a group homomorphism, where $F^* = F - \{ 0_F \}$ is a group under the multiplication operation.

b) Prove that if $char(F) = 2$, then $\varphi : F^* \to F^*$ is a group isomorphism and thus every element of F^* is a square; that is, for every $a \in F$, there exists $b \in F$ such that $a = b^2$.

c) Prove that if $char(F) \neq 2$, then $\varphi : F^* \to F^*$ has kernel of size two, and so exactly one half of the elements of F^* are squares.

d) Prove that for every $a \in F$, there exist $x, y \in F$ such that $a = x^2 + y^2$ (it is not required that $x \neq y$).
Hint: Let $S = \{ a \in F \mid$ there is $b \in F$ with $a = b^2 \}$. For any $a \in F$, let $a - S = \{ a - x \mid x \in S \}$. Prove that for any $a \in F$, $(a - S) \cap S \neq \varnothing$.

28. Determine all ring homomorphisms from $\mathbb{Z}_{15}$ to $\mathbb{Z}_{25}$.

29. Let R be a commutative ring. For any ideals I and J of R, let $IJ^{-1} = \{ x \in R \mid xJ \subseteq I \}$. Prove that for any ideals I and J of R, IJ^{-1} is an ideal of R and that $I \subseteq IJ^{-1}$.

30. Let R be a commutative unital ring. A subset M of R is said to be multiplicatively closed if M is closed under multiplication (that is, for any $a, b \in M$, $ab \in M$) and $1_R \in M$.

   a) Prove that if P is a prime ideal of R (see Problem 12 for the definition of a prime ideal), then $R - P$ is multiplicatively closed.

   b) Let M be a multiplicatively closed subset of R. On $R \times M$, define a relation $\sim$ by $(r, a) \sim (s, b)$ if there exists $x \in M$ such that $(rb - sa)x = 0_R$.

   (i) Prove that $\sim$ is an equivalence relation on $R \times M$.

   (ii) Denote the equivalence class of $(r, a) \in R \times M$ by $\frac{r}{a}$, and set $M^{-1}R = \{\, \frac{r}{a} \mid r \in R, \ a \in M \,\}$. Prove that for all $(r, a), (r_1, a_1), (s, b), (s_1, b_2) \in R \times M$, if $\frac{r}{a} = \frac{r_1}{a_1}$ and $\frac{s}{b} = \frac{s_1}{b_1}$, then $\frac{rb+sa}{ab} = \frac{r_1b_1+s_1a_1}{a_1b_1}$ and $\frac{rs}{ab} = \frac{r_1s_1}{a_1b_1}$. Thus $\frac{r}{a} + \frac{s}{b} = \frac{rb+sa}{ab}$ and $\frac{r}{a}\frac{s}{b} = \frac{rs}{ab}$ determine binary operations on $M^{-1}R$.

   (iii) Prove that $M^{-1}R$, with the two binary operations of addition and multiplication defined above, is a commutative unital ring.

   (iv) Prove that the map $f : R \to M^{-1}R$ for which $f(r) = \frac{r}{1_R}$ for each $r \in R$ is a ring homomorphism and that $f(M) \subseteq U(M^{-1}R)$. Further prove that f is injective if, and only if, M does not contain any divisors of 0_R.

   (v) Prove that if T is a commutative unital ring, then for any unital homomorphism $g : R \to T$ such that $g(M) \subseteq U(T)$, there exists a unique unital homomorphism $\hat{g} : M^{-1}R \to T$ such that $\hat{g} \circ f = g$.

   (vi) Prove that if $M = R - P$ for some prime ideal P of R, then $J = \{\, \frac{r}{a} \mid r \in P \,\}$ is a proper ideal of $M^{-1}R$ with the property that if I is any ideal of $M^{-1}R$, then $I \subseteq J$.

   In the case $M = R - P$, where P is a prime ideal of R, $M^{-1}R$ is typically denoted by R_P, and is called the localization of R at P, since a ring with a maximum proper ideal is called a local ring. Local rings arise quite naturally in algebraic geometry.

Chapter 9

Polynomial Rings

Introduction

To most students of elementary algebra, a polynomial is an expression of the form $a_0 + a_1 x + \cdots + a_n x^n$, where n is some nonnegative integer, and $a_0, a_1, \ldots, a_n$ are real numbers, called the coefficients of the polynomial, and x is thought of as a variable real number. In this chapter, we shall view polynomials in a somewhat different light, although the notation described above will be retained.

We shall define polynomials with coefficients from a given commutative unital ring R, and we shall show that the familar expressions for the sum and product of two polynomials give rise to binary operations on the set of all polynomials with coefficients from R, with the result that with these operations, the set of all polynomials with coefficients from R is a commutative unital ring.

For much of the chapter, we shall be concerned with the case when the coefficient ring is actually a field. As we shall see, the ring of polynomials with coefficients from a field shares many algebraic properties with the ring of integers. In particular, we shall show that in such a case, there is a form of Euclidean algorithm which is used to prove that the ring of polynomials with coefficients in field is a PID, and in fact, the greatest common divisor of two polynomials can be computed in much the same was as is the greatest common divisor of two integers. Furthermore, there is an analogue of the Fundamental Theorem of Algebra for the ring of polynomials over a field, and we shall prove the existence of maximal ideals and show how to find them. Once proven, this will provide a mechanism for the construction of new fields, as we have already established that the quotient ring of a commutative unital ring by a maximal ideal is a field.

9.2   The Ring of Power Series with Coefficients in a Commutative Unital Ring R

Throughout this section, R shall denote an arbitrary commutative unital ring. In Proposition 3.4.8, it was proven that for any nonempty set X, $\mathscr{F}(X, R)$, the set of all maps from X to R, has a binary operation that we shall call addition, where for $f, g \in \mathscr{F}(X, R)$, $f + g$ was defined to be the map from X to R for which $(f + g)(x) = f(x) + g(x)$ for every $x \in X$. It was also proven that since the addition operation on R is commutative and associative, and has identity O_R, the addition operation on $\mathscr{F}(X, R)$ is also commutative and associative, and has as its identity element the map $\mathbb{0} : X \mapsto R$ for which $\mathbb{0}(x) = 0_R$ for all $x \in X$. In this chapter, we shall be interested in the particular case of $X = \mathbb{N}$.

We point out that since the multiplication operation on R is commutative, we could construct a multiplication operation on $\mathscr{F}(\mathbb{N}, R)$ in an analogous fashion, but it turns out that there is another binary operation on $\mathscr{F}(\mathbb{N}, R)$ that is of far greater importance (and which we shall call multiplication). We shall return to this topic shortly, but first, we prove that the binary operation of addition on $\mathscr{F}(\mathbb{N}, R)$ makes $\mathscr{F}(\mathbb{N}, R)$ an abelian group. All that remains to be proven is the existence of additive inverses.

9.2.1 Lemma. *Let $f \in \mathscr{F}(\mathbb{N}, R)$, and let $g \in \mathscr{F}(\mathbb{N}, R)$ be the map for which $g(x) = -f(x)$ for each $x \in \mathbb{N}$. Then g is the additive inverse of f.*

Proof. We are to prove that $f + g = \mathbb{0}$ (since addition is commutative, this will also establish that $g + f = \mathbb{0}$). For any $n \in \mathbb{N}$, $(f + g)(n) = f(n) + g(n) = f(n) + (-f(n)) = 0_R$ and so $f + g = \mathbb{0}$. $\square$

We now introduce the binary operation of multiplication that will be so important in our work.

9.2.2 Definition. *For any $f, g \in \mathscr{F}(\mathbb{N}, R)$, let $fg \in \mathscr{F}(\mathbb{N}, R)$ be the map defined by $(fg)(n) = \sum_{i=0}^{n} f(i)g(n - i)$ for each $n \in \mathbb{N}$.*

Note that the product $f(i)g(n-i)$ is calculated using the multiplication operation of R, which is possible since $f(i)$ and $g(n - i)$ are elements of R.

9.2.3 Proposition. *Multiplication in $\mathscr{F}(\mathbb{N}, R)$ is associative; that is, for each $f, g, h \in \mathscr{F}(\mathbb{N}, R)$, $f(gh) = (fg)h$.*

Proof. Let $f, g, h \in \mathscr{F}(\mathbb{N}, R)$. To prove that $f(gh) = (fg)h$, we prove that for any $n \in \mathbb{N}$, $(f(gh))(n) = (f(gh))(n)$. Let $n \in \mathbb{N}$. Then

$$(f(gh))(n) = \sum_{i=0}^{n} f(i)(gh)(n-i) = \sum_{i=0}^{n} f(i)\Big(\sum_{j=0}^{n-i} g(j)h(i-j)\Big)$$

$$= \sum_{i=0}^{n}\sum_{j=0}^{n-i} f(i)\big(g(j)h(n-i-j)\big)$$

$$= \sum_{\substack{0 \le i,j \le n \\ j \le n-i}} f(i)\big(g(j)h(n-i-j)\big),$$

where the last equality above follows from the commutativity of addition in R. Let $W = \{(i,j) \in \mathbb{N}^2 \mid 0 \le i,j \le n, \ j \le n-i\}$. Recall that since multiplication in R is associative, we write $f(i)g(j)h(k)$ for either $f(i)\big(g(j)h(k)\big)$ or $\big(f(i)\big(g(j)\big)h(k)\big)$. Thus we have

$$(f(gh))(n) = \sum_{(i,j) \in W} f(i)g(j)h(n-i-j).$$

Similarly, we have

$$((fg)h)(n) = \sum_{j=0}^{n} (fg)(j)h(n-j) = \sum_{j=0}^{n}\Big(\sum_{i=0}^{j} f(i)g(j-i)\Big)h(n-j)$$

$$= \sum_{j=0}^{n}\sum_{i=0}^{j} f(i)g(j-i)\big)h(n-j)$$

Note that $n - j = n - i - (j - i)$, so for

$$U = \{(i,j) \in \mathbb{N}^2 \mid \text{ there exist } r, s \in \mathbb{N} \text{ such that }$$

$$0 \le r \le s \le n, i = r, \ j = s - r\},$$

we have (again, due to the commutativity of addition in R)

$$((fg)h)(n) = \sum_{(i,j) \in U} f(i)g(j)h(n-i-j).$$

Thus if we prove that $W = U$, it will follow that $((fg)h)(n) = (f(gh))(n)$, and thus $(fg)h = f(gh)$. Let $(i,j) \in W$, so $0 \le i \le n$ and $0 \le j \le n-i$. Then $0 \le i \le i+j \le n$, and so for $r = i$ and $s = j+i$ we have $0 \le r \le s \le n$, with $i = r$ and $j = s-r$. Thus $(i,j) \in U$, and so $W \subseteq U$. Now let $(i,j) \in U$, so there exist integers r and s such that $0 \le r \le s \le n$ with $i = r$ and $j = s-r$. But then $0 \le i \le n$, and $0 \le s-r \le n-r$; that is, $0 \le j \le n-i$, so $(i,j) \in W$. Thus $U \subseteq W$ and so $W = U$, as required. $\qquad\square$

9.2.4 Proposition. *For any $f, g, h \in \mathscr{F}(\mathbb{N}, R)$, $fg = gf$ and $f(g+h) = fg + fh$.*

Proof. Let $n \in \mathbb{N}$. Then $(fg)(n) = \sum_{i=0}^{n} f(i)g(n-i)$. Perform a change of summation variable by setting $j = n - i$. Then as i ranges from 0 up to n, j ranges from n down to 0, and so $(fg)(n) = \sum_{j=0}^{n} f(n-j)g(j)$. Since multiplication in R is commutative, we then have $(fg)(n) = \sum_{j=0}^{n} g(j)f(n-j) = (gf)(n)$. Thus $fg = gf$. As well,

$$
\begin{aligned}
(f(g+h))(n) &= \sum_{i=0}^{n} f(i)(g+h)(n-i) = \sum_{i=0}^{n} (f(i)g(n-i) + f(i)h(n-i)) \\
&= \sum_{i=0}^{n} f(i)g(n-i) + \sum_{i=0}^{n} f(i)h(n-i) = (fg)(n) + (fh)(n) \\
&= (fg + fh)(n)
\end{aligned}
$$

since multiplication in R distributes over addition and addition in R is commutative. Thus $f(g+h) = fg + fh$. $\qquad\square$

9.2.5 Lemma. *Let $\mathbb{1} \in \mathscr{F}(\mathbb{N}, R)$ be defined by $\mathbb{1}(n) = 1_R$ if $n = 0$, otherwise $\mathbb{1}(n) = 0_R$. Then $\mathbb{1}$ is the identity element for multiplication in $\mathscr{F}(\mathbb{N}, R)$.*

Proof. Let $f \in \mathscr{F}(\mathbb{N}, R)$, and let $n \in \mathbb{N}$. Then we have $(\mathbb{1}f)(n) = \sum_{i=0}^{n} \mathbb{1}(i)f(n-i)$. Since $\mathbb{1}(i) = 0_R$ for all $i > 0$, it follows that $(\mathbb{1}f)(n) = \mathbb{1}(0)f(n-0) = 1_R f(n) = f(n)$, and so $\mathbb{1}f = f$. Since multiplication in $\mathscr{F}(\mathbb{N}, R)$ is commutative, we conclude that $f\mathbb{1} = f$ as well, and so $\mathbb{1}$ is the identity for multiplication in $\mathscr{F}(\mathbb{N}, R)$. $\qquad\square$

The preceding results combine to prove that $\mathscr{F}(\mathbb{N}, R)$, with addition defined by $(f+g)(n) = f(n) + g(n)$ for $f, g \in \mathscr{F}(\mathbb{N}, R)$ and $n \in \mathbb{N}$, and multiplication defined by $(fg)(n) = \sum_{i=0}^{n} f(i)g(n-i)$ for $f, g \in \mathscr{F}(\mathbb{N}, R)$ and $n \in \mathbb{N}$, is a commutative unital ring, called the ring of power series with coefficients in the ring R.

9.3 The Ring of Polynomials with Coefficients in a Commutative Unital Ring

Again, throughout this section, R shall denote a commutative unital ring. As a special case of Definition 8.4.4, for $f \in \mathscr{F}(\mathbb{N}, R)$, the support of f, S_f, is the set $S_f = \{\, n \in \mathbb{N} \mid f(n) \neq 0_R \,\}$.

9.3.1 Definition. *Let $P(R) = \{ f \in \mathscr{F}(\mathbb{N}, R) \mid |S_f| \text{ is finite}\}$. An element $f \in P(R)$ is called a polynomial with coefficients from R.*

Note that $f \in P(R)$ if, and only if, there exists $m \in \mathbb{N}$ such that $f(n) = 0_R$ for all $n \in \mathbb{N}$ with $n \geq m$. For example, for any commutative unital ring R, the map $f : \mathbb{N} \to R$ for which $f(n) = 1_R$ for every $n \in \mathbb{N}$ is an element of $\mathscr{F}(\mathbb{N}, R)$ but not of $P(R)$. On the other hand, for any $m \in \mathbb{N}$, the map $g : \mathbb{N} \to R$ for which $g(n) = 1_R$ for every $n \in \mathbb{N}$ with $0 \leq n \leq m$, and $f(n) = 0_R$ for all $n \in \mathbb{N}$ with $n > m$ is an element of $P(R)$.

9.3.2 Proposition. *$P(R)$ is a subring of $\mathscr{F}(\mathbb{N}, R)$, and $\mathbb{1} \in P(R)$.*

Proof. Since $\mathbb{1}(i) = 0_R$ for all $i \geq 1$, $\mathbb{1} \in P(R)$. Let $f, g \in P(R)$. Then there exist $n, m \in \mathbb{N}$ such that $f(i) = 0_R$ for all $i \geq n$, and $g(i) = 0_R$ for all $i \geq m$. Let $i \in \mathbb{N}$ with $i \geq \max\{m, n\}$. Then $(f + g)(i) = f(i) + g(i) = 0_R + 0_R = 0_R$, so $f + g \in P(R)$. As well, $(-f)(i) = -(f(i)) = -0_R = 0_R$, so $-f \in P(R)$. Finally, let $i \geq m + n$. Then $(fg)(i) = \sum_{j=0}^{i} f(j)g(i - j)$. If $j \geq n$, then $f(j) = 0_R$. Otherwise, $j < n$, in which case $i - j > i - n \geq m + n - n = m$ and so $g(i - j) = 0_R$. Thus for $i \geq m + n$, $(fg)(i) = 0_R$, and so $fg \in P(R)$. $\square$

9.3.3 Definition. *For $f \in P(R)$ with $f \neq 0$, the degree of f, denoted by $\partial(f)$, is defined to be $\max\{n \in \mathbb{N} \mid f(n) \neq 0_R\}$. The degree of 0 is defined to be the symbol $-\infty$, and we adopt the following conventions regarding $-\infty$: for any $n \in \mathbb{Z}$, $n + (-\infty) = -\infty = (-\infty) + n < n$, and $(-\infty) + (-\infty) = -\infty$. If $f \neq 0$, then $f(\partial(f))$ is called the leading coefficient of f, and f is said to be monic if its leading coefficient is 1_R.*

9.3.4 Proposition. *Let $f, g \in P(R)$. Then $\partial(f + g) \leq \max\{\partial(f), \partial(g)\}$, and $\partial(fg) \leq \partial(f) + \partial(g)$. If R is an integral domain, then $\partial(fg) = \partial(f) + \partial(g)$.*

Proof. Let $i \in \mathbb{N}$ with $i > \max\{\partial(f), \partial(g)\}$, so $f(i) = 0_R = g(i)$ and thus $(f + g)(i) = 0_R$. Thus $\partial(f + g) \leq \max\{\partial(f), \partial(g)\}$ (note that we are implicitly relying on the conventions we introduced for $-\infty$). Now let $i \in \mathbb{N}$ with $i > \partial(f) + \partial(g)$. Then $(fg)(i) = \sum_{j=0}^{i} f(j)g(i - j)$. If $j > \partial(f)$, then $f(j) = 0_R$, while otherwise $j \leq \partial(f)$ and then $i - j \geq i - \partial(f) > \partial(f) + \partial(g) - \partial(f) = \partial(g)$ and so $g(i - j) = 0_R$. Thus $(fg)(i) = 0_R$, and so $\partial(fg) \leq \partial(f) + \partial(g)$. Note that if $\partial(f) + \partial(g) = -\infty$, then at least one of $\partial(f) = -\infty$ or $\partial(g) = -\infty$ holds; that is, either $f = 0$ or $g = 0$, and in either case, $fg = 0$, and then $\partial(fg) = -\infty = \partial f + \partial(g)$. Suppose now that $f \neq 0$ and $g \neq 0$. Then $\partial f, \partial g \in \mathbb{N}$. Let $i = \partial(f) + \partial(g)$. Then $(fg)(i) = \sum_{j=0}^{i} f(j)g(i-j)$. If $j > \partial(f)$, then $f(j) = 0_R$, while if $j < \partial(f)$, then $i - j > i - \partial(f) = \partial(f) + \partial(g) - \partial(f) = \partial(f)$, and then $g(i - j) = 0_R$. Thus $(fg)(i) = f(\partial(f))g(\partial(g))$, and by definition, $f(\partial(f)) \neq 0_R$ and $g(\partial(g)) \neq 0_R$. If R is an integral domain, then $f(\partial(f))g(\partial(g)) \neq 0_R$, and so $\partial(fg) = \partial(f) + \partial(g)$. $\square$

9.3.5 Corollary. *If R is an integral domain, then $P(R)$ is an integral domain.*

Proof. If $f, g \in P(R)$ are such that $f \neq 0$ and $g \neq 0$, then $\partial(f) > -\infty$ and $\partial(g) > -\infty$, so $\partial(fg) = \partial(f) + \partial(g) > -\infty$ and thus $fg \neq 0$. Since $P(R)$ is a commutative unital ring, it follows that $P(R)$ is an integral domain. $\square$

9.3.6 Proposition. *The map $\iota : R \to P(R)$ for which $\iota(r)(0) = r$, while for $n > 0$, $\iota(r)(n) = 0_R$, is an injective unital ring homomorphism with image $\{\, f \in P(R) \mid \partial(f) \leq 0 \,\}$.*

Proof. For any $r \in R$, it follows directly from the definition of $\iota(r)$ that $\partial(\iota(r)) \leq 0$. On the other hand, if $f \in P(R)$ has degree 0, then $r = f(0) \in R$ is not 0_R, while for any $i \in \mathbb{N}$ with $i > 0$, $f(i) = 0_R$, and thus $f = \iota(r)$. If $f = 0$, then $f = \iota(0_R)$. This proves that the image of ι is the set $\{\, f \in P(R) \mid \partial(f) \leq 0 \,\}$. For $r, s \in R$ with $r \neq s$, we have $\iota(r)(0) = r$, while $\iota(s)(0) = s$, so $\iota(r)(0) \neq \iota(s)(0)$ and therefore $\iota(r) \neq \iota(s)$. Thus ι is injective. It remains to prove that ι is a ring homomorphism. Let $r, s \in R$. Then for any $i \in \mathbb{N}$, $\iota(r + s)(i) = 0_R$ if $i > 0$, otherwise it equals $r + s$. Since $\iota(r)(0) = r$ and $\iota(s)(0) = s$, we have $(\iota(r) + \iota(s))(0) = \iota(r)(0) + \iota(s)(0) = r + s = \iota(r + s)(0)$. For $i \in \mathbb{N}$ with $i > 0$, $\iota(r + s)(i) = 0_R = 0_R + 0_R = \iota(r)(i) + \iota(s)(0) = (\iota(r) + \iota(s))(i)$, and so for every $i \in \mathbb{N}$, $\iota(r + s)(i) = (\iota(r) + \iota(s))(i)$. Thus $\iota(r + s) = \iota(r) + \iota(s)$. Finally, $\iota(rs)(0) = rs = \iota(r)(0)\iota(s)(0) = (\iota(r)\iota(s))(0)$ while for $i \in \mathbb{N}$ with $i > 0$, we have $\iota(rs)(i) = 0_R = \sum_{j=0}^{i} \iota(r)(j)\iota(s)(i - j)$ since either $j > 0$, in which case $\iota(r)(j) = 0_R$, or else $j = 0$ and then $\iota(s)(i) = 0_R$. Thus for all $i \in \mathbb{N}$, $\iota(rs)(i) = \iota(r)\iota(s)(i)$ and thus $\iota(rs) = \iota(r)\iota(s)$. Since $\iota(1_R) = 1$, ι is an injective unital ring homomorphism. $\square$

It is customary to identify $r \in R$ with $\iota(r) \in P(R)$, and in this way, we may think of R as being a subring of $P(R)$. We shall adopt this convention, so from here on, we consider R to be the subring of $P(R)$ whose elements are the polynomials of degree at most 0. Thus for any $r \in R$ and $f, g \in P(R)$,

$rf = fr$ and $r(f + g) = rf + rg$, and $1_R f = f$ (since we have identified 1_R with $\iota(1_R) = \mathbb{1}$). Moreover, for any $r, s \in R$ and $f \in P(R)$, we have $(r + s) = rf + sf$ and $(rs)f = r(sf)$.

9.3.7 Lemma. *Let $a \in R$ and $f \in P(R)$. Then for all $n \in \mathbb{N}$, $(af)(n) = a(f(n))$.*

Proof. Let $n \in \mathbb{N}$. Then $(af)(n) = \sum_{i=0}^{n} a(i)f(n - i) = a(0)f(n) = (a)(f(n))$ (recall that we are now treating $a \in R$ as the map from $\mathbb{N}$ to R for which $a(0) = a$, while $a(i) = 0_R$ for all $i \in \mathbb{N}$ with $i > 0$). $\square$

9.3.8 Definition. *Let $x \in P(R)$ be the polynomial of degree 1 for which $x(1) = 1_R$, while $x(i) = 0_R$ for all $i \in \mathbb{N}$ with $i \neq 1$.*

9.3.9 Proposition. *Let $n \geq 0$. Then $x^n(n) = 1_R$, while $x^n(i) = 0_R$ for all $i \in \mathbb{N}$ with $i \neq n$.*

Proof. The proof is by induction on $n \geq 0$. Since $x^0 = \mathbb{1}$, and $\mathbb{1}(0) = 1_R$, while $\mathbb{1}(i) = 0_R$ for all $i \in \mathbb{N}$ with $i > 0$, the assertion is valid for $n = 0$. Suppose now that $n \geq 0$ is an integer for which $x^n(n) = 1_R$ and $x^n(i) = 0_R$ for all $i \neq n$. Then

$$x^{n+1}(n + 1) = (x^n x)(n + 1) = \sum_{i=0}^{n+1} x^n(i)x(n + 1 - i)$$
$$= x^n(n)x(1) = 1_R^2 = 1_R,$$

while for $0 \leq i \leq n$, if $i = 0$, we have $x^{n+1}(i) = x^n(0)x(0) = 0_R$, while if $1 \leq i \leq n$, then since $x(t) = 0_R$ for all $t \neq 1$, and by hypothesis, $x^n(t) = 0_R$ for all $t < n$, we have

$$x^{n+1}(i) = x^n x(i) = \sum_{j=0}^{i} x^n(j)x(i - j) = x^n(i - 1)x(1) = 0_R.$$

On the other hand, by Proposition 9.3.4, $\partial(x^{n+1}) \leq n + 1$, and thus $x^{n+1}(i) = 0$ for all $i > n + 1$. The result follows now by induction. $\square$

9.3.10 Definition. *Let S be a commutative unital ring, and let R be a unital subring of S. For any subset $A \subseteq S$,*

$$R[A] = \cap\{\, T \mid T \text{ is a subring of } S \text{ and } R \cup A \subseteq T \,\}.$$

In the special case $A = \{\, a \,\}$, it is conventional to write $R[a]$ rather than $R[\{\, a \,\}]$.

It is immediate that $R[A]$ is a subring of S and $R \cup A \subseteq R[A]$. Furthermore, with respect to the subset ordering, $R[A]$ is the smallest subring of S that contains R and A. Our work below is focused on the case $A = \{\, x \,\} \subseteq P(R)$.

9.3.11 Proposition. *For $f \in P(R)$ with $f \neq 0$, and $n = \partial(f)$, $f = \sum_{i=0}^{n} f(i)x^i$.*

Proof. Let $f \in P(R)$ with $f \neq 0$, and let $n = \partial(f)$. By Lemma 9.3.7, for any $j \in \mathbb{N}$, we have $(\sum_{i=0}^{n} f(i)x^i)(j) = \sum_{i=0}^{n}(f(i)x^i)(j) = \sum_{i=0}^{n}(f(i))(x^i(j))$. Thus for $j \in \mathbb{N}$ with $j > n$, we have $(\sum_{i=0}^{n} f(i)x^i)(j) = 0_R = f(j)$. Suppose that $j \in \mathbb{N}$ with $j \leq n$. Then $(\sum_{i=0}^{n} f(i)x^i)(j) = (f(j))(x^j(j)) = f(j)$. Thus for all $j \in \mathbb{N}$, $f(j) = (\sum_{i=0}^{n} f(i)x^i))(j)$, and so $f = \sum_{i=0}^{n} f(i)x^i$. $\qquad\square$

9.3.12 Corollary. $R[x] = P(R)$.

Proof. Recall that we are regarding R to be a unital subring of $P(R)$, and thus $R[x] \subseteq P(R)$. If $f \in P(R)$ with $f \neq 0_R$, then by Proposition 9.3.11, $f \in T$ for every subring T of $P(R)$ that contains R and x, and thus $f \in R[x]$. Consequently, $P(R) \subseteq R[x]$, and so $P(R) = R[x]$, as required. $\qquad\square$

We shall take advantage of Corollary 9.3.12, and from now on, we shall refer to $P(R)$ as $R[x]$. Thus for any commutative unital ring R, the ring of polynomials with coefficients in R is

$$R[x] = \{\, f \in \mathscr{F}(\mathbb{N}, R) \mid \text{ there exists } n \in \mathbb{N} \\ \text{ and } a_0, a_1, \ldots, a_n \in R \text{ such that } f = \sum_{i=0}^{n} a_i x^i \,\}.$$

Note that if $f \in R[x]$ and $n \in \mathbb{N}$ is such that for some $a_0, a_1, \ldots, a_n \in R$, $f = \sum_{i=0}^{n} a_i x^i$, then it follows that for all $i \in \mathbb{N}$ with $i > n$, $f(i) = 0_R$, while for $i \leq n$, $f(i) = a_i$. Thus if $a_n \neq 0_R$, then $f(n) = a_n \neq 0_R$, and so $\partial f = n$. However, it is possible that $a_n = 0_R$. For example, $f = \sum_{i=0}^{n} a_i x^i = f = \sum_{i=0}^{n} a_i x^i + 0_R x^{n+1}$, since $0_R g = 0_R$ for any $g \in R[x]$ (recall that we consider R to be a subring of $R[x]$). In fact, given that $f = \sum_{i=0}^{n} a_i x^i$, then $\partial(f) = n$ if, and only if, $a_n \neq 0_R$.

9.3.13 Proposition. *Let R and S be commutative unital rings, and let $\varphi : R \to S$ be a unital ring homomorphism. Then for any $s \in S$, there is a unique ring homomorphism $\overline{\varphi} : R[x] \to S$ for which $\overline{\varphi}|_R = \varphi$ and $\overline{\varphi}(x) = s$ (note that $\overline{\varphi}$ is necessarily unital).*

Proof. Define $\overline{\varphi}$ as follows. First, define $\overline{\varphi}(0_R) = 0_S$. Then for $f \in R[x]$ with $f \neq 0_R$, define $\overline{\varphi}(f) = \sum_{i=0}^{\partial(f)} \varphi(f(i))s^i$. Then in particular, $\overline{\varphi}(1_R) = \varphi(1_R) = 1_S$. Let $f, g \in R[x]$ with $f, g \neq 0_R$. If $f + g = 0_R$, then $g = -f$, and we have $\overline{\varphi}(f) + \overline{\varphi}(g) = \sum_{i=0}^{\partial(f)} \varphi(f(i))s^i + \sum_{i=0}^{\partial(f)} \varphi(-f(i))s^i = \sum_{i=0}^{\partial(f)} \big(\varphi(f(i)) + \varphi(-f(i))\big)s^i = \sum_{i=0}^{\partial(f)} 0_S s^i = 0_S = \overline{\varphi}(0_R) = \overline{\varphi}(f + g)$. Suppose that $f + g \neq 0_R$. Since $\partial(f + g) \leq \max\{\partial(f), \partial(g)\}$, we may write $\overline{\varphi}(f + g) = \sum_{i=0}^{m} \varphi((f+g)(i))s^i$, where $m = \max\{\partial(f), \partial g)\}$ (since for any j with $\partial(f + g) < j$, $(f + g)(j) = 0_R$ and $\varphi(0_R) = 0_S$) Thus $\hat{\varphi}(f + g) = \sum_{i=0}^{m} \varphi((f(i) + g(i))s^i = \sum_{i=0}^{m} \varphi(f(i))s^i + \sum_{i=0}^{m} \varphi(g(i))s^i = \sum_{i=0}^{\partial(f)} \varphi(f(i))s^i + \sum_{i=0}^{\partial(g)} \varphi(g(i))s^i$ since $f(i) = 0_R$ if $i > \partial(f)$ and similarly, $g(i) = 0_R$ if $i > \partial(g)$. Thus $\overline{\varphi}(f + g) = \overline{\varphi}(f) + \overline{\varphi}(g)$. Since $f + 0_R = f = 0_R + f$, it follows that $\overline{\varphi}(f + 0_R) = \overline{\varphi}(f) = \overline{\varphi}(f) + 0_S = \overline{\varphi})(f) + \overline{\varphi}(0_R)$ as well. Thus for all $f, g \in P(R)$, $\overline{\varphi}(f + g) = \overline{\varphi}(f) + \overline{\varphi}(g)$. As well, since $0_R f = 0_R$ for all $f \in R[x]$, we have $\overline{\varphi}(0_R f) = \overline{\varphi}(0_R) = 0_S = 0_S \overline{\varphi}(f) = \overline{\varphi}(0_R)\overline{\varphi}(f)$, and as both $R[x]$ and S are commutative, it follows that $\overline{\varphi}(f 0_R) = \overline{\varphi}(f)\overline{\varphi}(0_R)$ as well. Finally, suppose that $f, g \in R[x]$ and $f, g \neq 0_R$. Then we have $\overline{\varphi}(f)\overline{\varphi}(g) = \big(\sum_{i=0}^{\partial(f)} \varphi(f(i))s^i\big)\big(\sum_{j=0}^{\partial(g)} \varphi(g(j))s^j\big) = \sum_{i=0}^{\partial(f)} \sum_{j=0}^{\partial(g)} \varphi(f(i)g(j))s^{i+j} = \sum_{k=0}^{\partial(f)+\partial(g)} \big(\varphi(\sum_{t=0}^{k} f(t)g(k - t))s^k\big) = \sum_{k=0}^{\partial(fg)} \varphi((fg)(k))s^k = \overline{\varphi}(fg)$. This completes the proof that $\overline{\varphi}$ is a unital ring homomorphism from $R[x]$ to S. For $a \in R$, we have $\overline{\varphi}(a) = \varphi(a)s^0 = \varphi(a)$, so $\overline{\varphi}\big|_R = \varphi$. Finally, $\overline{\varphi}(x) = \varphi(x(0))s^0 + \varphi(x(1))s^1 = \varphi(0_R)1_S + \varphi(1_R)s = 0_S + 1_S s = s$.

It remains to prove that $\overline{\varphi}$ is uniquely determined. Suppose that $\psi : R[x] \to S$ is a ring homomorphism for which $\psi\big|_R = \varphi$ and $\psi(x) = s$. Then for any nonzero $f \in R[x]$, we have $\psi(f) = \psi(\sum_{i=0}^{\partial(f)} f(i)x^i)^{\cdot} = \sum_{i=0}^{\partial(f)} \psi(f(i))\psi(x)^i = \sum_{i=0}^{\partial(f)} \varphi(f(i))s^i = \overline{\varphi}(f)$, and so $\psi = \overline{\varphi}$. $\square$

In the special case that R is a unital subring of S, then the inclusion map is a unital ring homomorphism from R to S, and in this case, for any $s \in S$, we shall denote $\overline{\varphi}(f)$ by $\hat{f}(s)$; that is, if $f = \sum_{i=0}^{n} a_i x^i$, then $\hat{f}(s) = \sum_{i=0}^{n} a_i s^i$.

9.3.14 Definition. *Let R and S be commutative unital rings with R a unital subring of S. If $f \in R[x]$ and $s \in S$ are such that $\hat{f}(s) = 0_R$, then s is called a root of f, or a zero of f.*

The construction of the polynomial ring can be iterated, since $R[x]$ is a commutative unital ring if R is a commutative unital ring. Thus we may consider the ring of polynomials with coefficients in $R[x]$. It would be very confusing to write $(R[x])[x]$, so it is customary to use a different symbol, y say, for the special map from $\mathbb{N}$ to $R[x]$ that sends 1 to $1_{R[x]}$ and sends all other $i \in \mathbb{N}$ to $0_{R[x]}$. As usual, we consider $R[x]$ to be a subring of $(R[x])[y]$, so we have $R \subseteq R[x] \subseteq (R[x])[y]$. In particular, with this convention, we have $1_R = 1_{R[x]}$.

Note that since $R \subseteq (R[x])[y]$ and $y \in (R[x])[y]$, $R[y] \subseteq (R[x])[y]$. As well, $x \in (R[x])[y]$, and so $(R[y])[x] \subseteq (R[x])[y]$. Similarly, we have $R \subseteq (R[y])[x]$, and $x \in (R[y])[x]$, so $R[x] \subseteq (R[y])[x]$. As $y \in (R[y])[x]$, we then obtain $(R[x])[y] \subseteq (R[y])[x]$. Thus $(R[x])[y] = (R[y])[x]$. We shall henceforth write simply $R[x, y]$ for either. Note that with this convention, we have $R[x, y] = R[y, x]$. Evidently, this process can be continued, thereby obtaining for any positive integer n the polynomial ring $R[x_1, x_2, \ldots, x_n]$ in the n variables $x_1, x_2, \ldots, x_n$.

For example, $2 - 3x^2y + xy^3 - 4xy \in \mathbb{Z}[x, y]$, and we may write it as $2 - (3x^2 + 4x)y + xy^3 \in (\mathbb{Z}[x])[y]$ or as $2 - 3yx^2 + (y^3 - 4y)x \in (\mathbb{Z}[y])[x]$.

Further, if R is an integral domain, then by Corollary 9.3.5, $R[x]$ is an integral domain, and thus for any positive integer n, $R[x_1, x_2, \ldots, x_n]$ is an integral domain.

9.4   The Division Algorithm and Applications

As we shall see, the ring of polynomials with coefficients in a field is a principal ideal domain, and so greatest common divisors of any two polynomials (not both zero) exist. In fact, there is an effective algorithm for finding greatest common divisors which is very similar to that for finding the greatest common divisor of two integers, where the role of the size of an integer is played by the degree of a polynomial. For these reasons, we shall now restrict our attention to polynomial rings whose coefficient rings are fields.

Throughout this section, F shall denote an arbitrary field.

9.4.1 Lemma. *Let $b, r \in F[x]$ with $b \neq 0_F$. If $\partial(r) \geq \partial(b)$, there exists $q \in F[x]$ such that $\partial(r - bq) < \partial(r)$.*

Proof. Suppose that $b, r \in F[x]$ are such that $b \neq 0$ and $m = \partial(b) < \partial(r) = n$, say. Let $r = \sum_{i=0}^{n} r_i x^i$ and $b = \sum_{i=0}^{m} b_i x^i$, so $b_m \neq 0$ and $r_n \neq 0$. Let $q = r_n b_m^{-1} x^{n-m} \in F[x]$. Then $\partial(bq) = \partial(b) + \partial(q) = m + n - m = n$. The leading term of bq is $b_m r_n b_m^{-1} x^n = r_n x^n$, the same as the leading term of r, and so $\partial(r - bq) < \partial(r)$.   □

9.4.2 Theorem (Division Theorem for Polynomials). *Let $a, b \in F[x]$ with $b \neq 0$. Then there exist unique $q, r \in F[x]$ such that*

(i) $a = bq + r$; and
(ii) $\partial(r) < \partial(b)$.

q is called the quotient and r the remainder when a is divided by q.

Proof. We first prove the existence of q and r with the two properties. Let $T = \{\, a - bt \mid t \in F[x] \,\}$. If $0_F \in T$, then for some $q \in F[x]$, we have $a = bq$, so for $r = 0_F$, we have $a = bq + r$ and $\partial(r) = -\infty < \partial(b)$. Suppose that $0_F \notin T$. Let $S = \{\, \partial(f) \mid f \in T \,\}$. Since $0_F \notin T$, $S \subseteq \mathbb{N}$, and since $a \in T$, $S \neq \varnothing$. By the Well Ordering Principle, S has a least element, n say, and so there exists $q \in F[x]$ such that $r = a - bq \in T$ has $\partial(r) = n$. If $n \geq \partial(b)$, then by Lemma 9.4.1, there exists $f \in T$ with $\partial(f) < \partial(r)$. Since $\partial(r) = n \leq \partial(h)$ for all $h \in T$, this is not possible and so we conclude that $\partial(r) = n < \partial(b)$. Thus we have obtained $q, r \in F[x]$ such that $a = bq + r$ and $\partial(r) < \partial(b)$, as required.

Suppose now that $q_1, r_1 \in F[x]$ are such that $bq + r = a = bq_1 + r_1$ and $\partial(r_1) < \partial(b)$. Then $b(q - q_1) = r_1 - r$. Suppose that $r_1 - r \neq 0_F$. Then $q - q_1 \neq 0$ and we have $\partial(r_1 - r) = \partial(b) + \partial(q - q_1) \geq \partial(b)$. But by Proposition 9.3.4, $\partial(r_1 - r) \leq \max\{\,\partial(r_1), \partial(r)\,\} < \partial(b)$, so $r_1 - r \neq 0_F$ is not possible. Thus $r_1 - r = 0_F$ and so $r_1 = r$. As well, since $F[x]$ is an integral domain and $b(q_1 - q) = 0_F$ and $b \neq 0_F$, it follows that $q_1 - q = 0$ and thus $q_1 = q$. $\qquad\square$

As an immediate application of the Division Theorem, we determine the group of units of $F[x]$. Recall that $\mathbb{Z}$ has only two units, ± 1. The situation is quite different in $F[x]$.

9.4.3 Proposition. *Let $f \in F[x]$. Then f is a unit of $F[x]$ if, and only if, $f = a$ for some $a \in F$ with $a \neq 0_F$.*

Proof. First, suppose that $f \in F[x]$ is a unit. Then there exists $g \in F[x]$ such that $fg = 1_F$, so $f \neq 0_F$, $g \neq 0_F$, and $\partial(f) + \partial(g) = \partial(1_F) = 0$. Thus $\partial(f) = 0$, so $f = a \in F$ and $a \neq 0_F$. Conversely, suppose that $f = a \in F$ with $a \neq 0_F$. Then $g = a^{-1} \in F$ and $fg = aa^{-1} = 1$, so f is a unit of $F[x]$. $\qquad\square$

9.4.4 Definition. *For $f, g \in F[x]$, we say that f divides g, denoted by $f \mid g$, if there exists $q \in F[x]$ such that $g = qf$. In such a case, we say that f is a divisor of g.*

9.4.5 Proposition. *Let $a, b, c \in F[x]$. Then the following hold:*

(i) *If $a \mid b$ and $b \mid c$, then $a \mid c$.*

(ii) *$a \mid b$ if, and only if, $a \mid rb$ for any unit r of $F[x]$.*

(iii) *$a \mid b$ and $b \mid a$ if, and only if, there exists a unit r of $F[x]$ such that $a = rb$. If $a \mid b$ and $b \mid a$ and both a and b are monic, then $a = b$.*

(iv) *If $a \mid b$ and $a \mid c$, then $a \mid (fb + gc)$ for all $f, g \in F[x]$.*

(v) *If $a \mid b$ and $a \mid (b + c)$, then $a \mid c$.*

Proof. For (i), suppose that $a \mid b$ and $b \mid c$. Then there exist $f, g \in F[x]$ such that $b = fa$ and $c = gb$, so $c = g(fa)$ and thus $a \mid c$.

For (ii), let r be a unit of $F[x]$. If $a \mid b$, then there exists $f \in F[x]$ such that $b = fa$, so $rb = rfa$ and thus $a \mid rb$. Conversely, suppose that $a \mid rb$. Then there exists $g \in F[x]$ such that $rb = ga$, from which we obtain $b = r^{-1}(rb) = r^{-1}(ga)$, and thus $a \mid b$.

For (iii), suppose first that $a \mid b$ and $b \mid a$. Then there exist $f, g \in F[x]$ such that $b = fa$ and $a = gb$, so $b = f(gb)$. Then $b(1_F - fg) = 0$. Since $F[x]$ is an integral domain, either $b = 0$ or else $1_F = fg$. If $b = 0_F$, then $a = gb = 0_F b$, so $a = b = 1_F b$. If $1_F = fg$, then f, g are units of $F[x]$, as required. If, in addition, both a and b are monic, then the leading coefficient of a and of b is 1_F, and since g is a unit with $a = gb$, it follows that $g = 1_F$ and so $a = b$. Conversely, suppose that there is a unit r of $F[x]$ such that $a = rb$. Then $b = r^{-1}a$, and then by (ii), we have $a \mid b$ and $b \mid a$.

For (iv), suppose that $a \mid b$ and $a \mid c$, and let $f, g \in F[x]$. There exist $p, q \in F[x]$ such that $b = pa$ and $c = qa$, so $fb + gc = f(pa) + g(qa) = (fp + gq)a$, proving that $a \mid (fb + gc)$.

Finally, for (v), suppose that $a \mid b$ and $a \mid (b + c)$. Then by (iv), a divides $(1_F)(b + c) + (-1_F)b = c$. $\square$

It is worth noting that any nonzero element of $F[x]$ can be expressed as a unit times a monic element of $F[x]$. For if $f \in F[x]$, $f \neq 0_F$, we have $f = a_0 + a_1 x + \cdots + a_n x^n$ for some $n \in \mathbb{N}$ and $a_0, a_1, \ldots, a_n \in F$, with $a_n \neq 0$. Then a_n is a unit of F and hence a unit of $F[x]$, and $f = a_n(a_n^{-1}a_0 + a_n^{-1}a_0 x + \cdots + x^n)$.

9.4.6 Proposition. *$F[x]$ is a PID. Moreover, if I is any nonzero ideal of $F[x]$, there is a unique monic $f \in I$ such that $I = (f)$.*

Proof. Let $I \triangleleft F[x]$. If $I = \{0_F\}$, then $I = (0_F)$, so we may suppose that $I \neq \{0_F\}$. Then $T = \{\partial(f) \mid f \in I, \; f \neq 0_F\} \neq \varnothing$, and $T \subseteq \mathbb{N}$, so by the Well Ordering Principle, T has a least element, m say. Let $p \in I$ have

$\partial(p) = m \geq 0$. We prove that $I = (p)$. Since $p \in I$, we have $(p) \subseteq I$. Let $a \in I$. Then since $p \neq 0_F$, we may apply Theorem 9.4.2 to a and p, thereby obtaining $q, r \in F[x]$ such that $a = pq + r$ with $\partial(r) < \partial(p)$. We wish to prove that $r = 0_F$. Suppose by way of contradiction that $r \neq 0_F$. Since $r = a - pq$ and $a, p \in I$, we find that $r \in I$. Now, p is an element of least degree in I, so $\partial(p) \leq \partial(r) < \partial(p)$. Since this is not possible, we conclude that $r = 0_F$ and $a = pq \in (p)$. Thus $I \subseteq (p)$, and so $I = (p)$. Let $p = rp_1$, where r is a unit of $F[x]$ and p_1 is monic. Then $I = (p) = (rp_1) = (p_1)$ by Proposition 9.4.5 (ii). Now suppose that $p_2 \in I$ is monic and that $I = (p_2)$. Since $p_1 \in I = (p_2)$, $p_2 \mid p_1$, and since $p_2 \in I = (p_1)$, $p_1 \mid p_2$. By Proposition 9.4.5 (iii), $p_1 = p_2$. □

9.4.7 Definition. *Let $a, b \in F[x]$, not both zero. Then $d \in F[x]$ is said to be a greatest common divisor of a and b if the following hold:*

(i) $d \mid a$ and $d \mid b$ (this declares that d is a common divisor of a and b).

(ii) If s is any common divisor of a and b (that is, if $s \mid a$ and $s \mid b$), then $s \mid d$.

(iii) d is monic.

9.4.8 Proposition. *For $a, b \in F[x]$, not both zero, there is one and only one greatest common divisor of a and b, and it is the unique monic generator of the ideal $aF[x] + bF[x]$.*

Proof. Since not both $a = 0_F$ and $b = 0_F$ hold, the ideal $aF[x] + bF[x]$ is not the zero ideal. By Proposition 9.4.6, there is a unique monic $d \in aF[x] + bF[x]$ such that $aF[x] + bF[x] = (d)$. Since $a, b \in (d)$, we have $d \mid a$ and $d \mid b$. Suppose that $c \in F[x]$ is such that $c \mid a$ and $c \mid b$. Since $d \in aF[x] + bF[x]$, we have $d = af + bg$ for some $f, g \in F[x]$. Since $c \mid a$ and $c \mid b$, there exist $p, q \in F[x]$ such that $a = cp$ and $b = cq$. Thus $d = cpf + cqg = c(pf + qg)$, and so $c \mid d$. Since d is monic, it follows that d is a greatest common divisor of a and b. If d_1 is any greatest common divisor of a and b, then $d \mid d_1$ and $d_1 \mid d$, and so, since both d and d_1 are monic, Proposition 9.4.5 (iii) implies that $d = d_1$. Thus d is the unique greatest common divisor of a and b. □

9.4.9 Notation. *For $a, b \in F[x]$, not both zero, the unique greatest common divisor of a and b shall be denoted by $\gcd(a, b)$.*

Note that for $a, b \in F[x]$, not both zero, $\gcd(a, b) \in aF[x] + bF[x]$, so there exist $p, q \in F[x]$ such that $\gcd(a, b) = ap + bq$.

9.4.10 Corollary. *Let E be a field with F a subfield of E. Then $F[x] \subseteq E[x]$. For $f, g \in F[x]$, the greatest common divisor of f and g in $E[x]$ is the same as the greatest common divisor of f and g in $F[x]$.*

Proof. Let d denote the greatest common divisor of f and g in $F[x]$, so $(f) + (g) = fF[x] + gF[x] = (d) = dF[x]$. Then $fE[x] + gE[x] = (fF[x])E[x] + (gF[x])E[x] = (fF[x] + gF[x])E[x] = dF[x]E[x] = dE[x]$, and so d is the greatest common divisor of f and g in $E[x]$.   □

One might find this corollary to be counter-intuitive, since it seems possible that $f, g \in F[x]$ might have more factors in $E[x]$ than they did in $F[x]$ (which is true) and thus it might seem possible that the greatest common divisor of f and g in $E[x]$ might be a nontrivial multiple of the greatest common divisor of f and g in $F[x]$. As the corollary shows, this is not the case.

We remark that the proof of Proposition 9.4.8 is an existence and uniqueness argument, with no indication as to how one might determine the greatest common divisor of $a, b \in F[x]$, not both zero. Fortunately, the Division Theorem (see Theorem 9.4.2) provides the foundation for a Euclidean algorithm for the calculation of the greatest common divisor of two polynomials, an algorithm that is entirely analogous to that for the computation of the greatest common divisor of two integers not both of which are zero (see Algorithm 4.3.7).

9.4.11 Lemma. *Let $a, b \in F[x]$, not both zero, and suppose that $q, r \in F[x]$ are such that $a = qb + r$. Then $\gcd(b, r)$ exists and $\gcd(a, b) = \gcd(b, r)$.*

Proof. If $b = 0$, then $r = a \neq 0$ since not both a and b were zero, while otherwise, $b \neq 0$. Thus by Proposition 9.4.8, $\gcd(b, r)$ exists. Since $\gcd(a, b)$ is a common divisor of a and b, it follows from Proposition 9.4.5 (iv) and the fact that $r = a - qb$ that $\gcd(a, b)$ divides r, and so $\gcd(a, b)$ is a common divisor of b and r. Thus $\gcd(a, b) \mid \gcd(b, r)$. Conversely, $\gcd(b, r)$ divides b and thus qb, and $\gcd(b, r)$ divides r, so by Proposition 9.4.5 (iv) and the fact that $a = qb + r$, we see that $\gcd(b, r)$ is a divisor of a and thus $\gcd(b, r)$ is a common divisor of a and b. This implies that $\gcd(b, r)$ divides $\gcd(a, b)$. We now know that each of $\gcd(a, b)$ and $\gcd(b, r)$ divides the other. Since they are both monic, it follows from Proposition 9.4.5 (iii) that $\gcd(a, b) = \gcd(b, r)$.   □

We can now formulate the Euclidean algorithm for the calculation of the greatest common divisor of two polynomials, not both of which are zero. Note that for any nonzero $p \in F[x]$, $gcd(p, 0_F) = up$, where u is the multiplicative inverse of the leading coefficient of p (thereby making up monic). Let $a, b \in F[x]$ with not both a, b zero. If $b = 0_F$, then $gcd(a, b) = ua$, where u is the multiplicative inverse of the leading coefficient of a. Otherwise, $b \neq 0_F$. In this case, we apply the Division Theorem to a and b, thereby obtaining $q, r \in F[x]$ such that $a = bq + r$ and $\partial(r) < \partial(b)$. By Proposition 9.4.11, $gcd(a, b) = gcd(b, r)$, and now $\partial(r) < \partial(b)$. If $r = 0_F$, then $gcd(a, b) = gcd(b, 0)_F = ub$ where u is the multiplicative inverse of the leading coefficient of b. Otherwise, neither b nor r is zero. Repeat the preceding step, with b in place of a, and r in place of b. After at most $\partial(b)$ iterations, we will reach a stage where $gcd(a, b) = gcd(g, 0)_F$ for some nonzero $g \in F[x]$, and thereby obtain $gcd(a, b) = ug$, where u is the multiplicative inverse of the leading coefficient of g.

9.4.12 Example. *We determine the greatest common divisor of* $3x^3 + 5x^2 + 6x, 4x^4 + 2x^3 + 6x^2 + 4x + 5 \in \mathbb{Z}_7[x]$. *We show the results of the sequence of applications of the Division Theorem, as described above. In our application of the division theorem, we make use of the facts that* $3^{-1} = 5$ *and* $(4)(5) = 6$ *in* $\mathbb{Z}_7$. *We calculate*

$$(4)(3^{-1})x(3x^3 + 5x^2 + 6x) = 6(3x^4 + 5x^3 + 6x^2) = 4x^4 + 2x^3 + x^2,$$

and thus $4x^4 + 2x^3 + 6x^2 + 4x + 5 - 6x(3x^3 + 5x^2 + 6x) = 5x^2 + 4x + 5$. *Our first application of the division theorem therefore yields*

$$4x^4 + 2x^3 + 6x^2 + 4x + 5 = 6x(3x^3 + 5x^2 + 6x) + (5x^2 + 4x + 5),$$

and thus $gcd(4x^4 + 2x^3 + 6x^2 + 4x + 5, 3x^3 + 5x^2 + 6x)$ *is equal to* $gcd(3x^3 + 5x^2 + 6x, 5x^2 + 4x + 5)$. *We have* $(3)(5^{-1})x(5x^2 + 4x + 5) = 2x(5x^2 + 4x + 5) = 3x^3 + x^2 + 3x$ *and*

$$3x^3 + 5x^2 + 6x - 2x(5x^2 + 4x + 5) = 3x^3 + 5x^2 + 6x - (3x^3 + x^2 + 3x) = 4x^2 + 3x,$$

while $(4)(5^{-1})(5x^2 + 4x + 5) = 5(5x^2 + 4x + 5) = 4x^2 + 6x + 4$, *so*

$$3x^3 + 5x^2 + 6x = 2x(5x^2 + 4x + 5) + 4x^2 + 3x$$
$$= 2x(5x^2 + 4x + 5) + 5(5x^2 + 4x + 5) - (3x + 4).$$

Thus the second application of the division theorem yields

$$3x^3 + 5x^2 + 6x = (2x + 5)(5x^2 + 4x + 5) + (4x + 3),$$

and so $gcd(3x^3 + 5x^2 + 6x, 5x^2 + 4x + 5) = gcd(5x^2 + 4x + 5, 4x + 3)$. *Now,* $5(4^{-1})x(4x+3) = 3x(4x+3) = 5x^2 + 2x$, *so* $5x^2 + 4x + 5 - 3x(4x+3) = 2x + 5$, *and* $2(4^{-1})(4x + 3) = 4(4x + 3) = 2x + 5$, *so the third application of the division theorem yields*

$$5x^2 + 4x + 5 = (3x + 4)(4x + 3) + 0_F.$$

Thus $gcd(5x^2 + 4x + 5, 4x + 3) = gcd(4x + 3, 0_F)$, *which is* $4x + 3$ *made monic; that is,* $gcd(4x + 3, 0_F) = 4^{-1}(4x + 3) = x + 6$. *Thus*

$$gcd(4x^4 + 2x^3 + 6x^2 + 4x + 5, 3x^3 + 5x^2 + 6x)$$
$$= gcd(3x^3 + 5x^2 + 6x, 5x^2 + 4x + 5)$$
$$= gcd(5x^2 + 4x + 5, 4x + 3) = 4^{-1}(4x + 3) = x + 6.$$

Moreover, we have

$$4x + 3 = (3x^3 + 5x^2 + 6x) - (2x + 5)(5x^2 + 4x + 5)$$
$$= (3x^3 + 5x^2 + 6x) - (2x + 5)((4x^4 + 2x^3 + 6x^2 + 4x + 5)$$
$$- 6x(3x^3 + 5x^2 + 6x))$$
$$= (5x^2 + 2x + 1)(3x^3 + 5x^2 + 6x)$$
$$+ (5x + 2)(4x^4 + 2x^3 + 6x^2 + 4x + 5).$$

Thus

$$gcd(4x^4 + 2x^3 + 6x^2 + 4x + 5, 3x^3 + 5x^2 + 6x)$$
$$= x + 6 = 2(4x + 3)$$
$$= 2(5x^2 + 2x + 1)(3x^3 + 5x^2 + 6x)$$
$$+ 2(5x + 2)(4x^4 + 2x^3 + 6x^2 + 4x + 5)$$
$$= (3x^2 + 4x + 2)(3x^3 + 5x^2 + 6x)$$
$$+ (3x + 4)(4x^4 + 2x^3 + 6x^2 + 4x + 5).$$

9.5   Irreducibility and Factorization of Polynomials

In this section, we prove the analogue of the Fundamental Theorem of Arithmetic (see Theorem 4.4.7). Througout this section, F shall denote an arbitrary field.

We have proven that $F[x]$ is an integral domain, and we recall that in an integral domain D, a nonzero, nonunit element p of D is irreducible in D if for all $x, y \in D$ with $p = xy$, either x or y is a unit of D. An element p that is not zero, not a unit of D, and not irreducible is said to reducible. In this section, we shall explore the notion of irreducibility in $F[x]$.

9.5.1 Example.

(i) $x^2+1 \in \mathbb{R}[x]$ *is irreducible over* $\mathbb{R}$. *For if not, there exist* $a, b, c, d \in \mathbb{R}$ *such that* $x^2 + 1 = (ax + b)(cx + d) = acx^2 + (ad + bc)x + bd$, *so* $bd = 1 = ac$, *and* $ad + bc = 0$. *But then* $bc = -ad$ *and so* $c = bcd = -ad^2$, *from which we obtain* $1 = ac = -a^2d^2$, *which is not possible. However,* $x^2+1 \in \mathbb{C}[x]$, *and over* $\mathbb{C}$, x^2+1 *is reducible since* $(x + i)(x - i) = x^2 + 1$.

(ii) $x^2 - 2 \in \mathbb{Q}[x]$ *is irreducible over* $\mathbb{Q}$, *for otherwise, there exist* $a, b, c, d \in \mathbb{Q}$ *such that* $x^2 - 2 = (ax + b)(cx + d) = acx^2 + (ad + bc)x + bd$, *which means that* $ac = 1$, $bd = -2$, *and* $ad + bc = 0$. *Then* $d + bc^2 = c(ad + bc) = 0$, *and thus* $-2 + b^2c^2 = b(d + bc^2) = 0$. *From this, we obtain* $(bc)^2 = 2$. *However, this means that* $\sqrt{2}$ *is rational, which is not the case. Thus* $x^2 - 2$ *is irreducible over* $\mathbb{Q}$. *However, over* $\mathbb{R}$ *it is reducible, since* $(x + \sqrt{2})(x - \sqrt{2}) = x^2 - 2$.

(iii) $x^2+1 \in \mathbb{Z}_2[x]$ *is reducible, since* $(x+1)^2 = x^2+(1+1)x+1 = x^2+1$.

The examples above show that in general, irreducibility of a polynomial depends on the field.

9.5.2 Proposition. *If* $f \in F[x]$ *and* $\partial(f) = 1$, *then* f *is irreducible.*

Proof. If $f \in F[x]$ is reducible, then there exist $g, h \in F[x]$ such that neither g nor h is a unit and $f = gh$. But then $\partial(g), \partial(h) \geq 1$, and so $\partial(f) = \partial(g) + \partial(h) \geq 2$. $\square$

9.5.3 Definition. *For* $f, g \in F[x]$, *we say that* f *and* g *are relatively prime if not both* f, g *are zero, and if* $gcd(f, g) = 1_F$.

9.5.4 Proposition. *Let* $a, b, c \in F[x]$. *If* $gcd(a, b) = 1_F$ *and* $a \mid bc$, *then* $a \mid c$.

Proof. Since $gcd(a, b) = 1_F$, there exist $f, g \in F[x]$ such that $1_F = fa + gb$, and thus $c = (cf)a + g(cb)$. Since $a \mid a$ and $a \mid cb$, it follows that $a \mid c$. $\square$

9.5.5 Corollary. *If* $p \in F[x]$ *is irreducible over* F, *and* $a, b \in F[x]$ *are such that* $p \mid ab$, *then either* $p \mid a$ *or* $p \mid b$.

Proof. Suppose that p does not divide a. Then $gcd(p, a) = 1_F$, and so there exist $r, s \in F[x]$ such that $rp + sa = 1_F$. Thus $b = rpb + sab$, and since $p \mid ab$, it follows that p divides $rpb + sab = b$. $\square$

9.5.6 Corollary. *If $p \in F[x]$ is irreducible over F, and for $n \in \mathbb{Z}^+$, $a_i \in F[x]$, $i = 1, 2, \ldots, n$, are such that $p \mid a_1 a_2 \cdots a_n$, then for some $i \in \{1, 2, \ldots, n\}$, $p \mid a_i$.*

Proof. The proof is an elementary inductive argument (on $n \geq 2$), where Corollary 9.5.5 provides the proof of the base case of the inductive argument.   □

9.5.7 Theorem. *Let $f \in F[x]$. If $\partial(f) > 0$, then there is a unit $u \in F$, and a positive integer k and monic irreducible $p_i \in F[x]$, $i = 1, 2, \ldots, k$ such that $f = u \prod_{i=1}^{k} p_i$. Moreover, up to the order of the factors, there is exactly one such factorization.*

Proof. We first prove the existence of such a factorization. The proof will be by induction on degree. If $f \in F[x]$ has degree 1, then $f = b + ax$ with $a \neq 0_F$, and thus a is a unit of $F[x]$, and $f = a(a^{-1}b + x)$. Suppose now that $n \geq 1$ is an integer for which every polynomial of degree at most n in $F[x]$ has such a factorization, and let $f \in F[x]$ with $\partial(f) = n + 1$. If f is irreducible, then for a the coefficient of x^{n+1} in f, $a \neq 0_F$ and so a is a unit of $F[x]$, and $f = a(a^{-1}f)$, where $a^{-1}f$ is monic and irreducible since f is irreducible. Suppose then that f is not irreducible. Then there exist $g, h \in F[x]$ with $\deg g > 0$, $\partial(h) > 0$ and $f = gh$. Thus $n + 1 = \partial(f) = \partial(g) + \partial(h)$, and so $0 < \partial(g), \partial(h) \leq n$. By hypothesis, there exist units u_1 and u_1, and positive integers k and l with monic irreducible p_i, q_j, $i = 1, 2, \ldots, k$, $j = 1, 2, \ldots, l$ such that $g = u_1 p_1 \ldots p_k$ and $h = u_2 q_1 \ldots q_l$. Thus $f = u_1 u_2 p_1 \ldots p_k q_1 \ldots q_l$, and since $u_1 u_2$ is a unit of $F[x]$, we have established that f has a factorization as the product of a unit of $F[x]$ and two or more monic irreducible elements of $F[x]$. The assertion follows now by induction.

We now turn to the proof of the uniqueness of the representation. We prove by induction on $n \geq 1$ that for any $f \in F[x]$ of degree n, if u_1 and u_2 are units, and k and l are positive integers for which there are monic irreducible $p_1, p_2, \ldots, p_k$ and $q_1, q_2, \ldots, q_l$ such that $f = u_1 p_1 p_2 \cdots p_k = u_2 q_1 q_2 \cdots q_l$, then $u_1 = u_2$, $k = l$, and there exists a relabelling of $q_1, q_2, \ldots, q_k$ such that for each $i = 1, 2, \ldots, k$, $p_i = q_i$. For $f \in F[x]$ of degree 1, say $f = ax + b$ with $a \neq 0_F$, $f = a(x + a^{-1}b)$ is the only possible representation (where $u = a$ and $p = x + a^{-1}b$), so the assertion holds for $n = 1$. Suppose now that $n > 1$ is an integer for which the assertion holds for every positive integer $m < n$. Let $f \in F[x]$ have degree n. Suppose that there are units u_1 and u_2, and positive integers k and l for which there are monic irreducible polynomials $p_1, p_2, \ldots, p_k$ and $q_1, q_2, \ldots, q_l$ such that $f = u_1 p_1 p_2 \cdots p_k = u_2 q_1 q_2 \cdots q_l$. Since $p_1 \mid f = q_1 q_2 \cdots q_l$ and p_1 is irreducible, it follows from Corollary 9.5.6 that for some i with $1 \leq i \leq l$, p_1 divides q_i. Then since both p_1 and q_i are monic and irreducible, $p_1 = q_i$. Relabel $q_1, q_2, \ldots, q_l$ if necessary so that $p_1 = q_1$. By consideration of degree, $k = 1$ if and only if $l = 1$, and so if $k = 1$, we are done. Otherwise, $k > 1$ and $l > 1$, and since $F[x]$ is an integral domain, we may cancel p_1 to obtain $p_2 \cdots, p_k = q_2 \cdots q_l$. Since $\partial(p_1) \geq 1$, $\partial(p_2 \cdots, p_k) < n$ and thus by hypothesis, $k = l$ and, relabelling $q_2, \ldots, q_k$ if necessary, $p_i = q_i$ for $i = 2, \ldots, k$. The result follows now by the second principle of mathematical induction. $\qquad \square$

To determine whether a given polynomial over an arbitrary field is irreducible or not is, in general, very hard. We shall see that there are easily applicable criteria when the degree of the polynomial is 2 or 3, and we shall present Eisenstein's irreducibility sufficiency condition for polynomials of arbitrary degree over the rational numbers. Also, we shall describe why, over the field of real numbers, every irreducible polynomial has degree either 1 or 2, and that over the field of complex numbers, the irreducible polynomials are precisely the polynomials of degree 1. For situations other than those described above, one must use ad hoc methods to establish irreducibility.

Our next result is a simple but useful fact with which readers may already be familiar from their studies of elementary algebra.

9.5.8 Proposition. *For $p \in F[x]$ and $b \in F$, the remainder when the Division Theorem is applied to p and $x - a$ is $\hat{p}(b)$. In particular, b is a root of p if, and only if, $x - b$ is a factor of p.*

Proof. By the Division Theorem, there exist $q, r \in F[x]$ such that $p = (x - b)q + r$ and $\partial(r) < \partial(x - b) = 1$, and so r is a constant. Now we have $\hat{p}(b) = (b - b)\hat{q}(b) + r$, and so $r = \hat{p}(b)$. If b is a root of p, then $\hat{p}(b) = 0_F$ and so $p = (x - a)q$. On the other hand, if $x - a$ is a factor of p, then by the uniqueness assertion of the divison theorem, $r = 0_F$ and so $\hat{p}(b) = r = 0_F$; that is, r is a root of p. $\qquad \square$

9.5.9 Corollary. *A polynomial over F of degree greater than 1 which has a root in F is reducible over F. Moreover, a polynomial of degree 2 or 3 over F is irreducible over F if, and only if, it has no root in F.*

Proof. Let $p \in F[x]$, and suppose first that p has $\partial(p) > 1$ and that p has a root in F, r say. Then by Proposition 9.5.8, $x - r$ is a factor of p, so there exists $q \in F[x]$ such that $p = (x - r)q$. Since $\partial(p) > 1$, $\partial(q) \geq 1$, and thus q is not a unit, so p is reducible.

Now suppose that $\partial(p)$ is either 2 or 3. Assume that p is reducible. Then $p = fg$ for some $f, g \in F[x]$ with $\partial(f) \geq 1$, $\partial(g) \geq 1$. Since $\partial(f) + \partial(g) \leq 3$, it follows that either $\partial(f) = 1$ or $\partial(g) = 1$. Now a polynomial of degree 1 is of the form $ax + b$ for some $a, b \in F$ with $a \neq 0_F$, and thus has $-a^{-1}b$ as a root. It follows that p has a root in F. We have proven now that if p has no root in F, then p is irreducible over F. As we proved in the first paragraph, if p has a root in F, then p is reducible over F; that is, if p is irreducible over F, then p has no root in F. $\qquad\square$

Note that if the degree of a polynomial is greater than 3, then it is quite possible for it to be reducible over F and yet have no root in F. For example, $x^4 + 2x^2 + 1 = (x^1 + 1)^2 \in \mathbb{R}[x]$ has no roots in $\mathbb{R}$.

For a polynomial of degree 2, there is a method called completing the square that may help in finding the roots of the polynomial.

9.5.10 Proposition. *Suppose that $char(F) \neq 2$. Then for any $a, b, c \in F$ with $a \neq 0_F$, $ax^2 + bx + c = a[(x + (2a)^{-1}b)^2 + (4a^2)^{-1}(4ac - b^2)]$. Then $ax^2 + bx + c$ is reducible if, and only if, there exists $r \in F$ such that $b^2 - 4ac = r^2$, in which case the roots of $ax^2 + bx + c$ in F are $-(2a)^{-1}(-b \pm r)$.*

Proof. Let $a, b, c \in F$ with $a \neq 0_F$. Then

$$a[(x + (2a)^{-1}b)^2 + (4a^2)^{-1}(4ac - b^2)]$$
$$= a[x^2 + 2(2a)^{-1}bx + (2a)^{-2}b^2 + (4a^2)^{-1}(4ac - b^2)]$$
$$= a[x^2 + a^{-1}bx + (4a^2)^{-1}(4ac)]$$
$$= ax^2 + bx + c.$$

Suppose now that $ax^2 + bx + c$ is reducible. Then by Corollary 9.5.9, there is $t \in F$ such that $at^2 + bt + c = 0_F$, and thus $a[(t + (2a)^{-1}b)^2 + (4a^2)^{-1}(4ac - b^2)] = 0_F$. Since $a \neq 0_F$, it follows that $(t + (2a)^{-1}b)^2 = (4a^2)^{-1}(b^2 - 4ac)$ and so $b^2 - 4ac = (2a(t + (2a)^{-1}b))^2$. Thus $r = 2at + b$ has the required property. Conversely, if there exists $r \in F$ such that $b^2 - 4ac = r^2$, then for $t = (2a)^{-1}(-b \pm r)$, we have

$$at^2 + bt + c = (4a)^{-1}(b^2 \mp 2br + r^2) + (2a)^{-1}(-b^2 \pm br) + c$$
$$= (4a)^{-1}(b^2 + (b^2 - 4ac) - 2b^2) + c = 0_F.$$

$\qquad\square$

For example, $x^2 + [2]_5 x + [3]_5 \in \mathbb{Z}_5[x]$ is irreducible since $[2]_5^2 - 4[1]_5[3]_5 = [4]_5 - [2]_5 = [2]_5$, and as $[1]_5^2 = [1]_5 = [4]_5^2$ and $[2]_5^2 = [3]_5^2 = [4]_5$, there is no $r \in \mathbb{Z}_5$ such that $r^2 = [2]_5$.

9.5.11 Definition. *If $p \in F[x]$ has a root $c \in F$, and $q \in F[x]$ and positive integer m are such that $p = (x - c)^m q$ and c is not a root of q, then c is said to be a root of p of multiplicity m.*

9.5.12 Corollary. *If $p \in F[x]$ is a nonzero polynomial of degree $n \geq 0$, then, counting multiplicities, p has at most n roots in F.*

Proof. The proof will be by induction on the degree of a polynomial. A nonzero polynomial of degree 0 is a nonzero constant and thus has 0 roots in F. Suppose now that $n \geq 1$ is an integer for which the hypothesis holds for all polynomials of degree d for any d with $0 \leq d \leq n$ and suppose that $f \in F[x]$ is nonzero with degree $n + 1$. If f has no roots, then there is nothing to prove, so suppose that f has a root $r \in F$. Further suppose that the multiplicity of r in f is m, so there exists $q \in F[x]$ such that $f = (x - r)^m q$ and r is not a root of q. Since f is not zero, neither is q, and so q is a nonzero polynomial of degree $d < n + 1$, and so by hypothesis, q has at most $\partial(q)$ roots in F, counting multiplicities. Since r is not a root of q, and any root s of f different from r must satisfy $0 = \hat{f}(s) = (s - r)^m \hat{q}(s)$ and $s - r \neq 0_F$, it follows that $\hat{q}(s) = 0_F$ and so s is a root of q. Thus the number of roots of f, counting multiplicities, is m plus the number of roots of q, counting multiplicities, and this is at most $m + \deg q = \partial(f)$. The result follows now by the second principal of mathematical induction. $\square$

Polynomials over $\mathbb{Q}$

Our goal in this section is to establish the necessary preliminaries for the proof of Eisenstein's criteria for irreducibility of $p \in \mathbb{Q}[x]$, and to prove Eisenstein's result itself. Recall that $\mathbb{Q}$ is the field of quotients of the integral domain $\mathbb{Z}$. It should come as no surprise then that the study of polynomials with coefficients in the field $\mathbb{Q}$ should rely on considerations of polynomials with coefficients in the integral domain $\mathbb{Z}$.

Our first result is a simple method for determining the rational roots of a polynomial with integer coefficients.

9.6.1 Proposition. *Let $f \in \mathbb{Z}[x]$ with f nonzero, and let $n = \partial(f)$. Then there exist integers $a_0, a_1, \ldots, a_n$ such that $f = \sum_{i=0}^{n} a_i x^i$. Suppose* that $r, s \in \mathbb{Z}$ are such that $\frac{r}{s}$ is a root of f, and that $(r, s) = 1$. Then r is a divisor of a_0, and s is a divisor of a_n.

Proof. Suppose that $r, s \in \mathbb{Z}$ are such that $(r, s) = 1$ and $\frac{r}{s}$ is a root of f. We multiply the equation $\hat{f}(\frac{r}{s}) = 0$ by s^n to obtain that $\sum_{i=0}^{n} a_i s^{n-i} r^i = 0$. Then $a_0 s^n + r(\sum_{i=1}^{n} a_i s^{n-i} r^{i-1}) = 0$, and so r divides $a_0 s^n$. Since $(r, s) = 1$, it follows that $(r, s^n) = 1$, and thus r divides a_0. Similarly, $a_n r^n + s(\sum_{i=0}^{n-1} a_i r^i s^{n-i-1}) = 0$, and so s divides $a_n r^n$. Since $(s, r^n) = 1$, it follows that s divides a_n, as claimed. $\square$

Thus every rational root of f will be found amongst the values $\pm\frac{c}{d}$ as c varies over all positive divisors of a_0 and d varies over all positive divisors of a_n. For example, to find the rational roots of $f = 6x^4 - x^3 + 4x^2 - x - 2 \in \mathbb{Z}[x]$, we need only check values in the set

$$\{\pm\frac{c}{d} \mid c \text{ is a positive divisor of } 2 \text{ and } d \text{ is a positive divisor of } 6 \}$$

$$= \{\pm 1, \pm\frac{1}{2}, \pm\frac{1}{3}, \pm\frac{1}{6}, \pm 2, \pm\frac{2}{3}\}.$$

We compute $f(1) = 6$, $f(-1) = 10$, $f(\frac{1}{2}) = -\frac{5}{4}$, and $f(-\frac{1}{2}) = 0$, so $-\frac{1}{2}$ is a root of f, and thus $x - (-\frac{1}{2})) = x + \frac{1}{2}$ is a factor of f. We find that

$$
\begin{array}{r}
6x^3 - 4x^2 + 6x - 4 \\
\hline
x + \tfrac{1}{2} \, \big) \, 6x^4 - x^3 + 4x^2 - x - 2 \\
6x^4 + 3x^3 \\
\hline
- 4x^3 + 4x^2 - x - 2 \\
- 4x^3 - 2x^2 \\
\hline
6x^2 - x - 2 \\
6x^2 + 3x \\
\hline
- 4x - 2 \\
- 4x - 2 \\
\hline
0
\end{array}
$$

and so $f = (x + \frac{1}{2})(6x^3 - 4x^2 + 6x - 4) = (x + \frac{1}{2})(2x^2(3x - 2) + 2(3x - 2)) = (x + \frac{1}{2})(3x - 2)(2x^2 + 2) = 6(x + \frac{1}{2})(x - \frac{2}{3})(x^2 + 1)$. Since $x^2 + 1 \geq 1 > 0$ for any $x \in \mathbb{R}$, it follows that we have found all rational (and indeed, all real) roots of f; namely $-\frac{1}{2}$ and $\frac{2}{3}$.

9.6.2 Definition. *If $f \in \mathbb{Z}[x]$ is not zero, then the content of f, denoted by $con(f)$, is the greatest common divisor of the coefficients of f. f is said to be primitive if $con(f) = 1$.*

Note that any $f \in \mathbb{Z}[x]$ can be written as the product of an integer and a primitive polynomial (simply factor out the greatest common divisor of the coefficients of the polynomial. For example, $15x^4 - 5x^3 + 10x + 30 = 5(3x^4 - x^3 + 2x + 6)$ and $3x^4 - x^3 + 2x + 6$ is primitive.

9.6.3 Proposition. *If $f, g \in \mathbb{Z}[x]$ are each primitive, then fg is primitive.*

Proof. Let $f, g \in \mathbb{Z}[x]$ be nonzero, and suppose that fg is not primitive. Then there exists a prime $p \in \mathbb{Z}$ such that p divides $con(fg)$. Let $\pi : \mathbb{Z} \to \mathbb{Z}_p$ denote the canonical ring homomorphism $(\pi(i) = [i]_p$ for each $i \in \mathbb{Z})$. By Proposition 9.3.13, there is a unique unital ring homomorphism $\overline{\pi} : \mathbb{Z}[x] \to \mathbb{Z}_p[x]$ for which $\overline{\pi}|_{\mathbb{Z}} = \pi$ and $\overline{\pi}(x) = x$. For any $\sum_{i=0}^{n} a_i x^i \in \mathbb{Z}[x]$, $\overline{\pi}(\sum_{i=0}^{n} a_i x^i) = \sum_{i=0}^{n} \pi(a_i) x^i$. Since π maps each coefficient of fg to 0, it follows that $\overline{\pi}(fg) = 0$. As $\overline{\pi}$ is a ring homomorphism, $\overline{\pi}(f)\overline{\pi}(g) = 0$. Finally, since $\mathbb{Z}_p$ is a field, $\mathbb{Z}_p[x]$ is an integral domain and thus either $\overline{\pi}(f) = 0$ or else $\overline{\pi}(g) = 0$. Consequently, either p divides each coefficient of f, or else p divides each coefficient of g, so at least one of f or g is not primitive. Thus if both f and g are primitive, then fg is primitive. $\square$

9.6.4 Lemma. *If $f, g \in \mathbb{Z}[x]$ are each primitive and $m, n \in \mathbb{Z}$ are such that $mf = ng$, then $f = \pm g$.*

Proof. Since $con(f) = 1$, $con(mf) = |m|$. Similarly, since $con(g) = 1$, $con(ng) = |n|$. If $mf = ng$, it follows that $|m| = |n|$ and thus $f = \pm g$. $\square$

The next result, Gauss's lemma, is the cornerstone of Eisenstein's irreducibility criterion (see Proposition 9.6.6). Gauss's lemma establishes that $f \in \mathbb{Z}[x]$ has a nontrivial factorization in $\mathbb{Q}[x]$ if, and only if, f has a nontrivial factorization in $\mathbb{Z}[x]$.

9.6.5 Proposition (Gauss's Lemma). *If $f \in \mathbb{Z}[x]$ is nonzero and $g, h \in \mathbb{Q}[x]$ are such that $f = gh$, then there exist $G, H \in \mathbb{Z}[x]$ and $a, b \in \mathbb{Q}$ such that $f = GH$ and $G = ag$ and $H = bh$.*

Proof. We may assume with no loss of generality that f is primitive. Let r denote the least common multiple of the denominators of the coefficients of g, and similarly, let s denote the least common multiple of the denominators of the coefficients of h. Then r and s are positive integers and $rg, sh \in \mathbb{Z}[x]$. If we let $m = con(rg)$ and $n = con(sh)$, then there exist primitive $G, H \in \mathbb{Z}[x]$ such that $rg = mG$, and $sh = nH$. Thus $g = \frac{m}{r}G$ and $h = \frac{n}{s}H$, and so $f = \frac{m}{r}\frac{n}{s}GH$; that is, $rsf = mnGH$. By Proposition 9.6.3, GH is primitive, and so by Lemma 9.6.4, $|rs| = |mn|$ and so $f = \pm GH$. We have $G = \frac{r}{m}g$ and $H = \frac{n}{s}h$, so we may choose $a = \frac{r}{m} \in \mathbb{Q}$ and $b = \frac{s}{n} \in \mathbb{Q}$ (if $f = -GH$, then replace H by the primitive polynomial $-H$, and replace b by $-b$). $\qquad\square$

9.6.6 Proposition (Eisenstein's irreducibility criterion). *Let $f = \sum_{i=0}^{n} a_i x^i \in \mathbb{Z}[x]$ be nonzero with degree $n > 0$. If there exists a prime $p \in \mathbb{Z}$ such that the following hold:*

    (i) p divides a_i for $i = 0, 1, \ldots, n-1$;

    (ii) p does not divide a_n;

    (iii) p^2 does not divide a_0 (so in particular, $a_0 \neq 0$),

then f is irreducible over $\mathbb{Q}$.

Proof. By Gauss's lemma, it will suffice to prove that f is irreducible over $\mathbb{Z}$. By way of contradiction, assume that f is reducible over $\mathbb{Z}$, and there exists a prime $p \in \mathbb{Z}$ such that the three conditions hold. Then there exist $g, h \in \mathbb{Z}[x]$ with $f = gh$, and neither g nor h are units (and since $f \neq 0$, we have $g \neq 0$ and $h \neq 0$ as well). Thus $m = \partial(g) \geq 1$ and $k = \partial(h) \geq 1$, say $g = \sum_{i=0}^{m} b_i x^i$ and $h = \sum_{i=0}^{k} c_i x^i$. Since $f = gh$, we have $a_n = b_m c_k$, and p does not divide a_n, so p does not divide either of b_m or c_k. Let $\pi : \mathbb{Z} \to \mathbb{Z}_p$ denote the canonical homomorphism ($\pi(i) = [i]_p$). By Proposition 9.3.13, there is a unique unital ring homomorphism $\overline{\pi} : \mathbb{Z}[x] \to \mathbb{Z}_p[x]$ for which $\overline{\pi}|_{\mathbb{Z}} = \pi$ and $\overline{\pi}(x) = x$. We have $\overline{\pi}(\sum_{i=0}^{n} a_i x^i) = \sum_{i=0}^{n} \pi(a_i)x^i$. Since p divides a_i for $i = 0, 1, \ldots, n-1$, $\pi(a_i) = [0]_p$ for $i = 0, 1, \ldots, n-1$. Thus $\overline{\pi}(f) = [a_n]_p x^n$, and $[a_n]_p \neq [0]_p$ since p does not divide a_n. As $\overline{\pi}$ is a ring homomorphism, $\overline{\pi}(g)\overline{\pi}(h) = [a_n]_p x^n$. Now, $\mathbb{Z}_p$ is a field, and so by Theorem 9.5.7, each of $\overline{\pi}(g)$ and $\overline{\pi}(h)$ can be written uniquely as the product of a unit and monic irreducible elements of $\mathbb{Z}_p[x]$. As each irreducible factor of either $\overline{\pi}(g)$ or $\overline{\pi}(h)$ is a factor of $[a_n]_p x^n$, it follows that $\overline{\pi}(g) = [r]_p x^k$ and $\overline{\pi}(h) = [s]_p x^l$ for some nonzero $[r]_p, [s]_p \in \mathbb{Z}_p$ and positive integers k and l for which $k+l = n$. But $\overline{\pi}(g) = \sum_{i=0}^{m} \pi(b_i)x^i$ and $\overline{\pi}(h) = \sum_{i=0}^{k} \pi(c_i)x^i$, and so $[0]_p \doteq \pi(b_0) = \pi(c_0)$. Thus p divides each of b_0 and c_0, and therefore p^2 divides $b_0 c_0 = a_0$. But by (iii), this is not the case, and so the assumption that f is reducible over $\mathbb{Z}$ has led to a contradiction. $\qquad\square$

9.6.7 Example.

    (i) For any prime $p \in \mathbb{Z}$, $x^2 - p \in \mathbb{Z}[x]$ is irreducible over $\mathbb{Q}$ by Eisenstein's irreducibility criterion. In particular, this proves that $x^2 - p$

has no rational roots, and so $\sqrt{p} \in \mathbb{R}$ but $\sqrt{p} \notin \mathbb{Q}$. More generally, for any integer $n \geq 2$, $x^n - p$ is irreducible over $\mathbb{Q}$ by Eisenstein's irreducibility criterion, and so $\sqrt[n]{p} \notin \mathbb{Q}$.

(ii) $x^3 - 6x^2 + 15x - 30 \in \mathbb{Z}[x]$ is irreducible over $\mathbb{Q}$ by Eisenstein's irreducibility criterion, by virtue of the prime 3, since 3 divides -30, 15, and -6, but 3 does not divide 1, the coefficient of x^3, nor does 3^2 divide -30.

Our next example is a classic, and it illustrates an interesting application of Eisenstein's irreducibility criterion. First, we make some initial observations pertaining to $x^p - 1 \in \mathbb{Z}[x]$, where p is a prime. For example, $x^2 - 1 = (x-1)(x+1)$, and $x^3 - 1 = (x-1)(x^2+x+1)$. It was shown in Corollary 9.5.9 that $x^2 + x + 1$ is irreducible over $\mathbb{Q}$ if, and only if, $x^2 + x + 1$ has no rational roots, and by Proposition 9.6.1, the only candidates for rational roots of $x^2 + x + 1$ are ± 1. Since neither 1 nor -1 are roots of $x^2 + x + 1$, we conclude that $x^2 + x + 1$ is irreducible over $\mathbb{Q}$. What about $x^5 - 1 = (x-1)(x^4+x^3+x^2+x+1)$? Should we expect $x^4+x^3+x^2+x+1$ to be irreducible over $\mathbb{Q}$? It is clear that neither 1 nor -1 is a root of $x^4 + x^3 + x^2 + x + 1$, but the situation is more complicated than it was for $p = 3$, since the irreducibility of a quartic over $\mathbb{Q}$ is no longer equivalent to not having a root in $\mathbb{Q}$. We note that Eisenstein's irreducibility criterion can't be directly applied to $x^4 + x^3 + x^2 + x + 1$. Our use of the phrase "directly applied" suggests that there is a work-around that will allow us to apply Eisenstein's criterion, and indeed, that is the case as we show next.

By Proposition 9.3.13, there is a unique ring homomorphism ϕ from $\mathbb{Z}[x]$ to $\mathbb{Z}[x]$ which is the identity mapping on $\mathbb{Z}$ and which sends x to $x + 1$. Similarly, there is a unique ring homomorphism θ from $\mathbb{Z}[x]$ to $\mathbb{Z}[x]$ which is the identity mapping on $\mathbb{Z}$, and which sends x to $x - 1$. Note that $\theta \circ \phi$ is then the unique ring homomorphism from $\mathbb{Z}[x]$ to $\mathbb{Z}[x]$ which is the identity on $\mathbb{Z}$ and which sends x to x; that is, $\theta \circ \phi$ is the identity mapping on $\mathbb{Z}[x]$. Similarly, $\phi \circ \theta$ is the identity mapping on $\mathbb{Z}[x]$, and so ϕ is a ring isomorphism from $\mathbb{Z}[x]$ to $\mathbb{Z}[x]$. Consider $\phi(x^5 - 1) = (x+1)^5 - 1 = x^5 + 5x^4 + 10x^3 + 10x^2 + 5x + 1 - 1 = x(x^4 + 5x^3 + 10x^2 + 10x + 5)$. As well, $\phi(x^5 - 1) = \phi((x-1)(x^4 + x^3 + x^2 + x + 1)) = x\phi(x^4 + x^3 + x^2 + x + 1)$ and so $\phi(x^4 + x^3 + x^2 + x + 1) = x^4 + 5x^3 + 10x^2 + 10x + 5$. We may apply Eisenstein's irreducibility criteria to $x^4 + 5x^3 + 10x^2 + 10x + 5$ with prime $p = 5$ to determine that $x^4 + 5x^3 + 10x^2 + 10x + 5$ is irreducible over $\mathbb{Q}$. Since θ is a ring isomorphism, it follows that $\theta(x^4 + 5x^3 + 10x^2 + 10x + 5) = \theta \circ \phi(x^4 + x^3 + x^2 + x + 1) = x^4 + x^3 + x^2 + x + 1$ is also irreducible over $\mathbb{Q}$.

9.6.8 Proposition. *For prime $p \in \mathbb{Z}$, $x^p - 1 = (x - 1)(\sum_{i=0}^{p-1} x^i)$ and $\sum_{i=0}^{p-1} x^i$ is irreducible over $\mathbb{Q}$.*

Proof. We shall follow the discussion in the paragraph that precedes this proposition, and make use of the isomorphism ϕ and θ introduced therein. Consider $\phi(x^p - 1) = (x + 1)^p - 1 = -1 + \sum_{i=0}^{p} \binom{p}{i} x^i = x(\sum_{i=1}^{p} \binom{p}{i} x^{i-1})$. As well, $(x - 1)(\sum_{i=0}^{p-1} x^i) = \sum_{i=0}^{p-1} x^{i+1} - \sum_{i=0}^{p-1} x^i = \sum_{i=1}^{p} x^i - \sum_{i=0}^{p-1} x^i = x^p - 1$, and so $\phi(x^p - 1) = \phi(x - 1)\phi(\sum_{i=0}^{p-1} x^i) = x\phi(\sum_{i=0}^{p-1} x^i)$. Thus $\phi(\sum_{i=0}^{p-1} x^i) = \sum_{i=1}^{p} \binom{p}{i} x^{i-1}$. We prove now that Eisenstein's irreducibility criterion can be applied to $\sum_{i=1}^{p} \binom{p}{i} x^{i-1}$. Note that $\binom{p}{p} = 1$, so this is a monic polynomial. Moreover, its constant term is $\binom{p}{1} = p$. Thus it suffices to prove that p divides $\binom{p}{i}$ for every integer i with $1 < i < p$. We observe that $\binom{p}{i} = \frac{p!}{i!(p-i)!}$, and p certainly divides $p!$. Since p is a prime, if n is any positive integer such that p divides $n!$, then p divides k for some k with $1 \leq k \leq n$. Thus for any integer i with $1 < i < p$, we have $0 < p - i < p - 1$ and so p does not divide $i!$ or $(p - i)!$. Thus p divides the integer $\binom{p}{i}$ for each integer i with $1 < i < p$. By Eisenstein's criterion, $\sum_{i=1}^{p} \binom{p}{i} x^{i-1}$ is irreducible over $\mathbb{Q}$, and thus $\theta(\sum_{i=1}^{p} \binom{p}{i} x^{i-1}) = \theta \circ \phi(\sum_{i=0}^{p-1} x^i) = \sum_{i=0}^{p-1} x^i$ is irreducible over $\mathbb{Q}$. □

9.7 Irreducible Polynomials over $\mathbb{R}$ and $\mathbb{C}$

The Fundamental Theorem of Algebra, which essentially states that the irreducible polynomials in $\mathbb{C}[x]$ are the polynomials of degree 1, has a long and interesting history. The first attempt to prove this theorem is usually credited to Jean le Rond d'Alembert (1717-1783), but Carl Friedrich Gauss is often considered to have offered its first satisfactory proof. Gauss returned to the elusive proof of this theorem time and time again. During his lifetime, he offered not less than four different proofs for the theorem, covering a timespan of fifty years. His first proof materialized in October of 1797 and was published in his doctoral dissertation in 1799, at which time he was 22 years of age. His thesis contains a critique of previous attempts to prove the theorem by d'Alembert, Euler, and others. As it turned out, his proof of 1799, using geometric notions, had its own unproven assertions (which were only completely proved in 1920), and Gauss was not satisfied with it either. In 1816, Gauss published two additional proofs, the first of which was very technical and almost strictly algebraic (the non-algebraic result needed was the fact that a polynomial of degree 3 with real coefficients had at least one real root). In 1849, just a few years before his death, Gauss offered a fourth proof.

While most modern texts on field theory offer what is essentially an algebraic proof (again, needing only the fact that a polynomial of degree 3 with real coefficients has at least one real root), the proof requires a considerable amount of algebraic machinery that is beyond the scope of the present text. Probably the shortest proof is provided in an elementary course on complex anaylsis, where it is tradiontally obtained as an immediate corollary to Liouville's theorem (which states that a map defined and differentiable at all points of $\mathbb{C}$ is constant if it is bounded).

Consequently, we propose to simply state the Fundamental Theorem of Algebra without proof, and then investigate its implications.

9.7.1 Theorem (Fundamental Theorem of Algebra). *Every polynomial of positive degree over $\mathbb{C}$ has a root in $\mathbb{C}$.*

9.7.2 Corollary. *Let $f \in \mathbb{C}[x]$. Then f is irreducible in $\mathbb{C}[x]$ if, and only if, $\partial(f) = 1$.*

Proof. By Proposition 9.5.2, if $f \in \mathbb{C}[x]$ and $\partial(f) = 1$, then f is irreducible in $\mathbb{C}[x]$. Conversely, suppose that $f \in \mathbb{C}[x]$ is irreducible. By Theorem 9.7.1, f has a root $r \in \mathbb{C}$, and so there exists $q \in \mathbb{C}[x]$ such that $f = (x - r)q$. But f is irreducible, and $x - r$ is not a unit of $\mathbb{C}[x]$, so it must be that q is a unit of $\mathbb{C}[x]$; that is, q is a nonzero constant, and so $\partial(q) = 0$. Thus $\partial(f) = \partial(x - r) + \partial(q) = 1 + 0 = 1$. $\qquad\square$

9.7.3 Corollary. *Let $f \in \mathbb{C}[x]$. Then the unique factorization of f as a product of a unit of $\mathbb{C}[x]$ and monic irreducible elements of $\mathbb{C}[x]$ is of the form*

$$f = a(x - r_1)^{n_1}(x - r_2)^{n_2} \cdots (x - r_k)^{n_k},$$

where $a \in \mathbb{C} - \{0\}$, $r_1, r_2, \ldots, r_k \in \mathbb{C}$ are the distinct roots of f, and $n_1, n_2, \ldots, n_k$ are positive integers for which $n_1 + n_2 + \cdots + n_k = \partial(f)$.

Proof. By Theorem 9.5.7, f has a factorization, unique up to the order of writing the factors, as a product of a unit of $\mathbb{C}[x]$ and monic irreducible elements of $\mathbb{C}[x]$. By Corollary 9.7.2, these are precisely the polynomials of the form $x - r$ for $r \in \mathbb{C}$. $\qquad\square$

We now turn our attention to the problem of determining the irreducible elements of $\mathbb{R}[x]$. It turns out that the Fundamental Theorem of Algebra will give us the whole story. This is made possible by considering $\mathbb{R}[x] \subseteq \mathbb{C}[x]$ (since $\mathbb{R}$ can be naturally considered to be a subfield of $\mathbb{C}$ via the injective unital ring homomorphism that maps $r \in \mathbb{R}$ to $r + 0i \in \mathbb{C}$).

We briefly recall the basic facts about $\mathbb{C} = \{\, a + bi \mid a, b \in \mathbb{R} \,\}$. Addition is defined by $(a + bi) + (c + di) = (a + c) + (b + d)i$, and multiplication is defined by $(a + bi)(c + di) = (ac - bd) + (ad + bc)i$. In particular, $i^2 = (0 + (1)i)(0 + (1)i) = -1 + 0i = -1$. For $a, b \in \mathbb{R}$, a is called the real part of $a + bi$, while b is called the imaginary part of $a + bi$. If $b = 0$, the complex number $a = a + (0)i$ is called real, while if $a = 0$, the complex number $0 + bi$ is called purely imaginary. The complex conjugate of $a + bi$, denoted by $\overline{a + bi}$, is $a - bi$.

9.7.4 Proposition. *Conjugation is a ring isomorphism; that is, the map that sends $a + bi \in \mathbb{C}$ to $a - bi$ is a ring isomorphism.*

Proof. Since conjugation is its own inverse, conjugation is a bijective map from $\mathbb{C}$ to $\mathbb{C}$. Let $a + bi, c + di \in \mathbb{C}$. Then $\overline{(a + bi)(c + di)} = \overline{ac - bd + (ad + bc)i} = ac - bd - (ad + bc)i$, while

$$\overline{a + bi}\,\overline{c + di} = (a - bi)(c - di) = ac - bd - (ad + bc)i,$$

so $\overline{(a + bi)(c + di)} = \overline{a + bi}\,\overline{c + di}$. As well, $\overline{a + bi} + \overline{c + di} = a - bi + c - di = a + c - (b + d)i = \overline{a + c + (b + d)i} = \overline{a + bi + c + di}$. $\qquad\square$

9.7.5 Proposition. *Let $f \in \mathbb{R}[x]$. Then $f \in \mathbb{C}[x]$, and if $z \in \mathbb{C}$ is a root of f, then $\overline{z}$ is also a root of f.*

Proof. Suppose that $f = \sum_{i=0}^{n} a_i x^i$, where $a_i \in \mathbb{R}$ for $i = 0, 1, \ldots, n$. Since $\hat{f}(z) = 0$, we have $\sum_{i=0}^{n} a_i z^i = 0$. By Proposition 9.7.4, $\overline{\sum_{i=0}^{n} a_i z^i} = \sum_{i=0}^{n} \overline{a_i z^i} = \sum_{i=0}^{n} \overline{a_i}\,\overline{z}^i = \sum_{i=0}^{n} a_i \overline{z}^i = \hat{f}(\overline{z})$. Since $\overline{0} = 0$, it follows that $0 = \overline{\hat{f}(z)} = \hat{f}(\overline{z})$, as claimed. $\qquad\square$

Thus the complex roots (nonreal) of $f \in \mathbb{R}[x]$ occur in pairs, with the two members of each pair being complex conjugates of each other. For example, the roots of $x^2 + 1 \in \mathbb{R}[x]$ are $\pm i$, and by Proposition 9.5.10, the roots of $x^2 + 2x + 3 \in \mathbb{R}[x]$ are $\dfrac{-2 \pm \sqrt{4 - 4(3)}}{2} = -1 \pm \sqrt{2}\sqrt{-1} = -1 + \sqrt{2}i, \overline{-1 + \sqrt{2}i}$.

9.7.6 Proposition. *Let $f \in \mathbb{R}[x]$. If f is irreducible over $\mathbb{R}$, then $\partial(f) = 1$ or 2; that is, f is either linear or quadratic.*

Proof. By Proposition 9.5.2, if f has degree 1, then f is irreducible over $\mathbb{R}$. Suppose that f is irreducible over $\mathbb{R}$. Then f has no roots in $\mathbb{R}$. However, since $\mathbb{R}[x] \subseteq \mathbb{C}[x]$, $f \in \mathbb{C}[x]$, and by Corollary 9.7.3, f factors completely as a product of a unit of $\mathbb{C}[x]$ and monic linear elements of $\mathbb{C}[x]$. Thus (counting multiplicities), f has $\partial(f)$ roots in $\mathbb{C}$. By Proposition 9.7.5, the complex roots of f occur in complex conjugate pairs, and so $\partial(f)$ must be even, say $\partial(f) = 2k$. Thus k denotes the number of complex conjugate pairs of roots of f. We prove that $k = 1$. Suppose that $a + bi$ is a root of f. Then $a - bi$ is also a root of f, and thus $(x - (a+bi))(x - (a-bi)) \in \mathbb{C}[x]$ is a factor of f. But $(x - (a+bi))(x - (a-bi)) = x^2 - 2ax + (a+bi)(a-bi) = x^2 + 2ax + (a^2 + b^2) \in \mathbb{R}[x]$. Thus each complex conjugate pair of roots of f provides a real quadratic factor of f. Since f is irreducible over $\mathbb{R}$, it follows that $k = 1$. $\qquad\square$

To complete our discussion of the irreducible polynomials of $\mathbb{R}[x]$, it remains to determine exactly which quadratic real polynomials are irreducible over $\mathbb{R}$.

9.7.7 Proposition. *For $a, b, c \in \mathbb{R}$ with $a \neq 0$, $ax^2 + bx + c$ is irreducible over $\mathbb{R}$ if, and only if, $b^2 - 4ac < 0$.*

Proof. We may appeal to Proposition 9.5.10 to conclude that f is reducible if, and only if, $b^2 - 4ac \geq 0$. $\qquad\square$

In summary, we have proven the following result.

9.7.8 Proposition. *Every real polynomial of positive degree is uniquely expressible as a nonzero real number times a product of monic linear or monic irreducible quadratic polynomials.*

In particular, we note that any polynomial of odd degree over $\mathbb{R}$ has at least one real root.

We remind the reader that the results of this section are largely existence theorems and are of little help in actually computing the factorization of a polynomial in $\mathbb{R}[x]$ or in $\mathbb{C}[x]$. Nevertheless, they are very powerful results, and are often used to answer some deep theoretical questions.

Quotient Rings of the Form $F[x]/(f)$, F a Field

Earlier on, we proved that $\mathbb{Z}/(n)$, the quotient ring $\mathbb{Z}$ modulo the principal ideal generated by $n \in \mathbb{Z}$ is equal to $\mathbb{Z}_n$, and in Chapter 4, it was proven that every nonzero element of $\mathbb{Z}_n$ is invertible if, and only if, n is a prime. More generally, it was proven that $[a]_n \in \mathbb{Z}/(n)$ is a unit if, and only if,

$(a, n) = 1$. In this section, we establish analogous results for the ring $F[x]$, where F is a field.

9.8.1 Proposition. *Let* $p \in F[x]$ *be nonzero. Then for* $a \in F[x]$, $a+(p) \in F[x]/(p)$ *is a unit of* $F[x]/(p)$ *if, and only if,* $\gcd(a, p) = 1$.

Proof. Let $a \in F[x]$. Suppose first that $\gcd(a, p) = 1$. Then by Proposition 9.4.8, there exist $r, s \in F[x]$ such that $1 = ra + sp$. Thus $1 + (p) = ra + (p) = (r + (p))(a + (p))$, and since multiplication in $F[x]/(p)$ is commutative, it follows that $a + (p)$ is a unit in $F[x]/(p)$. Conversely, suppose that $a + (p)$ is a unit in $F[x]/(p)$, and let $r \in F[x]$ be such that $r + (p)$ is the inverse of $a + (p)$ in $F[x]/(p)$. Then $1 + (p) = (a + (p))(r + (p)) = ar + (p)$, and so $1 - ar \in (p) = pF[x]$. Thus there exists $t \in F[x]$ with $1 = ar + tp$, and so $\gcd(a, p) = 1$. $\qquad\square$

9.8.2 Corollary. *Let* $p \in F[x]$ *be nonzero. Then* $F[x]/(p)$ *is a field if, and only if,* p *is irreducible over* F.

Proof. Suppose first of all that p is reducible. Then there exist $g, h \in F[x]$ such that $p = gh$ and neither g nor h is a unit (and since p is nonzero, both g and h are nonzero). Thus $\partial(p) = \partial(g) + \partial(h)$ and $\partial(g) \geq 1$, $\partial(h) \geq 1$, so $\partial(p) > \partial(g)$, $\partial(p) > \partial(h)$. Thus $g, h \notin (p)$, and so $g + (p) \neq 0 + (p)$, $h + (p) \neq 0 + (p)$, while $(g + (p))(h + (p)) = gh + (p) = p + (p) = 0 + (p)$. Thus $g + (p)$ is a nonzero divisor of zero in $F[x]/(p)$, and so $F[x]/(p)$ is not an integral domain and therefore not a field. Conversely, suppose that p is irreducible, and let $a \in F[x]$ with $a + (p) \neq 0 + (p)$. Since $\gcd(a, p)$ divides p, we must have either $\gcd(a, p) = up$ for some unit u of $F[x]$, or else $\gcd(a, p) = 1$. As $a \notin (p)$, it follows that $\gcd(a, p) = 1$, and so by Proposition 9.8.1, $a + (p)$ is a unit of $F[x]/(p)$. $\qquad\square$

9.8.3 Proposition. *If* $p \in F[x]$ *and* $n = \partial(p) \geq 0$, *then for every* $a \in F[x]$, *the coset* $a + (p)$ *has a unique representative of degree less than* n.

Proof. Since $\partial(p) \geq 0$, $p \neq 0$. Let $a \in F[x]$. Then by Theorem 9.4.2, there exist unique $q, r \in F[x]$ such that $a = qp + r$ and $\partial(r) < \partial(p)$, and so $r = a - qp \in a + pF[x] = a + (p)$. Suppose that $s \in a + (p)$ also satisfies $\partial(s) < \partial(p)$. Then $r - s \in (p)$, and so there exists $g \in F[x]$ such that $r - s = gp$. But then $a = qp + r = qp + gp + s = (q + g)p + s$. By the uniqueness assertion of the division theorem, $s = r$. $\qquad\square$

Note that Proposition 9.8.3 tells us that there is a one-to-one correspondence between $F[x]/(p)$ and $\{\, f \in F[x] \mid \partial(f) < \partial(p) \,\}$. If F is infinite, this tells us only that $F[x]/(p)$ is infinite (an alternative method to prove this is presented in Proposition 9.8.4). However, if F is finite, then while $F[x]$ is infinite, for nonzero $p \in F[x]$, the quotient ring $F[x]/(p)$ is finite, and we can compute its size exactly. Indeed, if F is a field with k elements, and $n \in \mathbb{N}$, then $\{\, f \in F[x] \mid \partial(f) < n \,\} = \{\, a_0 + a_1 x + \cdots + a_{n-1} x^{n-1} \mid a_0, a_1, \ldots, a_{n-1} \in F \,\}$. Since there are k ways to choose each coefficient a_i, $i = 0, 1, \ldots, n-1$, $|\{\, f \in F[x] \mid \partial(f) < n \,\}| = k^n$. Thus if $p \in F[x]$ is nonzero and $n = \partial(p)$, then $|F[x]/(p)| = k^n$.

9.8.4 Proposition. *Let F be a field, and let $f \in F[x]$ with $\partial(f) \geq 1$. Then the composition of the natural injective ring homomorphism from F into $F[x]$ followed by the natural projection homomorphism from $F[x]$ onto $F[x]/(f)$ is an injective ring homomorphism from F into the quotient ring $F[x]/(f)$.*

Proof. The composition of two ring homomorphisms is a ring homomorphism, and so the kernel of the composite homomorphism is an ideal I of F. Since F is a field, either $I = \{\, 0_F \,\}$ or $I = F$. By virtue of Proposition 9.8.3, the composite map is not trivial and so $I = \{\, 0_F \,\}$, as required. $\square$

We now proceed to work through a couple of examples, the first of which illustrates the ideas discussed in the preceding paragraph. We take as our field F the field $\mathbb{Z}_2 = \mathbb{Z}/(2) = \{\, [0]_2, [1]_2 \,\}$. To make it easier to follow this discussion, we shall simply write 0, respectively 1, for the cosets $[0]_2$ and $[[1]_2$. Note that in this field, $1 = -1$, as $1 + 1 = 0$. Consider $p = x^2 + x + 1 \in \mathbb{Z}_2[x]$. Since $0^2 + 0 + 1 = 1$, and $1^2 + 1 + 1 = 1$, $x^2 + x + 1$ has no roots in $\mathbb{Z}_2$ and is therefore irreducible over $\mathbb{Z}_2$. By Corollary 9.8.2, $\mathbb{Z}_2[x]/(p)$ is a field, and as discussed above, $\mathbb{Z}_2[x]/(p)$ has $2^2 = 4$ elements. For convenience, we shall simply write 0 for $0 + (p)$, 1 for $1 + (p)$, and we introduce a new symbol, t, for $x + (p)$. Then $\mathbb{Z}_2[x]/(p) = \{\, 0, 1, t, t+1 \,\}$. We give the Cayley tables for addition and multiplication in Tables 9.1 and 9.2, respectively.

To see how the entries in Table 9.2 were calculated, recall that $t = x + (p)$, and $p = x^2 + x + 1$. Since $0 + (p) = p + (p) = x^2 + x + 1 + (p) = (x + (p))^2 + (x + (p)) + (1 + (p))$, then since we are writing 0 and 1 for $0 + (p)$ and $1 + (p)$, respectively, we find that $t^2 + t + 1 = 0$. Thus for example, $t^2 = -t - 1 = t + 1$ (since $1 = -1$ in $\mathbb{Z}_2$), and $t(t + 1) = -1 = 1$. Also note that squaring in $F[x]/(p)$ is a simple operation since $(a + b)^2 =$

Table 9.1   Addition for $\mathbb{Z}_2[x]/(x^2 + x + 1)$

+	0	1	t	$t+1$
0	0	1	t	$t+1$
1	1	0	$t+1$	t
t	t	$t+1$	0	1
$t+1$	$t+1$	t	1	0

Table 9.2   Multiplication for $\mathbb{Z}_2[x]/(x^2 + x + 1)$

×	0	1	t	$t+1$
0	0	0	0	0
1	0	1	t	$t+1$
t	0	t	$t+1$	1
$t+1$	0	$t+1$	1	t

$a^2 + 2ab + b^2 = a^2 + b^2$ (every nonzero element has additive order 2). In particular, $(t + 1)^2 = t^2 + 1 = t + 1 + 1 = t$.

Note that, the additive group of $\mathbb{Z}_2[x]/(x^2 + x + 1)$ is isomorphic to $\mathbb{Z}_2 \oplus \mathbb{Z}_2$ under addition, and is thus another instance of the Klein 4-group (see Table 5.1), and the multiplicative group of units of $\mathbb{Z}_2[x]/(x^2 + x + 1)$ is a group of prime order 3, hence cyclic. We shall show shortly that these observations are typical of finite fields; that is, every finite field is isomorphic, as an additive group, to the additive group of a direct sum of n copies of the ring $\mathbb{Z}_p$, where p is the (prime) characteristic of the field and p^n is the size of the field, and the group of units of the field is a cyclic group.

For our second example, we consider $F = \mathbb{R}$, and $p = x^2 + 1$, which is irreducible over $\mathbb{R}$ as it has no real roots. Thus $\mathbb{R}[x]/(x^2 + 1)$ is a field, and $\mathbb{R}[x]/(x^2+1) = \{\, a+bx+(x^2+1) \mid a, b \in \mathbb{R} \,\}$. By Proposition 9.8.4, the map from $\mathbb{R}$ to $\mathbb{R}[x]/(x^2+1)$ is injective. Since $a \in \mathbb{R}$ is sent to $a+(x^2+1)$ by this mapping, we shall identify $a \in \mathbb{R}$ with $a + (x^2 + 1)$, and thus we shall write a in place of $a + (x^2 + 1)$. Furthermore, it is traditional in this particular example to let $i = x + (x^2 + 1)$ (we used the symbol t in the preceding example). Thus $a+bx+(x^2+1) = (a+(x^2+1))+(b+(x^2+1)(x+(x^2+1)) = a + bi$ with this notation, and we have $\mathbb{R}[x]/(x^2 + 1) = \{\, a + bi \mid a, b \in \mathbb{R} \,\}$. Note that $(x+(x^2+1))^2 = x^2+(x^2+1) = -1+x^2+1+(x^2+1) = -1+(x^2+1)$, and so $i^2 = -1$. For $a, b, c, d \in \mathbb{R}$, we have

$$(a + bi) + (c + di) = (a + c) + (b + d)i$$
$$(a + bi)(c + di) = ac + adi + bic + bidi = ac + (ad + bc)i + bdi^2$$
$$= (ac - bd) + (ad + bc)i$$

which should suggest to you that $\mathbb{R}[x]/(x^2 + 1)$ is isomorphic to $\mathbb{C}$ (which is indeed the case).

We have seen that when F is a field, the Euclidean algorithm for $F[x]$ suggests that computations in $F[x]$ should mimic similar computations in $\mathbb{Z}$. We have seen that when p is irreducible in $F[x]$, then $F[x]/(p)$ is a field, and more generally, that for arbitrary $f \in F[x]$, the invertible elements of $F[x]/(f)$ are given by $g \in F[x]$ for which $gcd(g, f) = 1$. We expect that we can use the Euclidean algorithm to find the inverse of $g \in F[x]/(f)$ when $gcd(g, f) = 1_F$. We give one example of such a computation. Let $F = \mathbb{Z}_5$, and $f = (x^2 + 1)(x + 1) = x^3 + x^2 + x + 1 \in \mathbb{Z}_5[x]$. Since f is not irreducible over $\mathbb{Z}_5$, $\mathbb{Z}_5[x]/(f)$ is not a field. Consider $g = x^2 + x + 1 \in \mathbb{Z}_5[x]$. In this simple case, the first application of the division theorem yields $f = xg + 1$, so $xg + 1 \in (f)$. This means that $xg + (f) = -1 + (f)$; that is, $(-x + (f))(g + (f)) = 1 + (f)$. Thus $(g + (f))^{-1} = -x + (f) = 4x + (f)$.

Our next result has far-reaching consequences, not the least of which is the fact that the group of units of a finite field is a cyclic group.

9.8.5 Definition. *For any finite group G, $\max\{ |g| \mid g \in G \}$ is called the exponent of G.*

9.8.6 Proposition. *If G is a finite abelian group with exponent m, then $|x|$ is a divisor of m for every $x \in G$.*

Proof. Suppose by way of contradiction that the result is false. Then there exists $x \in G$ for which $|x|$ is not a divisor of m. By the Fundamental Theorem of Arithmetic, there exists a prime p and integers i and j such that $|x| = p^i r$ for some positive integer r, $m = p^j n$ for some positive integer n, $i > j \geq 0$, and $(p, r) = 1 = (p, n)$. There exists $y \in G$ such that $|y| = m$, and since n is a divisor of m, there exists an element $z \in \langle y \rangle$ such that $|z| = n$. Similarly, since p^i divides $|x|$, there exists $a \in \langle x \rangle$ with $|a| = p^i$. Since $(p, n) = 1$, it follows that $(p^i, n) = 1$. Note that if $x \in \langle a \rangle \cap \langle z \rangle$, then $|x|$ is a divisor of $|a|$ and of $|z|$, and thus $|x| = 1$; that is, $x = e_G$. Thus $\langle a \rangle \cap \langle z \rangle = \{ e_G \}$. Consider now the order of az. For any positive integer k, if $(az)^k = e$, then $a^k = z^{-k}$, and thus $a^k, z^k \in \langle a \rangle \cap \langle z \rangle = \{ e \}$, so $a^k = z^k = e_G$. As a result, we see that any such integer k is a common multiple of $|a|$ and $|z|$ and hence is divisible by the least common multiple of $|a|$ and $|z|$. Since $|a|$ and $|z|$ are relatively prime, their least common multiple is their product. In particular, $|a||z|$ is a divisor of $|az|$. As $(az)^{|a||z|} = (a^{|a|})^{|z|}(z^{|z|})^{|a|} = e_G$, it follows that $|az| = |a||z|$. But then we have $|az| = |a||z| = p^i n > p^j n = m$, which contradicts the choice of m. $\qquad\square$

Here is a striking consequence of this result.

9.8.7 Proposition. *Let F be a field, and let G be a finite subgroup of F^*, the group of units of F. Then G is cyclic.*

Proof. Let m denote the exponent of G. Then by Proposition 9.8.6, $a^m = 1_F$, or $a^m - 1_F = 0_F$, for every $a \in G$. Thus G is a subset of the set of all roots of $x^m - 1_F$. Since a polynomial over F of degree m has at most m roots in F, it follows that $|G| \le m$. On the other hand, there exists $a \in G$ such that $|a| = m$, and since $\langle a \rangle \subseteq G$ we obtain that $m = |\langle a \rangle| \le |G|$. But then $|G| = m$, and so $\langle a \rangle = G$. $\qquad\square$

9.8.8 Corollary. *For any field F, and any positive integer m, the set of all m^{th} roots of unity in F (that is, the set of all roots of $x^m - 1_F$ in F) is a finite cyclic group.*

Proof. If x and y are m^{th} roots of unity in F, then $x^m = y^m = 1_F$ and thus $(xy)^m = x^m y^m = 1_F$, and $(x^{-1})^m = (x^m)^{-1} = 1_F^{-1} = 1_F$. Thus the set of all m^{th} roots of unity in F is closed under multiplication and taking of inverses and is therefore a subgroup of F^*. As there are at most m roots for $x^m - 1_F$ in F, the set of all m^{th} roots of unity in F is thus a finite subgroup of F^*, hence cyclic by Proposition 9.8.7. $\qquad\square$

9.8.9 Corollary. *For any finite field F, F^* is a cyclic group.*

Proof. F^* is finite, so the result follows immediately from Proposition 9.8.7. $\qquad\square$

We shall now see how this fact about the multiplicative group of units of a finite field also leads to the determination of its additive structure. Recall that by Proposition 8.7.2, every finite field has prime characteristic. Furthermore, it was shown that if F is a finite field of charactistic p, then the map $c: \mathbb{Z} \to F$ for which $c(n) = n 1_F$ is a ring homomorphism with kernel (p), and thus $\mathbb{Z}_p = \mathbb{Z}/(p)$ is isomorphic to the subring $\{ n 1_F \mid n \in \mathbb{Z} \}$ of F, and since $\mathbb{Z}_p$ is in fact a field, it follows that $\{ n 1_F \mid n \in \mathbb{Z} \}$ is a subfield of F, which we called its prime subfield. We may use this injective ring homomorphism to identify $\mathbb{Z}_p$ with its isomorphic image in F, and we shall therefore consider $\mathbb{Z}_p$ to be a subfield of F.

As we have just established, F^* is a cyclic group, so let $a \in F^*$ be a generator of the group; that is, $F^* = \langle a \rangle$. Now consider the ring homomorphism from $\mathbb{Z}_p[x]$ to F which is the identity on $\mathbb{Z}_p$ and which sends x to a. Since $F^* = \langle a \rangle$, this homomorphism is surjective. Since $\mathbb{Z}_p[x]$ is a principal ideal domain, there exists $f \in \mathbb{Z}_p[x]$ such that the kernel of this homomorphism is (f), and so $\mathbb{Z}_p[x]/(f)$ is isomorphic to F. Note that since $\mathbb{Z}_p[x]/(f)$ is a field, f is irreducible over $\mathbb{Z}_p$. Let $n = \partial(f)$. Then $\mathbb{Z}_p[x]/(f) = \{\, a_0 + a_1 x + \cdots + a_{n-1}x^{n-1} + (f) \mid a_0, a_1, \ldots, a_{n-1}) \in \mathbb{Z}_p \,\}$. Consider the mapping $\varphi : \mathbb{Z}_p[x]/(f) \to \mathbb{Z}_p \oplus \mathbb{Z}_p \oplus \cdots \oplus \mathbb{Z}_p$ (n summands) given by $\varphi(a_0 + a_1 x + \cdots + a_{n-1}x^{n-1}) = (a_0, a_1, \ldots, a_{n-1})$. This is obviously a bijective group homomorphism from the additive group of $\mathbb{Z}_p[x]/(f)$ onto the additive group of the direct sum of n copies of the ring (actually, field) $\mathbb{Z}_p$. Thus, as additive groups, F is isomorphic to $\mathbb{Z}_p[x]/(f)$, which is in turn isomorphic to $\oplus_{i=1}^n \mathbb{Z}_p$. We have proven the following result.

9.8.10 Proposition. *Let F be a finite field, and let p denote the characteristic of F. Then the additive group of F is isomorphic to the additive group of the direct sum of $n = \partial(f)$ copies of $\mathbb{Z}_p$. Furthermore, there exists $f \in \mathbb{Z}_p[x]$, irreducible over $\mathbb{Z}_p$, such that $\mathbb{Z}_p[x]/(f)$ is isomorphic to F, and so $|F| = p^{\partial(f)}$.*

In the discussion that preceded Proposition 9.8.10, it was shown that for every finite field F of characteristic p, there is $f \in \mathbb{Z}_p[x]$ irreducible over $\mathbb{Z}_p$ such that F is isomorphic to $\mathbb{Z}_p[x]/(f)$, and that $|F| = p^{\partial(f)}$. In the next chapter, we will prove that any two finite fields of the same size are isomorphic (note that this is quite different from the situation for finite groups, where in general, there are many isomorphism classes of finite groups of a given size). The striking consequence of this fact is that for any $f, g \in \mathbb{Z}_p[x]$ both irreducible over $\mathbb{Z}_p$, if $\partial(f) = \partial(g)$, then $\mathbb{Z}_p[x]/(f)$ is isomorphic to $\mathbb{Z}_p[x]/(g)$. It will also be shown that for any prime p, and any positive integer n, there is a field of order p^n.

Exercises

1. Apply the division algorithm to the following pairs of polynomials f, g over the indicated field F to find $q, r \in F[x]$ such that $f = qg + r$ and

$\partial(r) < \partial(g)$.

   a) $f = x^4 - 3x^3 + 5x^2 + x + 2$, $g = x^2 + 4x - 3$, $F = \mathbb{R}$.

   b) $f = x^4 - 3x^3 + 5x^2 + x + 2$, $g = x^2 + 4x - 3$, $F = \mathbb{Z}_7$.

   c) $f = x^5 + 4x^2 - 8$, $g = x^3 - 3x^2 + x + 10$, $F = \mathbb{R}$.

   d) $f = x^5 + 4x^2 - 8$, $g = x^3 - 3x^2 + x + 10$, $F = \mathbb{Z}_{11}$.

2. In each case, find the greatest common divisor of the given pair of polynomials over the indicated field and express it as a linear combination of the given polynomials.

   a) $x^3 + x^2 + x$ and $x^4 + 2x^2 + 2z + 1$ over $\mathbb{Z}_3$.

   b) $x^3 + x^2 + x$ and $x^4 + 2x^2 + 2x + 1$ over $\mathbb{Z}_7$.

   c) $x^3 - 6x^2 + x + 4$ and $x^5 - 6x + 1$ over $\mathbb{Z}_{11}$.

3. Give an example of a nonzero polynomial of degree 2 with coefficients in $\mathbb{Z}_4$ that has more than two roots in $\mathbb{Z}_4$ (of course, $\mathbb{Z}_4$ is not a field).

4. Find all polynomials of degree at most 3 in $\mathbb{Z}_2[x]$ that are irreducible over $\mathbb{Z}_2$. As well, find at least two quartic (degree 4) polynomials in $\mathbb{Z}_2[x]$ that are irreducible over $\mathbb{Z}_2$.

5. Let I be the ideal of $\mathbb{Z}[x]$ that is generated by 2 and x; that is, $I = 2\mathbb{Z}[x] + x\mathbb{Z}[x]$. Prove that I is not a principal ideal of $\mathbb{Z}[x]$.

6. Prove that the subring $\mathbb{Q}[\sqrt{2}]$ of $\mathbb{R}$ is actually a subfield of $\mathbb{R}$, and that $\mathbb{Q}[\sqrt{2}] = \{\, a + b\sqrt{2} \mid a, b \in \mathbb{Q} \,\}$. Hint: Consider the homomorphism from $\mathbb{Q}[x]$ into $\mathbb{R}$ that is the identity on $\mathbb{Q}$ and sends x to $\sqrt{2}$. Prove that its image is $\{\, a + b\sqrt{2} \mid a, b \in \mathbb{Q} \,\}$, and that the kernel of the homomorphism is $(x^2 - 2)$.

7. $p = x^3 + x + 1 \in \mathbb{Z}_2[x]$ is irreducible over $\mathbb{Z}_2$, so $\mathbb{Z}_2[x]/(p)$ is a field.

   a) Find the unique representative of the coset $x^7 + x^4 + x + 1 + (p)$ that has degree less than 3.

   b) Find the multiplicative inverse of $x^2 + 1 + (p)$ in $\mathbb{Z}_2[x]/(p)$.

8. a) Find $p, q \in \mathbb{Z}_3[x]$, each of degree 2 and irreducible over $\mathbb{Z}_3$. By Proposition 9.8.10, the fields $\mathbb{Z}_3[x]/(p)$ and $\mathbb{Z}_3[x]/(q)$ have the same size, 3^2, and it was remarked at the end of the section that these fields are isomorhic. Find an isomorphism between them.

   b) Prove that $p = x^3 - 2 \in \mathbb{Z}_7[x]$ and $q = x^3 + 2 \in \mathbb{Z}_7[x]$ are each irreducible over $\mathbb{Z}_7$. Find an isomorphism from $\mathbb{Z}_7[x]/(p)$ to $\mathbb{Z}_7[x]/(q)$.

9. Prove that if F is a field of characteristic p, then the map $\varphi : F \to F$ given by $\varphi(a) = a^p$ is an endomorphism of F that fixes each element of

the prime subfield of F (and thus φ is injective). If F is finite, then φ is an automorphism of F, called the Frobenius automorphism of F.

10. Prove that if $p \in \mathbb{Z}$ is a prime, then $x^p - x \in \mathbb{Z}_p[x]$ factors as a product of p linear factors over $\mathbb{Z}_p$.

11. Prove that if F is a field, and $p, q \in F[x]$ are relatively prime, and $f \in F[x]$ is divisible both by p and by q, then f is divisible by pq.

12. Find a field of order 8, and construct the Cayley tables for addition and multiplication.

13. Let F be a field, and let $f \in F[x]$. Prove that every element of the quotient ring $F[x]/(f)$ is either a unit or a zero divisor.

14. In each case below, determine whether or not the given polynomial $f \in F[x]$ is irreducible over the given field F.

    a) $3x^8 - 4x^3 + 6x^2 - 2x + 10$, $F = \mathbb{Q}$.

    b) $x^3 + 3x^2 + x + 1$, $F = \mathbb{Z}_5$.

    c) $x^5 + 15$, $F = \mathbb{Q}$.

    d) $x^4 - 6x^3 + 3x^2 + 9x + 12$, $F = \mathbb{R}$.

    e) $x^3 - 3x^2 + 4x - 2$, $F = \mathbb{C}$.

15. Find all polynomials of degree 3 in $\mathbb{Z}_3[x]$ that are irreducible over $\mathbb{Z}_3$.

16. List all polynomials of degree 4 over $\mathbb{Z}_2$, and write each as a product of polynomials irreducible over $\mathbb{Z}_2$.

17. For positive integers m and n, $n^{\frac{1}{m}}$ is defined to be the (unique) positive $r \in \mathbb{R}$ such that $r^m = n$. Prove that for any positive integers m and n, $n^{\frac{1}{m}} \in \mathbb{Q}$ if, and only if, $n = k^m$ for some $k \in \mathbb{Z}$. Hint: Use Gauss's lemma.

18. Let $p \in \mathbb{Z}$ be a prime, and let $\pi : \mathbb{Z} \to \mathbb{Z}_p$ denote the canonical projection mapping, so π is a unital ring homomorphism. By Proposition 9.3.13, there exists a unique ring homomrphism $\theta : \mathbb{Z}[x] \to \mathbb{Z}_p[x]$ for which $\theta|_{\mathbb{Z}} = \pi$ and $\theta(x) = x$.

    a) Prove that for any $f \in \mathbb{Z}[x]$, if $\partial(\theta(f)) = \partial(f)$ and $\theta(f)$ is irreducible over $\mathbb{Z}_p$, then f is irreducible over $\mathbb{Q}$. Find an example to illustrate that the condition $\partial(\theta(f)) = \partial(f)$ is necessary.

    b) Use the first part of this exercise to show that in each case below, the given polynomial is irreducible over $\mathbb{Q}$.

        (i) $3x^4 + 2x^3 - 6x^2 + 5x - 9$.

        (ii) $5x^4 - 15x^3 + 6x^2 - 4x + 25$.

(iii) $a_5x^5 + a_4x^4 + a_3x^3 + a_2x^2 + a_1x + a_0$, where each of $a_0, a_1, a_2, a_3, a_4, a_5$ is an odd integer.

19.   When the Division Theorem is applied to $x^{100} + x^{37} + 2$ and $x^2 - 1$ in $\mathbb{Q}[x]$, what is the remainder?

20.   Let F be a field, and let $I = \{\, f \in F[x] \mid \hat{f}(1_F) = 0_F \,\}$. Prove that I is an ideal of $F[x]$, and find a generator for I.

21.   Recall that an ideal I of a commutative ring R is said to be maximal if for any ideal J of R for which $I \subseteq J$, either $I = J$, or else $J = R$. Determine the maximal ideals of the following rings.

   a) $\mathbb{Q} \oplus \mathbb{Q}$.

   b) $\mathbb{Z}_6[x]/(x^2)$.

   c) $\mathbb{R}[x]/(x^2 - 3x - 2)$.

22.   Let $p, q \in \mathbb{Z}[x]$. Prove that p and q are relatively prime in $\mathbb{Q}[x]$ if, and only if, $p\mathbb{Z}[x] + q\mathbb{Z}[x]$ contains an integer.

23.   Express $x^5 + x + 1 \in \mathbb{Z}_2[x]$ as a product of irreducible elements of $\mathbb{Z}_2[x]$.

24.   Let $\theta : \mathbb{Z}[x] \to \mathbb{R}$ be the unique unital ring homomorphism whose restriction to $\mathbb{Z}$ is the inclusion mapping $\mathbb{Z} \subseteq \mathbb{R}$, and which sends x to $2 + \sqrt{3}$. Prove that $\ker(\theta)$ is a principal ideal of $\mathbb{Z}[x]$ and find a generator for $\ker(\theta)$.

25.   Let R be a commutative ring for which $a \in R$ and $a^2 = 0_R$ implies $a = 0_R$. Prove that if $p \in R[x]$ is a nonzero zero divisor of $R[x]$, then there exists $a \in R$ such that $ap = 0_R$.

26.   Let R be a commutative ring, and let $I \triangleleft R$. Recall that $I[x]$ denotes the subring of $R[x]$ that is generated by I and x; that is, $I[x]$ is the intersection of all subrings of $R[x]$ that contain I and x. Prove that $I[x]$ is an ideal of $R[x]$. Moreover, prove that $R[x]/I[x]$ is isomorphic to $(R/I)[x]$.

27.   Prove that $f \in \mathbb{Z}_2[x]$ is divisible by $x + 1$ if, and only if, f has an even number of nonzero terms.

28.   Let F be a field, and let n be a positive integer. For irreducible $p_1, p_2, \ldots, p_n \in F[x]$, prove that $F[x]/(p_1p_2 \cdots p_n)$ is isomorphic to $\oplus_{i=1}^{n} F[x]/(p_i)$.

29.   Let F be a field. Prove that if $\varphi : F[x] \to F[x]$ is an automorphism of $F[x]$ whose restriction to F is the identity on F, then $\partial(\varphi(x)) = 1$.

30.   Prove that $x^3 + 3x + 2 \in \mathbb{Z}[x]$ is irreducible over $\mathbb{Q}$.

31.   For each positive integer n, find an automorphism of $\mathbb{C}[x]$ of order n.

Chapter 10

Field Extensions

1 Introduction

In number theory, algebraic geometry, combinatorial mathematics, and other areas of mathematics, it often happens that the field we are operating in is too confining and we need to find a bigger field which will give us more scope. In this section, we shall discuss how to extend a field, and in particular, we shall show how our results can be applied to finite fields.

2 Definitions and Elementary Results

Throughout this chapter, F shall denote an arbitrary field.

10.2.1 Definition. *Let E be a field. We say that E is an extension of F if there exists an injective ring homomorphism $\varphi : F \to E$. The map φ is called an embedding monomorphism.*

Note that if E is a field, and F is a subfield of E, then E is an extension of F with embedding monomorphism the inclusion mapping; that is, the map that sends $a \in F$ to $a \in E$. For example, $\mathbb{R}$ is an extension of $\mathbb{Q}$, and $\mathbb{C}$ is an extension of $\mathbb{R}$, and of $\mathbb{Q}$.

Students sometimes have difficulties in dealing with the notion of an extension, particularly since mathematicians tend to think of F as actually being contained in the extension. Strictly speaking, an isomorphic copy of F, and not F itself, is contained in the extension. However, the following theorem shows that if $\varphi : F \to E$ determines E to be an extension of F, then we can construct a field that is isomorphic to the field E and which actually contains F.

10.2.2 Proposition. *Let E be an extension of F with embedding monomorphism φ. Then there exists a field E' containing F as a subfield, and E' is isomorphic to E by an isomorphism that extends φ.*

Proof. Let G be a set for which $G \cap F = \varnothing$ and there is a bijection α from $E - \varphi(F)$ onto G. Then define $E' = G \cup F$, and $\hat{\varphi} : E' \to E$ by $\hat{\varphi}(a) = \alpha(a)$ if $a \in G$, while if $a \in F$, let $\hat{\varphi}(a) = \varphi(a)$. Evidently, $\hat{\varphi}$ is a bijective mapping. Define addition and multiplication in E' by $a + b = \hat{\varphi}^{-1}(\hat{\varphi}(a) + \hat{\varphi}(b))$ and $ab = \hat{\varphi}^{-1}(\hat{\varphi}(a)\hat{\varphi}(b))$ for $a, b \in E'$. Then $\hat{\varphi}(a + b) = \hat{\varphi}(a) + \hat{\varphi}(b)$ and $\hat{\varphi}(ab) = \hat{\varphi}(a)\hat{\varphi}(b)$ for all $a, b \in E'$, and so the operations of addition and multiplication on E' make E' into a field and $\hat{\varphi} : E' \to E$ an isomorphism. By construction, $\hat{\varphi}\big|_F = \varphi$. $\qquad\square$

An irreducible polynomial of degree greater than 1 over F has no root in F, but it may have a root in an extension field of F. For example, $x^2 + 1 \in \mathbb{R}[x]$ is irreducible over $\mathbb{R}$, but has roots $i, -i \in \mathbb{C}$. A natural question which arises is: Given an irreducible polynomial f over F, does there exist an extension field E of F which contains a root of f? The following result, due to Kronecker, provides the definitive answer.

10.2.3 Proposition (Kronecker). *Let $p \in F[x]$ be irreducible over F. Then $F[x]/(p)$ is an extension field of F that contains a root of p.*

Proof. By Corollary 9.8.2, $F[x]/(p)$ is a field. Moreover, it was established in Proposition 9.8.4 that the map $\varphi : F \to F[x]/(p)$ that sends $a \in F$ to $a + (p) \in F[x]/(p)$ is an injective ring homomorphism; that is, an embedding monomorphism of F into $F[x]/(p)$. Thus $F[x]/(p)$ is an extension of F, and we shall consider F to be a subfield of $F[x]/(p)$ by identifying $a \in F$ with $a + (p) \in F[x]/(p)$. Let $t = x + (p) \in F[x]/(p)$. Then if $p = \sum_{i=0}^{n} a_i x^i$, we have $\hat{p}(t) = \sum_{i=0}^{n} a_i (x + (p))^i = \left(\sum_{i=0}^{n} a_i x^i \right) + (p) = p + (p) = 0_F + (p)$, so t is a root of p. $\qquad\square$

The construction described in Proposition 10.2.3 can be iterated to obtain an extension field that contains all roots of a given polynomial. We show precisely how this is done.

10.2.4 Definition. *Let $f \in F[x]$, and let E be an extension of F (and we shall regard F as a subfield of E). Then f is said to split over E if in $E[x]$, f can be written as a product with linear factors.*

10.2.5 Corollary. *For any $f \in F[x]$, there exists an extension of F over which f splits.*

Proof. We prove by induction on $n \geq 1$ that for any field K, and any polynomial $f \in K[x]$ of degree n, there is an extension E of K over which f splits. The base case holds trivially. Suppose that $n \geq 1$ is an integer for which the assertion holds, and let K be a field and $f \in K[x]$ have degree $n + 1$. By Proposition 10.2.3, there exists an extension E of K which contains a root r of f. Then $x - r \in E[x]$, and there exists $g \in E[x]$ such that $f = (x - r)g$. Since $\deg(f) = n + 1$, it follows that $\deg(g) = n$, and so by hypothesis, there exists an extension L of E over which g splits. Thus L is an extension of K and f splits over L. The result follows now by induction. $\qquad\square$

Suppose now that a field E is an extension of F. Then E is an abelian group under addition, and for any $a, b \in F$, and any $v \in E$, we have $(a+b)v = av + bv$ and $(ab)v = a(bv)$, while for any $a \in F$ and any $u, v \in E$, we have $a(u + v) = au + av$ and $1_F u = u$. Thus, using multiplication in E as scalar multiplication, all of the axioms of a vector space over F are satisfied, so in this way the field E can be considered as an F-vector space.

10.2.6 Definition. *For any extension E of F, the dimension of E as an F-vector space is denoted by $[E : F]$, and is called the degree of the extension. If $[E : F]$ is finite, E is said to be a finite-dimensional extension of F, or an extension of finite degree over F. If $B \subseteq E$, we write $span_F(B)$ for the F-vector subspace of E that is spanned by B; that is,*

$$span_F(B) = \cap\{\, U \mid U \text{ is an } F\text{-vector subspace of } E \text{ and } B \subseteq U \,\}.$$

If $B = \varnothing$, then $span_F(B) = \{\, 0_E \,\}$, while if $B \neq \varnothing$, then $span_F(B)$ is the set of all linear combinations of elements of B with coefficients from F.

Recall that for any $X, Y \subseteq F$, $XY = \{\, xy \mid x \in X,\ y \in Y \,\}$. With this binary operation, $\mathscr{P}(F)$ is a monoid with identity $\{\, 1_F \,\}$. We have defined exponentiation in any monoid, and so for $X \in \mathscr{P}(F)$, we may speak of X^n for any $n \in \mathbb{N}$. By definition, X^0 is the identity of the monoid; that is, $X^0 = \{\, 1_F \,\}$.

In preparation for the following proposition, the reader may wish to review the definition of $F[B]$ (see Definition 9.3.10).

10.2.7 Proposition. *Let E be an extension of F, and let $B \subseteq E$. Let $B^+ = \cup_{n=1}^{\infty} B^n$, and $B^* = B^+ \cup B^0$. Then $F[B] = span_F(B^*)$.*

Proof. First, note that $B \subseteq B^* \subseteq span_F(B^*)$, and F is the set of all F-linear combinations of $1_F \in B^*$, so $F \subseteq span_F(B^*)$. If we prove that

$span_F(B^*)$ is a subring of E, then $F[B] \subseteq span_F(B^*)$. Since $span_F(B^*)$ is a subgroup of E under addition, it only needs to be proven that $span_F(B^*)$ is closed under multiplication. But any element of $span_F(B^*)$ is a sum of elements from subsets of the form FB^n, and for any $m, n \in \mathbb{N}$, $FB^n FB^m = FB^{n+m} \subseteq span_F(B^*)$. Thus for any $a, b \in span_F(B^*)$, $ab \in span_F(B^*)$ and so $span_F(B^*)$ is a subring of E, proving that $F[B] \subseteq span_F(B^*)$. On the other hand, since any element of $span_F(B^*)$ is a sum of elements of the form FB^n for $n \in \mathbb{N}$, it suffices to prove that for each $n \in \mathbb{N}$, $FB^n \subseteq F[B]$. But $F \subseteq F[B]$ and $B \subseteq F[B]$, and $F[B]$ is closed under multiplication, so $FB^n \subseteq F[B]$ for any $n \in \mathbb{N}$, as required. $\qquad\square$

10.2.8 Definition. *For any extension E of F, and any subset B of E, let $F(B) = \cap\{T \mid T$ is a subfield of E and $F \cup B \subseteq T\}$. If B is finite, say $B = \{u_1, u_2, \ldots, u_k\}$, then we may write $F(u_1, u_2, \ldots, u_k)$ in place of $F(B)$.*

Since the intersection of any set of subrings of F is a subring of F, and since $a \in F(B)$, $a \neq 0_E$ implies that $a^{-1} \in F(B)$, it follows that $F(B)$ is a subfield of E. Note that if T is any subfield of E with $F \subseteq T$ and $B \subseteq T$, then $F(B) \subseteq T$. In this sense, $F(B)$ is the smallest subfield of E that contains both F and B. We note also that since $F[B]$ is the intersection of all subrings of E that contain $F \cup B$, while $F(B)$ is the intersection of all subfields of E that contain $F \cup B$, and each subfield of E is a subring of E, it follows that $F[B] \subseteq F(B)$.

10.2.9 Proposition. *Let E be an extension of F, and let B and C be subsets of E. Then $F(B \cup C) = F(B)(C)$. In particular, for $u_1, u_2, \ldots, u_k, u_{k+1} \in E$, $F(u_1, u_2, \ldots, u_{k+1}) = F(u_1, u_2, \ldots, u_k)(u_{k+1})$.*

Proof. The first assertion follows immediately from the fact that a subfield K of E contains F and $B \cup C$ if, and only if, it contains $F(B)$ and C. The second assertion then follows immediately from the first. $\qquad\square$

10.2.10 Proposition. *Let E be an extension of F, and let $B \subseteq E$. Then $F(B) = \{ab^{-1} \mid a, b \in F[B], b \neq 0_E\}$.*

Proof. Since $F[B]$ is an integral domain, we may form its field of quotients $Q(F[B])$. By Proposition 8.6.2, there is a unique ring homomorphism from $Q(F[B])$ into E extending the inclusion map of F into E, and the image of this homomorphism in E is the set $T = \{ab^{-1} \mid a, b \in F[B], b \neq 0_E\}$. Thus T is a field isomorphic to $Q(F[B])$, and $F[B] \subseteq T$, so $F(B) \subseteq T$.

Conversely, for any subfield L of E, if $F \subseteq L$ and $B \subseteq L$, then $F[B] \subseteq L$ and so for any $a, b \in F[B]$ with $b \neq 0_E$, $b^{-1} \in L$ and $a \in L$, so $ab^{-1} \in L$. Thus $T \subseteq F(B)$, and so $F(B) = T$, as required. □

Recall that in the case when E is an extension of F and $a \in E$, there is a unique ring homomorphism from $F[x]$ into E which acts as the identity mapping on F and sends x to a. This homomorphism sends $f \in F[x]$ to $\hat{f}(a)$, so its image is the set $\{\, \hat{f}(a) \mid f \in F[x] \,\}$, the subring $F[a]$ of E. It follows that $F(a) = \{\, \frac{\hat{p}(a)}{\hat{q}(a)} \mid p, q \in F[x],\ \hat{q}(a) \neq 0_e \,\}$. Of course, we have $F[a] \subseteq F(a)$, and in the next section, we investigate under what conditions equality holds.

3 Algebraic and Transcendental Elements

10.3.1 Definition. *Let E be an extension of F, and let $a \in F$. If the unique ring homomorphism from $F[x]$ into E which acts as the identity on F and which sends x to a has nontrivial kernel, then a is said to be algebraic over F, otherwise a is said to be transcendental over F. Equivalently, $a \in E$ is algebraic over F if there exists $f \in F[x]$ with $f \neq 0_F$ and $\hat{f}(a) = 0_F$, otherwise a is transcendental over F.*

For example, $\mathbb{R}$ is an extension of $\mathbb{Q}$ and $\sqrt{2} \in \mathbb{R}$ is algebraic over $\mathbb{Q}$ since $x^2 - 2 \in \mathbb{Q}[x]$ has $\sqrt{2}$ as a root. It is known (although not simple to prove) that e and π are transcendental over $\mathbb{Q}$.

10.3.2 Proposition. *Let E be an extension of F, and let $a \in E$. Then the following are equivalent:*

(i) *a is algebraic over F.*
(ii) *$F[a] = F(a)$.*
(iii) *There exists $p \in F[x]$ such that p is monic, irreducible over F, and a is a root of p.*

In such a case, the unique homomorphism $\theta : F[x] \to E$ that acts as the identity on F and sends x to a has kernel (p) and image $F(a)$, so $F[x]/(p) \simeq F(a)$.

Proof. Recall that $F[x]$ is a PID and every nontrivial ideal of $F[x]$ has a unique monic generator. Suppose that $a \in E$ is algebraic over F, and let $\theta : F[x] \to E$ be the unique homomorphism that acts as the identity on F and sends x to a. Let $p \in F[x]$ be the unique monic generator of the nontrivial ideal $\ker(\theta)$. Then every element of $F[x]$ that has a as a root is divisible by p. If p were reducible over F, then $p = fg$ for some nonunits $f, g \in F[x]$, and then $0_E = \widehat{fg}(a) = \hat{f}\hat{g}(a) = \hat{f}(a)\hat{g}(a)$, which would imply that $\hat{f}(a) = 0_E$ or $\hat{g}(a) = 0_E$; that is, either $f \in (p)$ or $g \in (p)$. Since $p = fg$ and p divides either f or g implies either that $\deg(p) = \deg(f)$ and g is a unit of F, or else that $\deg(p) = \deg(g)$ and f is a unit of F, neither of which is true, we conclude that p is irreducible over F. Thus $F[x]/(p)$ is a field, isomorphic to the image of θ, which is $F[a]$. This means that $F[a]$ is a field, and so $F[a] = F(a)$.

Next, suppose that $F[a] = F(a)$. Then the image of θ is the subfield $F(a)$ of E, which means that $F[x]/\ker(\theta)$ is a field. Since $\ker(\theta)$ is an ideal of $F[x]$, and $F[x]$ is not a field, it follows that $\ker(\theta)$ is nontrivial and so there exists a monic $p \in F[x]$ irreducible over F such that $F[x]/(p) \simeq F(a)$. Since $p \in \ker(\theta)$, $\hat{p}(a) = 0_E$; that is, a is a root of p.

Finally, suppose that there exists $p \in F[x]$ such that p is monic and irreducible over F, and that $\hat{p}(a) = 0_E$. Then $p \in \ker(\theta)$, so $\ker(\theta)$ is nontrivial and thus a is algebraic over F.   □

10.3.3 Corollary. *Let E be an extension of F, and let $a \in E$ be algebraic over F. Then there is a unique $p \in F[x]$ such that p is monic, irreducible over F, and a is a root of p.*

Proof. By Proposition 10.3.2, there is a monic irreducible $p \in F[x]$ that has a as a root, and the kernel of the unique homomorphism from $F[x]$ into E that acts as the identity on F and sends x to a has kernel (p). By Proposition 9.4.6, this uniquely determines p.   □

Note that every $a \in F$ is algebraic over F since $x - a \in F[x]$ is monic, irreducible over F, and has a as a root.

10.3.4 Definition. *If E is an extension of F, and $a \in E$ is algebraic over F, then the unique monic $p \in F[x]$ that is irreducible over F and has a as a root is called the* minimal polynomial *of a over F, denoted by $\min(a, F)$.*

It is important to realize that if $f \in F[x]$ has a as a root, then $\min(a, F)$ is a divisor of f (this does not depend on the extension field E to which a belongs).

10.3.5 Corollary. *Let E be an extension of F, and let $a \in E$. Then a is transcendental over F if, and only if, $F[x] \simeq F[a] \neq F(a)$, and for every nonzero $p \in F[x]$, $\hat{p}(a) \neq 0_E$.*

10.3.6 Example.

(i) $\mathbb{R}$ *is an extension of* $\mathbb{Q}$, *and* $\sqrt{2} \in \mathbb{R}$. $min(\sqrt{2}, \mathbb{Q}) = x^2 - 2$ *since* $x^2 - 2$ *is monic, irreducible over* $\mathbb{Q}$ *(by Eisenstein's criterion), and has* $\sqrt{2}$ *as a root.*

(ii) $\mathbb{C}$ *is an extension of* $\mathbb{R}$, *and* $i \in \mathbb{C}$. $min(i, \mathbb{R}) = x^2 + 1$ *as* $x^2 + 1$ *is monic and irreducible over* $\mathbb{R}$ *(since it has no real roots and has degree 2) and has* i *as a root.*

(iii) $\mathbb{C}$ *is an extension of* $\mathbb{R}$, *and* $\omega = \frac{1}{2}(-1 + \sqrt{3}i) \in \mathbb{C}$. $min(\omega, \mathbb{R}) = x^2 + x + 1$ *since* $x^2 + x + 1$ *is monic, irreducible since it has no real roots and has degree 2, and has* ω *as a root (by Proposition 9.5.10, the roots of* $x^2 + x + 1 \in \mathbb{R}[x] \subseteq \mathbb{C}[x]$ *in* $\mathbb{C}$ *are* ω *and* $\bar{\omega}$*).*

10.3.7 Proposition. *Let* E *be an extension of* F *and let* $a \in E$ *be algebraic over* F. *Let* $n = \deg(min(a, F))$. *Then* $F(a)$, *when considered as an* F-vector space, has basis $\{1_E, a, a^2, \ldots, a^{n-1}\}$. *Consequently,* $[F(a) : F] = n$.

Proof. By Proposition 10.3.2, $F(a) = F[a]$ and the unique homomorphism from $F[x]$ into E that acts as the identity on F and sends x to a has kernel (p) and image $F(a)$, so $F[x]/(p) \simeq F(a)$ by an isomorphism that sends $x + (p)$ to a, where $p = min(a, F)$. Let $n = \deg(p)$. By Proposition 9.8.3, each coset of (p) in $F[x]$ has a unique representative of degree less than n, which is to say that each element of $F[x]/(p)$ can be written uniquely as an F-linear combination of $1_F + (p)$, $x + (p), \ldots, x^{n-1} + (p)$, and thus each element of $F(a)$ can be written uniquely as an F-linear combination of $1_E, a, \ldots, a^{n-1}$. $\square$

For example, $[\mathbb{Q}(\sqrt{2}) : \mathbb{Q}] = 2$ and $\mathbb{Q}(\sqrt{2}) = \{a + b\sqrt{2} \in \mathbb{R} \mid a, b \in \mathbb{Q}\}$ since $min(\sqrt{2}, \mathbb{Q}) = x^2 - 2$.

We now have complete information about $F(a) \subseteq E$ when E is an extension of F and $a \in E$ is algebraic over F. Accordingly, we now turn our attention to the case when $a \in E$ is transcendental over F.

10.3.8 Proposition. *Let* E *be an extension of* F, *and let* $a \in E$ *be transcendental over* F. *Then* $F[a] \simeq F[x]$.

Proof. Let θ denote the unique homomorphism from $F[x]$ into E that acts as the identity on F and sends x to a. Since a is transcendental over F, $\ker(\theta) = \{0_F\}$, and thus θ is injective. By Proposition 10.2.7, $F[a] =$

$span_F(\{\, 1_E, a, a^2, \dots \})$, and so the image of θ is $F[a]$. Thus $F[x] \simeq F[a]$.
$\square$

10.3.9 Definition. . *The field of quotients of the integral domain $F[x]$ shall be denoted by $F(x)$.*

10.3.10 Corollary. *Let E be an extension of F, and let $a \in E$ be transcendental over F. Then $F(a) \simeq F(x)$.*

Proof. By Proposition 10.3.8, the injective ring homomorphism $\theta \colon F[x] \to E$ has image $F[a]$. The unique extension of θ to an injective ring homomorphism from $F(x)$ into E has image $\{\, \theta(f)\theta(g)^{-1} \mid f, g \in F[x],\ g \neq 0_F \,\}$. By Proposition 10.2.10, $F(a) = \{\, ab^{-1} \mid a, b \in F[a],\ b \neq 0_E \,\}$, and so the image of this extension of θ is $F(a)$. Thus $F(x) \simeq F(a)$. $\square$

$F(x)$ is often called the rational function field over F, and elements of $F(x)$ are usually written in the form f/g where $f, g \in F[x]$ and $g \neq 0_F$.

10.4   Algebraic Extensions

Extensions of a field F in which every element of the extension field is algebraic over F play important roles in many branches of mathematics, among them algebraic geometry and number theory. In this section, we shall derive basic properties of such extensions, and then we shall apply the theory to the study of finite fields. Once more, throughout this section, F shall denote an arbitrary field (unless otherwise specified).

10.4.1 Definition. *An extension field E of F is said to be algebraic over F, or that E is an algebraic extension of F, if every element of E is algebraic over F.*

Note that F is algebraic over itself, since every element of F is algebraic over F.

10.4.2 Proposition. *Every finite dimensional extension of F is algebraic over F.*

Proof. Suppose that E is an extension of F of finite dimension, and let $n = [E \colon F]$. Then every set of size greater than n in E is linearly dependent over F. Let $a \in E$. The set $\{\, 1_E, a, a^2, \dots, a^n \,\}$ has size $n + 1$ and is therefore linearly dependent. Thus there exist $c_0, c_1, \dots, c_n \in F$, not all zero, such that $c_0 + c_1 a + \cdots + c_n a^n = 0_E$. For $f = \sum_{i=0}^{n} c_i x^i \in F[x]$, we have $\hat{f}(a) = c_0 + c_1 a + \cdots + c_n a^n = 0_E$ and so a is a root of f. Thus a is algebraic over F. $\qquad\square$

The converse of the preceding result is not valid, as the following example illustrates.

10.4.3 Example. *For any positive integer n, $x^{2^n} - 2$ is irreducible over* $\mathbb{Q}$ *by Eisenstein's criterion, and so $2^{1/2^n} \in \mathbb{R}$ has $\min(2^{1/2^n}, \mathbb{Q}) = x^{2^n} - 2$. By Proposition 10.3.7, $[\mathbb{Q}(2^{1/2^n}) : \mathbb{Q}] = 2^n$, and $\{\, 2^{i/2^n} \mid i = 0, 1, \ldots, 2^n - 1 \,\}$ is a $\mathbb{Q}$-basis for $\mathbb{Q}(2^{1/2^n})$. Since $i/2^n = (2i)/2^{n+1}$ for any integer i, it follows that $\{\, 2^{i/2^n} \mid i = 0, 1, \ldots, 2^n - 1 \,\} \subseteq \{\, 2^{i/2^{n+1}} \mid i = 0, 1, \ldots, 2^{n+1} - 1 \,\}$, and thus $\mathbb{Q}(2^{1/2^n}) \subseteq \mathbb{Q}(2^{1/2^{n+1}})$ for every positive integer n. Let $K = \cup_{i=1}^{\infty} \mathbb{Q}(2^{1/2^n})$. Then $K \subseteq \mathbb{R}$, and K is closed under sums and products and the formation of additive and multiplicative inverses (in the latter case for nonzero elements of K), and so K is a subfield of $\mathbb{R}$, and $\mathbb{Q} \subseteq K$. For any $a \in K$, there exists a positive integer n such that $a \in \mathbb{Q}(2^{1/2^n})$, and so by Proposition 10.4.2, every element of K is algebraic over $\mathbb{Q}$. Thus K is an algebraic extension of $\mathbb{Q}$. Since the infinite set $\{\, 2^{1/2^n} \mid n$ a positive integer $\}$ is linearly independent over $\mathbb{Q}$, it follows that K is not a finite dimensional extension of $\mathbb{Q}$.*

10.4.4 Proposition. *Let E be an extension of F, and K be an extension of E, so K is an extension of F. Then $[K : F]$ is finite if, and only if, both $[E : F]$ and $[K : E]$ are finite, in which case $[K : F] = [K : E][E : F]$.*

Proof. Suppose first of all that $[K : F]$ is finite. Then there exists a finite subset B of K such that every element of K can be expressed as an F-linear combination of the elements of B. Since $F \subseteq E$, it follows that every element of K can be expressed as an E-linear combination of the elements of B and so $[K : E]$ is finite. As well, since E is an F-vector subspace of K, we have $[E : F] \leq [K : F]$ and so $[E : F]$ is finite.

Conversely, suppose that $[K : E]$ and $[E : F]$ are finite, and let $m = [K : E]$ and $n = [E : F]$. Let $\{\, u_1, u_2, \ldots, u_m \,\}$ be an E-basis for K, and let $\{\, v_1, v_2, \ldots, v_n \,\}$ be an F-basis for E. We prove that

$$S = \{\, u_i v_j \mid i = 1, 2, \ldots, m, \; j = 1, 2, \ldots, n \,\}$$

is an F-basis for K, from which we obtain $[K : F] = mn = [K : E][E : F]$. Let $u \in K$. Then there exist $a_1, a_2, \ldots, a_m \in E$ such that $u = \sum_{i=1}^{m} a_i u_i$. For each $i = 1, 2, \ldots, m$, there exist $b_{ij} \in F$, $j = 1, 2, \ldots, n$ such that $a_i = \sum_{j=1}^{n} b_{ij} v_j$, and thus $u = \sum_{i=1}^{m} (\sum_{j=1}^{n} b_{ij} v_j) u_i = \sum_{i=1}^{m} (\sum_{j=1}^{n} b_{ij} u_i v_j)$.

Thus S is an F-spanning set for K. It remains to prove that S is F-linearly independent. Suppose now that for each $i = 1, 2, \ldots, m$ and $j = 1, 2, \ldots, n$, we have $b_{ij} \in F$ such that $\sum_{i=1}^{m} \left(\sum_{j=1}^{n} b_{ij} u_i v_j \right) = 0_K$. Then $\sum_{i=1}^{m} \left(\sum_{j=1}^{n} b_{ij} v_j \right) u_i = 0_K$. Since $\sum_{j=1}^{n} b_{ij} v_j \in E$ for each $i = 1, 2, \ldots, m$, and $\{ u_1, u_2, \ldots, u_m \}$ is E-linearly independent, it follows that for each $i = 1, 2, \ldots, m$, $\sum_{j=1}^{n} b_{ij} v_j = 0_K = 0_E$. But then since $\{ v_1, v_2, \ldots, v_n \}$ is F-linearly independent and $b_{ij} \in F$ for all i and j, we obtain that $b_{ij} = 0_F$ for every $i = 1, 2, \ldots, m$ and $j = 1, 2, \ldots, n$. Thus S is F-linearly independent, and so S is an F-basis for K. $\square$

10.4.5 Corollary. *Let E be an extension of F, and let $a_1, a_2, \ldots, a_n \in E$ be algebraic over F. Then $F(a_1, a_2, \ldots, a_n)$ is a finite extension of F.*

Proof. The proof is by induction on n, with the base case given by Proposition 10.3.7. Suppose now that $n \geq 1$ is an integer for which for any choice of $a_1, a_2, \ldots, a_n \in E$, each algebraic over F, $F(a_1, a_2, \ldots, a_n)$ is a finite extension of F. Let $a_1, a_2, \ldots, a_{n+1} \in E$, each algebraic over F. We have $F(a_1, a_2, \ldots, a_{n+1}) = F(a_1, a_2, \ldots, a_n)(a_{n+1})$. By hypothesis, $F(a_1, a_2, \ldots, a_n)$ is a finite extension of F and thus of $F(a_{n+1})$, and by the base case, $F(a_{n+1})$ is a finite extension of F. Finally, by Proposition 10.4.4, $F(a_1, a_2, \ldots, a_{n+1})$ is a finite extension of F. The result follows now by induction. $\square$

10.4.6 Corollary. *If E is an algebraic extension of F, and K is an algebraic extension of E, then K is an algebraic extension of F.*

Proof. Let E and K be as described, and let $a \in K$. Then there exists $f = a_0 + a_1 x + \cdots + a_n x^n \in E[x]$ such that a is a root of f. Let $L = F(a_0, a_1, \ldots, a_n) \subseteq E$, so $f \in L[x]$. By Corollary 10.4.5, $[L : F]$ is finite. Since $f \in L[x]$ has a as a root, $[L(a) : L]$ is finite. By Proposition 10.4.4, $L(a)$ is a finite extension of F, and then by Proposition 10.4.2, $L(a)$ is an algebraic extension of F. Thus a is algebraic over F. This proves that every element of K is algebraic over F and so K is an algebraic extension of F. $\square$

10.4.7 Proposition. *Let E be an extension of F, and let*

$$Alg_F(E) = \{ a \in E \mid a \text{ is algebraic over } F \}.$$

Then $Alg_F(E)$ is a subfield of E containing F.

Proof. Since every element of F is algebraic over F, $F \subseteq Alg_F(E)$. Let $a, b \in Alg_F(E)$, and consider $F(a, b)$. By Corollary 10.4.5, $F(a, b)$ is a finite extension of F, and so by Proposition 10.4.2, $F(a, b) \subseteq Alg_F(E)$. Since $a + b$ and ab belong to $F(a, b)$, and if $a \neq 0_E$, $a^{-1} \in F(a, b)$, it follows that $a + b \in Alg_F(E)$, $ab \in Alg_F(E)$, and, if $a \neq 0_E$, $a^{-1} \in Alg_F(E)$. Thus $Alg_F(E)$ is a subfield of E. $\qquad \square$

As a particular example, we note that $\{\, a \in \mathbb{C} \mid a$ is algebraic over $\mathbb{Q} \,\}$ is a subfield of $\mathbb{C}$, and it is called the field of algebraic numbers. Every finite extension of $\mathbb{Q}$ is isomorphic to a subfield of the field of algebraic numbers, or in other words, up to isomorphism, every finite extension of $\mathbb{Q}$ is a subfield of the field of algebraic numbers. A finite extension of $\mathbb{Q}$ is called an algebraic number field.

10.4.8 Definition. *If E is an extension of F, $Alg_F(E)$ is called the algebraic closure of F in E.*

Recall that in Corollary 10.2.5, it was proven that given any $f \in F[x]$, there exists an extension E of F such that f splits in $E[x]$; that is, $f \in F[x] \subseteq E[x]$ can be written as a product of linear elements of $E[x]$.

10.4.9 Definition. *Let $f \in F[x]$. If E is an extension of F over which f splits, and $R_f \subseteq E$ is the set of roots of f in E, then $F(R_f)$ is called a splitting field for f over F.*

10.4.10 Proposition. *For any $f \in F[x]$, there exists a splitting field for f over F.*

Proof. By Corollary 10.2.5, for $f \in F[x]$, there is an extension E of F over which f splits. Let R_f denote the set of all roots of f in E. Then $F(R_f) \subseteq F'$ is a splitting field for f over F. $\qquad \square$

Our next major objective is to prove that any two splitting fields of $f \in F[x]$ are isomorphic (via an isomorphism that fixes the elements of F).

10.4.11 Lemma. *Let F' be a field for which there exists an isomorphism $\varphi : F \to F'$. Then the map $\psi : F[x] \to F'[x]$ given by $\psi(\sum_{i=0}^{n} a_i x^i) = \sum_{0}^{n} \varphi(a_i) x^i$ for each $\sum_{i=0}^{n} a_i x^i \in F[x]$ is the unique ring homomorphism from $F[x]$ to $F'[x]$ whose restriction to F is φ, and which sends x to x. Moreover, ψ is an isomorphism from $F[x]$ onto $F'[x]$.*

Proof. This is an immediate consequence of Proposition 9.3.13, which implies that there is a unique ring homomorphism $\psi: F[x] \to F'[x]$ whose restriction to F is φ, and which sends x to x. For any $\sum_{i=0}^{n} a_i x^i \in F[x]$, since ψ is a ring homomrphism, we have $\psi(\sum_{i=0}^{n} a_i x^i) = \sum_{i=0}^{n} \psi(a_i)\psi(x)^i = \sum_{i=0}^{n} \varphi(a_i)x^i$. Similarly, there exists a unique ring homomorphism $\gamma : F'[x] \to F[x]$ whose restriction to F' is φ^{-1} and which sends x to x. But then $\gamma \circ \psi$ is the unique extension of the identity map on F which sends x to x, and so $\gamma \circ \psi$ is the identity map on $F[x]$. Similarly, $\psi \circ \gamma$ is the identity map on $F'[x]$, and so ψ is an isomorphism from $F[x]$ to $F'[x]$. $\square$

10.4.12 Lemma. *Let F' be a field such that there exists an isomorphism $\varphi: F \to F'$, and let $\psi: F[x] \to F'[x]$ be the unique extension of φ to $F[x]$ that sends x to x. Let $f \in F[x]$ be irreducible over F, and let $f' = \psi(f)$. Let E be an extension of F in which f has a root r, and let E' be an extension of F' in which f' has a root r'. Then there exists an extension of φ to an isomorphism $\gamma: F(r) \to F'(r')$ for which $\gamma(r) = r'$.*

Proof. By Lemma 10.4.11, the map $\psi : F[x] \to F'[x]$ given by $\psi(\sum_{i=0}^{n} a_i x^i) = \sum_{i=0}^{n} \varphi(a_i)x^i$ is an isomorphism which extends φ and sends x to x. Since $f \in F[x]$ is irreducible over F, it follows that $\psi(f) \in F'[x]$ is irreducible over F'. Let E and E' be as described in the statement of the lemma. By Proposition 10.3.2, there is an isomorphism α from $F(r)$ onto $F[x]/(f)$ whose restriction to F (recall that we consider F to be a subfield of $F[x]/(f)$ by identifying $a \in F$ with $a + (f) \in F[x]/(f)$) is the identity mapping on F, and which maps r to $x + (f)$. Similarly, there is an isomorphism β from $F'[x]/(f')$ onto $F'(r')$ whose restriction to F' is the identity mapping on F', and which maps $x + (f')$ to r'. Since $f' = \psi(f)$, it follows that the homomorphism $\beta \circ \psi : F[x] \to F'[x]/(f')$ is surjective with kernel (f), and so we obtain an isomorphism $\delta: F[x]/(f) \to F'[x]/(f')$ whose restriction to F is φ, and which sends $x + (f)$ to $(x) + (f')$. Thus $\gamma = \beta \circ \delta \circ \alpha : F(r) \to F'(r')$ is an isomorphism whose restriction to F is φ and $\gamma(r) = r'$. $\square$

For convenience, from now on, for any isomorphism $\varphi: F \to F'$, F, F' fields, the unique extension of φ to an isomorphsim from $F[x]$ onto $F'[x]$ that sends x to x shall also be denoted by φ. Thus $\varphi(\sum_{i=0}^{n} a_i x^i) = \sum_{io=0}^{n} \varphi(a_i)x^i$ for any $\sum_{i=0}^{n} a_i x^i \in F[x]$.

10.4.13 Proposition. *For any fields F and F' for which there is an isomorphism $\varphi: F \to F'$, and any $f \in F[x]$, if E is any splitting field of* *f over F, and E′ is any splitting field for $\varphi(f)$ over F′, then there is an extension of φ to an isomorphism from E onto E′.*

Proof. The proof will be by induction on the degree of f. The result is immediate for polynomials f of degree 1, so suppose that $n \geq 1$ is an integer for which the result holds for all polynomials f of degree n. Let F and $F′$ be fields for which there is an isomorphism $\varphi: F \rightarrow F′$, and let $f \in F[x]$ be a polynomial of degree $n+1$. Let E be a splitting field for f over F, and let $E′$ be a splitting field for $\varphi(f)$ over $F′$. Let p be an irreducible factor of f. Then $\varphi(p)$ is an irreducible factor of $\varphi(f)$. Let $r \in E$ be a root of p, and let $r′ \in E′$ be a root of $\varphi(p)$. By Lemma 10.4.12, there is an extension of φ to an isomorphism ψ from $F(r)$ onto $F′(r′)$ Now, f factors in $F(r)$ as $(x-r)g$ for some $g \in F(r)[x]$, and then $\psi(f) = (x - \psi(r))\psi(g) = (x - r′)\psi(g)$, with $\psi(g) \in F′(r′)[x]$. Observe that E is a splitting field for g over $F(r)$, and $E′$ is a splitting field for $\varphi(g)$ over $F′(r′)$. Since $\deg(g) = n$, the induction hypothesis asserts that ψ has an extension to an isomorphism from E onto $E′$. Since ψ was an extension of φ, any extension of ψ is an extension of φ. The result follows now by induction. $\square$

10.4.14 Corollary. *Let $f \in F[x]$, and let E and $E′$ be splitting fields of f over F. Then there exists an isomorphism $\psi : E \rightarrow E′$ that fixes each element of F; that is, $\psi(a) = a$ for every $a \in F$.*

Proof. This is immediate from Proposition 10.4.13 by taking $F′ = F$ and φ the identity mapping on F. $\square$

As a result of Corollary 10.4.14, up to isomorphism fixing the elements of F, there is only one splitting field of $f \in F[x]$.

10.4.15 Notation. *For any $f \in F[x]$, we shall let $\mathrm{spl}_F(f)$ denote any splitting field of f over F.*

We have seen in Corollary 10.2.5 that for any $f \in F[x]$, there exists an extension of F in which f splits. It is natural to wonder whether there exists an extension of F in which every $f \in F[x]$ splits? It is far from obvious that such an extension exists, and in fact, the existence of such an extension of F cannot be proven just using the Zermelo-Fraenkel axioms of set theory. If one assumes that Axiom of Choice in addition to the Zermelo-Fraenkel axioms, then it is possible to prove that for every field F, there exists an extension $F′$ of F with the property that for every $f \in F[x]$, $F′$ contains a splitting field for f over F. Then $Alg_F(F′)$, the algebraic closure of F

in F', is called an algebraic closure of F. It can then be shown that any two algebraic closures of F are isomorphic by an isomorphism that fixes the elements of F. What we can prove is the following rather striking result.

10.4.16 Proposition. *Let F be a field, and let A be an algebraic closure of F. Then the only algebraic extension of A is A itself; that is, every $f \in A[x]$ splits over A.*

Proof. Suppose that E is an algebraic extension of A. Then by Proposition 10.4.6, E is algebraic over F. Let $a \in E$. Then there exists $f \in F[x]$ such that a is a root of f. But f splits over A, so A contains all roots of f. Thus $a \in A$, and so $E = A$. $\qquad\qquad\qquad\qquad\qquad\qquad\qquad\qquad\qquad$ $\square$

10.4.17 Definition. *A field F for which the only algebraic extension of F is F itself is called algebraically closed.*

Thus an algebraic closure of a field F is algebraically closed.

We have decided that the proof of the existence of an algebraic closure of a field F is beyond the intended level of this text, so we shall content ourselves with the above observations about algebraic closure. We should point out though that it is an immediate consequence of Liouville's theorem about functions of a complex variable that every $f \in \mathbb{C}[x]$ has a root in $\mathbb{C}$, which is equivalent to saying that $\mathbb{C}$ is algebraically closed. Since $\mathbb{C}$ is a finite extension of $\mathbb{R}$, it follows that $\mathbb{C}$ is an algebraic closure of $\mathbb{R}$.

10.5   Finite Fields

In Proposition 9.8.10, it was proven that if F is a finite field, then there exists a prime p and an irreducible $f \in \mathbb{Z}_p[x]$ such that F is isomorphic to $\mathbb{Z}_p[x]/(f)$, and as a result, $|F| = p^n$ for $n = \partial(f)$. Among other things, in this section we prove that for any prime p and any positive integer n, there exists a finite field of size p^n. Evidently, this is equivalent to proving that for every prime p, and every positive integer n, there exists an irreducible polynomial of degree n in $\mathbb{Z}_p[x]$. Moreover, we shall prove that any two finite fields of the same size are isomorphic.

In addressing these questions, the issue of repeated roots of a polynomial has to be addressed.

10.5.1 Definition. *Let $f \in F[x]$, and let E be any extension of F over which f splits. Then $r \in E$ is said to be a repeated root of f if $(x - r)^2$ is* *a divisor of f in $E[x]$. The maximum of $\{n \in \mathbb{Z}^+ \mid (x-r)^n \text{ divides } f\}$ is called the multiplicity of the root r in f.*

The notion of the formal derivative of a polynomial is very useful in the study of repeated roots.

10.5.2 Definition. *Let $f \in F[x]$. Then $f = \sum_{i=0}^{n} a_i x^i$ for unique $a_0, a_1, \ldots, a_n \in F$. The formal derivative of f is denoted by f' and is defined by $f' = \sum_{i=1}^{n} i a_i x^{i-1}$. Thus the formal derivative determines a mapping from $F[x]$ into $F[x]$.*

Although the derivative, as defined above, is independent of the notion of limit, the definition does coincide with that of the derivative of a polynomial function over the reals, and it should not come as a surprise to learn that the similarities are not just superficial.

10.5.3 Proposition. *The formal derivative is an F-linear endomorphism of $F[x]$. That is, the formal derivative has the following properties:*

(i) *For any $a \in F$ and $f \in F[x]$, $(af)' = a(f')$.*
(ii) *For any $f, g \in F[x]$, $(f+g)' = f' + g'$.*

Proof. Let $a, b_0, b_1, \ldots, b_n \in F$. Then $(a\sum_{i=0}^{n} b_i x^i)' = (\sum_{i=0}^{n} ab_i x^i)' = (\sum_{i=1}^{n} iab_i x^{i-1}) = a(\sum_{i=1}^{n} ib_i x^{i-1}) = a(\sum_{i=0}^{n} b_i x^i)'$.

Next, let $f, g \in F[x]$, and let $N = \max\{\partial f, \partial g\}$, and let $a_i \in F$, $i = 0, \ldots, N$ be such that $f = \sum_{i=0}^{N} a_i x^i$, and let $b_i \in F$, $i = 0, 1, \ldots, N$ be such that $g = \sum_{i=0}^{N} b_i x^i$. Then

$$(f+g)' = (\sum_{i=0}^{N} (a_i + b_i)x^i)' = \sum_{i=1}^{N} i(a_i + b_i)x^{i-1}$$

$$= \sum_{i=1}^{N} i a_i x^{i-1} + \sum_{i=1}^{N} i b_i x^{i-1} = f' + g'.$$

□

10.5.4 Corollary. *For any $f, g \in F[x]$, $(fg)' = fg' + f'g$.*

Proof. We prove first that for all $f \in F[x]$, $(fx^n)' = f(x^n)' + f'x^n$ by induction on n. Let $f \in F[x]$, so $f = \sum_{i=0}^{k} a_i x^i$ for some $a_0, a_i, \ldots, a_k \in F$.

We have

$$(fx)' = (\sum_{i=0}^{k} a_i x^{i+1})' = \sum_{i=0}^{k} (i+1)a_i x^i$$

$$= \sum_{i=1}^{k} ia_i x^i + \sum_{i=0}^{k} a_i x^i = x \sum_{i=1}^{k} ia_i x^{i-1} + f$$

$$= xf' + f(x')$$

since $x' = 1$. This proves that for any $f \in F[x]$, $(fx)' = fx' + f'x$. Suppose now that $n \geq 1$ is an integer for which the hypothesis holds, and let $f \in F[x]$. Then we have

$$(fx^{n+1})' = ((fx^n)x)' = fx^n + x(fx^n)'$$

$$= fx^n + xf'x^n + xfnx^{n-1}$$

$$= (n+1)x^n f + f'x^{n+1}$$

$$= (x^{n+1})'f + f'x^{n+1}$$

and it follows now by induction that for all $f \in F[x]$ and every positive integer n, $(fx^n)' = f(x^n)' + f'x^n$. Now let $f, g \in F[x]$, so $g = \sum_{i=0}^{m} b_i x^i$ for some $b_0, b_1, \ldots, b_m \in F$. By Proposition 10.5.3 and the preceding observation, we have

$$(fg)' = (\sum_{i=0}^{m} b_i f x^i)' = \sum_{i=0}^{m} (b_i f x^i)' = \sum_{i=0}^{m} b_i (f x^i)'$$

$$= b_0 f' + \sum_{i=1}^{m} b_i (f'x^i + f(x^i)') = (\sum_{i=0}^{m} b_i x^i)f' + f(\sum_{i=1}^{m} ib_i x^{i-1})$$

$$= gf' + fg'.$$

$\square$

10.5.5 Corollary. *For any $f \in F[x]$, and any positive integer n, $(f^n)' = nf^{n-1}f'$.*

Proof. The proof is by induction on n, and the base case of $n = 1$ is immediate. Suppose that $n \geq 1$ is an integer for which the result holds, and let $f \in F[x]$. Then $(f^{n+1})' = (ff^n)' = f(f^n)' + f'f^n = fnf^{n-1}f' + f'f^n = f'(nf^n + f^n) = f'(n+1)f^n$, and the result follows now by induction. $\square$

We now use the formal derivative to obtain a necessary and sufficient condition that a polynomial have a repeated root (in any extension field). The reader is reminded that in Proposition 9.4.10, it was proven that for

$f, g \in F[x]$ and E any extension field of F, the calculation of $gcd(f,g)$ in $E[x]$ produces the greatest common divisor of f and g in $F[x]$.

10.5.6 Proposition. *Let $f \in F[x]$ be a polynomial of degree at least 1, and let E be any extension of F over which f splits. Then f has a repeated root in E if, and only if, $gcd(f, f') \neq 1_F$.*

Proof. First, suppose that $f = (x - r)^2 g$ for some $r \in E$ and $g \in E[x]$. Then $f' = (x-r)^2 g' + 2(x-r)g = (x-r)((x-r)g' + 2g)$ and so $x - r$ is a divisor of $gcd(f, f')$. Thus $gcd(f, f') \neq 1_F$.

For the converse, suppose that $gcd(f, f') \neq 1_F$, and let p be an irreducible factor of $gcd(f, f')$. Let $r \in E$ be a root of p. Then r is a root of both f and f'. In particular, $f = (x - r)g$ for some $g \in E[x]$, so $f' = (x-r)g' + g$. But this means that r is a root of g, and so $g = (x-r)h$ for some $h \in E[x]$. Thus $f = (x-r)^2 h$ and so r is a repeated root of f. $\square$

10.5.7 Corollary. *If $f \in F[x]$ has degree $n \geq 1$ and $gcd(f, f') = 1_F$, then f has n distinct roots in $spl_F(f)$.*

Proof. Since f factors as a product of n linear factors over $spl_F(f)$, and by Proposition 10.5.6, no root is repeated, we see that f has n distinct roots in $spl_F(f)$. $\square$

10.5.8 Proposition. *Suppose that F has prime characteristic p. Then the mapping from F to F which sends $a \in F$ to $a^p \in F$ is an injective ring homomorphism. If F is finite, then it is an automorphism of F, called the Frobenius automorphism of F.*

Proof. Let $a, b \in F$. Certainly, $(ab)^p = a^p b^p$. The more interesting fact is that addition is preserved as well. We have $(a + b)^p = \sum_{i=0}^{p} \binom{p}{i} a^i b^{p-i}$. By Proposition 4.4.6, p is a divisor of $\binom{p}{i}$ for $i = 1, 2, \ldots, p - 1$ and so $\binom{p}{i} = 0_F$ for $i = 1, 2, \ldots, p-1$. Thus, as claimed, $(a + b)^p = a^p + b^p$. Since $1_F^p = 1_F \neq 0_F$, it follows that this is not the zero endomorphism of F, and thus has trivial kernel. It follows that the map is injective, and hence, if F is finite, bijective. $\square$

10.5.9 Corollary. *Suppose that F has prime characteristic p. Then for any positive integer n, and any $a, b \in F$, $(a + b)^{p^n} = a^{p^n} + b^{p^n}$.*

Proof. This is a straightforward proof by induction on n, with base case given by Proposition 10.5.8. Suppose $n \geq 1$ is an integer for which $(a +$

$b)^{p^n} = a^{p^n} + b^{p^n}$ for all $a, b \in F$. Then for any $a, b \in F$, we have $(a+b)^{p^{n+1}} = ((a + b)^{p^n})^p = (a^{p^n} + b^{p^n})^p = (a^{p^n})^p + (b^{p^n})^p = a^{p^{n+1}} + b^{p^{n+1}}$.   □

We are now ready to establish several important facts about finite fields.

10.5.10 Proposition. *For every prime p, and every positive integer n, there exists a field of order p^n. Moreover, any two finite fields of order p^n are isomorphic.*

Proof. Let E be a splitting field for $x^{p^n} - x \in \mathbb{Z}_p[x]$, and let

$$F = \{\, r \in E \mid r \text{ is a root of } x^{p^n} - x \,\}.$$

By Corollary 10.5.9, for any $a, b \in F$, $(a + b)^{p^n} = a^{p^n} + b^{p^n} = a + b$ and so $a + b \in F$. As well, $(ab)^{p^n} = a^{p^n} b^{p^n} = ab$ and so $ab \in F$. Since F is finite and closed under addition and multiplication, and $1_F, 0_F \in F$, F is a subfield of E. Moreover, since $(x^{p^n} - x)' = (x^{p^n})' - x' = p^n x^{p^n - 1} - 1 = -1$, $gcd(x^{p^n} - x, (x^{p^n} - x)']) = 1$ and so by Corollary 10.5.7, $|F| = p^n$. Note that $F = spl_{\mathbb{Z}_p}(x^{p^n} - x)$.

Now suppose that F' is any field of order p^n. Then under multiplication, $(F')^*$ is a group of order $p^n - 1$ and thus the order of each nonzero element of F' divides $p^n - 1$; that is, for $a \in F'$ with $a \neq 0_{F'}$, $a^{p^n - 1} = 1_{F'}$, and thus $a^{p^n} - a = 0_{F'}$. Thus every element of F', including $0_{F'}$, is a root of $x^{p^n} - x$, and so F' is a splitting field for $x^{p^n} - x$. By Corollary 10.4.14, F is isomorphic to F'.   □

Since any two fields of order p^n are isomorphic, we introduce the following notation, where the letters GF stand for "Galois field". The name is in honour of Évariste Galois (1811-1832).

10.5.11 Notation. $GF(p^n)$ *shall denote any field of order p^n.*

10.5.12 Corollary. *For any prime p, and any positive integer n, there exists at least one irreducible polynomial of degree n with coefficients in $\mathbb{Z}_p$.*

Proof. Let p be a prime and let n be a positive integer. By Proposition 10.5.10, there exists a field F of order p^n. By Proposition 9.8.7, the multiplicative group of units of F, F^*, is cyclic, and so there exists $a \in F$ such that $F = \{\, 0_F \,\} \cup \{\, a^i \mid i = 1, 2, \dots, p^n - 1 \,\}$. By Corollary 8.7.3, F contains a subfield isomorphic to $\mathbb{Z}_p$ which is a subfield of every subfield of F, and we shall identify this subfield with $\mathbb{Z}_p$. Thus F is an extension of $\mathbb{Z}_p$ and every subfield of F contains $\mathbb{Z}_p$. The smallest subfield of F that contains a therefore contains $\mathbb{Z}_p$, and so $F = \mathbb{Z}_p(a)$. By Proposition 10.3.2, F is isomorphic to $\mathbb{Z}_p[x]/(min(a, \mathbb{Z}_p))$, and by Proposition 10.3.7, $[F : \mathbb{Z}_p] = \partial(min(a, \mathbb{Z}_p))$. Since $[F : \mathbb{Z}_p]$ is the dimension of F as a $\mathbb{Z}_p$-vector space; that is, $[F : \mathbb{Z}_p]$ is the size of any $\mathbb{Z}_p$-basis for F, and every element of F can be uniquely written as a linear combination of the elements of a $\mathbb{Z}_p$-basis for F with coefficients from $\mathbb{Z}_p$, it follows that $p^n = |F| = p^{\partial(min(a, \mathbb{Z}_p))}$, and thus $\partial(min(a, \mathbb{Z}_p)) = n$. Since $min(a, \mathbb{Z}_p)$ is irreducible over $\mathbb{Z}_p$, we have proven that there exists an irreducible polynomial of degree n over $\mathbb{Z}_p$. $\qquad\square$

We close this section with one final observation. We have proven that for every prime p and positive integer n, there exists a field of order p^n, and that to construct such a field, one selects an irreducible polynomial f of degree n in $\mathbb{Z}_p[x]$ and forms $\mathbb{Z}_p[x]/(f)$, a field of order p^n. The selection of different irreducible polynomials of degree n will lead to different fields, but as we know from Proposition 10.5.10, any two such fields are isomorphic. How hard can it be to construct an isomorphism between two such fields? On the surface, it might seem to be a straightforward task. Let F and L be two fields of order p^n. Then we simply choose $a \in F$ such that $F = \mathbb{Z}_p(a)$ and construct $f = min(a, \mathbb{Z}_p)$. We then find a root $b \in L$ of f, so $f = min(b, \mathbb{Z}_p)$ and thus $\mathbb{Z}_p(b) \subseteq L$ has index $[\mathbb{Z}_p(b) : \mathbb{Z}_p] = \partial(f) = n$ and thus $\mathbb{Z}_p(b) = L$. There is a unique unital ring homomorphism from $\mathbb{Z}_p(a)$ to $\mathbb{Z}_p(b)$ that acts as the identity on $\mathbb{Z}_p$ and sends a to b. Such a homomorphism is necessarily bijective, and so is the sought-after isomorphism. So why did we say "On the surface" in the preceding discussion? The problem that we face is that of finding $a \in F$ for which $F = \mathbb{Z}_p(a)$. The job amounts to choosing a particular irreducible polynomial f of degree n over $\mathbb{Z}_p$, and then hunting for a root of f in F, and a root of f in L. It turns out that there is no simple way to do this.

In order to facilitate computations in $GF(p^n)$, it is important to use a standard construction of $GF(p^n)$. If this is done, then two different people will use the same representation of the elements of $GF(p^n)$ and thus can easily share the results of their work. One standard that is in widespread use is the choice of the so-called Conway polynomial of degree n to construct $GF(p^n)$. In order to describe the Conway polynomials, we first need to investigate the nature of the subfields of $GF(p^n)$.

10.5.13 Proposition. *Let p be a prime and n a positive integer. Then for every positive divisor m of n, there is a unique subfield of order p^m in $GF(p^n)$. Moreover, every subfield of $GF(p^n)$ has order p^m for some divisor m of n.*

Proof. First, suppose that m is a positive divisor of n, say $n = mk$. Then $p^n - 1 = p^{mk} - 1 = (p^m - 1)(\sum_{i=0}^{k-1}(p^m)^i)$, and so $p^m - 1$ is a divisor of $p^n - 1$. A similar computation then shows that $x^{p^m - 1} - 1$ is a divisor of $x^{p^n - 1} - 1$, and so every root of $x^{p^m} - x$ is a root of $x^{p^n} - x$. Since $GF(p^n)$ is a splitting field for $x^{p^n} - x$, it follows that $GF(p^n)$ contains a splitting field for $x^{p^m} - x$; that is, $GF(p^m) \subseteq GF(p^n)$. Since any subfield of $GF(p^n)$ of order p^m is the set of all roots of $x^{p^m} - x$ in $GF(p^n)$, it follows that there is only one such subfield. Note that if F is a subfield of $GF(p^n)$, then since $GF(p) = \mathbb{Z}_p$ is the prime subfield of $GF(p^n)$, we have $GF(p) \subseteq F$. Thus F is a finite-dimensional $GF(p)$-vector space, say $[F : GF(p)] = m$, and so $|F| = p^m$.

Now suppose that $GF(p^m) \subseteq GF(p^n)$. We are to prove that m is a divisor of n. We have $GF(p) = \mathbb{Z}_p \subseteq GF(p^m) \subseteq GF(p^n)$ and so by Proposition 10.4.4, we have

$$[GF(p^n) : GF(p^m)]\,[GF(p^m) : GF(p)] = [GF(p^n) : GF(p)] = n,$$

while $[GF(p^m) : GF(p)] = m$. Thus m is a divisor of n. $\qquad\square$

For any positive integer n, the multiplicative group $GF(p^n)^*$ of nonzero elements of $GF(p^n)$ is cyclic. Let $a \in GF(p^n)^*$ be a generator for this group. For any positive divisor m of n, $GF(p^m)^*$ is a subgroup (necessarily cyclic) of $GF(p^n)^*$, and $a^{\frac{p^n-1}{p^m-1}}$ is a generator of $GF(p^m)^*$. If $b \in GF(p^m)$ is a generator of $GF(p^m)^*$ and $a^{\frac{p^n-1}{p^m-1}}$ is a root of $min(b, \mathbb{Z}_p)$, then $min(a, \mathbb{Z}_p)$ and $min(b, \mathbb{Z}_p)$ are said to be compatible polynomials. This is equivalent to requiring that $min(a, \mathbb{Z}_p)$ be a divisor of $\widehat{min(b, \mathbb{Z}_p)}(x^{\frac{p^n-1}{p^m-1}})$ in $\mathbb{Z}_p[x]$; that is, if we let $h = min(b, \mathbb{Z}_p)$, then $min(a, \mathbb{Z}_p)$ is a divisor of $\hat{h}(x^{\frac{p^n-1}{p^m-1}})$ in $\mathbb{Z}_p[x]$. John H. Conway's suggestion was that for a given prime p, standard polynomials $f_1, f_2, \ldots$, be chosen so that for each n, f_n is a monic irreducible polynomial of degree n over $\mathbb{Z}_p$, and some root of f_n is a generator of $GF(p^n)^*$ (and therefore every root of f_n is a generator of $GF(p^n)^*$), and that for every n and m with m a divisor of n, f_n and f_m are compatible. The selection is made inductively, with $f_1 = x + 1$. Observe that for f_n and f_m to be compatible, for m a divisor of n, we require that for any root a of f_n, $a^{\frac{p^n-1}{p^m-1}}$ must be a root of f_m, which is equivalent to requiring that f_n be a divisor of $\hat{f}_m(x^{\frac{p^n-1}{p^m-1}})$. In general, there will be several monic irreducible polynomials of degree n over $\mathbb{Z}_p$ each of whose roots is a generator of $GF(p^n)^*$ and which is compatible with every f_m for m a positive proper divisor of n, so some convention must be introduced to ensure that the selection is uniquely determined (it was proven by Werner Nickel that such a selection is always possible). The method that has been adopted is to introduce a total ordering on the set of monic irreducible polynomials of each given degree over $\mathbb{Z}_p$, and then of all polynomials that meet the criteria, the least one (with respect to this ordering) is the one that is selected. Recall that we write the elements of $\mathbb{Z}_p$ as $0, 1, 2, \ldots, p-1$, and we order them as written, so $0 < 1 < \cdots < p-1$. Now for any positive integer n, the polynomials of degree n are of the form $a_n x^n + a_{n-1} x^{n-1} + \cdots + a_1 x + a_0$. We say that $a_n x^n + a_{n-1} x^{n-1} + \cdots + a_1 x + a_0 < b_n x^n + b_{n-1} x^{n-1} + \cdots + b_1 x + b_0$ if the sequence $(a_n, a_{n-1}, \ldots, a_1, a_0)$ is lexically less than the sequence $(b_n, b_{n-1}, \ldots, b_1, b_0)$. For example, for the prime $p = 3$, $f_1 = x + 1$, $f_2 = x^2 + 2x + 2$, $f_3 = x^3 + 2x + 1$, and $f_4 = x^4 + 2x^3 + 2$. Let us think about the addition and multiplication tables for $GF(9)$ using the Conway polynomial $f_2 = x^2 + 2x + 2$. Let $t = x + (f_2)$. The elements of $GF(9)$ are

$$0, 1, 2, t, t+1, t+2, 2t, 2t+1, 2t+2.$$

We compute the powers of t, using the fact that $t^2 + 2t + 2 = 0$, or $t^2 = -2t - 2 = t + 1$. We have t, $t^2 = t+1$, $t^3 = t(t+1) = t^2 + t = t + 1 + t = 2t+1$, $t^4 = 2t^2 + t = 2(t+1) + t = 2$, $t^5 = 2t$, $t^6 = 2t^2 = 2(t+1) = 2t+2$, $t^7 = 2t^2 + 2t = 2(t+1) + 2t = t+2$, and $t^8 = t^2 + 2t = t + 1 + 2t = 1$, as expected. Addition is quite straightforward if we represent each element as a linear combination of 1 and t, but not at all straightforward if we represent the elements as powers of t. On the other hand, multiplication of powers of t is immediate, but we have work to do to compute the product of two linear combinations of 1 and t. We choose to represent the nonzero elements of $GF(9)$ as powers of t, so there is no need to construct the multiplication table. Instead, we shall build the addition table. To compute $t^i + t^j$, we express each of t^i and t^j as linear combinations of 1 and t, add these linear combinations, and then find the result in the list of powers of t.

We complete our discussion with the verification that the second Conway polynomial for $p = 3$ does indeed divide the first Conway polynomial evaluated at $x^{\frac{3^2-1}{3^1-1}}$; that is, $x^4 + 1$. We evaluate $(x^2 + 2x + 2)(x^2 + x + 2) = x^4 + x^3 + 2x^2 + 2x^3 + 2x^2 + x + 2x^2 + 2x + 1 = x^4 + 1$, as required. As well, since 2 is a factor of 4, f_4 is to be a divisor of $\hat{f}_2(x^{\frac{3^4-1}{3^2-1}})$; that is, $x^4 + 2x^3 + 2$ must be a divisor of $(x^{10})^2 + 2(x^{10}) + 2$. We leave it to the reader to verify that $x^{20} + 2x^{10} + 2 = (x^4 + 2x^3 + 2)(x^{16} + x^{15} + x^{14} + x^{13} + 2x^{12} + x^{10} + 2x^9 + x^8 + x^7 + x^6 + x^4 + 2x^3 + 1)$.

Table 10.1   The addition table for $GF(9)$

+	0	1	t	t^2	t^3	t^4	t^5	t^6	t^7
0	0	1	t	t^2	t^3	t^4	t^5	t^6	t^7
1	1	t^4	t^2	t^7	t^6	0	t^3	t^5	t
t	t	t^2	t^5	t^3	1	t^7	0	t^4	t^6
t^2	t^2	t^7	t^3	t^6	t^4	t	1	0	t^5
t^3	t^3	t^6	1	t^4	t^7	t^5	t^2	t	0
t^4	t^4	0	t^7	t	t^5	1	t^6	t^3	t^2
t^5	t^5	t^3	0	1	t^2	t^6	t	t^7	t^4
t^6	t^6	t^5	t^4	0	t	t^3	t^7	t^2	1
t^7	t^7	t	t^6	t^5	0	t^2	t^4	1	t^3

10.6   Exercises

1. In the field $\mathbb{R}$, $\sqrt{2}$ and $\sqrt{3}$ are both algebraic over $\mathbb{Q}$.

   a) Since $\sqrt{2} + \sqrt{3} \in \mathbb{Q}(\sqrt{2}, \sqrt{3})$, an algebraic extension of $\mathbb{Q}$, $\sqrt{2} + \sqrt{3}$ is algebraic over $\mathbb{Q}$. Find a polynomial of degree 4 over $\mathbb{Q}$ which has $\sqrt{2} + \sqrt{3}$ as a root.

   b) What is $\partial(min(\sqrt{2} + \sqrt{3}, \mathbb{Q}))$? Justify your answer.

   c) Since $\sqrt{6} = \sqrt{2}\sqrt{3} \in \mathbb{Q}(\sqrt{2}, \sqrt{3})$, $\sqrt{6}$ is algebraic over $\mathbb{Q}$ as well. What is the degree of $min(\sqrt{6}, \mathbb{Q})$?

2. Prove that $\sqrt{2} + \sqrt[3]{5}$ is algebriac over $\mathbb{Q}$. Determine $[\mathbb{Q}(\sqrt{2} + \sqrt[3]{5}) : \mathbb{Q}]$.

3. Find $min(\sqrt{-3} + \sqrt{2}, \mathbb{Q})$ and $min(\sqrt[3]{2} + \sqrt[3]{4}, \mathbb{Q})$.

4. Suppose that F is a field, and E is an extension of F. Further suppose that $a, b \in E$ are both algebraic over F, and let $m = [F(a) : F]$ and $n = [F(b) : F]$. Prove that if $(m, n) = 1$, then $[F(a, b) : F] = mn$.

5. Let F be a field and $f \in F[x]$. Let E be a splitting field for f over F. Prove that $[E : F] \leq n!$.

6. For each $f \in \mathbb{Q}[x]$ below, determine $[spl_F(f) : F]$.

   a) $x^4 + x^2 + 1$.

   b) $x^4 + 1$.

   c) $x^6 + 1$.

   d) $x^4 - 2$.

   e) $x^5 - 1$.

   f) $x^9 + x^3 + 1$.

7. For any prime p, prove that $[spl_{\mathbb{Q}}(x^p - 1) : \mathbb{Q}] = p - 1$.

8. Prove that $\mathbb{Q}(\sqrt[3]{2})$ has no automorphisms other than the identity mapping.

9. Let m be a positive integer that is not a perfect square. Prove that for any $a, b \in \mathbb{Q}$, if $a + b\sqrt{m}$ is a root of a polynomial $f \in \mathbb{Q}[x]$, then $a - b\sqrt{m}$ is also a root of f.

10. Let $f = x^2 + x + 1 \in \mathbb{Z}_2[x]$, and let E be an extension of $\mathbb{Z}_2$ in which f splits. Let $b \in E$ be a root of f.

    a) What is $|\mathbb{Z}_2(b)|$? Express each element of $\mathbb{Z}_2(b)$ as a linear combination of powers of b with coefficients from $\mathbb{Z}_2$.

    b) Each of b^5, b^{-2}, and b^{100}, is an element in the list you provided for (a). In each case, identify the element in the list.

    c) Prove that $x^3 + x^2 + 1 \in \mathbb{Z}_2[x]$ splits in $\mathbb{Z}_2(b)$ by verifying that $1 + b$, $1 + b^2$, and $1 + b + b^2$ are roots of $x^3 + x^2 + 1$.

11. Let F be a field, and let E be an extension of F. For any $a, b \in F$ with $a \neq 0_F$, prove that for any $c \in E$, $F(c) = F(ac + b)$.

12. Let F be a field of characteristic 0. Prove that if $f \in F[x]$ is irreducible over F, then f has no repeated roots in any extension of F.

13. Let F be a field of characteristic 0, and let E be a finite extension of F. This exercise establishes a result due to Ernst Steinitz; namely that there exists $c \in E$ for which $E = F(c)$. The proof is inductive, and the base case does all of the work.
    Let $a, b \in E$. We wish to prove that there exists $c \in F(a, b)$ such that $F(a, b) = F(c)$. Let $f = min(a, F)$ and $g = min(b, F)$, and let $n = \partial(f)$ and $m = \partial(g)$. Enlarge E if necessary so that both f and g split over the extension, and let $a = a_1, a_2, \ldots, a_m \in E$ be the m distinct (by Problem 12) roots of f in E, and $b = b_1, b_2, \ldots, b_n$ the n distinct roots of g in E. Since $S = \{ (a - a_i)(b - b_j) \mid 1 \leq i \leq m, \ 2 \leq j \leq n \}$ is a finite subset of E, and F is infinite, there exists $d \in F - S$. Let $c = a + db$.

    a) Let $r = \hat{f}(c - dx) \in F(c)[x]$. Since $d \in F$, $r \in F(c)[x]$. Let $s = min(b, F(c))$. Prove that s divides $gcd(g, r)$.

    b) Prove that s splits over E (remember that both f and g split over E).

    c) Prove that b_i is a root of r if, and only if, $i = 1$.

    d) Deduce that $s = x - b$ and so $b \in F(c)$.

    e) Prove that $F(a, b) = F(c)$.

    f) Prove that if $r_1, r_2, \ldots, r_k \in E$ are algebraic over F, then there exists $c \in F(r_1, r_2, \ldots, r_k) \subseteq E$ such that $F(r_1, r_2, \ldots, r_k) = F(c)$.

14. Let F be a field, and let $f, g \in F[x]$ each be irreducible over F with $f \neq g$.

    a) Let E be an extension of F. Prove that if $b \in E$ is a root of f, then g is irreducible over $F(b)$.

    b) Prove that $gcd(f, g) = 1$.

15. Prove that $\mathbb{C}$ is the only finite extension of $\mathbb{R}$ that is different from $\mathbb{R}$.

16. Let E be an extension of $\mathbb{Q}$ for which $[E:\mathbb{Q}] = 2$. Prove that there exists $d \in \mathbb{Z}$ such that d is squarefree and $E = \mathbb{Q}(\sqrt{d})$.

17. Determine $[spl_{\mathbb{Z}_3}(x^4 - x^2 - 2):\mathbb{Z}_3]$.

18. Let p be a prime and let n be a positive integer.

    a) Prove that if f is any irreducible factor of $x^{p^n} - x \in \mathbb{Z}_p[x]$, then $\partial(f) \leq n$.

    b) Prove that $x^{p^n} - x$ has a factor of degree n that is irreducible over $\mathbb{Z}_p$.

19. Let $f = x^3 + 2x + 1 \in \mathbb{Z}_3[x]$ (note that $f = f_3$, the third Conway polynomial for $p = 3$).

    a) Prove that f is irreducible over $\mathbb{Z}_3$.

    b) Let $F = \mathbb{Z}_3[x]/(f)$, and let $t = x + (f) \in F$. Prove that t is a generator for the cyclic group F^* under multiplication.

    c) Prove that $g = x^3 + 2x + 2 \in \mathbb{Z}_3[x]$ is also irreducible over $\mathbb{Z}_3$, but $t = x + (g) \in K = \mathbb{Z}_3[x]/(g)$ is not a generator for the cyclic group K^* under multiplication. Find a generator for K^*.

20. Let $f = x^5 + x^3 + 1 \in \mathbb{Z}_2[x]$.

    a) Prove that f is irreducible over $\mathbb{Z}_2$.

    b) Let $t = x + (f) \in F = \mathbb{Z}_2[x]/(f)$. Without calculating the order of t explicitly in F^*, prove that t is a generator for the cyclic group F^* under multiplication.

21. Let p be a prime, and let b be a generator for the cyclic group $GF(p^{12})^*$ under multiplication. For each subfield K of $GF(p^{12})$, find a positive integer k such that b^k is a generator for the cyclic group K^* under multiplication.

22. Let p be a prime and n a positive integer. If K and L are subfields of $GF(p^n)$ of orders p^r and p^s, respectively, what is the order of $K \cap L$?

23. Let p be a prime and n a positive integer. For any positive divisor m of $p^n - 1$ and any $a \in GF(p^n)$, prove that $x^m - a \in GF(p^n)[x]$ has either m roots in $GF(p^n)$ or else it has no roots in $GF(p^n)$.

24. Let F be a field and let E be an algebraic extension of F. Prove that if R is a subring of E and $F \subseteq R$, then R is a subfield of E.

25. Let p be a prime and let n be a positive integer. Prove that the order of the Frobenius automorphism of $GF(p^n)$ in the automorphism group $\mathrm{Aut}(GF(p^n))$ is n. Recall that the Frobenius automorphism of $GF(p^n)$ is the map that sends a to a^p for every $a \in GF(p^n)$.

26. Let p be a prime, and let m, n be positive integers. Prove that $x^{p^m} + x + 1$ is a divisor of $x^{p^n} + x + 1$ in $\mathbb{Z}_p[x]$ if, and only if, m is an odd multiple of n.

Chapter 11

Latin Squares and Magic Squares

1 Introduction

We begin with a practical problem. Suppose a researcher has three different fertilizers which are to be tested on three different strains of wheat to determine which combination would give the maximum yield. How should the experiment be designed?

Obviously, the researcher would want to test each strain with each type of fertilizer so the test field should be divided up into a 3×3 matrix of nine plots, then each strain of wheat should be planted in three different locations, no two of which should receive the same fertilizer. To avoid skewing the results of the experiment because of the variability of the fertility of the soil in the nine different plots, the following constraints should be applied:

(i) the strains of wheat should be planted in such a way that in each row and each column of the matrix, each strain of wheat occurs once and only once; and

(ii) the fertilizers should be applied so that no two plots in the same row or column receive the same fertilizer.

We shall see that the solution to this problem leads quite naturally to the notion of *mutually orthogonal Latin squares*.

Suppose we denote the three strains of wheat by $1, 2, 3$, and the three types of fertilizer by a, b, c. For $s \in \{1, 2, 3\}$ and $t \in \{a, b, c\}$, we shall write (s, t) to convey that strain s of wheat is to be fertilized by the fertilizer of type t. Then our task is to place the nine pairs

$$(1, a), (1, b), (1, c), (2, a), (2, b), (2, c), (3, a), (3, b), (3, c)$$

as the entries of a 3×3 matrix in such a way that each type of fertilzer and each strain of wheat occurs once and only once in each row and column.

We try to solve the problem by first arranging $1, 2, 3$ in each of the three rows in such a way that for each $s = 1, 2, 3$, no column contains more than one copy of s. One such arrangement is shown in Figure 11.1 (i).

$$
\begin{bmatrix} 1\,2\,3 \\ 2\,3\,1 \\ 3\,1\,2 \end{bmatrix}
\qquad
\begin{bmatrix} a\ b\ c \\ c\ a\ b \\ b\ c\ a \end{bmatrix}
\qquad
\begin{bmatrix} (1,a)\ (2,b)\ (3,c) \\ (2,c)\ (3,a)\ (1,b) \\ (3,b)\ (1,c)\ (2,a) \end{bmatrix}
$$

(i)   (ii)   (iii)

Fig. 11.1

Next, we try to arrange the entries a, b, c in each of the three rows in such a way that for each $t = a, b, c$, no column contains more than one copy of t. This time, we are keeping an eye on the preceding arrangement, trying to place our letters a, b, c in such a way that at the end, if we superimpose the (i, j) entry of the first matrix with the (i, j) entry of the second matrix for each $i, j \in \{1, 2, 3\}$, we obtain all nine pairs. One such arrangement is shown in Figure 11.1 (ii). The resulting array of pairs is shown in Figure 11.1 (iii).

The considerations in the preceding example lead us quite naturally to the following definition. Recall that for any positive integer n, $J_n = \{1, 2, \ldots, n\}$.

11.1.1 Definition. *For any positive integer n, and any set X of size n, a Latin square over X is an $n \times n$ matrix L with entries from X such that each element of X appears exactly once in each row of L, and exactly once in each column of L.*

11.1.2 Notation. *For any positive integer n, an $n \times n$ Latin square L over J_n shall be written as a column vector of permutations, where each row is to be understood as an element of S_n, by means of the map notation decribed in Notation 3.4.6. Thus we write $L = [\sigma_i]$, where for each $i \in J_n$, σ_i is a bijective map from J_n to X given by $\sigma_i(j) = L_{ij}$ for each $i, j \in J_n$.*

Note that in this notation, for each $i, j, k \in J_n$ with $i \neq j$, $\sigma_i(k) \neq \sigma_j(k)$. To illustrate, consider the 3×3 Latin square shown in Figure 11.1 (i). This would be written as $\begin{bmatrix} \sigma_1 \\ \sigma_2 \\ \sigma_3 \end{bmatrix}$, where $\sigma_1 = [1\,2\,3]$, $\sigma_2 = [2\,3\,1]$, and $\sigma_3 = [3\,1\,2]$. The individual entries of σ_3 for example are accessed as $\sigma_3(1) = 3$, $\sigma_3(2) = 1$, and $\sigma_3(3) = 2$.

Unless there is some specific reason to choose otherwise, we shall normally take $X = J_n$ (that is, $X = \{1, 2, 3, \ldots, n\}$).

11.1.3 Definition. *For any positive integer n and set X of size n, and Latin squares $L_1 = [\sigma_i]$ and $L_2 = [\tau_i]$ over X, let*

$$L_1 \# L_2 = \{(\sigma_i(j), \tau_i(j)) \mid i, j \in J_n\}.$$

Then L_1 and L_2 are said to be orthogonal if $L_1 \# L_2 = X \times X$. A set $\mathscr{L}$ of Latin squares over X is said to be a set of mutually orthogonal Latin squares over X (typically abbreviated to say that $\mathscr{L}$ is a set of MOLS over X) if for any $L_1, L_2 \in \mathscr{L}$, L_1 and L_2 are orthogonal if $L_1 \neq L_2$.

11.1.4 Lemma. *For any positive integer n and any set X of size n, Latin squares L_1 and L_2 over X are orthogonal if, and only if, for every $(i, j), (k, l) \in J_n \times J_n$ with $(i, j) \neq (k, l)$, $(\sigma_i(j), \tau_i(j)) \neq (\sigma_k(l), \tau_k(l))$.*

Proof. Latin squares L_1 and L_2 over X are orthogonal if, and only if, $|L_1 \# L_2| = n^2$, which is equivalent to the requirement that for every $i, j, k, l \in J_n$, $(\sigma_i(j), \tau_i(j)) \neq (\sigma_k(l), \tau_k(l))$. $\qquad \square$

In the first example, we chose to represent the different types of fertilizer by the letters a, b, and c, but of course, we could have used the numbers 1, 2, and 3 just as easily. In that example, if we replace a by 1, b by 2, and c by 3, then the two resulting Latin squares over $\{1, 2, 3\}$ are orthogonal.

Our next result provides a nice way to construct a new pair of orthogonal Latin squares over J_n from a given pair of orthogonal Latin squares over J_n.

11.1.5 Proposition. *For any positive integer n and any set X of size n, if $L_1 = [\sigma_i]$ and $L_2 = [\tau_i]$ are orthogonal Latin squares over X, and $\pi, \lambda \in S_X$, the group of permutations of X, then $\pi(L_1) = [\pi \circ \sigma_i]$ and $\lambda(L_2) = [\lambda \circ \tau_i]$ are orthogonal Latin squares over X.*

Proof. For any $i, j, k, l \in J_n$, $(\pi \circ \sigma_i(j), \lambda \circ \tau_i(j)) = (\pi \circ \sigma_k(l), \lambda \circ \tau_k(l))$ if, and only if, $(\sigma_i(j), \tau_i(j)) = (\sigma_k(l), \tau_k(l))$, which by Lemma 11.1.4 is equivalent to the orthogonality of L_1 and L_2. $\qquad \square$

11.1.6 Proposition. *For any positive integer n, and any set X of size n, if $\mathscr{L}$ is a set of MOLS over X, then $|\mathscr{L}| \leq n - 1$.*

Proof. It suffices to prove the result for Latin squares over J_n. Let $\mathscr{L} = \{L_1, L_2, \ldots, L_k\}$ be a set of MOLS over J_n, and for each $i = 1, 2, \ldots, k$, write $L_i = [\sigma_j^i]$ (so $j \in J_n$). Now k applications of Proposition 11.1.5 tells us that $\{\, [(\sigma_1^i)^{-1} \circ \sigma_j^i] \mid i = 1, 2, \ldots, k \,\}$ is a set of MOLS of size k. For each i and j, let $\lambda_j^i = (\sigma_1^i)^{-1} \circ \sigma_j^i$, so our new set of MOLS is $\{\, [\lambda_j^i] \mid i = 1, 2, \ldots, k \,\}$. Note that the first row of each is $\lambda_1^i = \mathbb{1}_{J_n}$, which means that for any $i, j, l \in J_n$ with $i \neq j$, (l, l) in $L_i \# L_j$ is the pair $(\lambda_1^i(l), \lambda_1^j(l))$. As a consequence, for any $i, j \in J_n$ with $i \neq j$, $(\lambda_2^i(1), \lambda_2^j(1))$ is not a diagonal pair; that is, $\lambda_2^i(1) \neq \lambda_2^j(1)$. Thus

$$\{\, \lambda_2^i(1) \mid i = 1, 2, \ldots, k \,\}$$

is a set of size k. Moreover, for every $i \in J_n$, $1 = \lambda_1^i(1) \neq \lambda_2^i(1)$, and so $\{\, \lambda_2^i(1) \mid i = 1, 2, \ldots, k \,\} \subseteq J_n - \{1\}$. It follows now that $k \leq n - 1$, as required. $\square$

The preceding result suggests a very interesting question; namely for which values of n, if any, does there exist a set of MOLS over J_n of size $n - 1$? If n is a prime power, there exists a set of MOLS over J_n of size $n - 1$, as we shall shortly prove. For the particular value $n = 6$, it is known that there is no set of MOLS of size 2 over J_6, never mind a set of MOLS of size 5. As it turns out, 6 is peculiar in this regard, as it has been shown that for all $n \geq 2$ except for $n = 6$, there are at least two orthogonal Latin squares over J_n. However, it is still an open question as to whether or not there is a set of MOLS of size $n - 1$ for any value of n except for n a prime power (for which it is known that there is such a set of MOLS) or $n = 6$ (for which it is known that there is no such set of MOLS).

11.1.7 Proposition. *Let p be a prime, and let m be a positive integer. Then for $n = p^m$, there exists a set of MOLS of size $n - 1$ over J_n.*

Proof. Let $F = GF(p^m)$, and label the elements of F as $a_1, a_2, \ldots, a_n$, where $a_1 = 1_F$ and $a_n = 0_F$. For each $k = 1, 2, \ldots, n - 1$, let L_k denote the $n \times n$ matrix with entries from F as defined by $(L_k)_{ij} = a_k a_i + a_j$ for each $i, j \in J_n$. We prove first that for each k, L_k is a Latin square over F. Suppose first that in some row i, the entry in column j equals the entry in column l; that is, $a_k a_i + a_j = a_k a_i + a_l$. Then $a_j = a_l$, and so $j = l$. Thus for each i, if $j \neq l$, $(L_k)_{ij} \neq (L_k)_{il}$. Now suppose that in some column j, the entry in row i is equal to the entry in row l. Then $a_k a_i + a_j = a_k a_l + a_j$, and so $a_k a_i = a_k a_l$. Since $k \neq n$ and $a_n = 0_F$, $a_k \neq 0_F$ and so we may multiply by a_k^{-1} to obtain that $a_i = a_l$. But then $i = l$ and so we have proven that for $i \neq l$, then for any j, $(L_k)_{ij} \neq (L_k)_{lj}$. This proves that for each $k = 1, 2, \ldots, n - 1$, L_k is a Latin square over F.

Now let k, l be such that $1 \leq k, l \leq n - 1$ with $k \neq l$. We wish to prove that L_k and L_l are orthogonal. This requires we prove that $|L_k \# L_l| = n^2$, and for this, it suffices to prove that for any $i, j, r, s \in J_n$, if $((L_k)_{ij}, (L_l)_{ij}) = ((L_k)_{rs}, (L_l)_{rs})$, then $i = r$ and $j = s$. Suppose that $a_k a_i + a_j = a_k a_r + a_s$ and $a_l a_i + a_j = a_l a_r + a_s$. Then $a_k(a_i - a_r) = -(a_j - a_s)$ and $a_l(a_i - a_r) = -(a_j - a_s)$, from which we obtain $a_k(a_i - a_r) = a_l(a_i - a_r)$. If $a_i - a_r \neq 0_F$, then $a_k = a_l$, which implies $k = l$. Since we have chosen $k \neq l$, it follows that $a_i - a_r = 0_F$ and thus $a_i = a_r$, from which we deduce that $i = r$. But then $a_j - a_s = 0_F$, and so $a_j = a_s$, which proves that $j = s$, as required. $\qquad\square$

We illustrate the preceding proposition with an example.

11.1.8 Example. *Let $p = 2$ and $m = 2$, and construct a set of MOLS consisting of three 4×4 Latin squares with entries from $GF(2^2)$. To construct $GF(2^2)$, we use the Conway polynomial $f_{2,2} = x^2 + x + 1$, so that $t = x + (x^2 + x + 1)$ is a generator for the multiplicative group of units of $GF(2^2)$. Thus $GF(2^2) = \{1 = t^0, t, t^2, 0\}$. For convenience, we shall label these as*

$$1(= t^0), \ 2(= t^1), \ 3(= t^2), \ and \ 4(= 0).$$

We give the multiplication and addition tables for $GF(4)$ using this notation in Table 11.1.

Table 11.1   Multiplication and addition tables for $GF(4)$

×	1	2	3	4		+	1	2	3	4
1	1	2	3	4		1	4	3	2	1
2	2	3	1	4		2	3	4	1	2
3	3	1	2	4		3	2	1	4	3
4	4	4	4	4		4	1	2	3	4

We follow the prescription provided by Proposition 11.1.7 to construct the three Latin squares L_1, L_2, and L_3 formed by setting $a_1 = 1$, $a_2 = 2$, $a_3 = 3$, and $a_4 = 4$. The results are shown in Figure 11.2 (note that L_1 is simply the addition table for $GF(4)$ since $a_1 = 1$, whereas the other two are obtained by cyclically permuting the first three rows of L_1 – this is due to our choice of notation for the elements of F).

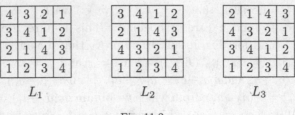

$$L_1 \qquad\qquad L_2 \qquad\qquad L_3$$

Fig. 11.2

11.2 Magic Squares

We end this chapter with a brief discussion of magic squares. This topic falls into the category of recreational mathematics, and it appears to have intrigued both amateur and professional mathematicians throughout the ages. The first recorded instance of a magic square is what is known as the Lo Shu square (shown in Figure 11.3 (i)), a 3 × 3 magic square with the property that every 3 × 3 magic square is obtained from the Lo Shu square by rotation or reflection. It dates back to around 2800 BCE as part of the legacy of Chinese mathematical traditions, and is an important emblem in Feng Shui, the art of geomancy concerned with the placement of objects in relation to the flow of "natural energy". Probably the most famous magic square is the 4 × 4 magic square which appeared in a woodcut created by Albrecht Dürer in 1514 AD, entitled "Melancholia". It is shown in Figure 11.3 (ii).

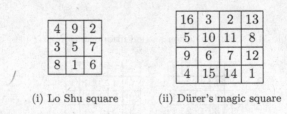

(i) Lo Shu square   (ii) Dürer's magic square

Fig. 11.3   Ancient magic squares

Observe the following: the entries of the Lo Shu square are the positive integers from 1 to $9 = 3^2$; the sum of the entries of each row is the same for each row (15) – the square is said to be row-magic; the sum of the entries of each column is the same for each column (also 15) – the square is said to be column-magic; and the sum of the entries on each of the two diagonals are the same (also 15) – the square is said to be diagonal-magic.

The magic square of the Melancholia woodcut is even more amazing than the Lo Shu square, as the sum 34 can be found in the rows, columns, diagonals, each of the quadrants, the center four squares, and the four corner squares, as well as the four corners of each of the four corner 3×3 squares. This sum can also be found in the four outer numbers clockwise from the corners $(3 + 8 + 14 + 9)$ and likewise the four counter-clockwise, the two sets of four symmetrical numbers $(2 + 8 + 9 + 15$ and $3 + 5 + 12 + 14)$, the sum of the middle two entries of the two outer columns and rows $(5 + 9 + 8 + 12$ and $3 + 2 + 15 + 14)$, and the four kite-shaped quartets $(3+5+11+15, 2+10+8+14, 3+9+7+15,$ and $2+6+12+14)$. The two numbers in the middle of the bottom row give the date of the engraving, which as we have mentioned above was 1514 AD.

11.2.1 Definition. *For any positive integer n, a semi-magic square of order n is an $n \times n$ array whose entries are the positive integers from 1 to n^2 (so each integer appearing once and only once) which is row and column-magic.*

Note that we may calculate the row-sum and the column-sum for an $n \times n$ magic square, since the sum of all n^2 values is $\sum_{i=1}^{n^2} i = \frac{n^2(n^2+1)}{2}$. In fact, in a row-magic square, since there are n rows and all row-sums are the same value, x say, then $nx = \frac{n^2(n^2+1)}{2}$ and so $x = \frac{n(n^2+1)}{2}$. Similarly, in a column-magic square, since all column-sums have the same value, each column sum is also equal to $\frac{n(n^2+1)}{2}$. Thus for an $n \times n$ semi-magic square, the row-sum and the column-sum are necessarily equal, since each is equal to $\frac{n(n^2+1)}{2}$.

11.2.2 Definition. *For any positive integer n, a (classical) magic square of order n is a semi-magic square of order n which is also diagonal magic, with the entries of each diagonal summing to $\frac{n(n^2+1)}{2}$ (the value of each row-sum and each column sum).*

For example, the Lo Shu square and Dürer's square are both magic squares, the first of order 3 and the second of order 4.

We describe now a procedure whereby a magic square of order $n \geq 1$ can be constructed from two orthogonal Latin squares over J_n. Let $I_n = \{0, 1, \ldots, n-1\}$, and define $f : I_n \times I_n \to I_{n^2-1}$ be the mapping defined by $f(r, s) = rn + s$ for each $(r, s) \in I_n \times I_n$. Suppose now that L_1 and L_2 are orthogonal Latin squares over J_n. Form orthogonal Latin squares H_1 and H_2 over I_n from L_1 and L_2, respectively, by subtracting 1 from each entry of L_1, respectively, L_2.

11.2.3 Proposition. *The $n \times n$ matrix $M = [f((H_1)_{ij}, (H_2)_{ij}) + 1]$ is a semi-magic square of order n.*

Proof. Since H_1 and H_2 are orthogonal Latin squares over I_n, each row (column) of H_1, respectively H_2, contains all entries of I_n. Thus for any $i \in J_n$, the sum of the entries in row i of M is

$$\sum_{j=1}^{n} (f((H_1)_{ij}, (H_2)_{ij}) + 1) = \sum_{j=1}^{n} ((H_1)_{ij} n + (H_2)_{ij} + 1)$$

$$= (\sum_{i=0}^{n-1} i) n + \sum_{i=0}^{n-1} i + n$$

$$= \frac{n^2(n-1)}{2} + \frac{n(n-1)}{2} + n$$

$$= \frac{n^3 - n^2 + n^2 - n + 2n}{2} = \frac{n^3 + n}{2}$$

$$= \frac{n(n^2+1)}{2},$$

as claimed. The argument that each column sum is $\frac{n(n^2+1)}{2}$ is essentially the same, and we therefore omit it. $\square$

In general, the semi-magic square that is obtained by this process is not a magic square. For example, if we use L_1 and L_2 from Example 11.1.8, the result is the non-magic semi-magic square shown in Figure 11.4 (i). However, if we use L_2 and L_3 from that example, we obtain the magic square shown in Figure 11.4 (ii).

15	12	5	2
10	13	4	7
8	3	14	9
1	6	11	16

10	13	4	7
7	3	14	9
15	12	5	2
1	6	11	16

(i) Semi-magic square   (ii) Magic square

Fig. 11.4   Semi-magic and magic squares

11.3   Exercises

1.   Construct four 8×8 MOLS, and use two of them to construct a magic square of order 8.

2. Arrange the four Jacks, the four Queens, the four Kings and the four Aces from a deck of cards in a 4×4 array so that each row and each column contains one card from each suit (Hearts, Diamonds, Clubs, Spades) and one card from each rank (Jack, Queen, King, Ace).

3. Find a magic square of order 5.

4. Quality control would like to find the best type of music to play to its assembly line workers in order to reduce the number of faulty products coming off the assembly line. As an experiment, a different type of music is played on four days in a week, while on the fifth day, no music is played. Design an experiment that will reduce the effect of the different days of the week.

5. Construct two MOLS of order 9.

6. Let m and n be positive integers, and suppose that $L = [\sigma_i]$ is a Latin square over J_n, and $M = [\tau_i]$ is a Latin square over J_m. Describe how to use L and M to construct a Latin square over $J_n \times J_m$.

7. Prove that a magic square of order 3 must have 5 at its centre.

8. A Latin square L is self-orthogonal if L and its transpose, L^t, are orthogonal.

   a) Prove that there is no 3×3 self-orthogonal Latin square.

   b) Give an example of a 4×4 self-orthogonal Latin square.

   c) Prove that the main diagonal in a self-orthogonal Latin square must have no duplicate entries (that is, if L is a self-orthogonal $n \times n$ Latin square, then for all $i, j \in J_n$, $L_{ii} = L_{jj}$ implies that $i = j$).

9. Let n be a positive integer, and suppose that n officers of n different ranks are chosen from each of n different regiments.

   a) Prove that if $n = 7$, it is possible to arrange the n^2 officers in an $n \times n$ array so that in each row and in each column, each rank and each regiment is represened exactly once.

   b) Prove that if $n = 6$, it is not possible to arrange the 36 chosen offices in such a manner.

10. A Latin square L over J_n is said to be in standard form, or in normal form, if the entries of the first row and the first column of L are in ascending order.

   a) Permute the rows and the columns of $L = \begin{bmatrix} 1\,3\,4\,2 \\ 3\,1\,2\,4 \\ 2\,4\,3\,1 \\ 4\,2\,1\,3 \end{bmatrix}$ to obtain a

Latin square M in standard form.

b) Find a 4×4 Latin square in standard form that is orthogonal to the Latin square M that you constructed in a).

11. This question refers to the mutually orthogonal Latin squares L_1, L_2, and L_3 that were constructed in Example 11.1.8. Let $\sigma, \tau, \lambda \in S_4$ be the permutations $\sigma = (1\,2\,3)$, $\tau = (1\,2)(3\,4)$, and $\lambda = (1\,3)(2\,4)$. Verify that $\{\sigma(L_1), \tau(L_2), \lambda(L_3)\}$ is a set of MOLS.

12. Find two 4×4 Latin squares each of which has $[1\,2\,3\,4]$ as its first row and $[2\,1\,4\,3]$ as its second row.

13. Prove that there is no 4×4 Latin square that is orthogonal to the Latin square shown at the right.
$$\begin{bmatrix} 1 & 2 & 3 & 4 \\ 2 & 3 & 4 & 1 \\ 3 & 4 & 1 & 2 \\ 4 & 1 & 2 & 3 \end{bmatrix}$$

Chapter 12

Group Actions, the Class Equation, and the Sylow Theorems

1 Group Actions

Examples of group actions abound in group theory. Conjugation, left (or right) multiplication by elements of a group, and multiplication of cosets by group elements are but a few of the specific examples of the general notion of a group action.

In this section, we define the concept of a group action and obtain some important results which will subsequently be used to prove basic results in the theory of groups.

Throughout this section, unless otherwise specified, X shall denote an arbitrary nonempty set, and G shall denote an arbitrary group. Recall that S_X denotes the symmetric group on X; that is, the group of all bijective mappings from X to itself with the group operation being composition of mappings.

12.1.1 Definition. *A (left) action of G on X is a homomorphism $\rho :$ $G \to S_X$. For each $g \in G$, the image of g under ρ shall be denoted by g^ρ. G is said to act on X via ρ if ρ is an action of G on X. Occasionally, we shall have need of the concept of a right action of G on X, by which is meant an antihomomorphism from G into S_X (for any groups G and H, an antihomomorphism $f : G \to H$ is a mapping for which $f(ab) = f(b)f(a)$ for all $a, b \in G$).*

As the need to use the notion of right action does not arise very often, we shall not introduce any special notation for a right action. Essentially every result that is proven below for left actions has its analogue for right actions, and we shall not remark on this any further.

Note that if ρ is an action of G on X, then $e_G^\rho = \mathbb{1}_X$, and for any

$g, h \in G$, $(gh)^\rho = g^\rho \circ h^\rho$.

12.1.2 Example.

(i) For any nonempty set X, the identity mapping from S_X to itself is an action of S_X on X.

(ii) Let G be a group, and let $X = Sub(G)$ denote the set of all subgroups of G. For each $g \in G$, let $g^\rho \in S_X$ denote the conjugation mapping defined by $g^\rho(H) = \{ gHg^{-1} \mid H \in X \}$. ρ is an action of G on $Sub(G)$, called the conjugation action.

(iii) For any group G, conjugation also provides an action of G on G, or on $\mathscr{P}(G)$.

(iv) Let G be a group, and let H be a subgroup of G. For $g \in G$, define $g^\rho : G/H \to G/H$ by $g^\rho(xH) = gxH$ for all $x \in G$. Then $g^\rho \in S_{G/H}$ for all $g \in G$, and $\rho : G \to S_{G/H}$ is an action of G on G/H.

We remark that it is not required that an action of a group G on a set X be injective; that is, it is entirely possible that the kernel of an action ρ of G on X may be nontrivial. For example, if G is a group, then the kernel of the conjugation action ρ of G on itself is $Z(G)$, the centre of G. Thus if G has nontrivial centre, the conjugation action of G on itself is not injective.

12.1.3 Definition. *Suppose that ρ is an action of G on X. If $\ker(\rho) = \{ e_G \}$, we say that ρ is a faithful action of G on X.*

Any action ρ of G on X can be made into a faithful action by an application of the first isomorphism theorem for groups. We obtain that $\hat{\rho} : G/\ker(\rho) \to S_X$ is a faithful action of $G/\ker \rho$ on X.

12.1.4 Notation. *Let ρ be an action of a group G on a set X. For $g \in G$ and $x \in X$, it is customary (when no ambiguity can result) to denote the result of applying g^ρ to x by gx, rather than $g^\rho(x)$. In fact, we shall often suppress any notation for the action homomorphism and simply say that the group G acts on the set X, with the result of applying $g \in G$ to $x \in X$ denoted by gx.*

12.1.5 Definition. *Let ρ be an action of a group G on a set X. Then for any $x \in X$, the set $Orb_G(x) = \{ gx \mid g \in G \}$ is called the orbit of x under the action. The set $O_\rho = \{ Orb_G(x) \mid x \in X \}$ is called the set of orbits of X under the action of G.*

12.1.6 Proposition. *Let ρ be an action of G on X. Then O_ρ is a partition of X.*

Proof. For every $x \in X$, $x \in Orb_G(x)$ since $e_G x = x$, and so $X = \cup_{x \in X} Orb_G(x)$, and for every $x \in X$, $Orb_G(x) \neq \varnothing$. It remains to prove that for any $x, y \in X$, if $Orb_G(x) \cap Orb_G(y) \neq \varnothing$, then $Orb_G(x) = Orb_G(y)$. Let $x \in X$, and suppose that $y \in Orb_G(x)$. Then there exists $g \in G$ such that $y = gx$. But then for every $h \in G$, $hy = h(gx) = (hg)x \in Orb_G(x)$, and thus $Orb_G(y) \subseteq Orb_G(x)$. Observe that $g^{-1}y = g^{-1}(gx) = (g^{-1}g)x = e_G x = x$, and thus $x \in Orb_G(y)$. As we have just shown, this implies that $Orb_G(x) \subseteq Orb_G(y)$, and thus $y \in Orb_G(x)$ implies that $Orb_G(y) = Orb_G(x)$. As a consequence, if $x, y \in X$ are such that $Orb_G(x) \cap Orb_G(y) \neq \varnothing$, say $z \in Orb_G(x) \cap Orb_G(y)$, then $Orb_G(x) = Orb_G(z) = Orb_G(y)$. $\qquad\square$

12.1.7 Definition. *Let ρ be an action of G on X. A subset Y of X is said to be closed under the action of ρ, or simply closed under the action, if for all $g \in G$, $gY \subseteq Y$ (where $gY = \{\, gy \mid y \in Y \,\}$).*

12.1.8 Proposition. *Let ρ be an action of G on X. A subset Y of X is closed under ρ if, and only if, Y is a union of orbits of X under ρ.*

Proof. It is evident that each orbit of X is closed under the action, and thus the union of orbits of X is closed under the action. For the converse, suppose that $Y \subset X$ is closed under the action. It suffices to prove that for every $y \in Y$, $Orb_G(y) \subseteq Y$. Let $y \in Y$, and let $x \in Orb_G(y)$. Then there exists $g \in G$ such that $gy = x$, and since Y is closed under the actin and $y \in Y$, it follows that $x \in Y$. Thus $Orb_G(y) \subseteq Y$. $\qquad\square$

If ρ is an action of G on X, and $Y \subseteq X$ is closed under the action of ρ, then ρ induces an action of G on Y; namely for each $g \in G$, $g^\rho\big|_Y \in S_Y$. To see this, observe that for any $g \in G$, both $gY \subseteq Y$ and $g^{-1}Y \subseteq Y$ hold. Thus $gY \subseteq Y \subseteq gY$ and so $g^\rho\big|_Y$ is surjective.

12.1.9 Definition. *An action ρ of G on X is said to be transitive if $|O_\rho| = 1$. We also say that G acts transitively on X (by ρ).*

In particular, if ρ is an action of G on X, then ρ induces a transitive action of G on $Orb_G(x)$ for each $x \in X$.

Up to this point, we have not placed any restrictions on either X or G. **For the rest of this section, we shall assume that X is finite.** Under this assumption, we are able to obtain some important information about the sizes of the orbits of G.

12.1.10 Definition. *Let ρ be an action of G on X. For any $x \in X$, the stabilizer of x, denoted by G_x, is the set of all elements of G that fix (or stabilize) x under ρ; that is, $G_x = \{ g \in G \mid gx = x \}$.*

12.1.11 Proposition. *If ρ is an action of G on X, then $[G : G_x] = |Orb_G(x)|$ for any $x \in X$.*

Proof. Let $x \in X$, and define $f : G/G_x \to Orb_G(x)$ by $f(gG_x) = gx$. That f is a mapping follows from the fact that if $gG_x = hG_x$, then $g^{-1}h \in G_x$ and so $g^{-1}hx = x$, giving $gx = hx$. Moreover, if $g_1, g_2 \in G$ are such that $f(g_1) = f(g_2)$, then $g_1 x = g_2 x$ and thus $g_1^{-1} g_2 \in G_x$; that is, $g_1 G_x = g_2 G_x$. As a result, f is injective. Finally, f is surjective, since for any $y \in Orb_G(x)$, there is $g \in G$ such that $y = gx$, and thus $f(gG_x) = gx = y$. This completes the proof that f is a bijective mapping from G/G_x onto $Orb_G(x)$. $\square$

12.1.12 Corollary. *If a group G acts transitively on the set X, then $[G : G_x] = |X|$ for every $x \in X$.*

Proof. This follows from Proposition 12.1.11 and the fact that since ρ is transitive, $Orb_G(x) = X$ for every $x \in X$. $\square$

12.1.13 Notation. *If ρ is an action of G on X, then for every $g \in G$, the fixed-point set of g under ρ is denoted by $Fix_\rho(g)$, or simply by $Fix(g)$ if no ambiguity will result, and is given by $Fix(g) = \{ x \in X \mid gx = x \}$.*

In Chapters 15 and 16, we shall use the following result, usually called Burnside's lemma, to great effect. The lemma appeared in William Burnside's foundational 1897 book entitled "On the Theory of Groups of Finite Order", in which he attributed the result to Ferdinand Georg Frobenius as the result had appeared in a paper published by Frobenius 10 years earlier. In fact, there is evidence that the result was known to Augustin-Louis Cauchy and others as early as 1845, and the lemma is sometimes humorously referred to as the "lemma that is not Burnside's".

12.1.14 Proposition (Burnside's lemma). *Let ρ be an action of a finite group G on X, and suppose that $n = |O_\rho|$. Then $|G| = \frac{1}{n} \sum_{g \in G} |Fix(g)|$.*

Proof. Let $S = \{(g, x) \mid g \in G,\ x \in X,\ \text{and}\ gx = x\}$. We determine $|S|$ via two different methods, from which the result will follow. First, $S = \cup_{x \in X} \{(g, x) \mid g \in G,\ gx = x\}$, and the union is disjoint, so

$$|S| = \sum_{x \in X} |\{(g, x) \mid g \in G,\ gx = x\}| = \sum_{x \in X} |G_x|.$$

By Proposition 12.1.11, for each $x \in X$, $|Orb_G(x)| = [G : G_x] = |G|/|G_x|$. Thus for each $x \in X$, and any $y \in Orb_G(x)$, since $Orb_G(y) = Orb_G(x)$, we have $|G_y| = |G|/|Orb_G(y)| = |G|/|Orb_G(x)|$, and so $\sum_{y \in Orb_G(x)} |G_y| = |G|$. Now, by Proposition 12.1.6, X is partitioned into the n orbits of the action of ρ, and so we obtain $|S| = \sum_{x \in X} |G_x| = n|G|$. On the other hand,

$$S = \cup_{g \in G} \{(g, x) \mid x \in X,\ gx = x\},$$

and this union is disjoint, so

$$|S| = \sum_{g \in G} |\{(g, x) \mid x \in X,\ gx = x\}| = \sum_{g \in G} |Fix(g)|.$$

$\square$

Observe that Burnside's lemma tells us that if a finite group G acts transitively on a finite set X (so $n = 1$ in Burnside's lemma), if $|Fix(g)| \geq 1$ for every $g \in G$, then $Fix(g)| = 1$ for every $g \in G$. In particular, in such a case we would have $|X| = |Fix(e_G)| = 1$. Thus if $|X| > 1$ and G acts transitively on X, then for at least one $g \in G$, $Fix(g) = \varnothing$.

2   The Class Equation of a Finite Group

The class equation of a finite group relates the order of the group to the orders and indices of certain subgroups of the group, often affording us useful information about the group. As we shall see, it is a result which is often used in inductive arguments.

Recall that for any group G, and $g \in G$, the centralizer of g in G is $C_G(g) = \{x \in G \mid gx = xg\}$.

12.2.1 Proposition (The Class Equation of a Group). *Let G be a finite group, and consider the conjugation action of G on itself (that is, the action $\rho : G \to S_G$ given by $g^\rho(x) = gxg^{-1}$ for all $g, x \in G$). Then the following hold.*

(i) For each $g \in G$, $G_g = C_G(g)$, and $|Orb_G(g)| = [G : G_g] = [G : C_G(g)]$.

(ii) Let $T \subseteq G$ be a set formed by choosing one element from each nontrivial orbit of the conjugation action. Then

$$|G| = |Z(G)| + \sum_{g \in T} [G : C_G(g)].$$

Proof. The first assertion follows immediately from the definition of G_g and Proposition 12.1.11. For the second assertion, recall that the set of orbits of an action is a partition of the set being acted upon, and the orbit of $g \in G$ under the conjugation action will be trivial if, and only if, $C_G(g) = G$; that is, $g \in Z(G)$. $\qquad \square$

12.2.2 Corollary. *A finite group of prime-power order has nontrivial centre.*

Proof. If G has trivial centre, then for each $g \in G$ different from the identity element, $C_G(g) \neq G$, and thus $[G : C_G(g)] \neq 1$. Since $[G : C_G(g)] = |G|/|C_G(g)|$, it follows from the fact that $|G| = p^k$ for some prime p and some positive integer k that p is a divisor of $[G : C_G(g)]$ for every $g \in G$ different from the identity. By the class equation for G, we have $|G| = 1 + pM$ for some positive integer M, and so p must divide 1. Since this is not the case, we conclude that G must have nontrivial centre. $\qquad \square$

12.3   The Sylow Theorems

The reader may recall that in Chapter 7, we proved Cauchy's theorem (Theorem 7.5.2), which stated that if G is any finite group whose order is divisible by a prime p, then G contains an element of order p; equivalently, G contains a subgroup of order p. Our first result in this section is often called the first Sylow theorem, and it provides the strongest possible extension of Cauchy's theorem. The class equation of the group plays a fundamental role in the proof.

12.3.1 Theorem (Sylow's First Theorem). *Let p be a prime, m a positive integer, and G a finite group whose order is divisible by p^m. Then G contains a subgroup of order p^m.*

Proof. The proof will be by induction on the order of the group, with base case the trivial group, for which the result holds vacuously. Suppose now that $n > 1$ is an integer for which the assertion holds for all finite groups of order less than n, and let G be a group of order n. If $|G|$ is not divisible by p, there is nothing to do, so suppose that $|G|$ is divisible by p^m for some positive integer m. By the class equation, we have $|G| = |Z(G)| + \sum_{g \in T} [G : C_G(g)]$ where T is obtained by choosing one element from each nontrivial conjugation orbit. Since p^m is a divisor of $|G|$, we have $|G| = p^m k$ for some positive integer k.

Case 1: p divides $|Z(G)|$. By Cauchy's theorem, $Z(G)$ contains a subgroup H of order p, and since every element of H commutes with each element of G, $H \triangleleft G$. Let $\pi : G \to G/H$ denote the canonical projection mapping (so $\pi(g) = gH$ for each $g \in G$). Now, $|G/H| = |G|/|H| = p^m k/p = p^{m-1} k < n$, so by hypothesis, G/H has a subgroup L of order p^{m-1}. But then $\pi^{-1}(L)$ is a subgroup of G of order $|L| \, |H| = (p^{m-1})(p) = p^m$.

Case 2: p does not divide $|Z(G)|$. If p were a factor of $[G : C_G(g)]$ for each $g \in T$, then it would follow that p is a factor of $|Z(G)|$, which is not the case. Thus for some $g \in T$, $[G : C_G(g)]$ is not divisible by p. Since $[G : C_G(g)] = |G|/|C_G(g)|$, we have p^m divides $|G| = [G : C_G(g)] \, |C_G(g)|$ and $(p^m, [G : C_G(g)]) = 1$, so it follows from Proposition 4.4.3 that p^m divides $|C_G(g)|$. As $|C_G(g)| < |G| = n$, our inductive hypothesis asserts that $C_G(g)$ (and hence G) contains a subgroup of order p^m. The result follows now by the second principle of mathematical induction. $\qquad \square$

Recall that the alternating group A_4 is a group of order 12 with no subgroup of order 6. Thus it is possible to have a finite group G and a positive divisor m of the order of the group, yet the group has no subgroup of order m. The first Sylow theorem is the closest thing to a converse of Lagrange's theorem that can be proved for arbitrary finite groups. There are stronger theorems which hold for special classes of groups, and we shall soon prove one such theorem for the class of finite abelian groups. In fact, the full converse of Lagrange's theorem holds for this class.

12.3.2 Definition. *Let G be a finite group, and let p be a prime divisor of $|G|$. A subgroup H of G is said to be a p-subgroup of G if $|H| = p^m$ for some positive integer m. If $|G| = p^m k$ and $(p, k) = 1$, then a subgroup of G of order p^m is called a Sylow p-subgroup of G.*

By Theorem 12.3.1, every finite group whose order is divisible by a prime p has at least one Sylow p-subgroup.

The reader may recall that in Chapter 6, it was shown (see Theorem 6.4.15) that for any group G and subgroup K of G and any subgroup H of $N_G(K)$, HK is a subgroup of $N_G(K)$ containing K as a normal subgroup, $H \cap K \triangleleft H$, and $HK/K \simeq H/(H \cap K)$, from which we may deduce that if

G is finite, then $|HK| = |H|\,|K|/|H \cap K|$.

12.3.3 Proposition. *Let G be a finite group whose order is divisible by the prime p, and let S be a Sylow p-subgroup of G. If P is a p-subgroup of G such that either $P \subseteq N_G(S)$ or $S \subseteq N_G(P)$, then $P \subseteq S$. In particular, if S_1 and S_2 are Sylow p-subgroups of G such that $S_1 \subseteq N_G(S_2)$, then $S_1 = S_2$.*

Proof. Suppose that P is a p-subgroup of G such that either $P \subseteq N_G(S)$ or $S \subseteq N_G(P)$. Then by Theorem 6.4.15, under either hypothesis, it follows that PS is a subgroup of G whose order is a divisor of $|P|\,|S|$, a power of p. Since both $P \subseteq PS$ and $S \subseteq PS$ hold, the maximality of the order of S asserts that $S = PS$, and thus $P \subseteq S$.

If S_1 and S_2 are p-Sylow subgroups of G such that $S_1 \subseteq N_G(S_2)$, then $S_1 \subseteq S_2$, and since $|S_1| = |S_2|$, it follows that $S_1 = S_2$. $\qquad\square$

12.3.4 Notation. *Let G be a finite group, and let p be a prime divisor of $|G|$. The number of Sylow p-subgroups of G shall be denoted by $n_p(G)$.*

12.3.5 Theorem (Sylow's Second Theorem). *Let G be a finite group and let p be a prime divisor of $|G|$. Then $n_p(G) \equiv 1 \mod p$.*

Proof. Let $\mathscr{S}$ denote the set of all Sylow p-subgroups of G, and choose $S \in \mathscr{S}$. S acts on $\mathscr{S}$ by conjugation, and since S is closed with respect to inverses and multiplication, $\{S\}$ is one of the orbits of the conjugation action of S on $\mathscr{S}$. Let $T \in \mathscr{S}$. The stabilizer of T under the conjugation action by S is $N_S(T)$. By Proposition 12.1.11, the size of the orbit of T is $[S : N_S(T)]$. If this is 1, then $S = N_S(T) \subseteq N_G(T)$, in which case Proposition 12.3.3 applies and we obtain $T = S$. If $T \neq S$, then we must have $[S : N_S(T)] > 1$, in which case p divides $[S : N_S(T)]$. Thus the sum over all orbits of the orbit size is an integer congruent to 1 modulo p. Since the set of all orbits of the conjugation action of S on $\mathscr{S}$ is a partition of $\mathscr{S}$, it follows that $n_p(G) = |\mathscr{S}| \equiv 1 \mod p$. $\qquad\square$

12.3.6 Corollary (Sylow). *Let G be a finite group, and let p be a prime divisor of $|G|$. Every p-subgroup of G is contained in a Sylow p-subgroup of G.*

Proof. Let $\mathscr{S}$ denote the set of all Sylow p-subgroups of G, and let P be a p-subgroup of G. Then P acts by conjugation on $\mathscr{S}$, and for any $S \in \mathscr{S}$, the stabilizer of S under the conjugation action is $N_P(S) = N_G(S) \cap P$. By

Proposition 12.1.11, the size of the orbit of S is $[P : N_P(S)]$, and since P is a p-subgroup, the size of the orbit of S is a power of p, possibly 1. Since $|\mathscr{S}|$ is the sum of the orbit sizes, they cannot all be greater than 1 since by Theorem 12.3.5, $|\mathscr{S}| = n_p(G) \equiv 1 \mod p$. Thus there is at least one Sylow p-subgroup S of G such that $P = N_P(S) \subseteq N_G(S)$. By Proposition 12.3.3, $P \subseteq S$. $\qquad\square$

The third theorem of Sylow establishes the somewhat surprising fact that any two Sylow p-subgroups of a finite group G are conjugate in G.

12.3.7 Theorem (Sylow's Third Theorem). *Let G be a finite group and p a prime divisor of $|G|$. Then any two Sylow p-subgroups of G are conjugate. Moreover, for any Sylow p-subgroup S of G, $n_p(G) = [G : N_G(S)]$, and so $n_p(G)$ is a divisor of $|G|/|S|$.*

Proof. Let $\mathscr{S}$ denote the set of all Sylow p-subgroups of G, and consider the conjugation action of G on $\mathscr{S}$. For any $S \in \mathscr{S}$, $Orb_G(S)$ has size $[G : N_G(S)]$, which is relatively prime to p since $S \subseteq N_G(S)$. Let $T \in \mathscr{S}$, and consider the conjugation action of T on $Orb_G(S)$. The orbit of S under the action of T on $Orb_G(S)$ has size $[T : N_T(S)]$, and since T is a p-subgroup, $[T : N_T(S)]$ is either 1 or a power of p. Since $|Orb_G(S)|$ is not a multiple of p, it follows that there exists $P \in Orb_G(S)$ whose orbit under the conjugation action of T has size 1; that is, $T = N_T(P) \subseteq N_G(P)$. By Proposition 12.3.3, $T = P$, and so $\mathscr{S} = Orb_G(S)$. Thus any two Sylow p-subgroups of G are conjugate in G. Moreover, $n_p(G) = |\mathscr{S}| = |Orb_G(S)| = [G : N_G(S)] = |G|/|N_G(S)|$, and since $S \subseteq N_G(S)$, it follows that $n_p(G)$ is a divisor of $|G|/|S|$. $\qquad\square$

4 Applications of the Sylow Theorems

In this section, we obtain structure theorems for some of the simpler finite groups. We remind the reader that in Proposition 5.5.8 (i), it was established that a finite cyclic group has subgroups of every possible size (that is, for each positive divisor of the order of the cyclic group, there is a subgroup of that size). Our first result of this section uses the Sylow theorems to extend this observation to all finite abelian groups.

12.4.1 Proposition. *If G is a finite abelian group, then for every positive divisor m of $|G|$, G has at least one subgroup of order m.*

Proof. We first observe that by Theorem 6.4.15, if H and K are normal subgroups of a finite group G such that $H \cap K = \{ e_G \}$, then HK is a subgroup of G of order $|H||K|$. Note that if $|H|$ and $|K|$ are relatively prime, then $H \cap K = \{ e_G \}$. As well, in an abelian group, every subgroup is normal. Suppose now that the prime factorization of $|G|$ is $\prod_{i=1}^{k} p_i^{r_i}$. By Theorem 12.3.1, G has at least one p_i-Sylow subgroup for each prime p_i, $i = 1, 2, \ldots, k$, and any p_i-Sylow subgroup of G has order $p_i^{r_i}$. By Theorem 12.3.7, any two p_i-Sylow subgroups of G are conjugate, and since G is abelian, it follows that G has a unique p_i-Sylow subgroup P_i for each $i = 1, 2, \ldots, k$. As well, by Theorem 12.3.1, for each i, P_i contains a subgroup of order $p_i^{s_i}$ for every integer s_i for which $0 \le s_i \le r_i$. Every positive divisor of $|G| = \prod_{i=1}^{k} p_i^{r_i}$ is of the form $\prod_{i=1}^{k} p_i^{s_i}$ where for each i, $0 \le s_i \le r_i$. Let m be a positive divisor of $|G|$, and write $m = \prod_{i=1}^{k} p_i^{s_i}$, where for each $i = 1, 2, \ldots, k$, $0 \le s_i \le r_i$. Now for each $i = 1, 2, \ldots, k$, let H_i be a subgroup of P_i of order $p_i^{s_i}$. We prove by induction on $l \ge 1$ that either $l > k$ or else $|H_1 H_2 \cdots H_l|$ has order $\prod_{i=1}^{l} p_i^{s_i}$, where the base case holds by choice of H_1. Suppose that $l \ge 1$ is such that the hypothesis holds. If $l \ge k$, there is nothing more to do, so suppose that $l < k$. We have $|H_1 H_2 \cdots H_l| = \prod_{i=1}^{l} p_i^{s_i}$, which is relatively prime to $p_{l+1}^{s_{l+1}}$, and so $(H_1 H_2 \cdots H_l) H_{l+1}$ has order $|H_1||H_2| \cdots |H_l||H_{l+1}|$, as required. The claim follows now by induction, and so $H_1 \cdots H_k$ is a subgroup of G of order $m = \prod_{i=1}^{k} p_i^{s_i}$. □

Recall that in Chapter 6, a group G was said to be simple if the only normal subgroups of G are the trivial subgroup and G itself. It was also observed there that the only finite abelian simple groups are the cyclic groups of prime order, so up to isomorphism, just the groups $\mathbb{Z}_p$ under addition, p a prime.

It should be pointed out that simple groups are, in fact, not simple at all (in structure)! If a group contains a nontrivial normal subgroup, then one can hope to obtain useful information about the group by studying the smaller, and likely less complicated group that results when the normal subgroup is factored out. On the other hand, in a simple group, this approach is not available to us.

Group theorists have shown that every finite nonabelian simple group is either alternating or else it belongs to one of 16 completely identified families of simple groups, or is one of 26 so-called sporadic simple groups.

We have already establised (see Section 7.5 of Chapter 7) that A_n is simple for every integer $n \ge 5$. One of the results of this section is to establish that A_5 is the smallest nonabelian simple group. The finite nonabelian simple groups other than the alternating groups are too difficult to investigate with just the tools that we have developed. One needs the machinery of group representations, a powerful tool in the study of groups. Simple groups often occur in connection with some geometric structure, and one is often able to deduce properties of such groups from the geometric structure.

12.4.2 Proposition. *Let G be a group (not necessarily finite), and let G act by left multiplication on the set of left cosets of a subgroup H of G. The kernel K of this action is $\bigcap_{g \in G} gHg^{-1}$. If $[G:H]$ is finite, say $n = [G:H]$, then G/K is isomorphic to a subgroup of S_n.*

Proof. $g \in K$ if, and only if, for every $x \in G$, $gxH = xH$; equivalently, $x^{-1}gx \in H$, or $g \in xHx^{-1}$. The second observation follows from the first isomorphism theorem and the fact that $|\{\, xH \mid x \in G \,\}| = [G:H]$. $\quad\square$

12.4.3 Corollary. *Let G be a finite group, and let p be the smallest prime divisor of $|G|$. If H is a subgroup of G with $[G:H] = p$, then $H \triangleleft G$.*

Proof. Let $K = \bigcap_{g \in G} gHg^{-1}$. Then $K \subseteq H$, $K \triangleleft G$, and by Proposition 12.4.2, G/K is isomorphic to a subgroup of S_p. Thus $|G|/|K|$ is a divisor of $p!$, and since p is the smallest prime that divides $|G|$, it follows that $|G|/|K|$ divides p. Since $K \subseteq H$ and H is a proper subgroup of G, $|G|/|K| > 1$ and thus $|G|/|K| = p = |G|/|H|$; that is, $|K| = |H|$. Thus $H = K \triangleleft G$. $\quad\square$

Note that 2 is the smallest prime, so if G contains a subgroup H of index 2, the preceding result applies and so H is normal in G. We established this fact earlier by elementary means (see Lemma 6.4.2).

We have now completed our development of the tools that will allow us to obtain structure theorems for some classes of groups.

12.4.4 Proposition. *Let p and q be primes with $p < q$, and let G be a group of order pq. Let P be a Sylow p-subgroup of G, and let Q be a Sylow q-subgroup of G. Then P and Q are cyclic, $Q \triangleleft G$, and $G = PQ$.*

  (i) If p does not divide $q - 1$, then $P \triangleleft G$ and G is a cyclic group.
  (ii) If p divides $q - 1$, then for any generator a of P and generator b of Q, there exists an integer m with $1 \leq m \leq q - 1$ such that $aba^{-1} = b^m$ and $m^p \equiv 1 \bmod q$. Moreover, in this case G is abelian if, and only if, $m = 1$, in which case G is cyclic.

Proof. Since $n_q(G)$ must divide p and be congruent to 1 modulo q, it follows that $n_q(G) = 1$ and thus $Q \triangleleft G$ and $|PQ| = |P|\,|Q|/|P \cap Q| = pq = |G|$.

As well, $n_p(G) \equiv 1 \mod p$ and $n_p(G)$ divides q. Thus for some positive integer m, $n_p(G) = 1 + mp$ and so $1 + mp$ divides q. Thus either $1 + mp = 1$ or $1 + mp = q$; that is, either $n_p(G) = 1$, or else p divides $q - 1$.

For (i), suppose that p does not divide $q - 1$. Then $n_p(G) = 1$ and so P is normal in G. Since $P \cap Q = \{e_G\}$, Lemma 6.4.12 asserts that each of P and Q centralizes the other, and thus $G = PQ$ is abelian, hence isomorphic to the direct product of P and Q, a cyclic group.

For (ii), suppose that p divides $q - 1$. In this case, $n_p(G) = 1$ or $n_p(G) = q$ are possible. Let a and b be generators for P and Q, respectively. If $n_p(G) = 1$, the argument for (i) applies and G is cyclic, so $aba^{-1} = b$ (that is, $m = 1$). Suppose that $n_p(G) = q$. Then P is not normal in G, and so G is not abelian. Since $aba^{-1} \in Q$, there exists a positive integer m such that $aba^{-1} = b^m$, and since $ab = ba$ would imply that G is abelian, it follows that $m > 1$. Now, $b = e_G b e_G = a^p b a^{-p} = b^{m^p}$, so $m^p \equiv 1 \mod q$, as required. $\qquad\square$

12.4.5 Proposition. *Every group of order 15 is cyclic, and every group G of order 30 contains a normal subgroup of order 15, and $n_3(G) = n_5(G) = 1$.*

Proof. For a group of order $15 = (3)(5)$, Proposition 12.4.4 (i) applies, and so G is cyclic. Suppose now that G is a group of order 30. Then $n_3(G) \equiv 1 \mod 3$ and is a divisor of 10, so $n_3(G) = 1$ and thus G contains a unique Sylow 3-subgroup $P \triangleleft G$. Let Q be a Sylow 5-subgroup of G. Then PQ is a subgroup of G of order $|P|\,|Q|/|P \cap Q| = 15$, and so PQ is cyclic. But then $PQ \subseteq N_G(Q)$, and so 15 is a divisor of $|N_G(Q)|$. Since $|G|/|N_G(Q)| = n_5 \equiv 1 \mod 5$, we see that $|N_G(Q)| = 30$, and so $N_G(Q) = G$; that is, $Q \triangleleft G$. It now follows that $PQ \triangleleft G$. $\qquad\square$

There is a very nice group-theoretic construction that should be viewed as a generalization of the direct product. This construction is called the semidirect product, and we discuss it next. It will be very useful for the construction of certain types of groups.

12.4.6 Definition. *Let G and H be groups and $\varphi : G \to \mathrm{Aut}(H)$ be a homomorphism. On the Cartesian product $H \times G$, define a binary operation by*

$$(h, g)(h_1, g_1) = (h\varphi_g(h_1), gg_1)$$

for all $h, h_1 \in H$ and $g, g_1 \in G$, where φ_g denotes $\varphi(g): H \to H$. It will be shown that $H \times G$ with this binary operation is a group, called the semidirect product of H by G (via φ), denoted by $H \rtimes_\varphi G$.

12.4.7 Proposition. *For any groups G and H, and any homomorphism $\varphi: G \to \mathrm{Aut}(H)$, $H \times G$, with the binary operation introduced in Definition 12.4.6, is a group.*

Proof. First, let us prove that the operation is associative. Let $(h, g), (h_1, g_1), (h_2, g_2) \in H \times G$. Then

$$(h, g)[(h_1, g_1)(h_2, g_2)] = (h, g)(h_1 \varphi_{g_1}(h_2), g_1 g_2) = (h \varphi_g(h_1 \varphi_{g_1}(h_2)), g(g_1 g_2))$$

$$= (h \varphi_g(h_1) \varphi_g \varphi_{g_1}(h_2), (gg_1)g_2)) \quad \text{since } \varphi_g \text{ is a homomorphism}$$

$$= (h \varphi_g(h_1) \varphi_{gg_1}(h_2), (gg_1)g_2)) \quad \text{since } \varphi \text{ is a homomorphism}$$

$$= (h \varphi_g(h_1), gg_1)(h_2, g_2)) = [(h, g)(h_1, g_1)](h_2, g_2)$$

Next, observe that for any $(h, g) \in H \times G$, we have $(e_H, e_G)(h, g) = (e_H \varphi_{e_g}(h), e_g g)$ and since φ is a homomorphism, it follows that φ_{e_G} is the identity of $\mathrm{Aut}(H)$ and so $\varphi_{e_G}(h) = h$. This establishes that $(e_H, e_G)(h, g) = (e_H \varphi_{e_g}(h), e_g g) = (h, g)$, and so (e_H, e_G) is a left identity for the binary operation.

Finally, note that for all $h \in H$, $g \in G$,

$$(\varphi_{g^{-1}}(h^{-1}), g^{-1})(h, g) = (\varphi_{g^{-1}}(h^{-1}) \varphi_{g^{-1}}(h), g^{-1} g)$$

$$= (\varphi_{g^{-1}}(h^{-1} h), e_G) \quad \text{since } \varphi_{g^{-1}} \text{ is a homomorphism}$$

$$= (\varphi_{g^{-1}}(e_H), e_G)$$

$$= (e_H, e_G) \quad \text{since } \varphi_{g^{-1}} \text{ is a homomorphism,}$$

and so every $(h, g) \in H \times G$ has a left inverse with respect to the left identity (e_H, e_G). By Lemma 5.3.1, with this binary operation, $H \times G$ is a group. $\square$

Let $\iota_H: H \to H \rtimes_\varphi G$ and $\iota_G: G \to H \rtimes_\varphi G$ be the mappings defined by $\iota_H(h) = (h, e_G)$ and $\iota_G(g) = (e_H, g)$ for all $h \in H$ and $g \in G$.

12.4.8 Lemma. *ι_H and ι_G are injective homomorphisms from H and G, respectively, into $H \rtimes_\varphi G$.*

Proof. Let $h_1, h_2 \in H$. Then $\iota_H(h_1) \iota_H(h_2) = (h_1, e_G)(h_2, e_G)$, and since φ is a homomorphism, φ_{e_G} is the identity automorphism of H, so $(h_1, e_G)(h_2, e_G) = (h_1 h_2, e_G) = \iota_H(h_1 h_2)$. Thus $\iota_H(h_1) \iota_H(h_2) =$

$\iota_H(h_1 h_2)$. Now let $g_1, g_2 \in G$. Then $\iota_G(g_1)\iota_G(g_2) = (e_H, g_1)(e_H, g_2) = (e_H \varphi_{g_1}(e_H), g_1 g_2) = (e_H, g_1 g_2) = \iota_G(g_1 g_2)$ since φ_{g_1} is an automorphism of H. Thus ι_G and ι_H are homomorphisms, each evidently injective. $\square$

Note that for any $h \in H$ and $g \in G$, we have $(e_H, g)(h, e_G)(e_H, g)^{-1} = (\varphi_g(h), e_G)$. Thus in $H \rtimes_\varphi G$, the action of G on H may be recovered as the action of $\iota_G(G)$ on $\iota_H(H)$ by conjugation. More precisely, let $H' = \iota_H(H)$ and $G' = \iota_G(G)$. Note that $(h, g) \in H' \cap G'$ implies that $h = e_H$ and $g = e_G$, so $H' \cap G' = \{(e_H, e_G)\}$. Further observe that for $(h, g) \in H \rtimes_\varphi G$, $(h, e_G)(e_H, g) = (h \varphi_{e_G}(e_H), e_G g) = (h, g)$, so $H'G' = H \rtimes_\varphi G$.

We shall identity $h \in H$ with $\iota_H(h) \in H \rtimes_\varphi G$ and $g \in G$ with $\iota_G(g) \in H \rtimes_\varphi G$, so that $(h, g) = (h, e_G)(e_H, g)$ shall be denoted by hg, and accordingly, we shall identify H with H' and G with G', so that $H \rtimes_\varphi G = HG$. Note that under this identification, both e_G and e_H are identified with (e_H, e_G), and we shall simply let e denote (e_H, e_G). Then $H \cap G = \{e\}$. Furthermore, note that for any $h, k \in H$ and any $g \in G$, we have $(gk)(h)(gk)^{-1} = g(khk^{-1})g^{-1} \in gHg^{-1}$, so it will follow that H is normal in $H \rtimes_\varphi G$ if we prove that for all $h \in H$ and $g \in G$, $ghg^{-1} \in H$. But as shown above, we have $ghg^{-1} = \varphi_g(h) \in H$. Thus $H \triangleleft H \rtimes_\varphi G$.

12.4.9 Example. *Recall that* $\mathrm{Aut}(C_3)$ *is a cyclic group of order 2 (let* $C_3 = \langle h \rangle$*, so that* $\mathrm{Aut}(C_3) = \left\langle \left(\begin{smallmatrix} h & h^2 & e \\ h^2 & h & e \end{smallmatrix} \right) \right\rangle$ *), so we obtain a surjective homomorphism* $\varphi : C_4 = \langle g \rangle \to \mathrm{Aut}(C_3)$ *by sending* g *to* $\left(\begin{smallmatrix} h & h^2 & e \\ h^2 & h & e \end{smallmatrix} \right)$*; that is,* $\varphi_g(h) = h^2$*. All 12 elements of* $C_3 \rtimes_\varphi C_4$ *can be uniquely represented in the form* $h^i g^j$ *for* $i = 0, 1, 2$ *and* $j = 0, 1, 2, 3$*, with multiplication completely determined by the rules* $g^4 = h^3 = e$ *and* $gh = h^2 g$*. Note that the latter can be written as* $ghg^{-1} = h^2$*, and so for any positive integer* i*,* $g^i h g^{-i} = h^{2^i}$*. We shall identify the central elements of* $C_3 \rtimes_\varphi C_4$*. Suppose that* $h^i g^j$ *is central, where* $0 \le i \le 2$*,* $0 \le j \le 3$*. Then in particular,* $h^i g^j h = h h^i g^j$*, so* $h^{2^j + i} g^j = h^{i+1} g^j$ *and thus 3 divides* $2^j - 1$*. This is only possible when* $j = 0$ *or 2. Consider first the case when* $j = 0$*. Suppose that* h^i *is central. Then* $gh^i g^{-1} = h^i$*, and so* $h^{2i} = h^i$*, which implies that 3 divides* i*. Since* $0 \le i \le 2$*, it follows that* $i = 0$*. Now,* $i = 0$ *and* $j = 0$ *yields the identity element, which is indeed central. Next, consider the case when* $j = 2$*. Suppose that* $h^i g^2$ *is central. Then* $gh^i g^2 = h^i g^2 g$ *and so* $h^{2i} g^3 = h^i g^3$*, which implies that* $h^i = e$*. It remains only to verify that* g^2 *is central, and it suffices to verify that* g^2 *commutes with* h*. We have* $g^2 h = h^4 g^2 = h g^2$*, so* g^2 *is central. Thus* $Z(C_3 \rtimes_\varphi C_4) = \{e, g^2\}$*.*

As one final calculation in this example, we determine the order of each *of the twelve elements. Of course, e has order 1, h and h^2 have order 3, g and g^3 have order 4, and g^2 has order 2. Let us examine the remaining 6 elements hg, hg^2, hg^3, h^2g, h^2g^2, and h^2g^3. We find that $hghg = h^3g^2 = g^2$. so $(hg)^4 = g^4 = e$ and thus hg has order 4, as does $(hg)^3 = g^2hg = hg^3$. Since h and g^2 commute and have relatively prime orders, it follows that hg^2 has order 6, and $(hg^2)^5 = h^2g^2$ also has order 6. Finally, $h^2gh^2g = h^6g^2 = g^2$, so h^2g has order 4 as does $(h^2g)^3 = g^2h^2g = h^2g^3$.*

12.4.10 Proposition. *Let H and K be groups, and let $\varphi, \psi : H \to \text{Aut}(K)$ be homomorphisms. If there is $\alpha \in \text{Aut}(H)$ such that $\psi = \varphi \circ \alpha$, then the bijective mapping f from the Cartesian product $K \times H$ to itself given by $f(k, h) = (k, \alpha(h))$ is an isomorphism from $K \rtimes_\psi H$ to $K \rtimes_\varphi H$.*

Proof. We have

$$f((k, h)(k_1, h_1)) = f(k\psi_h(k_1), hh_1) = (k\psi_h(k_1), \alpha(hh_1))$$
$$= (k\varphi_{\alpha(h)}(k_1), \alpha(h)\alpha(h_1))$$
$$= (k, \alpha(h))(k_1, \alpha(h_1)) = f(k, h)f(k_1, h_1).$$

$\square$

12.4.11 Example. *Since $\text{Aut}(C_3) \simeq C_2$, there is a one-to-one correspondence between the set of homomorphisms from $C_2 \times C_2$ to $\text{Aut}(C_3)$ and the set of homomorphisms from $C_2 \times C_2$ to C_2. Every homomorphism other than the trivial homomorphism (which provides us with the direct product $C_3 \times (C_2 \times C_2)$), is surjective, so will have a kernel of size 2. Moreover, since $C_2 \times C_2$ is abelian, every subgroup is normal. Since there are three subgroups of size 2, there are three surjective homomorphisms from $C_2 \times C_2$ onto C_2, each obtained by choosing one element of order 2 in $C_2 \times C_2$ to be mapped to the identity element of C_2, while the two remaining elements of order 2 in $C_2 \times C_2$ are mapped to the generator of C_2. However, since the product of any two of the three elements of order 2 in $C_2 \times C_2$ is the third element of order 2, it follows that every permutation of the three elements of order 2 gives an automorphism of $C_2 \times C_2$, and certainly, every automorphism of $C_2 \times C_2$ permutes the elements of order 2, so $\text{Aut}(C_2 \times C_2) \simeq S_3$. In particular, if $\varphi, \psi : C_2 \times C_2 \to \text{Aut}(C_3)$ are both surjective homomorphisms, then there exists an automorphism α of $C_2 \times C_2$ such that $\psi = \varphi \circ \alpha$. It follows that up to isomorphism, there are two groups that are a semidirect product of $C_2 \times C_2$ by C_3 (one of which is the direct product, while the other is in fact A_4).*

Note that the direct product of $C_2 \times C_2$ with C_3 is an abelian group, so the only nonabelian semidirect product of $C_2 \times C_2$ by C_3 is A_4. We may also characterize A_4 as follows.

12.4.12 Proposition. *Let G be a group of order 12 with four Sylow 3-subgroups. Then G is isomorphic to A_4.*

Proof. By Sylow's results, G must have either one or four Sylow 3-subgroups, and we are interested in the case that G has four of them, say P_1, P_2, P_3, P_4. Let φ denote the conjugation action of G on $\{P_1, P_2, P_3, P_4\}$. Since this action is transitive, and the stabilizer of a Sylow 3-subgroup P under this action is $N_G(P)$, we have $4 = |G|/|N_G(P)|$ and so $|N_G(P)| = 3 = |P|$. Since $P \subseteq N_G(P)$, it follows that $N_G(P) = P$ for each Sylow 3-subgroup P of G. But then if $g \in G$ is in the kernel of φ, $g \in N_G(P)$ for each $P \in X$, which would mean that g is the identity of G. Thus φ is injective. Now, as G contains eight elements of order 3, so does $\varphi(G)$. Use the bijection between $\{P_1, P_2, P_3, P_4\}$ and J_4 that maps P_i to i to consider φ as an injective homomorphism from G into S_4, and conclude that $\varphi(G)$ contains all eight 3-cycles of S_4. As A_4 is generated by the set of 3-cycles in S_4, it follows that $A_4 \subseteq \varphi(G)$. Finally, since $|A_4| = 12 = |G| = |\varphi(G)|$, $A_4 = \varphi(G)$, and so G is isomorphic to A_4. $\square$

12.4.13 Proposition. *If G is a group of order 24 with nonnormal Sylow 2-subgroups and nonnormal Sylow 3-subgroups, then G is isomorphic to S_4.*

Proof. Suppose that the Sylow 3-subgroups of G are not normal. Since the number of Sylow 3-subgroups is congruent to 1 modulo 3, and must divide $|G|/3 = 8$, it follows that G must have four Sylow 3-subgroups, so we let $X = \{P_1, P_2, P_3, P_4\}$ denote the set of the four Sylow 3-subgroups of G, and consider the conjugation action of G on X. As above, by identifying P_i with i, we may consider this action to be a homomorphism $\varphi : G \to S_4$. Since $|G| = |S_4|$, if we prove that $\ker(\varphi)$ is trivial, then φ is an isomorphism. We have $\ker(\varphi) = \cap_{i=1}^{4} N_G(P_i)$, and we have $|N_G(P_i)| = |G|/4 = 6$. It follows that $|\ker(\varphi)|$ is a divisor of 6. Moreover, if 3 divides $|\ker(\varphi)|$, then for some i, $P_i \subseteq N_G(P_j)$ for every $j = 1, 2, 3, 4$, and by Proposition 12.3.3, this would imply that $P_1 = P_2 = P_3 = P_4$, which is not the case. It remains to prove that $|\ker(\varphi)| \neq 2$. Suppose to the contrary that $|\ker(\varphi)| = 2$. Then for $K = \ker(\varphi)$, G/K is isomorphic to a subgroup of size $|G|/|K| = 24/2 = 12$ of S_4. Moreover, each subgroup of order 3 in G will map to a subgroup of order 3 in G/K, so $\varphi(G) \simeq G/K$ has four Sylow 3-subgroups. By Proposition

12.4.12, $\varphi(G)$ is isomorphic to A_4. But A_4 has a normal subgroup of order 4, $H = \{\mathbb{1}_{J_4}, (1\,2)(3\,4), (1\,3)(2\,4), (1\,4)(2\,3)\}$, and so $\varphi^{-1}(H)$ is a normal subgroup of G of order 8. But then $\varphi^{-1}(H)$ is a Sylow 2-subgroup of G, and since the Sylow 2-subgroups of G are not normal in G, this is not possible. Thus $|\ker(\varphi)| = 1$, and so G is isomorphic to S_4. $\qquad\square$

12.4.14 Proposition. *Let H and K be groups, $\alpha \in Aut(K)$, and $\varphi, \psi:$ $H \to Aut(K)$ be homomorphisms such that $\psi(h) = \alpha \circ \varphi_h \circ \alpha^{-1}$ for all $h \in H$; that is, $\psi = c_\alpha \circ \varphi$, where $c_\alpha(\beta) = \alpha \circ \beta \circ \alpha^{-1}$ for $\beta \in Aut(K)$. Then $K \rtimes_\varphi H \simeq K \rtimes_\psi H$ by the bijection $f : K \times H \to K \times H$ given by $f(k, h) = (\alpha(k), h)$.*

Proof. Let $(k, h), (k_1, h_1) \in K \rtimes_\varphi H$. Then

$$
\begin{aligned}
f((k, h)(k_1, h_1)) &= f(k\varphi_h(k_1), hh_1) = (\alpha(k)\alpha(\varphi_h(k_1)), hh_1) \\
&= (\alpha(k)(\alpha \circ \varphi_h)(k_1), hh_1) = (\alpha(k)(\alpha \circ \varphi_h)(\alpha^{-1}(\alpha(k_1))), hh_1) \\
&= (\alpha(k)(\alpha \circ \varphi_h \circ \alpha^{-1})(\alpha(k_1)), hh_1) = (\alpha(k)\psi_h(\alpha(k_1)), hh_1) \\
&= (\alpha(k), h)(\alpha(k_1), h_1) = f(k, h)f(k_1, h_1).
\end{aligned}
$$

$\qquad\square$

Of course, if $Aut(K)$ is abelian, then no information is provided by the preceding result.

12.4.15 Example. *Consider C_2 acting on $C_2 \times C_2$. We want a homomorphism from $C_2 \to Aut(C_2 \times C_2) \simeq S_3$. The nontrivial homomorphisms are in one-to-one correspondence with the elements of order 2 in S_3, of which there are three, so there are three nontrivial homomorphisms from C_2 to $Aut(C_2 \times C_2)$. However, the conjugation action of S_3 on itself is transitive on the set of elements of order 2, so for any nontrivial homomorphisms $\varphi, \psi : C_2 \to Aut(C_2 \times C_2)$, there exists $\alpha \in Aut(C_2 \times C_2)$ with $\psi = c_\alpha \circ \varphi$ and thus all three nontrivial homomorphisms result in isomorphic semidirect products of $(C_2 \times C_2) \rtimes C_2$.*

We now show how the results established in this chapter enable us to determine the structure of many finite groups.

12.4.16 Example.

(i) *Let G be a group of order $6 = (2)(3)$. Since 2 divides $3 - 1$, Proposition 12.4.4 (ii) applies. If G is abelian, then G is cyclic. Suppose that G is not abelian. Then the Sylow 3-subgroup $Q = \langle b \rangle$ of G is normal in G. Let $P = \langle a \rangle$ be a Sylow 2-subgroup of G. Then*

$aba^{-1} = b^m$ *for some positive integer* m *for which* $m^2 \equiv 1 \mod 3$. *Since* G *is not abelian,* $m > 1$ *and so* $m = 2$. *Thus* G *is generated by* a *and* b, *and* $a^2 = b^3 = 1$, *and* $aba^{-1} = b^2 = b^{-1}$. *The map* $f : G \to S_3$ *which sends* a *to* $\alpha = (1\,2)$ *and* b *to* $\beta = (1\,2\,3)$ *is in fact an isomorphism, since* S_3 *is generated by* α *and* β, $\alpha^2 = \mathbb{1}_{J_n} = \beta^3$, *and* $\alpha\beta\alpha^{-1} = (1\,2)\,(1\,2\,3)\,(1\,2) = (1\,3\,2) = \beta^{-1}$. *Thus up to isomorphism, there are two groups of order 6,* S_3 *and the cyclic group of order 6.*

(ii) *Let* G *be a group of order* $21 = (3)(7)$. *Once again, Proposition 12.4.4 (ii) applies. Let* P *be a Sylow 3-subgroup of* G, *say* $P = \langle a \rangle$, *and let* Q *be the unique Sylow 7-subgroup of* G, *say* $Q = \langle b \rangle$. *If* G *is abelian, then* G *is cyclic, so suppose that* G *is not abelian. There exists an integer* m *with* $1 \leq m \leq 6$ *such that* $aba^{-1} = b^m$ *and* $m^3 \equiv 1 \mod 7$, *and since* G *is not abelian,* $m > 1$. *To determine* m, *we examine* $\mathbb{Z}_7^*$ *for elements of order 3. Since* $\mathbb{Z}_7^*$ *is a cyclic group of order 6, it contains exactly two elements of order 3, namely* $[2]_7$ *and its inverse,* $[4]_7$. *Thus the only possible values for* m *are 2 or 4. Let* C_7 *denote a cyclic group of order 7, and let* b *be a generator of* C_7. $\mathrm{Aut}(C_7)$ *is isomorphic to* $\mathbb{Z}_7^*$, *a cyclic group of order 6, and thus* $\mathrm{Aut}(C_7)$ *has a cyclic subgroup of order 3. Since 2 has order 3 in* $\mathbb{Z}_7^*$, *it follows that the cyclic subgroup of order 3 in* $\mathrm{Aut}(C_7)$ *is generated by the automorphism which sends* b *to* b^2. *Let* C_3 *be a cyclic group of order 3 with generator* a. *Then the mapping* $\varphi : C_3 \to \mathrm{Aut}(C_7)$ *that is determined by sending* a *to the automorphism* α *of* C_7 *for which* $\alpha(b) = b^2$ *is an injective homomorphism. In the semidirect product* $C_3 \rtimes_\varphi C_7$, $a^3 = e = b^7$ *and* $aba^{-1} = \varphi_a(b) = b^2$. *Thus there exists a group of order 21 generated by elements* a *and* b *for which* $a^3 = e = b^7$ *and* $aba^{-1} = b^2$. *Note that the mapping* $\psi : C_3 \to \mathrm{Aut}(C_7)$ *that sends* a *to the automorphism that sends* b *to* b^4 *provides a semidirect product in which* $a^3 = e = b^7$ *and* $aba^{-1} = b^4$. *These two semidirect products are isomorphic by virtue of Proposition 12.4.14 as may be seen by choosing* $\alpha \in \mathrm{Aut}(C_7)$ *to be the automorphism determined by sending* b *to* b^2. *Then* $\alpha \circ \varphi_a \circ \alpha^{-1}(b) = \alpha \circ \varphi_a(b^4) = \alpha(b^8) = \alpha(b^2) = b^4 = \psi_a(b)$, *and so* $\alpha \circ \varphi_a \circ \alpha^{-1} = \psi_a$. *Since* C_3 *is cyclic generated by* a, *it follows that* $\alpha \circ \varphi_h \circ \alpha^{-1} = \psi_h$ *for all* $h \in C_3$. *We can see this internally in* G, *for* P *is a cyclic group of order 3 generated by* a, *and thus by* a^2. *Let* $a_1 = a^2$. *Then* G *is generated by* a_1 *and* b, $a_1^3 = e = b^7$, *and* $a_1 b a_1^{-1} = a^2 b a^{-2} = ab^2a^{-1} = b^4$.

12.4.17 Proposition. *If G is a group for which $G/Z(G)$ is cyclic, then G is abelian.*

Proof. Let $aZ(G)$ be a generator for $G/Z(G)$. Then every element of G is of the form $a^k g$ for some integer k and some $g \in Z(G)$. For any $m, n \in \mathbb{Z}$ and any $g, h \in Z(G)$, we have $(a^m g)(a^n h) = a^m (a^n h) g = a^n a^m h g = (a^n h)(a^m g)$, as required. $\square$

12.4.18 Corollary. *For any prime p, every group of order p^2 is abelian.*

Proof. By Corollary 12.2.2, a group of order p^2, p a prime, has nontrivial centre. If the centre is the entire group, the group is abelian, so suppose that the centre has order p. Then the quotient of the group by its centre has order p and is therefore cyclic. The result follows now from Proposition 12.4.17. $\square$

We give one more example of the semidirect product construction, wherein we consider three different actions of $C_8 = \langle a \rangle$, a cyclic group of order 8, on $C_{12} = \langle b \rangle$, a cyclic group of order 12. Recall that by Corollary 6.3.13 (iv), $\text{Aut}(C_{12})$ is isomorphic to U_{12}, which itself is isomorphic to the Klein 4-group. $\text{Aut}(C_{12})$ has three elements of order 2. As each automorphism of a cyclic group is completely determined by the image of a generator of the group, we see that the three nontrivial automorphisms of C_{12} are determined by sending b to b^5, b^7, and b^{11}, respectively. Since the homomorphic image of a cyclic group of order 8 must be a cyclic group of order a divisor of 8, we see that there are three nontrivial homomorphisms from C_8 into $\text{Aut}(C_{12})$, each with image a subgroup of order 2. Let φ_i denote the homomorphism from C_8 into $\text{Aut}(C_{12})$ that is determined by mapping a to the automorphism determined by sending b to b^i, $i = 5, 7, 11$. Thus we have three actions of C_8 on C_{12}, and we shall construct the semidirect product of C_{12} by C_8 for each action. Moreover, we shall prove that no two of the three constructed groups of order 96 are isomorphic.

Case 1: $G = C_{12} \rtimes_{\varphi_5} C_8$. The underlying set is $C_{12} \times C_8$, a set of size 96. We shall adopt the conventions described in the discussion that precedes Example 12.4.9 and consider C_{12} and C_8 to be subgroups of G, with $C_{12} \triangleleft G$. With this convention, the element (b^i, a^j) will be written simply as $b^i a^j$, and the action of C_8 on b is simply conjugation by b, which means that $ab = aba^{-1}a = b^5 a$. We determine the centre of G. An element $b^i a^j \in Z(G)$ if, and only if, it commutes with both a and b. Thus it suffices to determine i and j for which $b^i a = ab^i$ and $ba^j = a^j b$. We have

$ab^i = ab^i a^{-1} a = (aba^{-1})^i a = (b^5)^i a = b^{5i} a$, so we require that $i \equiv 5i$ mod 12; that is, i is a multiple of 3. Thus the values of i are $0, 3, 6, 9$. As well, $a^j b = a^j b a^{-j} a^j = b^{5^j} a^j$, so we require that $1 \equiv 5^j$ mod 12. Since 5 has order 2 in U_{12}, it follows that $j = 0, 2, 4, 6$, and so $Z(G) = \{ b^i a^j \mid i = 0, 3, 6, 9, \ j = 0, 2, 4, 6 \}$. Thus $|Z(G)| = 16$, and so $G/Z(G)$ is a group of order 6. By Example 12.4.16 (i), there are only two groups of order 6, the cyclic group of order 6 and S_3. If $G/Z(G)$ were cyclic, then by Proposition 12.4.17, G would be abelian. Since this is not the case, $G/Z(G)$ is isomorphic to S_3 in this case.

Case 2: $G = C_{12} \rtimes_{\varphi_7} C_8$. As in Case 1, $|G| = 96$, and we consider C_{12} and C_8 to be subgroups of G, with $C_{12} \lhd G$. The element (b^i, a^j) will be written simply as $b^i a^j$, and the action of C_8 on b is simply conjugation by b, which means that $ab = aba^{-1} a = b^7 a$. As in Case 1, we determine the centre of G. An element $b^i a^j \in Z(G)$ if, and only if, it commutes with both a and b. Thus it suffices to determine i and j for which $b^i a = ab^i$ and $ba^j = a^j b$. We have $ab^i = ab^i a^{-1} a = (aba^{-1})^i a = (b^7)^i a = b^{7i} a$, so we require that $i \equiv 7i$ mod 12; that is, i is a multiple of 2. Thus the values of i are $0, 2, 4, 6$. As well, $a^j b = a^j b a^{-j} a^j = b^{7^j} a^j$, so we require that $1 \equiv 7^j$ mod 12. Since 7 also has order 2 in U_{12}, it follows that $j = 0, 2, 4, 6$, and so $Z(G) = \{ b^i a^j \mid i = 0, 2, 4, 6, 8, 10, \ j = 0, 2, 4, 6 \}$. Thus $|Z(G)| = 24$, and so $G/Z(G)$ is a group of order 4. Up to isomorphism, there are only two groups of order 4, the cyclic group and the Klein 4-group. If $G/Z(G)$ were cyclic, then again by Proposition 12.4.17, G would be abelian. Since this is not the case, $G/Z(G)$ is isomorphic to the Klein 4-group.

Case 3: $G = C_{12} \rtimes_{\varphi_{11}} C_8$. Just as in the preceding two cases, $|G| = 96$, and we consider C_{12} and C_8 to be subgroups of G, with $C_{12} \lhd G$. The element (b^i, a^j) will be written simply as $b^i a^j$, and the action of C_8 on b is simply conjugation by b, which means that $ab = aba^{-1} a = b^{11} a$. Again, we determine the centre of G. An element $b^i a^j \in Z(G)$ if, and only if, it commutes with both a and b. Thus it suffices to determine i and j for which $b^i a = ab^i$ and $ba^j = a^j b$. We have $ab^i = ab^i a^{-1} a = (aba^{-1})^i a = (b^{11})^i a = b^{11i} a$, so we require that $i \equiv 11i$ mod 12; that is, i is a multiple of 6. Thus the values of i are $0, 6$. As well, $a^j b = a^j b a^{-j} a^j = b^{11^j} a^j$, so we require that $1 \equiv 11^j$ mod 12. Since 11 also has order 2 in U_{12}, it follows that $j = 0, 2, 4, 6$, and so $Z(G) = \{ b^i a^j \mid i = 0, 6 \ j = 0, 2, 4, 6 \}$. Thus $|Z(G)| = 8$, and so $G/Z(G)$ is a group of order 12. The reader is invited to prove that $G/Z(G)$ is isomorphic to $S_3 \times C_2$ (note that the image of C_{12} is a cyclic group of order 6, and the image of C_8 is a cyclic group of order 2 that acts on the cyclic group of order 6 by conjugation).

As we indicated in the discussion near the beginning of this section, our goal is to establish that A_5 is the smallest nonabelian simple group, by which we mean that a group of order less than 60 is either not simple or else is abelian, and that, up to isomorphism, A_5 is the only simple group of order 60.

We present one last result before we begin our examination of the groups of order at most 60.

12.4.19 Proposition. *Let G be a finite group. If there is a proper subgroup H of G such that $|G|$ does not divide $[G\!:\!H]!$, then G is not simple.*

Proof. By Proposition 12.4.2, if we set $K = \underset{g \in G}{\cap} gHg^{-1}$, then G/K is isomorphic to a subgroup of S_n where $n = [G\!:\!H]$. Thus $|G|/|K|$ is a divisor of $n!$. Since $|G|$ does not divide $n!$, it must be that $|K| \neq 1$, and since K is a subgroup of the proper subgroup H, K is a proper nontrivial normal subgroup of G. □

12.4.20 Corollary. *Let p be a prime with $p > 3$. Then any group of order $4p$ is not simple.*

Proof. Suppose that G is a group of order $4p$, where p is a prime greater than 3. Let P be a Sylow p-subgroup of G. Then $[G\!:\!P] = 4$, and $4! = 24$. Since $4p$ divides 24 if, and only if, p divides $(3)(2)$, we may apply Proposition 12.4.19 to conclude that G is not simple. □

12.4.21 Proposition. *Every group of order less than 60 is either cyclic of prime order, or is not simple.*

Proof. In Table 12.1, we show for each integer from 2 to 59 inclusive (except for the values 30, 40, and 56) why a group of that order is either of prime order or else fails to be simple. We shall write p to designate prime order, p^n to designate a finite p-group (which must have nontrivial centre and thus fail to be simple), pq to designate a group of order pq, where p and q are distinct primes (by Proposition 12.4.4, such a group fails to be simple), $4p$ to denote a group of order $4p$ where p is a prime greater than 3 (by Corollary 12.4.20, such a group is not simple), and $[G : n]$ where n is an integer to mean that G contains a (Sylow) subgroup of order n for which $|G|$ does not divide the factorial of $|G|/n$ (such a group fails to be simple by Proposition 12.4.19). The exceptional cases 30, 40, and 56 are indicated with a $*$.

Abstract Algebra

Table 12.1   Reasons for non-simplicity for groups of orders 2–59

2	3	4	5	6	7	8	9	10	11	12	13	14	15	16	17
p	p	p^2	p	pq	p	p^n	p^n	pq	p	$4p$	p	pq	pq	p^n	p
18	19	20	21	22	23	24	25	26	27	28	29	30	31	32	33
$[G\!:\!8]$	p	$4p$	pq	pq	p	$[G\!:\!8]$	p^n	pq	p^n	$4p$	p	$*$	p	p^n	pq
34	35	36	37	38	39	40	41	42	43	44	45	46	47	48	49
pq	pq	$[G\!:\!9]$	p	pq	pq	$*$	p	$[G\!:\!7]$	p	$4p$	pq	pq	p	$4p$	p^n
50	51	52	53	54	55	56	57	58	59						
$[G\!:\!25]$	pq	$4p$	p	$[G\!:\!27]$	pq	$*$	pq	pq	p						

Thus it remains to consider the groups of order 30, 40, or 56. In Proposition 12.4.5, it was established that any group of order 30 is not simple. For a group G of order 40, $n_5(G) \equiv 1 \mod 5$ and divides 8, so $n_5(G) = 1$, which means that no group of order 40 is simple. Finally, let G be a group of order $56 = (7)(2^3)$. Then $n_7(G) \equiv 1 \mod 7$ and divides 8, so $n_7(G) = 1$ or 8. If $n_7(G) = 1$, G has a normal Sylow 7-subgroup, so suppose that $n_7(G) = 8$. Then G contains 48 elements of order 7, so we have accounted for 49 elements of G, leaving 7, which together with the identity comprise the lone Sylow 2-subgroup. Thus no group of order 56 is simple.

$\square$

We finish up this section with an investigation of the groups of order 60. We shall prove that if G is a group of order 60, then G is simple if, and only if, $n_5(G) > 1$. We shall then use this fact to establish that a group G of order 60 is isomorphic to A_5 if, and only if, $n_5(G) > 1$.

12.4.22 Lemma. *Let G be a group of order 60 for which $n_5(G) > 1$. Then G has no subgroup of order 15, 20, or 30. Moreover, no subgroup of G whose order is divisible by 5 is normal in G.*

Proof. First, we know that $n_5(G)$ is congruent to 1 modulo 5 and divides 12, and by hypothesis, $n_5(G) > 1$, so $n_5(G) = 6$.

Suppose now that G has a subgroup H of order 15. By Proposition 12.4.5, H is cyclic, hence abelian, and H contains a Sylow 5-subgroup P of G. Thus $H \subseteq N_G(P)$, which is not possible as $|N_G(P)| = |G|/n_5(G) = 60/6 = 10$. By Proposition 12.4.5, G has no subgroup of order 30 either.

Now suppose that H is a subgroup of order 20 in G. Then H contains a Sylow 5-subgroup P of G. But $n_5(H) \equiv 1 \mod 5$ and divides $20/5 = 4$, so $n_5(H) = 1$. This means that P is normal in H, and so $H \subseteq N_G(P)$; that is, 20 is a divisor of 10. Since this is not the case, we conclude that G has no subgroup of order 20.

Finally, suppose that H is a proper normal subgroup of G and that 5 divides $|H|$. Then H contains a Sylow 5-subgroup of G, and since $H \triangleleft G$, H must contain all six Sylow 5-subgroups of G. Thus H contains all 24 elements of order 5, and so $|H| = 30$. But then by Proposition 12.4.5, G contain a subgroup of order 15, which we have proven is not the case. $\square$

12.4.23 Proposition. *Let G be a group of order* 60. *Then the following are equivalent:*

  (i) G is isomorphic to A_5.
  (ii) G is simple.
  (iii) $n_5(G) > 1$.

Proof. If G is isomorphic to A_5, then by Theorem 7.5.4, G is simple, and thus $n_5(G) > 1$. Suppose now that $n_5(G) > 1$. We prove first that G is simple, then that G is isomorphic to A_5. Suppose that G has a proper normal subgroup H. By Lemma 12.4.22, $|H|$ can only be one of 2, 3, 4, 6, or 12. If $|H| = 2$ or 4, then $|G/H| = 30$ or 15, and in either case, by Proposition 12.4.5, $n_5(G/H) = 1$ and so G contains a proper normal subgroup whose order is divisible by 5, which is not possible. If $|H| = 3$ or 6, then $|G/H| = 20$ or 10, and again, $n_5(G/H) = 1$, which we have seen is not possible. Suppose $|H| = 12$. If $n_3(H) = 1$, then H contains a unique subgroup of order 3, and since $H \triangleleft G$, every Sylow 3-subgroup of G is contained in H, so $n_3(G) = 1$. We have already seen that has no normal subgroup of order 3, so this is not possible. Thus $n_3(H)$ is a divisor of 4 and is congruent to 1 modulo 3, so it follows that $n_3(H) = 4$. But then H has 8 elements of order 3, so H contains a unique subgroup of order 4 which must therefore be normal in G. This is not possible since we have proven that G has no normal subgroup of order 4. Thus G is simple.

In particular, $n_2(G) > 1$. Since $n_2(G)$ divides $60/4 = 15$, $n_2(G) = 3, 5$, or 15. If $n_2(G) = 3$, then the normalizer of a Sylow 2-subgroup of G has order $60/3 = 20$, which we have proven is not possible. Thus $n_2(G)$ is either 5 or 15. If $n_2(G) = 5$, then $|N_G(K)| = 60/5 = 12$, and so G has a subgroup of order 12. Otherwise, $n_2(G) = 15$. In this case, if the intersection of each pair of distinct Sylow 2-subgroups is trivial, then we have $(3)(15) = 45$ elements of order 2 or 4, which is not possible as $n_5(G)$ must be equal to 6 and so G has $(6)(4) = 24$ elements of order 5, which would mean that $|G| \geq 70$. Thus we must have at least two Sylow 2-subgroups K_1 and K_2 whose intersection has size 2. Since every group of order 4 is abelian, it follows that both K_1 and K_2 are subgroups of $N_G(K_1 \cap K_2)$, and so we know that $|N_G(K_1 \cap K_2)|$ is a multiple of 4 and greater than 4, and must be a divisor of 60, so $|N_G(K_1 \cap K_2)| = 12, 20$, or 60. It can't be 20 as G has no subgroup of order 20, and it can't be 60 since G is simple.

In summary, either $n_2(G) = 5$, in which case the normalizer of any Sylow 2-subgroup of G has order 12, or else $n_2 = 15$, in which case there are distinct Sylow 2-subgroups K_1 and K_2 for which $N_G(K_1 \cap K_2)$ has order 12. But then by Proposition 12.4.2, there is an injective homomorphism $\rho: G \to S_5$. Since G has 24 elements of order 5, and G has no subgroup of order 30, it follows that G is generated by the 24 elements of order 5. Furthermore, S_5 contains exactly 24 elements of order 5, the 5-cycles, and so all 24 5-cycles are contained in $\rho(G)$. Since A_5 is simple, $n_5(A_5) > 1$ and so by Lemma 12.4.22, A_5 has no subgroup of order 30. Thus the 24 5-cycles generate A_5, and so we conclude that $\rho(G) = A_5$. □

12.5   Exercises

1. For each $\theta \in \mathbb{R}$, let $cis(\theta) = \cos(\theta) + i \sin(\theta) \in \mathbb{C}$.

    a) Prove De Moivre's theorem: for every $\theta \in \mathbb{R}$, and for every integer $n \geq 1$, $(cis(\theta))^n = cis(n\theta)$.

    b) Let $A = \begin{bmatrix} 0 & 1 \\ 1 & 0 \end{bmatrix}$ and for any positive integer k, let $B_k = \begin{bmatrix} \cos(\frac{2\pi}{k}) & -\sin(\frac{2\pi}{k}) \\ \sin(\frac{2\pi}{k}) & \cos(\frac{2\pi}{k}) \end{bmatrix}$. Prove that $A, B \in Gl_2(\mathbb{R})$, and that A has order 2 and B has order k.

    c) For each positive integer k, let D_k be the subgroup of $Gl_2(\mathbb{R})$ that is generated by A and B_k. Prove that $|D_k| = 2k$, and describe the conjugacy classes of D_k.

2. Prove that for each positive integer k, the group D_k of Exercise 1 is isomorphic to the dihedral group of order $2k$ that was introduced in Example 5.4.5.

3. Prove that if H and K are abelian subgroups of a group G, then $H \cap K$ is normal in the subgroup of G that is generated by H and K. Give an example to illustrate that the subgroup that is generated by H and K need not be the subset HK.

4. Prove that if H and K are groups, and $\varphi: H \to \text{Aut}(K)$ is a nontrivial homomorphism, then $K \rtimes_\varphi H$ is not abelian.

5. Let G be a finite group, and let H be a subgroup of G. If $H \cap gHg^{-1} =$

$\{e_G\}$ for every $g \in G - H$, then $|G - (\cup_{g \in G} gHg^{-1})| = \frac{|G|}{|H|} - 1$.

6. Prove that a finite nonabelian group in which every proper subgroup is abelian is not simple.

7. Let G be a group, and define normal subgroups $Z_k(G)$ inductively as follows: $Z_1(G) = Z(G)$, and for $k \geq 1$, define $Z_{k+1}(G) = \pi^{-1}(Z(G/Z_k(G)))$, where π denotes the canonical homomorphism from G onto $G/Z_k(G)$.

   a) Prove that for every positive integer k, $Z_k(G) \subseteq Z_{k+1}(G)$.

   The increasing series of normal subgroups $Z_k(G)$, $k \geq 1$, is called the upper central series of G.

   b) Prove that if G is a finite p-group for some prime p, then there exists a positive integer n such that $Z_n(G) = G$.

8. Let p be a prime and let G be a finite p-group.

   a) Prove that if H is a proper subgroup of G, then $H \neq N_G(H)$.

   b) Give an example of a finite group G that has a proper subgroup H for which $N_G(H) = H$.

9. Let G be a finite group. Prove that if $N_G(H) = C_G(H)$ for every abelian subgroup H of G, then G is abelian.

10. Let G be a group (not necessarily finite) which is generated by two elements of order 2.

    a) Prove that G has a cyclic subgroup of index 2.

    b) Prove that if G is finite (so necessarily of even order, say $|G| = 2k$), then G is isomorphic to the group D_k of Exercise 1.

11. Let G be a group, and let H and K be subgroups of G. For any $g \in G$, the set HgK is called a double coset of the pair (H, K). If both H and K are finite, prove that for any $g \in G$, $|HgK| = |H||K|/|gHg^{-1} \cap H|$. Hint: Consider the action of H on the left cosets of K in G.

12. a) Let n be a positive integer, and let G be a subgroup of S_n. Prove that if G contains an odd permutation, then G has a subgroup of index 2.

    b) Let m and k be positive integers with m odd, and let G be a group of order $2^k m$. Prove that if G has a cyclic Sylow 2-subgroup, then G has a subgroup of order m. Hint: Consider the left multiplication action of G on itself.

13. Prove that if G is a group of order 12 with no central element of order 2, then G is isomorphic to A_4.

14. Let p be a prime.
    a) Prove that a Sylow p-subgroup of S_p has order p.
    b) Prove that $n_p(S_p) = (p-2)!$.
    c) Use the Sylow theorems now to prove Wilson's theorem: $(p-1)! \equiv -1 \mod p$.

15. Let X be a nonempty set and $\rho: G \to S_X$ be an action of a group G on X, and let $K = \ker(\rho)$. Prove that if there are only finite many orbits of this action, say $X_1, X_2, \ldots, X_k$, then if we let ρ_i denote the restriction of the action to the orbit X_i and let $K_i = \ker(\rho_i)$ for each $i = 1, 2, \ldots, k$, then G/K is isomorphic to a subgroup of $\prod_{i=1}^{k} G/K_i$.

16. Let X be a nonempty set and let G be a group for which there is an action $\rho: G \to S_X$.
    a) Prove that for any $x \in X$, and any $g \in G$, $gG_xg^{-1} = G_{g^\rho x}$.
    b) Prove that if the action is transitive, then any two stabilizer subgroups are conjugate in G.

17. Let n be a positive integer and let G be a subgroup of S_n of prime power order, say $|G| = p^k$. Prove that if $n < p^2$, then G is abelian and every nonidentity element of G has order p.

18. Let p be a prime. Prove that if G is a finite p-group, then for each divisor m of $|G|$, G has a normal subgroup of order m.

19. Let p be a prime, and let G be a finite group for which p is a divisor of $|G|$. Let P be a Sylow p-subgroup of G. Prove that if H is a subgroup of G for which $N_G(P) \subseteq H$, then $N_G(H) = H$.

20. Let n be a positive integer, and let G be a subgroup of S_n. Prove that if the natural action of G on J_n is transitive, then G contains at least $n - 1$ elements g for which $Fix(g) = \varnothing$.

21. Let p be a prime, and let m and n be positive integers.
    a) Prove that if $(m, p) = 1$, then p is not a divisor of $\binom{p^n m}{p^n}$.
    b) Let G be a group of order $p^n m$. Let $\mathscr{S}$ be the set of all subsets of G of size p^n, and let G act on $\mathscr{S}$ by left multiplication (that is, for any $g \in G$ and $A \in \mathscr{S}$, $gA = \{ga \mid a \in A\}$). Prove that there is an orbit O of this action whose size is not a multiple of p.
    c) Let $A \in O$, and let G_A be the stabilizer of A; that is, $G_A = \{g \in G \mid gA = A\}$. Prove that $|G_A| = p^n$. Note that this provides

another proof of the fact that every finite group whose order is divisible by a prime p has a Sylow p-subgroup.

d) For any $A \in \mathscr{S}$, let $P = G_A$, so P is a Sylow p-subgroup of G. Let H be any p-subgroup of G, and consider the induced action of H on $\mathscr{S}$. Prove that H is contained in some conjugate of P (this provides another proof of the fact that every p-subgroup is contained in a Sylow p-subgroup and any two Sylow p-subgroups are conjugate in G).

22. Prove that a noncyclic group of order 21 must have exactly 14 elements of order 3.

23. Let p be a prime, and let G be a finite p-group.

a) If H is a nontrivial normal subgroup of G, prove that $|H \cap Z(G)| > 1$. Hint: Consider the conjugation action of G on H.

b) Suppose that $|G| = p^n$. Prove that for each $i = 1, 2, \ldots, n$, G has a normal subgroup P_i of order p^i such that $\{e_G\} \subseteq P_1 \subseteq P_2 \subseteq \cdots \subseteq P_n = G$.

24. Let G be a group of order 60, and suppose that G has a normal subgroup of order 2.

a) Prove that G has normal subgroups of orders 6, 10, and 30.

b) Prove that G has subgroups of orders 12 and 20.

c) Prove that G has a cyclic subgroup of order 30.

25. Let G be a group of order 60, and suppose that $n_3(G) = 1$. Prove that $n_5(G) = 1$.

26. Let p be a prime, and let G be a finite group for which p divides $|G|$. Let K be a normal subgroup of G, and let $\mathscr{P}$ denote the set of all Sylow p-subgroups of K.

a) Prove that the conjugation action of K on $\mathscr{P}$ is transitive.

b) Prove that if H is the normalizer subgroup of a Sylow p-subgroup of G, then $G = HK$.

27. Of all nonabelian finite groups of odd order, find one of smallest order.

28. Find the class equation for D_2, the group of symmetries of a square.

29. Find the class equation of D_3, the group of symmetries of a hexagon (a regular polygon with six sides).

30. Find a set of groups of order 30 such that each group of order 30 is isomorphic to exactly one group in the set.

Chapter 13

Finitely Generated Abelian Groups

1 Introduction and Preliminary Results

In this chapter, we shall obtain a complete classification of finitely generated abelian groups. In the course of our investigations, we shall have occasion to use results that are reminiscent of the use of matrices to describe linear transformations between finite dimensional vector spaces, and these results are the object of this section.

13.1.1 Definition. *Let G be an abelian group (with operation denoted additively), and let m and n be positive integers. Then for any $A \in M_{m \times n}(\mathbb{Z})$, define the map $L_A : G^n \to G^m$ as follows: for any $(x_1, x_2, \ldots, x_n) \in G^n$, let*

$$L_A(x_1, x_2, \ldots, x_n) = (\sum_{j=1}^{n} A_{1j} x_j, \sum_{j=1}^{n} A_{2j} x_j, \ldots, \sum_{j=1}^{n} A_{mj} x_j).$$

Note that L_A is actually a group homomorphism from G^n into G^m, but we shall not have need of this fact. Rather, the following results will be useful to us in the sequel.

13.1.2 Proposition. *Let G be an abelian group, and let m, n, k be positive integers. Then for any $A \in M_{m \times n}(\mathbb{Z})$ and any $B \in M_{n \times k}(\mathbb{Z})$, $L_A \circ L_B = L_{AB}$, and for any positive integer n, $L_{I_n} = \mathbb{1}_{G^n}$, where I_n is the $n \times n$ identity matrix in $M_{n \times n}(\mathbb{Z})$.*

Proof. Let $A \in M_{m \times n}(\mathbb{Z})$ and $B \in M_{n \times k}(\mathbb{Z})$. Then for any

$(x_1, x_2, \ldots, x_k) \in G^k$, we have

$$L_A \circ L_B(x_1, x_2, \ldots, x_k) = L_A(\sum_{j=1}^{k} B_{1j}x_j, \ldots, \sum_{j=1}^{k} B_{nj}x_j)$$

$$= (\sum_{t=1}^{n} A_{1t}(\sum_{j=1}^{k} B_{tj}x_j), \ldots, \sum_{t=1}^{n} A_{mt}(\sum_{j=1}^{k} B_{tj}x_j))$$

$$= (\sum_{j=1}^{k}(\sum_{t=1}^{n} A_{1t}B_{tj})x_j, \ldots, \sum_{t=1}^{n}(\sum_{j=1}^{k} A_{mt}B_{tj})x_j)$$

$$= (\sum_{j=1}^{k}(AB)_{1j}x_j, \ldots, \sum_{t=1}^{n}(AB)_{mj}x_j)$$

$$= L_{AB}(x_1, x_2, \ldots, x_k).$$

Thus $L_A \circ L_B = L_{AB}$. Note that we needed the fact that G was abelian in order to switch the order of summation. Now consider $L_{I_n}(x_1, x_2, \ldots, x_n) = (\sum_{j=1}^{n}(I_n)_{1j}x_j, \sum_{j=1}^{n}(I_n)_{2j}x_j, \ldots, \sum_{j=1}^{n}(I_n)_{nj}x_j) = (x_1, x_2, \ldots, x_n)$ as $(I_n)_{ij} = 1$ if $i = j$ and 0 otherwise. Thus $L_{I_n} = \mathbb{1}_{G^n}$. $\square$

13.1.3 Corollary. *Let G be an abelian group, and let n be a positive integer. If $A \in M_{n \times n}(\mathbb{Z})$ is invertible, then $L_A : G^n \to G^n$ is invertible and moreover, $(L_A)^{-1} = L_{A^{-1}}$.*

Proof. We use the fact that A is invertible in $M_{n \times n}(\mathbb{Z})$ and appeal to Proposition 13.1.2 to obtain $L_{A^{-1}} \circ L_A = L_{A^{-1}A} = L_{I_n} = \mathbb{1}_{G^n}$. A similar computation yields $L_A \circ L_{A^{-1}} = \mathbb{1}_{G^n}$, so L_A is invertible and $(L_A)^{-1} = L_{A^{-1}}$. $\square$

The next result provides a condition which ensures the existence of an invertible $A \in M_{n \times n}(\mathbb{Z})$ (see Corollary B.3.18) with properties that will be helpful to us in our investigation of finitely generated abelian groups.

13.1.4 Proposition. *Let $n \geq 2$ be an integer, and let $a_1, a_2, \ldots, a_n \in \mathbb{Z}$ satisfy $(a_1, a_2, \ldots, a_n) = 1$. Then there exists $A \in M_{n \times n}(\mathbb{Z})$ for which $\text{row}_1(A) = [a_1 \, a_2 \, \ldots \, a_n]$ and $\det(A) = 1$.*

Proof. The proof is by induction on $n \geq 2$. Suppose that a_1 and a_2 are relatively prime integers. Then there exist integers r, s such that $1 = ra_1 - sa_2$, and so $A = \left[\begin{smallmatrix} a_1 & a_2 \\ s & r \end{smallmatrix}\right]$ meets the requirements. Suppose now that $n \geq 2$ is an integer for which the hypothesis holds, and suppose that $a_1, a_2, \ldots, a_{n+1} \in \mathbb{Z}$ satisfy $(a_1, a_2, \ldots, a_{n+1}) = 1$. First, consider the case when $a_1 = \cdots a_n = 0$. Then $a_{n+1} = \pm 1$, and we form the $(n+1) \times (n+1)$ matrix with integer entries $A = \begin{bmatrix} 0_{1 \times n} & | & a_{n+1} \\ -I_n & | & 0_{n \times 1} \end{bmatrix}$. Then $\det(A) = (-1)^{1+n+1}(a_{n+1}) \det(-I_n) = a_{n+1}$. If $a_{n+1} = 1$, then A meets the requirements, and otherwise, $a_{n+1} = -1$, in which case, we change the sign of the entries in the second row of A to obtain a matrix with the same first row as A, but which now has determinant 1. We now consider the case when there exists i with $1 \leq i \leq n$ for which $a_i \neq 0$, and let $d = (a_1, a_2, \ldots, a_n)$. Then d divides each a_j, $j = 1, 2, \ldots, n$, and we set $b_j = a_j / d$, $j = 1, 2, \ldots, n$. Then $(b_1, b_2, \ldots, b_n) = 1$, and so by the inductive hypothesis, there exists $B \in M_{n \times n}(\mathbb{Z})$ such that $row_1(B) = [b_1 \, b_2 \, \cdots \, b_n]$ and $\det(B) = 1$. Scale the first row of B by d to obtain $C \in M_{n \times n}(\mathbb{Z})$ for which $row_1(C) = [a_1 \, a_2 \, \ldots \, a_n]$ and $\det(C) = d$. We must have $(a_{n+1}, d) = 1$, and so there exist $r, s \in \mathbb{Z}$ such that $r a_{n+1} - sd = 1$. Scale the first row of B by r to obtain $D \in M_{n \times n}(\mathbb{Z})$ with $\det(D) = r$, and then construct the $(n+1) \times (n+1)$ matrix

$$A = \begin{bmatrix} a_1 \cdots a_n & a_{n+1} \\ & s \\ D & 0 \\ & \vdots \\ & 0 \end{bmatrix}.$$

Expand the determinant of A along the last column to find that $\det(A) = (-1)^{1+n} a_{n+1} \det(D) + (-1)^{2+n} s \det(C) = (-1)^{n+1}(a_{n+1}r - sd) = (-1)^{n+1}$. If n is odd, then $\det(A) = 1$ and we have our desired matrix, while if n is even, then scale the last row of A to obtain $A' \in M_{(n+1) \times (n+1)}(\mathbb{Z})$ with $\det(A') = 1$ and $row_1(A') = [a_1 \, \ldots \, a_{n+1}]$. The result follows now by induction. $\qquad \square$

13.1.5 Proposition. *Let m and n be positive integers, and let $\varphi : \mathbb{Z}^m \to \mathbb{Z}^n$ be a group homomorphism from the additive group $(\mathbb{Z}^m, +)$ to the additive group $(\mathbb{Z}^n, +)$. Then there is a unique $\mathbb{Q}$-linear transformation $\hat{\varphi} : \mathbb{Q}^m \to \mathbb{Q}^n$ whose restriction to $\mathbb{Z}^m$ is φ. Moreover, $\hat{\varphi}$ is injective if, and only if, φ is injective, while if φ is surjective, then also $\hat{\varphi}$ is surjective.*

Proof. Let $\{ e_1, e_2, \ldots, e_m \}$ denote the standard basis for $\mathbb{Q}^m$, and observe that for each $i \in J_m$, $e_i \in \mathbb{Z}^m \subseteq \mathbb{Q}^m$. Let $\hat{\varphi}$ be the unique $\mathbb{Q}$-linear transformation from $\mathbb{Q}^m$ to $\mathbb{Q}^n$ for which $\hat{\varphi}(e_i) = \varphi(e_i)$ for each

$i \in J_m$. For any $(a_1, a_2, \ldots, a_m) \in \mathbb{Z}^m$, $(a_1, a_2, \ldots, a_n) = \sum_{i=1}^{m} a_i e_e$ and so $\hat{\varphi}(a_1, a_2, \ldots, a_m) = \hat{\varphi}(\sum_{i=1}^{m} a_i e_i) = \sum_{i=1}^{m} a_i \hat{\varphi}(e_i) = \sum_{i=1}^{m} \varphi(e_i) = \varphi(\sum_{i=1}^{m} a_i e_i) = \varphi(a_1, a_2, \ldots, a_m)$. Thus the restriction of $\hat{\varphi}$ to $\mathbb{Z}^m$ is equal to φ. Evidently, if $\hat{\varphi}$ is injective, then φ is injective. Suppose now that $\hat{\varphi}$ is not injective, and let u be in the null space of $\hat{\varphi}$ with $u \neq 0_{\mathbb{Q}^m}$. Let $k \in \mathbb{Z}$ be the product of the denominators of the nonzero entries of u, so that $ku \neq 0_{\mathbb{Q}^m}$ and $ku \in \mathbb{Z}^m$. Thus $0_{\mathbb{Z}^n} = 0_{\mathbb{Q}^n} = \hat{\varphi}(ku) = \varphi(ku)$ and so $ku \in \ker(\varphi)$, which proves that φ is not injective.

If φ is surjective, then the standard basis $\{ e_1, \ldots, e_n \}$ of $\mathbb{Z}^n$ is contained in $\hat{\varphi}(\mathbb{Q}^m)$ and thus $\hat{\varphi}$ is surjective. $\square$

We remark that in the preceding result, the surjectivity of $\hat{\varphi}$ does not imply that φ is surjective. For example, the additive group homomorphism $\hat{\varphi} : \mathbb{Z} \to \mathbb{Z}$ given by $f(n) = 2n$ for all $n \in \mathbb{Z}$ is not surjective, while $\hat{\varphi} : \mathbb{Q} \to \mathbb{Q}$ is given by $\hat{\varphi}(r) = 2r$ for all $r \in \mathbb{Q}$ is surjective.

13.1.6 Corollary. *Let m and n be positive integers such that the groups $(\mathbb{Z}^m, +)$ and $(\mathbb{Z}^n, +)$ are isomorphic. Then $m = n$.*

Proof. Suppose that there is an isomorphism $\varphi : \mathbb{Z}^m \to \mathbb{Z}^n$. Then by Proposition 13.1.5, $\hat{\varphi} : \mathbb{Q}^m \to \mathbb{Q}^n$ is a vector space isomorphism and thus $m = n$. $\square$

13.2   Direct Sum of Abelian Groups

In Chapter 6 (see Definition 6.5.1), the direct product of groups is defined. When all of the groups are abelian, there is a related construction, called the direct sum of abelian groups, that will be important to us in our classification of finitely generated abelian groups. We have already encountered the notion of direct sum in the context of rings (see Definition 8.4.4), and so we shall simply present the definition of direct sum of abelian groups and list the main properties of the direct sum, with the expectation that the reader can easily provide the proofs for these results by reviewing the corresponding results in Chapter 8.

13.2.1 Definition. *Let $\mathscr{I}$ be a nonempty set, and for each $\lambda \in \mathscr{I}$, let G_λ be an abelian group. For each $f \in \prod_{\lambda \in \mathscr{I}} G_\lambda$, let $S_f = \{ \lambda \in \mathscr{I} \mid f(\lambda) \neq 0_{G_\lambda} \}$. S_f is called the support of f. The subset*

$$\{ f \in \prod_{\lambda \in \mathscr{I}} G_\lambda \mid S_f \text{ is finite} \}$$

is called the direct sum of the abelian groups G_λ, $\lambda \in \mathscr{I}$, *and is denoted by*
$$\bigoplus_{\lambda \in \mathscr{I}} G_\lambda.$$

13.2.2 Proposition. *Let* $\mathscr{I}$ *be a nonempty set, and for each* $\lambda \in \mathscr{I}$, *let* G_λ *be an abelian group. Then* $\oplus_{\lambda \in \mathscr{I}} G_\lambda$ *is a subgroup of* $\prod_{\lambda \in \mathscr{I}} G_\lambda$.

It is apparent from the definition however that if $\mathscr{I}$ is a finite (nonempty) set, then $\oplus_{\lambda \in \mathscr{I}} G_\lambda = \prod_{\lambda \in \mathscr{I}} G_\lambda$, and it is this situation that most interests us.

13.2.3 Notation. *For any positive integer* n, *the direct sum of* n *abelian groups will always use* J_n *as its index set. For abelian groups* $G_1, G_2, \ldots, G_n$, *the direct sum* $\oplus_{i \in J_n} G_i$ *will usually be denoted by* $G_1 \oplus G_2 \oplus \cdots \oplus G_n$, *and its elements will be written as sequences of length* n *whose* i^{th} *entry belongs to* G_i *for* $i = 1, 2, \ldots, n$.

With the above notation, the operations of addition and multiplication in $G = G_1 \oplus G_2 \oplus \cdots \oplus G_n$ are given by:
$$(g_1, g_2, \ldots, g_n) + (h_1, h_2, \ldots, h_n) = (g_1 + h_1, g_2 + h_2, \ldots, g_n + h_n)$$
$$(g_1, g_2, \ldots, g_n)(h_1, h_2, \ldots, h_n) = (g_1 h_1, g_2 h_2, \ldots, g_n h_n)$$
for each $(g_1, g_2, \ldots, g_n), (h_1, h_2, \ldots, h_n) \in G$.

13.2.4 Proposition. *Let* $\mathscr{I}$ *be a nonempty set, and for each* $\lambda \in \mathscr{I}$, *let* G_λ *be an abelian group. Let* $\gamma \in \mathscr{I}$. *Then the mapping* $\iota_\gamma : G_\gamma \to G = \oplus_{\lambda \in \mathscr{I}} G_\lambda$, *where for each* $x \in G_\gamma$, $\iota_\gamma(x) \in \oplus_{\lambda \in \mathscr{I}} G_\lambda$ *is the mapping given by*
$$(\iota_\gamma(x))(\lambda) = \begin{cases} 0_{G_\lambda} & \text{if } \lambda \neq \gamma \\ x & \text{if } \lambda = \gamma \end{cases}$$
for each $\lambda \in \mathscr{I}$, *is an injective homomorphism with image*
$$\{ f \in G \mid f(\lambda) = 0_{G_\lambda} \text{ for all } \lambda \in \mathscr{I} \text{ with } \lambda \neq \gamma \}.$$

As well, for each $\gamma \in \mathscr{I}$, *the mapping* $\pi_\gamma : G \to G_\gamma$ *defined by* $\pi(x) = x(\gamma)$ *for each* $x \in G$ *is a surjective homomorphism with kernel*
$$\{ f \in G \mid f(\lambda) = 0_{G_\lambda} \text{ for all } \lambda \in \mathscr{I} \text{ with } \lambda \neq \gamma \}.$$

Finally, for each $\lambda \in \mathscr{I}$, $\pi_\lambda \circ \iota_\lambda$ *is the identity map on* G_λ, *while for any* $\delta \in \mathscr{I}$ *with* $\lambda \neq \delta$, $\pi_\delta \circ \iota_\lambda$ *is the zero map from* G_λ *into* G_δ.

13.2.5 Proposition. *Let* n *be a positive integer, and for each* $i \in J_n$, *let* G_i *be an abelian group. Then in the abelian group* $G = \oplus_{i \in J_n} G_i$, *the following hold:*

(i) For each $i \in J_n$, let $S_i = \iota_i(G_i)$, so S_i is a subgroup of G. Then $G = S_1 + S_2 + \cdots + S_n$, and in fact, for each $g \in G$, there exist unique $g_i \in S_i$, $i \in J_n$, for which $g = g_1 + g_2 + \cdots + g_n$.

(ii) For each $i \in J_n$, $S_i \cap (S_1 + S_2 + \cdots + S_{i-1} + S_{i+1} + \cdots + S_n) = \{0_G\}$ (where if $i = 1$, we mean that $S_1 \cap (S_2 + \cdots + S_n) = \{0_G\}$, and if $i = n$, we mean that $S_n \cap (S_1 + S_2 + \cdots + S_{n-1}) = \{0_G\}$).

13.2.6 Proposition. *Let G be an abelian group, and suppose that for some positive integer n, G has subgroups $S_1, S_2, \ldots, S_n$ for which the following hold:*

(i) $G = S_1 + S_2 + \cdots + S_n$.

(ii) *For each $i \in J_n$, $S_i \cap (S_1 + S_2 + \cdots + S_{i-1} + S_{i+1} + \cdots + S_n) = \{0_G\}$ (where if $i = 1$, we mean that $S_1 \cap (S_2 + \cdots + S_n) = \{0_G\}$, and if $i = n$, we mean that $S_n \cap (S_1 + S_2 + \cdots + S_{n-1}) = \{0_G\}$).*

Then $G \simeq S_1 \oplus S_2 \oplus \cdots \oplus S_n$. More precisely, for each element $g \in G$, there exist unique $g_i \in S_i$, $i \in J_n$, with $g = g_1 + g_2 + \cdots + g_n$, and the map $\theta : G \to \oplus_{i \in J_n} S_i$ for which $\theta(g) = (g_1, g_2, \ldots, g_n)$, where $g_i \in S_i$, $i \in J_n$, are the unique elements such that $g = g_1 + \cdots + g_n$, is a group isomorphism such that for each $i \in J_n$, $\theta\big|_{S_i} = \iota_i$.

13.3 Free Abelian Groups

In this section, we introduce the notion of a free abelian group. Finitely generated free abelian groups have a fundamental role in the classification of finitely generated abelian groups, and it is this classification theorem that is our objective in this chapter.

13.3.1 Definition. *Let G be an abelian group. If $a \in G$ has finite order, then a is said to be a torsion element of G. If every element of G is a torsion element, then G is said to be a torsion group, while if no element (except the identity element) of G has finite order, then G is said to be torsion-free. If G is generated by the subset $X = \{x_1, x_2, \ldots, x_n\}$, then G is said to be a free abelian group on X if for all $a_1, a_2, \ldots, a_n \in \mathbb{Z}$, $\sum_{i=1}^{n} a_i x_i = 0_G$ implies that $a_1 = a_2 = \cdots = a_n = 0$.*

13.3.2 Proposition. *Let G be an abelian group, and let $X = \{x_1, x_2, \ldots, x_n\} \subseteq G$. Then G is a free abelian group on X if, and only if, G is torsion-free, $0_G \notin X$, and $G = \oplus_{i=1}^{n} \langle x_i \rangle$.*

Proof. First, suppose that $X = \{x_1, x_2, \ldots, x_n\}$ and G is free abelian on X. If for some $i \in J_n$, $x_i = 0_G$, then for any $m \in \mathbb{Z}$, $ma_i = 0_G$, which contradicts the fact that G is free abelian on X. Thus $x_i \neq 0_G$ for every $i \in J_n$. Now, since X is a generating set for G, we have $G = \sum_{i=1}^{n} \langle x_i \rangle$. We prove that the sum is direct. For any $i \in J_n$, consider $g \in \langle x_i \rangle \cap \big(\sum_{j \in J_n,\ j \neq i} \langle x_j \rangle \big)$.

There exist integers $a_1, a_2, \ldots, a_n$ such that $x = -a_i x_i = \sum_{j \in J_n,\ j \neq i} a_j x_j$, and so $\sum_{i=1}^{n} a_j x_j = 0_G$. Since G is free abelian on X, we have $a_j = 0$ for every $j \in J_n$, and thus $x = 0_G$. This proves that the sum is direct. It remains to prove that G is torsion-free. Let $g \in G$ with $g \neq 0_G$. Then there exist $a_i \in \mathbb{Z}$, $i \in J_n$, such that $g = \sum_{i=1}^{n} a_i x_i$. Suppose by way of contradiction that g has finite order m. Then $0_G = mg = \sum_{i=1}^{n} (ma_i) x_i$, and since G is free abelian on X, it follows that $ma_i = 0$ for every $i \in J_n$. Since $g \neq 0_G$, $m > 0$ and thus $a_i = 0$ for all $i \in J_n$, which means that $g = 0_G$. Since g was chosen to be nonzero, we have a contradiction.

Conversely, suppose that $0_G \notin X = \{x_1, x_2, \ldots, x_n\} \subseteq G$ is a generating set for G such that $G = \oplus_{i=1}^{n} \langle x_i \rangle$, and that G is torsion-free. We are to prove that G is free abelian on X. Suppose that $a_1, a_2, \ldots, a_n \in \mathbb{Z}$ are such that $\sum_{i=1}^{n} a_i x_i = 0_G$. Since $a_i x_i \in \langle x_i \rangle$ for each $i \in J_n$, and the sum is direct, it follows that $a_i x_i = 0_G$ for each $i \in J_n$. Since $x_i \neq 0_G$ and G is torsion-free, it follows that $a_i = 0$ for every $i \in J_n$, as required. $\square$

13.3.3 Corollary. *An abelian group G is free abelian on a finite set X if, and only if, G is isomorphic to a direct sum of $|X|$ copies of $\mathbb{Z}$.*

Proof. Suppose first that $X = \{x_1, x_2, \ldots, x_n\} \subseteq G$ and that G is free abelian on X (so $0_G \notin X$). Then by Proposition 13.3.2, $G = \oplus_{i=1}^{n} \langle x_i \rangle$. For each $i \in J_n$, the cyclic subgroup $\langle x_i \rangle = \{mx_i \mid m \in \mathbb{Z}\} = \mathbb{Z}x_i$, and the mapping from $\mathbb{Z}x_i$ to $\mathbb{Z}$ given by $mx_i \mapsto m$ is an isomorphism. Thus G is isomorphic to $\oplus_{i=1}^{n} \mathbb{Z}$.

Conversely, suppose that for some positive integer n, there is an isomorphism $\varphi : \oplus_{i=1}^{n} \mathbb{Z} \to G$. For each $i \in J_n$, let $e_i \in \oplus_{i=1}^{n} \mathbb{Z}$ be the element for which $(e_i)_k = 0$ if $k \neq i$, while $(e_i)_i = 1$. Then $Y = \{e_1, e_2, \ldots, e_n\}$ is a generating set for $\oplus_{i=1}^{n} \mathbb{Z}$, and by Proposition 13.3.2, $\oplus_{i=1}^{n} \mathbb{Z}$ is free abelian on Y. Let $x_i = \varphi(e_i)$ for each $i \in J_n$. Then since φ is an isomorphism, it follows that $X = \{x_1, x_2, \ldots, x_n\}$ is a generating set for G, $0_G \notin X$, and $G = \oplus_{i=1}^{n} \langle x_i \rangle$, so by Proposition 13.3.2, G is free abelian on X. $\square$

13.3.4 Proposition. *Let G be a finitely generated abelian group. If G is torsion-free, and X is a generating set of G of least size, then G is free* *abelian on* X.

Proof. Since G is finitely generated, there is a finite generating set for G of least size n, say. Let $X = \{\, x_1, x_2, \ldots, x_n \,\}$ be a generating set of size n (note that $0_G \notin X$ by choice of n). We have $G = \sum_{i=1}^{n} \langle x_i \rangle$, and since G is torsion-free, $\langle x_i \rangle$ is isomorphic to $\mathbb{Z}$ for each $i \in J_n$. It suffices to prove that the sum is direct. Suppose to the contrary that there is $i \in J_n$ for which there exists nonzero $g \in \langle x_i \rangle \cap \sum_{j \in J_n, \ j \neq i} \langle x_j \rangle$. Then there exist $a_j \in \mathbb{Z}$ for each $j \in J_n$ such that $g = a_i x_i = \sum_{j \in J_n, \ j \neq i} a_j x_j$. We may assume that $(a_1, a_2, \ldots, a_n) = 1$ (otherwise one may cancel the greatest common divisor). By Proposition 13.1.4, there is $A \in M_{n \times n}(\mathbb{Z})$ with $\det(A) = 1$ having $[a_1 \, a_2 \, \cdots \, a_n]$ as its first row. Thus for some $y_2, y_2, \ldots, y_n \in G$, $L_A(g_1, g_2, \ldots, g_n) = (0_G, y_2, y_3, \ldots, y_n)$. By Corollary B.3.18, A is invertible in $M_{n \times n}(\mathbb{Z})$, and by Corollary 13.1.3, $L_{A^{-1}}(0, y_2, y_3, \ldots, y_n) = (g_1, g_2, \ldots, g_n)$. This tells us that $g_1, g_2, \ldots, g_n \in \langle y_2, y_3, \ldots, y_n \rangle$ and so $\{\, y_2, y_3, \ldots, y_n \,\}$ is a generating set of size $n - 1$ for G. But this contradicts the fact that G has no generating set of size $n - 1$, and so the sum is direct. By Proposition 13.3.2, G is free abelian on X. $\qquad\square$

13.3.5 Proposition. *Let G be an abelian group. The set T_G of all torsion elements of G is a subgroup of G, called the torsion subgroup of G, and the quotient group G/T is torsion-free.*

Proof. Since $0_G \in T$, $T \neq \varnothing$. Let $a, b \in T$. Then since $|b| = |-b|$, $-b \in T$, and $(|a||b|)(a + b) = |b|(|a|a) + |a|(|b|b) = 0_G + 0_G = 0_G$, so $|a + b|$ is a divisor of $|a||b|$. Thus T is a subgroup of G. Let $t = g + T \in G/T$, and suppose that m is a positive integer for which $mt = 0_G$. Then $mg \in T$, and so for $n = |mg| > 0$, $(nm)g = 0_G$. Since $nm \neq 0$, it follows that $|g|$ is a divisor of nm, and thus $g \in T$. But then $t = g + T = 0 + T$, the identity of G/T, and so G/T is torsion-free. $\qquad\square$

13.3.6 Theorem. *Let G be a finitely generated abelian group. Then there exists a finitely generated free abelian subgroup F of G such that $G = F \oplus T_G$. Moreover, T is finitely generated.*

Proof. Let $T = T_G$. By Proposition 13.3.5, G/T is torsion-free, and since it is a homomorphic image of a finitely generated abelian group, G/T is a finitely generated, torsion-free abelian group, hence by Proposition 13.3.4, there is a finite subset $X = \{\, x_1, x_2, \ldots, x_n \,\}$ such that $0_G + T \notin Y = \{\, x_1 + T, x_2 + T, \ldots, x_n + T \,\}$, and G/T is free abelian on Y. Let $F = \sum_{i=1}^{n} \langle x_i \rangle$. We prove that F is free abelian on X. Suppose that $a_1, a_2, \ldots, a_n \in \mathbb{Z}$ are such that $\sum_{i=1}^{n} a_i x_i = 0_G$. Then $\sum_{i=1}^{n} a_i(x_i + T) = 0_{G/T}$ and so $a_i = 0$ for every i since G/T is free abelian on $\{\, x_i + T \mid i \in J_n \,\}$. By Definition 13.3.1, G is free abelian on $\{\, x_i \mid i \in J_n \,\}$.

Now, for any $g \in G$, there exist $r_i \in \mathbb{Z}$, $i \in J_n$ such that $g + T = \sum_{i=1}^{n} r_i(x_i + T)$, and so $g \in (\sum_{i=1}^{n} r_i x_i) + T$. Thus $g \in F + T$, and so $G = F + T$. Finally, suppose that $g \in F \cap T$. Then there exist $r_i \in \mathbb{Z}$, $i \in J_n$ such that $g = \sum_{i=1}^{n} r_i x_i \in T$, and so $\sum_{i=1}^{n} r_i(x_i + T)$ is zero in G/T. But G/T is the direct sum of $\langle x_i + T \rangle$, $i \in J_n$, and so $r_i = 0$ for each $i \in J_n$. Thus $g = 0_G$, and so $F \cap T = \{\, 0_G \,\}$. We have now proven that $G = F \oplus T_G$. Since G/F is isomorphic to T_G and G/F is a homomorphic image of a finitely generated group, T_G is finitely generated. $\square$

13.3.7 Proposition. *Let G be a finitely generated free abelian group, and let n be the size of a smallest generating set for G. If Y is a subset of G with $|Y| > n$, then G is not a free abelian group on Y.*

Proof. Let $X = \{\, x_1, x_2, \ldots, x_n \,\}$ be a generating set of size n, so by Proposition 13.3.2, G is free abelian on X. Let $\{\, y_1, y_2, \ldots, y_m \,\} \subsetneq Y$ with $m > n$. For each $j \in J_m$, there exist $r_{ji} \in \mathbb{Z}$, $i \in J_n$, such that $y_j = \sum_{i=1}^{n} r_{ji} x_i$. We claim that there exist $a_j \in \mathbb{Z}$, $j \in J_m$, not all zero, such that $\sum_{j=1}^{m} a_j y_j = 0_G$; equivalently, that $\sum_{j=1}^{m} a_j (\sum_{i=1}^{n} r_{ji} x_i) = 0_G$. This in turn is equivalent to $\sum_{i=1}^{n} (\sum_{j=1}^{m} r_{ji} a_j) x_i = 0_G$. The latter equation, since G is free abelian over X, is equivalent to the requirement that $\sum_{j=1}^{m} r_{ji} a_j = 0$ for each $i \in J_n$; that is, that the homogeneous linear system $\sum_{j=1}^{m} r_{ji} t_j = 0$, $i \in J_n$, have a nontrivial solution in $\mathbb{Z}^n$. Since $m > n$, this system has a nontrivial solution $(t_1, t_2, \ldots, t_m) \in \mathbb{Q}^m$, and by clearing denominators, a nontrivial solution $(a_1, a_2, \ldots, a_m) \in \mathbb{Z}^m$ is obtained. Thus G is not free abelian over Y. $\square$

13.3.8 Proposition. *If G is a finitely generated abelian group and X_1, X_2 are subsets of G such that G is free abelian over X_1 and over X_2, then $|X_1| = |X_2|$.*

Proof. By Proposition 13.3.7, $|X_1| \le |X_2| \le |X_1|$. $\square$

13.3.9 Definition. *If G is a finitely generated free abelian group, then the rank of G, denoted by $rank(G)$, is the size of a smallest generating set for G.*

By Proposition 13.3.4, if G is a finitely generated torsion-free abelian group and X is a finite generating set for G of least size, then G is free abelian on X, and by Proposition 13.3.8, if Y is any subset of G such that G is free abelian on Y, then $|Y| = |X|$. Thus if X is a finite generating set of a finitely generated free abelian group G of rank n, and $|X| = n$, then G is free abelian on X.

13.3.10 Corollary. *Let G_1 and G_2 be finitely generated free abelian groups. Then G_1 and G_2 are isomorphic if, and only if, $rank(G_1) = rank(G_2)$.*

Proof. Let $m = rank(G_1)$ and $n = rank(G_2)$. Then G_1 is isomorphic to $\mathbb{Z}^m$ and G_2 is isomorphic to $\mathbb{Z}^n$, and so the result follows from Corollary 13.1.6.   □

Note that if G is a finitely generated abelian group, then there exists a finitely generated free abelian subgroup F of G such that $G = F \oplus T_G$, and thus G/T_G is isomorphic to F. If F_1 is a subgroup of G such that $G = F_1 \oplus T_G$, then G/T_G is isomorphic to F_1, and thus F_1 is isomorphic to F, and so F_1 is free abelian. By Corollary 13.3.10, $rank(F_1) = rank(F)$. The upshot of this observation is that while there are many choices for F such that $G = F \oplus T_G$, they will all have the same rank.

13.3.11 Definition. *Let G be a finitely generated abelian group, and let F be a free abelian subgroup of G such that $G = F \oplus T_G$. We define $rank(G) = rank(F)$.*

13.4   Finite Abelian Groups

We now turn our attention to the problem of describing the torsion subgroup of a finitely generated abelian group. Since a finitely generated abelian group for which all generators have finite order is simply a finite abelian group, the remainder of this section is devoted to identifying the structure of finite abelian groups.

Our first step is to prove that a finite abelian group is the direct sum of its Sylow subgroups. As every subgroup of an abelian group is normal, each Sylow subgroup of G is normal and thus for each prime divisor p of the order of a finite abelian group, there is one and only one Sylow p-subgroup of G.

13.4.1 Proposition. *Let G be a finite abelian group, and let $p_1, p_2, \ldots, p_n$ be the distinct prime divisors of $|G|$, and for each $i = 1, 2, \ldots, n$, let S_i*

denote the unique Sylow p_i-subgroup of G. Then $G = S_1 \oplus S_2 \oplus \cdots \oplus S_n$.

Proof. The proof is by induction on the number of distinct prime factors of $|G|$. There is nothing to prove if $|G|$ is a prime power, so suppose that the result holds for every finite abelian group whose order has n distinct prime factors, and let G be a finite abelian group for which $|G|$ has distinct prime factors p_i, $i = 1, 2, \ldots, n+1$. For each such i, let S_i denote the p_i-Sylow subgroup of G. Let $H = S_1 + S_2 + \cdots + S_n$. If $g \in H$, then there exist $g_i \in S_i$, $i = 1, 2, \ldots, n$, such that $g = g_1 + g_2 + \cdots + g_n$. Note that $|g|$ divides the least common multiple of the orders of g_i, $i = 1, 2, \ldots, n$. But for each i, $|g_i| = p_i^{r_i}$ for some nonnegative integer r_i, and so $|g|$ divides $p_1^{r_1} \cdots p_n^{r_n}$. It follows that $H \cap S_{n+1} = \{ e_G \}$. Moreover, for each $i = 1, 2, \ldots, n$, $S_i \subseteq H$ and so $|S_i|$ is a divisor of H, which means that $|S_1| |S_2| \cdots |S_n|$ divides H. But then $H + S_{n+1}$ has order $|G|$, and so $G = H + S_{n+1}$, and the sum is direct, with $|H| = |S_1| |S_2| \cdots |S_n|$. This means that $|H|$ has n distinct prime factors, and the Sylow subgroups of H are $S_1, S_2, \ldots, S_n$, so by hypothesis, $H = S_1 \oplus S_2 \oplus \cdots \oplus S_n$. Thus $G = S_1 \oplus \cdots \oplus S_n \oplus S_{n+1}$, as required. The result follows now by induction. $\square$

It remains to determine the structure of a finite abelian p-group, p a prime; that is, finite abelian groups of order a power of p (equivalently, by Cauchy's theorem (Theorem 7.5.2), finite abelian groups in which every element has order a power of p). We shall prove shortly that every finite abelian p-group is a direct sum of cyclic p-groups.

13.4.2 Proposition. *Let G be a finitely generated abelian group (written additively), and let $\{ x_1, x_2, \ldots, x_n \}$ be a generating set for G. Then for any $a_1, a_2, \ldots, a_n \in \mathbb{Z}$ for which $(a_1, a_2, \ldots, a_n) = 1$, there exists a generating set of size n for G that contains $y = \sum_{i=1}^n a_i x_i$.*

Proof. By Proposition 13.1.4, there exists $A \in M_{n \times n}(\mathbb{Z})$ with $row_1(A) = [a_1 \, a_2 \, \ldots \, a_n]$ and $\det(A) = 1$ (so by Corollary B.3.18, A is invertible in $M_{n \times n}(\mathbb{Z})$). For each $i = 1, 2, \ldots, n$, let $y_i = \sum_{j=1}^n A_{ij} x_j$. Then $(y_1, y_2, \ldots, y_n) = L_A(x_1, x_2, \ldots, x_n)$, and by Corollary 13.1.3, $(x_1, x_2, \ldots, x_n) = L_{A^{-1}}(y_1, y_2, \ldots, y_n)$, which establishes that $G = \langle x_1, x_2, \ldots, x_n \rangle \subseteq \langle y_1, y_2, \ldots, y_n \rangle \subseteq G$. Thus $\{ y_1, y_2, \ldots, y_n \}$ is a generating set for G, and $y_1 = \sum_{i=1}^n a_i x_i$. $\square$

13.4.3 Proposition. *Let p be a prime, and let G be a finite abelian p-group. Then for some positive integer n, there exist cyclic subgroups $\langle g_1 \rangle, \langle g_2 \rangle, \ldots, \langle g_n \rangle$ of G such that $G = \langle g_1 \rangle \oplus \langle g_2 \rangle \oplus \cdots \oplus \langle g_n \rangle$ and*

$|g_1| \leq |g_2| \leq \cdots \leq |g_n|$ *(equivalently, $|g_1| \mid |g_2| \mid \cdots \mid |g_n|$). More-over, if $h_1, h_2, \ldots, h_m \in G$ are such that $G = \langle h_1 \rangle \oplus \cdots \oplus \langle h_m \rangle$ and $|h_1| \leq |h_2| \leq \cdots \leq |h_m|$, then $m = n$ and $|g_i| = |h_i|$ for all $i = 1, 2, \ldots, n$.*

Proof. The proof is by induction on the size of a smallest set of generators for such a group, with hypothesis that if $\{g_1, g_2, \ldots, g_n\}$ is a generating set of least size for G, then $G = \langle g_1 \rangle \oplus \langle g_2 \rangle \oplus \cdots \oplus \langle g_n \rangle$. The base case is immediate, so suppose that $n \geq 1$ is an integer for which the hypothesis holds, and let G be a finite abelian p-group for which a smallest set of generators has size $n+1$. For any generating set $S = \{h_1, h_2, \ldots, h_{n+1}\}$ for G, let $\mu(S) = \min\{n \mid |h| = p^n, h \in S\}$. Then among all generating sets of size $n + 1$ for G, let $V = \{g_1, g_2, \ldots, g_{n+1}\}$ be one for which $\mu(V)$ is least, and further suppose that the elements have been labelled so that $|g_1| \geq |g_2| \geq \cdots \geq |g_{n+1}|$, so that $|g_{n+1}| = p^{\mu(V)}$. Let $H = \langle g_1 \rangle + \langle g_2 \rangle + \cdots + \langle g_n \rangle$. Then H is a finite abelian p-group that has a generating set of size n, and since $G = H + \langle g_{n+1} \rangle$, H can not have any generating set of size smaller than n. Thus we may apply the induction hypothesis to H to conclude that $H = \langle g_1 \rangle \oplus \cdots \oplus \langle g_n \rangle$ Let $z \in \langle g_{n+1} \rangle \cap H$, and suppose that $z \neq e_G$. Then there are integers $a_1, a_2, \ldots, a_{n+1}$ such that $z = a_{n+1} g_{n+1} = \sum_{i=1}^{n} a_i g_i$. Let $k = (a_1, a_2, \ldots, a_{n+1})$, and for each $i = 1, 2, \ldots, n + 1$, let $a_i = b_i k$. Then $(b_1, b_2, \ldots, b_{n+1}) = 1$. It then follows from Proposition 13.4.2 that for $y = (\sum_{i=1}^{n} b_i g_i) - b_{n+1} g_{n+1}$, there is a generating set U of size $n + 1$ for G that contains y. Since no generating set for G has fewer than $n+1$ elements, $y \neq 0$. As $ky = (\sum_{i=1}^{n} a_i g_i) - a_{n+1} g_{n+1} = 0$, we obtain that $|y|$ divides k. If $|g_{n+1}|$ was a divisor of $|y|$, then $|g_{n+1}|$ would divide a_{n+1} as k divides a_{n+1}, and this would mean that $z = a_{n+1} g_{n+1} = 0$, which is not the case. Thus $|g_{n+1}|$ does not divide $|y|$. Now, $|y| = p^t$ for some positive integer t, and $|g_{n+1}| = p^{\mu(V)}$, so since $|g_{n+1}|$ does not divide $|y|$, we conclude that $t < \mu(V)$ and thus $\mu(U) \leq t < \mu(V)$, which contradicts our choice of V. This contradiction has followed from the assumption that $z \neq e_G$, and so we conclude that $H \cap \langle g_{n+1} \rangle = \{e_G\}$. Since $G = H + \langle g_{n+1} \rangle$, it follows that $G = H \oplus \langle g_{n+1} \rangle = \langle g_1 \rangle \oplus \langle g_2 \rangle \oplus \cdots \oplus \langle g_{n+1} \rangle$. Suppose now that $h_1, h_2, \ldots, h_m \in G$ are such that $|h_1 \geq |h_2| \geq \cdots \geq |h_{n+1}|$ and $G = \langle h_1 \rangle \oplus \cdots \oplus \langle h_m \rangle$. Since no generating set for G has size less than $n+1$, it follows that $m \geq n+1$. We use induction on $i \geq 1$ to prove that if $i \leq n+1$, then for all integers j with $1 \leq j \leq i$, $|g_j| = |h_j|$. Since $|g_1| G = \{e_G\}$, it follows that $|g_1| \langle h_1 \rangle = \{0_G\}$ and thus $|h_1|$ divides $|g_1|$. Similarly, from $|h_1| G = \{e_G\}$ we obtain $|g_1|$ divides $|h_1|$, and so $|g_1| = |h_1|$. Suppose now that $k \geq 1$ is an integer for which the hypothesis holds. There is nothing to prove if $k + 1 > n + 1$, so suppose that $k + 1 \leq n + 1$. Then $k \leq n$ and we obtain that $|g_j| = |h_j|$ for all integers j with $1 \leq j \leq k$. Since $|g_i|$ divides $|g_{k+1}|$ if $i \geq k + 1$, it follows that $|g_{k+1}|G = |g_{k+1}|\langle g_1 \rangle \oplus \cdots \oplus |g_{k+1}|\langle g_k \rangle = |g_{k+1}|\langle h_1 \rangle \oplus \cdots \oplus |g_{k+1}|\langle h_k \rangle \oplus (|g_{k+1}|\langle h_{k+1} \rangle \oplus \cdots \oplus |g_{k+1}|\langle h_m \rangle)$. But

$$| \, |g_{k+1}|G \, | = | \, |g_{k+1}|\langle g_1 \rangle \oplus \cdots \oplus |g_{k+1}|\langle g_k \rangle \, | = | \, |g_{k+1}|\langle h_1 \rangle \oplus \cdots \oplus |g_{k+1}|\langle h_k \rangle \, |$$

and

$$|g_{k+1}|\langle g_1 \rangle \oplus \cdots \oplus |g_{k+1}|\langle g_k \rangle \supseteq |g_{k+1}|\langle h_1 \rangle \oplus \cdots \oplus |g_{k+1}|\langle h_k \rangle,$$

so $|g_{k+1}|\langle h_{k+1} \rangle \oplus \cdots \oplus |g_{k+1}|\langle h_m \rangle = \{ e_G \}$. In particular, $|h_{k+1}|$ divides $|g_{k+1}|$. By a symmetric argument, $|g_{k+1}|$ divides $|h_{k+1}|$, and so $|g_{k+1}| = |h_{k+1}|$. It follows now by induction that $|g_i| = |h_i|$ for all $i = 1, 2, \ldots, n$. If $m > n$, then $| \langle h_{n+1} \rangle \oplus \cdots \oplus \langle h_m \rangle | = 1$, which means that $h_m = e_G$. But then G would have a generating set of size less than $n + 1$, which is not possible. Thus $m = n$.

The result follows now by induction. $\qquad\square$

We summarize the results of this section in the following theorem, which is usually referred to as the fundamental theorem of finitely generated abelian groups.

13.4.4 Theorem. *If G is a finitely generated abelian group, then there exists a free abelian subgroup F of G of finite rank (called the rank of G) such that $G = F \oplus T_G$, where T_G is the torsion subgroup of G. If F_1 is any free abelian subgroup of G and T is any finite subgroup of G such that $G = F_1 \oplus T$, then $T = T_G$ and $rank(F_1) = rank(F)$. Moreover, there is a unique set of primes $S_G = \{ p_1, p_2, \ldots, p_n \}$ such that T_G is the direct sum of finite abelian p_i-subgroups $S_{p_i}(G)$ (the primes p_i are the prime divisors of $|T_G|$ and for each i, $S_{p_i}(G)$ is the Sylow p_i-subgroup of T_G), and for each i, there is a unique sequence of positive integers $(r_1, r_2, \ldots, r_{m_i})$ such that $r_1 \leq r_2 \leq \cdots \leq r_{m_i}$ and S_{p_i} is the direct sum of cyclic groups of order $p_i^{r_i}$ (so $|S_{p_i}| = r_1 + r_2 + \cdots + r_{m_i}$). Finally, if G and G_1 are finitely generated abelian groups, then G_1 is isomorphic to G if, and only if, $rank(G_1) = rank(G)$, $S_G = S_{G_1} = S$, and for each $p \in S$, $S_p(G)$ is isomorphic to $S_p(G_1)$, which holds if, and only if, the sequence of positive integers associated with p by G is the same as the sequence of positive integers associated with p by G_1.*

It is an immediate consequence of the fundamental theorem of finitely generated abelian groups that for each choice of rank and each positive integer n, we can determine (up to isomorphism) the exact number of finitely generated abelian groups with the chosen rank and with torsion subgroup of order n. For example, let us determine, up to isomorphism, the number of abelian groups of order $67,500$. Since the prime decomposition of $67,500$ is $(2^2)(3^3)(5^4)$, we need only determine the different abelian groups of order 2^2, of order 3^3, and of order 5^4. An abelian group of order 2^2 is either a direct sum of two cylic groups of order 2, or is a cyclic group of order 2^2 (corresponding to the two sequences $(1, 1)$ and (2). An abelian group of order 3^3 is isomorphic to $C_3 \oplus C_3 \oplus C_3$, or to $C_3 \oplus C_{3^2}$, or to C_{3^3} (corresponding to the three sequences $(1, 1, 1)$, $(1, 2)$, and (3)), and finally, an abelian group of order 5^4 is isomorphic to one of $C_5 \oplus C_5 \oplus C_5 \oplus C_5$ (with associated sequence $(1, 1, 1, 1)$), $C_5 \oplus C_5 \oplus C_{5^2}$ (with associated sequence $(1, 1, 2)$), $C_5 \oplus C_{5^3}$ (with associated sequence $(1, 3)$), $C_{5^2} \oplus C_{5^2}$ (with associated sequence $(2, 2)$), or to C_{5^4} (with associated sequence (4)). Thus there are in total $(2)(3)(5) = 30$ finite abelian groups of order $67,500$.

More generally, we observe that if p is a prime and n is a positive integer such that G is a finite abelian p-group of order p^n, then there exists a partition of n, by which is meant a sequence $(r_1, r_2, \ldots, r_k)$ such that $1 \le r_1 \le r_2 \le \cdots \le r_k$ and $\sum_{i=1}^{k} r_i = n$, such that G is isomorphic to $C_{p^{r_1}} \oplus C_{p^{r_2}} \oplus \cdots \oplus C_{p^{r_k}}$. By the Fundamental Theorem of Finitely Generated Abelian Groups, the size of a complete set of pairwise nonisomorphic finite abelian p-groups of order p^n is the number of partitions of n. For example, the partitions of 5 are $(1, 1, 1, 1, 1)$, $(1, 1, 1, 2)$, $(1, 1, 3)$, $(1, 4)$, $(1, 2, 2)$, $(2, 3)$, (5), so there are 7 partitions of 5. Thus for any prime p, there are 7 different (up to isomorphism) finite abelian groups of order p^5. In the preceding example, as part of our calculations, we determined the number of finite abelian 5-groups of order 5^4 and determined there were 5 of them, corresponding to the five partitions of 4; namely $(1, 1, 1, 1)$, $(1, 1, 2)$, $(1, 3)$, $(2, 2)$, and (5).

13.5   The Structure of the Group of Units of $\mathbb{Z}_n$

For $m, n \in \mathbb{Z}$ with $n \mid m$, $(m) \subseteq (n)$ and so by Theorem 8.3.4, the map $f : \mathbb{Z}_m \to \mathbb{Z}_n$ given by $f(i + (m)) = i + (n)$ is a surjective ring homomorphism with kernel $(n)/(m)$. Moreover, since $f(1 + (m)) = 1 + (n)$, f is unital. Let m be an integer greater than 1, and suppose the prime decomposition of m is given by $m = \prod_{i=1}^{k} p_i^{r_i}$. For each $i = 1, 2, \ldots, k$, let $\pi_i : \mathbb{Z}_m \to \mathbb{Z}_{p_i^{r_i}}$ be the surjective ring homomorphism given by $\pi_i(a + (m)) = a + (p_i^{r_i})$. Then the map $\pi : \mathbb{Z}_m \to \prod_{i=1}^{k} \mathbb{Z}_{p_i^{r_i}}$ given by

$$\pi(a + (m)) = (a + (p_1^{r_1}), a + (p_2^{r_2}), \ldots, a + (p_k^{r_k}))$$

for each $a \in \mathbb{Z}$ is a unital ring homomorphism. Since $|\mathbb{Z}_m| = m = \prod_{i=1}^{k} p_i^{r_i} = \prod_{i=1}^{k} |\mathbb{Z}_{p_i^{r_i}}| = |\prod_{i=1}^{k} \mathbb{Z}_{p_i^{r_i}}|$, if we prove that π is injective, then π is surjective and thus a ring isomorphism. Suppose that $a + (m) \in \ker(\pi)$. Then for each $i = 1, 2, \ldots, k$, $a + (p_i^{r_i}) = (p_i^{r_i})$, and so $a \in (p_i^{r_i})$; that is, $p_i^{r_i}$ divides a for each $i = 1, 2, \ldots, k$. But then the least common multiple of the pairwise relatively prime integers $p_1^{r_1}, \ldots, p_k^{r_k}$ divides a, and so $m = \prod_{i=1}^{k} p_i^{r_i}$ divides a. Thus $a + (m) = (m)$, and so $\ker(\pi) = \{0 + (m)\}$; that is, π is injective.

Note that since π is unital, units of $\mathbb{Z}_m$ map to units of $\prod_{i=1}^{k} \mathbb{Z}_{p_i^{r_i}}$; that is, $\pi(U_m)$ is contained in the group of units of $\prod_{i=1}^{k} \mathbb{Z}_{p_i^{r_i}}$, which is $\prod_{i=1}^{k} U_{p_i^{r_i}}$. Thus $\pi(U_m) \subseteq \prod_{i=1}^{k} U_{p_i^{r_i}}$. For any $x \in \prod_{i=1}^{k} U_{p_i^{r_i}}$, there exists $a + (m) \in \mathbb{Z}_m$ such that $\pi(a + (m)) = x$. As well, $x^{-1} \in \prod_{i=1}^{k} U_{p_i^{r_i}}$, and so there exists $b + (m) \in \mathbb{Z}_m$ such that $\pi(b + (m)) = x^{-1}$. But then $\pi(ab + (m)) = xx^{-1} = \pi(1 + (m))$. Since π is injective, we conclude that $(a + (m))(b + (m)) = 1 + (m)$ and so $a + (m) \in U_m$. This proves that $\pi(U_m) = \prod_{i=1}^{k} U_{p_i^{r_i}}$, and so the restriction of π gives an isomorphism from U_m onto $\prod_{i=1}^{k} U_{p_i^{r_i}}$. Thus the structure of the finite abelian group U_m is known once we determine the structure of U_{p^n} for any prime p and any positive integer n.

It was shown in Theorem 4.5.8 that for any positive integer n, $U_n = \{i + (n) \mid (i, n) = 1\}$. Since the Euler φ-function of n is defined to be

$$\varphi(n) = |\{i \in J_n \mid (i, n) = 1\}|,$$

we see that for any positive integer n, $|U_n| = \varphi(n)$. Note that we have just observed that from the prime decomposition of $m = \prod_{i=1}^{k} p_i^{r_i}$, we have $\varphi(m) = \prod_{i=1}^{k} \varphi(p_i^{r_i})$.

13.5.1 Proposition. *For any prime p, and any positive integer n,* $\varphi(p^n) = p^{n-1}(p - 1)$.

Proof. We have

$$\{i \in J_{p^n} \mid (i, p^n) = 1\} = \{i \in J_{p^n} \mid p \nmid i\}$$
$$= J_{p^n} - \{jp \mid j = 1, 2, \ldots, p^{n-1}\},$$

and so $\varphi(p^n) = p^n - p^{n-1} = p^{n-1}(p - 1)$. $\qquad\square$

13.5.2 Proposition. *Let p be an odd prime and n a positive integer. Then U_{p^n} is a cyclic group of order $p^n(p-1)$.*

Proof. Let p be an odd prime. Since $\mathbb{Z}_p$ is a field, it follows from Proposition 9.8.7 that U_p is cyclic of order $p-1$. Note that by Proposition 13.5.1, for any positive integer n, $|U_{p^n}| = \varphi(p^n) = p^{n-1}(p-1)$. We prove by induction on $n \geq 2$ that U_{p^n} is cyclic. For the base case, consider U_{p^2}, which is a finite abelian group of order $p(p-1)$. The surjective homomorphism $f : U_{p^2} \to U_p$ given by $f(a + (p^2)) = a + (p)$ has image the cyclic group U_p of order $p-1$, and since $|U_{p^2}| = p(p-1)$, it follows that the kernel of the homomorphism is a subgroup K of order p. Let $a + (p^2)$ be an element whose image is a generator of U_p. Then $p-1$ divides the order of $a + (p^2)$ and so $\langle a + (p^2) \rangle$ contains a cyclic subgroup H of order $p-1$. Since p and $p-1$ are relatively prime, HK is a subgroup of U_{p^2} of order $p(p-1)$ and so $U_{p^2} = HK$. Since H and K are each cyclic and their orders are relatively prime, $U_{p^2} = HK$ is cyclic. This completes the proof of the base case.

Suppose now that $n \geq 2$ is an integer for which the hypothesis holds. We consider the surjective homomorphisms $f : U_{p^{n+1}} \to U_{p^n}$ and $g : U_{p^n} \to U_{p^{n-1}}$ given by $f(a + (p^{n+1})) = a + (p^n)$ for all $a \in \mathbb{Z}$, and $g(a + (p^n)) = a + (p^{n-1})$ for all $a \in \mathbb{Z}$. Since U_{p^n} is cyclic, it follows that $U_{p^{n-1}}$, as the homomorphic image of a cyclic group, is also cyclic. Let $a + (p^n)$ be a generator for U_{p^n}, so the order of $a + (p^n)$ is $p^{n-1}(p-1)$. The order of $a + (p^{n+1})$ is a divisor of $|U_{p^{n+1}}| = p^n(p-1)$, and since $f(a + (p^{n+1})) = a + (p^n)$, which has order $p^{n-1}(p-1)$, it follows that $p^{n-1}(p-1)$ is a divisor of the order of $a + (p^{n+1})$. Thus the order of $a + (p^{n+1})$ is either $p^{n-1}(p-1)$ or $p^n(p-1)$. We shall prove that it is not $p^{n-1}(p-1)$, which then establishes that $U_{p^{n+1}}$ is a cyclic group of order $p^n(p-1)$ with generator $a + (p^{n+1})$. Since $a + (p^{n-1})$ is a generator for $U_{p^{n-1}}$, we have $a^{p^{n-2}(p-1)} = 1 + rp^{n-1}$ for some $r \in \mathbb{Z}$. Since the order of $a + (p^n)$ is $p^{n-1}(p-1)$, it follows that $p \nmid r$. But then $a^{p^{n-1}(p-1)} = (a^{p^{n-2}(p-1)})^p = (1 + rp^{n-1})^p = \sum_{i=0}^{p} \binom{p}{i}(rp^{n-1})^i$. Since $n \geq 2$, $n - 1 \geq 1$. As well, for each i with $1 \leq i \leq p-1$, p is a divisor of $\binom{p}{i}$ while p^2 is not a divisor of $\binom{p}{i}$. Thus the exponent of p in the summand with index i is $i(n-1) + 1$. For $1 < i < p$, the exponent of p is $n - 1 + (i-1)(n-1) + 1 \geq n - 1 + 1 + 1 = n + 1$, while for $i = p$, the exponent of p is $(n-1)p = n - 1 + (p-1)(n-1) \geq n - 1 + p - 1 \geq n - 1 + 2 = n + 1$ (here is where we need $p > 2$). Thus $\binom{p}{i}(rp^{n-1})^i \equiv 0 \mod p^{n+1}$ for $1 < i \leq p$, and so $a^{p^{n-1}(p-1)} \equiv 1 + \binom{p}{1}rp^{n-1} = 1 + rp^n \mod p^{n+1}$. Since $p \nmid r$, $a^{p^{n-1}(p-1)} + (p^{n+1}) \neq 1 + (p^{n+1})$, and so the order of $a + (p^{n+1})$ is not $p^{n-1}(p-1)$. As we have explained above, this means that the order of

$a + (p^{n+1})$ is $p^n(p-1) = |U_{p^{n+1}}|$, and so $U_{p^{n+1}}$ is cyclic, with generator $a + (p^{n+1})$. The result follows now by induction. □

The reader may be interested to note that what was actually proved in Proposition 13.5.2 was that if $a \in \mathbb{Z}$ is such that $a + (p^2)$ is a generator for U_{p^2}, then $a + (p^n)$ is a generator for U_{p^n} for every $n \geq 2$. It is interesting to consider the problem of finding a generator for U_{p^2}. We know that U_p is cyclic, but unfortunately, there is no algorithmic way to efficiently find a generator for U_p. However, suppose that we have found a generator for U_p, say $a \in \mathbb{Z}$ is such that $a + (p)$ is a generator for U_p. We prove that either $a + (p^2)$ has order $p(p-1)$ in U_{p^n}, or if not, then $(1+p)a + (p^2)$ has order $p(p-1)$ in U_{p^2}. Since $a + (p^2)$ maps to $a + (p)$ under the surjective unital ring homomorphism from $\mathbb{Z}_{p^2}$ onto $\mathbb{Z}_p$ that maps $i + (p^2)$ to $i + (p)$, it follows that the order of $a + (p^2)$ is a multiple of $p - 1$, the order of $a + (p)$, and is a divisor of $|U_{p^2}| = p(p-1)$. Thus if $a^{p-1} + (p^2) = 1 + (p^2)$, then $a + (p^2)$ has order $p - 1$. Suppose that this is the case. Then since $1 + p + (p^2)$ is in the kernel of this ring homomorphism, which must be a subgroup of order p, and $1 + p + (p^2) \neq 1 + (p^2)$, it follows that the order of $1 + p + (p^2)$ is p. Since p and $p - 1$ are relatively prime, the order of $(1+p)a + (p^2) = p(p-1)$; that is, $(1+p)a + (p^2)$ is a generator for U_{p^2}. On the other hand, if $a^{p-1} + (p^2) \neq 1 + (p^2)$, then the order of $a + (p^2)$ is $p(p-1)$. A natural question that arises now is whether or not every $a \in \mathbb{Z}$ for which $a + (p)$ is a generator for U_p is such that $a + (p^2)$ is a generator for U_{p^2}. Interestingly, it has been confirmed computationally that for every prime $p < 10^7$, if a is the least element of J_p for which $a + (p)$ is a generator of U_p, then also $a + (p^2)$ is a generator for U_{p^2}. The same does not hold true for every integer a for which $a + (p)$ is a generator of U_p. The smallest prime for which this occurs is $p = 29$, where $14 + (19)$ is a generator for U_{19}, but $14 + (19^2)$ is not a generator for U_{19^2}.

In summary, the major effort that is required to find a generator for U_{p^2} goes into finding a generator $a + (p)$ for U_p. After that, one need only check to see if $a^{p-1} + (p^2) = 1 + (p^2)$. If so, then $(1+p)a + (p^2)$ is a generator for U_{p^2}, while if not, then $a + (p^2)$ is a generator for U_{p^2}.

It remains to determine the structure of U_{2^n} for $n \geq 1$. Of course, U_2 is the trivial group, and U_{2^2} is a group of size $2^1(2-1) = 2$, so U_{2^2} is a cyclic group of order 2.

13.5.3 Proposition. *For $n \geq 3$, U_{2^n} is not cyclic. The cyclic subgroup $H = \langle 5 + (2^n) \rangle$ has order 2^{n-2} and does not contain $2^n - 1 + (2^n)$, an* *element of order 2. Let $K = \langle 2^n - 1 + (2^n) \rangle$, so K is a subgroup of U_{2^n} of order 2, and $H \cap K = \{1 + (2^n)\}$. Thus $U_{2^n} = KH \simeq \mathbb{Z}_2 \oplus \mathbb{Z}_{2^{n-2}}$.*

Proof. For any $n \geq 3$, $|U_{2^n}| = 2^{n-1}(2-1) = 2^{n-1}$, and by the fundamental theorem of finitely generated abelian groups, it is the direct sum of cyclic groups of prime power order, hence of cyclic groups each of order a power of 2. We may easily observe that U_{2^n} is not cyclic, since it does not have a unique subgroup of order 2. In fact, $2^n - 1 + (2^n)$, $2^{n-1} - 1 + (2^n)$, and $2^{n-1} + 1 + (2^n)$ are three distinct elements of order 2 (note that we need $n \geq 3$ for this observation). Next, we prove that for all $n \geq 3$, $5 + (2^n)$ has order 2^{n-2} in U_{2^n}. We shall manage this by using induction on $n \geq 2$ to prove that there exists an odd integer r such that $5^{2^{n-2}} = 1 + r2^n$. Since $5^{2^0} = 5 = 1 + (1)(2^2)$, the result holds when $n = 2$. Suppose now that $n \geq 2$ is an integer for which there exists an odd integer r with $5^{2^{n-2}} = 1 + r2^n$. Then $5^{2^{n+1-2}} = 5^{2^{n-1}} = (5^{2^{n-2}})^2 = (1 + r2^n)^2 = 1 + r2^{n+1} + r^2 2^{2n} = 1 + (r + r^2 2^{n-1})2^{n+1}$, and since $n \geq 2$, 2^{n-1} is even and thus $r + r^2 2^{n-1}$ is odd. The claim follows now by induction. Consider now any $n \geq 3$. Then for some odd integer r, $5^{2^{n-2}} = 1 + r2^n$, so the order of $5 + (2^n)$ in U_{2^n} divides 2^{n-2}. Suppose that $5^{2^{n-3}} \equiv 1 \bmod 2^n$. Then 2^n divides $5^{2^{n-3}} - 1$. But since $n - 1 \geq 2$, there exists an odd integer s such that $5^{2^{n-1-2}} = 1 + s2^{n-1}$, and so we conclude that 2^n divides $s2^{n-1}$; that is, s is even. Since this is not the case, we see that $5 + (2^n)$ has order 2^{n-2} in U_{2^n}. It follows now by induction that $5 + (2^n)$ has order 2^{n-2} in U_{2^n} for every $n \geq 3$.. At this point, we know that U_{2^n} is not cyclic, but contains a cyclic subgroup of order 2^{n-2}. Since $|U_{2^n}| = 2^{n-1}$, it follows from the fundamental theorem of finitely generated abelian groups that $U_{2^n} \simeq \mathbb{Z}_{2^{n-2}} \oplus \mathbb{Z}_2$, but we shall actually prove that $2^n - 1 + (2^n) \notin \langle 5 + (2^n) \rangle$, so $U_{2^n} = \langle 2^n - 1 + (2^n) \rangle \langle 5 + (2^n) \rangle \simeq \mathbb{Z}_2 \oplus \mathbb{Z}_{2^{n-2}}$. Suppose by way of contradiction that $2^n - 1 \in \langle 5 + (2^n) \rangle$. Since $|5 + (2^n)| = 2^{n-2}$, it follows that the unique element of order 2 in $\left\langle 5^{2^{n-2}} \right\rangle$ is $5^{2^{n-3}} + (2^n)$. As $n - 1 \geq 2$, there exists an odd integer r such that $5^{2^{n-3}} = 1 + r2^{n-1}$, and so we have $1 + r2^{n-1} \equiv 2^n - 1 \bmod 2^n$; that is, 2^n divides $2 + r2^{n-1}$, or 2^{n-1} divides $1 + r2^{n-2}$. Since $n \geq 3$, this is not possible. Thus $2^n - 1 + (2^n) \notin \langle 5 + (2^n) \rangle$, as claimed. $\square$

In summary, we have proven the following result.

13.5.4 Proposition. *Let m be a positive integer, and suppose the prime decomposition of m is $m = \prod_{i=1}^{k} p_i^{r_i}$. Then $U_m \simeq U_{p_1^{r_1}} \oplus U_{p_2^{r_2}} \oplus \cdots \oplus U_{p_k^{r_k}}$.*

For any positive integer r, U_{2^r} is trivial if $r = 1$, a cyclic group of order 2 if $r = 2$, while $U_{2^r} \simeq \mathbb{Z}_2 \oplus \mathbb{Z}_{2^{n-2}}$ for $r \geq 2$. For any odd prime p, and any positive integer r, $U_{p^r} \simeq \mathbb{Z}_{p^{n-1}(p-1)}$.

We remark that since every cyclic group is the direct sum of its Sylow subgroups, the representation of U_m as the direct sum of cyclic groups of prime power order is immediately obtained from Proposition 13.5.4. For example, consider the case of $m = 72$. We have $U_{72} \simeq U_{2^3} \oplus U_{3^2} \simeq \mathbb{Z}_2 \oplus \mathbb{Z}_2 \oplus \mathbb{Z}_3 \oplus \mathbb{Z}_2$.

6 Exercises

1. Let p and q be distinct primes. For each of the following values, determine the number of pairwise non-isomorphic abelian groups of that order, and express each as a sum of cyclic groups of prime power order.

   a) p^2.

   b) p^3.

   c) $p^4 q^3$.

2. a) Let G be a finite abelian group. Prove that there exist cyclic subgroups $C_1, C_2, \ldots, C_n$ of G, where n is the size of a smallest generating set for G, such that $G = C_1 \oplus C_2 \oplus \cdots \oplus C_n$, and if $n > 1$, $|C_i| \mid |C_{i+1}|$ for $i = 1, 2, \ldots, n - 1$.

   b) Express each abelian group of order 36 in the form described in (a).

3. What is the order of the automorphism group of $C_4 \oplus C_9$?

4. Prove that if G is a finite abelian but noncyclic group, then $\mathrm{Aut}(G)$ is not abelian.

5. Prove that the additive group of rational numbers $(\mathbb{Q}, +)$ cannot be written as the direct sum of two of its nontrivial subgroups.

6. a) Suppose that G is a finite group with the following property: for any nonidentity elements g, h of G, there is an automorphism φ of G such that $\varphi(g) = h$. Prove that there is a prime p such that G is a direct sum of cyclic groups of order p.

   b) Let p be a prime. If G is a direct sum of finitely many cyclic groups of order p, and $g, h \in G$, is there an automorphism φ of G such that $\varphi(g) = h$?

7. a) Let G be a finite abelian group with a generating set of size n.

Prove that every subgroup of G has a generating set of size at most n.

b)   Give an example of a finite abelian group with a generating set of size 2, and which contains a subgroup H such that every generating set for H has size at least 2.

8.   Let p be a prime and let G be an finite abelian p-group. Define the function $\varphi_p : G \to G$ by $\varphi_p(g) = pg$ for each $g \in G$.

a)   Prove that φ_p is a homomorphism.

b)   Prove that $\ker(\varphi_p) \simeq G/\varphi_p(G)$.

c)   Prove that the number of subgroups of G of order p is equal to the number of subgroups of G of index p.

d)   Let $G = C_{12} \times C_{36} \times C_{20}$. Find the number of elemens of order 2 in G (and thus the number of subgroups of index 2 in G).

9.   Let G be a free abelian group of finite rank n (so $G \simeq \mathbb{Z} \oplus \mathbb{Z} \oplus \cdots \oplus \mathbb{Z}$, where there are n summands). Let m be a positive integer. Prove that $G/mG \simeq \mathbb{Z}_m \oplus \mathbb{Z}_m \oplus \cdots \oplus \mathbb{Z}_m$ (n summands).

Chapter 14

Semigroups and Automata

1 Introduction

Although the idea of a semigroup appeared in the mathematical literature as early as 1904, it was not until 1928 that work in the theory of semigroups began in earnest with the publication of a paper by A. K. Suschkewitsch, in which he completely determined the structure of all finite simple semigroups; that is, semigroups which have no proper ideals. Since that time, the theory has been developed to a high degree of sophistication.

Apart from the intrinsic interest of the subject, semigroups play an important role in many other branches of mathematics, and in particular, they have a fundamental role in the study and classification of formal languages in theoretical computer science.

2 Semigroups

Recall that in Definition 3.4.20, a semigroup was defined as a set together with an associative binary operation on the set, and if the operation is commutative, we say that the semigroup is commutative. Further, if the binary operation has an identity, then the semigroup was called a monoid. Finally, it was observed that in a monoid, not every element need be invertible, but a monoid in which every element is invertible was called a group.

Recall that if $*$ is a binary operation on a set S, then a set $T \subseteq S$ was said to be closed under $*$ if for all $x, y \in T$, $x * y \in T$ (see Definition 3.4.11). Further, if $*$ is associative on S, then $*|_{T \times T}$ is associative on T.

14.2.1 Definition. *Let $(S, *)$ be a semigroup. If $\varnothing \neq T \subseteq S$ is closed with respect to the binary operation $*$ (so that $*|_{T \times T}$ is a binary operation* on T), then $(T, *|_{T \times T})$ is called a *subsemigroup* of $(S, *)$. Further, if S is a monoid and T is a subsemigroup of S containing the identity of S, then T is called a *submonoid* of S.

It will be convenient to have notation for the identity element of a monoid, and we shall use e_S to denote the identity element of a monoid S.

Just as we did for the case of groups, we shall typically write S in place of $(S, *)$ when the binary operation is unambiguously specified, and we shall typically employ multiplicative notation to denote the effect of applying $*$ to an input pair (x, y); that is, we shall write xy to mean $x * y$.

14.2.2 Definition. *Let S be a semigroup, and let $\theta \in S$. If $\theta a = \theta$ for all $a \in S$, then θ is called a* left zero *of S, while if $a\theta = \theta$ for all $a \in S$, then θ is called a* right zero *for S. If θ is both a left and a right zero for S, then θ is called a* zero *for S.*

It is possible that a semigroup may possess more than one right zero (respectively, left zero), and a semigroup may even possess both left zeroes and right zeroes. However, if a semigroup has a zero, then it has exactly one zero element, and then the zero element is the only left or right zero of the semigroup.

14.2.3 Proposition. *Let S be a semigroup, and let θ be a zero of S. Then θ is the unique left zero of S, and the unique right zero of S.*

Proof. Let $a \in S$ be a left zero of S. Then $a\theta = a$ since a is a left zero, but $a\theta = \theta$ since θ is a right zero. Thus $a = \theta$. Similarly, if a is a right zero of S, then $\theta a = a$ and $\theta a = \theta$, so $a = \theta$. $\qquad \square$

The notions of an identity element or a zero element are special cases of the more general notion of an idempotent element.

14.2.4 Definition. *Let S be a semigroup. An element $x \in S$ is said to be an* idempotent *of S if $x^2 = x$. E_S shall denote the set of all idempotents of S.*

Note that E_S may be empty. For example, the semigroup $\mathbb{Z}^+$ under addition has no idempotents.

14.2.5 Definition. *Let S be a semigroup, and let S^1 denote S if S is actually a monoid, otherwise let $S^1 = S \cup \{1\}$, where $1 \notin S$, and in the latter case, extend the binary operation on S to S^1 by defining $x1 = 1x = x$ for all $x \in S^1$. S^1 is said to be the result of* adjoining an identity *to S.*

The notation is potentially confusing if the discussion involves both the Cartesian product of some number of copies of S (in that we defined $S^1 = S$ and for $n \geq 1$, $S^{n+1} = S^n \times S$) and also the adjunction of an identity to S, but the context should always make things clear.

14.2.6 Proposition. *Let S be a semigroup. Then S^1 is a monoid and S is a subsemigroup of S^1.*

Proof. Let $x, y, z \in S^1$. If $x = 1$, then $x(yz) = yz$ and $(xy)z = yz$, while if $y = 1$, then $x(yz) = xz$ and $(xy)z = xz$. Finally, if $z = 1$, then $x(yz) = xy$ and $(xy)z = xy$. If $x, y, z \in S$, then $(xy)z = x(yz)$ by hypothesis. Thus the extension of $*$ to S^1 is an associative binary operation on S^1. It is immediate that S is closed under the binary operation on S^1, so S is a subsemigroup of the monoid S^1. $\qquad\square$

14.2.7 Definition. *Let S be a semigroup. A nonempty subset I of S is called a left ideal of S if $SI = \{ xa \mid x \in S, \ a \in I \} \subseteq I$, a right ideal of S if $IS = \{ ax \mid x \in S, \ a \in I \} \subseteq I$, and an ideal of S if it is both a left and a right ideal of S. If the only left ideal of S is S itself, then S is said to be a left-simple semigroup, and if the only right ideal of S is S itself, then S is said to be a right-simple semigroup. S is said to be simple if the only ideal of S is S itself. If S is a semigroup with zero θ, then S is said to be left (respectively right)-0-simple if $S^2 = \{ xy \mid x, y \in S \} \neq \{ \theta \}$ and the only left (respectively right) ideals of S are $\{ \theta \}$ and S, while S is said to be 0-simple· if $S^2 \neq \{ \theta \}$ and the only ideals of S are $\{ \theta \}$ and S.*

Note that each left ideal and each right ideal of S is a subsemigroup of S. Furthermore, for any $a \in S$, aS is a right ideal of S, and Sa is a left ideal of S. Evidently, S is left (right) simple if, and only if, $Sa = S$ $(aS = S)$ for all $a \in S$.

14.2.8 Proposition. *Let S be a semigroup. For any $a \in S$, $S^1 a = \{ a \} \cup aS$ is a left ideal of S, called the principal left ideal of S generated by a, and $aS^1 = \{ a \} \cup aS$ is a right ideal of S, called the principal right ideal of S generated by a. Finally, $S^1 a S^1 = \{ a \} \cup aS \cup Sa \cup SaS$ is an ideal of S, called the principal ideal of S that is generated by a.*

Proof. We note that for any $a \in S$, $S^1 a \subseteq S$, $aS^1 \subseteq S$, and $S^1 a S^1 \subseteq S$. Let $x, y \in S$ and $z, w \in S^1$. Then $x(za) = (xz)a \in S^1 a$, $(az)x = a(zx) \in aS^1$, and $x(zaw)y = (xz)a(wy) \in S^1 a S^1$, as required. $\qquad\square$

14.2.9 Example.

(i) $\mathbb{N}$ *is a monoid under each of the operations of addition and multiplication respectively. In the former case, the identity element is* 0, *while in the latter case, the identity element is* 1 *(and* 0 *is a zero of* $\mathbb{N}$ *under multiplication).* $\mathbb{Z}^+$ *is a subsemigroup of* $\mathbb{N}$ *in both cases. Under addition,* $\mathbb{Z}^+$ *does not have an identity element, while under multiplication,* $\mathbb{Z}^+$ *has identity* 1 *(but no zero element). Since each of addition and multiplication are commutative operations, every one-sided ideal is an ideal in both cases (addition and multiplication). If* I *is an ideal of* $\mathbb{N}$ *under addition, and* $n \in I$ *is the least element of* I, *then* $I = \{\, n + i \mid i \in \mathbb{N} \,\}$. *Under multiplication, the ideals of* $\mathbb{N}$ *are more difficult to describe.*

(ii) *Let* S *be a nonempty set, and define an operation on* S *by* $ab = b$ *for all* $a, b \in S$. *With this operation,* S *is a semigroup with no right ideals other than* S *itself; that is,* S *is right simple (and therefore simple). On the other hand, every nonempty subset of* S *is a left ideal of* S. *Every element of* S *is an idempotent of* S.

(iii) *Let* X *and* Y *be nonempty sets, and define an operation on the Cartesian product* $X \times Y$ *by* $(x, y)(x', y') = (x, y')$ *for all* $(x, y), (x', y') \in X \times Y$. *With this operation,* $X \times Y$ *is a semigroup in which every element is an idempotent. Moreover, for any* $x \in X$, $\{\, x \,\} \times Y$ *is a left ideal of* $X \times Y$, *and for any* $y \in Y$, $X \times \{\, y \,\}$ *is a right ideal of* $X \times Y$.

(iv) *For any positive integer* n, $M_{n \times n}(\mathbb{R})$, *with operation matrix multiplication, is a (noncommutative) monoid with identity element* I_n *and zero* $\mathbb{0}_n$. *The set of all* $A \in M_{n \times n}(\mathbb{R})$ *with at most one nonzero row is a right ideal, while the set of all* $A \in M_{n \times n}(\mathbb{R})$ *with at most one nonzero column is a left ideal. Any diagonal matrix whose diagonal elements are either* 0 *or* 1 *is an idempotent.*

(v) *For any set* X, *the set* $\mathscr{P}(X)$ *with the operation of set intersection is a commutative monoid with zero. The identity element is* X, *while the zero element is* $\varnothing$. *With the operation of set union,* $\mathscr{P}(X)$ *is a commutative monoid with zero, with identity element* $\varnothing$ *and zero* X. *In both cases, every element is an idempotent.*

(vi) $\mathbb{Z}^+$ *with binary operation defined by* $ab = (a, b)$ *for all* $a, b \in \mathbb{Z}^+$ *is a commutative semigroup with zero, but no identity. The zero element is* 1. *On the other hand,* $\mathbb{Z}^+$ *with binary operation given by* $ab = [a, b]$ *for all* $a, b \in \mathbb{Z}^+$ *is a commutative monoid without zero (the identity element is 1).*

(vii) *For any positive integer n, $\mathbb{Z}_n$ under multiplication is a commutative finite monoid with zero (of course, $\mathbb{Z}_n$ under addition is also a commutative monoid, in fact a group). There may be idempotents other than 1 or 0. For example, in $\mathbb{Z}_6$, $4^2 = 4$.*

(viii) *For any nonempty set X, the set T_X of all maps from X to X, with binary operation composition of maps, is a monoid with identity $\mathbb{1}_X$. For any $x \in X$, the map $r_x : X \to X$ defined by $r_x(t) = x$ for all $t \in X$, called a constant map, is an idempotent of T_X, and $\{r_x\}$ is a right ideal of T_X. The set of all constant maps, $\{r_x \mid x \in X\}$, is an ideal of T_X. Note that the symmetric group S_X is a submonoid of T_X. Since S_X is actually a group, we say that S_X is a subgroup of the monoid T_X. We also adopt the convention that if n is a positive integer and $X = J_n = \{1, 2, \ldots, n\}$, then we write T_n in place of T_{J_n}. Note that $|T_n| = n^n$.*

(ix) *For any semigroups P and T with $P \cap T = \varnothing$, let 0 denote a new element (that is, $0 \notin P \cup T$) and define an operation on $S = P \cup T \cup \{0\}$ as follows. Let $x, y \in S$. If $x, y \in P$, respectively T, define xy to be the product in P, respectively T, otherwise define $xy = 0$. With this binary operation, S is a semigroup with zero element 0. Each of $P \cup \{0\}$ and $T \cup \{0\}$ is an ideal of S.*

(x) *Let S and T be semigroups. On the Cartesian product $S \times T$, define the product operation by $(s, t)(s_1, t_1) = (ss_1, tt_1)$, where the operation in the first, respectively second, coordinate is computed in S, respectively T. This defines an associative binary operation on $S \times T$, and the resulting semigroup is called the direct product of S and T. Suppose that each of S and T are monoids with zero, and denote the identities and zeroes of S and T by 1_S, 1_T, 0_S, and 0_T respectively. Then $S \times T$ is a monoid with identity $(1_S, 1_T)$, and $S \times \{0_T\}$ and $\{0_S\} \times T$ are both ideals of $S \times T$, while $(0_S, 0_T)$ is the zero of $S \times T$.*

(xi) *Let S be a semigroup, and use the operation on S to define an operation on $\mathscr{P}(X)$ in a natural way by defining $XY = \{xy \mid x \in X, \ y \in Y\}$ for all $X, Y \in \mathscr{P}(S)$. With this operation, $\mathscr{P}(S)$ is a semigroup with zero $\varnothing$. Furthermore, if S is a monoid with identity 1_S, then $\mathscr{P}(S)$ is a monoid with identity $\{1_S\}$.*

We return now to a discussion of semigroups in general.

14.2.10 Proposition. *Let S be a semigroup, and let $\mathscr{S}$ be a nonempty set of subsemigroups of S. If $\cap \mathscr{S}$ is not empty, then $\cap \mathscr{S}$ is a subsemigroup of S.*

Proof. Let $x, y \in \cap \mathscr{S}$. Then for each $T \in \mathscr{S}$, $x, y \in T$ and thus $xy \in T$, which means that $xy \in T$ for every $T \in \mathscr{S}$. It follows now that either $\cap \mathscr{S} = \varnothing$ or else $\cap \mathscr{S}$ is nonempty and closed under the operation of S; hence a subsemigroup of S. $\qquad \square$

Unlike the situation for groups, it is entirely possible for $\cap \mathscr{S}$ to be empty. In fact, in Example 14.2.9, there are examples where $\mathscr{S}$ consists entirely of left (or right) ideals, yet $\cap \mathscr{S} = \varnothing$.

14.2.11 Definition. *Let S be a semigroup. For any nonempty subset A of S, let $\mathscr{S}_A = \{ T \subseteq S \mid T$ is a subsemigroup of S and $A \subseteq T \}$. Then $A \subseteq \cap \mathscr{S}_A$, and $\langle A \rangle$, the subsemigroup generated by A, is defined by $\langle A \rangle = \cap \mathscr{S}_A$. A is called a set of generators of $\langle A \rangle$, and if $A = \{ a \}$, then we write $\langle a \rangle$ rather than $\langle \{ a \} \rangle$.*

14.2.12 Proposition. *Let S be a semigroup and let A be a nonempty subset of S. Then $\langle A \rangle = \cup_{n \in \mathbb{Z}^+} A^n$, where $A^1 = A$ and for any positive integer n, $A^{n+1} = A^n A = \{ xy \mid x \in A^n, \ y \in A \}$.*

Proof. Since $\mathscr{P}(S)$ is a subsemigroup under the multiplication of subsets (see Example 14.2.9 (xi)), and by the exponential law, $A^m A^n = A^{m+n}$ for any positive integers m and n. Thus $T = \cup_{n \in \mathbb{Z}^+} A^n$ is closed under the operation on S, and since $A \subseteq T$, T is a subsemigroup of S and so $T \in \mathscr{S}_A$. By definition, $\langle A \rangle \subseteq T$. We prove now by induction on n that $A^n \subseteq \langle A \rangle$ for all $n \in \mathbb{Z}^+$. By definition, $A \subseteq \langle A \rangle$, so suppose that $n \in \mathbb{Z}^+$ is such that $A^n \subseteq \langle A \rangle$. Then $A^{n+1} = A^n A \subseteq \langle A \rangle \langle A \rangle \subseteq \langle A \rangle$ since by Proposition 14.2.10, $\langle A \rangle$ is a subsemigroup of S. It follows now by induction that $A^n \subseteq \langle A \rangle$ for every $n \in \mathbb{Z}^+$, and thus $T = \cup_{n \in \mathbb{Z}^+} A^n \subseteq \langle A \rangle$. $\qquad \square$

14.2.13 Definition. *Let S be a semigroup.*

    (i) *If there exists $a \in S$ such that $S = \langle a \rangle$, then S is called a cyclic, or monogenic, semigroup, and a is called a generator of S.*

    (ii) *If every cyclic subsemigroup of S is finite, then S is said to be periodic.*

Evidently, every finite semigroup is periodic, but there are infinite periodic semigroups. For example, in Example 14.2.9 (v), if X is infinite, then

$\mathscr{P}(X)$ is infinite, but under either intersection or union, every element is an idempotent and thus generates a subsemigroup of size one. For another example, consider the following.

14.2.14 Example. *Let X be an infinite set, and let $S = \{ f \in T_X \mid |f(X)| < \infty \}$. Since $f \circ g(X) = f(g(X)) \subseteq f(X)$ for any $f, g \in T_X$, it follows that if $f \in S$ and $g \in T_X$, then $f \circ g \in S$ and so S is a right ideal of T_X. Consequently, S is a subsemigroup of T_X. Furthermore, for any $f \in T_X$, and every positive integer n, $f^{n+1}(X) = f^n(f(X)) \subseteq f^n(X) \subseteq f(X)$, and if $f \in S$, then $f(X)$ is finite. Suppose that $f \in S$. Then $f(X)$ is finite, and so there exists a positive integer n such that $f^{n+1}(X) = f^n(X)$. But then $f|_{f^n(X)}$ is a surjective map from the finite set $f^n(X)$ to itself, and thus $f|_{f^n(X)}$ is a permutation of the finite set $f^n(X)$. It follows that there exists a positive integer m such that $(f|_{f^n(X)})^m = \mathbb{1}_{f^n(X)}$. Then for every $x \in X$, $f^m(f^n(x)) = f^n(x)$ and so $f^{m+n} = f^m \circ f^n = f^n$. It follows that $\langle f \rangle = \{ f^i \mid 1 \le i \le m + n \}$ and so S is periodic. As every constant map from X to X belongs to S, S is infinite.*

14.2.15 Definition. *Let S and T be semigroups. A homomorphism from S to T is a map $f : S \to T$ with the property that for all $x, y \in S$, $f(xy) = f(x)f(y)$. If S and T are monoids and $f(e_S) = e_T$, then we say that f is a unital homomorphism, or a monoid homomorphism. If f is a bijective homomorphism from S onto T, then f is said to be an isomorphism from S to T. An isomorphism from S to itself is called an automorphism of S.*

We remark that it is possible to have a semigroup homomorphism from a monoid S to a monoid T which is not a monoid homomorphism. However, in such a case the homomorphism is not surjective.

14.2.16 Proposition. *If S is a monoid, and T is a semigroup, and there is a surjective homomorphism $f : S \to T$, then T is a monoid and f is a unital homomorphism.*

Proof. It suffices to prove that $f(e_S)$ is an identity for T. Let $t \in T$. Since f is surjective, there exists $x \in S$ with $f(x) = t$, and so $f(e_S)t = f(e_S)f(x) = f(e_S x) = f(x) = t$, and $tf(e_S) = f(x)f(e_S) = f(x e_S) = f(x) = t$. Thus $f(e_S)$ is an identity of T. ∎

14.2.17 Proposition. *Let S and T be semigroups, and let $f : S \to T$ be an isomormphism from S to T. Then $f^{-1} : T \to S$ is an isomorphism from T to S.*

Proof. Since f is a bijective map from S onto T, f^{-1} is a bijective map from T onto S, so we need only prove that for all $x, y \in T$, $f^{-1}(xy) = f^{-1}(x)f^{-1}(y)$. Let $x, y \in T$. Then $f(f^{-1}(x)f^{-1}(y)) = f(f^{-1}(x))f(f^{-1}(y)) = xy = f(f^{-1}(xy))$. Since f is injective, we conclude that $f^{-1}(xy) = f^{-1}(x)f^{-1}(y)$, as required. $\qquad\square$

In Definition 5.2.4, we defined exponention in any group. The same definition can be offered for a semigroup, as the fundamental property of exponentiation relies only on the associativity of the binary operation.

14.2.18 Definition. *Let S be a semigroup, and let $a \in S$. Define $a^1 = a$, and for any $n \in \mathbb{Z}^+$, define $a^{n+1} = a^n a$. If S is a monoid with identity e, further define $a^0 = e$.*

14.2.19 Proposition. *Let S be a semigroup. Then for any $a \in S$ and any $m, n \in \mathbb{Z}^+$, $a^{m+n} = a^m a^n$ and $(a^m)^n = a^{mn}$. Moreover, if S is a monoid with identity e, then these equations hold for all $m, n \in \mathbb{N}$.*

Proof. Let $a \in S$. We prove the result by induction on n, with hypothesis that for all $m \in \mathbb{Z}^+$, $a^{m+n} = a^m a^n$. The result holds for $n = 1$ by definition of exponentiation, so suppose that $n \in \mathbb{Z}^+$ is such that the hypothesis holds. Then for any $m \in \mathbb{Z}^+$, $a^{m+(n+1)} = a^{(m+n)+1} = a^{m+n}a = (a^m a^n)a = a^m(a^n a) = a^m a^{n+1}$. It follows by induction now that for all $m, n \in \mathbb{Z}^+$, $a^{m+n} = a^m a^n$. Next, we prove by induction on n that for all $m \in \mathbb{Z}^+$, $(a^m)^n = a^{mn}$. Again, the result holds by definition for $n = 1$, so suppose that $n \in \mathbb{Z}^+$ is such that the hypothesis holds, and let $m \in \mathbb{Z}^+$. Then $(a^m)^{n+1} = (a^m)^n a^m = a^{mn}a^m = a^{mn+m} = a^{m(n+1)}$, and so the result follows by induction. The extension of the result in the case S is a monoid is immediate. $\qquad\square$

It is fruitful to interpret Proposition 14.2.19 in terms of homomorphisms. The mapping from the semigroup $\mathbb{Z}^+$ with the operation of addition to the semigroup S, with selected element $a \in S$, given by mapping $m \in \mathbb{Z}^+$ to $a^m \in S$ is a homomorphism from $\mathbb{Z}^+$ to S. Moreover, by Proposition 14.2.12, the image of this homomorphism is $\langle a \rangle$. It is worthwhile to investigate the differences between the situation when this mapping is injective and when it is not injective. If the mapping is injective, then $\mathbb{Z}^+$ (under addition) is isomorphic to $\langle a \rangle$, and in particular, $\langle a \rangle$ is infinite. What can be said about the situation when the mapping is not injective?

14.2.20 Proposition. *Let S be a semigroup, and let $a \in S$.*

(i) $\langle a \rangle$ *is finite if, and only if, the exponentiation mapping $m \mapsto a^m$ is not injective; that is, there exist $m, n \in \mathbb{Z}^+$ such that $m \neq n$ and $a^{m+n} = a^n$.*

(ii) *Suppose that $\langle a \rangle$ is finite, and let*

$$s = \min\{\, k \in \mathbb{Z}^+ \mid \text{ there exists } r \in \mathbb{Z}^+ \text{ with } r < k \text{ and } a^k = a^r \,\}.$$

Then there is a unique $r \in \mathbb{Z}^+$ such that $r < s$ and $a^r = a^s$. Furthermore, $K_a = \{\, a^t \mid t \in \mathbb{Z}^+, \ t \geq r \,\}$ is a finite cyclic group of order $s - r$, and K_a is an ideal of $\langle a \rangle$.

Proof. If $\langle a \rangle$ is finite, then there exist $m, n \in \mathbb{Z}^+$ such that $a^m = a^n$, so it suffices tó prove the converse. Suppose there exist $m, n \in \mathbb{Z}^+$ such that $a^m = a^n$ but $m \neq n$. Let s be as described above, and let $r \in \mathbb{Z}^+$ be such that $r < s$ and $a^r = a^s$. Suppose $r_1 \in \mathbb{Z}^+$ is such that $r_1 < s$ and $a^{r_1} = a^s = a^r$. If $r \neq r_1$, then by minimality of s, we must have $s = \max\{\, r, r_1 \,\}$, which is not possible, and so we conclude that $r = r_1$. Let $m = s - r$. By minimality of s, $a^k \neq a^n$ for all $k, n \in \mathbb{Z}^+$ with $k \neq n$ and $r \leq k, n < s$. Moreover, for any $k \in \mathbb{Z}^+$ with $r \leq k < s$, $a^{k+m} = a^k$. This follows by definition of m if $k = r$, while if $k > r$, we have $a^{k+m} = a^{k+s-r} = a^s a^{k-r} = a^r a^{k-r} = a^k$. It is now a simple induction on $n \in \mathbb{N}$ to prove that for any $k \in \mathbb{Z}^+$ with $r \leq k < s$, $a^{k+nm} = a^k$, based on the observation that $a^{k+(n+1)m} = a^{k+nm} a^m = a^k a^m = a^{k+m} = a^k$ if $z \in \mathbb{Z}^+$ and $a^{k+nm} = a^k$. Thus $K_a = \{\, a^t \mid t \in \mathbb{Z}^+, \ t \geq s \,\} = \{\, a^t \mid r \leq t < s \,\}$ is an ideal of $\langle a \rangle$ (and thus a subsemigroup of $\langle a \rangle$). Moreover, since $m = s - r$, there is a unique positive integer t such that $l = tm$ satisfies $r \leq l < s$. Then $a^l \in K_a$ and for any k with $r \leq k < s$, $a^k a^l = a^{k+tm} = a^k$; that is, K_a is a commutative monoid with identity a^l. Moreover, for any $k \in \mathbb{Z}^+$ with $r \leq k < s$, there is a unique $t \in \mathbb{Z}^+$ such that $r \leq t < s$ and $k + t \equiv l \mod m$, so $a^t \in K_a$ and since k, t, l all lie between r and $s - 1$ inclusive, and $k + t \equiv l \mod m$, we have $k + t \geq l$ and thus $a^k a^t = a^l$. Thus a^k is invertible with inverse a^t, which completes the proof that K_a is a finite abelian group. Finally, there exists unique $t \in \mathbb{Z}^+$ such that $r \leq t < s$ and $t \equiv 1 \mod m$. Then there exists $q \in \mathbb{Z}^+$ such that $t = 1 + qm$, and for each $k \in \mathbb{Z}^+$ with $r \leq k < s$, $a^k = a^{k+kqm} = a^{k(1+qm)} = (a^t)^k$, which means that $K_a = \langle a^t \rangle$. Thus K_a is a finite cyclic group of order m with identity a^l. $\square$

14.2.21 Corollary. *Every finite semigroup has idempotents.*

Proof. Let S be a finite semigroup, and let $a \in S$. Then $\langle a \rangle$ is finite, and by Proposition 14.2.19, K_a is a finite cyclic group. The identity of K_a is an idempotent of S. $\square$

14.2.22 Definition. *Let S be a finite cyclic semigroup, and let $a \in S$ be such that $\langle a \rangle = S$. Let $m = |K_a|$. Then S is said to have period m and index $|S| - m + 1$.*

Thus if $S = \langle a \rangle$ is finite, and r is the index of S and m is the period of S, then either $m = |S|$ and S is a finite cyclic group, or else $m < |S|$, $K_a = \{a^r, \ldots, a^{r+m-1}\}$, and $S - K_a = \{a, a^2, \ldots, a^{r-1}\}$. It follows from Proposition 14.2.20 that if S is a finite cyclic semigroup but not a group, then S has a unique generator. For suppose that $S = \langle a \rangle$ is a finite semigroup. By Proposition 14.2.20, there exist integers r and s with $1 \leq r < s$ such that $a^s = a^r$ with s minimal in this regard, and $K_a = \{a^i \mid i \geq r\}$ is an ideal of S. Suppose there exists $b \in S$ with $b \neq a$ and $S = \langle b \rangle$. Then there exist integers $i \geq 2$ and $j \geq 2$ such that $b = a^i$ and $a = b^j$. Thus we have $a = a^{ij}$ with $ij > 1$. Then by the minimality of s, we have $s \leq ij$. But then $a^{ij} \in K_a$, and thus $a \in K_a$. But this means that $S = K_a$ and so S is a finite cyclic group.

For an example of a cyclic semigroup of period m and index r, where m and r are arbitrary positive integers, consider the subsemigroup S of T_{m+r} that is generated by the map $\alpha : J_{m+r} \to J_{m+r}$ given by $\alpha(i) = i + 1$ for i such that $1 \leq i < m+r$, and $\alpha(m+r) = r+1$. Then $\alpha^r(J_{m+r}) = J_{m+r} - J_r$, and $\alpha|_{J_{m+r} - J_r}$ is a permutation of $J_{m+r} - J_r$ of order m (actually an m-cycle). Thus S is a cyclic semigroup of period m and index r.

Our next goal is to develop an analogue of the first isomorphism theorem. For this, we would need the notion of a quotient semigroup, so that if $f : S \to T$ is a surjective semigroup homomorphism, then somehow the data that comprises f can be used to construct a quotient S' of S in such a way that f induces an isomorphism from S' onto T. In the case of groups or rings, we were able to use normal subgroups or ideals, respectively, to construct the appropriate quotient object. That is no longer possible in the case of a general semigroup. However, all is not lost! The correct concept is that of a congruence relation.

14.2.23 Definition. *Let S be a semigroup. An equivalence relation τ on S is called a right congruence relation on S if for all $x, y, z \in S$, if $x\tau y$, then $xz\tau yz$ (that is, if $(x, y) \in \tau$, then $(xz, yz) \in \tau$), while it is called a left congruence if for all $x, y, z \in S$, if $x\tau y$, then $zx\tau zy$. If τ is both a left and a right congruence, then τ is called a congruence relation on S.*

14.2.24 Proposition. *Let S and T be semigroups, and let $f : S \to T$ be a homomorphism. Then the relation $\{(x, y) \in S \times S \mid f(x) = f(y)\}$ is a* *congruence relation on S.*

Proof. Let $x, y \in S$ be such that $f(x) = f(y)$. Then for any $z \in S$, $f(zx) = f(z)f(x) = f(z)f(y) = f(zy)$, and $f(xz) = f(x)f(z) = f(y)f(z) = f(yz)$. $\qquad \square$

14.2.25 Definition. *Let S and T be semigroups, and let $f : S \to T$ be a homomorphism. The congruence relation $\{ (x, y) \in S \times S \mid f(x) = f(y) \}$ on S is called the kernel relation of f, and is denoted by $\ker(f)$.*

14.2.26 Proposition. *Let S and T be semigroups, and let $f : S \to T$ be a homomorphism. For any $X, Y \in S/\ker(f)$, there exists a unique $Z \in S/\ker(f)$ such that $XY = \{ xy \mid x \in X, \ y \in Y \} \subseteq Z$.*

Proof. Let $x \in X$ and $y \in Y$. Since $S/\ker(f)$ is the set of all equivalence classes of $\ker(f)$ and is therefore a partition of S, there is a unique $Z \in S/\ker(f)$ for which $xy \in Z$. Let $x' \in X$ and $y' \in Y$. We must prove that $x'y' \in Z$. Since $\ker(f)$ is a congruence relation, we have $xy \equiv xy' \equiv x'y'$ mod $\ker(f)$, and so $x'y' \in Z$. $\qquad \square$

Recall that for any equivalence relation R on a set U, we have denoted the equivalence class of $u \in U$ by $[u]_R$. With this notation, we can rephrase Proposition 14.2.26 to say that for any $x, y \in S$, $[x]_{\ker(f)}[y]_{\ker(f)} = [xy]_{\ker(f)}$.

14.2.27 Notation. *Let S and T be semigroups, and let $f : S \to T$ be a homomorphism. For any $x \in S$, we shall write $[x]_f$ in place of $[x]_{\ker(f)}$.*

With this notation, Proposition 14.2.26 provides us with a binary operation on $S/\ker(f)$, namely $[x]_f[y]_f = [xy]_f$.

14.2.28 Proposition. *Let S and T be semigroups, and let $f : S \to T$ be a homomorphism. Then $S/\ker(f)$, with the binary operation described above, is a semigroup, and the mapping $\pi_f : S \to S/\ker(f)$ defined by $\pi_f(x) = [x]_f$ for $x \in S$ is a surjective homomorphism. Furthermore, if S is a monoid, then $S/\ker(f)$ is a monoid and π_f is a monoid homomoprhism.*

Proof. Let $x, y, z \in S$. Then $[x]_f([y]_f[z]_f) = [x]_f[yz]_f = [x(yz)]_f = [(xy)z]_f = ([x]_f[y]_f)[z]_f$, so the binary operation on $S/\ker(f)$ is associative. For $x, y \in S$, $\pi_f(xy) = [xy]_f = [x]_f[y]_f = \pi_f(x)\pi_f(y)$, so π_f is a homomorphism, evidently surjective. Suppose now that S is a monoid. Then for any $x \in X$, $[e_S]_f[x]_f = [e_S x]_f = [x]_f$, and $[x]_f[e_S]_f = [x e_S]_f = [x]_f$, so $S/\ker(f)$ is a monoid with identity $[e_S]_f$. By definition, $\pi_f(e_S) = [e_S]_f$, and so π_f is a monoid homomorphism if S is a monoid. $\qquad \square$

By definition of $\ker(f)$, for any $x \in S$, $f([x]_{\ker(f)}) = \{f(x)\}$. Thus in a natural way, we obtain a mapping $\overline{f} : S/\ker(f) \to T$; namely $\overline{f}([x]_f) = f(x)$ for every $x \in S$. Observe that $\overline{f} \circ \pi_f = f$, and since π_F is surjective, we see that the image of f is equal to the image of $\overline{f}$.

14.2.29 Proposition (First Semigroup Isomorphism Theorem). *Let S and T be semigroups, and let $f : S \to T$ be a homomorphism. Then $\overline{f} : S/\ker(f) \to T$ is an injective homomorphism. Moreover, $\overline{f}$ is an isomorphism if, and only if, f is surjective.*

Proof. By the preceding remarks, it suffices to prove that $\overline{f}$ is an injective homomorphism. Let $x, y \in S$, and suppose that $\overline{f}([x]_f) = \overline{f}([y]_f)$. Then $f(x) = f(y)$, and so $[x]_f = [y]_f$; that is, $\overline{f}$ is injective. As well, for $x, y \in S$, we have $\overline{f}([x]_f[y]_f) = \overline{f}([xy]_f) = f(xy) = f(x)f(y) = \overline{f}(x)\overline{f}(y)$, and so $\overline{f}$ is a homomorphism. $\qquad\square$

The Cayley representation theorem for semigroups is almost the same as the group-theoretic version. Recall that in Example 3.4.4, it was established that the set T_X of all maps of a set X to itself is a semigroup with the binary operation being composition of maps.

14.2.30 Theorem (Cayley Representation Theorem). *Let S be a semigroup. For each $x \in S$, let l_x denote the element of T_{S^1} that is defined by $l_x(t) = xt$ for all $t \in S$ (in particular, $l_x(1) = x$). Then the map $f : S \to T_{S^1}$ that maps $x \in S$ to l_x is an injective homomorphism.*

Proof. First, we prove that f is a homomorphism. Let $x, y \in S$. We are to prove that $l_{xy} = l_x \circ l_y$. Let $t \in S^1$. Then $l_{xy}(t) = (xy)t = x(yt) = l_x(yt) = l_x(l_y(t)) = l_x \circ l_y(t)$, and so $l_{xy} = l_x \circ l_y$. Thus f is a homomorphism. Now to prove that f is injective, let $x, y \in S$ and suppose that $f(x) = f(y)$; that is, $l_x = l_y$. Then $x = l_x(1) = l_y(1) = y$, and so f is injective. $\qquad\square$

Note that it was necessary to consider T_{S^1} and not just T_S. For example, if S is a cyclic semigroup with period 4 and index 2 (so $a^6 = a^2$), and for each $x \in S$, we define $l_x : S \to S$ by $l_x(t) = xt$ for all $t \in S$, then $l_a = \begin{pmatrix} a & a^2 & a^3 & a^4 & a^5 \\ a^2 & a^3 & a^4 & a^5 & a^2 \end{pmatrix} = l_{a^5}$, but $a \neq a^5$. Thus while the map from S into T_S is still a homomorphism, it is no longer injective.

In the study of the structure of a semigroup S, it is often important to be able to identify the subsemigroups of S that are in fact groups. Our next result describes the maximal subgroups of a semigroup S. Note that if G is a subgroup of a semigroup S, then G will have an identity, e_G, which will then be an idempotent of S. It is not necessarily the case that e_G is the identity of S, and in fact, it is not necessary that S have an identity in order that it may have subgroups. If e is an idempotent of S, then any subgroup of S that has e as its identity (and $\{e\}$ is one such subgroup) will be a subset of the subsemigroup eSe of S.

14.2.31 Proposition. *For any semigroup S, the following hold.*

   (i) *For any idempotent e of S, eSe is a monoid with identity e, and the set $G_e = \{a \in eSe \mid e \in aSa\} \subseteq eSe$ is the group of units of the monoid eSe and thus G_e is a subgroup of S.*
   (ii) *For any idempotents e and f of S, if $e \neq f$, then $G_e \cap G_f = \varnothing$.*
   (iii) *If G is a subgroup of S, then $G \subseteq G_{e_G}$.*

Thus the maximal subgroups of S are the subgroups G_e, where e is an idempotent of S.

Proof. (i) Since $G_e \subseteq eSe$, it suffices to prove that G_e is the set of all invertible elements of the monoid eSe. For $a \in G_e$, we have $e = axa = (ae)x(ae) = a(exae)$, with $exae \in eSe$. As well, $e = axa = (eaxe)a$ with $eaxe \in eSe$. Thus a has both a left and a right inverse in eSe, and so a is a unit of eSe. Conversely, let a be a unit of eSe, with inverse $b \in eSe$. Then $e = ab = ba = abba \in aSa$, and so $a \in G_e$. Thus G_e is the group of units of the monoid eSe.

   (ii) Suppose that e and f are idempotents of S such that $G_e \cap G_f \neq \varnothing$. Let $x \in G_e \cap G_f$. Then there exist $y \in G_e$ and $z \in G_f$ such that $yx = e = xy$ and $zx = f = xz$, and $ex = x = xe = fx = xf$. Thus $e = xy = fxy = fe$, and $f = zx = zxe = fe$, so $e = f$.

   (iii) Let G be a subgroup of S with identity e_G. Then for any $g \in G$, $g = e_G g e_G \in e_G S e_G$, and there exists $h \in G$ such that $e_G = gh = hg$, so $e_G = e_G^2 = ghhg \in gSg$. Thus $g \in G_{e_G}$. $\qquad\square$

14.2.32 Definition. *Let S be a semigroup.*

   (i) *S is said to be left (respectively, right) cancellative if for all $s, u, v \in S$, $su = sv$ (respectively, $us = vs$) implies $u = v$.*
   (ii) *S is said to be cancellative if it is both left and right cancellative.*
   (iii) *If S is left (respectively, right) simple and right (respectively, left) cancellative is called a left (respectively, right) group.*

We remark that if S is both left and right simple, then S is a group (see Corollary 5.3.3).

It was shown in Lemma 5.3.1 that a semigroup S with a left identity e such that for each $x \in S$, there exists $y \in S$ with $yx = e$ (every element of S has a left inverse with respect to the left identity e), then S is a group. The following example illustrates the fact that this need not be the case for a semigroup with left identity e in which every element has a right inverse with respect to e.

14.2.33 Example. *Let R be a set with more than one element, and define a binary operation on R by declaring $xy = y$ for all $x, y \in R$. Since $x(yz) = xz = z$, and $(xy)z = z$, $x(yz) = (xy)z$ for all $x, y, z \in R$ and so this binary operation is associative. Thus R, with this binary operation, is a semigroup in which every element is a left identity. Let G be a group and let $S = G \times R$ denote the product semigroup. Then for every $r \in R$, (e_G, r) is a left identity of S. Moreover, for any $r \in R$, for any $(g, x) \in S$, $(g, x)(g^{-1}, r) = (e_G, r)$, and so every element of S has a right inverse with respect to the left identity (e_G, r) of S. As a consequence, $(g, x)S = S$ for all $(g, x) \in S$, and so S is right simple. Moreover, for $(g, x), (h, y), (j, z) \in S$, if $(g, x)(h, y) = (g, x)(j, z)$, then $gh = gj$ and $y = xy = xz = z$, so $g^{-1}(gh) = g^{-1}(gj)$ and thus $(h, y) = (j, z)$. Thus S is left cancellative and therefore a right group. Since R has more than one element, S has more than one left identity and so S has no identity. In particular, this means that S is not a group.*

It turns out that every right group that is not a group is obtained via the construction presented in Example 14.2.33.

14.2.34 Proposition. *Let S be a right (respectively, left) group. Then E_S is a subsemigroup of S, and for any $e, f \in E_S$, $ef = f$ (respectively, $ef = e$). For any $e \in E_S$, $Se = eSe$ (respectively $eS = eSe$) is a maximal subgroup of S, and $S = \cup_{e \in E_S} eSe$ is a disjoint union of its maximal subgroups. Choose $e \in E_S$, and define the map $\varphi_e : S \to eSe \times E_S$ by $\varphi_e(x) = (xe, f)$ (respectively, $\varphi_e(x) = (ex, f)$), where f is the unique idempotent of S for which $x \in fSf$. Then φ_e is an isomorphism from S onto $eSe \times E_S$.*

Proof. We give only the argument for the case of a right group. Since S is right simple, for each $a \in S$, $aS = S$ and thus there exists $b \in S$ with $ab = a$. Then $ab^2 = ab$, and since S is left cancellative, $b^2 = b$. Thus $E_S \neq \varnothing$ and for every $a \in S$, there is $e \in E_S$ such that $a \in Se$. Let $e \in E_S$. Then for any $x \in S$, $ex = e^2 x$, and thus, since S is left cancellative, $x = ex$. In particular, for $f \in E_S$, $ef = f$ and $Se = eSe$. Thus E_S is a subsemigroup of S and for each $e \in E_S$, eSe is a subsemigroup of S with identity e. We claim that eSe is actually a subgroup of S. Let $exe \in eSe$. Since $exeS = S$, there exists $y \in S$ such that $exey = e$. But then $(exe)(eye) = e$, and so $(exe)(eye)(exe) = e(exe) = exe = (exe)e$. By left cancellation, $(eye)(exe) = e$ and thus every element of eSe has an inverse in eSe, which proves that eSe is a subgroup of S. But then, in the notation of Proposition 14.2.31, $G_e = \{ a \in S \mid a \in eSe \text{ and } e \in aSa \} = eSe$, so $eSe = Se$ is a maximal subgroup of S. We have shown that for each $a \in S$, there exists $e \in E_S$ with $a \in Se$, and thus $S = \cup_{e \in E_S} Se$. By Proposition 14.2.31, this union is disjoint since for each $e \in E_S$, $Se = eSe$ is a maximal subgroup of S. Fix $e \in E_S$. We prove that $\varphi_e : S \rightarrow eSe \times E_S$ as defined above is an isomorphism. Let $x, y \in S$, and let $f, f' \in E_S$ be such that $x \in Sf = fSf$ and $y \in Sf' = F'Sf'$. Then $xy \in Sf'$, and so $\varphi_e(xy) = (xye, f')$, while $\varphi_e(x) = (xe, f)$ and $\varphi_e(y) = (ye, f')$, so $\varphi_e(x)\varphi_e(y) = (xe, f)(ye, f') = (xeye, ff') = (xef'ye, f') = (xf'ye, f') = (xye, f') = \varphi_e(xy)$. Thus φ_e is a homomorphism. For any $x \in Se$ and $f \in E_S$, $xf \in Sf$ satisfies $xfe = xe = x$, so $\varphi_e(xf) = (xfe, f) = (xe, f) = (x, f)$, which establishes that φ_e is surjective. Finally, suppose that $x, y \in S$ are such that $\varphi_e(x) = \varphi_e(y)$. Then $xe = ye$ and there exists $f \in E$ such that $x, y \in Sf$, from which we obtain $x = xf = x(ef) = (xe)f = (ye)f = y(ef) = yf = f$. Thus φ_e is injective. $\qquad \square$

Note that if S is a right (respectively, left) group, then S is a group if, and only if $|E_S| = 1$. In fact, a right simple monoid must be a group. For suppose that S is right simple with identity e_S. Then every element of S has a right inverse; that is, for $x \in S$, $e_S \in S = xS$, and so there exists $y \in S$ with $xy = e_S$. But then $yx = e_S$ (for there is $z \in S$ with $yz = e_S$, and then $x = xe_S = x(yz) = (xy)z = e_Sz = z$), and so S is a group.

In a monoid S, an arbitrary $a \in S$ may fail to have an inverse. The inverse of $a \in S$ is an element $b \in S$ for which $ab = e_S = ba$, and the inverse is unique if it exists. We observe that if b is the inverse of a, then $aba = e_Sa = a$ and $bab = be_S = b$. This observation leads us to the more general notion of regularity.

14.2.35 Definition. *Let S be a semigroup. Then $a \in S$ is said to be a regular element if there exists $x \in S$ such that $axa = a$, and x is said to be an inverse of a if in addition, $xax = x$. S is said to be a regular semigroup if every element of S is a regular element, and a regular semigroup S is* *said to be an inverse semigroup if every element of S has a unique inverse. If S is an inverse semigroup and $a \in S$, then the unique inverse of a is denoted by a^{-1}.*

By Proposition 14.2.34, each right (or left) group is a union of its maximal subgroups and is therefore a regular semigroup. Every group is an inverse semigroup, but as we shall see later, not every inverse semigroup is a group.

Every idempotent of a semigroup is a regular element. We shall prove next that every regular element has an inverse, but in general, a regular element may have more than one inverse (that is, not every regular semigroup is an inverse semigroup). For example, in the right group $G \times E$ of Example 14.2.33, for any $e, f \in E$, $(e_G, e)(e_G, f) = (e_G, f)$, and so $(e_G, e)(e_G, f)(e_G, e) = (e_G, e)$. Switch the roles of e and f to obtain $(e_G, f)(e_G, e)(e_G, f) = (e_G, f)$, and so for each $f \in E$, (e_G, f) is an inverse for (e_G, e). If $|E| \geq 2$, then $G \times E$ is an example of a regular semigroup that is not an inverse semigroup, as for any $g \in G$, $e, f \in E$, we have $(g, e)(g^{-1}, f)(g, e) = (g, e)$ and $(g^{-1}, f)(g, e)(g^{-1}, f) = (g^{-1}, f)$, so (g^{-1}, f) is an inverse for (g, e), and $(g, e)(g^{-1}, e)(g, e) = (g, e)$, and $(g^{-1}, e)(g, e)(g^{-1}, e) = (g^{-1}, e)$, so (g^{-1}, e) is an inverse for (g, e).

14.2.36 Proposition. *Let S be a semigroup, and let a be a regular element. Then $aS^1 = aS$ and $S^1a = Sa$, and for any $x \in S$ such that $axa = a$, $ax, xa \in E_S$, and xax is an inverse of a.*

Proof. Let $a \in S$ be regular. If S is a monoid, then $S^1 = S$, so suppose that S is not a monoid, in which case $S^1 = S \cup \{1\}$. Then $aS \subseteq aS^1 = aS \cup \{a\}$ and $Sa \subseteq S^1a = Sa \cup \{a\}$. Since a is regular, there exists $x \in S$ such that $axa = a$, and thus $a \in aS \cap Sa$, which establishes that $aS^1 = aS$ and $S^1a = Sa$. Since $axa = a$, we obtain $axax = ax$ and $xaxa = xa$, which means that $ax, xa \in E_S$. Also from $axa = a$, we obtain that $a(xax)a = (axa)xa = axa = a$. We then compute $(xax)a(xax) = x(axaxa)x = xax$, and so xax is an inverse of a. $\square$

14.2.37 Proposition. *Let X be a nonempty set. Then T_X, the set of all maps from X to itself, with the binary operation composition of maps, is a regular semigroup.*

Proof. We know that composition of maps is associative, so T_X is a semigroup. We are to prove that for every $f \in T_X$, there exists $g \in T_X$ such that $fgf = f$. Fix $x_0 \in X$. Then for any $x \in X$, if $x \in f(X)$, choose

$y \in f^{-1}(x)$ and define $g(x) = y$ (so fg(x)=f(y)=x), while if $x \notin f(X)$, define $g(x) = x_0$. Then g is a map from X to X, and for any $x \in X$, $fgf(x) = fg(f(x)) = f(x)$, so $fgf = f$, as required. Note that if X is infinite, then the Axiom of Choice is required for the construction of g. $\square$

14.2.38 Proposition. *Let S be a semigroup. An element $a \in S$ is regular if, and only if, there exists $e \in E_S$ such that $aS^1 = eS$, which holds if, and only if there exists $f \in E_S$ such that $S^1 a = Sf$.*

Proof. Let $a \in S$. Suppose first that a is regular, and let $x \in S$ be such that $axa = a$. Then $aS^1 = axaS^1 \subseteq axS^1 \subseteq aS^1$ so $aS^1 = axS^1$ and $ax \in E_S$ (so $axS^1 = axS$). Similarly, $S^1 a = S^1 axa \subseteq S^1 xa \subseteq S^1 a$, so $S^1 a = S^1 xa = Sxa$ since $xa \in E_S$. Conversely, suppose that there exist $e, f \in E_S$ such that $aS^1 = eS$ and $S^1 a = Sf$. Then $a = ea$ and $e \in aS^1$, so $e = ax$ for some $x \in S^1$. If $e = a$, then a is regular, so suppose that $e \neq a$. Then $x \in S$ and we have $a = ea = axa$, and so a is regular. Similarly, if there exists $f \in E_S$ such that $S^1 a = Sf$, then either $f = a$ and so a is regular, or else $a = af$ and $f = ya$ for some $y \in S$, from which we obtain $a = af = aya$. $\square$

The following is a useful criterion for identifying the inverse semigroups within the class of regular semigroups.

14.2.39 Proposition. *Let S be a regular semigroup. Then S is an inverse semigroup if, and only if, $ef = fe$ for all $e, f \in E_S$. If S is an inverse semigroup, then E_S is a commutative subsemigroup of S.*

Proof. First, suppose that S is an inverse semigroup. Let $e, f \in E_S$, and let $a \in S$ be the unique inverse of ef, so $efaef = ef$ and $aefa = a$. Since $ef(fae)ef = ef$ and $(fae)ef(fae) = faefae = fae$, the idempotent fae is an inverse for ef, and thus $a = fae$ is an idempotent. But an idempotent is an inverse of itself, so the unique inverse of a is a, which means that $ef = a \in E_S$. Thus E_S is a subsemigroup of S, and for any $e, f \in E_S$, the inverse of ef is ef. Since $effeef = efef = ef$, and $feeffe = fefe = fe$, it follows that fe is an inverse for ef, and so the uniqueness of inverses gives $ef = fe$. Thus E_S is a commutative subsemigroup of S.

Conversely, suppose that $ef = fe$ for all $e, f \in E_S$, and let $a \in S$. Since S is regular, it follows from Proposition 14.2.36 that a has at least one inverse. Suppose that b and c are inverses of a, so $aba = a = aca$ and $bab = b$, $cac = c$. Then $abac = ac$, and $acab = ab$. Since $ab, ac \in E_S$,

$abac = acab$ and so $ac = ab$. Similarly, $baca = ba$ and $caba = ca$, so $ba = ca$. Thus $b = bab = bac = cac = c$. $\square$

14.2.40 Corollary. *Let S be an inverse semigroup. Then for all $a, b \in S$, $(ab)^{-1} = b^{-1}a^{-1}$.*

Proof. We have $ab(b^{-1}a^{-1})ab = a(bb^{-1})(a^{-1}a)b = a(a^{-1}a)(bb^{-1})b = ab$, and $b^{-1}a^{-1}abb^{-1}a^{-1} = b^{-1}bb^{-1}a^{-1}aa^{-1} = b^{-1}a^{-1}$, so $b^{-1}a^{-1}$ is an inverse of ab and thus $(ab)^{-1} = b^{-1}a^{-1}$. $\square$

14.2.41 Corollary. *A regular commutative semigroup is an inverse semigroup which is a union of groups.*

Proof. Let S be a regular commutative semigroup. By Proposition 14.2.39, S is an inverse semigroup. Moreover, for any $a \in S$, $aa^{-1} = a^{-1}a \in E_S$ since S is commutative, and so for $e = aa^{-1}$, $eae = a$ and $ea^{-1}e = a^{-1}$, which means that a is a unit of the monoid eSe; that is, a is an element of the group of units of the monoid eSe. Thus S is a union of subgroups of S. $\square$

14.3   The Semigroup of Relations on a Set

Our aim in this section is to construct a family $\mathscr{F}$ of inverse semigroups with the property that given any inverse semigroup S, there is $F \in \mathscr{F}$ and an injective semigroup homomorphism $f_F : S \to F$; that is, S can be embedded in F. This would provide a generalization of the result that states that every group is isomorphic to a subgroup of some permutation group. Instead of permutations of some set, we must work with relations on some set.

Recall that a map from a set X to a set Y is a special kind of a relation from X to Y, and if we have maps $f : X \to Y$ and $g : Y \to Z$, then we defined the composite map $g \circ f : X \to Z$ by
$$g \circ f = \{ (x, z) \in X \times Z \mid \exists \, y \in Y \text{ such that } (x, y) \in f \text{ and } (y, z) \in g \}.$$
This provided the result that $g \circ f(x) = g(f(x))$ for each $x \in X$. We define the composition of relations in general to be consistent with the definition of composition of maps.

14.3.1 Definition. *Let X, Y, and Z be sets, and let $R \subseteq X \times Y$, and $S \subseteq Y \times Z$. Then $S \circ R \subseteq X \times Z$ is defined by*
$$S \circ R = \{ (x, z) \in X \times Z \mid \exists \, y \in Y \text{ such that } (x, y) \in R \text{ and } (y, z) \in S \}.$$

14.3.2 Proposition. *Let* X, Y, Z, W *be sets, and let* $R \subseteq X \times Y$, $S \subseteq Y \times Z$, *and* $T \subseteq Z \times W$. *Then* $T \circ (S \circ R) = (T \circ S) \circ R$.

Proof. Let $(x, w) \in T \circ (S \circ R)$. Then there exists $z \in Z$ such that $(x, z) \in S \circ R$ and $(z, w) \in T$, and so there exists $y \in Y$ with $(x, y) \in R$ and $(y, z) \in S$. But from $(y, z) \in S$ and $(z, w) \in T$, we obtain that $(y, w) \in T \circ S$, and this, together with $(x, y) \in R$ yields $(x, w) \in (T \circ S) \circ R$. Thus $T \circ (S \circ R) \subseteq (T \circ S) \circ R$. The proof that $(T \circ S) \circ R \subseteq T \circ (S \circ R)$ is similar and is therefore omitted. $\square$

14.3.3 Corollary. *Let* X *be a nonempty set. Then* $\mathscr{R}(X)$, *the set of all relations from* X *to* X, *with the binary operation composition of relations, is a monoid with identity* $\mathbb{1}_X$, *the identity map on* X.

Proof. By Proposition 14.3.2, composition of relations is an associative binary operation on $\mathscr{R}(X)$. We prove that for each $R \in \mathscr{R}(X)$, $\mathbb{1}_X \circ R = R = R \circ \mathbb{1}_X$. Let $(a, b) \in \mathbb{1}_X \circ R$. Then there exists $x \in X$ such that $(a, x) \in R$ and $(x, b) \in \mathbb{1}_X$, in which case, $x = b$. Thus $(a, b) \in R$, and so $\mathbb{1}_X \circ R \subseteq R$. Next, if $(a, b) \in R$, then since $(b, b) \in \mathbb{1}_X$, we obtain $(a, b) \in \mathbb{1}_X \circ R$ and thus $R \subseteq \mathbb{1}_X \circ R$. This completes the proof that $R = \mathbb{1}_X \circ R$. The proof that $R \circ \mathbb{1}_X = R$ is similar and has been omitted. $\square$

14.3.4 Definition. *Let* X *be a nonempty set. Then* $\mathscr{I}(X)$ *is the subset of* $\mathscr{R}(X)$ *defined by*

$$\mathscr{I}(X) = \{\, R \in \mathscr{R}(X) \mid \text{ for all } a, b, c \in X, \ (a, b), (a, c) \in R \text{ implies } b = c,$$
$$\text{and for all } a, b, c \in X, \ (b, a), (c, a) \in R \text{ implies } b = c \,\}.$$

14.3.5 Proposition. *Let* X *be a nonempty set. Then* $\mathscr{I}(X)$ *is a submonoid of* $\mathscr{R}(X)$.

Proof. Let $R, S \in \mathscr{I}(X)$, and let $a, b, c \in X$ be such that $(a, b), (a, c) \in S \circ R$. Then there exist $x, y \in X$ such that $(a, x) \in R$, $(x, b) \in S$, $(a, y) \in R$, and $(y, c) \in S$. Since $R \in \mathscr{I}(X)$ and $(a, x), (a, y) \in R$, $x = y$. But then $(x, b), (x, c) \in S$, and since $S \in \mathscr{I}(X)$, we deduce that $b = c$. Next, suppose that $a, b, c \in X$ are such that $(c, a), (c, b) \in S \circ R$. Then there exist $x, y \in X$ such that $(c, x) \in R$, $(x, a) \in S$, $(c, y) \in R$, and $(y, b) \in S$. Since $R \in \mathscr{I}(X)$ and $(c, x), (c, y) \in R$, $x = y$. But then $(x, a), (x, b) \in S$, and since $S \in \mathscr{I}(X)$, we deduce that $a = b$. Thus $S \circ R \in \mathscr{I}(X)$, and so $\mathscr{I}(X)$ is a subsemigroup of $\mathscr{R}(X)$. It is obvious that $\mathbb{1}_X \in \mathscr{I}(X)$. $\square$

14.3.6 Proposition. *Let X be a nonempty set. Then $E_{\mathscr{I}(X)} = E_{\mathscr{R}(X)} \cap \mathscr{I}(X) = \{\, R \in \mathscr{R}(X) \mid R \subseteq \mathbb{1}_X \,\}$.*

Proof. Suppose first that $R \in E_{\mathscr{R}(X)} \cap \mathscr{I}(X)$. Then R is an idempotent, so $R \circ R = R$. Let $(a,b) \in R = R \circ R$. Then there exists $x \in X$ such that $(a,x), (x,b) \in R$. Now, $R \in \mathscr{I}(X)$ and $(a,b), (a,x) \in R$ imply that $b = x$, while $(a,b), (x,b) \in R$ imply that $a = x$ and thus $a = x = b$. This proves that $R \subseteq \mathbb{1}_X$. Conversely, suppose that $R \subseteq \mathbb{1}_X$. It is then immediate that $R \in \mathscr{I}(X)$, and $R = R \circ R$, so $R \in E_{\mathscr{R}(X)} \cap \mathscr{I}(X)$. $\qquad\square$

14.3.7 Proposition. *Let X be a nonempty set. Then $E_{\mathscr{I}(X)}$ is a commutative submonoid of $\mathscr{I}(X)$.*

Proof. Let $R, S \in E_{\mathscr{I}(X)}$. Then by Proposition 14.3.6, $R \cup S \subseteq \mathbb{1}_X$. Let $(a,b) \in S \circ R$. Then there is $x \in X$ such that $(a,x) \in R$, $(x,b) \in S$. But then $a = x = b$, and so $(a,b) \in \mathbb{1}_X$. Moreover, $(a,b) \in R \cap S$. Thus $S \circ R \subseteq R \cap S$. Let $(a,b) \in R \cap S \subseteq \mathbb{1}_X$. Then $a = b$, and we have $(a,a) \in R \cap S$, so $(a,a) \in S \circ R$ and thus $R \cap S \subseteq S \circ R$. Thus $S \circ R = R \cap S$ for any $R, S \in E_{\mathscr{I}(X)}$, which proves that $S \circ R = R \circ S$. By Proposition 14.3.6, $S \circ R \in E_{\mathscr{I}(X)}$, and so $E_{\mathscr{I}(X)}$ is a commutative subsemigroup of $\mathscr{I}(X)$. Finally, since $\mathbb{1}_X \in E_{\mathscr{I}(X)}$, $E_{\mathscr{I}(X)}$ is a submonoid of $\mathscr{I}(X)$. $\quad\square$

14.3.8 Proposition. *Let X be a nonempty set. Then $\mathscr{I}(X)$ is an inverse semigroup.*

Proof. By Proposition 14.2.39 and Proposition 14.3.7, it suffices to prove that $\mathscr{I}(X)$ is regular. For $R \in \mathscr{I}(X)$, let

$$S = \{\, (x,y) \in X \times X \mid (y,x) \in R \,\}.$$

Consider $(a,b) \in R \circ S \circ R$. There exist $x, y \in X$ such that $(a,x) \in R$, $(x,y) \in S$ and $(y,b) \in R$. Since $(x,y) \in S$, we have $(y,x) \in R$, and thus, since $R \in \mathscr{I}(X)$, $a = y$ and $x = b$. But then $(a,b) = (y,x) \in R$. Thus $R \circ S \circ R \subseteq R$. Now consider $(a,b) \in R$, so $(b,a) \in S$ and thus $(a,a) \in S \circ R$, which establishes that $(a,b) \in R \circ S \circ R$. Thus $R \subseteq R \circ S \circ R$, and so $R = R \circ S \circ R$. This proves that $\mathscr{I}(X)$ is a regular semigroup, as required. $\qquad\square$

Note that for $R \in \mathscr{I}(X)$, the unique inverse of R in $\mathscr{I}(X)$ is the relation $\{\, (b,a) \in X \times X \mid (a,b) \in R \,\}$. It is customary to refer to this relation as the inverse of R, and denote it by R^{-1}, which is in keeping with the notation for the inverse of an element in an inverse semigroup.

14.3.9 Theorem. *Let S be an inverse semigroup. For each $a \in S$, define $I(a) = \{\,(x,y) \in Sa^{-1}a \times Sa^{-1}a \mid y = xa^{-1}\,\}$. Then $I : S \to \mathscr{R}(S)$ is an injective homomorphism with $I(S) \subseteq \mathscr{I}(S)$.*

Proof. Let $a \in S$. If $x, y, z \in X$ are such that $(x,y), (x,z) \in I(a)$, then $y = xa^{-1} = z$, so suppose that $x, y, z \in X$ are such that $(y,x), (z,x) \in I(a)$. Then $y, z \in Sa^{-1}a$ and $ya^{-1} = x = za^{-1}$, so $y = ya^{-1}a$ and $z = a^{-1}a$, which yields $y = ya^{-1}a = za^{-1}a = z$. Thus $I(a) \in \mathscr{I}(S)$.

Next, suppose that $a, b \in S$ are such that $I(a) = I(b)$. Since $a^{-1}a \in Sa^{-1}a$ and $(a^{-1}a)a^{-1} = a^{-1}$, $(a^{-1}a, a^{-1}) \in I(a) = I(b)$, and thus $a^{-1}a \in Sb^{-1}b$ (which implies that $a^{-1}a = a^{-1}ab^{-1}b$), and $a^{-1}ab^{-1} = a^{-1}$. Similarly, $(b^{-1}b, b^{-1}) \in I(b) = I(a)$, so $b^{-1}ba^{-1} = b^{-1}$ and $b^{-1}b = b^{-1}ba^{-1}a$. Thus $b^{-1}b = b^{-1}ba^{-1}a = a^{-1}ab^{-1}b = a^{-1}a$ (recall that S is inverse, so idempotents commute). We now have $b^{-1} = b^{-1}bb^{-1} = a^{-1}ab^{-1} = a^{-1}$, and since $a^{-1} = b^{-1}$ implies that $a = b$ (since then a and b are each inverses of a^{-1} for example), it follows that I is injective.

Finally, let $a, b \in S$. We must prove that $I(ab) = I(a) \circ I(b)$. Consider $(x,y) \in I(a) \circ I(b)$. There exists $z \in S$ such that $(x,z) \in I(b)$ and $(z,y) \in I(a)$. Then $x \in Sb^{-1}b$, $z \in Sa^{-1}a$, $xb^{-1} = z$, and $za^{-1} = y$. Consequently, we have $y = za^{-1} = xb^{-1}a^{-1}$, and by Corollary 14.2.40, $b^{-1}a^{-1} = (ab)^{-1}$, so $y = x(ab)^{-1}$. Furthermore, $x = xb^{-1}b = zb = (za^{-1}a)b$, so $xb^{-1} = z(a^{-1}a)(bb^{-1}) = z(bb^{-1})(a^{-1}a) = zb(ab)^{-1}a \in S(ab)^{-1}a$ and thus $x = xb^{-1}b \in S(ab)^{-1}ab$. Thus $(x,y) \in I(ab)$, and so $I(a) \circ I(b) \subseteq I(ab)$. Conversely, suppose that $(x,y) \in I(ab)$. Then $x \in S(ab)^{-1}ab$ and $y = x(ab)^{-1} = xb^{-1}a^{-1}$. Let $z = xb^{-1} \in S(ab)^{-1}abb^{-1} = S(ab)^{-1}aa^{-1}abb^{-1} = S(ab)^{-1}abb^{-1}a^{-1}a \subseteq Sa^{-1}a$. Since $za^{-1} = xb^{-1}a^{-1} = x(ab)^{-1} = y$ and $z \in Sa^{-1}a$, $(z,y) \in I(a)$. Since $x \in S(ab)^{-1}ab \subseteq Sb^{-1}b$ and $xb^{-1} = z$, we have $(x,z) \in I(b)$, and so $(x,y) \in I(a) \circ I(b)$. Thus $I(ab) \subseteq I(a) \circ I(b)$, and so $I(ab) = I(a) \circ I(b)$. $\qquad\square$

4 Green's Relations

We introduce four fundamental equivalence relations, definable in any semigroup, which play a large role in the study of the structure of semigroups. In particular, they are indispensable in the investigation of the structure of regular or inverse semigroups.

14.4.1 Definition. *Let S be a semigroup. Define Green's relations*

$$\mathscr{L} = \{\, (x,y) \in S \times S \mid S^1 x = S^1 y \,\},$$
$$\mathscr{R} = \{\, (x,y) \in S \times S \mid x S^1 = y S^1 \,\},$$

Green's relation $\mathscr{H} = \mathscr{L} \cap \mathscr{R}$, *Green's relation* $\mathscr{D} = \mathscr{L} \circ \mathscr{R}$, *and Green's relation*

$$\mathscr{J} = \{\, (x,y) \in S \times S \mid S^1 x S^1 = S^1 y S^1 \,\}.$$

It is immediate that $\mathscr{L}$, $\mathscr{R}$, and $\mathscr{J}$ are equivalence relations since each is defined by an equality, and since the intersection of equivalence relations is again an equivalence relation, it follows that $\mathscr{H}$ is an equivalence relation. For $a \in S$, the equivalence class of a with respect to $\mathscr{L}$, $\mathscr{R}$, $\mathscr{H}$, and $\mathscr{J}$ shall be denoted respectively by L_a, R_a, H_a, and J_a. We shall prove shortly that $\mathscr{D}$ is an equivalence relation, and for $a \in S$, we shall then denote the equivalence class of a with respect to $\mathscr{D}$ by D_a. We shall have need of the following results.

14.4.2 Lemma. *Let X be a set, and let R, S be symmetric relations on X with $R \circ S \subseteq S \circ R$. Then $R \circ S = S \circ R$, and $R \circ S$ is a symmetric relation.*

Proof. By Definition 3.2.7 (ii), a relation T on X is symmetric if for all $(a,b) \in X \times X$, $(a,b) \in T$ implies that $(b,a) \in T$. Suppose that $(a,b) \in S \circ R$, so there exists $c \in X$ with $(a,c) \in R$ and $(c,b) \in S$. By the symmetry of R and S, $(c,a) \in R$ and $(b,c) \in S$, and thus $(b,a) \in R \circ S \subseteq S \circ R$ (which proves that $S \circ R$ is symmetric). But then there exists $d \in X$ with $(b,d) \in R$ and $(d,a) \in S$, and again by symmetry, we obtain $(d,b) \in R$ and $(a,d) \in S$, so $(a,b) \in R \circ S$. Thus $S \circ R \subseteq R \circ S$ and so $S \circ R = R \circ S$. $\square$

14.4.3 Proposition. *For any semigroup S, $\mathscr{L}$ is a right congruence relation on S, $\mathscr{R}$ is a left congruence relation on S, and $\mathscr{L} \circ \mathscr{R} = \mathscr{R} \circ \mathscr{L}$.*

Proof. Let $(x,y) \in \mathscr{L}$, so $S^1 x = S^1 y$. For any $z \in S$, $S^1 x z = S^1 y z$, and thus $(xz, yz) \in \mathscr{L}$. Thus $\mathscr{L}$ is a right congruence. The proof that $\mathscr{R}$ is a left congruence is similar and is therefore omitted. Now let $(x,y) \in \mathscr{L} \circ \mathscr{R}$, so there exists $z \in S$ with $(x,z) \in \mathscr{R}$ and $(z,y) \in \mathscr{L}$. Thus $x S^1 = z S^1$ and $S^1 z = S^1 y$, and so there exist $v, w \in S^1$ with $zv = x$ and $wz = y$. If $v = 1$, then $z = x$ and so $(x,y) \in \mathscr{L}$, $(y,y) \in \mathscr{R}$, from which we obtain $(x,y) \in \mathscr{R} \circ \mathscr{L}$. If $w = 1$, then $z = y$ and then $(x,y) \in \mathscr{R}$ and $(x,x) \in \mathscr{L}$, so $(x,y) \in \mathscr{R} \circ \mathscr{L}$. Suppose now that $v, w \in S$, and let $c = wx$. Since $\mathscr{R}$ is a left congruence and $(x,z) \in \mathscr{R}$, we obtain $(c,y) = (wx.wz) \in \mathscr{R}$. Now,

$c = wx = wzv = yv$, and since $\mathscr{L}$ is a right congruence and $(z,y) \in \mathscr{L}$, we have $(x,c) = (zv, yv) \in \mathscr{L}$ and thus $(x,y) \in \mathscr{R} \circ \mathscr{L}$. Thus $\mathscr{L} \circ \mathscr{R} \subseteq \mathscr{R} \circ \mathscr{L}$. Since each of $\mathscr{L}$ and $\mathscr{R}$ are equivalence relations and therefore symmetric relations, it follows from Lemma 14.4.2 that $\mathscr{L} \circ \mathscr{R} = \mathscr{R} \circ \mathscr{L}$. □

14.4.4 Lemma. *Let X be a set and R, S be relations on X such that S is reflexive. Then $R \subseteq R \circ S \cap S \circ R$. If S is reflexive and transitive, then $S = S \circ S$.*

Proof. Let $(a,b) \in R$. Since $(a,a) \in S$, we have $(a,b) \in R \circ S$, and since $(b,b) \in S$, we have $(a,b) \in S \circ R$. Suppose now that S is reflexive and transitive. Then as seen above, $S \subseteq S \circ S$ since S is reflexive, and we have $S \circ S \subseteq S$ since S is transitive. Thus $S = S \circ S$. □

14.4.5 Proposition. *Let S be a semigroup. Then $\mathscr{D} = \mathscr{L} \circ \mathscr{R}$ is the smallest equivalence relation on S that contains $\mathscr{L} \cup \mathscr{R}$; that is,*

$$\mathscr{D} = \cap \{ T \subseteq X \times X \mid T \text{ is an equivalence relation and } \mathscr{L} \cup \mathscr{R} \subseteq T \}.$$

Proof. By Proposition 14.4.3, $\mathscr{D} = \mathscr{L} \circ \mathscr{R} = \mathscr{R} \circ \mathscr{L}$, so $\mathscr{D} \circ \mathscr{D} = \mathscr{L} \circ \mathscr{R} \circ \mathscr{L} \circ \mathscr{R} = \mathscr{L} \circ \mathscr{L} \circ \mathscr{R} \circ \mathscr{R}$, and by Lemma 14.4.4, $\mathscr{L} \circ \mathscr{L} = \mathscr{L}$ and $\mathscr{R} \circ \mathscr{R} = \mathscr{R}$, so $\mathscr{D} \circ \mathscr{D} = \mathscr{L} \circ \mathscr{R} = \mathscr{D}$. Thus $\mathscr{D}$ is a transitive relation. Moreover, since $\mathscr{R}$ is reflexive, we have by Lemma 14.4.4 that $\mathscr{L} \subseteq \mathscr{D}$ and since $\mathscr{L} \circ \mathscr{R} = \mathscr{R} \circ \mathscr{L}$ and $\mathscr{L}$ is reflexive, we also have $\mathscr{R} \subseteq \mathscr{D}$. Thus $\mathscr{L} \cup \mathscr{R} \subseteq \mathscr{D}$, which in particular implies that $\mathscr{D}$ is reflexive. Finally, by Lemma 14.4.2, $\mathscr{D}$ is symmetric, and thus $\mathscr{D}$ is an equivalence relation on S containing $\mathscr{L} \cup \mathscr{R}$. Suppose now that T is an equivalence relation on S and that $\mathscr{L} \cup \mathscr{R} \subseteq T$. Let $(a,b) \in \mathscr{D}$. Then there exists $c \in S$ such that $(a,c) \in \mathscr{R}$ and $(c,b) \in \mathscr{L}$. But then $(a,c), (c,b) \in T$, and since T is transitive, this implies that $(a,b) \in T$. Thus $\mathscr{D} \subseteq T$, as required. □

14.4.6 Definition. *Let S be a semigroup. For $a \in S$, define maps $\rho_a, \lambda_a \colon S \to S$ by $\rho_a(x) = xa$ and $\lambda_a(x) = ax$ for all $x \in S$.*

14.4.7 Proposition (Green's Lemma). *Let S be a semigroup, and let $a \in S$. Then $H_a = L_a \cap R_a \subseteq L_a \cup R_a \subseteq D_a$, and the following hold.*

(i) *For $b \in R_a$ with $a \neq b$, there exist $u, v \in S$ such that $au = b$ and $bv = a$. Then $\rho_u|_{L_a}$ is a bijective map from L_a onto L_b with inverse $\rho_v|_{L_b}$. Moreover, for any $x \in L_a$, $\rho_u(H_x) = H_{\rho_u(x)}$.*

(ii) For $b \in R_a$ with $a \neq b$, there exist $u, v \in S$ such that $ua = b$ and $vb = a$. Then $\lambda_u\big|_{R_a}$ is a bijective map from R_a onto R_b with inverse $\lambda_v\big|_{R_b}$. Moreover, for any $x \in R_a$, $\lambda_u(H_x) = H_{\lambda_u(x)}$.

Proof. The containment assertions follow immediately from the fact that $\mathscr{H} = \mathscr{L} \cap \mathscr{R} \subseteq \mathscr{L} \cup \mathscr{R} \subseteq \mathscr{D}$. The proof of (ii) is similar to that of (i), and we shall prove only (i). Let $b \in R_a$ with $a \neq b$. Then there exist $u, v \in S$ with $b = au$ and $a = bv$. Consider ρ_u applied only to elements of L_a. For $x \in L_a$, $S^1 x = S^1 a$ and so $S^1 xu = S^1 au$, which implies that $\rho_u(x) = xu \in L_{au} = L_b$ and so $\rho_u(L_a) \subseteq L_b$. Similarly, $\rho_v(L_b) \subseteq L_a$. Moreover, for any $x \in L_a$, $x \in S^1 a$ and so $x = ta$ for some $t \in S^1$, which then yields $\rho_v(\rho_u(x)) = xuv = tauv = tbv = ta = x$. Thus $\rho_v \circ \rho_u$ is the identity mapping on L_a. Similarly, $\rho_u \circ \rho_v$ is the identity mapping on L_b, and so ρ_u is a bijective mapping from L_a onto L_b, with inverse ρ_v. Finally, we claim that for any $x \in L_a$, $\rho_u(x) \in R_x$. For let $x \in L_a$. There exists $y \in L_b$ such that $\rho_v(y) = x$, and $y = \rho_u \circ \rho_v(y) = xu$. Since $y = xy$ and $x = yv$, $(x, y) \in \mathscr{R}$, so $\rho_u(x) = xu = y \in R_x$. Thus for any $x \in L_a$, $\rho_u(H_x) = \rho_u(L_x \cap R_x) = \rho_u(L_a \cap R_x) \subseteq L_b \cap R_x = L_{\rho_u(x)} \cap R_{\rho_u(x)} = H_{\rho_u(x)}$. By essentially the same argument, $\rho_v(H_{\rho_u(x)}) \subseteq H_{\rho_v \circ \rho_u(x)} = H_x$, and thus the restrictions of ρ_u and ρ_v to H_x and $H_{\rho_u(x)}$, respectively, are inverses of each other. Thus ρ_u is a bijection from H_x onto $H_{\rho_u(x)}$. □

14.4.8 Lemma. *Let S be a semigroup, and let $a \in S$. For any $x, y \in D_a$, $L_x \cap R_y \neq \varnothing$.*

Proof. Let $x, y \in D_a$. Then $(x, y) \in \mathscr{D} = \mathscr{L} \circ \mathscr{R}$, and so there exists $z \in S$ with $(x, z) \in \mathscr{R}$ and $(z, y) \in \mathscr{L}$. We therefore obtain that $z \in L_y$ and $z \in R_x$, so $z \in L_x \cap R_y$. □

It is customary to represent a $\mathscr{D}$-class D of a semigroup S by a rectangle with equally spaced vertical lines and equally spaced horizontal lines drawn within the rectangle, with the understanding that the resulting vertical rectangles so formed represent the $\mathscr{L}$-classes contained within D, and the horizontal rectangles represent the $\mathscr{R}$-classes contained within D. The intersection of a vertical rectangle and a horizontal rectangle represents the corresponding $\mathscr{H}$-class. The resulting picture is called the egg-box picture of the $\mathscr{D}$-class (shown in Figure 14.1). For general discussions, one often places equal numbers of vertical and horizontal lines, but in general, the set of all $\mathscr{L}$-classes contained with a given $\mathscr{D}$-class will not have the same size as the set of all $\mathscr{R}$-classes contained within the $\mathscr{D}$-class. For example, if S is a semigroup with binary operation given by $xy = y$ for all $x, y \in S$, then S has a single $\mathscr{R}$-class (and thus a single $\mathscr{D}$-class), but every $\mathscr{L}$-class is a singleton. Note that every element of S is an idempotent.

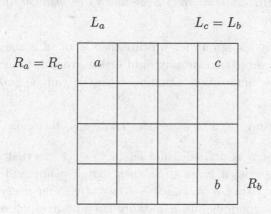

Fig. 14.1   An egg-box picture of a $\mathscr{D}$-class

14.4.9 Corollary. *Let S be a semigroup, and let $a \in S$. Then for any $x, y \in D_a$, there is a bijection from H_x to H_y. More precisely, if $z \in L_x \cap R_y$, and $u, w \in S$ are such that ρ_u is a bijection from L_x onto L_z, and λ_w is a bijection from R_z onto R_y, then $\lambda_w \circ \rho_u$ is a bijection from H_x onto H_y.*

Observe that if $x, y \in S$ satisfy either $S^1 x = S^1 y$ or $x S^1 = y S^1$, then $S^1 x S^1 = S^1 y S^1$, and so $\mathscr{D} \subseteq \mathscr{J}$. The poset diagram for Green's relations is shown in Figure 14.2.

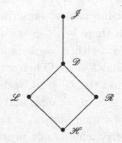

Fig. 14.2   The poset diagram for the poset of Green's relations

In general, the inclusion of $\mathscr{D}$ within $\mathscr{J}$ is proper, but in the special circumstance that $\mathscr{D}$ is the universal relation, then of course, $\mathscr{D} = \mathscr{J}$, and in fact, S must then be a simple semigroup.

14.4.10 Definition. *A semigroup S is said to be bisimple if $\mathscr{D}$ is the universal relation on S.*

For example, any semigroup for which either $\mathscr{L}$ or $\mathscr{R}$ is the universal relation is bisimple. In particular, any right or left group is bisimple.

There is another interesting situation where it can be proven that $\mathscr{D} = \mathscr{J}$.

14.4.11 Proposition. *If S is a periodic semigroup, then $\mathscr{D} = \mathscr{J}$.*

Proof. Suppose that S is periodic, and let $(a, b) \in \mathscr{J}$, so that $S^1 a S^1 = S^1 b S^1$. Thus there exist $u, v, r, s \in S^1$ such that $b = uav$ and $a = rbs$. Accordingly, we find that $a = ruavs = (ru)^n a(vs)^n$ for every positive integer n. Since S is periodic, for a suitably large positive integer n, $e = (ru)^n$ and $f = (vs)^n$ are idempotents, and $a = eaf$, which implies that $a = ea = af$. Thus $a = ea = (ru)^{n-1} rua$ and so $(a, ua) \in \mathscr{L}$. But then $(av, b) = (av, uav) \in \mathscr{L}$. As well, $a = af = a(vs)^n = av(sv)^{n-1} s$, and so $(av, a) \in \mathscr{R}$. Thus $(a, av) \in \mathscr{R}$ and $(av, b) \in \mathscr{L}$, so $(a, b) \in \mathscr{L} \circ \mathscr{R} = \mathscr{D}$. Thus $\mathscr{J} \subseteq \mathscr{D}$. As $\mathscr{D} \subseteq \mathscr{J}$ always holds, we have $\mathscr{D} = \mathscr{J}$. □

It is not always the case that $\mathscr{D} = \mathscr{J}$, and we shall soon present an example of a semigroup in which they are different. In this example, we will make use of the following fact.

14.4.12 Lemma. *If S is a cancellative semigroup without identity, then $\mathscr{D} = \mathbb{1}_S$.*

Proof. Suppose that S is a cancellative semigroup without identity, and let $a, b \in S$ be $\mathscr{D}$-related. Then there is $c \in S$ with $(a, c) \in \mathscr{L}$ and $(c, b) \in \mathscr{R}$, and so there exist $u, v \in S^1$ with $c = ua$ and $a = vc$, and there exist $r, s \in S^1$ such that $c = br$ and $b = cs$. Then $vua = vc = a$, so $avua = aa$. By cancellation, $avu = a$, and thus $e = vu$ satisfies $ae = ea = a$. Suppose that $e \neq 1$. Then for any $x \in S$, $ax = aex$, and so by cancellation, $x = ex$. Similarly, $xa = xea$ and so by cancellation, $x = xe$. Thus e is the identity for S. Since S has no identity, this is not possible, and we conclude that $e = 1$; that is, $u = v = 1$ and $a = c = b$, as required. □

14.4.13 Example. *We now construct a cancellative semigroup S without idempotents (so by Lemma 14.4.12, $\mathscr{D}$ is the identity relation) on which*

$\mathscr{J}$ *is the universal relation. Let* $S = \{ [\begin{smallmatrix} a & 0 \\ b & 1 \end{smallmatrix}] \mid a, b \in \mathbb{R}, a, b > 0 \}$. *Then*
S is closed under matrix multiplication, and so S is a subsemigroup of the
multiplicative semigroup of all 2×2 *real matrices. We claim that S is a*
cancellative semigroup and that $E_S = \varnothing$ *(so in particular, S is without an*
identity). For $A = [\begin{smallmatrix} a & 0 \\ b & 1 \end{smallmatrix}]$, $\det(A) = a > 0$ *and so A is invertible. Thus S*
is actually a subsemigroup of the group $Gl_2(\mathbb{R})$, *and thus S is cancellative.*
Suppose that $E \in S$ *is an idempotent of S, in which case it is an idempotent*
in the group $Gl_2(\mathbb{R})$. *Thus E must be the identity matrix, which is not*
possible since every matrix in S has nonzero entry in the second row, first
column. Thus S has no idempotents, and so by Lemma 14.4.12, $\mathscr{D}$ *is the*
identity relation on S. We prove now that $\mathscr{J}$ *is the universal relation on*
S. Let $A, B \in S$, *say* $A = [\begin{smallmatrix} a & 0 \\ b & 1 \end{smallmatrix}]$ *and* $B = [\begin{smallmatrix} c & 0 \\ d & 1 \end{smallmatrix}]$. *For any positive integer*
n, let $D_n = [\begin{smallmatrix} \frac{1}{n} & 0 \\ \frac{1}{n} & 1 \end{smallmatrix}]$, *so* $D_n \in S$. *Then* $AD_n = [\begin{smallmatrix} \frac{a}{n} & 0 \\ \frac{b}{n}+1 & 1 \end{smallmatrix}]$. *For* $x \in \mathbb{R}$ *with*
$x > 0$, *let* $C_n(x) = [\begin{smallmatrix} \frac{nc}{a} & 0 \\ x & 1 \end{smallmatrix}]$, *so* $C_n(x) \in S$. *Then*

$$C_n(x)AD_n = \left[\begin{smallmatrix} c & 0 \\ \frac{xa+b+1}{n} & 1 \end{smallmatrix} \right].$$

Choose and fix n large enough so that $\frac{b+1}{n} < d$; *that is,* $\frac{b+1}{d} < n$, *and so*
$nd - (b+1) > 0$. *Now solve the equation* $d = \frac{xa+b+1}{n}$ *for x. We obtain*
$x = \frac{nd-(b+1)}{a} > 0$, *and so* $C = C_n(\frac{nd-(b+1)}{a}) \in S$ *and* $B = CAD_n$. *Thus*
for any $A, B \in S$, $B \in S^1 A S^1$, *and so* $S^1 A S^1 = S^1 B S^1$ *for all* $A, B \in S$.
It follows that $\mathscr{J}$ *is the universal relation on S.*

Note that if S is a group, then $\mathscr{H}$ is in fact the universal relation, and
thus all five Green's relations are universal. More generally, in an $\mathscr{H}$-class
of a semigroup S, any two elements are both $\mathscr{L}$-related and $\mathscr{R}$-related, and
one might suspect that an $\mathscr{H}$-class that contained an idempotent would be
distinguished within the $\mathscr{D}$ class.

14.4.14 Proposition. *Let S be a semigroup, and let* $a \in S$. *Then* H_a *is*
a subgroup of S if, and only if, H_a *contains an idempotent. Moreover, for*
any $e \in E_S$, H_e *is a maximal subgroup of S.*

Proof. Suppose that H_a contains an idempotent e. Then for any $x \in H_a$,
$S^1 x = S^1 e$ and $eS^1 = xS^1$, so $x = xe = ex$. Thus $H_a \subseteq eSe$. Moreover,
there exist $u, v \in S^1$ such that $ux = e = xv$, so $e = e^2 = (xv)(ux) \in xSx$.
Thus $H_a \subseteq G_e = \{ x \in eSe \mid e \in xSx \}$, and by Proposition 14.2.31, G_e is a
maximal subgroup of S; namely the group of units of the monoid eSe. Let
$x \in G_e$, so $x \in eSe$ and $e \in xSx$. Then $x = ex = xe$ and $e = xyx$ for some
$y \in S$, so $x \in eS^1$ and $e \in xS^1$, so $xS^1 \subseteq eS^1 \subseteq xS^1$, giving $xS^1 = eS^1$;

that is, $(x,e) \in \mathscr{R}$. As well, $S^1 x \subseteq S^1 e \subseteq S^1 x$, and so $(x,e) \in \mathscr{L}$. Thus $(x,e) \in \mathscr{H}$, and so $x \in H_e = H_a$, which establishes that $G_e \subseteq H_a$. Thus $G_e = H_a$, and so H_a is a maximal subgroup of S.

Conversely, if H_a is a subgroup of S, then H_a contains an idempotent. $\qquad \square$

Thus for $a \in S$, D_a contains a subgroup of S if, and only if, $E_S \cap D_a \neq \varnothing$. Our next result establishes exactly when this happens.

14.4.15 Proposition. *Let S be a semigroup, and let $a \in S$. Then the following are equivalent.*

(i) *Every element of D_a is regular;*
(ii) *D_a contains a regular element;*
(iii) *D_a contains an idempotent.*

If a is regular, then every $\mathscr{L}$-class of D_a contains an idempotent, and every $\mathscr{R}$-class of D_a contains an idempotent.

Proof. That (i) implies (ii) is immediate. We prove that (ii) implies (iii) and (iii) implies (i). Suppose that D_a contains a regular element. Without any loss of generality, we may suppose that a is regular. Let $x \in S$ be such that $axa = a$. Let $e = ax$, so $e \in E_S$. Then $ea = a$, and so $eS^1 = aS^1$; that is, $(a,e) \in \mathscr{R}$, and so $e \in R_a \subseteq D_a$. Thus D_a contains an idempotent.

Finally, suppose that D_a contains an idempotent e, and let $x \in D_a$. Then there exists $c \in D_a$ such that $(e,c) \in \mathscr{R}$ and $(c,x) \in \mathscr{L}$, and so $cS^1 = eS^1$ and $S^1 c = S^1 x$. There exist $r,s \in S^1$ such that $c = er$ and $e = cs$, so $c = ec = csc$. Let $f = sc \in E_S$. Then $c = cf$ and we have $S^1 x = S^1 c = S^1 cf \subseteq S^1 f = S^1 sc \subseteq S^1 c$, so $S^1 x = S^1 f$. Therefore, there exist $u,v \in S^1$ with $x = uf$ and $f = vx$, from which we obtain $x = xf = xvx$. Thus x is a regular element.

Suppose that $a \in S$ is regular. Then for any $x \in D_a$, there exists $y \in S$ such that $x = xyx$. Let $e = xy$ and $f = yx$, so $e, f \in E_S$. Then $(x,e) \in \mathscr{R}$, and $(x,f) \in \mathscr{L}$, so $e \in R_x$ and $f \in L_x$. $\qquad \square$

14.4.16 Definition. *Let S be a semigroup, and let $a \in S$. If a is a regular element of S, then D_a is called a regular $\mathscr{D}$-class of S.*

By Proposition 14.4.15, every element of a regular $\mathscr{D}$-class is regular.

We remark that by Corollary 14.4.9, there is a bijection between any two $\mathscr{H}$-classes of a given $\mathscr{D}$-class. If D is a $\mathscr{D}$-class and e, f are idempotents contained in D, it is natural to wonder if this bijection allows us to compare the group structures of H_e and H_f (both are subgroups contained in D).

14.4.17 Proposition. *Let S be a semigroup, and let $a \in S$ be a regular element. Let $e, f \in E_S \cap D_a$. Then the bijection from H_e to H_f as described in Corollary 14.4.9 is a group isomorphism.*

Proof. There exists $c \in D_a$ with $(e, c) \in \mathscr{R}$ and $(c, f) \in \mathscr{L}$, and then there exist $u, v, r, s \in S^1$ such that $e = cu$, $c = ev$, $c = rf$, and $f = sc$. Then $\lambda_s \circ \rho_v$ is a bijective mapping from H_e onto H_f. Let $x, y \in H_e$. Then $\lambda_s \circ \rho_v(xy) = sxyv$. Since $x, y \in H_e$, $x = xe = ey$, so $sxyv = sxeyv = sxcuyv \doteq sx(ev)uyv = (sxv)(uyv)$. Since $sxv = \lambda_s \circ \rho_v(x) \in H_f$, we have $sxyv = (sxv)f(uyv)$. Now $fu = scu = se$, and so $sxyv = sxvseyv = sxvsyv$; that is, $\lambda_s \circ \rho_v(xy) = \lambda_s \circ \rho_v(x)\lambda_s \circ \rho_v(y)$, and so the bijective mapping $\lambda_s \circ \rho_v$ is a homomorphism. $\qquad \square$

We close this section with an application of the results that we have obtained to the case of the full transformation monoid T_X on a nonempty set X.

14.4.18 Proposition. *Let X be a nonempty set, and let $f, g \in T_X$. Then $(f, g) \in \mathscr{L}$ if, and only if, $\ker(f) = \ker(g)$, and $(f, g) \in \mathscr{R}$ if, and only if, $f(X) = g(X)$.*

Proof. Suppose that $(f, g) \in \mathscr{R}$, so there exist $r, s \in T_X$ such that $f = g \circ r$ and $g = f \circ s$. Then $f(X) = g(r(X)) \subseteq g(X) = f(s(X)) \subseteq f(X)$, and so $f(X) = g(X)$. Conversely, suppose that $f, g \in T_X$ and $f(X) = g(X)$. For each $x \in X$, $f(x) \in g(X)$, so there exists $y_x \in X$ with $f(x) = g(y_x)$. Define $s : X \to X$ by $s(x) = y_x$ for each $x \in S$. Then $g \circ s(x) = g(y_x) = f(x)$, and so $g \circ s = f$. A similar construction yields $r \in T_X$ such that $f \circ r = g$, and so $(f, g) \in \mathscr{R}$. Note that if X is infinite, then the Axiom of Choice is required for the construction of r and s.

Next, suppose that $f, g \in T_X$ and $\ker(f) = \ker(g)$. Then for each $x \in f(X)$, $f^{-1}(x)$ is an equivalence class of $\ker(g)$, and so $|g(f^{-1}(x))| = 1$. If $f(X) \neq X$, choose $y' \in X - f(X)$. Define $r \in T_X$ by $r(x) = g(f^{-1}(x))$ if $x \in f(X)$, otherwise set $r(x) = y'$. Then for each $x \in X$, $r \circ f(x) = g(f^{-1}(f(x))) = g(x)$, and so $r \circ f = g$. Similarly, we may construct $s \in T_X$ such that $s \circ g = f$ and so $(f, g) \in \mathscr{L}$. Conversely, suppose that $(f, g) \in \mathscr{L}$, so there exist $r, s \in T_X$ such that $f = s \circ g$ and $g = r \circ f$. Suppose that $x, y \in X$ are such that $f(x) = f(y)$. Then $g(x) = r \circ f(x) = r \circ f(y) = g(y)$, and so $\ker(f) \subseteq \ker(g)$. Similarly, if $x, y \in X$ are such that $g(x) = g(y)$, then $f(x) = s \circ g(x) = s \circ g(y) = f(y)$ and so $\ker(g) \subseteq \ker(f)$. $\qquad\square$

14.4.19 Corollary. *Let X be a nonempty set. Then for $f, g \in T_X$, $(f, g) \in \mathscr{D}$ if, and only if, $|f(X)| = |g(X)|$.*

Proof. Suppose first of all that $f, g \in T_X$ are such that $(f, g) \in \mathscr{D}$. Then there exists $h \in T_X$ such that $(f, h) \in \mathscr{L}$ and $(h, g) \in \mathscr{R}$. By Proposition 14.4.18, we obtain that $\ker(f) = \ker(h)$, and $h(X) = g(X)$. Thus $|h(X)| = |g(X)|$, and since $|X/\ker(f)| = |f(X)|$ and $|X/\ker(h)| = |h(X)|$, we obtain $|f(X)| = |h(X)| = |g(X)|$.

Conversely, suppose that $f, g \in T_X$ and $|f(X)| = |g(X)|$, so there exists a bijective map h' from $f(X)$ onto $g(X)$. Let $h : X \to X$ be any map for which $h\big|_{f(X)} = h'$. Then $h \circ f(X) = g(X)$, and so by Proposition 14.4.18, $h \circ f$ is $\mathscr{R}$-related to g. On the other hand, since $h\big|_{f(X)}$ is injective, $\ker(h \circ f) = \ker(f)$, and thus $h \circ f$ is $\mathscr{L}$-related to f. It follows now that $(f, g) \in \mathscr{D}$. $\qquad\square$

If X is a finite set, then T_X is finite and thus periodic, so by Proposition 14.4.11, $\mathscr{D} = \mathscr{J}$ on T_X. It is remarkable that this holds true as well when X is infinite.

14.4.20 Proposition. *Let X be a nonempty set. Then on T_X, $\mathscr{D} = \mathscr{J}$.*

Proof. We prove that if $f, g \in T_X$ and $(f, g) \in \mathscr{J}$, then $|f(X)| = |g(X)|$, and so by Proposition 14.4.19, $(f, g) \in \mathscr{D}$. Let $f, g \in T_X$, and suppose that $(f, g) \in \mathscr{J}$. Then there exist $r, s, u, v \in T_X$ such that $f = r \circ g \circ s$, and $g = u \circ f \circ v$. Thus $f(X) = r \circ g \circ s(X) \subseteq r \circ g(X)$, and since $|r \circ g(X)| \leq |g(X)|$, we have $|f(X)| \leq |g(X)|$. Similarly, $g(X) = u \circ f \circ v(X) \subseteq u \circ f(X)$, and since $|u \circ f(X)| \leq |f(X)|$, we have $|g(X)| \leq |f(X)|$. Thus $|f(X)| = |g(X)|$, and so $(f, g) \in \mathscr{D}$. $\qquad\square$

There is a natural partial order relation on the set of $\mathscr{J}$-classes of any semigroup S; namely for $a, b \in S$, we declare $J_a \leq J_b$ if $S^1 a S^1 \subseteq S^1 b S^1$. In the case of T_X, the result is a linear ordering, since for $f, g \in T_X$, $J_f \leq J_g$ if, and only if, $|f(X)| \leq |g(X)|$. One usually refers to $|f(X)|$ as $rank(f)$, the rank of f, so $J_f \leq J_g$ if, and only if, $rank(f) \leq rank(g)$.

14.4.21 Example. *Let $X = J_3 = \{1, 2, 3\}$. Then $|T_X| = 3^3 = 27$. The possible values of the rank of an element of T_X are 1,2,3, so there are three $\mathscr{D}$-classes, and since $\mathscr{D} = \mathscr{J}$, the set of $\mathscr{D}$-classes is partially ordered, as* *discussed above, with the $\mathscr{D}$-class of elements of rank 3 the largest element in the partial order, the $\mathscr{D}$-class of elements of rank 2 is in the middle, and the $\mathscr{D}$-class of elements of rank 1 (the constants) is the smallest $\mathscr{D}$ class.*

There are $3! = 6$ maps in the $\mathscr{D}$-class consisting of the maps of rank 3; namely the 6 permutations on J_3, which make up the symmetric group S_3. The maps of rank 2 have three possible image sets; namely $\{1,2\}$, $\{1,3\}$, and $\{2,3\}$. Each of these three sets of size 2 determines an $\mathscr{R}$-class. The six maps with image $\{1,2\}$ form one $\mathscr{L}$-class of the $\mathscr{D}$-class of rank 2 elements, and they are $[1\,1\,2]$, $[1\,2\,1]$, $[2\,1\,1]$, $[1\,2\,2]$, $[2\,1\,2]$, and $[2\,2\,1]$. The six maps with image $\{1,3\}$ form another $\mathscr{L}$-class, and they are $[1\,1\,3]$, $[1\,3\,1]$, $[3\,1\,1]$, $[1\,3\,3]$, $[3\,1\,3]$, and $[3\,3\,1]$. Finally, the six maps with image $\{2,3\}$ form an $\mathscr{L}$-class of the rank 2 $\mathscr{D}$-class, and they are $[2\,2\,3]$, $[2\,3\,2]$, $[3\,2\,2]$, $[2\,3\,3]$, $[3\,2\,3]$, and $[3\,3\,2]$. The kernel equivalence relation of each of these eighteen maps has two classes, one with two elements and one singleton, and each determines an $\mathscr{R}$-class of the rank 2 $\mathscr{D}$-class. They are $\{\{1,2\},\{3\}\}$, $\{\{1,3\},\{2\}\}$, and $\{\{2,3\},\{1\}\}$. Thus the $\mathscr{D}$-class of rank 2 elements has three $\mathscr{L}$-classes, and three $\mathscr{R}$-classes, and thus has nine $\mathscr{H}$-classes. There are six idempotents in this $\mathscr{D}$-class; namely $[1\,2\,2]$, $[1\,3\,3]$, $[1\,2\,1]$, $[3\,2\,3]$, $[1\,1\,3]$, and $[2\,2\,3]$. Each of the nine $\mathscr{H}$-classes of this $\mathscr{D}$-class has size 2, and six of the nine are cyclic groups of order 2.

Finally, there are three maps of rank 1; namely $[1\,1\,1]$, $[2\,2\,2]$, and $[3\,3\,3]$, the three constant maps. They form a single $\mathscr{R}$ class, since all three have the universal relation on J_3 as their kernel, so the three image sets $\{1\}$, $\{2\}$, and $\{3\}$ determine three $\mathscr{L}$-classes in the rank 1 $\mathscr{D}$-class, and so it has three $\mathscr{H}$-classes, each a trivial group. The egg-box diagram of each of the three $\mathscr{D}$-classes is shown in Figure 14.3.

.5 Semigroup Actions

In Chapter 12, the concept of a group action was introduced. More specifically, in Definition 12.1.1, the notion of a left group action on a set was presented. In this section, we introduce the analogous concept for semigroups. As our ultimate goal is to present some results in automata theory, where the tradional notation requires the concept of a right semigroup action, we shall work exclusively with right semigroup actions. In order to present this notion, we must first introduce the idea of the opposite semigroup.

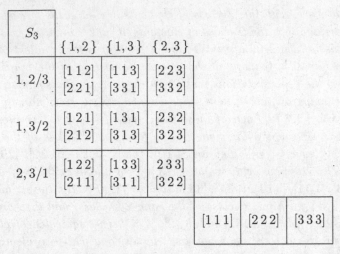

Fig. 14.3   The three $\mathscr{D}$-classes of T_3

14.5.1 Definition. *Let S be a semigroup. Define a binary operation $*$ on S by $a * b = ba$ for all $a, b \in S$. S, with this binary operation, shall be denoted by S^{op}.*

It is immediate that $*$ is associative, so S^{op} is a semigroup. Moreover, the idempotents of S and of S^{op} are the same, and if S has identity e_S, then S^{op} has e_S as its identity. If S is a group, then the map from S to S^{op} that sends a to its inverse is in fact an isomorphism. It is not true for semigroups in general that S is isomorphic to S^{op}. For example, under a semigroup isomorphism, a left ideal maps to a left ideal. In the right group E, with multiplication defined by $ef = f$ for all $e, f \in E$, every singleton subset is a left ideal of E, but the only right ideal of E is E itself. Thus E and E^{op} are not isomorphic semigroups.

We are now ready to introduce the notion of a monoid action. One might think that the appropriate idea would be to have a monoid homomorphism into the monoid of self-maps of a set, but it turns out that it is more convenient to generalize the notion further.

14.5.2 Definition. *Let X be a set and let M be a monoid. By a right action of M on X, we mean a monoid homomorphism $\varphi : M \to \mathscr{R}(X)^{op}$. For $U \subseteq X$ and $m \in M$, we shall denote $\varphi(m)(U)$, the image of U under $\varphi(m)$ by Um, so*

$$Um = \{ y \in X \mid \text{ there exists } u \in U \text{ with } (u, y) \in \varphi(m) \}.$$

In the particular case that U is a singleton, say $U = \{u\}$, we shall simply write um. If $|um| \leq 1$ for all $u \in X$ and all $m \in M$, then the action is said to be deterministic, and if $|um| = 1$ for all $u \in X$ and all $m \in M$, then the action is said to be complete (so a complete monoid action is deterministic).

We oberve that a complete monoid action on a set X is a monoid homomorphism into the monoid of self-maps of X.

Note that for $m, n \in M$, and $U \subseteq X$, we have $U(mn) = \varphi(mn)(U) = \varphi(m) * \varphi(n)(U) = \varphi(n) \circ \varphi(m)(U) = \varphi n(Um) = (Um)n$. Furthermore, since φ is required to be a monoid homomorphism, $\varphi(e_S)$ is the identity relation on X, and thus $Ue_M = U$ for all $U \subseteq X$.

14.5.3 Definition. *Let M be a monoid and X a set, and let $\varphi : M \to \mathscr{R}(X)^{op}$ be an action of M on X. If φ is injective, then the action is said to be a faithful action of M on X.*

Note that if φ is a faithful action, then $\mathscr{R}(X)^{op}$ contains a submonoid that is isomorphic to M.

14.5.4 Proposition. *Let M be a monoid acting on a set X by $\varphi : M \to \mathscr{R}(X)^{op}$. Then $\overline{\varphi} : M/\ker(\varphi) \to \mathscr{R}(X)^{op}$ is a faithful action of M on X such that for each $a \in M$ and each $U \subseteq X$, $\varphi(a)(U) = \overline{\varphi}(a)(\{\,[x]_{\ker(\varphi)} \mid x \in U\,\})$.*

Proof. This is a direct application of Proposition 14.2.29 to the homomorphism φ. $\square$

The construction of the action homomorphism $\overline{\varphi}$ is referred to as "making φ faithful".

14.5.5 Definition. *For a monoid M acting on a set X by $\varphi : M \to \mathscr{R}(X)^{op}$, the quotient monoid $M/\ker(\varphi)$ is called the transition monoid for the action.*

We now introduce some constructions that play an important role in the study of semigroup actions.

14.5.6 Definition. *Let M be a monoid. For $U, V \subseteq M$, let*

$$U^{-1}V = \{\,a \in M \mid Ua \cap V \neq \varnothing\,\}$$
$$= \{\,a \in M \mid \text{ there exists } u \in U \text{ such that } ua \in V\,\};$$

$$UV^{-1} = \{\,a \in M \mid aV \cap U \neq \varnothing\,\}$$
$$= \{\,a \in M \mid \text{ there exists } v \in V \text{ such that } av \in U\,\};$$

Furthermore, if X is a set upon which M acts, then for $Y, Z \subseteq X$, let

$$Y^{-1}Z = \{\, a \in M \mid Ya \cap Z \neq \varnothing \,\}$$
$$= \{\, a \in M \mid \text{ there exists } y \in Y \text{ and } z \in Z \text{ such that } z \in ya \,\};$$

and

$$YZ^{-1} = \{\, a \in M \mid aZ \cap Y \neq \varnothing \,\}$$
$$= \{\, a \in M \mid \text{ there exists } z \in Z \text{ and } y \in Y \text{ such that } y \in az \,\}.$$

14.5.7 Proposition. *Let M be a monoid acting on the set X. Let $Y, Z \subseteq X$, and let $A, B, C \subseteq M$. Then the following hold.*

    (i) $(AB)^{-1}C = B^{-1}(A^{-1}C)$ and $C(AB)^{-1} = (CB^{-1})A^{-1}$.

    (ii) $(ZA)^{-1}Y = A^{-1}(Z^{-1}Y)$ and $Y(BA)^{-1} = (YA^{-1})B^{-1}$.

Proof. (i) We have

$$(AB)^{-1}C = \{\, x \in M \mid \text{ there are } a \in A \text{ and } b \in B \text{ such that } abx \in C \,\}$$
$$= \{\, x \in M \mid \text{ there is } b \in B \text{ such that } bx \in A^{-1}C \,\}$$
$$= B^{-1}(A^{-1}C)$$

and

$$C(AB)^{-1} = \{\, x \in M \mid \text{ there are } a \in A \text{ and } b \in B \text{ such that } xab \in C \,\}$$
$$= \{\, x \in M \mid \text{ there is } a \in A \text{ such that } xa \in CB^{-1} \,\}$$
$$= (CB^{-1})A^{-1}.$$

For (ii), we have

$$(ZA)^{-1}Y = \{\, x \in M \mid \text{ there are } a \in A, z \in Z \text{ such that } zax \cap Y \neq \varnothing \,\}$$
$$= \{\, x \in M \mid \text{ there is } a \in A \text{ such that } ax \cap Z^{-1}Y \neq \varnothing \,\}$$
$$= A^{-1}(Z^{-1}Y),$$

and

$$Y(BA)^{-1} = \{\, x \in M \mid \text{ there are } a \in A \text{ and } b \in B \text{ such that } xba \cap Y \neq \varnothing \,\}$$
$$= \{\, x \in M \mid \text{ there is } b \in B \text{ such that } xb \cap YA^{-1} \neq \varnothing \,\}$$
$$= (YA^{-1})B^{-1}.$$

$\square$

For convenience, if any of the sets in Definition 14.5.6 is a singleton, say $\{\, x \,\}$, we shall simply write x rather than $\{\, x \,\}$. For example, for $a \in M$ and $B \subseteq M$, we shall write $a^{-1}B$ and Ba^{-1}, rather than $\{\, a \,\}^{-1}$ and $B\{\, a \,\}^{-1}$, respectively.

Moreover, we have the following associativity type of result.

14.5.8 Lemma. *Let M be a monoid, and let A, B, C be subsets of M. Then $(A^{-1}B)C^{-1} = A^{-1}(BC^{-1})$.*

Proof. We have

$$(A^{-1}B)C^{-1} = \{\, x \in M \mid \text{ there is } c \in C \text{ such that } xc \in A^{-1}B \,\}$$
$$= \{\, x \in M \mid \text{ there are } c \in C \text{ and } a \in A \text{ such that } axc \in B \,\}$$
$$= \{\, x \in M \mid \text{ there is } a \in A \text{ such that } ax \in BC^{-1} \,\}$$
$$= A^{-1}(BC^{-1}).$$

$\square$

In particular, for $a, c \in M$ and $B \subseteq M$, we have $a^{-1}(Bc^{-1}) = (a^{-1}B)c^{-1}$. As a result, we shall customarily omit the brackets and simply write $a^{-1}Bc^{-1}$.

14.5.9 Definition. *Let M be a monoid, and let $L \subseteq M$. Define relations ρ_L, λ_L, and P_L on M by:*

(i) $\rho_L = \{\, (x, y) \in M \times M \mid x^{-1}L = y^{-1}L \,\}$;

(ii) $\lambda_L = \{\, (x, y) \in M \times M \mid Lx^{-1} = Ly^{-1} \,\}$; and

(iii) $P_L = \{\, (x, y) \in M \times M \mid$ *for all* $u, v \in M$, $x \in u^{-1}Lv^{-1}$ *if,*

and only if, $y \in u^{-1}Lv^{-1}$ $\}$.

The notion of saturation is of importance in this work. If θ is an equivalence relation on a set X and $Y \subseteq X$, then Y is said to be saturated by θ (or θ saturates Y) if Y is a union of θ-classes.

14.5.10 Proposition. *Let M be a monoid, and let L be a subset of M. Then the following hold.*

(i) *ρ_L is a right congruence on M, and ρ_L saturates L. Moreover, if ρ is any right congruence on M that saturates L, then $\rho \subseteq \rho_L$.*

(ii) *λ_L is a left congruence on M, and λ_L saturates L. Moreover, if λ is any left congruence on M that saturates L, then $\lambda \subseteq \lambda_L$.*

(iii) *P_L is a congruence relation on M, and P_L saturates L, so $P_L \subseteq \rho_L \cap \lambda_L$. Moreover, if θ is any congruence on M that saturates L, then $\theta \subseteq P_L$.*

Proof. (i) Since ρ_L is the kernel of the map from M to $\mathscr{P}(M)$ that sends $x \in M$ to $x^{-1}L \in \mathscr{P}(M)$, ρ_L is an equivalence relation on M. Let $(x, y) \in \rho_L$, and $a \in M$. Then by Proposition 14.5.7, $(xa)^{-1}L = a^{-1}(x^{-1}L) = a^{-1}(y^{-1}L) = (ya)^{-1}L$, which proves that ρ_L is a right congruence. Furthermore, note that for $z \in M$, $z \in L$ if, and only if, $e_M \in z^{-1}L$.

Thus if $(x, y) \in \rho_L$, then $x^{-1}L = y^{-1}L$, and so $e_M \in x^{-1}L$ if, and only if, $e_M \in y^{-1}L$; that is, $x \in L$ if, and only if, $y \in L$. Thus ρ_L saturates L. Suppose now that ρ is a right congruence on M that saturates L, and let $(x, y) \in \rho$. Since ρ is a right congruence and $(x, y) \in \rho$, it follows that $(xa, ya) \in \rho$ for any $a \in M$, and since ρ saturates L, $xa \in L$ if, and only if, $ya \in L$. Thus $x^{-1}L = \{ a \in M \mid xa \in L \} = \{ a \in M \mid ya \in L \} = y^{-1}L$, so $(x, y) \in \rho_L$ and thus $\rho \subseteq \rho_L$.

(ii) Similarly, λ_L is an equivalence relation on M since it is the kernel of the map that sends $x \in M$ to $Lx^{-1} \in \mathscr{P}(M)$. Suppose that $(x, y) \in \lambda_L$, and let $a \in M$. Then by Proposition 14.5.7, $L(ax)^{-1} = (Lx^{-1})a^{-1} = (Ly^{-1})a^{-1} = L(ay)^{-1}$ and so $(ax, ay) \in \lambda_L$. Thus λ_L is a left congruence. Note that for $z \in M$, $z \in L$ if, and only if, $e_M \in Lz^{-1}$. Thus if $(x, y) \in \lambda_L$, then $Lx^{-1} = Ly^{-1}$, and so $e_M \in Lx^{-1}$ if, and only if, $e_M \in Ly^{-1}$; that is, $x \in L$ if, and only if, $y \in L$. Thus λ_L saturates L. Suppose now that λ is a left congruence on M that saturates L, and let $(x, y) \in \lambda$. Since λ is a left congruence and $(x, y) \in \lambda$, it follows that $(ax, ay) \in \lambda$ for any $a \in M$, and since λ saturates L, $ax \in L$ if, and only if, $ay \in L$. Thus $Lx^{-1} = \{ a \in M \mid ax \in L \} = \{ a \in M \mid ay \in L \} = Ly^{-1}$, so $(x, y) \in \lambda_L$ and thus $\lambda \subseteq \lambda_L$.

(iii) We observe that P_L is the kernel of the map from M to $\mathscr{F}(M \times M, \mathbb{Z}_2)$ that maps $x \in M$ to the map from $M \times M$ to $\mathbb{Z}_2$ that sends (u, v) to $1 \in \mathbb{Z}_2$ if $x \in u^{-1}Lv^{-1}$ or to $0 \in \mathbb{Z}_2$ if $x \notin u^{-1}Lv^{-1}$, and therefore P_L is an equivalence relation on M. Suppose that $(x, y) \in P_L$, and let $a \in M$. It follows from Proposition 14.5.7 that for $u, v \in M$, $a^{-1}(u^{-1}Lv^{-1}) = (ua)^{-1}Lv^{-1}$ and $(u^{-1}Lv^{-1})a^{-1} = u^{-1}L(av)^{-1}$. Thus for $(x, y) \in P_L$ and $a \in M$, $ax \in u^{-1}Lv^{-1}$ if, and only if, $x \in a^{-1}(u^{-1}Lv^{-1}) = (ua)^{-1}Lv^{-1}$, which in turn holds if, and only if, $y \in (ua)^{-1}Lv^{-1} = a^{-1}(u^{-1}Lv^{-1})$; that is, if, and only if $ay \in u^{-1}Lv^{-1}$, and so $(ax, ay) \in P_L$. The argument to establish that $(xa, ya) \in P_L$ is similar and so will be omitted. Suppose now that θ is a congruence on M that saturates L, and let $(x, y) \in \theta$. Let $u, v \in M$. Then $(uxv, uyv) \in \theta$, and we have $uxv \in L$ if, and only if, $uyv \in L$ since θ saturates L. If $x \in u^{-1}Lv^{-1}$, then $uxv \in L$, so $uyv \in L$ and thus $y \in u^{-1}Lv^{-1}$. Similarly, if $y \in u^{-1}Lv^{-1}$, then $x \in u^{-1}Lv^{-1}$ and so $(x, y) \in P_L$. Thus $\theta \subseteq P_L$.

Finally, since P_L is both a right congruence and a left congruence, we have $P_L \subseteq \rho_L \cap \lambda_L$, and since ρ_L (also λ_L) saturates L, it follows that P_L saturates L. $\qquad\square$

14.5.11 Notation. *For a monoid M and subset L of M, it is customary* to refer to the congruence P_L as the syntactic congruence on M that is determined by L.

In Chapter 12, it was shown that for a group G and subgroup H, there was a natural left action of G on the set of left cosets of H in G since left multiplication by $x \in G$ permutes the set of left cosets of H. Of course, right multiplication by $x \in G$ determines a right action of G on the set of right cosets of H, and we now describe the corresponding notion for semigroups. The analogue of the set of right cosets of a subgroup H of a group G will be the set of equivalence classes of a right congruence ρ on the monoid M.

14.5.12 Proposition. *Let M be a monoid and ρ be a right congruence on M. For $x \in M$, let $r_x : M \to M$ denote the map that sends $a \in M$ to $ax \in M$. Then for each ρ-class A, and each $x \in M$, there exists a unique ρ-class B such that $r_x(A) \subseteq B$. Specifically, if $A = [a]_\rho$, then $B = [ax]_\rho$. Thus r_x determines a map from M/ρ to M/ρ which sends $[a]_\rho$ to $[ax]_\rho$ for each $a \in M$. For simplicity, this map will also be denoted by r_x. The map $r : M \to T_{M/\rho}$ that sends $x \in M$ to $r_x \in T_{M/\rho}$ is a complete right action of M on M/ρ.*

Proof. Let $x \in M$. Then for any ρ-class $[a]_\rho$, for $y \in [a]_\rho$, $(a,y) \in \rho$ and so $(ax, yx) \in \rho$; that is, $yx \in [ax]_\rho$. Thus $r_x([a]_\rho) \subseteq [ax]_\rho$. It remains to prove that r is a homomorphism from M to $T_{M/\rho}^{op}$. Let $x,y \in M$. Then for $[a]_\rho \in M/\rho$, we have $r_{xy}([a]_\rho) = [axy]_\rho = r_y([ax]_\rho) = r_y(r_x([a]_\rho)) = r_y \circ r_x([a]_\rho) = r_x * r_y([a]_\rho)$, and so $r_{xy} = r_x * r_y$, as required. $\square$

14.5.13 Definition. *Let M be a monoid, and let X and Y be sets. Let $\varphi : M \to \mathscr{R}(X)^{op}$ and $\psi : M \to \mathscr{R}(Y)^{op}$ be actions of M on X and Y, respectively. Then a map $f : X \to Y$ is said to be an action homomorphism if for all $x \in X$ and $a \in M$, either both xa and $f(x)a$ are empty, or else neither is empty and then $f(xa) = f(x)a$, where we are using juxtaposition to denote both the action of a on an element of X and on an element of Y. If f is bijective, then f is said to be an action isomorphism.*

14.5.14 Proposition. *Let M be a monoid acting on sets X and Y by $\varphi : M \to \mathscr{R}(X)^{op}$ and $\psi : M \to \mathscr{R}(Y)^{op}$, respectively. If $f : X \to Y$ is an action isomorphism, then $f^{-1} : Y \to X$ is an action isomorphism.*

Proof. Let $y \in Y$ and $a \in M$. Let $x = f^{-1}(y)$, so $f(x) = y$. Since f is an action homomorphism, either $xa = \varnothing = f(x)a$, or else neither xa nor

$f(x)a$ is empty and then $f(xa) = f(x)a = ya$. Thus either both $f^{-1}(y)a$ and ya are empty, or neither is empty and then $f^{-1}(ya) = xa = f^{-1}(y)a$, as required. $\square$

14.5.15 Proposition.   *Let M be a monoid, and let ρ and τ be right congruences on M with $\rho \subseteq \tau$. Then the identity map on M induces a surjective action homomorphism $\pi : M/\rho \to M/\tau$.*

Proof. For $x \in M$, $[x]_\rho \subseteq [x]_\tau$, and we define $\pi([x]_\rho) = [x]_\tau$. It is immediate that π is a surjective map from M/ρ onto M/τ. Let $[x]_\rho \in M/\rho$ and $a \in M$. Then $\pi([x]_\rho a) = \pi([xa]_\rho) = [xa]_\tau = [x]_\tau a = \pi([x]_\rho)a$, and so π is an action homomorphism. $\square$

14.5.16 Corollary.   *Let M be a monoid and let L be a subset of M. If ρ is a right congruence that saturates L, then $\rho \subseteq \rho_L$ and $\pi : M/\rho \to M/\rho_L$ is a surjective action homomorphism.*

Proof. This is immediate from Propositions 14.5.10 and 14.5.15. $\square$

14.5.17 Proposition.   *Let M be a monoid, and let L be a subset of M. Let $X_L = \{\, u^{-1}L \mid u \in M \,\}$. For each $x \in M$, let $s_x : X_L \to X_L$ denote the map that sends $u^{-1}L$ to $(ux)^{-1}L$. Then the map $s : M \to T_{X_L}$ that sends $x \in M$ to $s_x \in T_{X_L}$ is a complete action of M on X_L. Moreover, the assignment $u^{-1}L \mapsto [u]_{\rho_L}$ determines an action isomorphism from X_L onto M/ρ_L.*

Proof. Our first step must be to verify that for $x \in M$, s_x is well-defined; that is, we must prove that if $u, v \in M$ are such that $u^{-1}L = v^{-1}L$, then $(ux)^{-1}L = (vx)^{-1}L$. Let $u, v \in M$ and suppose that $u^{-1}L = v^{-1}L$. Then by Proposition 14.5.7, $(ux)^{-1}L = x^{-1}(u^{-1}L) = x^{-1}(v^{-1}L) = (vu)^{-1}L$, and so s_x is indeed a well-defined map from X_L to itself. Next, we verify that s is an action of M on X_L. Let $x, y \in M$ and $u^{-1}L \in X_L$. Then $s_{xy}(u^{-1}L) = (uxy)^{-1}L = s_y((ux)^{-1}L) = s_y(s_x(u^{-1}L)) = s_y \circ s_x(u^{-1}L) = s_x * s_y(u^{-1}L)$, and so $s_{xy} = s_x * s_y$, as required.

By definition of ρ_L, for $u, v \in M$, $u^{-1}L = v^{-1}L$ if, and only if, $(u, v) \in \rho_L$; that is, if, and only if, $[u]_{\rho_L} = [v]_{\rho_L}$. Thus the assignment $u^{-1}L \mapsto [u]_{\rho_L}$ determines a bijective mapping from X_L onto M/ρ_L. That it is an action homomorphism follows immediately from the definition of the action of M on X_L and on M/ρ_L; namely that for $u^{-1}L \in X_L$ and $a \in M$, $(u^{-1}L)a = (ua)^{-1}L$ maps to $[ua]_{\rho_L} = [u]_{\rho_L} a$. $\square$

Our final result of this section is to describe the result of making the action of M on M/ρ_L faithful.

14.5.18 Proposition. *Let M be a monoid and let L be a subset of M. When the action of M on M/ρ_L is made faithful, the result is the complete action of M/P_L on M/ρ_L, where for $[a]_{P_L} \in M/P_L$ and $[x]_{\rho_L} \in M/\rho_L$, $[x]_{\rho_L}[a]_{P_L} = [xa]_{\rho_L}$.*

Proof. It suffices to verify that the kernel of the action of M on M/ρ_L is P_L. Let $x, y \in M$ be such that $[a]_{\rho_L} x = [a]_{\rho_L} y$ for all $a \in M$. Then for all $a \in M$, $[ax]_{\rho_L} = [ay]_{\rho_L}$, and so $(ax)^{-1}L = (ay)^{-1}$; equivalently, $axb \in L$ if, and only if, $ayb \in L$ for all $a, b \in M$. Thus for any $a, b \in M$, $x \in a^{-1}Lb^{-1}$ if, and only if, $y \in a^{-1}Lb^{-1}$, and so $(x, y) \in P_L$.

Conversely, suppose that $(x, y) \in P_L$. Then for all $u, v \in M$, $x \in u^{-1}Lv^{-1}$ if, and only if, $y \in u^{-1}Lv^{-1}$; that is, $uxv \in L$ if, and only if, $uyv \in L$. But this means that $v \in (ux)^{-1}L$ if, and only if, $v \in (uy)^{-1}L$, and so $(ux)^{-1}L = (uy)^{-1}L$, which establishes that $(ux, uy) \in \rho_L$. Thus $[u]_{\rho_L} x = [ux]_{\rho_L} = [uy]_{\rho_L} = [u]_{\rho_L} y$ for all $u \in M$ and so P_L is contained in the kernel of the action of M on M/ρ_L. $\square$

6 Automata Theory

The theory of automata is an application of the notion of semigroup actions to the special case where the monoid is the free monoid on a finite set (to be described shortly). An automaton is defined in such a way to model the operation of a machine which transitions from state to state according to instructions (inputs) given the machine. For example, a vending machine which dispenses candy bars, each costing 20¢, and which accepts only nickels or dimes (to keep the example simple), starts in its initial state (waiting for the first coin to be inserted), and moves from state to state as additional coins are inserted. Eventually, it reaches a terminal state when a total of 20¢ has been inserted, at which point it dispenses the selected candy bar and returns to its initial state. For example, it would reach its terminal state if four nickels have been inserted, or if two nickels and one dime have been inserted. Let the symbol a denote the insertion of a nickel, and let b denote the insertion of a dime. Then the machine will dispense a candy bar if it receives the input sequence (a, a, a, a), or any sequence of length three consisting of one b and two a's, or the sequence (b, b). The different internal states of the machine can be considered to be the proper prefixes of each of the sequences described above, which would give us as our set of states the set $\{aaa, aa, a, ab, ba, b\}$. We also require the initial state, which shall be denoted by i, and the terminal state, which shall be denoted by t. We can now draw a diagram, called a state diagram, for this machine. An arrow, labelled with either a or b, as appropriate, is drawn from one state to another if the recorded insertion causes the machine to change from the state at the tail of the arrow to the state at the head of the arrow.

Of course, there is the possibility that someone is not paying attention and begins to insert coins, then realizes that only one more nickel needs to be depositied to obtain the candy bar, but they have only a dime left. It is possible that they will follow through with the insertion of a dime, and then the machine should dispense the candy bar and return to its initial state (since it does not make change). They may also give up and leave in frustration, and then the machine will reward the next customer with a purchase after only inserting a nickel. We say that this machine recognizes the input sequences $aaaa$, aab, aba, baa, and bb, and also $aaab$, abb, and bab (as valid input sequences).

A state diagram of this machine is presented in Figure 14.4.

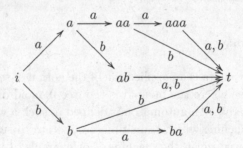

Fig. 14.4   A state diagram for the vending machine

Let us keep this simple example in mind as we develop the theory of automata. Our first step is to provide a home for the sequences that play an essential role in the theory (in the example, they were $aaaa$, aab, aba, baa, and bb, together with $aaab$, bab, and abb,).

14.6.1 Definition. *Let A be a nonempty set, and let A^+ denote the set of all finite sequences with entries from A. For any $u \in A^+$, the length of u is denoted by $|u|$. Define a binary operation, called concatenation, on A^+ as follows: for $u, v \in A^+$, the concatenation of u followed by v, is the sequence w of length $|u|+|v|$ for which $w(i) = u(i)$ for $i = 1, 2, \ldots, |u|$, while for i with*

$|u| < i \leq |u| + |v|$, *let* $w(i) = v(i - |u|)$. *For* $u = (a_1, a_2, \ldots, a_{|u|}) \in A^+$, *we shall adopt simpler notation and simply write* $u = a_1 a_2 \cdots a_{|u|}$. *Accordingly, the concatenation of u followed by v will be denoted by uv. A is called an alphabet, and the elements of* A^+ *are called words on the alphabet A.*

For example, if $A = \{a, b, c\}$, then $u = a \in A^+$ and $|u| = 1$, and $v = abaa \in A^+$ with $|v| = 4$. Then $uv = aabaa$ has length 5. We shall also find it convenient to denote a constant subsequence with exponential notation, so for example, we would write $aabaa = a^2 ba^2$.

14.6.2 Proposition. *Let A be a nonempty set. Then* A^+, *with the operation of concatenation, is a semigroup. Moreover, for any semigroup S, and any map* $f : A \to S$, *there is a unique homomorphism from* A^+ *to S whose restriction to A is equal to f (for convenience, we shall denote this unique homomorphism also by f).*

Proof. For $u, v, w \in A^*$, both $u(vw)$ and $(uv)w$ are equal to y of length $n = |u| + |v| + |w|$, where for i with $1 \leq i \leq |u|$, $y(i) = u(i)$, for i with $|u| < i \leq |u| + v|$, $y(i) = v(i - |u|)$, and for i with $|u| + |v| < i \leq n$, $y(i) = w(i - |u| - |v|)$, and so $u(vw) = (uv)w$. Thus concatenation is an associative operation on A^+.

Let S be a semigroup and let $f : A \to S$ be a map. Define $g : A^+ \to S$ by induction on length as follows. For $u \in A^+$ with $|u| = 1$, $u = a$ for some $a \in A$ and we define $g(u) = f(a)$. Suppose now that $n \geq 1$ is an integer for which $g(u)$ is defined for all $u \in A^+$ with $|u| \leq n$. Let $v \in A^+$ be such that $|v| = n + 1$. Then $v = ua$ for a unique $u \in A^+$ with $|u| = n$ and a unique $a \in A$, and we define $g(v) = g(u)f(a)$. Thus we have inductively defined a map $g : A^+ \to S$ whose restriction to A is f. We prove now that g is a homomorphism. Let $u \in A^+$. We prove by induction on $|v|$ that for all $v \in A^+$, $g(uv) = g(u)g(v)$. For $v \in A^+$ with $|v| = 1$, $v = a$ for some $a \in A$ and we have $g(ua) = g(u)f(a) = g(u)g(a)$. Suppose now that $n \geq 1$ is an integer for which $g(uv) = g(u)g(v)$ for all $v \in A^+$ with $|v| \leq n$, and let $v \in A^+$ be such that $|v| = n + 1$. Then $v = wa$ for a unique $w \in A^+$ with $|w| = n$, and a unique $a \in A$, so $g(uv) = g(uwa) = g(uw)f(a) = g(u)g(w)g(a) = g(u)g(wa) = g(u)g(v)$. It follows now by induction that $g(uv) = g(u)g(v)$ for all $v \in A^+$, and since u was arbitrary in A^+, this proves that g is a homomorphism. It remains to prove the uniqueness of g. Let $h : A^+ \to S$ be a homomorphism for which $h(a) = f(a)$ for every $a \in A$. We must prove that $h = g$. Again, we use induction on $|u|$. For $u \in A^+$ with $|u| = 1$, $u = a$ for some $a \in A$ and so

$h(u) = h(a) = f(a) = g(a) = g(u)$. Suppose now that $n \geq 1$ is an integer such that for all $u \in A^+$ with $|u| \leq n$, $h(u) = g(u)$, and let $u \in A^+$ be such that $|u| = n + 1$. Then $u = va$ for a unique $v \in A^+$ with $|v| = n$ and a unique $a \in A$, and so $h(u) = h(va) = h(v)h(a) = g(v)g(a) = g(va) = g(u)$. It follows now by induction that $h = g$. $\qquad\square$

14.6.3 Definition. *Let A be a nonempty set. Then the semigroup A^+, with the operation of concatenation, is called a free semigroup on the set A. The monoid $A^* = (A^+)^1$ obtained by adjoining an identity to A^+ is called the free monoid on the set A. The adjoined identity element 1 is often called the empty word, and we define $|1| = 0$.*

14.6.4 Corollary. *For any nonempty set A, and any monoid S and map $f : A \to S$, there is a unique monoid homomorphism from A^* to S whose restriction to A is f.*

Proof. Since S is a semigroup, there is a unique semigroup homomorphism $g : A^+ \to S$ whose restriction to A is f. Extend g to A^* by defining $g(1) = e_S$ to obtain a monoid homomorphism from A^* to S whose restriction to A is f. If $h : A^* \to S$ is a monoid homomorphism whose restriction to A is f, then the restriction of h to A^+ must be g, and since $h(1) = e_M = g(1)$, it follows that $h = g$. $\qquad\square$

Consider the state diagram shown in Figure 14.4 again. For convenience, let us relabel the states as $q_1 = a$, $q_2 = b$, $q_3 = aa$, $q_4 = ab$, $q_5 = aaa$, and $q_6 = ba$. Then the complete set of states is $Q = \{\, i, q_1, q_2, q_3, q_4, q_5, q_6, t \,\}$. Note that the arrows with the label a on them determine a relation on Q; namely

$$R_a = \{\, (i, q_1), (q_1, q_3), (q_3, q_5), (q_5, t), (q_2, q_6), (q_4, t), (q_6, t) \,\},$$

and similarly, the arrows with the label b determine a relation on Q; namely

$$R_b = \{\, (i, q_2), (q_1, q_4), (q_2, t), (q_3, t), (q_4, t), (q_5, t), (q_6, t) \,\}.$$

Set $A = \{\, a, b \,\}$ and consider the action of A^* on Q that is determined by mapping a to R_a and b to R_b. The reader can easily verify that under this action, $L = \{\, aaaa, aab, aba, baa, bb, aaab, bab, abb \,\}$ is equal to $i^{-1}t$.

This brief discussion is intended to provide motivation for the definitions and terminology we are about to introduce.

14.6.5 Definition. *Let A be a finite set, Q be a set, $f : A \to \mathscr{R}(Q)$ be a map, and let $\varphi : A^* \to \mathscr{R}(Q)^{op}$ be the unique monoid homomorphism for*

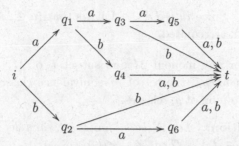

Fig. 14.5   The relabelled state diagram for the vending machine

which $\varphi|_A = f$. Then φ is a monoid action of A^* on Q. For $I \subseteq Q$ and $T \subseteq Q$, the data $\mathscr{A} = (Q, A, f, I, T)$ is called an A^*-automaton with set of initial states I and set of terminal states T. The subset $L = I^{-1}T$ of A^* is called the behaviour of $\mathscr{A}$, and we say that L is recognized by $\mathscr{A}$, or that $\mathscr{A}$ recognizes L. $\mathscr{A}$ is said to be deterministic (respectively, complete) if $|I| = 1$ and the action φ of A^* on Q is deterministic (respectively, complete). The elements of Q are called the states of $\mathscr{A}$, and the elements of

$$IA^* = \{\, q \in Q \mid \text{ there exists } m \in A^* \text{ and } i \in I \text{ with } q \in im \,\}$$

are said to be the states reachable from I. $\mathscr{A}$ is said to be accessible if $IA^* = Q$; that is, if every state is reachable from i. Finally, $\mathscr{A}$ is said to be finite if Q is a finite set.

It is customary to suppress the explicit mention of f if no uncertainty can arise. Thus we would typically write $\mathscr{A} = (Q, A, i, T)$ in place of (Q, A, f, I, T). Also, in a state diagram for $\mathscr{A}$, each initial state is typically identified by a short arrow pointing towards the state, while each element of T is identified by a short arrow pointing out from the state. As well, in the case of a deterministic automata, there is only one initial state, so $I = \{\, i \,\}$ and then we would write $\mathscr{A} = (Q, A, i, T)$.

14.6.6 Proposition. Let $\mathscr{A} = (Q, A, I, T)$ be an automaton, and let $L = I^{-1}T$. Then $(IA^*, A, I, T \cap IA^*)$ is an accessible automaton that recognizes L.

Proof. Note that $IA^* \subseteq Q$, and for $u, v \in A^*$ and $i \in I$, we have $(iu)v = i(uv) \in IA^*$, so the restriction of the action of $a \in A$ to IA^* determines an action of A^* on IA^*. As well, $I \subseteq IA^*$, and $IA^* \cap T \subseteq IA^*$. It remains to verify that $I^{-1}(IA^* \cap T) = L$. It is immediate that $I^{-1}(IA^* \cap T) \subseteq L$, so we prove that $L \subseteq I^{-1}(IA^* \cap T)$. Let $u \in L$, so there exists $i \in I$

such that $iu \in T$. Since $iu \in IA^*$, it follows that $iu \in IA^* \cap T$, and thus $u \in I^{-1}(IA^* \cap T)$, as required. $\qquad\square$

Recall that for any monoid M and subset L of M, M acts by right multiplication on M/ρ_L, the set of ρ_L-equivalence classes. Furthermore, recall that L is a union of ρ_L-classes.

14.6.7 Proposition. *Let A be a finite set, and let $L \subseteq A^*$. Let $\overline{L} = \{ [u]_{\rho_L} \mid u \in L \}$. Then $\mathscr{M}_L = (A^*/\rho_L, A, [1]_{\rho_L}, \overline{L})$ is a complete accessible A^*-automaton that recognizes L. Moreover, if $\mathscr{D} = (Q, A, i, T)$ is any accessible deterministic A^*-automaton that recognizes L, then there is a surjective action homomorphism from Q onto A^*/ρ_L.*

Proof. By Proposition 14.5.12, the action of right multiplication by A^* on A^*/ρ_L is a complete action. As $[1]_{\rho_L} A^* = \{ [x]_{\rho_L} \mid x \in A^* \} = A^*/\rho_L$, the action is accessible. Furthermore, by Proposition 14.5.10, ρ_L satuates L and thus $[1]_{\rho_L}^{-1}\overline{L} = \{ x \in A^* \mid [x]_{\rho_L} \in \overline{L} \} = L$; that is, $\mathscr{M}_L$ recognizes L. Suppose now that $\mathscr{D} = (Q, A, I, T)$ is an accessible A^*-automaton that recognizes L. We claim that the assignment of $q \in Q$ to $[u]_{\rho_L}$, where $u \in i^{-1}q$ ($i^{-1}q$ is nonempty since $\mathscr{D}$ is accessible), determines a map from Q to A^*/ρ_L. For suppose that $q \in Q$ and $u, v \in i^{-1}q$. Then $iu = q = iv$ and so $v^{-1}L = v^{-1}(i^{-1}T) = (iv)^{-1}T = (iu)^{-1}T = u^{-1}(i^{-1}T) = u^{-1}L$. But this means that $(u, v) \in \rho_L$; that is, $[u]_{\rho_L} = [v]_{\rho_L}$ and so the result of the assignment does not depend on our choice of element from $i^{-1}q$. Thus the assignment does determine a map from Q to A^*/ρ_L. Moreover, for any $u \in A^*$, if we let $q = iu$, then $u \in i^{-1}q$ and so $q \mapsto [u]_{\rho_L}$, and so the map is surjective. Finally, for $q \in Q$ and $v \in A^*$, q maps to $[u]_{\rho_L}$ where $iu = q$, and thus $iuv = qv$, which means that qv maps to $[uv]_{\rho_L} = [u]_{\rho_L} v$, so the map is a surjective action homomorphism. $\qquad\square$

14.6.8 Definition. *Let A be a finite set, and let $L \subseteq A^*$. Then $\mathscr{M}_L$ is called the minimal automaton recognizing L.*

The minimality is in the sense of homomorphic image. The set of all accessible, deterministic A^*-automata that recognize L is preordered by $\mathscr{A} \geq \mathscr{B}$ if there is a surjective action homomorphism from $\mathscr{A}$ to $\mathscr{B}$, and so $\mathscr{A} \geq \mathscr{M}_L$ for every $\mathscr{A}$ in this set. We note that if $\mathscr{M}_L$ is finite and $\mathscr{A}$ in this set is such that $\mathscr{M}_L \geq \mathscr{A}$, then $\mathscr{A}$ is isomorphic to $\mathscr{M}_L$.

14.6.9 Corollary. *Let A be a finite set, and let $L \subseteq A^*$. Then $\mathscr{M}_L$ is isomorphic to the complete automaton $(X_L, A, 1^{-1}L, \{ u^{-1}L \mid u \in L \})$,*

Table 14.1 $u^{-1}L$ for $u \in \{a,b\}^*$

u	$u^{-1}L$
1	L
a	$\{a^3, a^2b, ba, b^2\}$
b	$\{a^2, ab, b\}$
a^2	$\{a^2, ab, b\}$
ab	$\{a, b\}$
ba	$\{a, b\}$
b^2	$\{1\}$
a^3	$\{a, b\}$
a^2b	$\{1\}$
aba	$\{1\}$
ab^2	$\{1\}$
ba^2	$\{1\}$
bab	$\{1\}$
b^2a	$\varnothing$
b^3	$\varnothing$
a^4	$\{1\}$
a^3b	$\{1\}$.

where $X_L = \{u^{-1}L \mid u \in A^*\}$, where for $v \in A^*$ and $u^{-1}L \in X_L$, the action of v on $u^{-1}L$ results in $(uv)^{-1}L$.

Proof. This follows immediately from Propositions 14.6.7 and 14.5.17, wherein the latter result provided the isomorphism from X_L to A^*/ρ_L that maps $u^{-1}L$ to $[u]_{\rho_L}$. $\qquad\square$

14.6.10 Example. *Let us construct* $\mathscr{M}_L$, *where*

$$L = \{a^4, a^2b, aba, ba^2, b^2, a^3b, ab^2, bab\} \subseteq A^* = \{a, b\}^*$$

is recognized by the automata for the vending machine whose state diagram was shown in Figure 14.4. In order to determine the ρ_L-classes, we must first calculate $u^{-1}L$ for each $u \in A^*$. The results are displayed in Table 14.1, where if $u \in A^*$ does not appear in the righthand column, $u^{-1}L = \varnothing$.

Thus the ρ_L-classes are $i = \{1\}$, $q_1 = \{a\}$, $q_2 = \{a^2, b\}$, $q_3 = \{ab, ba, a^3\}$, $t = [a^4]_{\rho_L} = L$, *and*

$$q_5 = \{b^2a, b^3\} \cup \{u \mid |u| \geq 4,\ u \neq a^4, a^3b\}.$$

The state diagram for $\mathscr{M}_L$ *is shown in Figure 14.6.*

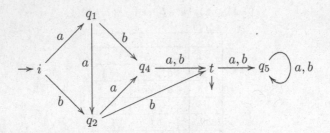

Fig. 14.6 The state diagram for the minimal automata for the vending machine example

Table 14.2 The Cayley table of the transition monoid for $\mathcal{M}_L$

	1	α	α^2	α^3	α^4	β	β^2	$\alpha\beta$	0
1	1	α	α^2	α^3	α^4	β	β^2	$\alpha\beta$	0
α	α	α^2	α^3	α^4	0	$\alpha\beta$	α^4	β^2	0
α^2	α^2	α^3	α^4	0	0	β^2	0	α^4	0
α^3	α^3	α^4	0	0	0	α^4	0	0	0
α^4	α^4	0	0	0	0	0	0	0	0
β	β	α^3	α^4	0	0	β^2	0	α^4	0
β^2	β^2	0	0	0	0	0	0	0	0
$\alpha\beta$	$\alpha\beta$	α^4	0	0	0	α^4	0	0	0
0	0	0	0	0	0	0	0	0	0

Let us continue with the above example and compute the transition monoid, A^*/P_L, for $\mathcal{M}_L$. Since $P_L \subseteq \rho_L$, each ρ_L-class is a union of P_L-classes. Thus we search for the coarsest refinement of the partition A^*/ρ_L that determines a congruence relation on A^*. It is sufficient to ensure that the cells of our refinement are compatible under multiplication on both the left and the right by a and b since these two elements are the generators of A^*. The reader is invited to verify that by splitting the ρ_L-class $\{\,ab, ba, a^3\,\}$ into $\{\,ab\,\}$, $\{\,a^3, ba\,\}$, and the ρ_L-class L into $\{\,a^2b, b^2\,\}$, $\{\,aba, ab^2, ba^2, bab, a^4, a^3b\,\}$, we obtain the desired partition. Let $\alpha = \{\,a\,\}$ and $\beta = \{\,b\,\}$. Then α and β generate A^*/P_L, and the Cayley table for the operation on A^*/P_L is given in Table 14.2 (in which we have substituted 0 for q_5).

Given a finite set A, a subset L of A^* is called a language over the alphabet A. The development of automata theory was carried out in large part due to the desire to try to carry out natural language processing by machine, which helps to explain the terminology of this section. Recall that if $\mathscr{A} = (Q, A, I, T)$ is an automaton (where we have suppressed explicit mention of the action of the elements of A on Q), then the language that is recognized by $\mathscr{A}$ is $L = I^{-1}T$, and L is called the behaviour of $\mathscr{A}$. We have seen that if $\mathscr{A}$ is deterministic, then there is an accessible deterministic automaton that recognizes the behaviour of $\mathscr{A}$, and that there is an accessible complete automaton, $\mathscr{M}_L$, that recognizes L, and this automata is simplest possible, in the sense that it is a homomorphic image of any deterministic automaton that recognizes L. It is natural to wonder about those languages that are recognized by nondeterministic automata: do they differ in some interesting way from those recognized by deterministic automata? The next result answers this question in full.

14.6.11 Proposition. *Let $\mathscr{A} = (Q, A, I, T)$ be an automaton, and let $L = I^{-1}T$. For $\hat{T} = \{ X \in \mathscr{P}(Q) \mid X \cap T \neq \varnothing \}$, $\hat{\mathscr{A}} = (\mathscr{P}(Q), A, I, \hat{T})$ is a complete automaton whose behaviour is L.*

Proof. Let $f : A \to \mathscr{R}(Q)$ determine the action of A^* on Q. We wish to extend f to a map from A to $(T_{\mathscr{P}(Q)})^{op}$. For $U \subseteq Q$ and $a \in A$, we have defined $Ua = \{ ua \mid u \in U \}$, and so we map $a \in A$ to

$$\{ (U, Ua) \mid U \in \mathscr{P}(Q) \} \in T_{\mathscr{P}(Q)}.$$

This map from A to $(T_{\mathscr{P}(Q)})^{op}$ determines the complete automaton $\hat{\mathscr{A}} = (\mathscr{P}(Q), A, I, \hat{T})$. Note that $I \in \mathscr{P}(Q)$ is the single initial state of $\hat{\mathscr{A}}$. It remains to prove that the behaviour of $\hat{\mathscr{A}}$ is L. Let $u \in I^{-1}\hat{T}$, so $Iu \in \hat{T}$; that is, $Iu \cap T \neq \varnothing$. Thus there exists $i \in I$ such that $iu \in T$, and so $u \in I^{-1}T = L$. Thus $I^{-1}\hat{T} \subseteq L$. Conversely, let $u \in L$, so there exists $i \in I$ such that $iu \in T$. But then $Iu \cap T \neq \varnothing$, and so $Iu \in \hat{T}$, which means that $u \in I^{-1}\hat{T}$. $\qquad\square$

We illustrate the preceding result with an example.

14.6.12 Example. *Let $A = \{ a, b \}$, $Q = \{ 1, 2, 3 \}$, and $f : A \to \mathscr{R}(Q)$ be given by $f(a) = \{ (1, 2), (1, 3) \}$ and $f(b) = \{ (3, 1), (3, 2) \}$. The state diagram for the automaton $\mathscr{A} = (Q, A, 1, \{ 2 \})$ that is determined by f is shown in Figure 14.7.*

It is evident that $1^{-1}2 = \{ a, ab \} \cup \{ (ab)^n a, a(ba)^n b \mid n \in \mathbb{Z}^+ \}$, which we can write as $1^{-1}2 = \{ (ab)^n a \mid n \in \mathbb{Z}, \ n \geq 0 \} \cup \{ a(ba)^n b \mid n \in \mathbb{Z}, \ n \geq 0 \}$ by using the convention that $u^0 = 1$ for $u \in A^*$. We can then regard $\langle ab \rangle \cup \{ 1 \} = \{ (ab)^n \mid n \in \mathbb{Z}, \ n \geq 0 \}$ as the free monoid on one generator ab, which we write as $\{ ab \}^*$, or $(ab)^*$. With this convention, we have $L = 1^{-1}2 = (ab)^*a \cup a(ba)^*b$. Since $|Q| = 3$, $\mathscr{P}(Q)$ contains 8 elements;

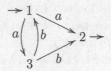

Fig. 14.7   The state diagram for the example of the subset construction automaton $\mathscr{A}$

namely $\varnothing, \{1\}, \{2\}, \{3\}, \{1,2\}, \{1,3\}, \{2,3\}$, and $\{1,2,3\}$. Note that $I = \{1\}$, and $T = \{2\}$, so $\hat{T} = \{\{2\}, \{1,2\}, \{2,3\}, \{1,2,3\}\}$. The extension of the action of A on Q is determined by

$$a \mapsto \begin{pmatrix} \varnothing & \{1\} & \{2\} & \{3\} & \{1,2\} & \{1,3\} & \{2,3\} & \{1,2,3\} \\ \varnothing & \{2,3\} & \varnothing & \varnothing & \{2,3\} & \{2,3\} & \varnothing & \{2,3\} \end{pmatrix}$$

and

$$b \mapsto \begin{pmatrix} \varnothing & \{1\} & \{2\} & \{3\} & \{1,2\} & \{1,3\} & \{2,3\} & \{1,2,3\} \\ \varnothing & \varnothing & \varnothing & \{1,2\} & \varnothing & \{1,2\} & \{1,2\} & \{1,2\} \end{pmatrix}$$

The state diagram for $D(\mathscr{A})$ is shown in Figure 14.8.

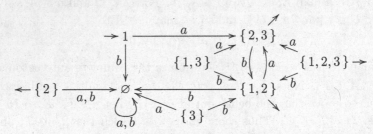

Fig. 14.8   The state diagram for $D(\mathscr{A})$

We may apply Proposition 14.6.6 to the automaton shown in Figure 14.8 to obtain an accessible deterministic automaton that accepts L, and the state diagram for it is shown in Figure 14.9.

Once we are able to convert a given automaton $\mathscr{A}$ into an accessible complete automaton, there is another procedure for the construction of the minimal automaton that recognizes the same language as $\mathscr{A}$.

14.6.13 Proposition. *Let A be a finite set, and let $\mathscr{A} = (Q, A, i, T)$ be an accessible, complete automaton. Let $\kappa = \{(q, r) \in Q \times Q \mid q^{-1}T = r^{-1}T\}$. Then κ is an equivalence relation on Q that saturates T and $\mathscr{B} = (Q/\kappa, A, [i]_\kappa, \overline{T})$, where $\overline{T} = \{[t]_\kappa \mid t \in T\}$ and the action of $a \in A$ on $q^{-1}T$ results in $(qa)^{-1}T$, is an automaton isomorphic to $\mathscr{M}_{i^{-1}T}$.*

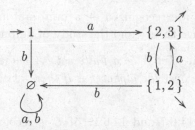

Fig. 14.9   The state diagram for $D(\mathscr{A})$ made accessible

Proof. Since κ is the kernel of the map from Q to $\mathscr{P}(Q)$ that sends $q \in Q$ to $q^{-1}T \in \mathscr{P}(Q)$, κ is an equivalence relation on Q. Moreover, by Proposition 14.5.7, if $q^{-1}T = r^{-1}T$, then for any $u \in A^*$, $(qu)^{-1}T = u^{-1}(q^{-1}T) = u^{-1}(r^{-1}T) = (ru)^{-1}$, and so the action of A^* on Q is compatible with κ; that is, that action of $u \in A^*$ sends κ-related elements to κ-related elements. Thus the action of A^* on Q induces an action of Q^* on Q/κ, where the action of $u \in A^*$ is to map $q^{-1}T \in Q/\kappa$ to $(qu)^{-1}T \in Q/\kappa$. Since $\mathscr{A}$ is accessible and complete, it follows that $\mathscr{B} = (Q/\kappa, A, [i]_\kappa, \overline{T})$ is also accessible and complete. To see that κ saturates T, let $q \in T$ and suppose that $r \in Q$ is such that $(q, r) \in \kappa$. Then $1 \in q^{-1}T = r^{-1}T$, and so $r = r1 \in T$. We have $([i]_\kappa)^{-1}\overline{T} = \{\, u \in A^* \mid [iu]_\kappa \in \overline{T} \,\} = \{\, u \in A^* \mid iu \in T \,\} = i^{-1}T$, as claimed. It remains to prove that $\mathscr{B}$ is isomoporphic to $\mathscr{M}_{i^{-1}T}$. Let $L = i^{-1}T$. By Corollary 14.6.9, $\mathscr{M}_L$ is isomorphic to $(X_L, A, 1^{-1}L, \{\, u^{-1}L \mid u \in L \,\})$, where the action of $v \in A^*$ on $u^{-1}L \in X_L$ results in $(uv)^{-1}L$. Observe that for $u, v \in A^*$ such that $(iu, iv) \in \kappa$, we have $(iu)^{-1}T = (iv)^{-1}T$ and thus

$$
\begin{aligned}
u^{-1}L &= \{\, w \in A^* \mid uw \in L \,\} = \{\, w \in A^* \mid iuw \in T \,\} \\
&= (iu)^{-1}T = (iv)^{-1}T = \{\, w \in A^* \mid ivw \in T \,\} \\
&= \{\, w \in A^* \mid vw \in L \,\} = v^{-1}L.
\end{aligned}
$$

It follows that α defined by $\alpha([iu]_\kappa) = u^{-1}L$ defines a map from Q/κ onto X_L. Moreover, for $[iu]_\kappa \in Q/\kappa$ and $v \in A^*$, we have $\alpha([iu]_\kappa)v = (u^{-1}L)v = (uv)^{-1}L = \alpha([iuv]_\kappa) = \alpha([iu]_\kappa v)$, and so α is a surjective action homomorphism. It remains to prove that α is injective. Suppose that $[iu]_\kappa, [iv]_\kappa \in Q/\kappa$ are such that $u^{-1}L = \alpha([iu]_\kappa) = \alpha([iv]_\kappa) = v^{-1}L$. Then $(iu)^{-1}T = \{\, w \in A^* \mid iuw \in T \,\} = \{\, w \in A^* \mid uw \in L \,\} = u^{-1}L = v^{-1}L = \{\, w \in A^* \mid vw \in L \,\} = \{\, w \in A^* \mid ivw \in T \,\} = (iv)^{-1}T$, and so $(iu, iv) \in \kappa$; that is, $[iu]_\kappa = [iv]_\kappa$. Thus α is injective, and so α is an action isomorphism from $\mathscr{B}$ to $(X_L, A, 1^{-1}L, \{\, u^{-1}L \mid u \in L \,\})$. $\qquad\square$

Languages which are recognized by a finite automaton play a special role in formal language theory and the study of automata.

14.6.14 Definition. *Let A be a finite set. A subset L of A^* is said to be a regular language over the alphabet A if there exists a finite automaton that recognizes L.*

By Propositions 14.6.7 and 14.6.11, if L is recognized by some finite A^*-automaton, then the minimal automaton $\mathcal{M}_L$ is a finite automaton. Thus, given a language L, to determine whether it is regular or not, one need only construct the minimal automaton recognizing L to obtain the answer. As it turns out, there is another approach to this question. There is a famous theorem due to S. C. Kleene (and so referred to as Kleene's Theorem) whose proof is bejond the scope of this short introduction to the theory of automata theory and formal languages, but which is quoted below. It enables one to identify a regular language without having to produce a finite automaton at all!

Before we can present Kleene's result, we need some additional terminology. The binary operation on A^* can be extended to a binary operation on $\mathscr{P}(A)$ in the obvious way; that is, for $U, V \subseteq A^*$, $UV = \{\, uv \mid u \in U,\ v \in V \,\}$. We shall refer to this binary operation on A^* as product, as in UV is the product of U and V. Product is an associative operation on $\mathscr{P}(A^*)$, and thus the exponentiation laws hold; that is, for positive integers i and j, $U^i U^j = U^{i+j}$, and $(U^i)^j = U^{ij}$. Then for $U \subseteq A^*$, we define $U^+ = \cup_{i \in \mathbb{Z}^+} U^i$, and $U^* = U^+ \cup \{\, 1 \,\}$. We say that U^* is obtained by an application of the star operation to U.

14.6.15 Definition. *Let A be a finite set. The set of rational languages over the alphabet A is the smallest subset of $\mathscr{P}(A^*)$ that contains all finite subsets of A^* and is closed with respect to the star operation, union, and product. A language $L \subseteq A^*$ is said to be rational (over the alphabet A) if it belongs to the set of rational languages over A.*

For example if $A = \{\, a, b \,\}$, then $L_1 = a^* = \{\, 1 \,\} \cup \{\, a^i \mid i \in \mathbb{Z}^+ \,\}$, $L_2 = a^* \cup b^*$, $L_3 = a^* \{\, b \,\}^2$, which we simply write as $a^* b^2$, and $L_4 = a^* b^2 \cup \{\, ab, b^2 a \,\}^*$ are all rational languages over A.

14.6.16 Proposition (Kleene's Theorem). *Let A be a finite set, and let $L \subseteq A^*$. Then L is rational if, and only if, it is regular.*

14.6.17 Example. *Let $A = \{\, a, b \,\}$ and $L = a^* b \cup b^* a$. L is a rational language, and thus by Kleene's theorem, L is a regular language. In this* *relatively simple example, it is quite easy to see how to construct a finite automaton that recognizes L. We use two initial states. i_1 and i_2, and one final state t, so $I = \{i_1, i_2\}$ and $T = \{t\}$. The action of a is given by $\{(i_1, i_1), (i_2, t)\}$, and the action of b by $\{(i_2, i_2), (i_1, t)\}$. The state diagram is shown in Figure 14.10.*

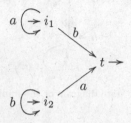

Fig. 14.10   The state diagram for the example illustrating Kleene's theorem

Let us also compute the minimal automata that recognizes L. This time, rather than using Proposition 14.6.7 for the computation of $\mathcal{M}_L$, we shall use Corollary 14.6.9. The set of states is $X_L = \{u^{-1}L \mid u \in A^*\}$, and the action of $u \in A^*$ on state $v^{-1}L$ results in $(vu)^{-1}L$. We observe that $1^{-1}L = L$, $a^{-1}L = a^*b \cup \{1\}$, $b^{-1}L = b^*a \cup \{1\}$, and for any integer $i \geq 2$, $(a^i)^{-1}L = a^*b$, $(b^i)^{-1}L = b^*a$, while for any integer $i \geq 1$, $(a^ib)^{-1}L = \{1\} = (b^ia)^{-1}L$. For any positive integer i and $u \in A^+$, $(a^ibu)^{-1}L = \varnothing = (b^iau)^{-1}L$. Now, any $v \in A^*$ is either 1, or else of the form a^iw, $w = 1$ or $w = by$, or else of the form b^iw, $w = 1$ or $w = ay$ for some $y \in A^*$. We have therefore determined all elements of X_L, so

$$X_L = \{1^{-1}L, a^{-1}L, b^{-1}L, (a^2)^{-1}L, (b^2)^{-1}L, (ab)^{-1}L, (aba)^{-1}L\}.$$

For notational convenience, let us relabel the elements of X_L as $i = 1^{-1}L$, $1 = a^{-1}L$, $2 = b^{-1}L$, $3 = (a^2)^{-1}L$, $4 = (b^2)^{-1}L$, $t = (ab)^{-1}L$, and $0 = (aba)^{-1}L$. The state diagram for $\mathcal{M}_L$ is shown in Figure 14.11.

The following result, known as the Pumping Lemma, gives a necessary condition for an infinite language to be regular. Since it is a necessary condition, it is typically used to show that a given language is not regular.

14.6.18 Proposition (Pumping Lemma). *Let A be a finite set and let L be an infinite regular language on the alphabet A. Then there is a positive integer n such that for all $z \in L$ with $|z| > n$, there exist $u, v, w \in A^*$ with $z = uvw$, $|uv| < n$, $v \neq 1$, and $uv^kw \in L$ for $k \in \mathbb{Z}^+$.*

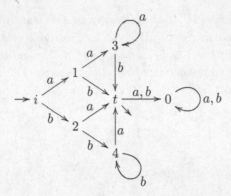

Fig. 14.11   The state diagram of the minimal automaton

Proof. Since L is regular, there exists a finite accessible complete automaton $\mathscr{A} = (Q, A, i, T)$ that recognizes L. Let $n = |Q|$. Now suppose that $z \in L$ and $m = |z| > n$. Then there exist $a_1, a_2, \ldots, a_m \in A$ such that $z = a_1 a_2 \ldots, a_m$. Let $q_0 = i$, and for $j = 1, 2, \ldots, m$, let $q_j = i a_1 a_2 \ldots, a_j$, so that in the state diagram for $\mathscr{A}$, there is a labelled path

$$i = q_0 \xrightarrow{a_1} q_1 \xrightarrow{a_2} q_2 \cdots q_{m-1} \xrightarrow{a_m} q_m \in T.$$

Consider the initial segment

$$i = q_0 \xrightarrow{a_1} q_1 \xrightarrow{a_2} q_2 \cdots q_n \xrightarrow{a_n} q_n.$$

Since there are $n + 1 > |Q|$ states that appear in this path, there must be a state that appears more than once in the path. Let $0 \le r < s \le n < m$ be integers such that $q_r = q_s$. Then

$$q_r \xrightarrow{a_{r+1}} q_{r+1} \xrightarrow{a_{r+2}} q_{r+2} \cdots \xrightarrow{a_s} q_s = q_r$$

is a cycle in the state diagram. If $r = 0$, then set $u = 1$, while if $r > 0$, set $u = a_1 a_2 \cdots a_r$. In either case, set $v = a_{r+1} a_{r+2} \cdots a_s$, so $v \ne 1$, and set $w = a_{s+1} a_{s+2} \cdots a_m$. Then $z = uvw$, and for any $j \ge 1$, $uv^j w$ is in $i^{-1} T = L$. Observe that since $s > r$, $|uv| = n - (s - r) < n$. $\square$

14.6.19 Example.   *Let $A = \{a, b\}$. We show how the pumping lemma can be used to prove that $L = \{a^i b^i \mid i \ge 1\}$ is not a regular language over A. For if it were, then by the pumping lemma, there exists a positive integer n such that for every integer i with $2i > n$, there exist $u, v, w \in A^*$ with $v \ne 1$, $a^i b^i = uvw$, and for every positive integer j, $uv^j w = a^k b^k$ for some positive integer k. Suppose such an n exists, and let $z = a^n b^n$. Then there exist $u, v, w \in A^*$ such that $a^n b^n = uvw$, $|uv| < n$, $v \ne 1$, and $uv^j w \in L$ for* *every positive integer j. Observe that since $|uv| < n$, both u and v must be powers of a, say $u = a^i$ and $v = a^k$, where i may be 0, but $k > 0$. Finally, $w = a^t b^n$, where $t = n - (i + k) > 0$. Since $uv^2w \in L$, we have $a^{i+2k+t}b^n \in L$, which is not possible since $i + 2k + t = (i + k + t) + k = n + k > n$.*

Another way that one might attempt to prove that the language L of Example 14.6.19 is not regular would be to construct $\mathcal{M}_L$ and demonstrate that this is not a finite automaton. We shall illustrate this approach, using Corollary 14.6.9 to construct $\mathcal{M}_L$, whereby the set of states of $\mathcal{M}_L$ can be taken to be $X_L = \{ u^{-1}L \mid u \in A^* \}$. Since for any positive integer i, $(a^i)^{-1}L = \{ a^j b^{i+j} \mid j \geq 0 \}$, we see that for $i \neq k$ $(a^i)^{-1}L \neq (a^k)^{-1}L$ and thus X_L is infinite.

7   Exercises

1. For any semigroup S, let $\mathscr{C}_S$ denote the set of all congruences on S. $\mathscr{C}_S$ is partially ordered by inclusion. Let ω_S denote the universal congruence on S. For any $\rho, \tau \in \mathscr{C}_S$ with $\rho \subseteq \tau$, let $[\rho, \tau] = \{ \sigma \in \mathscr{C}_S \mid \rho \subseteq \sigma \subseteq \tau \}$. Let $\rho, \tau \in \mathscr{C}_S$ with $\rho \subseteq \tau$.

   a) Let $\sigma \in [\rho, \tau]$, so that ρ saturates σ. Define the relation σ' on S/ρ by
   $$\sigma' = \{ ([a]_\rho, [b]_\rho) \in S/\rho \times S/\rho \mid (a, b) \in \sigma \}.$$
   Prove that $\sigma' \in \mathscr{C}_{S/\rho}$.

   b) Prove that the map from $[\rho, \tau]$ to $[\mathbb{1}_{S/\rho}, \omega_{S/\rho}]$ that is determined by mapping $\sigma \in [\rho, \tau]$ to $\sigma' \in [\mathbb{1}_{S/\rho}, \omega_{S/\rho}]$ is an order isomorphism; that is, it is bijective and for $\sigma_1 \subseteq \sigma_2 \in [\rho, \tau]$, $\sigma_1 \subseteq \sigma_2$ if, and only if, $\sigma_1' \subseteq \sigma_2'$.

2. Let S be a semigroup. If the intersection of all ideals of S is not empty, then it is an ideal of S called the kernel of S, and denoted by $\ker(S)$, while if the intersection is empty, then S is said not to have a kernel.

   a) Let S be a semigroup with kernel K. Prove that K is a simple semigroup.

   b) Give an example of a semigroup without a kernel.

3. Let S be a semigroup.

   a) Prove that if $\mathscr{C}$ is a set of congruences on S, then
   $$\cap \mathscr{C} = \{ (a, b) \in S \times S \mid (a, b) \in \rho \text{ for each } \rho \in \mathscr{C} \}$$
   is a congruence on S.

    b)   By the first part, for congruences $\rho_1, \rho_2, \ldots, \rho_n$ on S, $\rho = \cap_{i=1}^{n} \rho_i$ is a congruence on S. Prove S/ρ is isomorphic to a subsemigroup of $S/\rho_1 \times S/\rho_2 \times \cdots \times S/\rho_n$.

    c)   Let ρ and σ be congruences on S, and let

$$\rho \vee \sigma = \cap\{\,\tau \mid \tau \text{ is a congruence on } S \text{ and } \rho \cup \sigma \subseteq \tau\,\}.$$

By the first part, $\rho \vee \sigma$ is a congruence on S. Prove that $(a, b) \in \rho \vee \sigma$ if, and only if, there exists an odd positive integer $n = 2m + 1$ and $a = a_0, a_1, a_2, \ldots, a_n = b \in S$ with $(a_{2i}, a_{2i+1}) \in \rho$ and $(a_{2i+1}, a_{2i+2}) \in \sigma$ for all i with $0 \le i < m$. Note that the apparent asymmetry between ρ and σ is illusory (since $\rho \vee \sigma = \sigma \vee \rho$).

4.   Consider the monoid $M_{2\times2}(\mathbb{R})$ with the operation of matrix multiplication. Let S and T denote the subsets of $M_{2\times2}(\mathbb{R})$ consisting of those matrices with zero first row, respectively, those with zero first column.

    a)   Prove that S is a subsemigroup of $M_{2\times2}(\mathbb{R})$, and determine all idempotents of S.

    b)   Prove that T is a subsemigroup of $M_{2\times2}(\mathbb{R})$, and determine all idempotents of T.

    c)   Determine whether or not S and T are isomorphic.

    d)   Give an example of a subsemigroup of $M_{2\times2}(\mathbb{R})$ under multiplication which does not have any idempotents.

5.   a)   Let S, T, and R be semigroups, and let $f : S \to T$ and $g : T \to R$ be homomorphisms. Prove that $g \circ f : S \to R$ is a homomorphism.

    b)   Let S be a semigroup. Prove that $\mathrm{End}(S)$ is a monoid under the operation composition of maps.

    c)   Prove that $\mathrm{End}(\mathbb{N}, +)$ is isomorphic to $(\mathbb{N}, \times)$.

6.   Let X be a nonempty set.

    a)   Determine all $g \in T_X$ for which $\{\, g \circ f \mid f \in T_X \,\} = T_X$.

    b)   Determine all $g \in T_X$ for which $\{\, f \circ g \mid f \in T_X \,\} = T_X$.

    c)   Determine all $g \in T_X$ for which $g \circ f = g$ for every $f \in T_X$.

    d)   Determine all $g \in T_X$ for which $f \circ g = g$ for every $f \in T_X$.

7.   Prove that a cancellative semigroup can contain at most one idempotent.

8.   A semigroup S is said to be a left (respectively, right) zero semigroup if every element of S is a left (respectively, right) zero. Prove that if S is a semigroup with the property that for all $a, b, c, d \in S$, if $ab = cd$, then either $a = c$ or $b = d$, then S is either a left zero semigroup or else S is a right zero semigroup.

9. Let X be a finite set. For $\alpha \in T_X$, define $rank(\alpha) = |\alpha(X)|$ and $defect(\alpha) = |X| - rank(\alpha)$.

   a) Let $\alpha \in T_X$. Prove that if $r = rank(\alpha) < |X|$, then there exist $\beta, \gamma \in T_X$ such that $\alpha = \gamma \circ \beta$, $rank(\beta) = r+1$, and $defect(\gamma) = 1$. Moreover, prove that β and γ may be chosen so that β and α agree at all but one point, and γ is idempotent.

   b) Prove by induction on k, $0 \le k \le |X| - 1$, that every element of defect k can be written as the composite of an element of S_X, followed by k idempotents each of defect 1. Hint: The base case is immediate, and for the inductive step, suppose that α has defect $k + 1 \le |X| - 1$ and let $X/\ker(\alpha) = \{X_1, X_2, \ldots, X_r\}$, where $r = rank(\alpha)$. For each $i = 1, 2, \ldots, r$, let $a_i = \alpha(X_i)$. Since $r = rank(\alpha) = |X| - defect(\alpha) \le |X| - 1$, at least one of the cells of $X/\ker(\alpha)$ has more than one element. Without loss of generality, suppose that $|A_1| > 1$. Choose $a \in A_1$ and let $A_1' = A_1 - \{a\}$. Define $\beta \in T_X$ by $\beta(x) = \alpha(x)$ for all $x \in X - \{a\}$, and let $\beta(a) = b$, where b is chosen from $X - \alpha(X)$. Then $rank(\beta) = r + 1$ and so $defect(\beta) = k$. Define $\gamma \in T_X$ by $\gamma(x) = x$ for all $x \in X - \{b\}$, while $\gamma(b) = a_1$. Then $defect(\gamma) = 1$, $\gamma^2 = \gamma$, and $\gamma \circ \beta = \alpha$.

10. Let S be a semigroup. For each $a \in S$, the map $\lambda_a : S \to S$ defined by $\lambda_a(x) = ax$ for all $x \in S$ is called the inner left translation of S by a. Similarly, the map $\rho_a : S \to S$ defined by $\rho_a(x) = xa$ for all $x \in S$ is called the inner right translation of S by a. Note that $\lambda_a(xy) = \lambda_a(x)y$ and $\rho_a(xy) = x\rho_a(y)$ for all $x, y \in S$. A map $\lambda : S \to S$ for which $\lambda(xy) = \lambda(x)y$ for all $x, y \in S$ is called a left translation of S, and a map $\rho : S \to S$ for which $\rho(xy) = x\rho(y)$ for all $x, y \in S$ is called a right translation of S. Prove that the set T of all left (respectively, right) translations of S is a monoid under composition of maps, and that the set of all inner left (respectively, right) translations of S is a left (respectively, right) ideal of T.

11. Let S be a periodic commutative semigroup. For each $e \in E_S$, let $S_e = \{x \in S \mid e \in \langle x \rangle\}$ (see Definition 14.2.11).

    a) Prove that for each $e \in E_S$, S_e is a subsemigroup of S.

    b) Prove that for any $e, f \in E_S$, if $e \ne f$, then $S_e \cap S_f = \varnothing$.

    c) Prove that $S = \cup_{e \in E_S} S_e$.

    d) Prove that for any $e, f \in E_S$, $S_e S_f \subseteq S_{ef}$.

12. Prove that a regular semigroup with exactly one idempotent is a group.

13. Prove that a semigroup S is regular if, and only if, $AB = A \cap B$ for all right ideals A and all left ideals B of S.

14. Let S be a semigroup. Let $e \in E_S$ and $a, b \in S$ be such that $ab = e$ but $ba \neq e$, yet e is the identity of $\langle a, b \rangle$. Observe that every element of $\langle a, b \rangle$ can be written in the form $b^m a^n$, where m and n are nonnegative integers (in $\langle a, b \rangle$, $a^0 = e = b^0$).

   a) Prove that each element of $\langle a, b \rangle$ can be uniquely represented in that way as follows.

      (i)   Prove that a and b are each elements of infinite order.

      (ii) Prove that $\langle a \rangle \cap \langle b \rangle = \{\, e \,\}$.

      (iii) Prove that if $r \geq 0$ and $s \geq 0$ are such that $b^r a^s = e$, then $r = s = 0$.

      (iv) Prove that if r, s, m, n are nonnegative integers such that $b^r a^s = b^m a^n$, then $r = m$ and $s = n$.

   b) Define $\alpha, \beta \in T_{\mathbb{N}}$ as follows: $\alpha(n) = n + 1$ for all $n \in \mathbb{N}$, and $\beta(n) = n - 1$ for all $n \in \mathbb{N}$ with $n > 0$, while $\beta(0) = 0$. The subsemigroup $B = \langle \alpha, \beta \rangle$ of $T_{\mathbb{N}}$ is called the bicyclic semigroup.

      (i)   Prove that $\alpha\beta = \mathbb{1}_{\mathbb{N}}$ but $\beta\alpha \neq \mathbb{1}_{\mathbb{N}}$.

      (ii) Prove that $E_B = \{\, \beta^i \alpha^i \mid i \geq 0 \,\}$.

      (iii) Prove that B is bisimple; that is, B has only one $\mathscr{D}$ class.

      (iv) Prove that the egg-box picture of B is the infinite array

$$
\begin{array}{ccccc}
\mathbb{1}_{\mathbb{N}} & \alpha & \alpha^2 & \alpha^3 & \cdots \\
\beta & \beta\alpha & \beta\alpha^2 & \beta\alpha^3 & \cdots \\
\beta^2 & \beta^2\alpha & \beta^2\alpha^2 & \beta^2\alpha^3 & \cdots \\
\vdots & \vdots & \vdots & \vdots & \ddots
\end{array}
$$

   so that each row of the diagram is an $\mathscr{R}$-class of B, and each column is an $\mathscr{L}$-class of B.

15. Draw the egg-box picture for T_4, and for each $\mathscr{D}$-class of T_4, indicate the number of $\mathscr{R}$ and $\mathscr{L}$-classes in it. Furthermore, determine the size of each $\mathscr{H}$-class.

16. Prove that for any positive integer n, every ideal of T_n is principal; that is, for each ideal I of T_n, there exists $\alpha \in T_n$ such that $I = T_n \alpha T_n$.

17. Prove that a commutative semigroup S can be embedded in a group if, and only if, S is cancellative.

18. a) [Lallement's Lemma] Let S be a regular semigroup, and let $\varphi : S \to T$ be a surjective homomorphism from S onto a semigroup T. Prove that for every $f \in E_T$, $E_S \cap \varphi^{-1}(f) \neq \varnothing$. Hint: Let $f \in E_T$ and $a \in \varphi^{-1}(f)$. Let x be an inverse of a^2 (so $a^2 x a^2 = a^2$ and $x a^2 x = x$), and consider axa.

   b) Prove that a homomorphic image of an inverse semigroup is an inverse semigroup.

19. a) Let S be a semigroup, and for $e, f \in E_S$, write $e \leq f$ if $ef = fe = e$. Prove that this defines a partial order relation on E_S.

   b) This problem describes the order relation on E_{T_n} for any positive integer n.

   (i) Let $\alpha \in T_n$. Prove that $\alpha \in E_{T_n}$ if, and only if, $\alpha|_{\alpha(J_n)} = \mathbb{1}_{\alpha(J_n)}$.

   (ii) Prove that for $\alpha, \beta \in T_n$ (and so in particular for $\alpha, \beta \in E_{T_n}$), if $\alpha\beta = \beta\alpha = \alpha$, then $\ker(\beta) \subseteq \ker(\alpha)$ and $\alpha(J_n) \subseteq \beta(J_n)$.

   (iii) Prove that if $\alpha, \beta \in E_{T_n}$ are such that $\ker(\beta) \subseteq \ker(\alpha)$ and $\alpha(J_n) \subseteq \beta(J_n)$, then $\alpha\beta = \beta\alpha = \alpha$.

   c) Let $\alpha = \left(\begin{smallmatrix} 1\,2\,3\,4\,5\,6\,7\,8\,9 \\ 2\,2\,5\,2\,5\,2\,8\,8\,8 \end{smallmatrix} \right) \in T_9$, and observe that $\alpha \in E_{T_9}$. Find $\beta, \gamma \in E_{T_9}$ such that $\beta \neq \gamma$ and $\alpha \leq \beta$, $\alpha \leq \gamma$.

20. List the elements of $\mathscr{I}(J_2)$ (since each element of $\mathscr{I}(X)$ can be thought of as a bijective map from a subset of X onto a subset of X, we may use the two row map notation to describe the entries of $\mathscr{I}(X)$ for X finite, so for example, $\left(\begin{smallmatrix} 1 \\ 2 \end{smallmatrix} \right)$, and $\left(\begin{smallmatrix} 1\,2 \\ 2\,1 \end{smallmatrix} \right)$ are elements of $\mathscr{I}(J_2)$). For each element of $\mathscr{I}(J_2)$, find its inverse.

21. Let A be a finite set.

   a) Prove that if $L \subseteq A^*$ is regular, then $A^* - L$ is regular.

   b) Prove that if $K, L \subseteq A^*$ are both regular, then $K \cap L$ is regular.

22. Let A be a finite set, and let $L \subseteq A^*$. We say that L is recognized by a monoid M if there is a homomorphism $\varphi : A^* \to M$ and a subset L' of M such that $\varphi^{-1}(L') = L$.

   a) Prove that the transition monoid A^*/P_L recognizes L.

   b) Prove that the following are equivalent:

   (i) L is regular.

   (ii) A^*/P_L is finite.

   (iii) L is recognized by some finite monoid.

   c) A monoid M is said to divide a monoid N if there is a submonoid P of N such that M is a homomorphic image of P. For any finite set A and $L \subseteq A^*$, prove that the following are equivalent:

      (i) M recognizes L.

      (ii) A^*/P_L divides M.

23. Draw a state diagram for a finite automaton that recognizes $L = a^*b^* \subseteq \{a, b\}^*$.

24. Let $A = \{a, b\}$ and $L = A^*abA^*$.

   a) Draw a state diagram for a finite automaton that recognizes L.

   b) Draw a state diagram for $\mathscr{M}_L$.

   c) Construct the Cayley table for the monoid A^*/P_L.

25. Let $A = \{a, b\}$. For each of the following languages, draw a state diagram for a finite deterministic automaton that recognizes the language.

   a) $L = A^*aa$.

   b) $L = A^*aaA^*$.

   c) $L = A^*ab$.

26. Let $A = \{a\}$. Prove that $L = \{a^{n^2} \mid n \in \mathbb{Z}^+\}$ is not regular.

27. Let A be a finite set, and let $\mathscr{R}_A = \{L \subseteq A^* \mid L \text{ is regular}\}$. Consider any map $f : A \to \mathscr{R}_A \subseteq \mathscr{P}(A^*)$. Observe that $\mathscr{P}(A^*)$ is a monoid with operation multiplication of subsets of A^*; that is, for $X, Y \in \mathscr{P}(A^*)$, $XY = \{uv \mid u \in X,\ v \in Y\} \in \mathscr{P}(A^*)$. The identity element of $\mathscr{P}(A^*)$ is $\{1\}$. Let $\hat{f} : A^* \to \mathscr{P}(A^*)$ denote the unique monoid homomorphism from A^* to $\mathscr{P}(A^*)$ whose restriction to A is f. Prove that for each $L \in \mathscr{R}_A$, $\hat{f}(L) \in \mathscr{R}_A$.

28. Consider $B = \mathscr{P}(\{\varnothing\}) = \{\varnothing, \{\varnothing\}\}$. For convenience, let us label the elements of B as $0 = \varnothing$ and $1 = \{\varnothing\}$. Then set union and set intersection are associative, commutative operations on B, which we shall denote by $+$ and juxtaposition, respectively, and refer to as addition and multiplication. Furthermore, addition distributes over multiplication (and in fact, multiplication distributes over addition, but we shall not have need of this fact). For positive integers m, n, let $M_{m \times n}(B)$ denote the set of all $m \times n$ matrices with entries from B. The usual definition of matrix multiplication provides us with a map from $M_{m \times n}(B) \times M_{n \times k}(B)$ to $M_{m \times k}(B)$, and the associativity result is still valid; that is, if $A \in M_{m \times n}(B)$, $B \in M_{n \times k}(B)$, and $C \in M_{k \times p}(B)$, then $AB, BC, (AB)C$, and $A(BC)$ are all defined and $A(BC) = (AB)C$. In

particular, $M_{n \times n}(B)$ is a (noncommutative) monoid with identity I_n, the $n \times n$ diagonal matrix with each diagonal entry equal to 1.

Let A be a finite set, and let $n = |A|$. Let $f : A \to M_{n \times n}(B)$ be any map, and let $\hat{f}$ denote the unique monoid homomorphism from A^* to $M_{n \times n}(B)$ whose restriction to A is f. Let $\alpha \in M_{1 \times n}(B)$ and $\beta \in M_{n \times 1}(B)$. Then for any $X \in M_{n \times n}(B)$, $\alpha X \beta \in M_{1 \times 1}(B)$. Let $L_{(\alpha, \beta, f)} = \{ w \in A^* \mid \alpha \hat{f}(w) \beta = [1] \}$. We say that $L_{(\alpha, \beta, f)}$ is the language determined by (α, β, f). Further, we say that $L \subseteq A^*$ has a matrix representation if there exist $\alpha \in M_{1 \times n}(B)$, $\beta \in M_{n \times 1}(B)$, and $f : A \to M_{n \times n}(B)$ such that $L = L_{(\alpha, \beta, f)}$.

a) Let $A = \{a, b\}$, $f : A \to M_{2 \times 2}(B)$ be given by $f(a) = \begin{bmatrix} 0 & 1 & 0 \\ 0 & 0 & 1 \\ 0 & 0 & 1 \end{bmatrix}$ and $f(b) = \begin{bmatrix} 1 & 0 & 0 \\ 0 & 1 & 1 \\ 0 & 0 & 1 \end{bmatrix}$, $\alpha = [1\,0\,0]$ and $\beta = \begin{bmatrix} 0 \\ 1 \\ 0 \end{bmatrix}$. Let $L = L_{(\alpha, \beta, f)}$.

   (i) Describe the elements of L.

   (ii) Determine the elements of X_L (see Corollary 14.6.9). Is L regular?

b) Prove that $L \subseteq A^*$ is regular if, and only if, L has a matrix representation. Hint: To prove that a regular language L has a matrix representation, let $\mathscr{A} = (Q, A, i, T)$ be a finite deterministic automaton that recognizes L. Let $|Q| = n$, and label the elements of Q as $q_1, q_2, \ldots, q_n$, with $q_1 = i$. Then for $k = |T|$, there are indices $i_1, i_2, \ldots, i_k$ such that $T = \{ q_{i_1}, q_{i_2}, \ldots, q_{i_k} \}$. Let $\alpha \in M_{1 \times n}(B)$ be the matrix with nonzero entries only in columns $i_1, i_2, \ldots, i_k$, and let $\beta \in M_{n \times 1}(B)$ have nonzero entry only in the first row. Finally, define $f : A \to M_{n \times n}(B)$ by setting $f(a)$ equal to the $n \times n$ matrix with nonzero entries only in positions (r, s) where there is an arrow with the label a from q_s to q_r in the state diagram for $\mathscr{A}$.

29. [H. J. Shyr] Let A be a finite set. An automaton $\mathscr{A} = (Q, A, i, T)$ is said to be commutative if $qxy = qyx$ for all $q \in Q$ and all $x, y \in A^*$, while $\mathscr{A}$ is said to be quasi commutative if $ixy = iyx$ for all $x, y \in A^*$. $L \subseteq A^*$ is said to be commutative (respectively, quasi commutative) if $\mathscr{M}_L$ is commutative (respectively, quasi commutative). Prove that the following are equivalent:

a) L is quasi commutative.

b) For all $u, v \in A^*$, $(uv)^{-1}L = (vu)^{-1}L$.

c) $\{ (x, y) \in A^* \times A^* \mid xuvy \in L \} = \{ (x, y) \in A^* \times A^* \mid xvuy \in L \}$ for all $u, v \in A^*$.

d) A^*/P_L is commutative.

e) Define $\pi : A^* \to \mathscr{P}(A^*)$ by $\pi(1) = 1$ and for $u \in A^+$, we have $u = a_1 a_2 \cdots a_k$ for unique $a_1, a_2, \ldots, a_k \in A$, and we set $\pi(u) = \{\, a_{\tau(1)} a_{\tau(2)} \cdots a_{\tau(k)} \mid \tau \in S_{|u|} \,\}$. Then L has the property that $\pi(u) \subseteq L$ if, and only if, $u \in L$.

f) L is commutative.

Chapter 15

Isometries

1 Isometries of $\mathbb{R}^n$

In this chapter, we shall briefly study certain subgroups of the isometry group of $\mathbb{R}^n$, n a positive integer, and we shall then focus on finite subgroups of the isometry group of $\mathbb{R}^2$ and finite groups of rotations in $\mathbb{R}^3$.

We begin with a quick review of concepts from linear algebra that will be needed in this study. The reader who feels unsure about any elementary concepts of linear algebra should consult Appendix B. The inner product will be the usual dot product in $\mathbb{R}^n$, so for $u, v \in \mathbb{R}^n$ with $u = (u_1, u_2, \ldots, u_n)$ and $v = (v_1, v_2, \ldots, v_n)$, $u \cdot v = \sum_{i=1}^{n} u_i v_i$. The length of u, denoted by $||u||$, is defined by $||u|| = \sqrt{u \cdot u}$, and the distance between u and v is $||u - v||$. The standard basis for $\mathbb{R}^n$ is the set $\{e_1, e_2, \ldots, e_n\}$, where for each $i = 1, 2, \ldots, n$, e_i is the element of $\mathbb{R}^n$ for which every coordinate except for the ith is 0, and the ith coordinate is 1. For example, the standard basis for $\mathbb{R}^3$ is $\{e_1, e_2, e_3\}$, where $e_1 = (1, 0, 0)$, $e_2 = (0, 1, 0)$, and $e_3 = (0, 0, 1)$. The reader should be aware that the same notation is used for each $\mathbb{R}^n$, so for example, $e_1 = (1)$ in $\mathbb{R}^1 = \mathbb{R}$, $e_1 = (1, 0)$ in $\mathbb{R}^2$, $e_1 = (1, 0, 0)$ in $\mathbb{R}^3$, and so on. The standard basis is orthonormal with respect to the dot product; that is, $e_i \cdot e_j = 0$ if $i \neq j$, while $e_i \cdot e_i = 1$.

15.1.1 Definition. *Let n be a positive integer. A map $T : \mathbb{R}^n \to \mathbb{R}^n$ is called an isometry (or rigid motion) of $\mathbb{R}^n$ if T preserves distances; that is, for all $u, v \in \mathbb{R}^n$, $||u - v|| = ||T(u) - T(v)||$. If T is an isometry, then T is said to be orthogonal if $T(0) = 0$.*

15.1.2 Definition. *For any $\alpha \in \mathbb{R}^n$, the map $T_\alpha : \mathbb{R}^n \to \mathbb{R}^n$ defined by $T_\alpha(u) = u + \alpha$ for all $u \in \mathbb{R}^n$ is called a translation.*

15.1.3 Proposition. *Each translation of $\mathbb{R}^n$ is an isometry of $\mathbb{R}^n$.*

Proof. Let $\alpha \in \mathbb{R}^n$. To prove that T_α is an isometry, we observe that for $u, v \in \mathbb{R}^n$, $||T_\alpha(u) - T_\alpha(v)|| = ||u + \alpha - (v + \alpha)|| = ||u - v||$, as required. $\square$

15.1.4 Proposition. *Let S and T be isometries. Then $S \circ T$ is an isometry.*

Proof. Let $u, v \in \mathbb{R}^n$. Then

$$||S \circ T(u) - S \circ T(v)|| = ||S(T(u)) - S(T(v))|| = ||T(u) - T(v)||$$
$$= ||u - v||,$$

and so $S \circ T$ is an isometry. $\square$

At this point, while it is true that an isometry is a bijective map, the fact is not entirely obvious. That an isometry must be injective is immediate, since distinct elements u and v map to elements u' and v' with $||u' - v'|| = ||u - v|| \neq 0$, and thus $u' \neq v'$. It is apparent that any translation is surjective (and thus bijective) since for $\alpha \in \mathbb{R}^n$, and any $v \in \mathbb{R}^n$, $T_\alpha(v - \alpha) = v$. The next result will be most helpful in our treatment of isometries in general.

15.1.5 Proposition. *If T is orthogonal, then T is a nonsingular linear transformation; that is, an invertible linear transformation. Moreover, for each $u, v \in \mathbb{R}^n$, $||T(u)|| = ||u||$, and $T(u) \cdot T(v) = u \cdot v$.*

Proof. Let $T : \mathbb{R}^n \to \mathbb{R}^n$ be an isometry for which $T(0) = 0$. If we prove that T is a linear transformation, then by the dimension theorem for finite dimensional vector spaces and the fact that T is injective (hence the dimension of the null space of T is 0), the dimension of the image of T will be n and so T is surjective. We prove now that T is a linear transformation. Note that since T is orthogonal, $T(0) = 0$ and so for any $w \in R^n$, $||w|| = ||w - 0|| = ||T(w) - T(0)|| = ||T(w) - 0|| = ||T(w)||$. Moreover, for any $u, v \in \mathbb{R}^n$, we have $||u||^2 - 2u \cdot v + ||v||^2 = ||u - v||^2 = ||T(u) - T(v)||^2 = ||T(u)||^2 + ||T(v)||^2 - 2T(u) \cdot T(v)$, and thus $u \cdot v = T(u) \cdot T(v)$. We wish to prove that $T(u + v) = T(u) + T(v)$, or equivalently,

$||T(u+v) - T(u) - T(v)|| = 0$. We have

$$||T(u+v) - (T(u) + T(v))||^2$$
$$= (T(u+v) - (T(u) + T(v))) \cdot (T(u+v) - (T(u) + T(v)))$$
$$= T(u+v) \cdot T(u+v) - 2T(u+v) \cdot T(u) - 2T(u+v) \cdot T(v))$$
$$+ T(u) \cdot T(u) + 2T(u) \cdot T(v)) + T(v) \cdot T(v)$$
$$= (u+v) \cdot (u+v) - 2(u+v) \cdot u - 2(u+v) \cdot v$$
$$+ u \cdot u + 2u \cdot v + v \cdot v = 0,$$

proving that $T(u+v) = T(u) + T(v)$. As well, for any $r \in \mathbb{R}$, we have

$$||T(ru) - rT(u)||^2 = ||T(ru)||^2 - 2T(ru) \cdot rT(u) + ||rT(u)||^2$$
$$= ||ru||^2 - 2r(ru) \cdot u + r^2||u||^2$$
$$= r^2||u||^2 - 2r^2||u||^2 + r^2||u||^2 = 0,$$

and so $T(ru) - rT(u) = \mathbb{0}$, or $T(ru) = rT(u)$. Thus T is linear and the result follows. $\qquad\square$

15.1.6 Corollary. *Let n be a positive integer, and let $T \colon \mathbb{R}^n \to \mathbb{R}^n$ be an orthogonal map. For any (ordered) orthonormal basis $\alpha = \{\, u_1, u_2, \ldots, u_n \,\}$ for $\mathbb{R}^n$, let $c_\alpha \colon \mathbb{R}^n \to \mathbb{R}^n$ be the mapping defined by $c_\alpha(v) = (a_1, a_2, \ldots, a_n)$ where $v = \sum_{i=1}^n a_i u_i$ is the (unique) representation of $v \in \mathbb{R}^n$ as a linear combination of the elements of α. Then c_α is an orthogonal map.*

Proof. Let $v, w \in \mathbb{R}^n$, and suppose that $v = \sum_{i=1}^n a_i u_i$ and $w = \sum_{i=1}^n b_i u_i$. Then $v - w = \sum_{i=1}^n (a_i - b_i) u_i$, and so $||v - w||^2 = (\sum_{i=1}^n (a_i - b_i) u_i) \cdot (\sum_{j=1}^n (a_j - b_j) u_j) = \sum_{i=1}^n (a_i - b_i)^2 = ||(a_1, a_2, \ldots, a_n) - (b_1, b_2, \ldots, b_n)||^2 = ||c_\alpha(v) - c_\alpha(w)||^2$. Thus c_α is an isometry. Moreover, $c_\alpha(\mathbb{0}) = \mathbb{0}$, so c_α is orthogonal. $\qquad\square$

15.1.7 Corollary. *Every isometry of $\mathbb{R}^n$ can be expressed as a nonsingular linear isometry (a linear transformation that is also an isometry) followed by a translation. Moreover, the nonsingular linear isometry and the translation are uniquely determined by this requirement.*

Proof. Let T be an isometry, and let $\alpha = T(\mathbb{0})$. Now let $S = T_{-\alpha} \circ T$, and observe that $S(\mathbb{0}) = -\alpha + T(\mathbb{0}) = \mathbb{0}$. By Proposition 15.1.4, S is an isometry, and thus by Proposition 15.1.5, S is a nonsingular linear transformation. Since $T_\alpha \circ T_{-\alpha} = I_{\mathbb{R}^n}$, the identity transformation on $\mathbb{R}^n$, we obtain that $T_\alpha \circ S = T_\alpha \circ T_{-\alpha} \circ T = T$.

Suppose now that R and S are nonsingular linear isometries and $\alpha, \beta \in \mathbb{R}^n$ are such that $T_\alpha \circ R = T_\beta \circ S$. Then $\alpha = \alpha + 0 = \alpha + R(0) = T_\alpha \circ R(0) = T_\beta \circ S(0) = \beta + S(0) = \beta + 0 = \beta$. But then $T_\alpha = T_\beta$, and thus $R = S$. $\square$

15.1.8 Corollary. *Every isometry is a bijective mapping, and the inverse of an isometry is an isometry.*

Proof. Let T be an isometry. By Corollary 15.1.7, together with the fact that every translation is bijective, we obtain that T is a bijective mapping. For $u, v \in \mathbb{R}^n$, there exist $x, y \in \mathbb{R}^n$ such that $u = T(x)$ and $v = T(y)$. Thus $\|T^{-1}(u) - T^{-1}(v)\| = \|x - y\| = \|T(x) - T(y)\| = \|u - v\|$ and so T^{-1} is an isometry. $\square$

15.1.9 Definition. *For any positive integer n, $E(\mathbb{R}^n)$ denotes the set of all isometries of $\mathbb{R}^n$, $Gl(\mathbb{R}^n)$ denotes the set of all nonsingular linear transformations from $\mathbb{R}^n$ to itself, and $O(\mathbb{R}^n)$ denotes the subset of $Gl(\mathbb{R}^n)$ that consists of all linear isometries of $\mathbb{R}^n$.*

By Proposition 15.1.4 and Corollary 15.1.8, $E(\mathbb{R}^n)$ is a subgroup of the group $S_{\mathbb{R}^n}$ of all bijective mappings from $\mathbb{R}^n$ to itself, as is $Gl(\mathbb{R}^n)$. $E(\mathbb{R}^n)$ is called the Euclidean group in dimension n. Furthermore, $O(\mathbb{R}^n)$ is a subgroup of $Gl(\mathbb{R}^n)$, called the orthogonal group in dimension n. The natural action of $S_{\mathbb{R}^n}$ on $\mathbb{R}^n$ provides an action of $O(\mathbb{R}^n)$ on $\mathbb{R}^n$. It follows that we may use this action to form the semidirect product of the additive group $\mathbb{R}^n$ by the group $O(\mathbb{R}^n)$, and we denote this by $\mathbb{R}^n \rtimes O(\mathbb{R}^n)$. The semidirect product operation is given by $(\alpha, R)(\beta, S) = (\alpha + R(\beta), R \circ S)$ (see Definition 12.4.6).

Observe that Corollary 15.1.7 establishes that the mapping

$$\psi : E(\mathbb{R}^n) \to \mathbb{R}^n \times O(\mathbb{R}^n)$$

given by $\psi(T) = (\alpha, S)$ for $T \in E(\mathbb{R}^n)$, where $T = T_\alpha \circ S$, is injective. Moreover, by Proposition 15.1.4 and Corollary 15.1.7, ψ is surjective. Let T and S be isometries, and suppose that $T = T_\alpha \circ R$ and $S = T_\beta \circ Q$ where R and Q are nonsingular linear isometries. By Proposition 15.1.4, $T \circ S$ is an isometry, and so there exists $\gamma \in \mathbb{R}^n$ and a nonsingular linear isometry P such that $T \circ S = T_\gamma \circ P$. As $T \circ S = T_\alpha \circ R \circ T_\beta \circ Q$, we have for every $u \in \mathbb{R}^n$ that $T \circ S(u) = \alpha + R(\beta + Q(u)) = \alpha + R(\beta) + R \circ Q(u)$. Since R and Q are nonsingular linear isometries, $R \circ Q$ is a nonsingular linear isometry, and we have $T \circ S = T_{\alpha + R(\beta)} \circ (R \circ Q)$. Thus by Corollary 15.1.7, $\gamma = \alpha + R(\beta)$, and $P = R \circ Q$. We summarize this discussion in the following proposition.

15.1.10 Proposition. *The mapping* $\psi : E(\mathbb{R}^n) \to \mathbb{R}^n \times O(\mathbb{R}^n)$ *is an isomorphism from the group* $E(\mathbb{R}^n)$ *to the semidirect product* $\mathbb{R}^n \rtimes O(\mathbb{R}^n)$.

From our earlier work on semidirect products, we know that $\mathbb{R}^n$ is a normal subgroup of $\mathbb{R}^n \rtimes O(\mathbb{R}^n)$, and since the subset of all translations of $\mathbb{R}^n$ is a subgroup of $E(\mathbb{R}^n)$ ($T_\alpha \circ T_\beta = T_{\alpha+\beta}$ and $T_\alpha^{-1} = T_{-\alpha}$), and the fact that $\psi(T_\alpha) = (\alpha, I_{\mathbb{R}^n})$, we see that the subgroup of translations of $\mathbb{R}^n$ is a normal subgroup of $E(\mathbb{R}^n)$. We shall denote the set of all translations of $\mathbb{R}^n$ by $\mathrm{Trans}(\mathbb{R}^n)$. Thus $\mathrm{Trans}(\mathbb{R}^n) \lhd E(\mathbb{R}^n)$ and $E(\mathbb{R}^n)/\mathrm{Trans}(\mathbb{R}^n) \simeq O(\mathbb{R}^n)$.

Recall that in Example 5.2.9 (ix), the orthogonal matrix group $O(n)$ was defined by $O(n) = \{ A \in Gl_n(\mathbb{R}) \mid A^{-1} = A^t \}$. For $A \in O(n)$ and $X, Y \in M_{n \times 1}(\mathbb{R})$, we have AX and AY are elements of $M_{n \times 1}(\mathbb{R})$, and $||AX - AY||^2 = (AX - AY)^t(AX - AY) = X^t A^t AX - Y^t A^t AX - X^t A^t AY + Y^t A^t AX$. Since $A^t = A^{-1}$, we obtain that $||AX - AY||^2 = X^t X - Y^t X - X^t Y + Y^t Y = (X - Y)^t(X - Y) = ||X - Y||^2$, so if we make the usual identification of $\mathbb{R}^n$ with $M_{n \times 1}(\mathbb{R})$, then we can think of left multiplication by A as giving a mapping from $\mathbb{R}^n$ to itself, and the properties of matrix multiplication tell us that this mapping is linear, while the computation we have just completed establishes that the mapping is an isometry. It is shown in Appendix B that, relative to an orthonormal basis for $\mathbb{R}^n$, the matrix representative A of a linear transformation $T : \mathbb{R}^n \to \mathbb{R}^n$ is an orthogonal matrix (that is, an element of $O(n)$) if, and only if, T is a linear isometry. If bases are chosen in a coherent manner, the matrix representative of a composition of two linear transformations from $\mathbb{R}^n$ to $\mathbb{R}^n$ is the product of the matrix representatives of the two linear transformations, and thus the map from $O(\mathbb{R}^n)$ to $O(n)$ that sends a linear isometry to its matrix representative relative to a chosen orthonormal basis for $\mathbb{R}^n$ is a group isomorphism. Thus the study of the algebraic properties of $O(\mathbb{R}^n)$ amounts to the study of the same properties in $O(n)$, and we shall largely focus on $O(n)$ from now on.

The determinant mapping from $Gl_n(\mathbb{R})$ to $\mathbb{R}^* = \mathbb{R} - \{0\}$ is a group homomorphism (where the operation on $\mathbb{R}^*$ is multiplication), and for $A \in O(n)$, $A^{-1} = A^t$ and so $1 = \det(I_n) = \det(A^{-1}A) = \det(A^t A) = \det(A^t)\det(A) = \det(A)^2$, and so $\det(A) = \pm 1$. Thus the map that sends $A \in O(n)$ to the multiplicative subgroup $\{1, -1\}$ of $\mathbb{R}^*$ is a group homomorphism. It is perhaps not immediately obvious that it is surjective, and so we shall prove this now. Let n be a positive integer, and let E denote the elementary $n \times n$ matrix obtained by scaling row n of I_n by -1, so $\det(E) = -1$. As well, $E^{-1} = E = E^t$, so E is orthogonal, and so the map from $O(n)$ to $\{1, -1\}$ is surjective.

15.1.11 Definition. *The kernel of the homomorphism from $O(n)$ onto $\{1, -1\}$ is called the special orthogonal group of degree n, denoted by $SO(n)$.*

Thus $SO(n)$ has index 2 in $O(n)$.

15.2   Finite Subgroups of the Isometry Group of $\mathbb{R}^2$

We prove that the finite subgroups of $E(\mathbb{R}^2)$ are either cyclic or dihedral. Our first result shows that it suffices to prove that every finite subgroup of $O(\mathbb{R}^2)$ is either cyclic or dihedral.

15.2.1 Proposition. *Let n be a positive integer. If G is a finite subgroup of $E(\mathbb{R}^n)$, then there exists a translation T_β such that the conjugate of G by T_β is a subgroup of $O(\mathbb{R}^n)$.*

Proof. Let $y \in \mathbb{R}^n$, and set $x = \frac{1}{|G|} \sum_{g \in G} g(y)$. For $h \in G$, $h = T_\alpha \circ S$ for some $\alpha \in \mathbb{R}^n$ and $S \in O(\mathbb{R}^n)$. Since S is linear, we have $h(x) = \alpha + S(\frac{1}{|G|} \sum_{g \in G} g(y)) = \alpha + \frac{1}{|G|} \sum_{g \in G} S(g(y)) = \frac{1}{|G|} \sum_{g \in G} (\alpha + S(g(y))) = \frac{1}{|G|} \sum_{g \in G} h \circ g(y) = \frac{1}{|G|} \sum_{g \in G} g(y) = x$ since for all $h \in G$, $hG = G$. Thus for all $h \in G$, $h(x) = x$, and so $T_{-x} \circ h \circ T_x(0) = T_{-x} \circ h(x) = T_{-x}(x) = 0$. By Proposition 15.1.5, $T_{-x} \circ h \circ T_x \in O(\mathbb{R}^n)$, and thus $T_x^{-1} \circ h \circ T_x \in O(\mathbb{R}^n)$ for each $h \in G$; that is, the conjugate of G by T_x is a subgroup of $O(\mathbb{R}^n)$. $\quad\square$

As indicated above, we are interested in the case $n = 2$, and Proposition 15.2.1 tells us that it suffices to focus on finite subgroups of $O(\mathbb{R}^2)$, or equivalently, finite subgroups of $O(2)$. Recall that $SO(2)$ has index 2 in $O(2)$, and in fact, if $E \in O(2)$ has determinant -1, then $O(2)$ is the disjoint union of $SO(2)$ and its coset $E\,SO(2)$. We begin by identifying the elements of $SO(2)$.

15.2.2 Proposition. *For any $A \in O(2)$, $A \in SO(2)$ if, and only if, there exists $\theta \in \mathbb{R}$ with $0 \le \theta < 2\pi$ such that $A = \begin{bmatrix} \cos(\theta) & -\sin(\theta) \\ \sin(\theta) & \cos(\theta) \end{bmatrix}$, while $A \notin SO(2)$ if, and only if, there exists $\theta \in \mathbb{R}$ such that $A = \begin{bmatrix} \cos(\theta) & \sin(\theta) \\ \sin(\theta) & -\cos(\theta) \end{bmatrix}$. If $A \in O(2) - SO(2)$, then $A^2 = I_2$ (and since $A \ne I_2$, A has order 2).*

Proof. If A is of the first form, then $\det(A) = \cos^2(\theta) + \sin^2(\theta) = 1$, and so $A^{-1} = \begin{bmatrix} \cos(\theta) & \sin(\theta) \\ -\sin(\theta) & \cos(\theta) \end{bmatrix} = A^t$; that is, $A \in SO(2)$. If A is of the second form, then $\det(A) = -1$, and $A^{-1} = -\begin{bmatrix} -\cos(\theta) & -\sin(\theta) \\ -\sin(\theta) & \cos(\theta) \end{bmatrix} = A = A^t$, so $A \in O(2) - SO(2)$, as required. Conversely, suppose that $A \in O(2)$. By Proposition 15.1.5, Ae_1 has length 1 and so lies on the unit circle. Thus for some $\theta \in \mathbb{R}$, Ae_1, the first column of A, is $\begin{bmatrix} \cos(\theta) \\ \sin(\theta) \end{bmatrix}$. Moreover, also by Proposition 15.1.5, since $e_1 \cdot e_2 = 0$, $Ae_1 \cdot Ae_2 = 0$ and thus Ae_2 is on the unit circle, orthogonal to $\begin{bmatrix} \cos(\theta) \\ \sin(\theta) \end{bmatrix}$, and so Ae_2 is either $\begin{bmatrix} -\sin(\theta) \\ \cos(\theta) \end{bmatrix}$ or $\begin{bmatrix} \sin(\theta) \\ -\cos(\theta) \end{bmatrix}$. Since Ae_2 is the second column of A, the first case results in a matrix of determinant 1; hence an element of $SO(2)$, while the second case results in a matrix of determinant -1; hence an element of $O(2) - SO(2)$. Direct computation proves that any matrix A of this form satisfies $A^2 = I_2$.

Finally, the periodicity of the sine and cosine functions ensures that we may choose θ such that $0 \leq \theta < 2\pi$. $\qquad\square$

We have seen that an element of $SO(2)$ is in fact a rotation of the plane about the origin by some angle (if $A = \begin{bmatrix} \cos(\theta) & -\sin(\theta) \\ \sin(\theta) & \cos(\theta) \end{bmatrix}$, then the effect of left multiplication by A is to rotate the plane about the origin by the angle of radian measure θ). What is the geometric interpretation of the elements of $O(2) - SO(2)$? The answer is to be found by determining the eigenvectors and eigenvalues of a matrix $A = \begin{bmatrix} \cos(\theta) & \sin(\theta) \\ \sin(\theta) & -\cos(\theta) \end{bmatrix}$. The characteristic polynomial $c_A(\lambda)$ of A is $c_A(\lambda) = \det(A - \lambda I_2) = \begin{vmatrix} \cos(\theta)-\lambda & \sin(\theta) \\ \sin(\theta) & -\cos(\theta)-\lambda \end{vmatrix} = \lambda^2 - (\sin^2(\theta) + \cos^2(\theta)) = \lambda^2 - 1$, and so the eigenvalues of A are 1 and -1. The elements of the one-dimensional eigenspace associated with eigenvalue 1 are fixed by T_A, while each v in the one-dimensional eigenspace associated with eigenvalue -1 is sent to $-v$ by T_A.

We note that since A is symmetric, the two eigenspaces are orthogonal to each other. To see this, suppose that v and w are eigenvectors with associated eigenvalues 1 and -1, respectively. Then $v \cdot w = v^t w = (Av)^t w = v^t(A^t w) = v^t(Aw) = -v^t w = -v \cdot w$, and so $v \cdot w = 0$. Thus T_A fixes a line l through the origin, and reflects $l^\perp$, the line through the origin that is orthogonal to l, across the origin; that is, T_A is the reflection of the plane in the line l. We show that $(\cos(\frac{\theta}{2}), \sin(\frac{\theta}{2}))$ is a unit direction vector for l; that is, $(\cos(\frac{\theta}{2}), \sin(\frac{\theta}{2}))$ is a unit eigenvector for A with associated eigenvalue 1. Let $v = \begin{bmatrix} a \\ b \end{bmatrix}$ be an eigenvector of A with associated eigenvalue 1. Then $Av = v$, and so (a, b) is a solution to the system of linear equations $a(\cos(\theta) - 1) + b\sin(\theta) = 0$ and $a\sin(\theta) + b(-\cos(\theta) - 1) = 0$. We take advantage of the double-angle formulas for cosine and sine; namely $\cos(\theta) = \cos^2(\frac{\theta}{2}) - \sin^2(\frac{\theta}{2})$ and $\sin(\theta) = 2\sin(\frac{\theta}{2})\cos(\frac{\theta}{2})$, to obtain that $a(-2\sin^2(\frac{\theta}{2})) + b(2\sin(\frac{\theta}{2})\cos(\frac{\theta}{2})) = 0$ and $a(2\sin(\frac{\theta}{2})\cos(\frac{\theta}{2})) + b(-2\cos^2(\frac{\theta}{2})) = 0$; that is,

$$\sin(\frac{\theta}{2})(a\sin(\frac{\theta}{2}) - b\cos(\frac{\theta}{2})) = 0$$
$$\cos(\frac{\theta}{2})(a\sin(\frac{\theta}{2}) - b\cos(\frac{\theta}{2})) = 0.$$

Since $\sin(\frac{\theta}{2}) = 0$ and $\cos(\frac{\theta}{2}) = 0$ cannot both hold, it follows that

$$a\sin(\frac{\theta}{2}) - b\cos(\frac{\theta}{2}) = 0,$$

and so the eigenspace associated with eigenvalue 1 is the orthogonal complement of the one-dimensional subspace spanned by $(\sin(\frac{\theta}{2}), -\cos(\frac{\theta}{2}))$. Since $(\cos(\frac{\theta}{2}), \sin(\frac{\theta}{2}))$ is orthogonal to $(\sin(\frac{\theta}{2}), -\cos(\frac{\theta}{2}))$, it follows that the eigenspace of A that is associated with eigenvalue 1 is the span of $(\cos(\frac{\theta}{2}), \sin(\frac{\theta}{2}))$ (which is a unit vector). We shall write $M_{\frac{\theta}{2}}$ to denote the map which is the reflection of the plane in the line l through the origin with direction vector $(\cos(\frac{\theta}{2}), \sin(\frac{\theta}{2}))$.

Thus each element of $SO(2)$ corresponds to a rotation of the plane about the origin, while each element of $O(2) - SO(2)$ corresponds to the reflection of the plane in a line that passes through the origin. As a result of these observations, we shall refer to an element of $SO(2)$ as a rotation, while each element of $O(2) - SO(2)$ shall be called a reflection.

15.2.3 Proposition. *Any finite subgroup of $SO(2)$ is cyclic.*

Proof. Let G be a finite subgroup of $SO(2)$. If G is trivial, then G is cyclic, so we may assume that G is not trivial. By Proposition 15.2.2, each element of $SO(2)$ may be described as $R_\theta = \begin{bmatrix} \cos(\theta) & -\sin(\theta) \\ \sin(\theta) & \cos(\theta) \end{bmatrix}$ for some θ with $0 \le \theta < 2\pi$. Note that $R_\theta = I_2$ if, and only if, $\theta = 0$. Let

$$\alpha = \min\{\, \theta \in (0, 2\pi) \mid R_\theta \in G \,\}.$$

Then $\langle R_\alpha \rangle \subseteq G$, and $0 < \alpha < 2\pi$. Suppose that $G \ne \langle R_\alpha \rangle$. Then there exists β with $0 < \beta < 2\pi$ such that $R_\beta \in G$ but $R_\beta \ne R_\alpha^n$ for any positive integer n. Note that R_α provides a rotation of the plane by the angle α, and thus R_α^n provides a rotation of the plane by the angle $n\alpha$ for any $n \ge 1$, so $R_\alpha^n = R_{n\alpha}$. Let k be the largest nonnegative integer such that $\beta > k\alpha$ (if $\beta = m\alpha$ for any nonnegative integer m, then $R_\beta \in \langle R_\alpha \rangle$, and we have assumed this not to be the case). Thus $k\alpha < \beta < (k+1)\alpha$, and so $0 < \beta - k\alpha < \alpha$. By our choice of α, $R_{\beta-k\alpha} \notin G$. However, $R_{\beta-k\alpha}$

provides a rotation first by $-k\alpha$, then by β; that is, $R_{\beta-k\alpha} = R_\beta R_\alpha^{-k} \in G$. Since this contradiction was based on the assumption that $G \neq \langle R_\alpha \rangle$, we conclude that $G = \langle R_\alpha \rangle$. $\qquad\qquad\square$

Recall that in Example 5.4.5, the dihedral group D_n was defined as the group of rigid motions of the plane that map a regular n-gon centered at the origin to itself, and it was shown there that D_n consists of the n rotations in the subgroup $\left\langle R_{\frac{2\pi}{n}} \right\rangle$, together with n reflections, determined as follows: if n is even, the vertices of the n-gon occur in diametrically opposite pairs, and if $n = 2m$, then we have m pairs of diametrically opposite vertices, and reflection in the line through a diametrically opposite pair is an element of D_n, and furthermore, the $n = 2m$ sides of the n-gon can be partitioned into opposing pairs, and for each of these m pairs of opposite sides, reflection in the line joining their midpoints is an element of D_n. When n is odd, the n reflections that belong to D_n are simply the reflections in the n lines that join the origin to a vertex of the n-gon.

15.2.4 Proposition. *If G is a finite subgroup of $O(2)$ and G contains a reflection, then G has even order $2m$, G contains a cyclic normal subgroup N of order m generated by a rotation R_θ, and G contains m reflections. Furthermore, if M_β is any reflection in G, then $H = \langle M_\beta \rangle$ acts on N by conjugation, with action φ determined by $M_\beta^{-1} R_\theta M_\beta = R_\theta^{-1}$, and G is isomorphic to the semidirect product $H \rtimes_\varphi N$. Finally, $G = D_m$.*

Proof. The determinant homomorphism from $O(2)$ onto $\{1, -1\}$, restricted to G is still surjective since G contains a reflection. Moreover, $N = \ker(\det) = G \cap SO(2)$ is a finite subgroup of $SO(2)$, and so by Proposition 15.2.3, there exists $\theta \in [0, 2\pi)$ such that $N = \langle R_\theta \rangle$. Since $G/\ker(\det)$ is isomorphic to $\{1, -1\}$, it follows that $|G/N| = 2$ and thus $|G| = 2|N|$. Let $m = |N|$. Then $|G - N| = m$, and so G consists of a cyclic group of order m, generated by the rotation R_θ, and m reflections. Let $M_\beta = \begin{bmatrix} \cos(2\beta) & \sin(2\beta) \\ \sin(2\beta) & -\cos(2\beta) \end{bmatrix} \in G$, so M is an orthogonal reflection across the line l through the orgin making angle β with the positive x-axis. Then $G = N \cup N M_\beta$, so each element of G has a unique representation in the form $R_\theta^i M_\beta^j$, where $i = 0, 1, \ldots, m-1$, and $j = 0, 1$. Let us compute $M_\beta^{-1} R_\theta M_\beta$. Since M_β has order 2, it is its own inverse. Moreover, $M_\beta R_\theta$ has determinant -1, so it is a reflection and thus is its own inverse. It follows that $M_\beta R_\theta = (M_\beta R_\theta)^{-1} = R_\theta^{-1} M_\beta^{-1} = R_\theta^{-1} M_\beta$, and thus $M_\beta R_\theta M_\beta = R_\theta^{-1}$, or equivalently, $M_\beta^{-1} R_\theta M_\beta = R_\theta^{-1}$. Thus $H = \langle M_\beta \rangle$ acts on N by conjugation, with the action determined by $M_\beta^{-1} R_\theta M_\beta = R_\theta^{-1}$, and so $G \simeq H \rtimes_\varphi N$. It remains to prove that G is isomorphic to D_m. We consider the regular m-gon centered at the origin for which some vertex makes angle β with the positive x-axis. D_m is the group of m rotations of this regular m-gon together with the m reflections as described above. It follows that $H \subseteq D_m$ and $N \subseteq D_m$, so $G = HN \subseteq D_m$. Since $|G| = 2m = |D_m|$, it follows that $G = D_m$. $\qquad\square$

15.3 The Classification of the Finite Subgroups of $SO(3)$

15.3.1 Definition. *For positive integers m and n, and $A \in M_{m\times n}(\mathbb{R})$, let $T_A : \mathbb{R}^n \to \mathbb{R}^m$ denote the linear mapping defined by $T_A(x) = Ax$ for each $x \in \mathbb{R}^n$ (where we are treating an element of $\mathbb{R}^n$ as a column vector).*

15.3.2 Proposition. *For any $A \in O(3)$, $\det(A)$ is an eigenvalue of A. Furthermore, if $A \neq \det(A)I_3$, then the algebraic multiplicity of the eigenvalue $\det(A)$ is 1.*

Proof. Let $A \in O(3)$. Then $A^t = A^{-1}$ and so $1/\det A = \det(A^{-1}) = \det(A^t) = \det(A)$. Let $\lambda = \det(A)$ (so $\lambda = \pm 1$ and $\frac{1}{\lambda} = \lambda$). Then

$$\det(A - \lambda I_3) = \det(AI_3 - \lambda AA^t) = \det(A)\det(I_3 - \lambda A^t)$$

$$= \lambda \det(I_3 - \lambda A) = (-1)^3 \lambda^4 \det(A - \frac{1}{\lambda}I_3) = -\det(A - \lambda I_3).$$

Thus $\det(A - \lambda I_3) = -\det(A - \lambda I_3)$ and so $\det(A - \lambda I_3) = 0$. Thus $\lambda = \det(A)$ is an eigenvalue of A. Since the characteristic polynomial of A has degree 3, it either has one real root and two complex roots, in which case the eigenspace of $\det(A)$ has dimension 1, or else it has three real roots, all of magnitude 1. In the latter case, if all three have the same sign, then $A = \pm I_3$. Otherwise, exactly two will have the same sign, and those two will have to be $-\det(A)$ since the product of all three is $\det(A)$. Thus in this last case as well, the algebraic multiplicity of the eigenvalue $\det(A)$ is 1. $\qquad\square$

15.3.3 Proposition. *Let $A \in O(3)$. Then there exists $\theta \in \mathbb{R}$ and $P \in O(3)$ such that $A = P^{-1} \begin{bmatrix} \det(A) & 0 & 0 \\ 0 & \cos(\theta) & -\sin(\theta) \\ 0 & \sin(\theta) & \cos(\theta) \end{bmatrix} P$.*

Proof. Let $A \in O(3)$. By Proposition 15.3.2, A has $\det(A)$ as an eigenvalue. If $A = I_3$, then $\theta = 0$ and $P = I_3$ have the required property, while if $A =$

$-I_3$, then $\theta = \pi$ and $P = I_3$ have the required property. Thus we may suppose that $A \neq \pm I_3$, in which case by Proposition 15.3.2, the eigenspace with associated eigenvalue $\det(A)$ has dimension 1; that is, it is a line l through the origin. Let u be a unit length eigenvector for A with eigenvalue $\det(A)$. We may extend $\{u\}$ to an (ordered) orthonormal basis $\alpha = \{u, v, w\}$ for $\mathbb{R}^3$. By Corollary 15.1.6, c_α is an orthogonal linear mapping, and so $c_\alpha \circ T$ is orthogonal. It follows that $\{c_\alpha(T(u)), c_\alpha(T(v)), c_\alpha(T(w))\}$ is an orthonormal basis for $\mathbb{R}^3$, and so the matrix B that represents T_A relative to α is an orthogonal matrix; that is, $B = [\,c_\alpha(T(u)) \mid c_\alpha(T(v)) \mid c_\alpha(T(w))\,]$ is orthogonal. Let P denote the matrix that represents the identity mapping relative to the standard basis for the domain and the orthonomal basis α in the codomain; so $P = [\,c_\alpha(u) \mid c_\alpha(v) \mid c_\alpha(w)\,]$. Then by Corollary 15.1.6, $\{c_\alpha(u), c_\alpha(v), c_\alpha(w)\}$ is an orthonormal basis for $\mathbb{R}^3$ and so P is orthogonal. Moreover, since $T \circ 1_{\mathbb{R}^3} = 1_{\mathbb{R}^3} \circ T (= T)$, we have by Proposition B.2.31 (found in Appendix B) that $BP = PA$, or $B = PAP^{-1}$. Since $A \in O(3)$, we know that for all $x, y \in \mathbb{R}^n$, $Ax \cdot Ay = x \cdot y$, and $||Ax|| = ||x||$. Now, u is an eigenvector of A with eigenvalue $\det(A)$, so $T(u) = \det(A)u = (\det(A))u + (0)v + (0)w$ and thus the first column of B is $\det(A)e_1$. Suppose that $a, b, c \in \mathbb{R}$ are such that $T(v) = au + bv + cw$. Then $u \cdot T(v) = a(u \cdot u) + b(u \cdot v) + c(u \cdot w) = a$ since $\{u, v, w\}$ is an orthonormal set. Moreover, since $T(u) = \det(A)u$, we have $u \cdot T(v) = \frac{1}{\det(A)}T(u) \cdot T(v) = \frac{1}{\det(A)}u \cdot v = 0$, so $a = 0$. Similarly, if $d, e, f \in \mathbb{R}$ are such that $T(w) = du + ev + fw$, then $d = u \cdot T(w) = \frac{1}{\det(A)}T(u) \cdot T(w) = 0$. Thus $B = \begin{bmatrix} \det(A) & 0 & 0 \\ 0 & b & e \\ 0 & c & f \end{bmatrix}$ and so it follows that $\begin{bmatrix} b & e \\ c & f \end{bmatrix}$ is an orthogonal 2×2 matrix. Furthermore, $\det(A) = \det(PAP^{-1}) = \det(B) = \det(A)\det(\begin{bmatrix} b & e \\ c & f \end{bmatrix})$, and so $\begin{bmatrix} b & e \\ c & f \end{bmatrix} \in SO(2)$. By Proposition 15.2.2, there exists θ with $0 \le \theta < 2\pi$ such that $b = \cos(\theta)$, $e = -\sin(\theta)$, $c = \sin(\theta)$, and $f = \cos(\theta)$. $\qquad\square$

15.3.4 Corollary. *Let $A \in O(3)$ with $A \neq I_3$. Then there exists a line l in $\mathbb{R}^3$ and $\theta \in \mathbb{R}$ with $0 < \theta < 2\pi$ such that if $\det(A) = 1$, then T_A is a rotation of $\mathbb{R}^3$ about the axis l by the angle θ, while if $\det(A) = -1$, then T_A is a rotation of $\mathbb{R}^3$ about the axis l by the angle θ, followed by the reflection across the plane that is orthogonal to l and passes through the origin.*

Proof. By Proposition 15.3.3, there exists an orthogonal matrix P such that $B = PAP^{-1}$ is $\begin{bmatrix} \det(A) & 0 & 0 \\ 0 & \cos(\theta) & -\sin(\theta) \\ 0 & \sin(\theta) & \cos(\theta) \end{bmatrix}$, and so there exists an orthonormal basis $\{u, v, w\}$ for $\mathbb{R}^n$ such that u is an eigenvector with eigenvalue $\det(A)$, and the eigenspace for eigenvalue $\det(A)$ is $l = span(u)$, a line through the origin. If $\det(A) = 1$, then T_A is rotation about the line l by the angle θ, while if $\det(A) = -1$, then T_A is rotation about the line l by the angle θ, followed by reflection across the plane orthogonal to l that passes through the origin. For in this case, we have

$$B = \begin{bmatrix} -1 & 0 & 0 \\ 0 & \cos(\theta) & -\sin(\theta) \\ 0 & \sin(\theta) & \cos(\theta) \end{bmatrix} = \begin{bmatrix} -1 & 0 & 0 \\ 0 & 1 & 0 \\ 0 & 0 & 1 \end{bmatrix} \begin{bmatrix} 1 & 0 & 0 \\ 0 & \cos(\theta) & -\sin(\theta) \\ 0 & \sin(\theta) & \cos(\theta) \end{bmatrix}.$$

$\square$

We remark that a reflection across a plane in $\mathbb{R}^3$ cannot be viewed as a motion in $\mathbb{R}^3$, whereas the rotation of $\mathbb{R}^3$ about a line is indeed a transformation that can be carried out in $\mathbb{R}^3$. For this reason, we shall now focus our efforts on describing the finite groups of rotations of $\mathbb{R}^3$.

15.3.5 Definition. *For any $A \in SO(3)$ with $A \neq I_3$, the eigenspace E_1 of T_A that is associated with eigenvalue 1 is called the axis of rotation, and a unit vector $v \in E_1$ is called a pole of A (since the eigenspace E_1 of T_A has dimension 1, there are exactly two poles of A). If G is a finite subgroup of $SO(3)$, let $\Pi_G = \{ v \in \mathbb{R}^n \mid \text{there exists } A \in G \text{ such that } v \text{ is a pole of } A \}$, the set of poles of G.*

15.3.6 Proposition. *Let G be a finite subgroup of $SO(3)$. Then the natural action of G on $\mathbb{R}^3$ induces an action of G on Π_G.*

Proof. Let $x \in \Pi_G$, and let $A \in G$. Since T_A preserves lengths, it follows that $||Ax|| = ||x|| = 1$, and so it suffices to prove that there exists $B \in G$ such that Ax is an eigenvector for B. Since $x \in \Pi_G$, there exists $C \in G$ such that $Cx = x$. Then for $B = ACA^{-1}$, $B \in G$ and $BAx = ACA^{-1}(Ax) = ACx = Ax$, as required. $\square$

15.3.7 Proposition. *Let G be a finite subgroup of $SO(3)$. Then G is either cyclic or dihedral, or is isomorphic to A_4, A_5, or S_4.*

Proof. If $|G| \leq 3$, then G is cyclic, so we may assume that $|G| \geq 4$. Let $T = \{ (A, x) \mid A \in G, \ A \neq I_3, \ x \in \Pi_G, Ax = x \}$. We count the number of elements of T in two different ways. First, for $x \in \Pi_G$, let $G_x = \{ A \in G \mid Ax = x \}$. Then T is the disjoint union of the sets $(G_x - \{ I_3 \}) \times \{ x \}$ for $x \in \Pi_G$, and thus $|T| = \sum_{x \in \Pi_G}(|G_x| - 1)$. On the other hand, for each $A \in G$ with $A \neq I_3$, Π_G contains exactly two poles of A and thus $|T| = 2(|G| - 1)$. Thus $\sum_{x \in \Pi_G}(|G_x| - 1) = 2(|G| - 1)$, or equivalently, $\sum_{x \in \Pi_G} |G_x| - |\Pi_G| =$

$2(|G| - 1)$. Suppose that there are s orbits $O_1, O_2, \ldots, O_s$ of Π_G under the action of G. For each $i = 1, 2, \ldots, s$, $|O_i| = |G|/|G_x|$ for any $x \in O_i$, and for $x \in O_i$ we shall let $p_i = |G_x|$, so $|O_i| = |G|/p_i$ for $i = 1, 2, \ldots, s$. Then $\sum_{x \in \Pi_G} |G_x| = \sum_{i=1}^{s} \frac{|G|}{p_i} p_i = s|G|$ and so $s|G| - |\Pi_G| = 2(|G| - 1)$, or $|\Pi_G| = |G|(s - 2) + 2$. Since $|G| \geq 4$, it follows that $s \geq 2$. Also, we have $\sum_{x \in \Pi_G} (|G_x| - 1) = \sum_{i=1}^{s} \frac{|G|}{p_i}(p_i - 1) = |G| \sum_{i=1}^{s}(1 - \frac{1}{p_i})$, and so $|G| \sum_{i=1}^{s}(1 - \frac{1}{p_i}) = 2(|G| - 1)$. Divide through by $|G|$ to obtain that

$$\sum_{i=1}^{s}(1 - \frac{1}{p_i}) = 2 - \frac{2}{|G|}.$$

Furthermore, since $|G| \geq 4$, we obtain $\frac{3}{2} \leq 2 - \frac{2}{|G|} < 2$, and as a result, $\sum_{i=1}^{s}(1 - \frac{1}{p_i}) < 2$. By definition of Π_G, for $x \in \Pi_G$, there exists $A \in G$ with $A \neq I_3$ and $Ax = x$; that is, $A \in G_x$. Thus for every $i = 1, 2, \ldots, s$, $p_i \geq 2$ and thus $1 - \frac{1}{p_i} \geq \frac{1}{2}$. It follows now that $2 > \sum_{i=1}^{s}(1 - \frac{1}{p_i}) \geq \frac{s}{2}$; equivalently, $s < 4$. Thus $2 \leq s < 4$, and so s is either 2 or 3. We now consider the various possibilities.

Case 1: $s = 2$. Then $|\Pi_G| = |G|(s - 2) + 2 = 2$, and so every element of G has the same axis of rotation. But then G is isomorphic to a finite subgroup of rotations of $\mathbb{R}^2$ (the plane orthogonal to the axis of rotation that passes through the origin), and by Proposition 15.2.3, G is cyclic.

Case 2: $s = 3$. Assume that p_1, p_2, and p_3 were labelled so that $p_1 \leq p_2 \leq p_3$. We have $|\Pi_G| = |G|(s - 2) + 2 = |G| + 2$, and so there are $\frac{|\Pi_G|}{2} = \frac{(|G|+2)}{2}$ lines through the origin that occur as axes of rotation for elements of G. Now, from $\sum_{i=1}^{s}(1 - \frac{1}{p_i}) = 2 - \frac{2}{|G|}$ we obtain that $(1 - \frac{1}{p_1}) + (1 - \frac{1}{p_2}) + (1 - \frac{1}{p_3}) = 2 - \frac{2}{|G|}$ and so $1 + \frac{2}{|G|} = \frac{1}{p_1} + \frac{1}{p_2} + \frac{1}{p_3}$. If $p_1 \geq 3$, we would have $1 + \frac{2}{|G|} \leq 1$, which is not possible. Thus $p_1 = 2$, and so

$$\frac{1}{p_2} + \frac{1}{p_3} = \frac{1}{2} + \frac{2}{|G|}.$$

Furthermore, if $p_2 > 3$, then $\frac{1}{2} + \frac{2}{|G|} = \frac{1}{p_2} + \frac{1}{p_3} \leq \frac{1}{4} + \frac{1}{4} = \frac{1}{2}$, which is also not possible. Thus $p_2 \leq 3$, so $p_2 = 2$ or $p_2 = 3$.

Case 2 (i): $p_2 = 2$. Then $p_3 = \frac{|G|}{2}$. For $x \in O_3$, $|G_x| = \frac{|G|}{2}$, and G_x is isomorphic to a group of rotations of the plane orthogonal to the line through x and the origin. By Proposition 15.2.3, G_x is a cyclic group, and since it has index 2 in G, $G_x \triangleleft G$. Since $p_2 = 2$, for $y \in O_2$, $|G_y| = 2$, so G_y is a cyclic group of order 2, and $G_y \cap G_x = \{I_3\}$. Thus G is a semidirect product of the finite cyclic group G_x by the cyclic group G_y of order 2, where the action of G_y on G_x is the conjugation action. As shown in Proposition 15.2.4, such a group is isomorphic to the dihedral group D_n, where $|G| = 2n$.

Case 2 (ii): $p_2 = 3$. Then $\frac{1}{2} + \frac{2}{|G|} = \frac{1}{p_2} + \frac{1}{p_3} = \frac{1}{3} + \frac{1}{p_3}$ and so $\frac{1}{p_3} = \frac{1}{6} + \frac{2}{|G|} > \frac{1}{6}$. But then $3 = p_2 \le p_3 < 6$, and so $p_3 = 3, 4$, or 5.

Suppose first of all that $p_3 = 3$, in which case $\frac{1}{3} + \frac{1}{3} = \frac{1}{2} + \frac{2}{|G|}$, and thus $|G| = 12$. Then $|O_1| = |G|/p_1 = 6$, and this determines 3 axes of rotation for elements of G, and for $x \in O_1$, $|G_x| = p_1 = 2$ (which means that G has at least three elements of order 2), while $|O_2| = 4$, and this determines an additional 2 axes of rotation for elements of G, and for $x \in O_2$, $|G_x| = 3$. Finally, $|O_3| = 4$, and this determines the final 2 axes of rotation for elements of G, and for $x \in O_3$, $|G_x| = 3$. Now, G has at least two subgroups of order 3, which means that the Sylow 3-subgroups of G are not normal. By Proposition 12.4.12, G is isomorphic to A_4. Note that in this case, the elements of G determine $3 + 2 + 2 = 7 = (12 + 2)/2$ axes of rotation.

Next, consider the possibility that $p_3 = 4$. Then $\frac{1}{3} + \frac{1}{4} = \frac{1}{2} + \frac{2}{|G|}$, and thus $|G| = 24$. This time, we find that $|O_1| = |G|/p_1 = 12$, and this determines 6 axes of rotation for elements of G, and for $x \in O_1$, $|G_x| = p_1 = 2$, while $|O_2| = 24/3 = 8$, and this determines 4 axes of rotation for elements of G, and for $x \in O_2$, $|G_x| = 3$. Finally, $|O_3| = 24/4 = 6$, and this determines the final 3 axes of rotation for elements of G, and for $x \in O_3$, $|G_x| = 4$. Now, G has at least four subgroups of order 3, and since a subgroup of order 3 is a Sylow 3-subgroup of G, it follows that G has exactly four subgroups of order 3. G contains at least 6 elements of order 2 from the stabilizers of elements of O_1. As well, each stabilizer of an element in O_3 is isomorphic to a subgroup of order 4 of $SO(2)$, and hence G has at least 3 cyclic subgroups of order 4. But then G has at least 6 elements of order 4, and at least 6 elements of order 2, while a Sylow 2-subgroup of G has order 8. Thus the Sylow 2-subgroups of G are not normal, and by Proposition 12.4.13, the only group of order 24 with nonnormal Sylow 2 and 3 subgroups is S_4. Thus G is isomorphic to S_4. Note that in this case, the elements of G determine $6 + 4 + 3 = 13 = (24 + 2)/2$ axes of rotation.

Finally, we consider the possiblity that $p_3 = 5$. Then $\frac{1}{3} + \frac{1}{5} = \frac{1}{2} + \frac{2}{|G|}$, and thus $|G| = 60$. Now, we find that $|O_1| = |G|/p_1 = 30$, and this determines 15 axes of rotation for elements of G, and for $x \in O_1$, $|G_x| = p_1 = 2$, while $|O_2| = 60/3 = 20$, and this determines 10 axes of rotation for elements of G, and for $x \in O_2$, $|G_x| = 3$. Finally, $|O_3| = 60/5 = 12$, and this determines the final 6 axes of rotation for elements of G, and for $x \in O_3$, $|G_x| = 5$.

Since G is a group of order 60 with (at least) 6 Sylow 5-subgroups, it follows from Proposition 12.4.23 that G is isomorphic to A_5. Note that in this final case, the elements of G determine $15 + 10 + 6 = 31$ axes of rotation. $\qquad\square$

We finish this discussion with a proof that every group identified in Proposition 15.3.7 actually does occur as a subgroup of $SO(3)$. It is evident that for every positive integer n, the cyclic subgroup of $SO(3)$ that is generated by $A = \begin{bmatrix} 1 & 0 & 0 \\ 0 & \cos(\frac{2\pi}{n}) & -\sin(\frac{2\pi}{n}) \\ 0 & \sin(\frac{2\pi}{n}) & \cos(\frac{2\pi}{n}) \end{bmatrix}$ has order n.

Next, for any integer $n \geq 3$, let P be a regular n-gon in the plane, and construct $H = [-1, 1] \times P \subseteq \mathbb{R}^3$. Then the rotation group of H has a cyclic subgroup of order n, with axis $[-1, 1] \times o$, where o is the center of the regular n-gon, and n subgroups of order 2, generated by the n rotations by π about an axis through the origin and one side of H. Thus the symmetry group of H is isomorphic to D_n.

The question of the existence of A_4, A_5, and S_4 as subgroups of $SO(3)$ will be dealt with in the next section, wherein it will be shown that they arise as rotation groups of the Platonic solids.

4 The Platonic Solids

What should we consider to be the analogue in $\mathbb{R}^3$ of the closed n-gons in $\mathbb{R}^2$? One reasonable family of objects would be those regions in $\mathbb{R}^3$ that can be formed by the intersection of planes (playing the role of the lines in the construction of a polygon). If four or more planes in $\mathbb{R}^3$ intersect in such a way as to form exactly one bounded region, that region is called a polyhedron (many faces) in $\mathbb{R}^3$. The portion of a plane that forms one boundary side of the enclosed region is called a face of the polyhedron. A line segment that is the intersection of two faces of the polyhedron is called an edge of the polyhedron and is said to be incident to each of the two faces, while the endpoints of an edge are called vertices of the polyhedron, said to be incident to the edge and to those faces that contain the vertices. Furthermore, a polyhedron with n faces is called an n-hedron.

15.4.1 Definition. *For any integer $n \geq 4$, an n-hedron is said to be regular if it is convex (for any two points u and v in the n-hedron, every point on the line segment joining u and v is also in the n-hedron), its n faces are congruent regular polygons, and the number of edges incident to a vertex does not depend on the vertex. The dual of a convex n-hedron A is formed*

by taking as its vertices the midpoint of each face of A, and drawing an
edge between two vertices if, and only if, the two faces whose midpoints are
being joined are adjacent (share a boundary edge).

If A is a convex n-hedron, with say V vertices and E edges, then Euler's polyhedron formula states that $V - E + n = 2$. Since the dual of a convex polyhedron is again a convex polyhedron, we deduce from Euler's formula that if A is a convex n-hedron, then A^*, the dual of A, is a convex polyhedron with V^* vertices, E^* edges, and n' faces, so $V^* - E^* + n' = 2$. We know that $V^* = n$, and since each edge of A separates two faces of A, it follows that each edge of A gives rise to an edge of A^*, so $E^* = E$. Thus $V - E + n = 2 = V^* - E^* + n' = n - E + n'$, and so n', the number of vertices of A^*, is equal to V. It is evident that if A is a regular n-hedron with V vertices, then A^* is a regular V-hedron with n vertices, inscribed inside A. Moreover, the dual of A^* will be a copy of A, scaled so as to be inscribed in A^*. Thus the dual of A is similar to A, by which we mean that if A is centred at the origin, so that A^* is also centred at the origin, then $(A^*)^*$ can be scaled up to equal A. In this sense, we say that $(A^*)^* = A$.

We have seen that in the plane, there is a regular n-gon for each integer $n \geq 3$, and so it is natural to wonder for what values of n (necessarily, $n \geq 4$) there exists a regular n-hedron. As it turns out, the situation is radically different in three dimensions than it is for two dimensions. We shall see that there are regular n-hedra only for $n = 4, 6, 8, 12$, and 20.

To see this, suppose that $n \geq 4$ is an integer for which there exists a regular n-hedron, say with each face a regular m-gon, and with k edges incident to each vertex. It follows that k faces meet at each vertex. At a vertex v then, the sum of the interior angles at that vertex in each of the k incident faces must be less than 2π (since a total angle of 2π would correspond to all incident faces lying in the same plane), and since the interior angle at any vertex of a regular m-gon is $\frac{(m-2)\pi}{m}$, it follows that $\frac{(m-2)k\pi}{m} < 2\pi$, and so $\frac{(m-2)k}{m} < 2$. Since $m \geq 3$ (and $k \geq 3$), we consider the cases $m = 3, 4, 5, \ldots$ and determine for each what the possible values of k are. When $m = 3$, we must have $k < 6$, so the possible values of k are $3, 4, 5$. When $m = 4$, we must have $k < 4$, so the only possibility is $k = 3$. When $m = 5$, we must have $k < \frac{10}{3}$, so $k = 3$ is the only possibility. For $m \geq 6$, we have $k < 3$, so there are no regular polyhedra for which a face is a regular m-gon with $m \geq 6$.

Now, in a regular n-hedron, with V vertices, E edges, and each face a regular m-gon, each edge is in the boundary of two faces. Thus in the product mn, each edge has been counted twice and so we obtain the equation $mn = 2E$. As well, at each vertex, there are exactly k incident edges and so in the product kV, each edge has been counted twice (each edge has two distinct endpoints). Thus $kV = 2E = mn$ and so we obtain $E = \frac{mn}{2}$ and $V = \frac{mn}{k}$. From Euler's polyhedron formula, we find that $\frac{mn}{k} - \frac{mn}{2} + n = 2$, and so $n(\frac{m}{k} - \frac{m}{2} + 1) = 2$.

Let us consider the cases determined above.

Case 1: $m = 3 = k$. Then $2 = n(1 - \frac{3}{2} + 1) = n(\frac{1}{2})$ and so $n = 4$. As well, $V = \frac{mn}{k} = 4$, and $E = \frac{mn}{2} = 6$. A regular 4-hedron is called a regular tetrahedron, and each face is an equilateral triangle, it has four vertices and six edges, with three edges incident to each vertex.

Case 2: $m = 3$, $k = 4$. Then $2 = n(\frac{3}{4} - \frac{3}{2} + 1) = n(\frac{1}{4})$ and so $n = 8$. As well, $V = \frac{mn}{k} = 6$ and $E = \frac{mn}{2} = 12$. A regular 8-hedron is called a regular octahedron. Each face of a regular octahedron would be an equilateral triangle, and the octahedron would have six vertices and twelve edges, with each vertex having four incident edges.

Case 3: $m = 3$, $k = 5$. Then $2 = n(\frac{3}{5} - \frac{3}{2} + 1) = n(\frac{1}{10})$ and so $n = 20$. As well, $V = \frac{mn}{k} = 12$ and $E = \frac{mn}{2} = 30$. A regular 20-hedron is called a regular icosahedron. Each face of a regular icosahedron wouold be an equilateral triangle, and a regular icosahedron would have twelve vertices and thirty edges, with five edges incident to each vertex.

Case 4: $m = 4$, $k = 3$. Then $2 = n(\frac{4}{3} - \frac{4}{2} + 1) = n(\frac{1}{3})$ and so $n = 6$. As well, $V = \frac{mn}{k} = 8$ and $E = \frac{mn}{2} = 12$. A regular 6-hedron is called a regular hexahedron, or simply a cube, with each face a square, and it has eight vertices and twelve edges, with three edges incident to each vertex.

Case 5: $m = 5$, $k = 3$. Then $2 = n(\frac{5}{3} - \frac{5}{2} + 1) = n(\frac{1}{6})$ and so $n = 12$. As well, $V = \frac{mn}{k} = 20$ and $E = \frac{mn}{2} = 30$. A regular 12-hedron is called a regular dodecahedron. It would have each face a regular pentagon, and would have twenty vertices and thirty edges, with three edges incident to each vertex.

Of course, we have not yet established that any of the above actually exist. It is evident that a regular tetrahedron exists (and is self-dual), since if we take three equilateral triangles of the same size and make a "cone" shape out of them by choosing a vertex on each and mating the three chosen vertices, with the sides identified in the obvious manner, the base of the "cone" is an equilateral triangle of the same size as the other three; hence we have built a regular tetrahedron.

Of course, the cube exists, and its dual is the octahedron. We shall show how to construct a regular icosahedron, and that will complete the demon-

424 *Abstract Algebra*

stration of existence since the dual of a regular icosahedron is a regular dodecahedron. One way to build an icosahedron is to mimic our construction
of the regular tetrahedron: take five equilateral triangles of the same size
and build a 5-sided "cone" out of them. Then take another five triangles of
the same shape and size, and build a second 5-sided "cone". Now invert one
and position it below the other, and join each vertex on the pentagonal base
of one to two adjacent vertices on the pentagonal base of the other. The
result is an icosahedron. However, it may be difficult to convince oneself
that the triangles that join the two "cones" can be made equilateral of the
same size as the original ten, so we offer another construction. Construct
a cube with vertices at the eight points $(\pm 1, \pm 1, \pm 1)$.

Fig. 15.1   The construction of a regular icosahedron

Then on each face of the cube, select two points equidistant from the
centre of the face, situated on a line through the midpoint of the face that
runs parallel to one of the sides as shown in Figure 15.1. Both of the
selected points are vertices of our icosahedron. Draw a line segment joining
the two selected points – this will be one edge in the regular icosahedron
under construction. Let λ denote the distance from the midpoint of the
face to either selected point. The distance between the two selected points
on a face is therefore 2λ, and so our goal is to have every edge have length
2λ. This will require that we make a judicious choice for the value of λ, and
the following discussion establishes that this is possible (and that there is
exactly one value of λ with this property). Due to the way the orientation
of the lines on which the selected points lie have been chosen (which is such
that as one scans around the cube, starting at any face, the orientation of
the lines on which the chosen points lie alternates), each edge e that we have
constructed so far lies on a face which has two adjacent faces on which the
orientation of the edges drawn on them has one endpoint closer to e than
the other. Join the endpoints of e to the closer endpoint on each of these
two adjacent edges (one such pair of edges is shown as dotted lines in Figure

15.1). Since there are 6 faces of the cube and each contributes 2 vertices, we have a total of 12 vertices. Furthermore, since we initially had created 6 edges, and each of these six edges leads to the construction of 4 more edges, we have contructed a polyhedron with 12 vertices and $6 + 24 = 30$ edges. By Euler's polyhedron formula, such a polyhedron has $2 - 12 + 30 = 20$ faces, and thus is an icosahedron. By the construction, all of the first 6 constructed edges have the same length, namely 2λ, and also all of the subsequently constructed 24 edges will have the same length, and we require that all 30 of them have the same length. It suffices to choose one of the 24 edges and determine its length. Choose one face and look at one of the edges drawn from an endpoint of the edge in that face to a point on an adjacent face. The coordinates of the two points on our chosen face are of the form $(1, \pm\lambda, 0)$, or on the face opposite, $(-1, \pm\lambda, 0)$, or $(0, 1, \pm\lambda)$ or on the face opposite $(0, -1, \pm\lambda)$, or else $(\pm\lambda, 0, 1)$ or on the face opposite $(\pm\lambda, 0, -1)$. If we consider the edge with endpoints $(1, \pm\lambda, 0)$, then two of the edges would be drawn from the two endpoints to $(\lambda, 0, 1)$, and so the length of either edge will be $\sqrt{(1-\lambda)^2 + \lambda^2 + 1} = \sqrt{2\lambda^2 - 2\lambda + 2} = \sqrt{2}\sqrt{\lambda^2 - \lambda + 1}$. Thus we require λ to make $2\lambda = \sqrt{2}\sqrt{\lambda^2 - \lambda + 1}$, which will occur if $2\lambda^2 = \lambda^2 - \lambda + 1$. We are therefore looking for a root of the quadratic $\lambda^2 + \lambda - 1$. Since the two roots of this quadratic are $\frac{-1 \pm \sqrt{5}}{2}$, one of which is negative, it follows that we require $\lambda = \frac{\sqrt{5}-1}{2}$. The reader might be intrigued to note that λ is the reciprocal of the golden ratio $\frac{1+\sqrt{5}}{2}$.

We are now ready to complete the classification of the finite subgroups of $SO(3)$.

15.4.2 Proposition. *The rotation group of the regular tetrahedron is the alternating group A_4, and the rotation group of both the regular dodecahedron and its dual, the regular icosahedron, is the alternating group A_5. The rotation group of the cube and its dual, the regular octahedron, is S_4.*

Proof. We shall make repeated use of Proposition 12.1.11, which states that if G acts on a set X and $x \in X$, then $[G:G_x] = |Orb_G(x)|$, or equivalently, $|G| = |G_x||Orb_G(x)|$. We begin with the rotation group G of the regular tetrahedron T, where the tetrahedron has been positioned so that its centre of mass is the origin. The natural action of G on $\mathbb{R}^3$ induces a transitive action of G on the set of vertices of T, and if $g \in G$ fixed every vertex of T, then by convexity, g would fix every element of T. Thus the restriction of the action of G to the set of vertices of T, a set of size 4, provides an injective homomorphism from G into S_4. We may therefore regard G as a

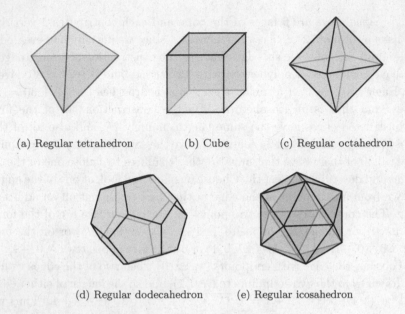

(a) Regular tetrahedron   (b) Cube   (c) Regular octahedron

(d) Regular dodecahedron   (e) Regular icosahedron

Fig. 15.2   The five Platonic solids

subgroup of S_4. Since the stabilizer of any vertex v has size 3, it follows from Proposition 12.1.11 that $|G| = 12$. Moreover, G_v contains two 3-cycles, ρ and ρ^2, and since this holds for each vertex of T, it follows that G contains all 8 3-cycles of S_4. As the 3-cycles of S_4 generate A_4, we obtain that $A_4 \subseteq G$ and so $G = A_4$.

Next, consider the rotation group G of the cube. Again, the natural action of G on the cube induces a transitive action of G on the 8 vertices of the cube and the stabilizer of a vertex under this induced action has size 3, so by Proposition 12.1.11, $|G| = 24$. The action of G on the cube also induces an action of G on the set of four sets of diametrically opposite vertices, which identify the four diagonals of the cube. Label the vertices in a clockwise traversal of the boundary of one face of the cube as 1, 2, 3, and 4, and then label the vertices at the other end of the diagonals incident to 1, 2, 3, and 4 as $1'$, $2'$, $3'$, and $4'$, respectively (see Figure 15.3).

Suppose $g \in G$ acts as the identity mapping on the four diagonals. Then $g(\{i, i'\}) = \{i, i'\}$ for each $i = 1, 2, 3, 4$. If, for some i, $g(i) = i$ and $g(i') = i'$, then g is rotation of $\mathbb{R}^3$ about the line through i and i', and so the orbit under the action of $\langle g \rangle$ of a vertex adjacent to i is the set of all three vertices adjacent to i, and similarly, the orbit of a vertex adjacent to

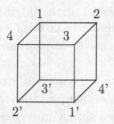

Fig. 15.3

i' is the set of all three vertices adjacent to i'. Since neither of these two orbits contain a vertex j and its matching vertex j', it follows that g fixes all eight vertices of the cube; that is, g acts as the identity mapping on the set of vertices of the cube. Thus if g acts as the identity mapping on the set of four sets of two diametrically opposite vertices, and g does not act as the identity on the set of vertices of the cube, then g acts as the permutation $(1\,1')(2\,2')(3\,3')(4\,4')$ on the set of the four diagonals. Note that a rotation preserves angles (including their sign). The triple of directed line segments $(\{\,2',1'\,\},\{\,2',3'\,\},\{\,2',4\,\})$ determines a right-handed coordinate system for $\mathbb{R}^3$, while its image under g is $(\{\,2,1\,\},\{\,2,3\,\},\{\,2,4'\,\})$, which determines a left-handed coordinate system for $\mathbb{R}^3$. Since this is not possible, we conclude that the induced action of G on the set of 4 diagonal pairs of vertices is faithful; that is, if $g \in G$ fixes each diagonal pair of vertices, then $g = e_G$. Thus we may consider G to be a subgroup of the group of permutations of the set of 4 diagonal pairs of vertices, and hence we may regard G as a subgroup of S_4. As $|G| = 24 = |S_4|$, we have obtained that $G = S_4$.

Finally, we consider the rotation group G of the regular icosahedron. The natural action of G on the regular icosahedron induces a transitive action on its set of 12 vertices, and for any vertex v, the stabilizer of v under the action of G is a cyclic group of order 5, since 5 triangles, each an equilateral triangle, meet at v. Thus by Proposition 12.1.11, $|G| = |G_v||Orb_G(v)| = (5)(12) = 60$. Furthermore, since G has order 60, the order of a Sylow 5-subgroup of G is 5. We have seen that the stabilizer of each vertex is a cyclic group of order 5, so G has at least 12 Sylow 5-subgroups. Thus by Proposition 12.4.23, G is isomorphic to A_5. □

We remark that each of the eleven nonidentity rotations of the regular tetrahedron is a rotation of $\mathbb{R}^3$ about one of the following seven axes: four of them obtained as the line through a vertex and the centre of the tetrahedron (which we consider to be placed at the origin), while the other three are obtained by joining the midpoint of an edge e to the midpoint of the edge opposite e (the unique edge not incident to an endpoint of e), there being three such pairs of edges.

In the case of the cube, the 23 nonidentity rotations are about 13 different axes; namely the three lines that pass through the midpoints of opposing faces, the four lines that pass through a vertex and the one diametrically opposite, and the six lines each of which passes through the midpoints of a pair of opposite edges.

Finally, the 59 nonidentity rotations of the regular icosahedron require 31 different axes. Each triangular face has a face opposite (that is, the line through the midpoint of a face and the origin passes through the midpoint of exactly one other face), and the axes obtained by joining the midpoints of opposite pairs of faces provide 10 of these 31. Each edge has an opposite edge (again, the line which passes through the midpoint of an edge and the origin passes through the midpoint of exactly one other edge), and so the 15 pairs of opposite edges provide another 15 axes of rotation. Each vertex has an opposite, and the 6 pairs of opposing vertices determine the final 6 axes of rotation of the regular icosahedron.

15.5 Exercises

1. Let $\{u, v, w\}$ be an orthonormal basis for $\mathbb{R}^3$. Prove that the triple (u, v, w) determines a right-handed coordinate system (by which is meant $w = u \times v$) if, and only if, $w \cdot (u \times v) = 1$ (where $u \times v$ is the cross product in $\mathbb{R}^3$).

2. Let $T \in O(\mathbb{R}^3)$ be a rotation with axis of rotation a line l (through the origin). Let $\{v_1, v_2, v_3\}$ be an orthonormal basis for $\mathbb{R}^3$, and let $v_i' = T(v_i)$ for $i = 1, 2, 3$.

    a) Prove that the three vectors $v_i - v_i'$, $i = 1, 2, 3$, are coplanar and that l is orthogonal to the plane determined by $v_i - v_i'$, $i = 1, 2, 3$.

    b) If $v_i = e_i$, $i = 1, 2, 3$, and $T = T_A$, where $A = \begin{bmatrix} \frac{1}{3} & -\frac{2}{3} & \frac{2}{3} \\ \frac{2}{3} & \frac{2}{3} & \frac{1}{3} \\ -\frac{2}{3} & \frac{1}{3} & \frac{2}{3} \end{bmatrix}$, use (a) to find a direction vector u for l, the axis of rotation of T.

    c) Let $S \in O(\mathbb{R}^3)$ and suppose that S is not a rotation (so $\det(S) = -1$). Let $v_i' = S(v_i)$, $i = 1, 2, 3$, and prove that the vectors $v_i + v_i'$, $i = 1, 2, 3$, are coplanar, and that for any vector u on the line through the origin that is orthogonal to the plane determined by $v_i - v'_i$, $i = 1, 2, 3$, $S(u) = -u$.

d) If $v_i = e_i$, $i = 1, 2, 3$, and $S = T_B$, where $B = \begin{bmatrix} \frac{1}{3} & -\frac{2}{3} & \frac{2}{3} \\ \frac{2}{3} & \frac{2}{3} & \frac{1}{3} \\ \frac{2}{3} & -\frac{1}{3} & -\frac{2}{3} \end{bmatrix}$, use (c) to find a vector u on the line through the origin that is orthogonal to the plane determined by $e_i - e'_i$, $i = 1, 2, 3$.

3. Consider the cube with vertices at $(\pm 1, \pm 1, \pm 1)$. Find the matrices that represent an element of the rotation group of the cube of orders 2, 3, and 4, respectively (relative to the standard basis). As well, find the matrices (relative to the standard basis) that represent the elements of the unique normal subgroup of the rotation group of the cube.

4. Determine the rotation group of each of the following (find their matrices relative to the standard basis for $\mathbb{R}^3$):

a) A box with vertices at $(\pm 1, \pm 2, \pm 1)$.

b) A box with vertices at $(\pm 1, \pm 2, \pm 3)$.

5. a) The rotation group G of the cube induces an action on the set of three line segments formed by joining the centroids of pairs of opposite faces of the cube. What is the kernel K of this action? Can you identify the quotient group G/K?

b) Now suppose that we have three colours, and we colour each of the three lines segments formed in (a), using different colours for different line segments, so that we have three coloured line segments, say R, G, B. Let $H = \{ g \in G \mid g(R) = R, \ g(G) = G, \ g(B) = B \}$. Can you identity H (that is, can you find a familiar group that is isomorphic to H)?

6. For any positive integer n, find two elements of $Gl_2(\mathbb{R})$ that generate a dihedral group of order $2n$.

7. Let n be a positive integer. Prove that the set of elements of order 2 in D_n form a single conjugacy class if n is odd, while they form two classes of size $n/2$ when n is even.

8. Let n be any positive integer. For any subspace W of $\mathbb{R}^n$, $\mathbb{R}^n = W \oplus W^\perp$, where $W^\perp$, the orthogonal complement of W, is the subspace $\{ v \in \mathbb{R}^n \mid v \cdot w = 0 \text{ for all } w \in W \}$. A subspace of $\mathbb{R}^n$ of dimension $n - 1$ is called a hyperplane. It follows that if H is a hyperplane of $\mathbb{R}^n$, then $H^\perp$ has dimension 1.

a) Let $u \in \mathbb{R}^n$, $u \neq 0$. Define a mapping $T_u : \mathbb{R}^n \to \mathbb{R}^n$ by $T_u(v) = v - 2\frac{v \cdot u}{u \cdot u} u$ for each $v \in \mathbb{R}^n$. Prove that T_v is an orthogonal linear transformation which fixes every element of the hyperplane $span(u)^{\perp}$ and maps u to $-u$. Can you find a basis for $\mathbb{R}^n$ such that the matrix for T_u relative to your basis has a very simple form?

b) A finite group G is called a reflection group if there exist a positive integers n and k and nonzero vectors $v_1, v_2, \ldots, v_k \in \mathbb{R}^n$ such that G is isomorphic to the subgroup of $O(\mathbb{R}^n)$ that is generated by $T_{v_1}, \ldots, T_{v_k}$.

   (i) Prove that every dihedral group is a reflection group.

   (ii) For any positive integer n, define the mapping $\Gamma : S_n \to O(n)$ by $\Gamma(\sigma) = [\, e_{\sigma(1)} | e_{\sigma(2)} | \cdots | e_{\sigma(n)} \,]$ for each $\sigma \in S_n$. Prove that Γ is an injective homomorphism.

   (iii) Prove that for each positive integer n, $\Gamma(\tau)$ is a reflection if τ is a transposition. Specifically, if $\tau = (i\, j)$, find a vector $u \in \mathbb{R}^n$ such that $\Gamma(\tau) = T_u$. Note that since S_n is generated by transpositions, this will prove that S_n is a reflection group.

   (iv) Prove that for any positive integer n, the only reflections in $\Gamma(S_n)$ are the images of the transpositions.

   (v) Prove that for any positive integer n, $\Gamma(S_n)$ fixes each element of the line through the origin of $\mathbb{R}^n$ with direction vector $u = \sum_{i=1}^{n} e_i$, and maps the hyperplane $W = (span(u))^{\perp}$ to itself. Further, prove that the only element of W that is fixed by $\Gamma(S_n)$ is the zero vector.

Chapter 16

Pólya-Burnside Enumeration

1 Introduction

We begin this chapter with a simple example which shows one type of problem this particular method of enumeration deals with. The method is based on what is often called Burnside's lemma, Proposition 12.1.14, which states that if a finite group G acts on a finite set X and the action has n orbits, then $|G| = \frac{1}{n} \sum_{g \in G} |Fix(g)|$.

In our example, suppose that we have three colours, (R)ed, (B)lue, and (Y)ellow, and we have at least four medallions of each colour. We wish to string 4 medallions on a silver chain, and we would like to determine the number of distinguishable ways in which this can be done.

As the problem stands, it is rather vague. The first thing to determine is whether or not one can distinguish between two medallions of the same colour. Suppose we stipulate that the answer to that question is no, one cannot tell the difference between two medallions of the same colour. Let us further stipulate that the medallions are symmetric front to back, so that it does not matter which way the medallion is being held when it is threaded onto the chain, and the clasp is symmetric so that if the necklace is turned over, the clasp looks the same. Let us further specify that the clasp is large enough to prevent a medallion from slipping over it. It appears then that we are trying to determine the number of ways to place four marks of up to three different colours at four equally spaced locations around a circle that has a reference mark on it (the clasp location) midway between one pair of adjacent mark locations. However, we recognize that a necklace is to be considered the same as the one obtained by turning the necklace over (keeping the reference mark in the same position), which is to say that two colour assignments for which one is a mirror image of the other relative to the line through the clasp location and the centre of the circle are to be considered the same.

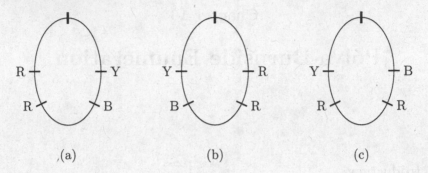

(a)   (b)   (c)

Fig. 16.1

In Figure 16.1, we have shown three examples of the necklace construction. The necklaces (a) and (b) are actually the same, since turning (a) over (reflecting it across the line that passes through the clasp and the bottom of the chain) results in (b). Necklaces (a) and (c) are different (note that if we were able to thread the medallions past the clasp, then (a) and (c) would be considered the same). As a result of these observations, we see that a necklace construction can be thought of as a mapping from the set J_4 to the set $\{R, B, Y\}$, where the elements of J_4 mark the locations at which a medallion is to be placed on the chain (see Figure 16.2). For example, necklace (a) of Figure 16.1 would correspond to the map $f = \left(\begin{smallmatrix} 1 & 2 & 3 & 4 \\ Y & B & R & R \end{smallmatrix}\right)$.

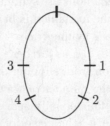

Fig. 16.2

Now, it is a simple matter to count the number of maps from J_4 to $\{R, B, Y\}$. For each $i \in J_4$, there are three possible values that can be assigned to i, and so there are 3^4 maps from J_4 to $\{R, B, Y\}$. However, distinct maps do not necessarily correspond to different necklaces. For example, map $g = \begin{pmatrix} 1 & 2 & 3 & 4 \\ R & R & B & Y \end{pmatrix}$ corresponds to necklace (b) of Figure 16.1, which we have observed to be the same as necklace (a).

How could we recognize that f and g both correspond to the same necklace without drawing the necklaces? Well, we note that given any necklace, if we exchange the medallions at positions 1 and 4, and as well, exchange the medallions at positions 2 and 3, the same necklace is produced. In terms of the mappings from J_4 to $\{R, B, Y\}$, we would say that for any map f, f and $f \circ (1\,4)(2\,3)$ result in the same necklace. For example, for the maps f and g given above, and $\sigma = (1\,4)(2\,3) \in S_4$,

$$f \circ \sigma = f \circ (1\,4)(2\,3) = \begin{pmatrix} 1 & 2 & 3 & 4 \\ Y & B & R & R \end{pmatrix} \circ \begin{pmatrix} 1 & 2 & 3 & 4 \\ 4 & 3 & 2 & 1 \end{pmatrix} = \begin{pmatrix} 1 & 2 & 3 & 4 \\ R & R & B & Y \end{pmatrix} = g.$$

In view of this observation, we see that we should utilize the (right) action of the group $G = \langle \sigma \rangle \subseteq S_4$ on the set of all maps from J_4 to $\{R, B, Y\}$ whereby $f : J_4 \to \{R, B, Y\}$ is mapped to $f \circ \pi$, where $\pi \in G = \{\sigma, \mathbb{1}_{J_4}\}$, since $\sigma^2 = \mathbb{1}_{J_4}$. Due to our specifications, two maps correspond to the same necklace if, and only if, they belong to the same orbit of this action, and so the number of different necklaces that can be made is equal to n, the number of orbits of this action. Since $|G| = 2$, we find by Burnside's lemma that n is given by

$$n = \frac{1}{|G|} \sum_{\pi \in G} |Fix(\pi)| = \frac{1}{2}(|Fix(\mathbb{1}_{J_4})| + |Fix(\sigma)|).$$

As well, $|Fix(\mathbb{1}_{J_4})| = 81$, the total number of maps from J_4 to $\{R, B, Y\}$. For a map f to be in $Fix(\sigma)$, f must be constant on the cycles of σ (the orbits of J_4 under the natural action of G on J_4), so $f \in Fix(\sigma)$ if, and only if, $f(1) = f(4)$ and $f(2) = f(3)$. Since we have 3 choices each for $f(1)$ and $f(2)$, it follows that $|Fix(\sigma)| = 9$, and so $n = \frac{1}{2}(81 + 9) = 45$. Thus there are 45 different necklaces that can be made.

Let us now modify our specifications and see what that does to the number of distinct necklaces that can be made. Suppose that there is no clasp (once the medallions are threaded on, the necklace has a final link put in to close the chain), and that the medallions have a "front" side (perhaps a coloured stone set in a silver holder), so that the necklace cannot be turned over. Under these specifications, necklaces (a) and (c) of Figure 16.1 are the same, while (a) and (b) are different. This can be dealt with via an approach similar to that above, but this time, the group G in question is $\{\mathbb{1}_{J_4}, (1\,2\,3\,4), (1\,3)(2\,4), (1\,4\,3\,2)\}$, with the action on the

set of all maps from J_4 to $\{R, B, Y\}$ defined in the same way as above.
Since the same colour must be assigned to all elements of each cycle in
the cycle decomposition of an element of G, we find that $|Fix(\mathbb{1}_{J_4})| = 81$,
$|Fix((1\,2\,3\,4))| = 3$, $|Fix((1\,3)(2\,4))| = 9$, and $|Fix((1\,4\,3\,2))| = 3$. For
example,

$$Fix((1\,2\,3\,4)) = \left\{ \begin{pmatrix} 1 & 2 & 3 & 4 \\ R & R & R & R \end{pmatrix}, \begin{pmatrix} 1 & 2 & 3 & 4 \\ B & B & B & B \end{pmatrix}, \begin{pmatrix} 1 & 2 & 3 & 4 \\ Y & Y & Y & Y \end{pmatrix} \right\}.$$

Since $|G| = 4$, the number of orbits of the action of G on the set of all maps
from J_4 to $\{R, B, Y\}$ is $\frac{1}{4}(81 + 3 + 9 + 3) = \frac{96}{4} = 24$. Thus we can make
24 different necklaces (according to our new conventions).

Let us make one final modification to this example. Suppose now that
there is no clasp, but the medallions are reversible, so that the necklace
can be turned over. Now (a), (b), and (c) of Figure 16.1 are all the same
necklace. This time, we require the dihedral group D_4 to act on our set of
maps from J_4 to $\{R, B, Y\}$, where our copy of D_4 is

$$D_4 = \{\mathbb{1}_{J_4}, (1\,3), (2\,4), (1,2)(3\,4), (1\,4)(2\,3), (1\,2\,3\,4), (1\,3)(2\,4), (1\,4\,3\,2)\}.$$

As before, for any $\pi \in D_4$, two maps belong to $Fix(\pi)$ if, and only if, they
are each constant on every cycle of π. Thus $|Fix(\mathbb{1}_{J_4})| = 3^4$, while if π
is a 2-cycle, $|Fix(\pi)| = 3^3$, and if π is a product of two disjoint 2-cycles,
$|Fix(\pi)| = 3^2$. Finally, if π is a 4-cycle, $|Fix(\pi)| = 3$. Thus the number
of orbits of the action of D_4 on the set of all maps from J_4 to $\{R, B, Y\}$
is $\frac{1}{8}(81 + 2(27) + 3(9) + 2(3)) = \frac{168}{8} = 21$, and so there are 21 different
necklaces that can be made according to these specifications.

16.2   A Theorem of Pólya

In this section, we present a result due to George Pólya that may be regarded as a generalization of Burnside's lemma, and which has application
to a wide variety of enumeration problems.

16.2.1 Definition. *Let n be a positive integer, and let G be a subgroup
of S_n. For any positive integer m, in this context, we shall consider the
elements of J_m to represent m different "colours", and so we refer to a
map from J_n to J_m as a colouring of J_n with m colours (even if the map is
not surjective). Let $C(m) = \mathscr{F}(J_n, J_m)$, the set of all mappings from J_n to
J_m. Define a (right) action of G on $C(m)$ by the requirement that $\sigma \in G$
sends $f \in C(m)$ to $f \circ \sigma \in C(m)$.*

The problem that we now pursue is the determination of the number of orbits of this action of G on $C(m)$ for every positive integer m.

16.2.2 Definition. *Let n be a positive integer, and let $\mathbb{Z}[x_1, x_2, \ldots, x_n]$ denote the polynomial ring in n (commuting) variables $x_1, x_2, \ldots, x_n$ with coefficients from $\mathbb{Z}$. Define a map $Z : S_n \to \mathbb{Z}[x_1, x_2, \ldots, x_n]$ by $Z(\sigma) = x_1^{s_1} x_2^{s_2} \cdots x_n^{s_n}$, where for each $i = 1, 2, \ldots, n$, $s_i \geq 0$ is the number of i-cycles (note that the 1-cycles are included) in the cycle decomposition of σ. Then define Z on the set of all subgroups of S_n by $Z_G = \frac{1}{|G|} \sum_{\sigma \in X} Z(\sigma)$ for each subgroup G of S_n. For any $(k_1, k_2, \ldots, k_n) \in \mathbb{Z}^n$, and any $\sigma \in S_n$, let $Z(\sigma)(k_1, k_2, \ldots, k_n) = k_1^{s_1} k_2^{s_2} \cdots k_n^{s_n}$ if $Z(\sigma) = x_1^{s_1} x_2^{s_2} \cdots x_n^{s_n}$, and for any subgroup G of S_n, let $Z_G(k_1, k_2, \ldots, k_n) = \frac{1}{|G|} \sum_{\sigma \in G} Z(\sigma)(k_1, k_2, \ldots, k_n)$.*

For example, if $\sigma = (1\,2\,3)(4\,5\,6)(7\,8\,9\,10) \in S_{10}$, then $Z(\sigma) = x_1^0 x_2^0 x_3^2 x_4^1 x_5^0 x_6^0 x_7^0 x_8^0 x_9^0 x_{10}^0 = x_3^2 x_4$. For another example, consider the subgroup $G = S_3 = \{\, \mathbb{1}_{J_4}, (1\,2\,3), (1\,3\,2), (1\,2), (1\,3), (2\,4) \,\}$. Since $\mathbb{1}_{J_3} = (1)(2)(3)$, $Z(\mathbb{1}_{J_3}) = x_1^3$, while $Z((1\,2\,3)) = Z((1\,3\,2)) = x_3$, and $Z((1\,2)) = Z((1\,3)) = Z((2\,3)) = x_1 x_2$, so $Z_G = \frac{1}{6}(x_1^3 + 3 x_1 x_2 + 2 x_3)$.

For a final example, the nonidentity elements of the subgroup D_4 of S_4 consist of two 2-cycles, three products of two disjoint 2-cycles, and two 4-cycles, so $Z_{D_4} = \frac{1}{8}(x_1^4 + 2 x_1^2 x_2 + 3 x_2^2 + 2 x_4)$.

The alert reader may have noticed that in our action of D_4 on $\mathscr{F}(J_4, \{\, R, B, Y \,\})$ in the first section, the number of necklaces was $\frac{1}{8}(3^4 + 2(3^3) + 3(9) + 2(3))$, which we point out is the value $Z_{D_4}(3, 3, 3, 3)$, where 3 is the number of colours; that is, the number of elements in the set $\{\, R, B, Y \,\}$. This is no coincidence, as our next result shows.

16.2.3 Theorem (Pólya). *Let n be a positive integer, and let G be a subgroup of S_n. Then for any positive integer m, the number of orbits of the action of G on $C = \mathscr{F}(J_n, J_m)$, as defined in Definition 16.2.1, is $Z_G(i_1, i_2, \ldots, i_n)$, where $i_1 = i_2 = \cdots = i_n = m$.*

Proof. Let m be a positive integer, and consider $\sigma \in G$. Suppose that $Z(\sigma) = x_1^{s_1} x_2^{s_2} \cdots x_n^{s_n}$. We compute $|Fix(\sigma)|$. We have

$$Fix(\sigma) = \{\, f \in C(m) \mid f \circ \sigma = f \,\}$$
$$= \{\, f \in C(m) \mid f(\sigma(i)) = f(i) \text{ for all } i = 1, 2, \ldots, n \,\}$$

and so $f \in Fix(\sigma)$ if, and only if, f is constant on each orbit of the action of σ on J_n (since the above shows that $f(\sigma^j(i)) = f(i)$ for all $i \in J_n$ and for all integers $j \geq 1$ if $f \in Fix(\sigma)$). Hence

$$|Fix(\sigma)| = m^{s_1} m^{s_2} \cdots m^{s_n} = Z(\sigma)(m, m, \ldots, m).$$

It now follows from Burnside's lemma that the number of orbits is

$$\frac{1}{|G|} \sum_{\sigma \in G} |Fix(\sigma)| = \frac{1}{|G|} \sum_{\sigma \in G} Z(\sigma)(m, m, \ldots, m) = Z_G(m, m, \ldots, m).$$

$\square$

As our first application of Pólya's theorem, we determine the number of necklaces that can be made if we are to place five medallions (available in each of the familiar three colours, R, B, and Y) on the silver chain (with no clasp), and the medallions are the same front and back (so the pattern of a necklace, and the pattern obtained by turning the necklace over are to be considered the same). Thus given any pattern for a necklace, any rotation of the pattern results in a pattern that we consider to produce the same necklace, and the reflection of the pattern across the vertical line that bisects the chain results in a pattern that is considered to make the same necklace. It follows that we need to consider the action of the dihedral group $G = D_5 \subseteq S_5$ on the set $\mathscr{F}(J_5, J_3)$. We need to determine $|Fix(\sigma)|$ for each $\sigma \in D_5$. Our copy of D_5 in S_5 is generated by $\sigma = (1\,2\,3\,4\,5)$ and $\tau = (1\,4)(2\,3)$ (imagine that we have put the fifth medallion at the original marker for the clasp). The elements are the five powers of σ, all but one of which are 5-cycles, while the other is the identity mapping, and the composites $\tau \circ \sigma^i$, $i = 1, 2, 3, 4, 5$, each of which is a product of two disjoint 2-cycles. Thus $Z_{D_5} = \frac{1}{10}(x_1^5 + 4x_5 + 5x_1 x_2^2)$ and so the number of necklaces is $Z_{D_4}(3, 3, 3, 3, 3) = \frac{1}{10}(3^5 + 4(3) + 5(3)(3^2)) = \frac{1}{10}(243 + 12 + 135) = 39$.

16.3 Enumeration Examples

We finish up this chapter with a variety of enumeration problems that we are now equipped to analyze and solve.

For our first example in this section, let us determine the number of necklaces with five medallions, where the necklace has no clasp, but we must use exactly three (R)ed and two (B)lue medallions on each necklace. Such a necklace is obtained from a pattern which consists of a surjective mapping $f : J_5 \to J_2$ wherein $|f^{-1}(1)| = 3$ and $|f^{-1}(2)| = 2$ (where we have relabelled R as 1 and B as 2). Since a rotation of a pattern by one-fifth of a full rotation produces a pattern that we have declared will produce the same necklace as the original pattern, and as well, if we agree to place the fifth medallion at the position the clasp formerly occupied, we have decreed that reflecting the pattern across the line through the clasp location that bisects the chain will result in a pattern that produces the same necklace as the original pattern. Thus we should consider the action of the dihedral group $D_5 = \langle (1\,2\,3\,4\,5), (1\,4)(2\,3) \rangle \subseteq S_5$ on the set

$$X = \{\, f : J_5 \to J_2 \mid |f^{-1}(1)| = 3,\ |f^{-1}(2)| = 2 \,\}.$$

Note that for any $\sigma \in S_5$, if $f \in X$, then $f \circ \sigma \in X$ as well, so D_5 does indeed act on X. First, note that $|X| = \binom{5}{2} = 10$. The number of different necklaces we can make is the number of orbits of X under the action of D_5. D_5 has five 5-cycles, four products of two disjoint 2-cycles, and the identity map, and for $\sigma \in D_5$, $f \in Fix(\sigma)$ if, and only if, f is constant on the orbits of σ; that is, on the elements of each cycle in σ. For a product of two disjoint 2-cycles (and so with one 1-cycle), the elements it fixes are those elements of X which are formed by choosing one of the 2-cycles to map to 2, while the other 3 elements are mapped to 1. Thus if $\sigma \in D_5$ is a product of two disjoint 2-cycles, then $|Fix(\sigma)| = 2$. If $\sigma \in D_5$ is a 5-cycle, then no $f \in X$ will belong to $Fix(\sigma)$, and so $|Fix(\sigma)| = 0$. Since $|Fix(\mathbb{1}_{J_5})| = |X| = 10$, the number of orbits of the action of D_5 on X is $\frac{1}{10}(10 + 5(2) + 4(0)) = 2$. Thus there are only two necklaces that meet these requirements. Actually, these are easily seen to be the one where the two blue medallions are separated by red medallions, and the one where the two blue medallions are adjacent. A pattern for the first would be $f = \left(\begin{smallmatrix} 1 & 2 & 3 & 4 & 5 \\ R & B & R & B & R \end{smallmatrix} \right)$, and a pattern for the second would be $g = \left(\begin{smallmatrix} 1 & 2 & 3 & 4 & 5 \\ B & B & R & R & R \end{smallmatrix} \right)$.

In our next example, we consider the group G of rotations of the cube (recall that G is isomorphic to S_4). The natural action of G on $\mathbb{R}^n$ induces an action of G on the set of the six faces of the cube. We wish to determine the number of different ways to paint the faces of the cube so that two faces are painted (R)ed, two faces are painted (B)lue, and one face (Y)ellow and one face (G)reen. Let us label the six faces of the cube with the labels 1,2,3,4,5,6, so that we can think of the set of faces of the cube as the set J_6. Furthermore, an element of G that fixes all six faces must then fix all eight vertices of the cube, and the only element of G with this property is the identity mapping. Thus we may consider G to be a subgroup of size 24 of S_6. Each colouring is a surjective mapping from J_6 to J_4 (where we identity R with 1, B with 2, Y with 3, and G with 4) for which $|f^{-1}(1)| = |f^{-1}(2)| = 2$ and $|f^{-1}(3)| = |f^{-1}(4)| = 1$. $|X| = \binom{6}{2}\binom{4}{2}2 = \frac{6!}{4} = 180$, and so $|Fix(\mathbb{1}_{J_6})| = 180$. Each rotation of the cube by $\frac{\pi}{2}$ about an axis joining the midpoints of two opposite faces is a 4-cycle, and there will be six of these, while each of the three squares of these is a product of two disjoint 2-cycles. Then we have the rotations by $\frac{2\pi}{3}$ about each of the four diagonals, and such a rotation is a product of two disjoint 3-cycles, so we obtain six such rotations. Finally, we have the rotations by π about an axis joining the midpoint of two opposite edges, and so we obtain 6 of these, each a product of three pairwise disjoint 2-cycles. The fixed point set of a 4-cycle has size 0, since any of its elements would require that 4 faces get the same colour. The fixed point set of a product of two disjoint 2-cycles has size $(2)(2)$ since we must choose one of the 2-cycles and paint its faces with colour 1, then colour the faces in the other 2-cycle with colour 2, then choose one of the remaining faces to colour with colour 3. The fixed point set of a product of two disjoint 3-cycles is empty, as is the fixed point set of a product of three pairwise disjoint 2-cycles. Thus the number of orbits is $\frac{1}{24}(180 + 3(4)) = \frac{192}{24} = 8$, and so there are 8 ways to paint the cube according to these specifications.

Suppose now that the faces of the cube are to be painted with the same four colours, but now there is no stipulation as to the number of faces that may be painted with a given colour. In this case, we can apply Pólya's theorem. We have $Z_G = \frac{1}{24}(x_1^6 + 6x_1^2x_4 + 3x_1^2x_2^2 + 8x_3^2 + 6x_2^3)$ and so the number of orbits (and thus the number of ways of painting the cube according the our new specifications) is

$$\frac{1}{24}(4^6 + 6(4)^2(4) + 3(4^2)(4^2) + +8(4^2) + 6(4^3)) = 240.$$

Incidentally, as a byproduct of Pólya's theorem, we see that for any subgroup G of S_n acting on $\mathscr{F}(J_n, J_m)$ we see that $|G|$ is a factor of the integer $Z_G(m, m, m \ldots, m)$. In the example just completed, we see that 24 is a factor of $m^6 + 6m^3 + 3m^4 + 8m^2 + 6m^3 = m^6 + 3m^4 + 12m^3 + 8m^2$.

As our final example of this section, we show how Burnside's lemma can be used to compute the number of different so-called switching functions. We consider, for any positive integer n, the set $\mathscr{F}(\mathbb{Z}_2^n, \mathbb{Z}_2)$. Of course, this set has size $|\mathbb{Z}_2|^{|\mathbb{Z}_2^n|} = 2^{2^n}$, a large number. For example, for $n = 3$, there are $2^8 = 256$ mappings from $\mathbb{Z}_2^3$ to $\mathbb{Z}_2$.

16.3.1 Definition. *Let n be a positive integer, and consider the permutation action of S_n on $\mathbb{Z}_2^n$, whereby $\sigma \in S_n$ maps $(a_1, a_2, \ldots, a_n) \in \mathbb{Z}_2^n$ to $(a_{\sigma(1)}, a_{\sigma(2)}, \ldots, a_{\sigma(n)})$. Then define a right action of S_n on $\mathscr{F}(\mathbb{Z}_2^n, \mathbb{Z}_2)$ by $\sigma \in S_n$ maps $f \in \mathscr{F}(\mathbb{Z}_2^n, \mathbb{Z}_2)$ to $f \circ \sigma$. Thus for $(a_1, a_2, \ldots, a_n) \in \mathbb{Z}_2^n$, $f \circ \sigma(a_1, a_2, \ldots, a_n) = f(a_{\sigma(1)}, a_{\sigma(2)}, \ldots, a_{\sigma(n)})$.*

Note that for $\sigma, \tau \in S_n$ and $f \in \mathscr{F}(\mathbb{Z}_2^n, \mathbb{Z}_2)$, $f \circ (\sigma \circ \tau) = (f \circ \sigma) \circ \tau$, so as indicated, this is a right action.

For example, if $n = 3$ and we represent an element of $\mathbb{Z}_2^3$ as a sequence of three entries from $\mathbb{Z}_2$, with no commas or spaces between the successive entries, then $\mathbb{Z}_2^3 = \{\, 000, 001, 010, 011, 100, 101, 110, 111 \,\}$. For $\sigma = (1\,2\,3) \in S_3$ and

$$f = \begin{pmatrix} 000\ 001\ 010\ 011\ 100\ 101\ 110\ 111 \\ 1 \quad 1 \quad 0 \quad 0 \quad 1 \quad 1 \quad 0 \quad 1 \end{pmatrix}$$

the action of σ on f results in

$$f \circ \sigma = \begin{pmatrix} 000\ 001\ 010\ 011\ 100\ 101\ 110\ 111 \\ 1 \quad 1 \quad 1 \quad 1 \quad 0 \quad 0 \quad 0 \quad 1 \end{pmatrix}.$$

The name "switching function" stems from the fact that a mapping from $\mathbb{Z}_2^n$ to $\mathbb{Z}_2$ accepts n binary inputs, and outputs a binary value, so it represents a logic circuit (from which all present day computing devices are constructed). If f and g are two switching functions, and we know how to compute the map f (imagine that we have physically built a device that computes f), then it is not necessary to figure out how to compute g if we have $\sigma \in S_n$ such that $g = f \circ \sigma$, we simply feed the n inputs intended for g into f after permuting them by σ. The result is the value for g at the original inputs. From this point of view, it is natural to ask how many switching functions we would have to know how to construct in order to be able to evaluate every switching function; that is, we would like to know the number of orbits of the action of S_n on $\mathscr{F}(\mathbb{Z}_2^n, \mathbb{Z}_2)$.

16.3.2 Proposition. *Let n be a positive integer. The number of orbits of the action of S_n on $\mathscr{F}(\mathbb{Z}_2^n, \mathbb{Z}_2)$ is*

$$\frac{1}{n!} \sum_{\sigma \in S_n} 2^{Z_{\langle \sigma \rangle}(2,2,\ldots,2)}.$$

Proof. By Burnside's lemma, the number of orbits is $\frac{1}{|S_n|} \sum_{\sigma \in S_n} |Fix(\sigma)| = \frac{1}{n!} \sum_{\sigma \in S_n} |Fix(\sigma)|$. Observe that for $\sigma \in S_n$, $f \in Fix(\sigma)$ if, and only if, f is constant on the orbits of the action of $\langle \sigma \rangle$ on $\mathbb{Z}_2^n$, and so $|Fix(\sigma)| = 2^k$, where k is the number of orbits of the action of $\langle \sigma \rangle$ on $\mathbb{Z}_2^n$. By Pólya's result (Theorem 16.2.3), $k = Z_{\langle \sigma \rangle}(2, 2, \ldots, 2)$. □

As an example, let us determine the number of orbits of the action of S_3 on $\mathscr{F}(\mathbb{Z}_2^3, \mathbb{Z}_2)$. Since $S_3 = \{\, \mathbb{1}_{J_3}, (1\,2), (2\,3), (1\,3), (1\,2\,3), (1\,3\,2) \,\}$, and the number of orbits of $\sigma \in S_3$ depends only on the cycle structure of σ, we have three computations to perform, one of which is $Z_{\mathbb{1}_{J_3}}(2, 2, 2) = 2^3 = 8$. For any 2-cycle, say $\sigma = (1\,2)$, $Z_\sigma(2, 2, 2) = 2^2 = 4$, as positions 1 and 2 must be equal, either 0 or 1, and position 3 can be either 0 or 1. For any 3-cycle, say $\sigma = (1\,2\,3)$, $Z_\sigma(2,2,2) = 2$, as all three entries must be equal, either to 0 or to 1. Thus $Z_{\langle \mathbb{1}_{J_3} \rangle}(2,2,2) = \frac{1}{1}(8) = 8$, while for any 2-cycle σ, $Z_{\langle \sigma \rangle}(2,2,2) = \frac{1}{2}(8+4) = 6$, and for any 3-cycle σ, $Z_{\langle \sigma \rangle}(2,2,2) = \frac{1}{3}(8+2+2) = 4$. Putting it all together, the number of orbits of the action of S_3 on $\mathscr{F}(\mathbb{Z}_2^3, \mathbb{Z}_2)$ is

$$\frac{1}{3!}(2^8 + 3(2^6) + 2(2^4)) = \frac{1}{6}(15)(2^5) = (5)(16) = 80.$$

16.4   Exercises

1. Given n colours, what is the number of different colourings of the vertices of a cube (under the assumption that if one colouring can be obtained by a rotation of the cube after the cube has been coloured, then the two colourings are the same).

2. Suppose that you are given at least 10 red beads and at least 10 blue beads.

   a) In how many ways can 10 beads be arranged on a chain (with no clasp, and with the understanding that if one arrangement could be obtained by simply turning over another arrangement, the two are to be considered the same arrangement).

   b) Under the same assumption as (a), but now the necklace is to have a clasp past which beads cannot be threaded, how many different arrangements are there?

   c) Under the same assumptions as in (a), but with the requirement that an arrangement is to consist of exactly 7 beads, 4 of which are to be red, what is the number of possible arrangments?

3. Consider a regular tetrahedron (see Chapter 15), which has 4 faces, four vertices, and 6 edges.

   a) Suppose we wish to colour the edges of the regular tetrahedron, and that we have n available colours (n a positive integer). How many different colourings are possible, if we consider two colourings to be the same if one can be obtained from the other by a rotation of the tetrahedron?

   b) In how many ways can we colour the four faces of the tetrahedron, if we have 4 colours available, and two colourings are to be considered the same if one can be obtained from the other by a rotation of the tetrahedron?

4. We consider colourings of the faces of a cube, wherein two colourings are to be considered the same if one can be obtained from the other by a rotation of the cube.

   a) In how many ways can the faces of a cube be coloured if there are 6 available colours, and we require that no two faces are to be coloured with the same colour?

   b) In how many ways can the faces of a cube be coloured with colours blue, white, and yellow, if exactly two of the faces are to be coloured blue, one white, and the remaining three yellow?

5. For any positive integer n, $\varphi(n) = |\{\, k \in \mathbb{Z} \mid 1 \leq k \leq n,\ (k, n) = 1 \,\}|$ (that is, φ is the Euler φ-function, see Exercise 14 (e) of Chapter 4)).

   a) Suppose that G is a cyclic group of order n, with generator g, and that G acts on a finite set S. Prove that the number of orbits of this action is $\frac{1}{n} \sum_{d|n} |Fix(g)|^{\frac{n}{d}} \varphi(d)$,

   b) Prove that for any positive integers k and n, n divides $\sum_{d|n} k^{\frac{n}{d}} \varphi(d)$.

6. Suppose that a rectangular sheet of fabric of size 4 units by 5 units is to be dyed in such a way that the result is a grid of 20 squares, each 1 unit by 1 unit. Further suppose that the dye job is to result in 6 red squares and 14 yellow squares.

   a) How many different patterns are there, if the dye does not penetrate to show up on the other side of fabric (so there is no question of being able to turn the sheet over and consider the back side as a pattern)?

   b) Now suppose that the dye does penetrate the fabric perfectly, so that the fabric can be turned over (so a colouring pattern of the fabric is to be considered the same as the pattern on the reverse side of the pattern). Under this assumption, how many colourings are possible?

7. Suppose that we have a square sheet of fabric that has been partitioned into 3 rows and 3 columns, and we wish to colour the 9 squares so formed using colours red and blue, subject to the requirement that two of the squares are coloured red and the remaining seven squares are coloured blue. We assume that the colouring penetrates the fabric perfectly, and the colouring that is obtained by turning the sheet over is a valid colouring. If two colourings are to be considered the same if one can be obtained from the other by applying an element of D_4 (the symmetries of the square) to the sheet, what is the number of different possible

colourings?

8. Suppose that we have three boxes, two of which have been labelled with the same label, while the third box has been assigned a different label. Further suppose that we have two identical white balls and two identifical black balls. In how many ways can the four balls be deposited in the three boxes (so the only thing that matters is the number of each colour that are deposited in each box, and we do not distinguish between the two boxes with the same label)?

9. What is the number of orbits of the action of S_4 on $\mathcal{F}(\mathbb{Z}_2^4, \mathbb{Z}_2)$ (see Definition 16.3.1)?

10. Consider a regular pentagon centered at the origin of the plane. In how many ways may we colour its vertices with 3 available colours, subject to the specification that two colourings are to be considered the same if one can be obtained from the other by an application of an element of D_5?

11. Let G be a finite group acting on a finite set X, and suppose that the action has s orbits $O_1, O_2, \ldots, O_s$. Suppose further that A is an abelian group, and that additive notation is used to denote the binary operation of A. Let $\omega : X \to A$ be any map which is constant on each of the s orbits O_i, and use ω to define a map $W : G \to A$ by $W(g) = \sum_{x \in Fix(g)} \omega(x)$. Finally, for each $i = 1, 2, \ldots, s$, let $\alpha_i \in O_i$. Prove that $\sum_{g \in G} W(g) = |G| \sum_{i=1}^{s} \omega(\alpha_i)$. Observe that if $A = \mathbb{Z}$ and ω is the constant function $\omega(x) = 1$ for all $x \in X$, then the equality reduces to Burnside's lemma, Proposition 12.1.14.

12. Let n be a positive integer, and let $G = \langle (1\,2\,\ldots, n) \rangle \subseteq S_n$.

     a) Prove that for any positive integer k, $(1\,2\,\ldots\,n)^k$ is the product of d pairwise disjoint cycles, each of length $\frac{n}{d}$, where $d = (k, n)$.

     b) Prove that $Z_G(x_1, x_2, \ldots, x_n) = \frac{1}{n} \sum_{d|n} \varphi(d) x_d^{\frac{n}{d}}$, where φ is the Euler φ-function (see Exercise 14 (e) of Chapter 4).

13. Let G be the rotation group of the cube.

     a) For the induced action of G on the vertices of the cube, find $Z_G(x_1, x_2, \ldots, x_8)$.

     b) For the induced action of G on the faces of the cube, find $Z_G(x_1, x_2, \ldots, x_6)$.

14. Suppose that X and Y are disjoint sets, and that G is a group acting on X and H is a group acting on Y. Define a mapping from the product

group $G \times H$ into $S_{X \cup Y}$ by

$$(g,h)z = \begin{cases} gz & \text{if } z \in X \\ hz & \text{if } z \in Y \end{cases}$$

for all $z \in X \cup Y$, for all $(g,h) \in G \times H$.

a) Prove that in fact, for each $(g,h) \in G \times H$, this determines a permutation of $X \cup Y$, and then prove that this mapping is an action of $G \times H$ on $X \cup Y$.

b) If G, H, X, and Y are finite, and $n = |X|$, $m = |Y|$, determine $Z_{G \times H}(x_1, x_2, \ldots, x_n)$ in terms of $Z_G(x_1, \ldots, x_n)$ and $Z_H(x_{n+1}, \ldots, x_{n+m})$.

c) In the particular case when $G = S_3$, acting naturally on J_3, and D_4, acting on the set $V = \{\, a, b, c, d \,\}$ of vertices of a square, where the vertices have been labelled during a clockwise tour of the square, determine $Z_{S_3 \times D_4}(x_1, x_2, \ldots, x_7)$.

15. For any finite group G, consider the action of G on itself given by left multplication $g \mapsto \lambda_g : G \to G$, where for each $x \in G$, $\lambda_g(x) = gx$. Let

$$G = \left\{ \begin{bmatrix} 1 & a & b \\ 0 & 1 & c \\ 0 & 0 & 1 \end{bmatrix} \in Gl_3(\mathbb{Z}_3) \;\middle|\; a, b, c \in \mathbb{Z}_3 \right\}.$$

Prove that G is a subgroup of $Gl_3(\mathbb{Z}_3)$ of order 27 and then for the action of G on itself by left multiplication, compute $Z_G(x_1, x_2, \ldots, x_{27})$.

Chapter 17

Group Codes

1 Introduction

Messages consisting of sequences of binary digits (such a sequence is called a word) transmitted over noisy channels are subject to error. Although we cannot entirely prevent the noisy channel from causing such errors, we can minimize the probability of them being misinterpreted at the receiving end of the channel by encoding the words, by which we mean that redundant information is built into the data being transmitted, as this will enable the receiving end to detect and possibly reduce or even eliminate errors that may have occurred during transmission.

For example, suppose we want to send a message that consists of a sequence of two binary digits. The possible message words, each of length two, are 00, 10, 01, and 11. If the probability of receiving a 1 as a 0, or a 0 as a 1 is 0.01, then the probability of receiving an incorrect word is $1 - (0.99)^2$, approximately 0.02. It is clear that the probability of receiving an incorrect word increases with the length of the word and the length of the message.

The basic idea is to adjoin check digits to the words which carry the information (in this case, 00, 10, 01, and 11), resulting in an encoded word. These check digits are redundant in the sense that they do not carry any new information, but their presence permits the detection, and possibly even the correction of errors. We think of the encoded words as consisting of two parts: the message digits come first, followed by the check digits as we read from left to right. Any collection of such encoded words is called a code.

In the example above, let us adjoin a 0 or a 1 to each of the four words 00, 10, 01, and 11 to obtain the four words 000, 101, 110, and 011. The

445

property that we have bestowed upon the words is that each word now contains an even number of 1's. Note that our original four words, each of length 2, constituted the entire collection of binary words of length two on the alphabet $\{0, 1\}$. On the other hand, the set of encoded words is a proper subset of the set of all words of length three on the same alphabet. This is in a sense analogous to the situation in English where not all 26^4 "words" with four letters are words of the English language.

Suppose now that in the transmission of a message written in the code above, a single error occurs in one of the words, so the received word will have an odd number of 1's. As a result, the receiver can detect that the word is not a word of the "language" and can then ask for a re-transmission. This code is therefore called a single-error-detecting code. In fact, the code will detect any odd number of errors but no even number of errors will be detected. Note that once the receiver decides that no error has occurred, then the original message word is recovered by simply truncating the last character of the code word.

The process that we have described above is illustrated schematically in Figure 17.1.

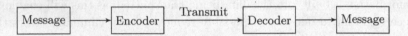

Fig. 17.1   The encoding, transmission, and decoding process

Let us now compute the probability that a given transmitted word will be incorrectly interpreted at the receiving end. This will happen only when exactly two errors occur and the probability of this event is $\binom{3}{2}(0.1)^2(0.99) = 0.000297$, a considerable improvement over 0.02, the probability of error if the uncoded words are transmitted. As is always the case, we don't get something for nothing: for the sake of accuracy, we have increased transmission time. A moment's thought will convince the reader that this will always be the case: as accuracy increases, efficiency decreases.

It is often the case that a message can only be sent once. For example, transmissions between a space probe and a ground station can take days, months, or even years to receive, and there is no opportunity to arrange for a repeat transmission if some error is detected. In such a situation, the error-detection capability is of very little value. Instead, it would be important to be able to detect and correct errors. Once again, we illustrate with a simple example.

Suppose we want our language to consist of the two words 0 and 1, where the probability of error is again 0.01. We encode the information words as 000 and 111 and stipulate the following decoding scheme: the received word is decoded according to majority rule; that is, the received word will be decoded as 1 if it has more 1's than 0's, otherwise it will be decoded as 0. What is the probability that a decoding error will occur? Suppose that xxx is transmitted. It will be incorrectly decoded if there is an error in at least two of the coordinate positions. This will happen with probability $\binom{3}{2}(0.01)^2(0.99) + (0.01)^3$, which is approximately equal to 0.000298. This is an improvement over 0.01, the probability of error if the uncoded words are transmitted. Once again, we have paid for accuracy by tripling the transmission time.

.We remark that the English language has considerable error correcting capacity. For example, if we come across the word "depcribe", we would have no difficulty in recognizing that there is a typographical error in the third letter and we would correct it to read "describe". On the other hand, if the word "cup" mistakenly appears as "cap", then unless the context tells us, we do not even detect the error.

We take our cue from these observations and conjecture: the more dissimilar the words of a code are, the more likely it will be that we shall be able to recognize and even correct errors. We shall prove a precise version of this conjecture after we have introduced the concept of Hamming distance.

17.1.1 Definition. *Let n be a positive integer. An element of $\mathbb{Z}_2^n$ is called a word of length n. A subset C of $\mathbb{Z}_2^n$ is called a code of length n. If $u \in \mathbb{Z}_2^n$, say $u = a_1 a_2 \cdots a_n$, then we write $u_i = a_i$ for each $i \in J_n$.*

We shall write an element of $\mathbb{Z}_2^n$ as a block of n contiguous characters, each either 0 or 1, with no comma or other separator involved. Thus for example, $\{\,0100, 1100, 1110, 1111\,\}$ is a code of length 4 (and also of size 4).

Example 8.4.7 established that the support map $S : \mathbb{Z}_2^n \to \mathscr{P}(J_n)$ defined in Definition 8.4.4 is a ring isomorphism. Thus for any $u, v \in \mathbb{Z}_2^n$, $S(u + v) = S(u) \triangle S(v)$, and $S(uv) = S(u) \cap S(v)$ (recall that the ring operations in $\mathbb{Z}_2^n$ are defined coordinatewise). For example, in $\mathbb{Z}_2^5$, $01110 + 11011 = 10101$, $(01110)(11011) = 01010$, $S(01110) = \{\,2,3,4\,\}$, $S(11011) = \{\,1,2,4,5\,\}$, $S(10101) = \{\,1,3,5\,\}$, and $S(01010) = \{\,2,4\,\}$. Note that $S(01110 + 11011) = S(10101) = \{\,1,3,5\,\}$, and $S(01110) \triangle S(11011) = \{\,2,3,4\,\} \triangle \{\,1,2,4,5\,\} = \{\,1,3,5\,\}$.

17.1.2 Definition. *Let n be a positive integer. For any $u, v \in \mathbb{Z}_2^n$, the*

Hamming distance between u and v is $h(u,v) = |\{i \in J_n \mid u_i \neq v_i\}|$. For any $u \in \mathbb{Z}_2^n$, the weight of u, denoted by $w(u)$, is $h(u,0)$.

Note that for any $u, v \in \mathbb{Z}_2^n$, $\{i \in J_n \mid u_i \neq v_i\} = \{i \in J_n \mid u_i = 1 \text{ and } v_i = 0\} \triangle \{i \in J_n \mid u_i = 0 \text{ and } v_i = 1\} = S(u) \triangle S(v) = S(u+v)$, so $h(u,v) = |S(u+v)|$.

17.1.3 Proposition. *Let n be a positive integer. Then for any $u, v, w \in \mathbb{Z}_2^n$, $h(u,v) = h(u+w, v+w)$.*

Proof. Let $u, v, w \in \mathbb{Z}_2^n$. Then $S(u+w+v+w) = S(u+v)$ and so $h(u+w, v+w) = |S(u+w+v+w)| = |S(u,v)| = h(u,v)$. $\square$

17.1.4 Corollary. *Let n be a positive integer. Then for any $u, v, w \in \mathbb{Z}_2^n$, $h(u,v) = w(u+v)$.*

Proof. By Proposition 17.1.3, for $u, v \in \mathbb{Z}_2^n$. $h(u,v) = h(u+v, v+v) = h(u+w, 0) = w(u+v)$. $\square$

17.1.5 Lemma. *Let n be a positive integer. For any $u, v \in \mathbb{Z}_2^n$, $h(u,v) = w(u) + w(v) - 2w(uv)$.*

Proof. We have $h(u,v) = |S(u+v)| = |S(u) \triangle S(v)| = |S(u) - S(v)| + |S(v) - S(u)| = |S(u) - (S(u) \cap S(v))| + |S(v) - (S(v) \cap S(u))| = |S(u)| - |S(uv)| + |S(v)| - |S(uv)| = w(u) + w(v) - 2w(uv)$. $\square$

17.1.6 Proposition. *Let n be a positive integer. The Hamming distance $h: \mathbb{Z}_2^n \times \mathbb{Z}_2^n \to \mathbb{Z} \subseteq \mathbb{R}$ is a metric; that is, it satisfies the following three requirements:*

    (i) For all $u, v \in \mathbb{Z}_2^n$, $h(u,v) \geq 0$, and $h(u,v) = 0$ if, and only if, $u = v$.
    (ii) For all $u, v \in \mathbb{Z}_2^n$, $h(u,v) = h(v,u)$.
    (iii) For all $u, v, w \in \mathbb{Z}_2^n$, $h(u,v) \leq h(u,w) + h(w,v)$ (the triangle inequality).

Proof. Let $u, v, w \in \mathbb{Z}_2^n$. For (i), $h(u,v) = |S(u+v)| \geq 0$. If $u = v$, then $u + v = 0$, and $S(0) = \varnothing$, giving $h(u,u) = 0$. On the other hand, suppose that $h(u,v) = 0$. Then $|S(u+v)| = 0$ and so $S(u+v) = \varnothing$. But then $S(u) \triangle S(v) = \varnothing$ and so $S(u) = S(v)$, and since S is bijective, this implies that $u = v$.

    (ii) is immediate since $u + v = v + u$. For (iii), $h(u,v) = |S(u+v)| = |S(u+w+w+v)| = |S(u+w) \triangle S(v+w)| \leq |S(u+w)| + |S(v+w)| = h(u,w) + h(w,v)$. $\square$

Suppose that we would like to design a code C that would allow for the correction of any pattern of k or fewer errors. We start with the set of all 2^n words of length n (where the value of n is to be determined) over the alphabet $\mathbb{Z}_2 = \{0, 1\}$, and to the tail-end of each of word, we adjoin k elements of $\mathbb{Z}_2$ (an element of $\mathbb{Z}_2$ shall be referred to as a digit, and these k elements that we adjoin to a word are called the check-digits for the word). Of course, the process by which we determine the value of the check bits for each word is at this time unspecified. The resulting set C of size 2^n is then a subset of $\mathbb{Z}_2^{n+k}$ and is called a code of length $n + k$.

Once a word of C has been transmitted through a noisy channel, how does the receiver decode the word, given that errors may have occurred during transmission? If we make the simplifying assumption that the noisy channel is such that the probability of error in any one digit is independent of what happens to any other digit, then for any nonnegative integer m, it is less likely for $m + 1$ errors to occur than it is for m errors to occur.

In view of this assumption, we adopt the following algorithm for the receiver to use upon receiving a word. If the word u is sent and word v is received, a search of C is conducted to determine the word $w \in C$ that is closest (in terms of Hamming distance) to v. The received word v is then to be decoded as w. Suppose that the code has the property that no word is within a distance at most k from more than one word of the code. If, in fact, not more than k digits of v are in error, then w will be u. On the other hand, if more than k errors have occurred, then the decoding scheme may yield a word different from u. Such a decoding scheme is known as "nearest neighbour decoding".

Suppose that v was transmitted and corrected to u as described above. Then the error word for this received word v is $\epsilon = v + u$. Evidently, $S(\epsilon)$ contains the coordinate positions of the entries that are in error. In view of this observation, we may describe the nearest-neighbor scheme as follows: for a received word v, find $u \in C$ for which $w(v + u)$ is a minimum, and correct v to u.

17.1.7 Definition. *Let n be a positive integer, and let $C \subseteq \mathbb{Z}_2^n$. For any integer d with $0 \leq d \leq n$, C is said to be d-error-detecting if it recognizes all error words of weight at most d; it is said to be d-error-correcting if, using the nearest neighbor decoding scheme, it corrects all errors with error word of weight at most d. The minimum Hamming distance of the code C is $\mu(C) = \min\{ h(u,v) \mid u,v \in C, \ u \neq v \}$.*

For example, if $C = \{ 0000, 0110, 1001, 1111 \}$, then $\mu(C) = 2$.

We shall now show precisely how the minimum Hamming distance of a code is related to its error-detecting and error-correctng capacities.

17.1.8 Proposition. *Let n be a positive integer, and $C \subseteq \mathbb{Z}_2^n$. For any integer d with $0 \le d \le n$, C is d-error-detecting if, and only if, $\mu(C) \ge d+1$, while C is d-error-correcting if, and only if, $\mu(C) \ge 2d + 1$.*

Proof. Let d be an integer with $0 \le d \le n$. First, suppose that $\mu(C) \le d$. Let $u, v \in C$ be such that $\mu(C) = h(u, v) = |S(u + v)|$, and let $\epsilon = u + v$. Then $w(\epsilon) = \mu(C) \le d$, but if u is transmitted and received as $u + \epsilon = v$, then we would not recognize that any errors had occurred since $v \in C$. Thus there is at least one occurrence of at most d errors that will not be detected, and so C is not d-error-detecting. Conversely, suppose that $\mu(C) \ge d + 1$. Then for any $\epsilon \in \mathbb{Z}_2^n$ with $w(\epsilon) \le d$, if $u \in C$ is transmitted and $u + \epsilon$ is received, the error(s) will be detected since $u + \epsilon \notin C$ (if $u + \epsilon \in C$, then $\mu(C) \le h(u, u + \epsilon) = |S(u + u + \epsilon)| = |S(\epsilon)| \le d < \mu(C)$, which is not possible). Thus C will detect every pattern of d or fewer errors.

Suppose now that C is d-error-correcting, but that $\mu(C) \le 2d$. Let $u, v \in C$ be such that $h(u, v) = \mu(C) \le 2d$. Let $\epsilon = u + v$, so $w(\epsilon) = \mu(C)$. Partition $S(\epsilon)$ into two subsets U and V, each of equal size $\mu(C)/2$ if $\mu(C)$ is even, while if $\mu(C)$ is odd, then arrange it so that $|U| = |V| + 1$. Since S is bijective, there exist $\epsilon_1, \epsilon_2 \in \mathbb{Z}_2^n$ such that $S(\epsilon_1) = U$ and $S(\epsilon_2) = V$. Then $S(\epsilon_1 + \epsilon_2) = S(\epsilon_1) \triangle S(\epsilon_2) = U \triangle V = U \cup V = S(\epsilon)$, and so $\epsilon = \epsilon_1 + \epsilon_2$. Thus $u + v = \epsilon_1 + \epsilon_2$, and so $u + \epsilon_1 = v + \epsilon_2 = w$, say, and $h(u, w) = |S(u + w)| = w(\epsilon_1) \ge w(\epsilon_2) = |S(v + w)| = h(v, w)$. Suppose that u is transmitted and received as w. We know that $w(\epsilon_2) \le w(\epsilon_1)$. If $w(\epsilon_2) < w(\epsilon_1)$, then the nearest neighbor decoding scheme would "correct" w to v, while if $w(\epsilon_1) = w(\epsilon_2)$ (that is, $h(u, w) = h(v, w)$), then we would have two elements of C equidistant from w. In this latter case, suppose that there is $y \in C$ with $h(y, w) < h(u, w)$. Then $h(y, u) \le h(y, w) + h(w, u) = h(y, w) + h(w, v) < h(u, w) + h(v, w) = w(\epsilon_1) + w(\epsilon_2) = w(\epsilon) = \mu(C)$; that is, $h(y, u) < \mu(C)$, which is not possible as $y, u \in C$. Thus the two choices of u and v are not only equidistant from w, but no element of C is closer to w than either u or v, and we have found an instance of at most d errors occurring which is either incorrectly decoded by the nearest neighbor decoding scheme, or is unable to be decoded by the nearest neighbor decoding scheme. This contradicts the fact that C was a d-error-correcting code, and so we conclude that if C is d-error-detecting code, then $\mu(C) \ge 2d + 1$.

Now suppose that $\mu(C) \geq 2d + 1$, and let $\epsilon \in \mathbb{Z}_2^n$ have $w(\epsilon) \leq d$. Let $v = u + \epsilon$. We claim that $\{ w \in C \mid h(v, w) \leq d \} = \{ u \}$. Since $h(v, u) = w(\epsilon) \leq d$, $u \in \{ w \in C \mid h(v, w) \leq d \}$. Let $w \in C$ be such that $h(v, w) \leq d$. Then $h(u, w) \leq h(u, v) + h(v, w) \leq d + d < \mu(C)$, which is not possible if $u \neq w$ and so we obtain that $w = u$. $\qquad\square$

17.1.9 Proposition. *Let n be a positive integer, and let C be a subgroup of $\mathbb{Z}_2^n$. Then $\mu(C) = \min\{ w(u) \mid u \in C, \ u \neq 0 \}$.*

Proof. Since C is a subgroup of $\mathbb{Z}_2^n$, we have $\{ u + v \mid u, v \in C, u \neq v \} = \{ u \in C \mid u \neq 0 \}$, and thus $\mu(C) = \min\{ h(u, v) \mid u, v \in C, \ u \neq v \} = \min\{ w(u + v) \mid u, v \in C, \ u \neq v \} = \min\{ w(u) \mid u \in C, \ u \neq 0 \}$. $\qquad\square$

Thus for example, $C = \langle \{ 101100, 011010, 110001 \} \rangle$ is the subgroup of $\mathbb{Z}_2^6$ of order 8 whose elements are:

$$C = \{ 000000, 101100, 011010, 110001, 110110, 101011, 011010, 000111 \}.$$

By Proposition 17.1.9, $\mu(C) = 3$ and so by Proposition 17.1.8, C is a 2-error-detecting code and a single error-correcting code.

Two problems of practical importance which we should address are those of designing codes with high error-correcting capacity, but which are efficient in the sense that the decoding scheme is easy to implement and transmission time is not seriously compromised. With the nearest neighbor decoding scheme (and that is the scheme we shall employ), if we encode our message words at random, we must have a list of codewords at hand against which we must compare each received word to determine whether or not it is in the code. This is obviously a time-consuming procedure. If, however, we choose a code with some mathematical structure or whose words have a certain pattern, then the encoding and decoding algorithms can be quite simple and efficient. We have already seen in Proposition 17.1.9 that if a code C is a subgroup of $\mathbb{Z}_2^n$, then the task of determining $\mu(C)$ is reduced from that of the calculation and comparison of $\binom{|C|}{2}$ values to that of the calculation and comparison of $|C|$ values. It is reasonable to expect that other gains can be made by considering codes that are subgroups of $\mathbb{Z}_2^n$, and this we proceed to do in the next section.

.2 Group Codes

In this section, we shall apply results from group theory to the problem of code design. The notions of homomorphism and subgroup will play prominent roles in the theoretical development, while it will be seen that cosets can be effectively used in the practical process of decoding.

17.2.1 Definition. *For any positive integer n, a subgroup of $\mathbb{Z}_2^n$ (under addition) is called a* group code *(of length n).*

17.2.2 Definition. *Let n and k be positive integers with $k \leq n$. Then for any injective mapping $f : J_k \to J_n$, let $\Pi_f : \mathbb{Z}_2^n \to \mathbb{Z}_2^k$ be the map defined by $\Pi_f(a_1 a_2 \cdots a_n) = a_{f(1)} a_{f(2)} \cdots a_{f(k)}$ for each word $a_1 a_2 \cdots a_n \in \mathbb{Z}_2^n$.*

For example, when $n = 4$ and $k = 2$, we might consider $f = \left(\begin{smallmatrix} 1 & 2 \\ 1 & 3 \end{smallmatrix}\right)$. Then $\Pi_f(0101) = 00$ and $\Pi_f(1011) = 11 = \Pi_f(1010)$.

17.2.3 Proposition. *Let n and k be positive integers with $k \leq n$, and let $f : J_k \to J_n$ be an injective mapping. Then $\Pi_f : \mathbb{Z}_2^n \to \mathbb{Z}_2^k$ is a surjective homomorphism.*

Proof. For the purposes of this demonstration, we will return to the representation of an element of $\mathbb{Z}_2^n$ as $(a_1, a_2, \ldots, a_n)$. For any $u = (a_1, a_2, \ldots, a_n)$ and $v = (b_1, b_2, \ldots, b_n)$ in $\mathbb{Z}_2^n$,

$$
\begin{aligned}
\Pi_f(u + v) &= \Pi_f(a_1 + b_1, a_2 + b_2, \ldots, a_n + b_n) \\
&= (a_{f(1)} + b_{f(1)}, \ldots, a_{f(k)} + b_{f(k)}) \\
&= (a_{f(1)}, \ldots, a_{f(k)}) + (b_{f(1)}, \ldots, b_{f(k)}) \\
&= \Pi_f(a_1, \ldots, a_n) + \Pi_f(b_1, \ldots, b_n),
\end{aligned}
$$

and so Π_f is a homomorphism. For any $(a_1, \ldots, a_k) \in \mathbb{Z}_2^k$, let $(b_1, b_2, \ldots, b_n) \in \mathbb{Z}_2^n$ be defined by $b_i = a_{f^{-1}(i)}$ for $i \in f(J_k)$, while for $i \in J_n - f(J_k)$, let $b_i = 0$. Then

$$
\begin{aligned}
\Pi_f(b_1, b_2, \ldots, b_n) &= (b_{f(1)}, \ldots, b_{f(k)}) \\
&= (a_{f^{-1}(f(1))}, \ldots, a_{f^{-1}(f(k))}) \\
&= (a_1, \ldots, a_k),
\end{aligned}
$$

and so Π_f is surjective. $\qquad\square$

Note that if $f \in S_n$, then $\Pi_f : \mathbb{Z}_2^n \to \mathbb{Z}_2^n$ is a group isomorphism. In such a case, for any group code $C \subseteq \mathbb{Z}_2^n$, we say that C and $\Pi_f(C)$ are *equivalent codes*.

17.2.4 Proposition. *Let n be a positive integer, and let $\sigma \in S_n$. For any group code $C \subseteq \mathbb{Z}_2^n$, let $C' = \Pi_\sigma(C)$. Then C' is a group code with $\mu(C) = \mu(C')$ and $|C| = |C'|$.*

Proof. Certainly, C' is a group code and $|C| = |C'|$ since Π_σ is an isomorphism. Moreover, since the effect of Π_σ on a word u is to permute the entries of u, it follows that $w(u) = w(\Pi_\sigma(u))$ for all $u \in \mathbb{Z}_2^n$. Consequently, for any $u \in C$, $w(u) = w(\Pi_\sigma(u))$ and so $\mu(C) = \mu(C')$. $\qquad\square$

Observe that since $|\mathbb{Z}_2^n| = 2^n$, Lagrange's theorem asserts that a group code will have order 2^k for some $k \le n$. Our next result shows for any group code $C \subseteq \mathbb{Z}_2^n$ with $|C| = 2^k$, there is $\sigma \in S_n$ such that $\Pi_\sigma : \mathbb{Z}_2^n \to \mathbb{Z}_2^n$ maps C to a group code C' with $\mu(C) = \mu(C')$ and for any $u \in \mathbb{Z}_2^k$, there is a unique $v \in \mathbb{Z}_2^{n-k}$ with $uv \in C'$ (where by uv we mean the juxtaposition of the two words u and v). Once we have established this fact, we shall henceforth restrict our attention to group codes with this property.

Before we present this result, we offer an example. Consider the group code $C = \{\,000, 001, 100, 101\,\}$. With $\sigma = (2\,3)$, we obtain $\Pi_\sigma(C) = \{\,000, 010, 100, 110\,\}$. Note that each of $00, 01, 10, 11$ appears exactly once as the first two characters of a code word.

17.2.5 Proposition. *Let n be a positive integer, and let $C \subseteq \mathbb{Z}_2^n$ be a group code of size 2^k. Then there exists $\sigma \in S_n$ such that $C' = \Pi_\sigma(C)$ satisfies $\Pi_f(C') = \mathbb{Z}_2^k$, where $f : J_k \to J_n$ is the injective map $f(i) = i$ for $1 \le i \le k$.*

Proof. The proof will be by induction on $n \ge 1$, with hypothesis that for all k with $1 \le k \le n$, if $C \subseteq \mathbb{Z}_2^n$ is a group code of size 2^k, then there exists $\sigma \in S_n$ such that $C' = \Pi_\sigma(C)$ satisfies $\Pi_f(C') = \mathbb{Z}_2^k$, where $f(i) = i$ for all $i \in J_k$. For $n = 1$, the only possible value for k is 1, and $C = \{\,0, 1\,\}$ is the only subgroup of $\mathbb{Z}_2^1$ of size 2^1, and $\sigma = \mathbb{1}_n \in S_1$ has the required property. Suppose now that $n \ge 1$ is an integer for which the hypothesis holds. Let C be a subgroup of $\mathbb{Z}_2^{n+1}$ of order 2^k, where $1 \le k \le n + 1$. If $k = n + 1$, then $\sigma = \mathbb{1}_{J_{n+1}}$ has the required property, so we may assume that $k \le n$. Let $g : J_n \to J_{n+1}$ be the map $g(i) = i$ for each $i \in J_n$, and consider $\Pi_g(C)$. We consider two cases.

Case 1: $|\Pi_g(C)| = |C|$. Then $C_1 = \Pi_g(C)$ is a subgroup of order 2^k in $\mathbb{Z}_2^n$, where $k \le n$, and so by hypothesis, there exists $\sigma \in S_n$ such that $C_1' = \Pi_\sigma(C_1)$ satisfies $\Pi_f(C_1') = \mathbb{Z}_2^k$, where $f(i) = i$ for all i with $1 \le i \le k$. We may consider $\sigma \in S_{n+1}$ if we declare $\sigma(n + 1) = n + 1$, and then $C' = \Pi_\sigma(C)$ satisfies $\Pi_f(C') = \mathbb{Z}_2^k$, where we consider $f : J_k \to J_n \subseteq J_{n+1}$.

Case 2: $|\Pi_g(C)| \ne |C|$. By Proposition 17.2.3, $\Pi_g : \mathbb{Z}_2^{n+1} \to \mathbb{Z}_2^n$ is surjective and so $|\ker(\Pi_g)| = 2^{n+1}/2^n = 2$. Since $|\Pi_g(C)| \ne |C|$, it follows that $\ker(\Pi_g) \cap C \ne \{\,0\,\}$ and so $\ker(\Pi_g) \subseteq C$, giving $|\Pi_g(C)| = |C|/2 = 2^{k-1}$, with $\ker(\Pi_g) = \{\,00\ldots0, 00\ldots01\,\}$. If $k = 1$, then $C = \ker(\Pi_g)$, and if so, $\sigma = (1\,n + 1)$ produces $C' = \Pi_\sigma(C) = \{\,000\ldots0, 100\ldots0\,\}$, and so $\Pi_f(C) = \{0,1\} = \mathbb{Z}_2^1 = \mathbb{Z}_2^k$ where $f(1) = 1$. Assume therefore that $k \geq 2$. Set $C_1 = \Pi_g(C)$. Let us write $C_10 = \{\,u0 \mid u \in C_1\,\} \subseteq \mathbb{Z}_2^{n+1}$ and $C_11 = \{\,u1 \mid u \in C_1\,\} \subseteq \mathbb{Z}_2^{n+1}$. Then $C = C_10 \cup C_11$. Since $C_1 \subseteq \mathbb{Z}_2^n$ and $|C_1| = 2^{k-1}$ with $1 \leq k - 1 \leq n - 1 < n$, by our induction hypothesis there exists $\sigma \in S_n$ such that $C_1' = \Pi_\sigma(C_1)$ satisfies $\Pi_f(C_1') = \mathbb{Z}_2^{k-1}$, where $f : J_{k-1} \to J_n$ is the mapping $f(i) = i$ for $i \in J_{k-1}$. Consider $\sigma \in S_{n+1}$ as before, by prescribing $\sigma(n + 1) = n + 1$, and let τ denote the 2-cycle $(k\,n{+}1)$. Then $\tau \circ \sigma \in S_{n+1}$ and if we let $C' = \tau \circ \sigma(C)$, then we have

$$C' = \tau \circ \sigma(C) = \tau \circ \sigma(C_10 \cup C_11)$$
$$= \tau(\sigma(C_10) \cup \sigma(C_11)),$$

and so $\Pi_f(C') = \mathbb{Z}_2^{k-1} \times \{0,1\} = \mathbb{Z}_2^k$, where $f : J_k \to J_{n+1}$ is the mapping $f(i) = i$ for $i \in J_k$.

Thus the hypothesis is valid for $n + 1$. The result now follows by induction. $\qquad\square$

17.2.6 Definition. *Let n and k be positive integers with $k \leq n$. An (n, k) group code C is a subgroup C of $\mathbb{Z}_2^n$ such that $\Pi_f(C) = \mathbb{Z}_2^k$ where $f : J_k \to J_n$ is given by $f(i) = i$ for $i \in J_k$.*

Thus an (n, k) code is a group code in which for each $u \in \mathbb{Z}_2^k$, there exists one, and only one, $w \in \mathbb{Z}_2^{n-k}$ such that uw, the concatenation of u and w, is an element of C.

Proposition 17.2.5 has established that every group code of length n and order 2^k is equivalent to an (n, k) group code, and by Proposition 17.2.4, equivalent group codes have the same size and equal error-detection and error-correction capabilities.

17.3 Construction of Group Codes

It is apparent that we should not employ trial and error methods in the construction of our group code due to the difficulty of predicting $\mu(C)$ for a code C constructed in such a manner. In this section, we show how to construct all (n, k) group codes. Our methods rely on the fact that $\mathbb{Z}_2^n$ is a finite dimensional vector space (of dimension n in fact) over the finite field $\mathbb{Z}_2$. The fact that a subgroup of $\mathbb{Z}_2^n$ is a $\mathbb{Z}_2$-subspace of $\mathbb{Z}_2^n$ and vice-versa is what makes it all work. A group homomorphism from $\mathbb{Z}_2^n$ to $\mathbb{Z}_2^m$ is nothing but a $\mathbb{Z}_2$-linear transformation, and as $\mathbb{Z}_2^n$ is a $\mathbb{Z}_2$-vector space of dimension n and $\mathbb{Z}_2^m$ is a $\mathbb{Z}_2$-vector space of dimension m, a linear transformation $T:\mathbb{Z}_2^n \to \mathbb{Z}_2^m$ is uniquely determined by its effect on a $\mathbb{Z}_2$-basis for $\mathbb{Z}_2^n$, and of course, one basis would be the standard basis $\{e_1, e_2, \ldots, e_n\}$, where for each $i \in J_n$, e_i is the n-bit sequence with every entry 0 except for the ith entry, which is 1. Just as we did for linear transformations between real finite dimensional vector spaces, we are able to represent every linear transformation from $\mathbb{Z}_2^n$ to $\mathbb{Z}_2^m$ by an $m \times n$ matrix with entries from $\mathbb{Z}_2$. If $T:\mathbb{Z}_2^n \to \mathbb{Z}_2^m$ is a $\mathbb{Z}_2$-linear transformation, then the matrix that represents T relative to the standard ordered basis for $\mathbb{Z}_2^n$ and for $\mathbb{Z}_2^m$ is given by

$$[T] = \begin{bmatrix} c_{st}(T(e_1)) \mid c_{st}(T(e_2)) \mid \cdots \mid c_{st}(T(e_n)) \end{bmatrix},$$

where $c_{st}(a_1 e_1 + \cdots + a_m e_m) = \begin{bmatrix} a_1 \\ a_2 \\ \vdots \\ a_m \end{bmatrix}$ for each $a_1 e_1 + \cdots + a_m e_m \in \mathbb{Z}_2^m$.

Again, just as expected, for each $u \in \mathbb{Z}_2^n$, $[T]u = T(u)$, where in the expression $[T]u$, we are considering $u \in \mathbb{Z}_2^n$ to be a column vector. Conversely, for any $A \in M_{m \times n}(\mathbb{Z}_2)$, let $T_A:\mathbb{Z}_2^n \to \mathbb{Z}_2^m$ denote the linear transformation given by $T_A(u) = Au$ for each $u \in \mathbb{Z}_2^n$.

17.3.1 Definition. *Let n and k be positive integers, with $k \leq n$, and let $A \in M_{(n-k) \times k}(\mathbb{Z}_2)$. Then $M_A = \begin{bmatrix} I_k \\ \cdots \\ A \end{bmatrix} \in M_{n \times k}(\mathbb{Z}_2)$ where $I_k \in M_{k \times k}(\mathbb{Z}_2)$ is the identity matrix, is called an (n, k) generator matrix. The group code $C = T_{M_A}(\mathbb{Z}_2^k) \subseteq \mathbb{Z}_2^n$ is called the code associated with the matrix M_A.*

Observe that the structure of M_A ensures that $C = T_{M_A}(\mathbb{Z}_2^k)$ is an (n, k) code, since for any $u \in \mathbb{Z}_2^k$, regarded as a column vector when required for matrix multiplication, we have $T_{M_A}(u) = \begin{bmatrix} I_k \\ \cdots \\ A \end{bmatrix} u = \begin{bmatrix} I_k u \\ \cdots \\ Au \end{bmatrix} = \begin{bmatrix} u \\ \cdots \\ Au \end{bmatrix}$. Thus the first k characters of each code word make up u, the data word that was encoded by appending so-called check digits in the form of the $n - k$ vector Au. Thus each code word in C has the data word it encodes in the first k characters, and this is followed by $n - k$ check digits given by Au. Note that since for each $i \in J_n$, $M_A e_i$ is the ith column of M_A, C is the column space of M_A; that is, C is the subspace spanned by the n columns of M_A, or equivalently, C is the subgroup of $\mathbb{Z}_2^n$ that is generated by the set of n columns of M_A.

For example, consider $A = \left[\begin{smallmatrix} 1 & 1 & 0 \\ 1 & 0 & 1 \end{smallmatrix}\right]$, so $M_A = \begin{bmatrix} 1 & 0 & 0 \\ 0 & 1 & 0 \\ 0 & 0 & 1 \\ 1 & 1 & 0 \\ 1 & 0 & 1 \end{bmatrix}$. Let us compute

$$
\begin{aligned}
C &= T_{M_A}(\mathbb{Z}_2^3) \\
&= \left\{ M_A\begin{bmatrix} 0 \\ 0 \\ 0 \end{bmatrix}, M_A\begin{bmatrix} 1 \\ 0 \\ 0 \end{bmatrix}, M_A\begin{bmatrix} 0 \\ 1 \\ 0 \end{bmatrix}, M_A\begin{bmatrix} 0 \\ 0 \\ 1 \end{bmatrix}, M_A\begin{bmatrix} 1 \\ 1 \\ 0 \end{bmatrix}, M_A\begin{bmatrix} 1 \\ 0 \\ 1 \end{bmatrix}, M_A\begin{bmatrix} 0 \\ 1 \\ 1 \end{bmatrix}, M_A\begin{bmatrix} 1 \\ 1 \\ 1 \end{bmatrix} \right\} \\
&= \left\{ \begin{bmatrix} 0 \\ 0 \\ 0 \\ 0 \\ 0 \end{bmatrix}, \begin{bmatrix} 1 \\ 0 \\ 0 \\ 1 \\ 1 \end{bmatrix}, \begin{bmatrix} 0 \\ 1 \\ 0 \\ 1 \\ 0 \end{bmatrix}, \begin{bmatrix} 0 \\ 0 \\ 1 \\ 0 \\ 1 \end{bmatrix}, \begin{bmatrix} 1 \\ 1 \\ 0 \\ 0 \\ 1 \end{bmatrix}, \begin{bmatrix} 1 \\ 0 \\ 1 \\ 1 \\ 0 \end{bmatrix}, \begin{bmatrix} 0 \\ 1 \\ 1 \\ 1 \\ 1 \end{bmatrix}, \begin{bmatrix} 1 \\ 1 \\ 1 \\ 0 \\ 9 \end{bmatrix} \right\}.
\end{aligned}
$$

Note that $\mu(C) = \min\{\, w(u) \mid u \in C,\ u \neq \mathbb{0} \,\} = 2$, so C is only a single error-detecting code.

Our next result establishes that every (n, k) group code arises as the code associated with an (n, k) generator matrix.

17.3.2 Proposition. *Let n and k be positive integers with $k \leq n$. Then for every (n, k) group code C, there exists $A \in M_{(n-k) \times k}(\mathbb{Z}_2)$ such that $C = T_{M_A}(\mathbb{Z}_2^k)$.*

Proof. Let C be an (n, k) group code. Then for each $i \in J_k$, there exists one, and only one, word in C of the form $e_i | u_i$, where $u_i \in \mathbb{Z}_2^{n-k}$. Let A be the $(n - k) \times k$ matrix whose columns are, respectively, $u_1, u_2, \ldots, u_k$. and consider M_A. For each $i \in J_k$, $T_{M_A}(e_i) = \operatorname{col}_i(M_A) = e_i | u_i \in C$, and so $T_{M_A}(\mathbb{Z}_2^k) \subseteq C$. Since the k columns of M_A are linearly independent, T_{M_A} is injective and so $|T_{M_A}(\mathbb{Z}_2^k)| = 2^k = |C|$. Consequently, $C = T_{M_A}(\mathbb{Z}_2^k)$, as required. $\qquad\square$

From a theoretical standpoint, it is unnecessary to think of the words of an (n, k) code as consisting of a message word of length k to which have been adjoined $n - k$ check digits. One merely considers C as a subgroup of size 2^k in $\mathbb{Z}_2^n$. In practice, however, it is useful to think of the set of all words of length k as the vocabulary in terms of which the messages are written and the generator matrix as the device which adjoins the check digits to produce the code words. The generator matrix affords an efficient way of producing the code without having to store the code words. We merely store the matrix in some electronic device and, to transmit a message, we input the length k data words to be encoded by the device.

4   At the Receiving End

We have just seen that the generator matrix is a useful tool at the transmitting end. At the receiving end, it turns out that we may use another matrix, called a parity-check matrix for the code, to determine whether or not the received word is a code word.

17.4.1 Definition. *Let n and k be positive integers with $k \leq n$, and let $A \in M_{(n-k) \times k}(\mathbb{Z}_2)$. The parity-check matrix P_A associated with M_A is the $(n-k) \times n$ matrix $P_A = [\, A \; I_{n-k} \,]$.*

For example, if $A = \left[\begin{smallmatrix} 1 & 1 & 0 \\ 1 & 0 & 1 \end{smallmatrix}\right]$, then $P_A = \left[\begin{smallmatrix} 1 & 1 & 0 & 1 & 0 \\ 1 & 0 & 1 & 0 & 1 \end{smallmatrix}\right]$. Note that $P_A M_A = 0$.

17.4.2 Proposition. *Let n and k be positive integers with $k \leq n$, and let $A \in M_{(n-k) \times k}(\mathbb{Z}_2)$. Then $T_{M_A} : \mathbb{Z}_2^k \to \mathbb{Z}_2^n$ is injective, $T_{P_A} : \mathbb{Z}_2^n \to \mathbb{Z}_2^{n-k}$ is surjective, and $\ker(T_{P_A}) = T_{M_A}(\mathbb{Z}_2^k)$.*

Proof. Let $u \in \ker(T_{P_A})$. Then $u = v|w$ for some $v \in \mathbb{Z}_2^k$ and some $w \in \mathbb{Z}_2^{n-k}$. We therefore obtain $0 = T_{P_A}(u) = T_A(v) + w$, which means that $w = T_A(v)$. But then $T_{M_A}(v) = v|T_A(v) = v|w = u$, and so $u \in T_{M_A}(\mathbb{Z}_2^k)$. Thus $\ker(T_{P_A}) \subseteq T_{M_A}(\mathbb{Z}_2^k)$. By construction, $rank(M_A) = k$ and so T_{M_A} is injective, and $rank(P_A) = n - k$, so T_{M_P} is surjective. Thus $\dim(\ker(T_{P_A})) = k = \dim(T_{M_A}(\mathbb{Z}_2^k))$. Since $\ker(T_{P_A}) \subseteq T_{M_A}(\mathbb{Z}_2^k)$, we conclude that $\ker(T_{P_A}) = T_{M_A}(\mathbb{Z}_2^k)$. $\qquad\square$

Thus if we choose $A \in M_{(n-k) \times k}(\mathbb{Z}_2)$ and set $M = \left[\begin{smallmatrix} I_k \\ A \end{smallmatrix}\right]$ and $P = [\, A \; I_{n-k} \,]$, then Proposition 17.4.2 tells us that the (n, k) group code $C = T_{M_A}(\mathbb{Z}_2^k)$ can also be viewed as $\ker(T_{P_A})$, and we will see shortly how to use this to our advantage.

5   Nearest Neighbor Decoding for Group Codes

So far, the discussion has centred on the detection of errors. We now show how the notion of the cosets of a subgroup can be employed to help with the correction of received words.

An (n, k) group code C is a subgroup of $\mathbb{Z}_2^n$ of order 2^k and as such, the set of all (left) cosets of C is a partition of $\mathbb{Z}_2^n$. The number of cosets of C in $\mathbb{Z}_2^n$ is equal to $[\mathbb{Z}_2^n : C] = |\mathbb{Z}_2^n|/|C| = 2^{n-k}$. Suppose now that $u \in C$ is transmitted, and let v denote the received word. Then $\epsilon = u + v$ is the so-called error word for the transmission, since adding ϵ to v will provide

u. But note that $\epsilon + u \in \epsilon + C$, so $v \in \epsilon + C$, or equivalently, $v + C = \epsilon + C$. We have just proven that the error word that we seek is in the same coset of C as the received word v. We therefore obtain the following description of the nearest neighbor decoding scheme for a group code C:

(i) Compute the coset of C that contains v (this is $v + C$).

(ii) Search in $v + C$ to determine the word of least weight and call this ϵ. Then return $v + \epsilon$ as the decoded result.

Note that it is possible that some coset will have more than one word of least weight. In other words, not every error can be correctly decoded by the nearest neighbor scheme. In such a case, the algorithm must refuse to "decode" the received word, although it can, and should, return the fact that an error has occurred. The word of minimum weight in a coset, if unique, is called the coset leader for that coset.

In order to implement the nearest neighbor decoding scheme, we may precompute each coset and store the results. Then when a word is received, we could search each coset, looking for a match. When we identify the coset that contains the received word, we then retrieve the coset leader (the word of least weight) for that coset and add it to the received word to complete the decoding process. However, this is where we can utilize the fact that C is also equal to $\ker(T_{P_A})$. All of the elements of a given coset are mapped to the same element of $\mathbb{Z}_2^{n-k}$, and different cosets map to different elements of $\mathbb{Z}_2^{n-k}$. Thus we could construct an array of size 2^{n-k}, and in position i of the array, we store the coset leader for the coset that T_{P_A} maps to i (or if a coset leader did not exist for the coset, then we store an error flag). With this data structure, which is called a syndrome-coset leader table, our nearest neighbor decoding scheme is implemented as follows: apply T_{P_A} to v, the received word, to obtain an element $i \in \mathbb{Z}_2^{n-k}$. Look up the coset leader (error word) ϵ that is stored in position i of the syndrome-coset leader table, and add ϵ to v. The result is the code word that we declare to have been sent. The value $T_{P_A}(v)$ is called the syndrome of the error, hence the name syndrome-coset leader table. Let us work through an example of this process.

17.5.1 Example. *Consider the* $(4, 2)$ *group code with generator matrix* $M = \begin{bmatrix} 1 & 0 \\ 0 & 1 \\ 1 & 1 \\ 0 & 1 \end{bmatrix}$ *and parity check matrix* $P = \begin{bmatrix} 1 & 1 & 1 & 0 \\ 0 & 1 & 0 & 1 \end{bmatrix}$. *We find that*

$$C = \{\, 0000, 1010, 0111, 1101 \,\}$$

Table 17.1   A syndrome-coset leader table

Syndrome	Coset Leader
00	0000
01	0001
10	ERROR
11	0100

and the cosets of C in $\mathbb{Z}_2^4$, other than C itself, are:

$$1000 + C = \{\, 1000, 0010, 1111, 0101 \,\}$$
$$0100 + C = \{\, 0100, 1110, 0011, 1001 \,\}$$
$$0001 + C = \{\, 0001, 1011, 0110, 1100 \,\}.$$

The coset leaders, in order that the cosets are listed immediately above, are 0000, ERROR *(since the coset* $1000 + C$ *has two words of least weight),* 0100, *and* 0001.s *The syndromes of the four cosets are, respectively,* $T_{P_A}(0000) = 00$, $T_{P_A}(1000) = 10$, $T_{P_A}(0100) = 11$, *and* $T_{P_A}(0001) = 01$. *The resulting syndrome-coset leader table is shown in Table 17.1.*

For example, suppose that 1111 *is received. We compute* $T_{P_A}(1111) =$ 10, *look up* 10 *in the syndrome-coset leader table to find that the coset leader with that sydrome is Error. Thus we have determined that an error did occur, but we are unable to determine which code word was sent. It could have been either* $1000 + 1111 = 0111$, *or it could have been* $0010 + 1111 =$ 1101. *On the other hand, if* 0110 *is received, we calculate its syndrome,* $T_{P_A}(0110) = 01$, *look up* 01 *in the syndrome-coset leader table, and find that the coset leader for that syndrome is* 0001. *We decode the received word* 0110 *as* $0110 + 0001 = 0111$.

We remark that of course, it is possible that for example, code word 1101 is sent, but received as 1010 (three characters were changed during transmission). In this case, we fail entirely to detect that anything has gone wrong, and we happily accept 1010 as the code word that was transmitted. The point is that it is presumably the case that it is extremely unlikely that as many as three characters will be in error (otherwise we would have designed a more robust code for this application). In our present case, $\mu(C) = 2$, and so C is not capable of correcting every single error, although it will detect every single error.

We also point out that the errors that will be correctly decoded are precisely the coset leaders. For if $u \in C$ is transmitted, and error word ϵ occurs, so that $u + \epsilon$ is the received word, then the coset leader of the coset

$u + \epsilon + C = \epsilon + C$ is our choice of error according to the nearest neighbor decoding scheme, and so if ϵ is not the unique word of least weight in $\epsilon + C$, then the coset leader is $\delta \neq \epsilon$, and we decode $u + \epsilon$ as $u + \epsilon + \delta \neq u$ since $\epsilon + \delta = \mathbb{O}$ if, and only if, $\epsilon = \delta$.

17.6   Hamming Codes

Up to this point, we have been working with predetermined n and k, and considering various (n, k) group codes. In this section, we consider the difference $r = n - k$ to be given and we investigate the properties of (m, l) group codes for which m and l are positive integers with $l \leq m$ and $m - l = r$. The idea is of course to maximize the data payload for the fixed cost of transmitting r check bits.

We should also realize that the properties of an (n, k) group code constructed with generating matrix $M = \left[\begin{smallmatrix} I_k \\ A \end{smallmatrix} \right]$ depend entirely on our choice of the $(n - k) \times k$ matrix A. What do we have to know about A in order to conclude that the resulting code C has predetermined error-correction capabilities? In general, that is quite difficult to say, but if we decide that we want to construct an (n, k) group code that is single-error-correcting, the requirements on A are quite easily determined. We require that $\mu(C) \geq 3$, so let us now proceed to determine what requirements A must meet in order to ensure that $\mu(C) \geq 3$. It is easier to conduct this analysis if we exploit Proposition 17.4.2 and think of C as $\ker(T_{P_A})$, where P is the parity-check matrix $[\, A \; I_{n-k} \,]$.

In summary, give a positive integer r (which is to be the number of check bits we append to a data word to construct a code word), we consider a positive integer k and an $r \times k$ matrix A and ask what we need to know about A in order to ensure that the code contructed with generator matrix $M = \left[\begin{smallmatrix} I_k \\ A \end{smallmatrix} \right]$ is a single error-correcting code. Our goal is to find the maximum value of k for which it is possible to construct an $r \times k$ matrix A for which the resulting $(r + k, k)$ group code is single error-correcting.

17.6.1 Proposition. *Let r and k be positive integers, and let $A \in M_{r \times k}(\mathbb{Z}_2)$. Then $C = T_{M_A}(\mathbb{Z}_2^k)$ is a single error-correcting code if, and only if, each column of A has weight at least 2, and no two columns of A are equal.*

Proof. By Proposition 17.1.8, C is single error-correcting if, and only if, $\mu(C) \geq 3$, and by Proposition 17.1.9, $\mu(C)$ is the weight of the lightest nonzero word in C. Furthermore, by Proposition 17.4.2, $C = \ker(T_{P_A})$. $\ker(T_{P_A})$ will contain no word of weight 1 if, and only if, A has no zero column, and C will contain no word of weight 2 if, and only if, A has no two identical columns, nor a column of weight 1, since $P_A = [A\, I_r]$ has each of the r standard basis vectors for $\mathbb{Z}_p^r$ as columns. $\qquad\square$

Since $|\mathbb{Z}_2^r - \{\mathbb{0}, e_1, e_2, \dots, e_r\}| = 2^r - 1 - r$, it follows that we must have $k \leq 2^r - 1 - r$ if $A \in M_{r \times k}(\mathbb{Z}_2)$ meets the requirements of Proposition 17.6.1. Since we were looking to maximize k, we have achieved our goal: the maximum value is $k = 2^r - 1 - r$.

17.6.2 Definition. *For a positive integer r, a $(2^r - 1, 2^r - 1 - r)$ group code is called a Hamming (n, k) code, where $n = 2^r - 1$ and $k = 2^r - 1 - r$.*

17.6.3 Proposition. *For any positive integer $r \geq 2$, there exists a Hamming $(2^r - 1, 2^r - r - 1)$ group code with $\mu(C) = 3$.*

Proof. Let $n = 2^r - 1$ and $k = n - r$. Choose A to be an $r \times k$ matrix whose $k = 2^r - r - 1$ columns are the elements of $\mathbb{Z}_2^r$ which have weight at least 2, as there are exactly $2^r - 1 - r$ such columns. Then by Proposition 17.6.1, C is single error-correcting and thus by Proposition 17.1.8, $\mu(C) \geq 3$. Since A has at least one column of weight 2, it follows that M_A has at least one column of weight 3 and so $\mu(C) \leq 3$. Thus $\mu(C) = 3$. $\qquad\square$

For $r = 2$, the $(3, 1)$ Hamming code is the triple repetition code, with generator matrix $\begin{bmatrix} 1 \\ 1 \\ 1 \end{bmatrix}$. The code is $C = \{000, 111\}$, and there are $2^r = 4$ cosets in total, of which C is one. The others are $100 + C = \{100, 011\}$, $010 + C = \{010, 101\}$, and $001 + C = \{001, 110\}$. In general, a $(2^r - 1, 2^r - r - 1)$ Hamming code has size $2^{2^r - r - 1}$, and $2^{2^r - 1}/2^{2^r - r - 1} = 2^r$ cosets. Moreover, there are $2^r - 1$ words of weight 1, no two in the same coset, and of course, none in the code. Thus the code, plus the $2^r - 1$ cosets of the words of weight 1 account for all 2^r cosets of the Hamming code. As some cosets will contain words of weight 2, it follows that if we have decided to correct all errors of weight 1, we will not be able to detect errors of weight 2. It is often desirable to be able to correct every occurrence of a single error, and as well, be able to detect every occurrence of two errors. A very common way to accomplish this is to extend the Hamming code by an additional check bit, this time a simple parity check bit.

Suppose that $M_A = \begin{bmatrix} I_k \\ A \end{bmatrix}$ is the generator matrix for a (n, k) Hamming code, where $n = 2^r - 1$ and $k = 2^r - r - 1$. Let $\mathbb{1}$ denote the $1 \times k$ matrix with every entry equal to 1, and set $A' = \begin{bmatrix} A \\ 1 \end{bmatrix} \in M_{(n-k+1)\times k}(\mathbb{Z}_2)$. Then $M_{A'} = \begin{bmatrix} I_k \\ A' \end{bmatrix} \in M_{(n+1)\times k}(\mathbb{Z}_2)$, and for $P_{A'} = \begin{bmatrix} A' & I_{(n+1)-k} \end{bmatrix} \in M_{(n+1-k)\times(n+1)}(\mathbb{Z}_2)$, we have $P_{A'}M_{A'} = A'I_k + I_{n+1-k}A' = \mathbb{0}$. Therefore we have $C' = T_{M_{A'}}(\mathbb{Z}_2^k) \subseteq \ker(T_{P_{A'}})$, and by construction of $M_{A'}$, $T_{M_{A'}}$ is injective, so $|C| = 2^k$. As well, $T_{P_{A'}}(\mathbb{Z}_2^{n+1})$ contains all $n+1-k$ weight 1 vectors, which form a basis for $\mathbb{Z}_2^{n+1-k}$, and so $T_{P_{A'}}$ is surjective. It follows that $|\ker(T_{P_{A'}})| = 2^{n+1}/2^{n+1-k} = 2^k = |C'|$, and so $C' = \ker(T_{P_{A'}})$. Since $\mu(C) = 3$, and we have added a parity bit to each word of C to form the elements of C', it follows that C' contains words of weight 4, and every nonzero word of C' has weight at least 4, so $\mu(C') = 4$. This means that no coset of C' will contain both a word of weight 1 and a word of weight 2, and thus we may correct every single error, and as well, detect every occurrence of two errors.

We shall work out the process for the triple repetition code (the Hamming $(3,1)$ code), which when extended becomes a $(4,2)$ group code. We have $A = \begin{bmatrix} 1 \end{bmatrix}$, so $M_A = \begin{bmatrix} 1 \\ 1 \\ 1 \end{bmatrix}$ and $P_A = \begin{bmatrix} 1 & 1 & 0 \\ 1 & 0 & 1 \end{bmatrix}$. The code is $C = \{\,000, 111\,\}$, and there are 4 cosets, which are, in addition to C,

$$100 + C = \{\,100, 011\,\}$$
$$010 + C = \{\,010, 101\,\}$$
$$001 + C = \{\,001, 110\,\}.$$

We compute $M_{A'} = \begin{bmatrix} 1 \\ 1 \\ 1 \\ 1 \end{bmatrix}$, and $P_{A'} = \begin{bmatrix} A' & I_3 \end{bmatrix} = \begin{bmatrix} 1 & 1 & 0 & 0 \\ 1 & 0 & 1 & 0 \\ 1 & 0 & 0 & 1 \end{bmatrix}$. We have $C' = \{\,0000, 1111\,\}$, and there are now 8 cosets, C' and seven others which we compute below.

$$1000 + C' = \{\,1000, 0111\,\}$$
$$0100 + C' = \{\,0100, 1011\,\}$$
$$0010 + C' = \{\,0010, 1101\,\}$$
$$0001 + C' = \{\,0001, 1110\,\}$$
$$1100 + C' = \{\,1100, 0011\,\}$$
$$0110 + C' = \{\,0110, 1001\,\}$$
$$1010 + C' = \{\,1010, 0101\,\}.$$

Our syndrome-coset leader table for decoding the extended Hamming $(4,2)$ code is shown in Table 17.2.

In computer memory, where bit errors are quite rare, Hamming codes are widely used. In particular, for ECC (error correcting code) memory,

Table 17.2 A syndrome-coset leader table for
the extended Hamming $(4, 2)$ code

Syndrome	Coset Leader
000	0000
100	0100
010	0010
001	0001
110	ERROR
101	ERROR
011	ERROR
111	1000

an extended Hamming code is quite popular. Since the Hamming code for $r = 6$ is a $(63, 57)$ code, it is common to take $r = 7$ to obtain the $(127, 120)$ Hamming code, but only use enough linearly independent columns in A to encode 64 bit data words with 7 check bits; that is, a $(71, 64)$ group code. With the addition of the parity check bit as described above to construct a single error-correcting, double error detecting code, we obtain a $(72, 64)$ group code.

7 Exercises

1. Suppose a word of length n is transmitted across a noisy channel, and let p denote the probability that an individual bit will be in error upon receipt. For any nonnegative integer i, let P_i denote the probability that exactly i bits of the n are in error. Prove that for each nonnegative integer i, $P_{i+1} < P_i$ if, and only if, $p < \frac{i+1}{n+1}$.

2. For positive integers n and d with $d \leq n$, let $A(n, d)$ denote the number of words in a largest possible code of length n with minimum Hamming distance d. There is no known formula for $A(n, d)$ in general, but some of its values can be computed. In each case below, determine the value of $A(n, d)$ for any positive integer n, and d as given.

   a) $A(n, n)$.

   b) $A(n, 1)$.

   c) $A(n, 2)$, where $n \geq 2$.

3. Let n be a positive integer, and let $\alpha \in \mathbb{Z}_2^n$. Prove that for any code $C \subseteq \mathbb{Z}_2^n$, $\mu(C) = \mu(\alpha + C)$.

4. a) Either construct, or else prove that it is impossible to construct, a

(5,3) code which will detect up to two errors.

b)   Either construct, or else prove that it is impossible to construct, a (6,3) code which will detect up to two errors.

5. Let n be a positive integer. For any $\alpha \in \mathbb{Z}_2^n$ and any nonnnegative real number r, define $S(\alpha;r)$, the sphere with centre α and radius r, by $S(\alpha;r) = \{\gamma \in \mathbb{Z}_2^n \mid h(\alpha,\gamma) \leq r\}$. Prove that for $\alpha, \beta \in \mathbb{Z}_2^n$ and m, k nonnegative integers, $S(\alpha;m) \cap S(\beta;k) \neq \varnothing$ if, and only if, $h(\alpha,\beta) \leq m + k$.

6. Let n be a positive integer. Prove that for any positive integer m, and any $\beta_0, \beta_1, \ldots, \beta_n \in \mathbb{Z}_2^m \in \mathbb{Z}_2^n$, $\sum_{i=1}^{m} h(\beta_{i-1}, \beta_i) \geq h(\beta_0, \beta_m)$.

7. Let n be a positive integer. For any $\alpha, \beta, \gamma \in \mathbb{Z}_2^n$, we say that β lies between α and γ if $h(\alpha,\beta) + h(\beta,\gamma) = h(\alpha,\gamma)$.

a)   Prove that β lies between α and γ if, and only if, $S(\alpha) \cap S(\gamma) \subseteq S(\beta) \subseteq S(\alpha) \cup S(\gamma)$.

b)   Prove or disprove the following statement: if β lies between α and γ, and δ lies between α and β, then δ lies between α and γ.

8. Let n be a positive integer and r a nonnegative real number. Prove that for any $\alpha, \beta \in \mathbb{Z}_2^n$, $S(\alpha;r) + \beta = S(\alpha + \beta;r)$ (see Exercise 5 for the definition of $S(\alpha;r)$).

9. Given positive integers m and n, and $\alpha \in \mathbb{Z}_2^n$, find a formula for $|S(\alpha;m)|$. Use this formula to calculate $|S(0110101;4)|$.

10. For any positive integer n, the subgroup of $\mathbb{Z}_2^n$ that consists of all words of even weight is called a parity-check code. Find the generator and parity check matrices for these codes.

11. For any positive integer n, the n-repetition code consists of the two words of length n in which all entries are equal; either 0 or 1. Find the generator and parity-check matrices for an n-repetition code.

12. Prove that in a group code, either all of the words have even weight, or else half have even weight and half have odd weight.

13. Let n be a positive integer, and let C be a group code of length n. Let α be the syndrome of a word $\beta \in \mathbb{Z}_2^n$. Prove that if $w(\beta) = m$, then there are m columns of P whose sum is α.

14. Let $A = \begin{bmatrix} 1 & 0 & 0 & 1 & 1 \\ 1 & 1 & 1 & 0 & 1 \\ 0 & 1 & 0 & 0 & 0 \\ 1 & 1 & 1 & 1 & 0 \end{bmatrix}$.

a)   If $C = T_{M_A}(\mathbb{Z}_2^5)$, what is $|C|$?

b)   Construct a syndrome-coset leader table for C.

c) Use your table to decode each of the three words: 110101011, 00101001, and 110011001.

d) Find $\mu(C)$.

15. Suppose that P is the parity-check matrix for a group code C, and no column of P is zero, and no column of P is the sum of fewer than d other columns of P.

a) Prove that $\mu(C) \geq d+1$.

b) Is the converse true?

16. Let n be a positive integer, and let $\alpha, \beta, \gamma \in \mathbb{Z}_2^n$.

a) Prove that if $w(\alpha)$ is odd and $\alpha = \beta + \gamma$, then exactly one of β or γ is of odd weight.

b) Use Exercise 15 and the first part of this exercise to prove that the group code C whose parity-check matrix is $P = \begin{bmatrix} 1 & 1 & 1 & 0 & 1 & 0 & 0 & 0 \\ 1 & 1 & 0 & 1 & 0 & 1 & 0 & 0 \\ 1 & 0 & 1 & 1 & 0 & 0 & 1 & 0 \\ 0 & 1 & 1 & 1 & 0 & 0 & 0 & 1 \end{bmatrix}$ has $\mu(C) = 4$.

c) How many cosets of C are there in $\mathbb{Z}_2^8$?

d) Construct a syndrome-coset leader table, and use it to decode each of the following words: 1110100, 01110000, and 11001100.

17. Let n be a positive integer. For any words $\alpha, \beta \in \mathbb{Z}_2^n$, say $\alpha = a_1 a_2 \ldots a_n$ and $\beta = b_1 b_2 \ldots b_n$, define the dot product of α and β to be $\alpha \cdot \beta = \sum_{i=1}^n a_i b_i$. For $C \subseteq \mathbb{Z}_2^n$, define the dual $C^\perp$ of C by $C^\perp = \{\alpha \in \mathbb{Z}_2^n \mid \alpha \cdot \beta = 0 \text{ for all } \beta \in C\}$.

a) Prove that if C is a group code, then $C^\perp$ is a group code.

b) Prove that if M is the generator matrix for C, then $C^\perp = \ker(T_{M^t})$, where M^t denotes the transpose of M. Conclude that if $|C| = 2^k$, then $|C^\perp| = 2^{n-k}$.

18. Let m be a positive integer. For any $i, j \in J_m$ with $i \neq j$, the elementary row operations $R_{i \leftrightarrow j}$ and $R_i \to R_i + R_j$ on any matrix with m rows, with entries from $\mathbb{Z}_2$ are defined in the usual way: for any positive integer n, if $M \in M_{m \times n}(\mathbb{Z}_2)$, then $R_{i \leftrightarrow j}$ applied to M simply exchanges rows i and j of M, while $R_i \to R_i + R_j$ applied to M replaces row i of M by the sum of rows i and j of M. Let E_{ij} denote the result of applying $R_{i \leftrightarrow j}$ to I_m, and let $E_i(j)$ denote the result of applying $R_i \to R_i + R_j$ to I_m. We call E_{ij} and $E_i(j)$ elementary matrices of types I and II, respectively.

a) Prove that for any positive integer n, and any $M \in M_{m \times n}(\mathbb{Z}_2)$, if L is the matrix obtained by applying $R_{i \leftrightarrow j}$ to M, then $L = E_{ij} M$, while if L is the matrix obtained by applying $R_i \to R_i + R_j$ to M, then $L = E_i(j)M$. In either case, we write $M \to L$.

b) Prove that for any $i, j \in J_m$ with $i \neq j$, $E_{ij}, E_i(j) \in Gl_m(\mathbb{Z}_2)$, and each has order 2. In particular, this means that for $M, L \in M_{m \times n}(\mathbb{Z}_2)$ with $L = EM$ for some elementary matrix E, then $EL = M$.

For any positive integer n, and M and L in $M_{m \times n}(\mathbb{Z}_2)$, we say that M is row equivalent to L, written $M \sim L$, if there exist finitely many matrices $M_0 = M, M_1, M_2, \ldots, M_t = L$ with $M_i \to M_{i+1}$ for each $i = 0, 1, \ldots, t-1$.

c) Prove that row equivalence is an equivalence relation on $M_{m \times n}(\mathbb{Z}_2)$.

d) Prove that if $M, L \in M_{m \times n}(\mathbb{Z}_2)$ and $M \sim L$, then $\ker(T_M) = \ker(T_L)$.

e) Let $\sigma \in S_n$. For $P \in M_{m \times n}(\mathbb{Z}_2)$, let $P' \in M_{m \times n}(\mathbb{Z}_2)$ denote the matrix whose i^{th} column is the $\sigma(i)^{th}$ column of P, $i \in J_n$. Let $C = \ker(T_P)$ and $C' = \ker(T_{P'})$. Prove that $C' = \Pi_\sigma(C)$ (see Definition 17.2.2).

f) Let $P = \begin{bmatrix} 0 & 1 & 0 & 1 & 0 & 1 & 0 & 1 & 1 & 0 \\ 1 & 1 & 0 & 0 & 1 & 1 & 0 & 0 & 1 & 1 \\ 1 & 0 & 1 & 0 & 0 & 0 & 1 & 0 & 0 & 0 \\ 0 & 0 & 1 & 1 & 1 & 0 & 1 & 1 & 1 & 1 \end{bmatrix}$ and let $C = \ker(T_P)$. Determine $|C|$ and $\mu(C)$. Hint: Find a matrix L that is row-equivalent to P and which has e_1, e_2, e_3, and e_4 among its columns. Then determine a permutation σ of J_n such that when the columns of P are rearranged according to σ, the resulting matrix $P' = [\, I_4 \; A \,]$ for some 4×6 matrix A. Now use the preceding parts of this exercise.

19. Let n and k be positive integers with $k < n$, and let $A \in M_{(n-k) \times k}(\mathbb{Z}_2)$. Suppose that $C = \ker(T_{P_A})$ is such that $\mu(C)$ is odd. Prove that if $Q = \begin{bmatrix} 1 & 1 & \cdots & 1 \\ 0 & & & \\ \vdots & & P_A & \\ 0 & & & \end{bmatrix}$ and $C' = \ker(T_Q)$, then $\mu(C') = \mu(C) + 1$.

Chapter 18

Polynomial Codes

1 Definitions and Elementary Results

In Chapter 9, it was shown that for any field F, the ring $F[x]$ of polynomials with coefficients in F is a principal ideal domain. Moreover, it was proven there (Proposition 9.3.11) that every $p \in F[x]$ can be uniquely written as a linear combination of (finitely many) elements of $B = \{\, x^i \mid i \in \mathbb{N} \,\}$, which tells us that $F[x]$ is an F-vector space with countably infinite basis B. We are now interested in the particular case when $F = \mathbb{Z}_2$.

18.1.1 Definition. *For any positive integer n, let*

$$(\mathbb{Z}_2[x]; n) = \{\, p \in \mathbb{Z}_2[x] \mid \partial(p) < n \,\} = \{\, \sum_{i=0}^{n-1} a_i x^i \mid a_0, a_1, \ldots, a_{n-1} \in \mathbb{Z}_2 \,\}.$$

Observe that $(\mathbb{Z}_2[x]; n)$ is a vector subspace of dimension n in $\mathbb{Z}_2[x]$, with $\mathbb{Z}_2$ basis $\{\, 1, x, x^2, \ldots, x^{n-1} \,\}$. We therefore obtain a natural linear isomorphism between the $\mathbb{Z}_2$-vector space $\mathbb{Z}_2^n$ of dimension n and $(\mathbb{Z}_2[x]; n)$ which we shall denote by π_n. For $v = a_0 a_1 \cdots a_{n-1} \in \mathbb{Z}_2^n$, define $\pi_n(v) \in \mathbb{Z}_2[x]$ by $\pi_n(v) = \sum_{i=0}^{n-1} a_i x^i$. For example, if $n = 4$ and $v = 1011$, we have $\pi_4(v) = 1 + x^2 + x^3$.

18.1.2 Proposition. *For each positive integer n, the map $\pi_n : \mathbb{Z}_2^n \to \mathbb{Z}_2[x]$ is an injective additive group homomorphism (equivalently, $\mathbb{Z}_2$-linear map) whose image is $(\mathbb{Z}_2[x]; n)$.*

Proof. It is evident that π_n is a $\mathbb{Z}_2$-linear map from $\mathbb{Z}_2^n$ into $\mathbb{Z}_2[x]$ with image the set of all polynomials of degree less than n, which is a subspace of dimension n with basis $\{\, 1, x, x^2, \ldots, x^{n-1} \,\}$. Since $\{\, 1, x, x^2, \ldots, x^{n-1} \,\}$ is a $\mathbb{Z}_2$ basis for $(\mathbb{Z}_2[x]; n)$, π_n is injective. $\qquad \square$

Thus π_n is an isomorphism from $\mathbb{Z}_2^n$ onto $(\mathbb{Z}_2[x]; n)$. Among other things, this implies that if C is a subgroup of $\mathbb{Z}_2^n$, then $\pi_n(C)$ is a subgroup of $(\mathbb{Z}_2[x]; n)$ isomorphic to C.

18.1.3 Definition. *For $p \in \mathbb{Z}_2[x]$, let $M_p : \mathbb{Z}_2[x] \to \mathbb{Z}_2[x]$ be the map given by $M_p(f) = pf$ for each $f \in \mathbb{Z}_2[x]$.*

18.1.4 Proposition. *If $p \in \mathbb{Z}_2[x]$ is nonzero, then $M_p : \mathbb{Z}_2[x] \to \mathbb{Z}_2[x]$ is an injective $\mathbb{Z}_2$-linear map.*

Proof. Suppose that $p \in \mathbb{Z}_2[x]$ is nonzero. Since multiplication in $\mathbb{Z}_2[x]$ distributes across addition, M_p is a $\mathbb{Z}_2$-linear map. Since $\mathbb{Z}_2$ is a field, $\mathbb{Z}_2[x]$ is an integral domain and thus $pf = 0$ implies that $f = 0$. Thus M_p is injective.   □

18.1.5 Corollary. *For any positive integer j, and any nonzero polynomial p, $M_p((\mathbb{Z}_2[x]; j))$ is a subgroup of $(\mathbb{Z}_2[x]; \partial(p) + j)$ of order 2^j.*

18.1.6 Definition. *For positive integers n and k with $k < n$, and $p \in \mathbb{Z}_2[x]$ of degree $n - k$, $P_{n,k,p}$, the polynomial code generated by p, is defined by $P_{n,k,p} = \pi_n^{-1} \circ M_p \circ \pi_k(\mathbb{Z}_2^k) \subseteq \mathbb{Z}_2^n$. Moreover, p is said to be the generator polynomial for the code.*

By Corollary 18.1.5, $|P_{n,k,p}| = 2^k$.

18.1.7 Example. *We construct the code $P_{5,3,p}$ with generator polynomial $p = 1 + x + x^2$. Since*

$$\pi_3(\mathbb{Z}_2^3) = (\mathbb{Z}_2[x]; 3) = \{\, 0, 1, x, 1 + x, x^2, 1 + x^2, 1 + x + x^2, x + x^2 \,\},$$

we obtain

$$\begin{aligned}
P_{5,3,p} &= \pi_5^{-1}\{\, 0, p, xp, (1 + x)p, x^2 p, (1 + x^2)p, (1 + x + x^2)p, (x + x^2)p \,\} \\
&= \pi_5^{-1}\{\, 0, 1 + x + x^2, x + x^2 + x^3, 1 + x^3, \\
&\qquad\qquad x^2 + x^3 + x^4, 1 + x + x^3 + x^4, 1 + x^2 + x^4, x + x^4 \,\} \\
&= \{\, 00000, 11100, 01110, 10010, 00111, 11011, 10101, 01001 \,\}
\end{aligned}$$

Note that $\mu(P_{5,3,p}) = 2$, so $P_{5,3,p}$ is not able to correct even a single error, but can detect every single error. Thus $P_{5,3,p}$ is a $(5,3)$ single error-detecting group code.

We observe that if p has zero constant term, then every word of $P_{n,k,p}$ will have zero first entry. Since this is clearly wasteful, we shall only consider generating polynomials with nonzero constant term.

18.1.8 Proposition. *Let n and k be positive integers with $k < n$, and let $p \in \mathbb{Z}_2[x]$ have degree $n - k$ and nonzero constant term. If p does not divide $x^m + 1$ for each integer m with $1 \leq m < n$, then $\mu(P_{n,k,p}) \geq 3$.*

Proof. It suffices to prove that $\pi_n(P_{n,k,p})$ does not contain any polynomial which is either a monomial or a sum of two monomials. Suppose by way of contradiction that $\pi_n(P_{n,k,p})$ contains a monomial x^i. But then p is a factor of x^i, and thus $p = x^{n-k}$, contradicting our choice of p as having nonzero constant term. Suppose now that $\pi_n(P_{n,k,p})$ contains a sum of two monomials, say $x^i + x^j$ with $0 \leq i < j < n$. Then p divides $x^i(1 + x^{j-i})$, and since p has nonzero constant term, $gcd(p, x^i) = 1$ and thus p divides $1 + x^{j-i}$. But $1 \leq j - i < n$, so this contradicts the hypothesis of the proposition. Therefore $\pi_n(P_{n,k,p})$ does not contain any monomial or a sum of two monomials, and so $\mu(P_{n,k,p}) \geq 3$. $\qquad\square$

Certain polynomial codes have the property that if the entries in a code word are shifted one character to the right, while the rightmost character moves into the first position, the result is another code word. Such codes are important as they facilitate simplified encoding and decoding procedures. For example, every Hamming code has this property.

18.1.9 Definition. *For any positive integer n, a code $C \subseteq \mathbb{Z}$ is cyclic if for each $v = a_0 a_1 \cdots a_{n-1} \in C$, $a_{n-1} a_0 a_1 \cdot a_{n-2} \in C$.*

18.1.10 Proposition. *Let n and k be positive integers with $k < n$, and let $p \in \mathbb{Z}_2[x]$ have degree $n - k$ and nonzero constant term. Then $P_{n,k,p}$ is cyclic if, and only if, p divides $x^n - 1$.*

Proof. First, suppose that $P_{n,k,p}$ is cyclic. Since $\partial(x^k p) = n$, the cyclic shift of $x^k p$ is $x^k p - x^n + 1$ since the coefficient of the leading term of $x^k p$ is 1. As $P_{n,k,p}$ is cyclic and $x^k p \in P_{n,k,p}$, it follows that $x^k p - x^n + 1 \in P_{n,k,p}$. But then p divides $x^k p - (x^n - 1)$, and thus p divides $x^n - 1$.

Conversely, suppose that p divides $x^n - 1$, and let $c = a_0 + a_1 x + \cdots + a_{n-1} x^{n-1} \in P_{n,k,p}$; that is, p divides c. We want to prove that p divides $xc - a_{n-1} x^n + a_{n-1} = xc - a_{n-1}(x^n - 1)$. But this is immediate since p divides c and p divides $x^n - 1$. $\qquad\square$

In Chapter 10, it was shown that if $p \in \mathbb{Z}_2[x]$ is irreducible over $\mathbb{Z}_2$ of degree $r > 0$, then the quotient ring $\mathbb{Z}_2[x]/(p)$ is a field of order 2^r. Moreover, the multiplicative group of units of a finite field is cyclic.

18.1.11 Definition. *If $p \in \mathbb{Z}_2[x]$ is irreducible over $\mathbb{Z}_2$ and $x + (p)$ generates the cyclic group of units of $\mathbb{Z}_2[x]/(p)$, then p is called a primitive polynomial.*

The Conway polynomials discussed in Chapter 10 are particularly nice examples of primitive polynomials. For example, $1 + x + x^2$, $1 + x + x^3$, and $1 + x + x^4$ are Conway polynomials.

18.1.12 Corollary. *Let r be a positive integer and set $n = 2^r - 1$ and $k = n - r = 2^r - r - 1$. If $p \in \mathbb{Z}_2[x]$ is primitive of degree r, then $P_{n,k,p}$ is a single error-correcting group code.*

Proof. Since $x^n \equiv 1 \mod (p)$, but $x^t \not\equiv 1 \mod (p)$ for every positive integer $t < n$, the result follows from Proposition 18.1.8. $\qquad\qquad\square$

In our earlier discussion about group codes, it was considered convenient not to have to perform further computations to extract the encoded data word from a code word, once error-correction has been completed. We arranged it so that our encoding scheme embedded the data word as the first k characters of a code word, and thus the data word is extracted from the code word by truncating the code word after k characters. We now establish that there is an encoding homomorphism from $(\mathbb{Z}_2[x]; k)$ to $(\mathbb{Z}_2[x]; n)$ that achieves an equivalent result for the polynomial codes. Due to the polynomial algebra involved, it is more convenient to embed the data word as the last k characters of a code word, rather than the first k characters.

18.1.13 Proposition. *Let n and k be positive integers with $m = n - k \geq 1$, and let $p \in \mathbb{Z}_2[x]$ have $\partial(p) = m$. Define the map $E : (\mathbb{Z}_2[x]; k) \to (\mathbb{Z}_2[x]; n)$ as follows: for $f \in (\mathbb{Z}_2[x]; k)$, let $q, r \in \mathbb{Z}_2[x]$ be the unique polynomials for which $x^m f = qp + r$ and $\partial(r) < \partial(p) = m$. Set $E(f) = r + x^m f \in (\mathbb{Z}_2[x]; n)$. Then E is an injective group homomorphism with image $(\mathbb{Z}[x]; n) \cap (p)$.*

Proof. We may recover f and r from $E(f)$ by an application of the division theorem to $E(f)$ and x^m, as f and r are the unique polynomials for which $E(f) = r + x^m f$ and $\partial(r) < m = \partial(x^m)$. Thus E is injective. To see that E is a homomorphism, suppose that $f_1, f_2 \in (\mathbb{Z}_2[x]; k)$, and that $x^m f_1 = q_1 p + r_1$ and $x^m f_2 = q_2 p + r_2$ where $\partial(r_1) < m$ and $\partial(r_2) < m$. Then $E(f_1) = r_1 + x^m f_1$ and $E(f_2) = r_2 + x^m f_2$. Moreover, $x^m (f_1 + f_2) = x^m f_1 +$

$x^m f_2 = (r_1 + r_2) + p(q_1 + q_2)$, with $\partial(r_1 + r_2) \leq \max\{\partial(r_1), \partial(r_2)\} < m$, so $E(f_1 + f_2) = (r_1 + r_2) + x^m(f_1 + f_2) = E(f_1) + E(f_2)$, as required.

Now, for each $f \in (\mathbb{Z}_2[x]; k)$, $E(f) = r + x^m f = qp$, so $E(f) \in (p)$ for every $f \in (\mathbb{Z}_2[x]; k)$. Thus the image of E is contained in $(\mathbb{Z}_2[x]; n) \cap (p)$. Let $g \in (\mathbb{Z}_2[x]; n) \cap (p)$, so $g = hp$ for some $h \in (\mathbb{Z}_2[x]; k)$. Apply the division theorem to g and x^m to obtain unique polynomials r and q such that $g = r + x^m q$ and $\partial(r) < m$. Note that $m + \partial(q) = \partial(g) < n$, so $\partial(q) < n - m = k$ and thus $q \in (\mathbb{Z}_2[x]; k)$. But then $E(q) = x^m q + r = g$ and so E is surjective; that is, the image of E is $(\mathbb{Z}[x]; n) \cap (p)$. $\square$

Now, observe that $\pi_n(P_{n,k,p}) \subseteq (\mathbb{Z}_2[x]; n) \cap (p) = E((\mathbb{Z}_2[x]; k))$. Since π_n and E are injective and $|P_{n,k,p}| = 2^k$, it follows that $|\pi_n(P_{n,k,p})| = 2^k = |E((\mathbb{Z}_2[x]; k))|$ and thus $\pi_n(P_{n,k,p}) = E((\mathbb{Z}_2[x]; k))$. We have established that $P_{n,k,p} = \pi_n^{-1} \circ E \circ \pi_k((\mathbb{Z}_2[x]; k))$, and so $\pi_n^{-1} \circ E \circ \pi_k$ is an injective homomorphism from $(\mathbb{Z}_2[x]; k)$ into $\mathbb{Z}_2^n$ with image $P_{n,k,p}$.

Let us work through this encoding for the word $v = 110 \in \mathbb{Z}_2^3$ in Example 18.1.7, where the generating polynomial was $p = 1 + x + x^2$. In this example, we have $n = 5$ and $k = 3$, and we are computing $\pi_5^{-1} \circ E \circ \pi_3(v)$. $\pi_3(v) = 1 + x$, and to evaluate $E(1+x)$, we compute $x^{5-3}(1+x) = x^2 + x^3$, then apply the division theorem to $x^2 + x^3$ and p, whereby we have $x^2 + x^3 = xp + x$, so $E(x^2 + x^3) = x + x^2(1 + x) = x + x^2 + x^3$ and $\pi_5^{-1}(x + x^2 + x^3) = 01110$. As expected, the data word 110 occupies the last three character positions of the code word 01110.

We now continue with our consideration of the general polynomial code $P_{n,k,p}$. As $\pi_n^{-1} \circ E \circ \pi_k$ is a homomorphism and thus a $\mathbb{Z}_2$-linear map from $\mathbb{Z}_2^k$ into $\mathbb{Z}_2^n$ with image $P_{n,k,p}$, and as every linear map from $\mathbb{Z}_2^k$ into $\mathbb{Z}_2^n$ is of the form T_M for some $M \in M_{n \times k}(\mathbb{Z}_2)$, we wish to determine M such that $T_M = \pi_n^{-1} \circ E \circ \pi_k$. M is the matrix representative of the encoding homomorphism relative to the standard ordered basis for both $\mathbb{Z}_2^k$ and $\mathbb{Z}_2^n$; that is, the columns of M are the images $\pi_n^{-1} \circ E \circ \pi_k(e_i)$, $i = 1, 2, \ldots, k$. As e_i itself will be the last k characters of $\pi_n^{-1} \circ E \circ \pi_k(e_i)$, M will have I_k as its lower $k \times k$ block, while the upper $(n-k) \times k$ block, which we denote by A, has as its columns $\pi_n^{-1}(r_i)$, where r_i is the remainder when $x^{n-k} x^{i-1}$ is divided by p, $i = 1, 2, \ldots, k$. Thus $M = \begin{bmatrix} A \\ I_i \end{bmatrix}$. The parity-check matrix is then $P = \begin{bmatrix} I_{n-k} & A \end{bmatrix}$.

18.1.14 Example. *We find the encoding matrix for the code presented in Example 18.1.7, where the generator polynomial was $p = 1 + x + x^2$ and $n = 5$, $k = 3$. We compute $\pi_3(100) = 1$, $\pi_3(010) = x$, and $\pi_3(001) = x^2$,* so we apply the division theorem to $x^2(1)$, $x^2(x)$, and $x^2(x^2)$, respectively, and p to obtain $x^2 = 1(1 + x + x^2) + (1 + x)$, $x^3 = (1 + x)(1 + x + x^2) + 1$, and $x^4 = (x + x^2)(1 + x + x^2) + x$. Thus $E(1) = 1 + x + x^2(1) = 1 + x + x^2$, $E(x) = 1 + x^2(x) = 1 + x^3$, and $E(x^2) = x + x^2(x^2) = x + x^4$, and so $\pi_5^{-1} \circ E \circ \pi_3$ maps $100, 010, 001$, respectively, to $11100, 10010$, and 01001, so

$$M = \begin{bmatrix} 1 & 1 & 0 \\ 1 & 0 & 1 \\ 1 & 0 & 0 \\ 0 & 1 & 1 \\ 0 & 0 & 1 \end{bmatrix}. \text{ Thus } A = \begin{bmatrix} 1 & 1 & 0 \\ 1 & 0 & 1 \end{bmatrix}, \text{ and so } P = \begin{bmatrix} 1 & 0 & 1 & 1 & 0 \\ 0 & 1 & 1 & 0 & 1 \end{bmatrix}.$$

18.2 BCH Codes

BCH codes are named after their inventors, Alexis Hocquenghem (published in September of 1959), and independently of Hocquenghem's work, Raj Chandra Bose and his student, Dwijendra Kumar Ray-Chaudhuri (published in March of 1960).

The most general form of a BCH code can be described as a polynomial code with coefficients from a finite field $GF(q)$, where q is a prime power. However, the BCH codes that are widely used in applications are binary codes; that is, they are polynomial codes with coefficients from $GF(2) = \mathbb{Z}_2$, and there are some real advantages to working over $\mathbb{Z}_2$, as we shall see. Accordingly, we shall restrict our attention to BCH codes over $\mathbb{Z}_2$.

For a positive integer m, we consider the field $GF(2^m)$, which is an extension of degree m over $\mathbb{Z}_2$. To construct $GF(2^m)$, we choose an irreducible polynomial $p \in \mathbb{Z}_2[x]$ of degree m and construct the quotient ring (which is then a field) $\mathbb{Z}_2[x]/(p)$. By Proposition 10.5.10, if p_1 and p_2 are each irreducible of degree m over $\mathbb{Z}_2$, then $\mathbb{Z}_2[x]/(p_1)$ is isomorphic to $\mathbb{Z}_2[x]/(p_2)$, which leads one to wonder: is there a preferred choice of irreducible p of degree m with which to construct $GF(2^m) \simeq \mathbb{Z}_2[x]/(p)$? Basically, the answer is yes: we will find it easier to work with the elements of $\mathbb{Z}_2[x]/(p)$ if the coset $x + (p)$ is a generator of the multiplicative (cyclic) group $GF(2^m)^*$.

18.2.1 Definition. *Let m be a positive integer. A generator of $GF(2^m)^*$ is called a* primitive element *of $GF(2^m)$.*

Thus we seek $p \in \mathbb{Z}_2[x]$, irreducible of degree m, for which $x + (p)$ is a primitive element of $\mathbb{Z}_2[x]/(p)$; that is, a primitive polynomial of degree m. It is known that for every positive integer m, there exists a primitive polynomial of degree m over $\mathbb{Z}_2$, and in fact, the Conway polynomials discussed at the end of Chapter 10 are particularly nice examples of primitive polynomials. For example, $1 + x + x^4$ is a Conway polynomial. Let $t = x + (1 + x + x^4) \in \mathbb{Z}_2[x]/(1 + x + x^4)$. We verify that $\langle t \rangle = GF(2^4)^*$, a cyclic group of order 15. We are to verify that $|t| = 15$. It suffices to verify that $|t| > 5$, and so we compute the sequence of the first five powers of t in $\mathbb{Z}_2[x]/(1 + x + x^4)$: t, t^2, t^3, $t^4 = 1 + t$, and $t^5 = t + t^2$. Since $t^5 \neq 1$, it follows that $|t| > 5$ and is a divisor of 15, so $|t| = 15$, as asserted.

It is of interest to note that not every irreducible polynomial over $\mathbb{Z}_2$ is primitive. For example, $p = 1 + x + x^2 + x^3 + x^4$ is irreducible over $\mathbb{Z}_2$ (neither 0 nor 1 is a root, so if it is not irreducible, it must be a product of irreducible polynomials of degree 2, and the only irreducible polynomial of degree 2 over $\mathbb{Z}_2$ is $1 + x + x^2$, whose square is $1 + x^2 + x^4$), but not primitive, as we now demonstrate. Let $t = x + (p)$. We compute the sequence of the first five powers of t (since $\mathbb{Z}_2[x]/(p)$ is a field of order $2^4 = 16$), which is t, t^2, t^3, $t^4 = 1 + t + t^2 + t^3$, and $t^5 = t + t^2 + t^3 + t^4 = 1$. Thus $|t| = 5$ in this case.

We proceed now to describe the construction of binary BCH codes.

18.2.2 Proposition. *Let $m \geq 2$ and $d \geq 3$ be integers with $d \leq n = 2^m - 1$, and let t be a primitive element for $GF(2^m)$. For each $i = 1, 2, \ldots, d-1$, let $m_i = min(t^i, \mathbb{Z}_2) \in \mathbb{Z}_2[x]$, the minimal polynomial for t^i over $\mathbb{Z}_2$, and let $p = [m_1, m_2, \ldots, m_{d-1}]$ and $r = \partial(p)$. Then $r \leq min\{n, (d-1)m\}$ and $P_{n,n-r,p}$ is a cyclic code with $\mu(P_{n,n-r,p}) \geq d$.*

Proof. First, note that p is the polynomial of lowest degree having each of $t, t^2, \ldots, t^{d-1}$ as roots. Secondly, observe that $|t| = 2^m - 1 = n$, so for any positive integer i, t^i is a root of $x^n - 1$. Thus for each $i = 1, 2, \ldots, d-1$, m_i divides $x^n - 1$, and so p, the least common multiple of the $m_i's$, divides $x^n - 1$ (it is possible that $p = x^n - 1$, in which case all code characters will be check characters; that is, we would not be able to encode any data with $P_{n,n-r,p}$), and so $r \leq n$. Moreover, since p divides the product $m_1 m_2 \ldots m_{d-1}$, $r = \partial(p) \leq \partial(m_1 m_2 \cdots m_{d-1})$. For each $i = 1, 2, \ldots, d-1$, $t^i \in GF(2^m)$ and thus by Proposition 10.3.7 and Proposition 10.4.4, $\partial(m_i) = [\mathbb{Z}_2(t^i) : \mathbb{Z}_2]$ is a divisor of $[GF(2^m) : \mathbb{Z}_2] = m$. It follows that $\partial(t^i) \leq m$, and so $\partial(p) \leq (d-1)m$. Let $k = n - r$, and note that by Proposition 18.1.10, $P_{n,k,p}$ is cyclic. We prove that $\mu(P_{n,k,p}) \geq d$. Suppose to the contrary that $P_{n,k,p}$ contains a nonzero code word c of weight less than d. If $c = c_0 c_1 \cdots c_{n-1}$, then $\pi_n(c) = \sum_{i=0}^{n-1} c_i x^i \in (\mathbb{Z}_2[x]; n)$ is divisible by p and has at most $d-1$ nonzero coefficients. Suppose that $0 \leq n_1 < n_2 < \cdots < n_{d-1} < n$ are indices such that $\{c_{n_j} \mid j = 1, 2, \ldots, d-1\}$ contains all nonzero entries of c. Then $\sum_{j=1}^{d-1} c_{n_j} x^{n_j}$ is divisible by p, which means that each of t^i, $i = 1, 2, \ldots, d - 1$, is a root of $\sum_{j=1}^{d-1} c_{n_j} x^{n_j}$. But then $(c_{n_1}, \ldots, c_{n_{d-1}})$ is a nontrivial solution to the following homogeneous system of $d - 1$ linear equations in the $d - 1$ unknowns $y_1, y_2, \ldots, y_{d-1}$:

$$t^{n_1} y_1 + t^{n_2} y_2 + \cdots + t^{n_{d-1}} y_{d-1} = 0$$
$$t^{2n_1} y_1 + t^{2n_2} y_2 + \cdots + t^{2n_{d-1}} y_{d-1} = 0$$
$$\vdots$$
$$t^{(d-1)n_1} y_1 + t^{(d-1)n_2} y_2 + \cdots + t^{(d-1)n_{d-1}} y_{d-1} = 0.$$

The coefficient matrix of this system is the $(d-1) \times (d-1)$ Vandermonde matrix

$$\begin{bmatrix} t^{n_1} & t^{n_2} & \cdots & t^{n_{d-1}} \\ t^{2n_1} & t^{2n_2} & \cdots & t^{2n_{d-1}} \\ \vdots & \vdots & \ddots & \vdots \\ t^{(d-1)n_1} & t^{(d-1)n_2} & \cdots & t^{(d-1)n_{d-1}} \end{bmatrix}$$

which has determinant

$$t^{n_1} t^{n_2} \cdots t^{n_{d-1}} \begin{vmatrix} 1 & 1 & \cdots & 1 \\ t^{n_1} & t^{n_2} & \cdots & t^{n_{d-1}} \\ \vdots & \vdots & \ddots & \vdots \\ t^{(d-2)n_1} & t^{(d-2)n_2} & \cdots & t^{(d-2)n_{d-1}} \end{vmatrix} = \prod_{i=1}^{d-1} t^{n_i} \prod_{1 \le j < i \le d-1} (t^{n_i} - t^{n_j}).$$

Since $n_i \ne n_j$ if $i \ne j$, and since $n_i < n$ for every $i = 1, 2, \ldots, d - 1$, it follows that $t^{n_i} - t^{n_j} \ne 0$ for every $1 \le i < j \le d - 1$, and so the coefficient matrix is invertible. But then the only solution to the homogeneous system is the trivial solution, and so we have obtained a contradiction. As this has followed from our assumption that $P_{n,k,p}$ contains a nonzero word of weight less than d, we conclude that $\mu(P_{n,k,p}) \ge d$. $\qquad\square$

We remark that the least common multiple described in Proposition 18.2.2 is simply the product of the distinct minimal polynomials that appear in the list $m_1, m_2, \ldots, m_{d-1}$, as each is irreducible over $\mathbb{Z}_2$. It is always the case that at least two elements in the list t^i, $i = 1, 2, \ldots, d - 1$ will have the same minimal polynomial, and the following lemma presents a useful observation in this regard.

18.2.3 Lemma. *Let m be a positive integer, and let t be a primitive element of $GF(2^m)$. For any l with $1 \leq l \leq 2^m - 1$, write $l = 2^k j$ where j is an odd integer, and $k \geq 0$. Then the minimal polynomial for t^l over $\mathbb{Z}_2$ is the minimal polynomial for t^j over $\mathbb{Z}_2$.*

Proof. The Frobenius automorphism of $\mathbb{Z}_2$ extends to an automorphism of the polynomial ring $\mathbb{Z}_2[x]$. As a result, for any $\sum_{i=0}^{n} a_i x^i \in \mathbb{Z}_2[x]$, $(\sum_{i=0}^{n} a_i x^i)^2 = \sum_{i=0}^{n} a_i x^{2i}$ and so by induction, for any positive integer k, $(\sum_{i=0}^{n} a_i x^i)^{2^k} = \sum_{i=0}^{n} a_i x^{2^k i}$. It follows that if $f \in \mathbb{Z}_2[x]$ is the minimal polynomial for t^j, and $l = 2^k j$ for nonnegative k and odd j, then $\hat{f}(t^l) = \hat{f}(t^{2^k j}) = (\hat{f}(t^j))^{2^k} = (0)^{2^k} = 0$; that is, t^l is a root of f, and so the minimal polynomial for t^l is a divisor of f. Since f is irreducible over $\mathbb{Z}_2$, we obtain that f is the minimal polynomial for t^l over $\mathbb{Z}_2$. $\qquad\square$

In Proposition 18.2.2, it was proven that the generator polynomial for a BCH code in $\mathbb{Z}_2^m$ had degree bounded by $(d-1)m$. As a consequence of the preceding lemma, we are able to obtain an improved estimate for the bound on the degree of the generator polynomial.

18.2.4 Proposition. *Let $m \geq 2$ and $d \geq 3$ be integers with d odd and $d \leq n = 2^m - 1$, and let t be a primitive element for $GF(2^m)$. Then p, as defined in Proposition 18.2.2, is equal to $[m_1, m_3, \ldots, m_{d-2}]$; that is, the least common multiple of the minimal polynomials of t^{2i-1}, $i = 1, 2, \ldots, \frac{d-1}{2}$, and so $\partial(p) \leq (d-1)m/2$.*

Proof. By Lemma 18.2.3, $m_{2i} = m_i$ for each odd exponent i, and since d is odd and $1 \leq i \leq d-1$, the odd exponents range from 1 up to $d-2$, hence the result. $\qquad\square$

Let us work through an example of the construction of a binary BCH code.

18.2.5 Example. *The polynomial $q = 1 + x + x^4 \in \mathbb{Z}_2[x]$ is a Conway polynomial, so $t = x + (q)$ is a generator for the cyclic group of units of $\mathbb{Z}_2[x]/(q) = GF(2^4)*$ and thus $|t| = 15$. For $d = 5$, we obtain $p = [m_1, m_3]$, with $m_1 = q$. To compute $m_3 = \min(t^3, \mathbb{Z}_2)$, we note that the Frobenius automorphism of $GF(2^4)$ will map roots of m_3 to roots of m_3, and so t^3, t^6, t^{12}, and $t^{24} = t^9$ (and after this, we repeat the cycle) are all roots of m_3. Thus $(x-t^3)(x-t^6)(x-t^{12})(x-t^9)$ divides m_3 and so $\partial(m_3) \geq 4$. Moreover, $(t^3)^5 = t^{15} = 1$, and so t^3 is a root of $x^5 - 1 = (x-1)(x^4+x^3+x^2+x+1)$, and since $t^3 \neq 1$, we obtain that t^3 is a root of $x^4 + x^3 + x^2 + x + 1$. But*

Table 18.1   The BCH code $P_{15,7,p}$, where $p = 1 + x^4 + 6 + x^7 + x^8$

```
00000000 0000000   11001110 0100000   10001011 1000000   01000101 1100000
00010111 0000001   11011001 0100001   10011100 1000001   01010010 1100001
00101110 0000010   11100000 0100010   10100101 1000010   01101011 1100010
00111001 0000011   11110111 0100011   10110010 1000011   01111100 1100011
01011100 0000100   10010010 0100100   11010111 1000100   00011001 1100100
01001011 0000101   10000101 0100101   11000000 1000101   00001110 1100101
01110010 0000110   10111100 0100110   11111001 1000110   00110111 1100110
01100101 0000111   10101011 0100111   11101110 1000111   00100000 1100111
10111000 0001000   01110110 0101000   00110011 1001000   11111101 1101000
10101111 0001001   01100001 0101001   00100100 1001001   11101010 1101001
10010110 0001010   01011000 0101010   00011101 1001010   11010011 1101010
10000001 0001011   01001111 0101011   00001010 1001011   11000100 1101011
11100100 0001100   00101010 0101100   01101111 1001100   10100001 1101100
11110011 0001101   00111101 0101101   01111000 1001101   10110110 1101101
11001010 0001110   00000100 0101110   01000001 1001110   10001111 1101110
11011101 0001111   00010011 0101111   01010110 1001111   10011000 1101111
01100111 0010000   10101001 0110000   11101100 1010000   00100010 1110000
01110000 0010001   10111110 0110001   11111011 1010001   00110101 1110001
01001001 0010010   10000111 0110010   11000010 1010010   00001100 1110010
01011110 0010011   10010000 0110011   11010101 1010011   00011011 1110011
00111011 0010100   11110101 0110100   10110000 1010100   01111110 1110100
00101100 0010101   11100010 0110101   10100101 1010101   01101001 1110101
00010101 0010110   11011011 0110110   10011110 1010110   01010000 1110110
00000010 0010111   11001100 0110111   10001001 1010111   01000111 1110111
11011111 0011000   00010001 0111000   01010100 1011000   10011010 1111000
11001000 0011001   00000110 0111001   01000011 1011001   10001101 1111001
11110001 0011010   00111111 0111010   01111010 1011010   10110100 1111010
11100110 0011011   00101000 0111011   01101101 1011011   10100011 1111011
10000011 0011100   01001101 0111100   00001000 1011100   11000110 1111100
10010100 0011101   01011010 0111101   00011111 1011101   11010001 1111101
10101101 0011110   01100011 0111110   00100110 1011110   11101000 1111110
10111010 0011111   01110100 0111111   00110001 1011111   11111111 1111111
```

then m_3 divides $x^4 + x^3 + x^2 + x + 1$, and $\partial(m_3) = 4$, which means that $m_3 = x^4 + x^3 + x^2 + x + 1$. Thus $p = [1 + x + x^4, 1 + x + x^2 + x^3 + x^4]$. Since $1 + x + x^4$ and $1 + x + x^2 + x^3 + x^4$ are distinct polynomials irreducible over $\mathbb{Z}_2$, they are relatively prime and thus their least common multiple is their product; that is, $p = (1+x+x^4)(1+x+x^2+x^3+x^4) = 1+x^4+x^6+x^7+x^8$. We have determined that $r = \partial(p) = 8$, while the length of the code words is $n = 2^4 - 1 = 15$, so $k = n - r = 7$. Thus we have constructed $P_{15,7,p}$, with $\mu(P_{15,7,p}) \geq 5$, which makes $P_{15,7,p}$ a 2-error-correcting code. Note that $(1)(p) \in \pi_{15}(P_{15,7,p})$, and so $\pi_{15}^{-1}(p) = 10001\,01110\,00000 \in P_{15,7,p}$, which means that $\mu(P_{15,7,p}) \leq 5$, and so $\mu(P_{15,7,p}) = 5$. We follow the procedure that was described in Proposition 18.1.13 to encode the elements of $\mathbb{Z}_2^7$ as elements of $\mathbb{Z}_2^{15}$. Each element u of $\mathbb{Z}_2^7$ is first encoded as an element of $(\mathbb{Z}_2[x]; 7)$, which is then multiplied by x^{n-k} to form $f \in (\mathbb{Z}_2[x]; n)$ and the remainder of this product upon division by p is then added to f to obtain the code word that encodes u. We list the code $P_{15,7,p}$ in Table 18.1. The data word appears as the final 7 characters of the code word, and we have provided a minor separation between the eighth and ninth charcters to make it easier to identify the data word.

3 Decoding for a BCH Code

A BCH code is a group code, and we could use a parity check matrix to compute the syndrome of a received word and consult a syndrome-coset leader table, just as was done in Chapter 17. However, in the case of a BCH code, it is frequently the case that the number of syndromes is approximately the same as the size of the code, or even larger. For example, the BCH code constructed in Example 18.2.5 encodes 7-character data words, using 8 check characters. The size of the code $P_{15,7,p}$ is 128, while there are $2^8 = 256$ syndromes. Thus we should like to find some computational scheme to identify and correct the errors for which the code is designed to handle without using such large table-lookups.

There are several error-correction schemes currently employed for the decoding of a BCH code. All are based on arithmetic in the finite field $\mathbb{Z}_2[x]/(q)$, where q is the primitive polynomial used in the construction of the encoding polynomial p. The general situation is outside of the scope of our work in this text, but we shall describe the error-correction procedure for the particular code that we presented in Example 18.2.5. In that example, we used the Conway polynomial $q = 1 + x + x^4$ to construct $p = m_1 m_3 = (1 + x + x^4)(1 + x + x^2 + x^3 + x^4)$, the encoding polynomial for $P_{15,7,p}$.

The general idea is that a code word c is transmitted, and received as r. If no errors occurred in the transmission, $r = c$ and no error correction is required. Of course, the receiver is not privy to c, and so must use some computation to determine whether or not the received word is in error. The parity check matrix is suitable for this purpose. To form the generator matrix for $P_{15,7,p}$, a 15×7 matrix G, we take a $\mathbb{Z}_2$ basis for $P_{15,7,p}$ and use each basis element as one of the columns of G.

It is conventional to take the encoded forms of the standard basis of $\mathbb{Z}_2^7$; namely

{ 100010111000000, 110011100100000,

011001110010000, 101110000001000,

010111000000100, 001011100000010,

000101110000001 },

arranged in order so that the lower 7×7 submatrix of G is I_7. If we denote the upper 8×7 block as A, then $G = \begin{bmatrix} A \\ I_7 \end{bmatrix}$ and $P = \begin{bmatrix} I_8 & A \end{bmatrix}$. Thus

$$
G = \left[\begin{array}{ccccccc}
1 & 1 & 0 & 1 & 0 & 0 & 0 \\
0 & 1 & 1 & 0 & 1 & 0 & 0 \\
0 & 0 & 1 & 1 & 0 & 1 & 0 \\
0 & 0 & 0 & 1 & 1 & 0 & 1 \\
1 & 1 & 0 & 1 & 1 & 1 & 1 \\
0 & 1 & 1 & 0 & 1 & 1 & 1 \\
1 & 1 & 1 & 0 & 0 & 1 & 0 \\
1 & 0 & 1 & 0 & 0 & 0 & 0 \\
\hline
1 & 0 & 0 & 0 & 0 & 0 & 0 \\
0 & 1 & 0 & 0 & 0 & 0 & 0 \\
0 & 0 & 1 & 0 & 0 & 0 & 0 \\
0 & 0 & 0 & 1 & 0 & 0 & 0 \\
0 & 0 & 0 & 0 & 1 & 0 & 0 \\
0 & 0 & 0 & 0 & 0 & 1 & 0 \\
0 & 0 & 0 & 0 & 0 & 0 & 1
\end{array}\right]
\quad \text{and} \quad
P = \left[\begin{array}{cccccccc|ccccccc}
1 & 0 & 0 & 0 & 0 & 0 & 0 & 0 & 1 & 1 & 0 & 1 & 0 & 0 & 0 \\
0 & 1 & 0 & 0 & 0 & 0 & 0 & 0 & 0 & 1 & 1 & 0 & 1 & 0 & 0 \\
0 & 0 & 1 & 0 & 0 & 0 & 0 & 0 & 0 & 0 & 1 & 1 & 0 & 1 & 0 \\
0 & 0 & 0 & 1 & 0 & 0 & 0 & 0 & 0 & 0 & 0 & 1 & 1 & 0 & 1 \\
0 & 0 & 0 & 0 & 1 & 0 & 0 & 0 & 1 & 1 & 0 & 1 & 1 & 1 & 1 \\
0 & 0 & 0 & 0 & 0 & 1 & 0 & 0 & 0 & 1 & 1 & 0 & 1 & 1 & 1 \\
0 & 0 & 0 & 0 & 0 & 0 & 1 & 0 & 1 & 1 & 1 & 0 & 0 & 1 & 0 \\
0 & 0 & 0 & 0 & 0 & 0 & 0 & 1 & 1 & 0 & 1 & 0 & 0 & 0 & 0
\end{array}\right].
$$

To detect whether or not a detectable error has occurred (one that does not transform one code word into another), it suffices to multiply the final 7 characters of the received word by A and compare to the first 8 characters of the received word. If the two agree, we declare that no error has occurred, while if they differ, then a detectable error has occurred and we proceed to try to correct it.

For the BCH code $P_{15,7,p}$, we have determined that $\mu(P_{15,7,p}) = 5$, so the nearest-neighbor-decoding scheme will correct all patterns of 2 or fewer errors. Suppose that a code word c is transmitted, and during the transmission, exactly two errors occur. Let r denote the received word, so c and r differ at exactly two locations, say index $0 \leq i_1 < i_2 \leq 14$. Let us move the discussion over to $\mathbb{Z}_2[x]$ by means of the injective map π_{15}. Let $f = \pi_{15}(c)$ and $g = \pi_{15}(r)$, so $f, g \in (\mathbb{Z}_2[x]; 15)$, and $f - g = e$ is a polynomial which is a sum of two monomials, $x^{i_1} + x^{i_2}$. Our objective is

Table 18.2 $GF(2^4) = \mathbb{Z}_2[x]/(1+x+x^4)$, with $t = x + (1+x+x^4)$

t	t
t^2	t^2
t^3	t^3
t^4	$1+t$
t^5	$t+t^2$
t^6	t^2+t^3
t^7	$1+t+t^3$
t^8	$1+t^2$
t^9	$t+t^3$
t^{10}	$1+t+t^2$
t^{11}	$t+t^2+t^3$
t^{12}	$1+t+t^2+t^3$
t^{13}	$1+t^2+t^3$
t^{14}	$1+t^3$
t^{15}	1

to figure how to identify the indices i_1 and i_2 from the polynomial g that corresponds to the received word r. The computational work will take place in the finite field $\mathbb{Z}_2[x]/(q) = \mathbb{Z}_2[x]/(1+x+x^4)$, for which $t = x + (q)$ is a generator of the cyclic group of units of the field since $1+x+x^4$ is a Conway polynomial. We list the elements of $GF(2^4) = \mathbb{Z}_2[x]/(q)$ in Table 18.2. Each element of $GF(2^4)$ can be written as a $\mathbb{Z}_2$-linear combination of 1, t, t^2, and t^3.

The idea is to construct what is called the error-locator polynomial in $GF(2^4)[x]$, which, in the case of two errors in position i_1 and i_2, has the form $(1+t^{i_1}x)(1+t^{i_2}x) = 1 + (t^{i_1}+t^{i_2})x + t^{i_1}t^{i_2}x^2$. Of course, we don't know the values t^{i_1} and t^{i_2}; these are the values that we wish to determine. The key observation is that $g+e = f$ is divisible by $p = (1+x+x^4)(1+x+x^2+x^3+x^4)$ since c is a code word. Recall that t is a root of $m_1 = q$, and t^3 is a root of $m_3 = 1+x+x^2+x^3+x^4 = m_3$, and $p = m_1m_3$, so t and t^3 are roots of p and thus of f. It follows that $\hat{f}(t) = \hat{f}(t^3) = 0$ and so $\hat{e}(t) = \hat{g}(t)$ and $\hat{e}(t^3) = \hat{g}(t^3)$. But $\hat{e}(t) = t^{i_1} + t^{i_2}$, and $\hat{e}(t^3) = t^{3i_1} + t^{3i_2}$. Thus $t^{i_1} + t^{i_2}$, the coefficient of x in the error-locator polynomial, can be computed as $\hat{g}(t)$, and $t^{i_1}t^{i_2}$, the coefficient of x^2 in the error-locator polynomial, can be computed as follows:

$$\hat{g}(t^3) = (t^{i_1})^3 + (t^{i_2})^3 = (t^{i_1} + t^{i_2})^3 + 3(t^{i_1})^2 t^{i_2} + 3t^{i_1}(t^{i_2})^2$$

$$= \hat{g}(t)^3 + t^{i_1} t^{i_2}(t^{i_1} + t^{i_2}) = \hat{g}(t)^3 + t^{i_1} t^{i_2} \hat{g}(t),$$

and so $t^{i_1} t^{i_2} = (\hat{g}(t^3) + \hat{g}(t)^3)(\hat{g}(t))^{-1}$. Thus in the event of two errors, the error-locator polynomial is

$$1 + \hat{g}(t)x + (\hat{g}(t))^{-1}(\hat{g}(t^3) + \hat{g}(t)^3)x^2.$$

We know that the two roots of this polynomial are t^{-i_1} and t^{-i_2}, so we examine the 15 nonzero elements of $GF(2^4)$ to find the two roots of the error-locator polynomial, at which point we will know the value of $(t^{i_1})^{-1}$ and $(t^{i_2})^{-1}$. We then consult Table 18.2 to find these values (in the event that either is not found as a monomial power of t), thereby determining the exponents i_1 and i_2. Once we are in possession of i_1 and i_2, we change the bits at those locations in r to recover c. Note that the quadratic formula is not available to us in a field of characteristic 2, and so we must indeed search for the roots of the error-locator polynomial.

Before we work through an example of this process, we should consider the case of a single error. Suppose that upon transmission of code word c, a single error occurs, with the result that the received word r differs from c in a single location, say with index i. We move to $(\mathbb{Z}_2[x]; 15)$ by means of π_{15}, with $f = \pi_{15}(c)$ and $g = \pi_{15}(r)$, so $f = g + e$, where $e = x^i$. Then as in the discussion of the case of two errors, we evaluate $\hat{g}(t) = \hat{f}(t) + \hat{e}(t) = \hat{e}(t) = t^i$, and $\hat{g}(t^3) = \hat{f}(t^3) + \hat{e}(t^3) = \hat{e}(t^3) = t^{3i}$. In this case, the error-locator polynomial in $GF(2^4)[x]$ is $1 - t^i x = 1 - \hat{g}(t)x = 1 - t^i x$, which has root $x = t^{-i}$. Thus in this case, finding the root is quite simple: one computes $\hat{g}(t)$ and then, if not a monomial in t, search Table 18.2 to find it, thereby determining i, which enables the correction of the error to be carried out.

We should point out that of course, we do not know in advance whether one error has occurred or two errors have occurred; we only know that at least one error has occurred (by multiplying the ostensible data word as extracted from the received word r by A and not obtaining the check bits that form the first 8 characters of the received word). However, observe that in the case of a single error, we have $\hat{g}(t^3) = t^{3i}$, so $(\hat{g}(t^3) + \hat{g}(t)^3)(\hat{g}(t))^{-1} = (t^{3i} + t^{3i})t^{-i} = 0$. Thus the error-locator polynomial for both cases can be taken to be

$$1 + \hat{g}(t)x + (\hat{g}(t))^{-1}(\hat{g}(t^3) + \hat{g}(t)^3)x^2.$$

We should also point out that the evaluation of $\hat{g}(t)$ and $\hat{g}(t^3)$ could involve many powers of t. There is a very nice observation that will greatly simplify this part of the computation. Recall that t is a root of $1 + x + x^4$, while t^3 is a root of $1 + x + x^2 + x^3 + x^4$. If we determine the remainder r_1 when g is divided by $1 + x + x^4$, and the remainder r_2 when g is divided by $1 + x + x^2 + x^3 + x^4$, say $g = (1 + x + x^4)q_1 + r_1 = (1 + x + x^2 + x^3 + x^4)q_2 + r_2$ for some quotients $q_1, q_2 \in \mathbb{Z}_2[x]$, then $\hat{g}(t) = \hat{r_1}(t)$ and $\hat{g}(t^3) = \hat{r_2}(t^3)$. Since $\partial(r_1)$ and $\partial(r_2)$ are both less than 4, the evaluation of $\hat{g}(t)$ and of $\hat{g}(t^3)$ need not involve any exponents greater than 3.

Let us now work through an example. Suppose that we transmit code word 11111001 1011001 and it is received as 11111011 1010001. Let us first check to see whether a detectable error has occurred. We extract the data portion of the received word, 1010001, and multiply by A to obtain

$$
\begin{bmatrix}
1 & 1 & 0 & 1 & 0 & 0 & 0 \\
0 & 1 & 1 & 0 & 1 & 0 & 0 \\
0 & 0 & 1 & 1 & 0 & 1 & 0 \\
0 & 0 & 0 & 1 & 1 & 0 & 1 \\
1 & 1 & 0 & 1 & 1 & 1 & 1 \\
0 & 1 & 1 & 0 & 1 & 1 & 1 \\
1 & 1 & 1 & 0 & 0 & 1 & 0 \\
1 & 0 & 1 & 0 & 0 & 0 & 0
\end{bmatrix}
\begin{bmatrix}
1 \\ 0 \\ 1 \\ 0 \\ 0 \\ 0 \\ 1
\end{bmatrix}
=
\begin{bmatrix}
1 \\ 1 \\ 1 \\ 1 \\ 0 \\ 0 \\ 0 \\ 0
\end{bmatrix}
\neq
\begin{bmatrix}
1 \\ 1 \\ 1 \\ 1 \\ 1 \\ 0 \\ 1 \\ 1
\end{bmatrix},
$$

the parity check bit portion of the received word, and so we conclude that at least one error has occurred. We set $g = \pi_{15}(111110111010001) = 1 + x + x^2 + x^3 + x^4 + x^6 + x^8 + x^{10} + x^{12} + x^{14}$. The remainder r_1, when g is divided by $1 + x + x^4$, is x^2. The reader might wish to verify that $g = (1 + x^3 + x^4 + x^5 + x^7 + x^8 + x^{10})(1 + x + x^4) + x^2$. As well, the remainder r_2, when g is divided by $1 + x + x^2 + x^3 + x^4$, is 0. Actually, if we let $h = 1 + x + x^2 + x^3 + x^4$, we have $g = h(1 + x^6 h)$. Thus for the error-locator polynomial, the coefficient of x is $\hat{r_1}(t) = t^2$, and the coefficient of x^2 is $(\hat{g}(t))^{-1}(\hat{g}(t^3) + \hat{g}(t)^3) = t^{-2}(0 + t^6) = t^4$, so the error-locator polynomial is

$$
1 + t^2 x + t^4 x^2.
$$

We work our way through Table 18.2, searching for a root of the error-locator polynomial. t is not a root since $1 + t^3 + t^6 = 1 + t^3 + t^2 + t^3 = 1 + t^2 \neq 0$, and t^2 gives $1 + t^4 + t^8 = 1 + 1 + t + 1 + t^2 = 1 + t + t^2 \neq 0$. Then $1 + t^2(t^3) + t^4(t^3)^2 = 1 + t^5 + t^{10} = 1 + t + t^2 + 1 + t + t^2 = 0$, so t^3 is a root of the error-locator polynomial. Once we have found one root, the other root may be immediately recovered by observing that $ax^2 + bx + c = a(x^2 + a^{-1}bx + a^{-1}c)$ and $a^{-1}c$ is the product of the two roots. Since

$(t^4)^{-1} = t^{11}$, we obtain $1 + t^2 x + t^4 x^2 = t^4(x^2 + t^{-2}x + t^{11})$, so t^{11} is the product of the two roots of the error-locator polynomial, and we now know one of the roots to be t^3. Thus the other root is t^8, and so we calculate $t^{-3} = t^{12}$ and $t^{-8} = t^7$ to obtain that the indices of the two errors (for there were indeed two errors) are 7 and 12. and we recover $c = 11111001\,1011001$ from r by toggling the bits at indices 7 and 12 in the received word.

18.4   Exercises

1. Find generator and parity-check matrices for the following polynomial codes:

    a)   the $(4,1)$ code generated by $1 + x + x^2 + x^3$.

    b)   the $(7,3)$ code generated by $(1+x)(1+x+x^3)$.

    c)   the $(9,4)$ code generated by $1 + x^2 + x^5$.

2. Construct a syndrome-coset leader table for each of the following codes:

    a)   the $(3,1)$ code generated by $1 + x + x^2$.

    b)   the $(7,4)$ code generated by $1 + x + x^3$.

    c)   the $(9,3)$ code generated by $1 + x^3 + x^6$.

3. Consider the $(63,56)$ code generated by $(1+x)(1+x+x^6)$.

    a)   What is the number of check-digits?

    b)   How many different syndromes are there?

    c)   How many errors can we be sure to correct with this code?

    d)   How many errors can we be sure to detect with this code?

4. Using the primitive polynomial $p = 1 + x^2 + x^5$, construct the field of order 32 as $\mathbb{Z}_2[x]/(p)$. Each nonzero element is of the form t^i for some $i = 0, 1, \ldots, 31$, where $t = x + (p)$. Each element can also be written as a linear combination of $\{1, t, t^2, t^3, t^4\}$. Produce a table similar to Table 18.2 for this field.

5. Let $P_{15,7,p}$ be the BCH code given in Example 18.2.5. Decode each of the following received words:

    a)   10101001 1110000.

    b)   00001100 1110010.

    c)   11000011 0011100.

    d)   01111010 1011100.

6. Let n be a positive integer.

   a) Let $\pi : \mathbb{Z}_2[x] \to \mathbb{Z}_2[x]/(x^n + 1)$ denote the canonical projection homomorphism, and let $\varphi = \pi \circ \pi_n : \mathbb{Z}_2^n \to \mathbb{Z}_2[x]/(x^n + 1)$. Prove that φ is an isomorphism (with respect to addition).

   b) Let C be a cyclic code of length n, and let $C' = \varphi(C)$, so C' is a $\mathbb{Z}_2$-subspace of $\mathbb{Z}_2[x]/(x^n + 1)$. Prove that C' is an ideal of the quotient ring $\mathbb{Z}_2[x]/(x^n + 1)$.

   c) Prove that if $I \lhd \mathbb{Z}_2[x]/(x^n + 1)$, then $\varphi^{-1}(I)$ is a cyclic code of length n.

   d) Prove that every ideal I of $\mathbb{Z}_2[x]/(x^n + 1)$ is principal, generated by the coset of some divisor g of $x^n + 1$. Prove that the cyclic code $\varphi^{-1}(I)$ is a polynomial code generated by g.

   e) In $\mathbb{Z}_2[x]$, $x^6 + 1 = (x^3)^2 + 1 = (x^3 + 1)^2 = (x + 1)^2(x^2 + x + 1)^2$, and so $g - (x+1)(x^2 + x + 1)$ is a divisor of $x^6 + 1$. Determine the cyclic code of length 9 that is generated by g.

7. In $\mathbb{Z}_2[x]$, $x^9 + 1$ can be factored as

$$(x^3)^3 + 1 = (x^3 + 1)((x^3)^2 + x^3 + 1)$$
$$= (x + 1)(x^2 + x + 1)(x^6 + x^3 + 1).$$

   a) Prove that each of these three factors of $x^9 + 1$ is irreducible over $\mathbb{Z}_2$.

   b) Determine the number of cyclic codes of length 9.

8. a) Prove that $x^7 + 1 \in \mathbb{Z}_2[x]$ has two irreducible factors of degree 3.

   b) Construct two $(7, 4)$ cyclic codes.

   c) How many cyclic codes of length 7 are there? For each cyclic code C of length 7, determine $\mu(C)$.

Appendix A

Rational, Real, and Complex Numbers

1 Introduction

This appendix is included as a brief reminder to the reader of some of the elementary properties of the three number systems of the title.

In Chapter 8 (see Proposition 8.6.2), an algebraic construction was given that when applied to $\mathbb{Z}$, the ring of integers, constructed $\mathbb{Q}$, the field of rational numbers, as an extension ring of $\mathbb{Z}$. Elements of $\mathbb{Q}$ were represented by symbols of the form $\frac{a}{b}$ where $a, b \in \mathbb{Z}$ with $b \neq 0$. Addition and multiplication were defined by:

$$\frac{a}{b} + \frac{c}{d} = \frac{ad + bc}{bd} \quad \text{and} \quad \frac{a}{b}\frac{c}{d} = \frac{ac}{bd}$$

for every $\frac{a}{b}, \frac{c}{d} \in \mathbb{Q}$.

In $\mathbb{Q}$, every nonzero integer has a multiplicative inverse, and so we can solve equations of the form $ax = b$ where $a, b \in \mathbb{Z}$ with $a \neq 0$ and more generally, we can solve equations of the form $rx = s$ where $r, s \in \mathbb{Q}$ with $r \neq 0$. However, even a simple equation of the form $x^2 = 2$ has no solution in $\mathbb{Q}$. The need for a number system that extends $\mathbb{Q}$ in which we would be able to solve such equations led to the contruction of the field $\mathbb{R}$ of real numbers. This construction is not entirely an algebraic construction, but rather a combination of algebraic and metric concepts (the latter relying on the order relation on $\mathbb{Q}$), and we shall not attempt to describe it here. The set of rational numbers is dense in $\mathbb{R}$, by which is meant that between any two distinct real numbers there is a rational number (and thus between any two distinct real numbers there are infinitely many rational numbers). The elements of $\mathbb{R} - \mathbb{Q}$ are called the irrational real numbers. For example, $\sqrt{2}$ and $-\sqrt{2}$, the two solutions to the equation $x^2 = 2$, are irrational numbers. The order relation on $\mathbb{Q}$ extends to a linear ordering of $\mathbb{R}$, by which we mean that for any $a, b \in \mathbb{R}$, exactly one of $a < b$, $a = b$, or $a > b$ holds.

However, although the extension of $\mathbb{Q}$ to $\mathbb{R}$ allows us to solve equations that have no solution in $\mathbb{Q}$, but for which it was not unreasonable from the point of view of applications to expect that a solution should exist (for example, it is quite reasonable to want to know the length of a diagonal of a square of side 1, but this requires the existence of $\sqrt{2}$; an analog of this would be the need to find a solution for the quadratic equation $x^2 + x + 1 = 0$: since $x^2 + x + 1 = (x + \frac{1}{2})^2 + \frac{3}{4}$, this equation has no real solution), we are led to extend the field of real numbers. This last extension is algebraic, and can be thought of as the set $\mathbb{R} \times \mathbb{R}$ with the ordinary addition of ordered pairs, but with a multiplication operation defined by $(a, b)(c, d) = (ac - bd, ad + bc)$. The pair $(1, 0)$ is the multiplicative identity element, and $i = (0, 1)$ has the property that $i^2 = (-1, 0)$. This number field is called the field of complex numbers, and is denoted by $\mathbb{C}$. $\mathbb{R} \times \mathbb{R}$ is a real vector space, and $(a, b) = a(1, 0) + b(0, 1)$. It is customary to think of $\mathbb{R}$ as a subfield of $\mathbb{C}$ by identifying $r \in \mathbb{R}$ with $(r, 0) \in \mathbb{C}$, and $(0, b)$ with $b(0, 1) = bi$, so that with this convention, $(a, b) = a + bi$. Thus

$$\mathbb{C} = \{ a + bi \mid a, b \in \mathbb{R} \}.$$

Note that $i^2 = (-1, 0) = -(1, 0)$, so that with our conventions, $i^2 = -1$. Thus the equation $x^2 + 1 = 0$ has solutions in $\mathbb{C}$; namely $x = i$ and $x = -i$. More generally, $\mathbb{C}$ is an example of an algebraically closed field, by which is meant that for any $f \in \mathbb{C}[x]$, $\hat{f}(z) = 0$ has a solution $z \in \mathbb{C}$.

A.2   The Order Relation on the Real Number System

In this section, we discuss the properties of the familiar order relation on both $\mathbb{Q}$ and on $\mathbb{R}$. As we mentioned in the first section, the order relation on $\mathbb{R}$ is an extension of the order relation on $\mathbb{Q}$, and is a linear ordering of $\mathbb{R}$. Moreover, it is compatible with the operations of addition and multiplication on $\mathbb{R}$, in the sense that for any $a, b, c \in \mathbb{R}$, if $a < b$, then $a + c < b + c$, and if $c > 0$, then $ac < bc$, while if $c < 0$, then $ac > bc$.

A.2.1 Proposition. *For any $a, b \in \mathbb{R}$ with $0 < a < b$, and any positive integer n, $0 < a^n < b^n$.*

Proof. Let $a, b \in \mathbb{R}$ be such that $0 < a < b$. We shall use induction on n, and the base case is immediate. Suppose that $n \geq 1$ is an integer for which $0 < a^n < b^n$. Then since $a > 0$, we have $0 < a^{n+1} < ab^n$, and since $b^n > 0$, we have $ab^n < b^{n+1}$, so $0 < a^{n+1} < b^{n+1}$. The result follows now by induction. $\square$

A.2.2 Corollary. *Let $a, b \in \mathbb{R}$ be such that $0 < a < b$, and let n be a positive integer. Then $0 < \sqrt[n]{a} < \sqrt[n]{b}$.*

Proof. Since a and b are both positive, $a_1 = \sqrt[n]{a}$ and $b_1 = \sqrt[n]{b}$ exist and are positive. Suppose that $a_1 \geq b_1$. Then by Proposition A.2.1, $a = a_1^n \geq b_1^n = b$, which is not the case. Thus $0 < a_1 < b_1$, as required. $\square$

A.2.3 Definition. *Let $S \subseteq \mathbb{R}$ be nonempty. Then S is bounded above if there exists $r \in \mathbb{R}$ such that for all $x \in S$, $x \leq r$. Such a number r is called an upper bound for S. Further, S is said to have a least upper bound if the set of all upper bounds of S is not empty and has a least element b, in which case b is called a least upper bound of S. In other words, $b \in \mathbb{R}$ is a least upper bound for S if b is an upper bound for S and for any upper bound r of S, $b \leq r$.*

It is immediate that if S has a least upper bound, then S has exactly one least upper bound. It is a consequence of the construction of the set of real numbers that the following result holds (and since we have not discussed the construction, we are not in a position to offer a proof–we wish to simply record it as a fact).

A.2.4 Proposition. *Every nonempty subset of $\mathbb{R}$ which is bounded above has a least upper bound.*

In view of the uniqueness of the least upper bound, we shall speak of the least upper bound of a set S, and denote it by $lub(S)$. The least upper bound of a set S is also often referred to as the supremum of the set, in which case we write $sup(S)$.

We point out that Proposition A.2.4 does not apply to $\mathbb{Q}$. For example, by Example 9.6.7 (i), $\sqrt{2} \notin \mathbb{Q}$. Suppose that $A = \{\, x \in \mathbb{Q} \mid x < \sqrt{2} \,\}$ has a least upper bound $r \in \mathbb{Q}$. Then $r \leq \sqrt{2}$, and $r \in \mathbb{Q}$ while $\sqrt{2} \notin \mathbb{Q}$, so $r < \sqrt{2}$. Thus $r^2 < 2$. Since $r^2 \in \mathbb{Q}$, $e = 2 - r^2 \in \mathbb{Q}$. For a positive integer n, consider $(r + \frac{e}{n})^2 = r^2 + 2\frac{er}{n} + \frac{e^2}{n^2} = r^2 + e(\frac{2r}{n} + \frac{e}{n^2})$. Choose n large enough so that $\frac{2r}{n} + \frac{e}{n^2} < 1$, and set $s = r + \frac{e}{n}$. Then $s \in \mathbb{Q}$ and $s^2 < r^2 + e = 2$, so $s \in A$. But then $s \leq r$ which is not possible since $\frac{e}{n} > 0$. Thus A does not have a least upper bound in $\mathbb{Q}$.

However, we may use Proposition A.2.4 to prove that $\mathbb{Q}$ is a dense subset of $\mathbb{R}$.

A.2.5 Proposition. *For any positive real numbers a and b (not necessarily distinct), there exists a positive integer n such that $na > b$.*

Proof. Let a and b be positive real numbers, and suppose to the contrary that no such positive integer exists. Let $S = \{\, na \mid n \in \mathbb{Z},\ n > 0 \,\}$. By supposition, for every $x \in S$, $x \leq b$ (since $x = na$ for some positive integer n), so S is bounded above. By Proposition A.2.4, $lub(S)$ exists. Now, $a \in S$ and $a > 0$, so $lub(S) - a < lub(S)$, which means that $lub(S) - a$ is not an upper bound for S. Thus there exists $x \in S$ with $x > lub(S) - a$. But then $x = ma$ for some positive integer m, and so $lub(S) < (m+1)a$. Since $(m+1)$ is a positive integer, $(m+1)a \in S$, which means that $lub(S) < y$ for $y = (m+1)a \in S$. As this is a contradiction of the least upper bound definition, our supposition must be false and so the result follows. $\square$

The property described in Proposition A.2.5 is called the Archimedean property, and by virtue of Proposition A.2.5, we refer to $\mathbb{R}$ as an Archimedean field.

A.2.6 Proposition. *Let $a, b \in \mathbb{R}$ with $a < b$. Then there exists $r \in \mathbb{Q}$ such that $a < r < b$.*

Proof. We apply Proposition A.2.5 to the positive real numbers $b - a$ and 1 to obtain the existence of a positive integer n for which $n(b-a) > 1$, and so $b - a > \frac{1}{n}$. Suppose first that $b > 0$. By Proposition A.2.5 applied to $\frac{1}{n}$ and b, there is a positive integer m such that $\frac{m}{n} > b$, and so the set $T = \{\, k \in \mathbb{N} \mid \frac{k}{n} \geq b \,\}$ is a nonempty subset of $\mathbb{N}$. By the Well Ordering Property for $\mathbb{N}$, T has a least element, t say. Since $t \in T$ and $0 \notin T$ (since $b > 0$), t is a positive integer. But then $t - 1 \in \mathbb{N}$, and $t - 1 < t$, so $t - 1 \notin T$, which means that $\frac{t-1}{n} < b$. Also, $\frac{1}{n} < b - a$ implies that $-\frac{1}{n} > a - b$, and so $a < b - \frac{1}{n} < \frac{t}{n} - \frac{1}{n} = \frac{t-1}{n} < b$. Since $r = \frac{t-1}{n} \in \mathbb{Q}$, we have found $r \in \mathbb{Q}$ for which $a < r < b$.

Now suppose that $b \leq 0$. From $a < b \leq 0$, we obtain that $0 \leq -b < -a$. We may therefore apply the argument of the first paragraph to determine that there exists $r \in \mathbb{Q}$ with $-b < r < -a$, and thus $a < -r < b$. Since $r \in \mathbb{Q}$ implies that $-r \in \mathbb{Q}$, the proof is complete. $\square$

We are now able to present an example of a set of rational numbers that is bounded above but which has no least upper bound in $\mathbb{Q}$.

A.2.7 Example. *For any integer $n > 1$ and any prime p, let $S = \{\, r \in \mathbb{Q} \mid r > 0,\ r^n < p \,\}$. Note that $1 \in S$, so S is not empty. By Corollary A.2.2, if $r \in S$, then $0 < r^n < p$ and thus $r < \sqrt[n]{p}$. Thus in $\mathbb{R}$, S is bounded above by $\sqrt[n]{p}$, and so in $\mathbb{R}$, $lub(S)$ exists. We prove that $lub(S) = \sqrt[n]{p}$ (which will complete the proof since $\sqrt[n]{p} \notin \mathbb{Q}$ by virtue of Example 9.6.7). Suppose by* *way of contradiction that $\sqrt[n]{p} \neq lub(S)$ and let $l = lub(S)$. Since $\sqrt[n]{p}$ is an upper bound for S, it follows that $0 < l < \sqrt[n]{p}$, and then by Proposition A.2.6, there exists a rational number r such that $l < r < \sqrt[n]{p}$. But then $0 < r^n < p$, which means that $r \in S$ and this contradicts the definition of least upper bound (since $l < r$). Thus our supposition that $\sqrt[n]{p} \neq lub(S)$ must be false, and so $\sqrt[n]{p} = lub(S)$.*

3 Decimal Representation of Rational Numbers

The construction of the set of real numbers builds every real number r as the limit of an infinite sequence of rational numbers. There are always (infinitely) many sequences of rational numbers that converge to the same real number r. Among these, we find infinite series that sum to r; that is, $r = \sum_{i=0}^{\infty} q_i = \lim_{n=0}^{\infty} S_n$, where for each $n \geq 0$, $S_n = \sum_{i=0}^{n} q_i$, and $q_i \in \mathbb{Q}$ for each i. But even then, we do not obtain a unique representation of r. Let us introduce some conventions. We shall restrict our attention to nonnegative real numbers, as the representation of a negative real number r is simply -1 times the representation of $|r|$. Further, we shall require that $q_0 \in \mathbb{Z}$, and that $q_i \geq 0$ for each i. Our sequence of partial sums, as the sequence $\{ S_n \}$ is called, has initial term an integer, and is increasing. Even these conventions are not enough, as we may still construct (infinitely) many series that meet these requirements and converge to r. We now make an additional requirement: $q_i = \frac{d_i}{10^i}$, where $d_i \in \{ 0, 1, 2, \ldots, 9 \}$ for each $i \geq 1$. Every real number r can be represented as the sum of an infinite series that meets these requirements, and such a series is called a decimal representation of r. If $r = \sum_{i=0}^{\infty} \frac{d_i}{10^i}$ where $d_0 \in \mathbb{Z}$ and $d_i \in \{ 0, 1, \ldots, 9 \}$ for each $i \geq 1$, we might write this decimal representation of r as $r = d_0.d_1 d_2 d_3 \cdots$. Every real number either has a unique decimal representation, or two decimal representations. If r has a decimal representation for which there exists a positive integer n such that $d_i = 9$ for all $i \geq n$, then r also has a decimal representation for which $d_i = 0$ for all $i \geq n$, and vice-versa. For example, $2 = 2 + \sum_{i=1}^{\infty} \frac{0}{10^i}$ and $2 = 1 + \sum_{i=1}^{\infty} \frac{9}{10^i}$. As we mentioned above, we might write $2.000 \cdots = 1.9999 \cdots$, but of course, this notation is somewhat vague.

A.3.1 Definition. *A decimal representation $\sum_{i=0}^{\infty} \frac{d_i}{10^i}$ for which there exist nonnegative integers m and n such that $d_{m+i} = d_{m+n+i}$ for all $i \geq 1$ is called a repeating decimal representation, and if $r = \sum_{i=0}^{\infty} \frac{d_i}{10^i}$, we write*

$$r = d_0.d_1 d_2 \cdots d_m \overline{d_{m+1} d_{m+1} \cdots d_{m+n}}$$

to denote the fact that the block $d_{m+1}d_{m+1}\cdots d_{m+n}$ *repeats forever.*

For example, we write $2 = 1.\overline{9}$, or $\frac{1}{3} = 0.\overline{3}$.

A.3.2 Proposition. *A real number r is rational if, and only if, it has a repeating decimal representation.*

Proof. Let $r \in \mathbb{R}$, and suppose that r has the repeating decimal representation $r = d_0.d_1d_2\cdots d_m\overline{d_{m+1}d_{m+1}\cdots d_{m+n}} = \sum_{i=0}^{\infty}\frac{d_i}{10^i}$ where $d_{m+i} = d_{m+n+i}$ for all $i \geq 1$. Then $10^m r = \sum_{i=0}^{m} d_i 10^{m-i} + 10^m \sum_{i=1}^{\infty}\frac{d_{m+i}}{10^{m+i}}$ and

$$10^{m+n}r = \sum_{i=0}^{m+n} d_i 10^{m+n-i} + 10^{m+n}\sum_{i=1}^{\infty}\frac{d_{m+n+i}}{10^{m+n+i}}$$

$$= \sum_{i=0}^{m+n} d_i 10^{m+n-i} + 10^m \sum_{i=1}^{\infty}\frac{d_{m+i}}{10^{m+i}}.$$

Set $k = \sum_{i=0}^{m+n} d_i 10^{m+n-i} \in \mathbb{Z}$ and $l = \sum_{i=0}^{m} d_i 10^{m-i} \in \mathbb{Z}$ to obtain that $10^{m+n}r = k + (10^m r - l)$. Thus $(10^{m+n} - 10^m)r = k - l$ and so $r = \frac{k-l}{10^{m+n}-10^m} \in \mathbb{Q}$.

Conversely, suppose that there exist integers m and n with $r = \frac{m}{n}$. We may assume that $n > 0$. Apply the division theorem to obtain quotient q and remainder s such that $m = qn + s$ and $0 \leq s < n$. It follows that $\frac{m}{n} = q + \frac{s}{n}$, and $0 \leq \frac{s}{n} < 1$. It therefore suffices to prove that for any positive integers m and n with $0 \leq m < n$, $\frac{m}{n}$ has a repeating decimal representation. If $m = 0$, then we have $\frac{m}{n} = \sum_{i=0}^{\infty}\frac{d_i}{10^i}$ where $d_i = 0$ for $i \geq 0$; that is, $\frac{m}{n} = d_0\overline{0}$. We now consider the case when $m > 0$. For any integer k, let $\mathrm{Mod}_n(k)$ denote the remainder when k is divided by n; that is, by the division theorem, there exist $q, t \in \mathbb{Z}$ with $0 \leq t < n$ and $k = qn + t$, and then $\mathrm{Mod}_n(k) = t$. Consider the sequence $\mathrm{Mod}_n(m), \mathrm{Mod}_n(10m), \ldots, \mathrm{Mod}(10^n m)$ of length $n+1$. Since $\{\,k \in \mathbb{Z} \mid 0 \leq k < n\,\}$ has size k and $\mathrm{Mod}_n(10^i m)$ is in this set for every $i \geq 0$, it follows that there exist $0 \leq i < j \leq n$ such that $\mathrm{Mod}_n(10^i m) = \mathrm{Mod}_n(10^j m)$. Let $t = j - i$, so $t \geq 1$ and $10^i 10^t = 10^j$. There are integers q_1, q_2, r such that $10^i m = q_1 n + r$, $10^j m = q_2 n + r$, and $0 \leq r < n$. Observe that $10^i \frac{m}{n} = q_1 + \frac{r}{n} \geq q_1$, and $\frac{m}{n} < 1$, so $0 \leq q_1 < 10^i$. We also obtain $10^j m = 10^t 10^i m = 10^t q_1 n + 10^t r$. There exist $q_3, s \in \mathbb{Z}$ with $0 \leq s < n$ and $10^t r = q_3 n + s$. Note that $q_3 < q_3 + \frac{s}{n} = 10^t \frac{r}{n} < 10^t$ since $r < n$. Since $q_2 n + r = 10^j m = 10^t q_1 n + q_3 n + s = (10^t q_1 + q_3)n + s$, the uniqueness clause of the division theorem tells us that $s = r$ and $q_2 = 10^t q_1 + q_3$. We therefore have $10^t r = q_3 n + r$ and so $r(10^t - 1) = q_3 n$; equivalently,

$\frac{r}{n} = \frac{q_3}{10^t-1}$. From $10^j m = 10^t q_1 n + q_3 n + r = (10^t q_1 + q_3)n + r$, we obtain (since $10^j = 10^t 10^i$)

$$\begin{aligned}
\frac{m}{n} &= \frac{10^t q_1}{10^j} + \frac{q_3}{10^j} + \frac{r}{n 10^j} \\
&= \frac{q_1}{10^i} + \frac{1}{10^i}\frac{q_3}{10^t} + \frac{1}{10^j}\frac{q_3}{10^t - 1} \\
&= \frac{q_1}{10^i} + \frac{1}{10^i}\frac{q_3}{10^t} + \frac{1}{10^j}\frac{q_3}{10^t}\frac{1}{1 - 10^{-t}} \\
&= \frac{q_1}{10^i} + \frac{1}{10^i}\frac{q_3}{10^t} + \frac{1}{10^j}\frac{q_3}{10^t}\sum_{i=0}^{\infty}(\frac{1}{10^t})^i \\
&= \frac{q_1}{10^i} + \frac{1}{10^i}\frac{q_3}{10^t} + \frac{1}{10^i}\frac{q_3}{10^t}\sum_{i=1}^{\infty}(\frac{1}{10^t})^i \\
&= \frac{q_1}{10^i} + \frac{1}{10^i}\frac{q_3}{10^t}\sum_{i=0}^{\infty}(\frac{1}{10^t})^i.
\end{aligned}$$

Recall that $0 \le q_1 < 10^i$, so we may write q_1 as an i-digit integer (where we will use 0 for high order digits if $q_1 < 10^{i-1}$ (for example, $2 < 10^3$ so we would write 002 to represent 2 as a 3-digit integer), say $q_1 = d_1 d_2 \cdots d_i$. Thus $\frac{q_1}{10^i} = 0.d_1 d_2 \cdots d_i$. As well, $0 \le q_3 < 10^t$, so we can write q_3 as a t-digit integer, say $q_3 = d_{i+1} d_{i+2} \cdots d_{i+t}$, and thus $\frac{q_3}{10^t} = 0.d_{i+1} d_{i+2} \cdots d_{i+t}$. We have now established that

$$\frac{m}{n} = 0.d_1 d_2 \cdots d_i + 0.\underbrace{00 \cdots 0}_{i} d_{i+1} d_{i+2} \cdots d_{i+t} \sum_{i=0}^{\infty}(\frac{1}{10^t})^i$$

$$= 0.d_1 d_2 \cdots d_i \overline{d_{i+1} d_{i+2} \cdots d_{i+t}}.$$

and so $\frac{m}{n}$ has a repeating decimal representation, as claimed. $\quad\square$

For example, $1.3\overline{45}$ is a repeating decimal representation, and so it represents a rational number. Thus there exist positive integers m and n such that $\frac{m}{n} = 1.3\overline{45}$. In the proof of Proposition A.3.2, we actually saw how to find such integers. We have $10(1.3\overline{45}) = 13 + 0.\overline{45}$, and $10^3(1.3\overline{45}) = 1345 + 0.\overline{45}$, so $(1000 - 10)1.3\overline{45} = 1345 - 13 = 1332$. Thus $1.3\overline{45} = \frac{1332}{990} = \frac{666}{495} = \frac{222}{165}$.

To determine a repeating decimal representation for a rational number, the proof above is essentially performing long division to find remainders repeatedly, and one watches to see if either a remainder of 0 turns up (in which case the repeating part of the decimal representation is $\overline{0}$), or a remainder is calculated and it has appeared earlier, in which case we have just determined the repeating portion of the decimal representation. For example, $\frac{1}{3} = 0.\overline{3}$, and $\frac{11}{14} = 0.7\overline{857142}$. These representations are found by the long division process (in the second long division, we show the remainders at each step off to the right):

$$
\begin{array}{r}
0.3 \\
3\,\overline{)\,1.0} \\
\underline{9} \\
1
\end{array}
\qquad\qquad
\begin{array}{r}
0.7857142 \\
14\overline{)\,11.0} \\
\underline{98} \\
120 \quad 12 \\
\underline{112} \\
80 \quad 8 \\
\underline{70} \\
100 \quad 10 \\
\underline{98} \\
20 \quad 2 \\
\underline{14} \\
60 \quad 6 \\
\underline{56} \\
40 \quad 4 \\
\underline{28} \\
12 \;\; 12
\end{array}
$$

For $\frac{1}{3}$, the remainder after the first application of the division theorem was the same as the initial numerator, and so the repeating part starts with the first decimal digit, whereas for $\frac{11}{14}$, the repetition starts with the second decimal digit, as the seventh remainder (12) is the same as the first remainder rather than the original numerator.

A.4   Complex Numbers

We have seen that the set $\mathbb{C}$ of complex numbers is $\mathbb{R}^2$ with the usual real vector space structure and a multiplication operation defined by $(a, b)(c, d) = (ac - bd, ad + bc)$. For $r \in \mathbb{R}$, we agreed to denote $(r, 0)$ by simply r, and $(0, r)$ by $r(0, 1)$, with $i = (0, 1)$. We have determined that $i^2 = -1$, and so, writing $(a, b) = a + bi$ for any $a, b \in \mathbb{R}$, we obtain $(a + bi)(c + di) = ac + bdi^2 + bci + adi = ac - bd + (ad + bc)i$ and $(a + bi) + (c + di) = (a + c) + (b + d)i$. Note that we have relied on the fact that the multiplication operation is commutative and associative, and that multiplication distributes over addition. All of these facts can be simply established by observing that the map λ from $\mathbb{R}^2$ to $M_{2\times 2}(\mathbb{R})$ given by $\lambda(a + bi) = \begin{bmatrix} a & -b \\ b & a \end{bmatrix}$ is an injective $\mathbb{R}$-linear map, and under λ, multiplication in $\mathbb{C}$ corresponds to matrix multiplication in $M_{2\times 2}(\mathbb{R})$. We verify this last assertion, and leave the others for the reader to work through.

Let $a + bi, c + di \in \mathbb{C}$. Then $(a + bi)(c + di) = ac - bd + (ad + bc)i$, so

$$\lambda((a + bi)(c + di)) = \begin{bmatrix} ac-bd & -(ad+bc) \\ ad+bc & ac-bd \end{bmatrix},$$

while

$$\lambda(a + bi)\lambda(c + di) = \begin{bmatrix} a & -b \\ b & a \end{bmatrix} \begin{bmatrix} c & -d \\ d & c \end{bmatrix}$$
$$= \begin{bmatrix} ac-bd & -ad-bc \\ bc+ad & -bd+ac \end{bmatrix}$$
$$= \lambda((a + bi)(c + di)).$$

The real part of $a + bi$ is $Re(a + bi) = a$, and the imaginary part of $a + bi$ is $Im(a + bi) = b$. For $z \in \mathbb{C}$, if $Im(z) = 0$, we say that z is real, while if $Re(z) = 0$, we say that z is pure imaginary. The set $\{\, a + 0i \mid a \in \mathbb{R} \,\}$ is called the real axis, while the set $\{\, 0 + bi \mid b \in \mathbb{R} \,\}$ is called the imaginary axis.

A.4.1 Definition. *For $z = a + bi \in \mathbb{C}$, the complex conjugate of z, denoted by $\overline{z}$, is $a - bi$.*

Note that when considered as a map from $\mathbb{R}^2$ to $\mathbb{R}^2$, complex conjugation is simply reflection in the real axis.

A.4.2 Proposition. *Let $z, w \in \mathbb{C}$. Then the following hold:*

(i) $\overline{z + w} = \overline{z} + \overline{w}$.

(ii) $\overline{zw} = (\overline{z})(\overline{w})$.

(iii) $\overline{(\overline{z})} = z$.

(iv) $z\overline{z}$ is a nonnegative real number, and is 0 if, and only if, $z = 0$. More precisely, if $z = a + bi$, then $z\overline{z} = a^2 + b^2$.

(v) If $z \neq 0$, then z^{-1} exists and is given by $z^{-1} = \frac{1}{z\overline{z}}\overline{z}$. It is customary to write $\frac{1}{z}$ for z^{-1}.

(vi) z is real if, and only if, $z = \overline{z}$.

(vii) If $r \in \mathbb{R}$, then $\overline{rz} = r\overline{z}$.

(viii) $z + \overline{z} = 2Re(z)$, and $z - \overline{z} = 2Im(z)$.

(ix) If $z \neq 0$, then $\overline{z^{-1}} = (\overline{z})^{-1}$, and so $\overline{\left(\frac{w}{z}\right)} = \frac{\overline{w}}{\overline{z}}$ if $z \neq 0$.

Proof. Let $z = a + bi$ and $w = c + di$. Then $\overline{z + w} = \overline{(a + c) + (b + d)i} = a + c - (b + d)i = a - bi + c - di = \overline{z} + \overline{w}$, and $\overline{zw} = \overline{ac - bd + (ad + bc)i} = ac - bd - (ad + bc)i$, while $\overline{z}\,\overline{w} = (a - bi)(c - di) = ac - bd - (ad + bc)i = \overline{zw}$. For (iii), we have $\overline{(\overline{z})} = \overline{(\overline{(a + bi)})} = \overline{a - bi} = a + bi = z$. We compute $z\overline{z} = (a + bi)(a - bi) = a^2 + b^2 + (-ab + ab)i = a^2 + b^2 \geq 0$ and note that $a^2 + b^2 = 0$ if, and only if, $a = 0 = b$. Suppose now that $z \neq 0$, so $a^2 + b^2 \neq 0$. Then $z\overline{z} = a^2 + b^2$ and thus $z(\frac{1}{a^2+b^2}\overline{z}) = 1$. Since complex multiplication is commutative, it follows that the multiplicative inverse of z is $z^{-1} = \frac{1}{z\overline{z}}\overline{z}$. If $z = a + bi$ is real, then $b = 0$ and we have $\overline{z} = a - 0i = a = a + 0i = z$, while if $\overline{z} = z$, then $a + bi = a - bi$ and so $2bi = 0$ which implies that $b = 0$ and so z is real. Now, suppose that $r \in \mathbb{R}$. Then $\overline{rz} = \overline{r}\,\overline{z} = r\overline{z}$. Finally, for (ix), if $z \neq 0$, we have $\overline{z}\overline{z^{-1}} = \overline{zz^{-1}} = \overline{1} = 1$, so $(\overline{z})^{-1} = \overline{z^{-1}}$, and so $\overline{\left(\frac{w}{z}\right)} = \overline{wz^{-1}} = \overline{w}\,\overline{z^{-1}} = \overline{w}(\overline{z})^{-1} = \frac{\overline{w}}{\overline{z}}$. $\square$

By Proposition A.4.2 (iv), for every $z \in \mathbb{C}$, $z\overline{z}$ is a nonnegative real number, and so $\sqrt{z\overline{z}}$ is a nonnegative real number.

A.4.3 Definition. *The modulus of the complex number z, denoted by $|z|$, is given by $|z| = \sqrt{z\overline{z}}$.*

Note that $|a + bi|$ is simply the length of the vector (a, b).

A.4.4 Proposition. *For any $z, w \in \mathbb{C}$, the following hold:*

(i) $|z|^2 = z\overline{z}$, and $|\overline{z}| = |z|$.
(ii) $|zw| = |z|\,|w|$.
(iii) If $w \neq 0$, then $\left|\frac{z}{w}\right| = \frac{|z|}{|w|}$.

Proof. Let $z, w \in \mathbb{C}$. That $|z|^2 = z\overline{z}$ holds follows immediately from the definition, and $|z| = \sqrt{z\overline{z}} = \sqrt{\overline{z}\,\overline{\overline{z}}} = |\overline{z}|$. As for the second assertion, we have $|zw| = \sqrt{zw\overline{zw}} = \sqrt{z\overline{z}w\overline{w}} = \sqrt{z\overline{z}}\sqrt{w\overline{w}} = |z|\,|w|$. Finally, if $w \neq 0$, then $|w| \neq 0$, and so, by Proposition A.4.2 (iv), from $\left|\frac{z}{w}\right||w| = \left|\frac{z}{w}w\right| = |z|$ we obtain that $\left|\frac{z}{w}\right| = \frac{|z|}{|w|}$. $\square$

A.4.5 Proposition. *For any $z \in \mathbb{C}$,*

$$\max\{\,|Re(z)|, |Im(z)|\,\} \leq |z| \leq |Re(z)| + |Im(z)|.$$

Proof. Let $z \in \mathbb{C}$, so $z = x + yi$ for some $x, y \in \mathbb{R}$. Then $Re(z) = x$ and $Im(z) = y$, and we have

$$\max\{\,|Re(z)|, |Im(z)|\,\} = \max\{\,|x|, |y|\,\} = \max\{\,\sqrt{x^2}, \sqrt{y^2}\,\}$$
$$\leq \sqrt{x^2 + y^2} = |z| \leq \sqrt{x^2 + 2|x|\,|y| + y^2}$$
$$= \sqrt{(|x| + |y|)^2} = |x| + |y|.$$

$\square$

A.4.6 Corollary (The Triangle Inequality). *For $z, w \in \mathbb{C}$, $|z + w| \leq |z| + |w|$.*

Proof. Let $z, w \in \mathbb{C}$. Then

$$|z + w|^2 = (z + w)\overline{(z + w)} = (z + w)(\overline{z} + \overline{w})$$
$$= z\overline{z} + z\overline{w} + w\overline{z} + w\overline{w}$$
$$= |z|^2 + z\overline{w} + \overline{z\overline{w}} + |w|^2$$
$$= |z|^2 + 2Re(z\overline{w}) + |w|^2.$$

By Proposition A.4.5, $Re(z\overline{w}) \leq |z\overline{w}| = |z| |\overline{w}|$, and by Proposition A.4.4 (i), $|\overline{w}| = |w|$. Thus $|z + w|^2 \leq |z|^2 + 2|z| |w| + |w|^2 = (|z| + |w|)^2$, and so $|z + w| \leq |z| + |w|$. □

5 The Polar Form of a Complex Number

Recall that the definition of the trigonometric functions $\sin(\theta)$ and $\cos(\theta)$ for $\theta \in \mathbb{R}$ involves measuring distance along the circumference of the unit circle centred at the origin. The reference point is taken to be $(1,0)$, and we measure a distance $|\theta|$ on the circumference of the circle from the reference point $(1,0)$, in the counterclockwise direction if $\theta > 0$, otherwise in the clockwise direction. The point on the circle that is obtained by this procedure is defined to have coordinates $(\cos(\theta), \sin(\theta))$. For example, since the circumference of a circle of radius 1 is 2π, it follows that one quarter of a circle has measure $\frac{\pi}{2}$, so $(\cos(\frac{\pi}{2}), \sin(\frac{\pi}{2})) = (0, 1)$. If we construct the ray from the origin through the point $(\cos(\theta), \sin(\theta))$, then we can uniquely identify points on this ray in terms of their distance from the origin. Since the point $(\cos(\theta), \sin(\theta))$ is on this ray and is a distance 1 from the origin, the point on this ray that is a distance $r > 0$ from the origin is $r(\cos(\theta), \sin(\theta)) = (r \cos(\theta), r \sin(\theta))$. Thus for a point (a, b) on this ray, we have $(a, b) = (r \cos(\theta), r \sin(\theta))$, so $a = r \cos(\theta)$ and $b = r \sin(\theta)$. Since $(\cos(\theta), \sin(\theta))$ is on the unit circle, as we have observed above, $\cos^2(\theta) + \sin^2(\theta) = 1$, and so $a^2 + b^2 = r^2(\cos^2(\theta) + \sin^2(\theta)) = r^2$. Thus $r = \sqrt{a^2 + b^2}$. Now, in the context of the complex numbers, $(a, b) = a + bi$, and so we have $r = |a + bi|$. As well, although the sine and cosine functions are periodic ($\sin(\theta) = \sin(\theta + 2\pi)$ and $\cos(\theta) = \cos(\theta + 2\pi)$ for every $\theta \in \mathbb{R}$), we can recover these values from the known coordinate values a and b, for $\cos(\theta) = \frac{a}{r}$ and $\sin(\theta) = \frac{b}{r}$, where $-1 \leq \frac{a}{r} \leq 1$ and $-1 \leq \frac{b}{r} \leq 1$.

A.5.1 Definition. *For $z \in \mathbb{C}$ with $z \neq 0$, the set of all values of $\theta \in \mathbb{R}$ for which $\cos(\theta) = \frac{Re(z)}{|z|}$ and $\sin(\theta) = \frac{Im(z)}{|z|}$ is called the argument of z, denoted by $arg(z)$, while the unique value of $\theta \in arg(z)$ for which $-\pi < \theta \leq \pi$ is*

called the principal value of the argument of z, denoted by $Arg(z)$. Thus
$arg(z) = \{ Arg(z) + 2n\pi \mid n \in \mathbb{Z} \}.$

Thus for $z \in \mathbb{C}$ with $z \neq 0$, $z = |z|(cos(\theta) + i \sin(\theta)$ for any $\theta \in arg(z)$. It is conventional to write $\text{cis}(\theta)$ for $\cos(\theta) + i\sin(\theta)$. With this notation, $z = |z| \text{cis}(\theta)$. This is called the polar form of z.

A.5.2 Proposition.   *For any $\alpha, \beta \in \mathbb{R}$, $cis(\alpha + \beta) = cis(\alpha)cis(\beta)$.*

Proof. The addition formulas for the sine and cosine functions state that $\cos(\alpha + \beta) = \cos(\alpha)\cos(\beta) - \sin(\alpha)\sin(\beta)$ and $\sin(\alpha + \beta) = \sin(\alpha)\cos(\beta) + \cos(\alpha)\sin(\beta)$, so $\text{cis}(\alpha + \beta) = \cos(\alpha + \beta) + i\sin(\alpha + \beta) = \cos(\alpha)\cos(\beta) - \sin(\alpha)\sin(\beta) + i(\sin(\alpha)\cos(\beta) + \cos(\alpha)\sin(\beta))$, while

$$\text{cis}(\alpha)\text{cis}(\beta) = (\cos(\alpha) + i\sin(\alpha))(\cos(\beta) + i\sin(\beta))$$
$$= \cos(\alpha)\cos(\beta) - \sin(\alpha)\sin(\beta) + i(\cos(\alpha)\sin(\beta) + \sin(\alpha)\cos(\beta))$$
$$= \text{cis}(\alpha + \beta).$$

$\square$

We remark that the multiplication of two complex numbers takes a particularly pleasing form in polar coordinates: for $z, w \in \mathbb{C}$, neither 0, we have $z = |z|\text{cis}(\theta)$ and $w = |w|\text{cis}(\omega)$ for some $\theta, \omega \in \mathbb{R}$, and then $zw = |z||w|\text{cis}(\theta)\text{cis}(\omega) = |z||w|\text{cis}(\theta + \omega)$. In particular, for any $z \in \mathbb{C}$, $z \neq 0$, $zi = |z|\text{cis}(\theta)\text{cis}(\frac{\pi}{2}) = |z|\text{cis}(\theta + \frac{\pi}{2})$. Thus multiplication by i amounts to a rotation in the counterclockwise direction through an angle of $\frac{\pi}{2}$.

A.5.3 Corollary (De Moivre's Theorem).   *For any $\theta \in \mathbb{R}$ and any positive integer n, $(cis(\theta))^n = cis(n\theta)$.*

Proof. This follows from Proposition A.5.2 via a straightforward argument by induction. Let $\theta \in \mathbb{R}$. For $n \geq 1$ an integer for which the hypothesis holds, we have $(\text{cis}(\theta))^{n+1} = (\text{cis}(\theta))^n\text{cis}(\theta) = \text{cis}(n\theta)\text{cis}(\theta) = \text{cis}(n\theta + \theta) = \text{cis}((n+1)\theta)$, and so the result follows by induction. $\square$

In fact, De Moivre's theorem holds for integers $n \leq 0$ as well. For $n = 0$, we observe that $(\text{cis}(\theta))^0 = 1 = \text{cis}(0) = \text{cis}(0\theta)$, while for $n < 0$, the result follows from the fact that $\text{cis}(n\theta)\text{cis}(|n|\theta) = \text{cis}((n + |n|)\theta) = \text{cis}(0) = 1$, and so $\text{cis}(n\theta) = (\text{cis}(|n|\theta))^{-1} = ((\text{cis}(\theta))^{|n|})^{-1} = (\text{cis}(\theta))^{-|n|} = (\text{cis}(\theta))^n$ when $n < 0$.

The periodicity of the trigonometric functions comes into play when we must determine whether or not two nonzero complex numbers given in polar form are equal.

A.5.4 Proposition. *Let r, s be positive real numbers, and let $\theta, \omega \in \mathbb{R}$. Then $r\, cis(\theta) = s\, cis(\omega)$ if, and only if, $r = s$ and $\theta - \omega = 2n\pi$ for some integer n.*

Proof. Suppose first of all that $r\, \mathrm{cis}(\theta) = s\, \mathrm{cis}(\omega)$. Then $r = |r\, \mathrm{cis}(\theta)| = |s\, \mathrm{cis}(\omega)| = s$, and thus $\mathrm{cis}(\theta) = \mathrm{cis}(\omega)$, which means that $\mathrm{cis}(\theta)\mathrm{cis}(-\omega) = \mathrm{cis}(\theta)(\mathrm{cis}(\omega))^{-1} = 1$ and so $\mathrm{cis}(\theta - \omega) = 1$. It follows that $\cos(\theta - \omega) = 1$ and $\sin(\theta - \omega) = 0$. The zeroes of the sine function are $m\pi$, $m \in \mathbb{Z}$, and for m odd, $\cos(m\pi) = -1$, while for m even, $\cos(m\pi) = 1$. Thus $\theta - \omega = 2n\pi$ for some $n \in \mathbb{Z}$.

The converse follows immediately from the periodicity of the trigonometric functions. $\qquad\square$

One immediate consequence of De Moivre's theorem is the fact that for any positive integer n, there is $z \in \mathbb{C}$ with $z \neq 1$ and for which $z^n = 1$; namely $(\mathrm{cis}(\frac{2\pi}{n}))^n = \mathrm{cis}(2\pi) = 1$. Also note that if $z^n = 1$ and $w^n = 1$, then $(zw)^n = 1$ and $(z^{-1})^n = (z^n)^{-1} = 1$, so the set $\{\, z \in \mathbb{C} \mid z^n = 1 \,\}$ is a subgroup of the multiplicative group of nonzero elements of $\mathbb{C}$.

A.5.5 Definition. *For any positive integer n, the subgroup $\{\, z \in \mathbb{C} \mid z^n = 1 \,\}$ is called the group of nth roots of unity.*

We remark that it is a finite group, since it consists of the roots of $x^n - 1 \in \mathbb{C}[x]$ and since $\mathbb{C}$ is a field, there are at most (and in the case of $\mathbb{C}$, exactly) n roots of the equation $x^n - 1 = 0$ (see Corollary 9.7.3). Since every finite subgroup of the multiplicative group of a field is a cyclic group, it follows that the group of nth roots of unity is cyclic.

A.5.6 Proposition. *Let n be a positive integer. Then the group of nth roots of unity is $\{\, cis(\frac{2k\pi}{n}) \mid k = 0, 1, \ldots, n - 1 \,\} = \langle cis(\frac{2\pi}{n}) \rangle$.*

Proof. By De Moivre's theorem, $(\mathrm{cis}(\frac{2k\pi}{n}))^n = \mathrm{cis}(2k\pi) = 1$, so for each $k \in \mathbb{Z}$, $\mathrm{cis}(\frac{2k\pi}{n})$ is an nth root of unity. By Proposition A.5.4, the set $\{\, \mathrm{cis}(\frac{2k\pi}{n}) \mid k = 0, 1, \ldots, n - 1 \,\}$ has size n, and thus must be the complete set of roots of $x^n - 1$. $\qquad\square$

Any generator of the cyclic group of nth roots of unity is called a primitive n^{th} root of unity. In particular, $\mathrm{cis}(\frac{2\pi}{n})$ is a primitive n^{th} root of unity, and for $k = 0, 1, \ldots, n - 1$, $\text{cis}(\frac{2k\pi}{n})$ is a primitive n^{th} root of unity if, and only if, k and n are relatively prime.

When plotted in the plane, the set of nth roots of unity forms a regular n-gon with vertices on the unit circle. For example, the six 6th roots of unity are the vertices of the regular 6-gon displayed in Figure A.1.

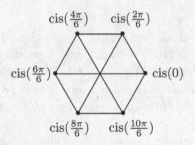

Fig. A.1   The 6th-roots of unity

Our next objective is to prove that for any positive integer n and any nonzero $w \in \mathbb{C}$, the equation $z^n = w$ has exactly n solutions. Since $w \neq 0$, we may write $w = |w|\text{cis}(\theta)$ for some $\theta \in \mathbb{R}$. Certainly, if z is a solution, then $z \neq 0$, so we search for solutions of the form $z = r\,\text{cis}(\beta)$, $r > 0$, $\beta \in \mathbb{R}$. By De Moivre's theorem, this requires that $r^n \text{cis}(n\beta) = |w|\,\text{cis}(\theta)$, and by Proposition A.5.4, it is necessary and sufficient that $r^n = |w|$ and $n\beta = \theta + 2k\pi$ for $k \in \mathbb{Z}$. Thus $r = |w|^{\frac{1}{n}}$ and $\beta = \frac{\theta}{n} + k\frac{2\pi}{n}$, for $k \in \mathbb{Z}$. By the periodicity of the sine and cosine functions, we obtain as our full set of solutions

$$\{\, |w|^{\frac{1}{n}} \text{cis}(\tfrac{\theta}{n} + k\tfrac{2\pi}{n}) \mid k = 0, 1, \ldots, n - 1 \,\}$$
$$= \{\, |w|^{\frac{1}{n}} \text{cis}(\tfrac{\theta}{n})(\text{cis}(\tfrac{2\pi}{n}))^k \mid k = 0, 1, \ldots, n - 1 \,\}.$$

A.5.7 Example. *Find all sixth roots of* $1 - i$. *We calculate* $|1 - i| = \sqrt{2}$, *so for* $\theta = Arg(1 - i)$, *we have* $\cos(\theta) = \frac{1}{\sqrt{2}}$ *and* $\sin(\theta) = -\frac{1}{\sqrt{2}}$, *which means that* $\theta = -\frac{\pi}{4}$. *Thus the* 6th-*roots of* $1 - i$ *are* $2^{\frac{1}{12}} cis((-\frac{\pi}{4} + 2k\pi)/6)$, $k = 0, 1, 2, 3, 4, 5$; *that is, the* 6th-*roots of* $1 - i$ *are* $2^{\frac{1}{12}} cis(-\frac{\pi}{24}) cis(\frac{k\pi}{3})$, $k = 0, 1, 2, 3, 4, 5.$

A.5.8 Example. *Compute* $(-\frac{1}{2} + \frac{\sqrt{3}}{2})^{12}$. *Since* $|-\frac{1}{2} + \frac{\sqrt{3}}{2}| = 1$, *we see that if* $\theta = Arg(-\frac{1}{2} + \frac{\sqrt{3}}{2})$, *then* $\cos(\theta) = (-1/2)/1 = -\frac{1}{2}$ *and* $\sin(\theta) = \frac{\sqrt{3}}{2}$, *so* $\theta = \frac{2\pi}{3}$. *Now by De Moivre's theorem,* $(-\frac{1}{2} + \frac{\sqrt{3}}{2})^{12} = (1)^{12} cis(12\frac{2\pi}{3})| =$

$cis(8\pi) = 1.$

Our next example shows how to determine the square roots of $z \in \mathbb{C}$, expressed in terms $Re(z)$ and $Im(z)$.

A.5.9 Example. *For $z = a + ib \in \mathbb{C}$ with $b \neq 0$,*

$$z^{1/2} = \pm\frac{1}{\sqrt{2}}\left(\sqrt{|z| + a} + i\frac{b}{|b|}\sqrt{|z| - a}\right).$$

Suppose that $x, y \in \mathbb{R}$ are such that $(x + yi)^2 = a + bi$. Since $(x + yi)^2 = x^2 - y^2 + 2xyi$, we obtain $x^2 - y^2 = a$ and $2xy = b$. Since $b \neq 0$, it follows that $x \neq 0$ and thus $y = b/2x$. This gives $a = x^2 - (\frac{b}{2x})^2$, or $(x^2)^2 - ax^2 - \frac{b^2}{4} = 0$, which is a quadratic equation in the unknown x^2. We solve this and find that $x^2 = \frac{1}{2}(a \pm \sqrt{a^2 + 4\frac{b^2}{4}}) = \frac{1}{2}(a \pm \sqrt{a^2 + b^2})$. Since $x \in \mathbb{R}$ and $x^2 \geq 0$, we must have $x^2 = \frac{1}{2}(a + |z|)$, and so $x = \pm\frac{1}{\sqrt{2}}\sqrt{a + |z|}$. Write $x = \delta\frac{1}{\sqrt{2}}\sqrt{a + |z|}$, where $\delta = \pm 1$. It now follows that $y = \frac{b}{2(\delta\frac{1}{\sqrt{2}}\sqrt{a+|z|})} = \frac{b\sqrt{|z|-a}}{\delta\sqrt{2}\sqrt{(|z|+a)(|z|-a)}} = \delta\frac{b\sqrt{|z|-a}}{\sqrt{2}\sqrt{b^2}} = \delta\frac{b}{|b|}\frac{\sqrt{|z|-a}}{\sqrt{2}}$. Thus $x + yi = \delta\frac{1}{\sqrt{2}}(\sqrt{|z| + a} + i\frac{b}{|b|}\sqrt{|z| - a}) = \pm\frac{1}{\sqrt{2}}(\sqrt{|z| + a} + i\frac{b}{|b|}\sqrt{|z| - a}).$

6 Exercises

1. a) Prove that if $a \in \mathbb{R}$ is rational and $b \in \mathbb{R}$ is irrational, then $a + b$, ab and a/b are all irrational.

   b) Prove that if $a, b \in \mathbb{R}$ are such that $a + b$ is rational and $a - b$ is irrational, then both a and b are irrational. Find examples of irrational a, b such that $a + b$ is rational and $a - b$ is irrational.

   c) Give examples to show that it is possible for the sum, product, and quotient of two irrational numbers to be rational.

2. Prove that between any two real numbers there is an irrational number (note that we have proved that between any two real numbers, there is a rational number).

3. Find a decimal representation for each of the following rational numbers:

   a) $\frac{22}{7}$;

   b) $\frac{5}{11}$;

   c) $\frac{7}{12}$;

   d) $\frac{5}{8}$;

    e) $\frac{22}{9}$.

4. In each case below, find suitable integers m and n:

    a) $\frac{m}{n} = 1.2\overline{34}$;

    b) $\frac{m}{n} = 0.012\overline{501}$;

    c) $\frac{m}{n} = 2.2\overline{4}$;

    d) $\frac{m}{n} = 0.\overline{9}$.

5. In each case below, either provide a proof of the assertion or else provide a counterexample for the assertion.

    a) For any $a, b \in \mathbb{R}$, if $a + b$ and ab are rational, then both a and b are rational.

    b) For any nonzero $a, b \in \mathbb{R}$, if $a + b$ and $\frac{a}{b}$ are rational, then both a and b are rational.

6. Prove that for any $\epsilon > 0$, there exists a positive integer n such that $\frac{1}{n} < \epsilon$.

7. Prove that if $f \in \mathbb{R}[x]$ and $r = a + bi \in \mathbb{C}$ is a root of f with $b \neq 0$, then $x^2 + 2ax + (a^2 + b^2)$ is a divisor of f in $\mathbb{C}[x]$.

8. In each case below, find all $z \in \mathbb{C}$ that satisfy the given equation:

    a) $z^3 + 1 = 0$;

    b) $z^4 + 16 = 0$;

    c) $z^3 + 1 + i = 0$;

    d) $z^6 + 1 = 0$;

    e) $z^2 + z + 1 = 0$.

9. Express $z^4 + 16$ as a product of two quadratic polynomials each with real coefficients.

10. a) Prove that

$$U = \{\, z \in \mathbb{C} \mid \text{there exists a positive integer } n \text{ such that } z^n = 1 \,\}$$

        is a subgroup of the multiplicative group $\mathbb{C}^* = \mathbb{C} - \{\, 0 \,\}$.

    b) Prove that for every positive integer n, the cyclic group of nth-roots of unity is a subgroup of U.

    c) Prove that U is not a cyclic group.

11. a) Use De Moivre's theorem to prove that for every positive integer n and every $\theta \in \mathbb{R}$, $\cos(n\theta) = \sum_{i=0}^{m}(-1)^i \binom{n}{2i} \cos^{n-2i}(\theta) \sin^{2i}(\theta)$, where $m = \frac{n}{2}$ if n is even, otherwise $m = (n-1)/2$.

b) 'For $\theta \in \mathbb{R}$, express $\cos(4\theta)$ in terms of powers of $\cos(\theta)$.

12. Let $a, b \in \mathbb{C}$ with $a \neq 0$.

   a) Prove that the parametric equation $z = b + ta$, $t \in \mathbb{R}$, represents a straight line in the plane passing through b and parallel to a.

   b) Prove that a and ia represent two directions perpendicular to each other (that is, prove that the angle between a and ia is an odd multiple of $\frac{\pi}{2}$).

   c) Prove that a line perpendicular to the direction a can be written as the set of $z \in \mathbb{C}$ which satisfy $\overline{a}z + a\overline{z} = c$ for some $c \in \mathbb{R}$.

   d) Let $z \in \mathbb{C}$. Prove that if z' is the reflection of z in the line $\overline{a}z + a\overline{z} = c$, where $c \in \mathbb{R}$, then $\overline{a}z' + a\overline{z} = c$. Then find the reflection of $(1, 1)$ (that is, $z = 1 + i$) in the line with equation $2x + y = 1$.

13. Let $a \in \mathbb{C}$. Prove that any circle with center $-a$ can be represented by an equation of the form $z\overline{z} + \overline{a}z + a\overline{z} + b = 0$ for some $b \in \mathbb{R}$ with $b < |a|^2$.

14. Let $S \subseteq \mathbb{R}$ be nonempty. Then S is bounded below if there exists $r \in \mathbb{R}$ such that for all $x \in S$, $r \leq x$. Such a number r is called a lower bound for S. Further, S is said to have a greatest lower bound if the set of all lower bounds of S is not empty and has a greatest element b, in which case b is called a greatest lower bound of S.

   a) Let $S \subseteq \mathbb{R}$ be nonempty. Prove that if S has a lower bound r, then $-S = \{-x \mid x \in S\}$ has upper bound $-r$.

   b) Let $S \subseteq \mathbb{R}$ be nonempty and bounded below. By (a), $lub(-S)$ exists. Prove that $-lub(-S)$ is the unique greatest lower bound of S.

Appendix B

Linear Algebra

1 Vector Spaces

It is assumed that the reader is familiar with the basic ideas of linear algebra and this appendix is provided not as a detailed account of the subject, but rather as a resource for the reader to consult for a quick verification of some of the facts. The pace will be fast and some of the proofs may differ from the proofs encountered by the reader in more elementary courses. However, given that the reader will probably not have to consult this appendix until the later chapters, it is hoped that by that time, the reader will have acquired sufficient mathematical maturity to find the exposition digestible.

B.1.1 Definition. *Let F be a field. A vector space over F consists of:*

(i) *an abelian group $(V, +)$, whose elements are called the vectors of the vector space;*

(ii) *an action of F on V (that is, a unital ring homomorphism from F into $\text{End}(V)$, the ring of homomorphisms from V to itself, where the multiplication operation is composition of homomorphisms. The effect of the action of $r \in F$ on $v \in V$ is denoted by rv, and the action of F on V is called scalar multiplication.*

For the reader's convenience, we shall explicitly list the properties of scalar multiplication (these follow immediately from the definition).

B.1.2 Proposition. *Let V be a vector space over the field F. Then the following hold:*

(i) *For every $r, s \in F$ and every $v \in V$, $(r + s)v = rv + sv$.*

(ii) *For every $r, s \in F$ and every $v \in V$, $(rs)v = r(sv)$.*

503

(iii) For every $r \in F$ and every $v, w \in V$, $r(v + w) = rv + rw$.
(iv) For every $v \in V$, $1_F v = v$.

For every field F, there are vector spaces over F.

B.1.3 Proposition. *For any field F, and any nonempty set X, the abelian group $V = \mathscr{F}(X, F)$ under addition can be made into an F-vector space as follows. For each $r \in F$, let $\theta_r : V \to V$ be the map given by $\theta_r(f)(x) = (r)f(x)$ for each $x \in X$, and then define $\theta : F \to \mathscr{F}(V, V)$ by $\theta(r) = \theta_r$ for all $r \in F$. Then θ is a unital ring homomorphism from F into End(V).*

Proof. First, we prove that $\theta_r \in \text{End}(V)$ for each $r \in F$. Let $r \in F$, $f, g \in V$. Then for each $x \in X$, $\theta_r(f + g)(x) = (r)((f + g)(x)) = r(f(x) + g(x)) = (r)(f(x)) + (r)(g(x)) = \theta_r(f)(x) + \theta_r(g)(x) = (\theta_r(f) + \theta_r(g))(x)$, and so $\theta_r(f + g) = \theta_r(f) + \theta_r(g)$. Thus $\theta_r \in \text{End}(V)$ for each $r \in F$. Next, we prove that θ is a ring homomorphism. Let $r, s \in F$, and $f \in V$. Then for each $x \in X$, $\theta_{rs}(f)(x) = (rs)(f(x)) = r(s(f(x)) = (r)(\theta_s(f)(x)) = \theta_r(\theta_s(f))(x)$, and so $\theta_{rs}(f) = \theta_r(\theta_s(f)) = \theta_r \circ \theta_x(f)$. Since f was arbitrary in V, it follows that $\theta_{rs} = \theta_r \circ \theta_s$. As well, $\theta_{r+s}(f)(x) = (r + s)(f(x)) = (r)(f(x)) + (x)(f(x)) = \theta_r(f)(x) + \theta_s(f)(x) = (\theta_r(f) + \theta_s(f))(x)$, and so $\theta_{r+s}(f) = \theta_r(f) + \theta_s(f) = (\theta_r + \theta_s)(f)$. Thus $\theta_{r+s} = \theta_r + \theta_s$, and so θ is a ring homomorphism. Finally, let $f \in V$ and $x \in X$. Then $\theta_{1_F}(f)(x) = (1_F)(f(x)) = f(x)$, and so $\theta_{1_F}(f) = f$ for all $f \in V$. But this means that $\theta_{1_F} = \mathbb{1}_V$, and so θ is a unital ring homomorphism. $\quad\square$

B.1.4 Corollary. *For any field F, and any positive integer n, the abelian group $V = F^n$, the direct sum of n copies of $(F, +)$, is a vector space over F with scalar multiplication given by $r(a_1, a_2, \ldots, a_n) = (ra_1, ra_2, \ldots, ra_n)$ for each $r \in F$ and $(a_1, a_2, \ldots, a_n) \in F^n$.*

Proof. This follows immediately from Proposition B.1.3 since $F^n = \mathscr{F}(J_n, F)$. Recall that we have adopted the convention that for $f \in F^n$, if for each $i \in J_n$, we let a_i denote $f(i) \in F$, then we represent f by the n-tuple $(a_1, a_2, \ldots, a_n)$. $\quad\square$

In Chapter 3, we discussed matrices with entries from $\mathbb{Z}$, $\mathbb{Q}$, $\mathbb{R}$, or $\mathbb{C}$, and defined matrix addition and multiplication in Definition 3.4.25 for matrices with entries from any of these rings. The arguments presented there are valid for $M_{m \times n}(R)$, where R is any commutative unital ring, and thus in particular for $M_{m \times n}(F)$. We may define scalar multiplication on $M_{m \times n}(F)$ as follows: for any $A \in M_{m \times n}(F)$ and $r \in F$, $rA \in M_{m \times n}(F)$ is the $m \times n$

matrix for which $(rA)_{ij} = r(A_{ij})$ for each $(i,j) \in J_m \times J_n$; that is, the entries of rA are those of A, each multiplied by r using the multiplication operation of the field F. By an argument very much like that used to prove Proposition B.1.3, it may be proven that $M_{m \times n}(F)$ is a vector space over F.

B.1.5 Proposition. *Let V be a vector space over the field F. Then the following hold:*

(i) *For any $v \in V$, $0_F v = 0_V$.*

(ii) *For any $v \in V$, $(-1_F)v = -v$*

(iii) *For any $v \in V$ and $r \in F$, $rv = 0_V$ if, and only if, either $r = 0_F$ or $v = 0_V$.*

Proof. Let $v \in V$. Then $0_F v = (0_F + 0_F)v = 0_F v + 0_F v$, and thus $0_V = 0_F v + (-0_F v) = 0_F v + 0_F v + (-0_F v) = 0_F v$. Then for (ii), we have $v + (-1_F)v = 1_F v + (-1_F)v = (1_F - 1_F)v = 0_F v = 0_V$, so $-v = (-1_F)v$. Finally, suppose that $r \in F$. Then $r 0_V = r(0_V + 0_V) = r 0_V + r 0_V$, and thus $r 0_V = 0_V$, and this, together with (i), proves that if $r = 0_F$ or $v = 0_V$, then $rv = 0_V$. Now suppose that $rv = 0_V$. If $r \neq 0_F$, then $v = 1_F v = (r^{-1} r)v = r^{-1}(rv) = r^{-1} 0_V = 0_V$. $\quad\square$

In the study of a vector space V over a field F, it is very common to encounter subsets which are nonempty, closed under addition, and every scalar multiple of an element of the subset is again an element of the subset.

B.1.6 Definition. *Let V be a vector space over a field F, and let $W \subseteq V$.*

(i) *If for all $v, w \in W$, $v + w \in W$, W is said to be closed with respect to addition.*

(ii) *If for all $v \in W$ and $r \in F$, $rv \in W$, then W is said to be closed with respect to scalar multiplication.*

We note that the empty set is both closed with respect to addition and with respect to scalar multiplication. If W is a nonempty subset of a vector space V over the field F, and W is closed with respect to addition, then the addition operation of V, when restricted to apply only to elements of W, provides an associative, commutative binary operation on W. Furthermore, if W is also closed under scalar multiplication, then W is closed under additive inverse, and so is a subgroup of V, and the action of F on V, when restricted to W, provides an action of F on W. Thus if W is nonempty and closed under both addition and scalar multiplication, then W is a vector space over the field F. Since its vector space structure is obtained from that of V, such a subset W is called a subspace of V.

The notion of linear combination plays a central role in linear algebra.

B.1.7 Definition. *Let V be a vector space over a field F, and let $S \subseteq V$ be nonempty. Then $v \in V$ is said to be a linear combination of elements of S if for some positive integer n there exist $v_1, v_2, \ldots, v_n \in S$ and $r_1, r_2, \ldots, r_n \in F$ such that $v = \sum_{i=1}^{n} r_i v_i$. Furthermore, the span of S, denoted by $sp(S)$, is the set of all linear combinations of elements of S. Finally, we define the span of the empty set by $sp(\varnothing) = \{0_V\}$.*

B.1.8 Example. *If $F = \mathbb{R}$, $V = \mathbb{R}^3$, and $S = \{(1,0,1),(1,1,0),(0,1,1)\}$, then*

$$sp(S) = \{a(1,0,1) + b(1,1,0) + c(0,1,1) \mid a,b,c \in \mathbb{R}\}$$
$$= \{(a+b, b+c, a+c) \mid a,b,c \in \mathbb{R}\}.$$

Of course, in this example, we can construct individual elements of $sp(S)$, but given $v \in \mathbb{R}^3$, it is not immediately clear how one would determine whether or not $v \in sp(S)$.

B.1.9 Proposition. *Let V be a vector space over the field F, and let S be a subset of V. Then $sp(S)$ is a subspace of V.*

Proof. We must show that $W = sp(S)$ is nonempty and closed under addition and scalar multiplication. If $S = \varnothing$, then $W = \{0_V\}$, which is nonempty and closed under both addition and scalar multiplication by virtue of Proposition B.1.5. Suppose that $S \neq \varnothing$. Since $S \subseteq sp(S)$, it follows that $sp(S) \neq \varnothing$. Let $v, w \in sp(S)$. Then there exist positive integers m and n, and $v_1, v_2, \ldots, v_m, w_1, w_2, \ldots, w_n \in S$ and $r_1, r_2, \ldots, r_m, s_1, s_2, \ldots, s_n \in F$ such that $v = \sum_{i=1}^{m} r_i v_i$ and $w = \sum_{j=1}^{n} s_j w_j$. Then $v + w = \sum_{i=1}^{m} r_i v_i + \sum_{j=1}^{n} s_j w_j \in sp(S)$, and for any $r \in F$, $rv = r(\sum_{i=1}^{m} r_i v_i) = \sum_{i=1}^{m} r(r_i v_i) = \sum_{i=1}^{m} (rr_i)v_i \in sp(S)$. Thus $sp(S)$ is nonempty, and closed under both addition and scalar multiplication. $\square$

B.1.10 Proposition. *Let V be a vector space over the field F, and let $S \subseteq V$. Then*

$$sp(S) = \cap\{W \subseteq V \mid W \text{ is a subspace of } V \text{ and } S \subseteq W\}.$$

Proof. Let W be a subspace of V with $S \subseteq W$. Then since W is closed under addition and scalar multiplication, W is closed under linear combinations and since $S \subseteq W$, it follows that $sp(S) \subseteq W$. Thus $sp(S) \subseteq \cap\{W \subseteq V \mid W$ is a subspace of V and $S \subseteq W\}$. On the other hand, by Proposition B.1.9, $sp(S)$ is a subspace of V and $S \subseteq sp(S)$, so $\cap\{W \subseteq V \mid W$ is a subspace of V and $S \subseteq W\} \subseteq sp(S)$. □

B.1.11 Corollary. *Let V be a vector space over the field F. Then for any $S \subseteq T \subseteq V$, $sp(S) \subseteq sp(T)$. As well, if W is a subspace of V, then $sp(W) = W$.*

Proof. Suppose that $S \subseteq T \subseteq V$. Then by Proposition B.1.9, $sp(T)$ is a subspace of V, and $S \subseteq T \subseteq sp(T)$, so by Proposition B.1.10, $sp(S) \subseteq sp(T)$. For any subspace W of V, we have $sp(W) \subseteq W \subseteq sp(W)$ and thus $sp(W) = W$. □

B.1.12 Definition. *Let V be a vector space over F. A set $S \subseteq V$ is linearly independent if for all $v \in S$, $v \notin sp(S - \{v\})$. If S is not linearly independent, then S is said to be linearly dependent.*

Note that $\varnothing$ is linearly independent, and any set that contains 0_V is linearly independent as $o_V \in sp(S)$ for any subset S of V.

B.1.13 Proposition. *Let V be a vector space over a field F, and let S be a nonempty subset of V. Then S is linearly independent if, and only if, for every positive integer n and every choice of n distinct elements $v_1, v_2, \ldots, v_n \in S$ and $r_1, r_2, \ldots, r_n \in F$, $\sum_{i=1}^{n} r_i v_i = 0_V$ implies that $r_i = 0_F$ for each $i = 1, 2, \ldots, n$.*

Proof. Suppose first of all that S is linearly independent, and suppose by way of contradiction that there is a positive integer n and $v_1, v_2, \ldots, v_n \in S$ and $r_1, r_2, \ldots, r_n$, such that $\sum_{i=1}^{n} r_i v_i = 0_V$, and that there exists $i \in J_n$ with $r_i \neq 0_F$. Let k denote the maximum of $i = 1, 2, \ldots, n$ for which $r_i \neq 0_F$, so $\sum_{i=1}^{k} r_i v_i = 0_V$. If $k = 1$, then $r_1 v_1 = 0_V$ and since $0_V \notin S$ because S is linearly independent, it follows that $v_1 \neq 0_V$, so by Proposition B.1.5 (iii), $r_1 = 0$. As this contradicts the choice of k, we conclude that $k > 1$. But then we have $-r_k v_k = \sum_{i=1}^{k-1} r_i v_i$, and so $v_k = \sum_{i=1}^{k-1} (r_k^{-1} r_i) v_i \in sp(S - \{v_k\})$, contradicting the fact that S is linearly independent. Thus $\sum_{i=1}^{n} r_i v_i = 0_V$ implies that $r_i = 0_F$ for every $i = 1, 2, \ldots, n$. Conversely, suppose that S is a nonempty subset of V with the indicated property, and suppose that S is linearly dependent. Then there exists $v \in S$ such that $v \in$

$sp(S - \{v\})$, and so there exists a positive integer n and n distinct elements $v_1, v_2, \ldots, v_n \in S - \{v\}$ and $r_1, r_2, \ldots, r_n \in F$ such that $v = \sum_{i=1}^{n} r_i v_i$. Then for $r = -1_F$, we have $rv + \sum_{i=1}^{n} r_i v_i = 0_V$, with $\{v, v_1, v_2, \ldots, v_n\}$ a set of $n+1$ distinct elements of S, and so by hypothesis, $r_i = 0_F$ for $i = 1, 2, \ldots, n$ and $r = 0_F$. As this contradicts the fact that $r = -1_F \neq 0_F$, we conclude that S is linearly independent. $\qquad\square$

B.1.14 Proposition. *Let V be a vector space over the field F, and let S be a linearly independent subset of V. Then every subset T of S is linearly independent.*

Proof. Let $T \subseteq S$ and suppose by way of contradiction that T is linearly dependent. Then there exists $v \in T$ such that $v \in sp(T - \{v\}) \subseteq sp(S - \{v\})$, which contradicts the fact that S is linearly independent. $\qquad\square$

B.1.15 Proposition. *Let V be a vector space over a field F. A subset S of V is linearly independent if, and only if, every finite subset of S is linearly independent.*

Proof. Let S be a subset of V. Suppose first of all that S is linearly independent. Then by Proposition B.1.14, every subset of S is linearly independent, and so in particular, every finite subset of S is linearly independent.

Suppose now that every finite subset of S is linearly independent. We are to prove that S itself is linearly independent. Suppose to the contrary that there exists $v \in S$ such that $v \in sp(S - \{v\})$. Then there exists a finite subset T of $S - \{v\}$ such that $v \in sp(T)$. But then $T' = T \cup \{v\}$ is a finite subset of S and thus by assumption is linearly independent. But $v \in T'$ satisfies $v \in sp(T' - \{v\})$, which contradicts the linear independence of T'. Thus S is linearly independent. $\qquad\square$

B.1.16 Proposition. *Let V be a vector space over F, and let S be a finite subset of V. Then there exists $T \subseteq S$ such that $sp(T) = sp(S)$ and T is linearly independent.*

Proof. Since S is finite, the set $A = \{\, |T| \mid T \subseteq S, \text{ and } sp(T) = sp(S) \,\}$ is not empty, so by the Well Ordering Principle, A has a least element m. Let $T \subseteq S$ be such that $|T| = m$ and $sp(T) = sp(S)$. If T is not linearly independent, then there exists $v \in T$ such that $v \in sp(T - \{v\})$. But then $sp(S) = sp(T) = sp((T - \{v\}) \cup \{v\}) \subseteq sp(sp(T - \{v\})) = sp(T - \{v\}) \subseteq sp(T) = sp(S)$, so $sp(S) = sp(T - \{v\})$. As this contradicts our choice of T, we conclude that T is linearly independent. $\qquad\square$

B.1.17 Definition. *Let V be a vector space over a field F. If there exists a subset S of V such that S is linearly independent over F and $sp(S) = V$ (that is, S is a spanning set for V), then S is called an F-basis for V, or simply a basis for V when the field is understood.*

By Proposition B.1.16, if V has a finite spanning set, then V has a basis; in fact, V has a finite basis.

B.1.18 Proposition. *Let V be a vector space over F, and let S be a subset of V. If $v, w \in V$ are such that $v \notin sp(S)$ but $v \in sp(S \cup \{w\})$, then $w \in sp(S \cup \{v\})$.*

Proof. Suppose that $v, w \in V$ are such that $v \in sp(S \cup \{w\})$ but $v \notin sp(S)$. Then for some positive integer n, there exist $v_1, v_2, \ldots, v_n \in S$ and $r, r_1, r_2, \ldots, r_n \in F$ such that $v = rw + \sum_{i=1}^{n} r_i v_i$. If $r = 0_F$, then $v \in sp(S)$, which is not the case, so $r \neq 0_F$. But then $(-r)w = (-v) + \sum_{i=1}^{n} r_i v_i \in sp(S \cup \{v\})$. Since $-r \neq 0_F$ and $sp(S \cup \{v\})$ is a subspace of V, $w = (-r)^{-1}(rw) = (-r)^{-1}(-1_F)v + \sum_{i=1}^{n}((-r)^{-1}r_i)v_i \in sp(S \cup \{v\})$, as claimed. $\square$

B.1.19 Proposition. *Let V be a vector space over F, and let S be a linearly independent subset of V. If $v \in V - sp(S)$, then $S \cup \{v\}$ is linearly independent.*

Proof. Let $v \in V - sp(S)$, and suppose that $S \cup \{v\}$ is linearly dependent. Since $v \notin sp(S) = sp((S \cup \{v\}) - \{v\})$, there must be $w \in S \cup \{v\}$ with $w \in sp((S \cup \{v\}) - \{w\})$. Necessarily, $w \neq v$ and so $w \in S$ and $w \in sp((S - \{w\}) \cup \{v\})$, but $w \notin S - \{w\}$ since S is linearly independent. By Proposition B.1.18, $v \in sp((S - \{w\}) \cup \{w\}) = sp(S)$. But we had chosen $v \notin sp(S)$, so we have obtained a contradiction as a result of our assumption that $S \cup \{v\}$ is linearly dependent. Thus $S \cup \{v\}$ is linearly independent. $\square$

B.1.20 Proposition. *Let V be a vector space over the field F, and suppose that $V = sp(A)$ for some finite set $A \subseteq V$. Then every subset $S \subseteq V$ with $|S| > |A|$ is linearly dependent.*

Proof. By Proposition B.1.16, we may assume that A is linearly independent. Suppose by way of contradiction that there exists a linearly independent subset S for which $|S| > |A|$. Then the set

$$\{ |A \cap S| \mid S \text{ linearly independent and } |S| > |A| \}$$

is nonempty, and since A is finite, this set has a maximum element, m say. Let S be linearly independent with $|A \cap S| = m$, and let $B = A \cap S$. If $A \subseteq S$, then since $|S| > |A|$, $A \neq S$, while each $v \in S - A$ is in $sp(A) \subseteq sp(S - \{v\})$, which means that S is not linearly independent. Since S is linearly independent, it follow that A is not a subset of S; that is, $B \neq A$. We prove that there exists $T \subseteq V$ such that T is linearly independent and $|A \cap T| > |B|$ (which will contradict our choice of S). Let $v \in A - B$ (so $v \notin S$). If $v \notin sp(S)$, then by Proposition B.1.19, $T = S \cup \{v\}$ is linearly independent and $T \cap S = B \cup \{v\}$, and so in this case, we obtain a contradiction immediately. Suppose now that $v \in sp(S)$. Then there is a finite subset of S with v in its span, and of all such finite subsets of S, let C be one of smallest size. Then $v \in sp(C)$ and for any $w \in C$, $v \notin sp(C - \{w\})$ by choice of C. If $C \subseteq B$, then $v \in sp(B)$. However, $v \in A - B$, so $B \cup \{v\} \subseteq A$ is linearly independent and thus $v \notin sp((B \cup \{v\}) - \{v\}) = sp(B)$. As a result, we see that C is not a subset of B. Choose $w \in C - B$. Then $v \notin sp(C - \{w\})$, but $v \in sp(C)$, so by Proposition B.1.18, $w \in sp((C - \{w\}) \cup \{v\}) \subseteq sp((S - \{w\}) \cup \{v\})$. Let $T = (S - \{w\}) \cup \{v\}$. If $v \in sp(S - \{w\})$, then $sp((S - \{w\}) \cup \{v\}) = sp(S - \{w\})$ and so $w \in sp(S - \{w\})$. But this contradicts the fact that S is linearly independent. Thus $v \notin sp(S - \{w\})$, and so $T = (S - \{w\}) \cup \{v\}$ is linearly independent. But $T \cap A = ((S - \{w\}) \cap A) \cup (\{v\} \cap A) = (S \cap A) \cup \{v\}$ as $w \in S - B = S - (S \cap A) = S - A$ and thus $w \notin A$, while $v \in A$. We have therefore proven that T is linearly independent and $|T \cap A| > |S \cap A|$, which contradicts our choice of S. Thus there can be no linearly independent subset S with $|S| > |A|$. $\qquad\square$

As we have mentioned above, due to Proposition B.1.16, every vector space that has a finite spanning set has a basis; in fact, every finite spanning set contains a basis for V within it. Evidently, a vector space has a finite spanning set if, and only if, it has a finite basis.

B.1.21 Proposition. *Let V be a vector space over a field F and suppose that V has a finite basis B. Then every basis for V is finite of the same size as B; that is, if C is a basis for V, then $|C| = |B|$.*

Proof. Let C be a basis for V. Since C is a linearly independent subset of V and B is a finite spanning set for V, Proposition B.1.20 asserts that $|C| \leq |B|$. Thus C is a finite basis for V, hence a finite spanning set for V. As B is linearly independent, Proposition B.1.20 also tells us that $|B| \leq |C|$, and thus $|C| = |B|$. $\qquad\square$

B.1.22 Definition. *Let V be a vector space over a field F and suppose that V has a finite basis B. Then $|B|$ is called the F-dimension, or simply dimension, of V, and denoted by $\dim_F V$ or simply $\dim V$.*

For example, if F is a field, then for any positive integer n, and each integer $i \in J_n$, define $e_i \in F^n$ to be the n-tuple with every entry 0_F except for the ith entry, which is 1_F. Then the set $\{\, e_i \mid i \in J_n \,\}$ is a basis for F^n, called the standard basis for F^n, and so $\dim_F F^n = n$. In particular, $\{\, e_1, e_2 \,\}$ is a basis for F^2, where $e_1 = (1_F, 0_F)$ and $e_2 = (0_F, 1_F)$. In the case of $F = \mathbb{R}$, we have $\dim_{\mathbb{R}}(\mathbb{R}^2) = 2$. However, recall that elements of $\mathbb{C}$ are just the elements of $\mathbb{R}^2$, and $\mathbb{C}$ is a field, so $\dim_{\mathbb{C}}(\mathbb{C}) = 1$, while $\dim_{\mathbb{R}}(\mathbb{C}) = 2$. The set $\{\, 1 \,\}$ is the standard $\mathbb{C}$-basis for $\mathbb{C}$ over $\mathbb{C}$, while the set $\{\, (1,0), (0,1) \,\} = \{\, 1, i \,\}$ is the standard basis for $\mathbb{C}$ as an $\mathbb{R}$-vector space.

Our next result establishes that in a finite dimensional vector space V, every linearly independent subset is contained in at least one basis for V.

B.1.23 Proposition. *Let V be a vector space of finite dimension over the field F, and let S be a linearly independent subset of V. Then there exists a basis B for V with $S \subseteq B$.*

Proof. Every linearly independent subset of V has size at most $n = \dim_F(V)$. Let B be a linearly independent subset of V of greatest size, subject to the requirement that $S \subseteq B$. If $sp(B) \neq V$, then there exists $v \in V - sp(B)$, and then by Proposition B.1.19, $B \cup \{\, v \,\}$ is linearly independent and $S \subseteq B \subseteq B \cup \{\, v \,\}$, contradicting our choice of B. Thus $sp(B) = V$ and so B is a basis for V. $\qquad\square$

B.1.24 Corollary. *Let V be a vector space of finite dimension n over the field F, and let W be a subspace of V. Then W has finite dimension, and $\dim_F(W) \leq \dim_F(V)$, with equality holding if, and only if, $W = V$.*

Proof. Every linearly independent subset of W is a linearly independent subset of V and so has size at most n. Let B be a linearly independent subset of W of greatest size. Suppose that $sp(B) \neq W$. Then there exists $v \in W = sp(B)$, and so $B \cup \{\, v \,\} \subseteq W$ is linearly independent, contradicting the choice of B. Thus B is a basis for W. Now by Proposition B.1.23, there exists a basis A for V with $B \subseteq A$. Thus $\dim_F(W) = |B| \leq |A| = n$, and of course, if $W = V$, then B is a basis for V and thus $\dim_F(W) = n$. On the other hand, if $\dim_F(W) = n$, then $B = A$ and so we obtain that $W = sp(B) = sp(A) = V$. $\qquad\square$

Moreover, the next result provides a useful means for finding a basis for a finite dimensional vector space.

B.1.25 Proposition. *Let V be a vector space over the field F of finite dimension n. Then the following hold for any subset S of V:*

(i) *If S is linearly independent, then S is a basis for V if, and only if, $|S| = n$.*

(ii) *If S is a spanning set for V, then S is a basis for V if, and only if, $|S| = n$.*

Proof. Suppose first that S is linearly independent. By Proposition B.1.23, there exists a basis B for V with $S \subseteq B$. Since $\dim_F(V) = n$, it follows that $|B| = n$. If S is a basis for V, then by Proposition B.1.21, $|S| = n$. On the other hand, if $|S| = n$, then $S = B$ and so S is a basis for V.

Now suppose that S is a spanning set for V. If S is a basis for V, then by Proposition B.1.21, $|S| = n$. Suppose now that S is a spanning set of size n. Then by Proposition B.1.16, there is a basis B for V with $B \subseteq S$. By Proposition B.1.21, $|B| = n = |S|$, and so $S = B$. $\square$

If V is a vector space of dimension n, then we may construct a basis for V by starting with any linearly independent subset S of V (including the possibility that we start with the empty set) and if $|S| < n$, choose arbitrarily $v \in V - sp(S)$ and form $S' = S \cup \{v\}$. Now S' is a linearly independent subset of V of size $|S| + 1$. If $|S'| < n$, repeat this step. After at most n repetitions of this step, the result will be a linearly independent subset T of V of size n, which is therefore a basis for V.

Alternatively, if we establish that a set of size $\dim_F(V)$ is a spanning set, then it is a basis for V. Let us review the Example B.1.8 in this context. There we had $F = \mathbb{R}$ and $V = \mathbb{R}^3$, so $\dim_{\mathbb{R}}(V) = 3$ as we have the standard basis $B = \{e_1, e_2, e_3\}$ for $\mathbb{R}^3$ over $\mathbb{R}$. The set $S = \{(1,0,1),(1,1,0),(0,1,1)\}$ has size 3, and we claim that it is a spanning set for V, in which case it is a basis for V. Recall that $sp(S)$ is a subspace of $\mathbb{R}^3$ and is therefore closed under the formation of linear combinations. In particular, $(1,0,1) - (1,1,0) = (0,-1,1) \in sp(S)$ and so $(0,0,2) = (0,-1,1) + (0,1,1) \in sp(S)$. But then $e_3 = \frac{1}{2}(0,0,2) \in sp(S)$. Now $e_2 = (0,1,1) - e_3 \in sp(S)$ and $e_1 = (1,0,1) - e_3 \in sp(S)$, so $B = \{e_1, e_2, e_3\} \subseteq sp(S)$. By Proposition B.1.11, $\mathbb{R}^3 = sp(B) \subseteq sp(sp(S)) = sp(S) \subseteq \mathbb{R}^3$ and thus $\mathbb{R}^3 = sp(S)$.

B.1.26 Proposition. *Let V be a vector space of finite dimension n over* the field F, and let $B = \{v_1, v_2, \ldots, v_n\}$ be a basis for V. Then every $v \in V$ has a unique representation as a linear combination of the elements of B; that is, for every $v \in V$, there exist unique $r_1, r_2, \ldots, r_n \in F$ such that $v = \sum_{i=1}^{n} r_i v_i$.

Proof. Let $v \in V$. As $V = sp(B)$, there exist $r_1, r_2, \ldots, r_n \in F$ such that $v = \sum_{i=1}^{n} r_i v_i$; the question is: are they unique? Suppose that $a_1, a_2, \ldots, a_n \in F$ are such that $v = \sum_{i=1}^{n} a_i v_i$. Then $0_V = v - v = \sum_{i=1}^{n} (r_i - a_i) v_i$. Since B is linearly independent, by Proposition B.1.13, this implies that $r_i - a_i = 0_F$ for each $i = 1, 2, \ldots, n$; that is, $r_i = a_i$ for each $i = 1, 2, \ldots, n$. □

2 Linear Transformations

B.2.1 Definition. *Let V and W be vector spaces over the field F. A map $T : V \to W$ is called a linear transformation if T is a homomorphism from the abelian group $(V, +)$ to the abelian group $(W, +)$ for which $T(rv) = rT(v)$ for every $r \in F$ and every $v \in V$. In the event that $W = V$, we also say that T is a linear operator on V.*

B.2.2 Proposition. *Let V and W be vector spaces over the field F. A map $T : V \to W$ is a linear transformation if, and only if, $T(rv + sw) = rT(v) + sT(w)$ for every $r, s \in F$.*

Proof. Let $T : V \to W$ be a map, and let $r, s \in F$ and $v, w \in V$. If T is a linear transformation, then $T(rv + sw) = T(rv) + T(sw) = rT(v) + sT(w)$, as required. Conversely, if $T(rv + sw) = rT(v) + sT(w)$ for all $r, s \in F$ and all $v, w \in V$, then for any $v, w \in V$, $T(v + w) = T(1_F v + 1_F w) = 1_F T(v) + 1_F T(w) = T(v) + T)w)$, and so T is a homomorphism from $(V, +)$ to $(W, +)$. As well, for any $r \in F$ and $v \in V$,

$$T(rv) = T(rv + 0_V) = T(rv + 0_F 0_V) = rT(v) + 0_F T(0_V) = rT(v),$$

and so T is a linear transformation from V to W. □

One very important linear transformation is the trace mapping from $M_{n \times n}(F)$ to F.

B.2.3 Definition. *Let F be a field and n be a positive integer. Then define the trace mapping $tr : M_{n \times n}(F) \to F$ by $tr(A) = \sum_{i=1}^{n} A_{ii}$ for each $A \in M_{n \times n}(F)$.*

B.2.4 Proposition. *Let F be a field and n be a positive integer. Then the trace mapping $tr \colon M_{n \times n}(F) \to F$ is a linear mapping.*

Proof. Let $r, s \in F$ and $A, B \in M_{n \times n}(F)$. Then

$$tr(rA + sB) = \sum_{i=1}^{n}(rA + sB)_{ii} = \sum_{i=1}^{n} rA_{ii} + sB_{ii}$$

$$= r\sum_{i=1}^{n} A_{ii} + s\sum_{i=1}^{n} B_{ii} = r\,tr(A) + s\,tr(B),$$

and so the trace mapping is linear. $\qquad\square$

The transpose mapping is another important example of a linear mapping.

B.2.5 Definition. *For any field F and positive integers m and n, the transpose mapping from $M_{m \times n}(F)$ to $M_{n \times m}(F)$ is defined as follows: for $A \in M_{m \times n}(F)$, the transpose of A, denoted by A^t, is the $n \times m$ matrix with entries from F for which $(A^t)_{ij} = A_{ji}$ for each $(i, j) \in J_n \times J_m$.*

B.2.6 Proposition. *For any positive integers m, n, the transpose mapping from $M_{m \times n}(F)$ to $M_{n \times m}(F)$ is a bijective linear transformation. In fact, for every $A \in M_{m \times n}(F)$, $(A^t)^t = A$.*

Proof. Let $A, B \in M_{m \times n}(F)$ and $r \in F$. Then for any $(i, j) \in J_n \times J_m$, $((A + B)^t)_{ij} = (A + B)_{ji} = A_{ji} + B_{ji} = (A^t)_{ij} + (B^t)_{ij} = (A^t + B^t)_{ij}$, and so $(A + B)^t = A^t + B^t$. Similarly, for any $(i, j) \in J_n \times J_m$, $((rA)^t)_{ij} = (rA)_{ji} = r(A_{ji}) = r(A^t_{ij}) = (rA^t)_{ij}$, and so $(rA)^t = rA^t$. For any $A \in M_{m \times n}(F)$, $A^t \in M_{n \times m}(F)$ and so $(A^t)^t \in M_{m \times n}(F)$. For any $(i, j) \in J_m \times J_n$, $((A^t)^t)_{ij} = (A^t)_{ji} = A_{ij}$, and thus $A = (A^t)^t$. Now, suppose that $A, B \in M_{m \times n}(F)$ are such that $A^t = B^t$. Then $A = (A^t)^t = (B^t)^t = B$, which proves that the transpose mapping is injective. As well, for any $B \in M_{n \times m}(F)$, let $A = B^t \in M_{m \times n}(F)$. Then $A^t = (B^t)^t = B$ and thus the transpose mapping is surjective. $\qquad\square$

B.2.7 Proposition. *Let m, n, k be positive integers and let F be a field. For any $A \in M_{n \times n}(F)$ and $B \in M_{n \times k}(F)$, $(AB)^t = B^t A^t$.*

Proof. $AB \in M_{m \times k}(F)$, so $(AB)^t \in M_{k \times m}(F)$, as is $B^t A^t$. For any $(i, j) \in J_k \times J_m$, we have $((AB)^t)_{ij} = (AB)_{ji} = \sum_{t=1}^{n} A_{jt} B_{ti} = \sum_{t=1}^{n}(B^t)_{it}(A^t)_{tj} = (B^t A^t)_{ij}$. $\qquad\square$

B.2.8 Proposition. *Let m and n be positive integers, and let F be a field. For any $A \in M_{m \times n}(F)$, the mapping $T_A : M_{n \times 1}(F) \to M_{m \times 1}(F)$ defined by $T_A(v) = Av$ for all $v \in M_{n \times 1}(F)$ is a linear transformation.*

Proof. This is an immediate consequence of the properties of matrix multiplication. Let $r, s \in F$, and let $v, w \in M_{n \times 1}(F)$. Then $T_A(rv + sw) = A(rv + sw) = r(Av) + s(Aw) = rT(v) + sT(w)$. □

B.2.9 Corollary. *Let m, n, k be positive integers, and let F be a field. For any $A \in M_{n \times n}(F)$ and $B \in M_{n \times k}(F)$, $T_B \circ T_A = T_{BA}$.*

Proof. Let $v \in F^n$. Then $T_B \circ T_A(v) = T_B(Av) = B(Av) = (BA)v = T_{BA}(v)$. □

The reader will have noticed that we have chosen to emphasize the notion of linear transformations because they have a more geometric flavour which often leads to a more intuitive grasp of the subject. However, both matrices and linear transformations have important roles to play: the former when we need to compute, and the latter when we want to develop the theoretical aspects of linear algebra.

B.2.10 Proposition. *Let V and W be vector spaces over a field F, and $T : V \to W$ be a linear transformation. Then the following hold:*

(i) *$T(0_V) = 0_W$.*

(ii) *If U is a subspace of V, then $T(U)$ is a subspace of W.*

(iii) *If U is a subspace of W, then $T^{-1}(W) = \{ v \in V \mid T(v) \in W \}$ is a subspace of V.*

(iv) *If Y is a vector space over the field F and $S : V \to W$ is a linear transformation, then $S \circ T : V \to Y$ is a linear transformation.*

Proof. Since T is a group homomorphism from $(V, +)$ to $(W, +)$, we need only verify that if U is a subspace of V, then $T(U)$ is closed under scalar multiplication in W, and if U is a subspace of W, then $T^{-1}(U)$ is closed under scalar multiplication in V. Suppose that U is a subspace of V, and let $r \in F$ and $w \in T(U)$. Then there exists $u \in U$ such that $T(u) = w$ and so $ru = rT(v) = T(rv) \in T(U)$ since U is closed under scalar multiplication. Similarly, if U is a subspace of W, then for any $r \in F$ and $v \in T^{-1}(U)$, $T(v) \in U$ and so $T(rv) = rT(v) \in U$ since U is closed under scalar multiplication. Thus $rv \in T^{-1}(U)$, which proves that $T^{-1}(U)$ is closed under scalar multiplication.

Finally, let $v, w \in V$ and $r, s \in F$. Then $S \circ T(rv + sw) = S(T(rv + sw)) = S(rT(v) + sT(w)) = r(S(T(v)) + sS(T(w)) = r\, S \circ T(v) + s\, S \circ T(w)$, and so $S \circ T$ is linear.   □

B.2.11 Definition. *Let V and W be vector spaces over a field F, and let $T : V \to W$ be a linear transformation. The null space of T, denoted by null(T), is the kernel of T (when regarded just as a group homomorphism from $(V, +)$ to $(W, +)$). That is, null$(T) = T^{-1}(0_W)$. Moreover, $T(V)$ is called the image of T.*

B.2.12 Corollary. *Let V and W be vector spaces over a field F, and let $T : V \to W$ be a linear transformation. Then null(T) is a subspace of V and $T(V)$ is a subspace of W.*

Proof. Since $\{\, 0_W \,\}$ is a subspace of W, null$(T) = T^{-1}(0_W)$ is a subspace of V by Proposition B.2.10 (iii), and since V is a subspace of itself, $T(V)$ is a subspace of W.   □

B.2.13 Proposition. *Let V and W be vector spaces over the field F, and let $T : V \to W$ be a linear transformation. Then T is injective if, and only if, null$(T) = \{\, 0_V \,\}$.*

Proof. This is simply Proposition 6.4.7 stated in the context of vector spaces.   □

B.2.14 Proposition (Dimension Theorem). *Let V and W be vector spaces over F, and suppose that V is finite dimensional. Let $T : V \to W$ be a linear transformation. Then $T(V)$ is a finite dimensional subspace of W, and $\dim_F(T(V)) = \dim_F(V) - \dim_F(\text{null}(T))$.*

Proof. By Proposition B.2.12, null(T) is a subspace of V, and by Proposition B.1.24, $\dim_F(\text{null}(T)) \le n$, with equality holding if, and only if, null$(T) = V$. In the event that null$(T) = V$, $T(V) = \{\, 0_W \,\}$, which has $\varnothing$ as a basis, so $\dim_F(T(V)) = 0 = n - \dim_F(\text{null}(T))$. Suppose that null$(T) \ne V$, so $\dim_F(\text{null}(T)) < n$. Let B be a basis for null(T), say $B = \{\, v_1, v_2, \ldots, v_k \,\}$. By Proposition B.1.23, there exists a basis A for V with $B \subseteq A$. Let $A - B = \{\, v_{k+1}, \ldots, v_n \,\}$. We claim that

$$S = \{\, T(v_{k+i}) \mid i \in J_{n-k} \,\}$$

is a basis for $T(V)$. Let $w \in T(V)$. Then there exists $v \in V$ such that $T(v) = w$, and so there exist $r_i \in F$, $i \in J_n$, with $v = \sum_{i=1}^{n} r_i v_i$ and thus $w = T(v) = T(\sum_{i=1}^{n} r_i v_i) = \sum_{i=1}^{n} r_i T(v_i)$. Since $v_i \in$ null(T) for

$i = 1, 2, \ldots, k$, we have $T(v_i) = 0_W$ for $i = 1, 2, \ldots, k$, and thus $w = \sum_{i=k+1}^{n} r_i T(v_i) \in sp(S)$. This establishes that $T(V) = sp(S)$.

It remains to prove that S is linearly independent. Suppose to the contrary that S is linearly dependent. Then there exists $j \in J_{n-k}$ such that $T(v_{k+j}) \in sp(\{T(v_{k+i}) \mid i \in J_{n-k} - \{j\}\})$, and so for some $r_{k+i} \in F$, $i \in J_{n-k} - \{j\}$, $T(v_j) = \sum_{i \in J_{n-k} - \{j\}} r_i T(v_{k+i})$. But then $(-1_F)v_j + \sum_{i \in J_{n-k} - \{j\}} r_i v_{k+i} \in null(T)$, and so there exist $r_i \in F$, $i \in J_k$, such that $(-1_F)v_j + \sum_{i \in J_{n-k} - \{j\}} r_i v_{k+i} = \sum_{i=1}^{k} r_i v_i$. Then $(-1_F)v_j + \sum_{i \in J_n - \{j\}} r_i v_i = 0_V$. Since $\{v_i \mid i \in J_n\}$ is linearly independent, all coefficients of v_i, $i \in J_n$ must therefore be 0_F. As the coefficient of v_j is -1_F, this is not possible and thus S is linearly independent. We have now proven that S is a basis for $T(V)$. Since $|S| = n - k$, we have $\dim_F(T(V)) = n - k = n - \dim_F(null(T))$. $\qquad\square$

B.2.15 Definition. *Let V and W be vector spaces over F with V of finite dimension. Let $T : V \to W$ be a linear transformation. The rank of T, denoted by $rank(T)$, is the dimension of $T(V)$, and the nullity of T, denoted by $nullity(T)$, is the dimension of $null(T)$.*

Thus the dimension theorem asserts that for V finite dimensional and $T : V \to W$ a linear transformation, $\dim_F(V) = rank(T) + nullity(T)$.

B.2.16 Corollary. *Let V and W be vector spaces of the same finite dimension over the field F, and let $T : V \to W$ be linear. Then T is surjective if, and only if, T is injective.*

Proof. T is surjective if, and only if, $\dim_F(T(V)) = \dim_F(W)$, and by hypothesis, $\dim_F(W) = \dim_F(V)$, so T is surjective if, and only if, $\dim_F(T(V)) = \dim_F(V)$. By the dimension theorem, $\dim_F(V) = \dim_F(T(V)) + \dim_F(null(T))$, so we see that T is surjective if, and only if, $\dim_F(null(T)) = 0$; that is, if, and only if, $null(T) = \{0_V\}$. Finally, by Proposition B.2.13, T is injective if, and only if, $null(T) = \{0_V\}$. $\qquad\square$

We observe that for any vector space V over a field F, $(V, +)$ is an abelian group and every subspace of V is an additive subgroup of $(V, +)$, so every subspace W of V is a normal subgroup of $(V, +)$ and we may consider the quotient group $(V/W, +)$, where the addition is defined by $(u + W) + (v + W) = (u + v) + W$. We next prove that there is a natural way to define scalar multiplication on V/W by using the scalar multiplication of V.

B.2.17 Proposition. *Let V be a vector space over a field F, and let W be a subspace of V. Then for any $u, v \in V$ and any $r \in F$, if $u + W = v + W$, then $ru + W = rv + W$.*

Proof. Let $u, v \in V$ be such that $u + W = v + W$; that is, $u - v \in W$. Let $r \in F$. Since W is closed under scalar multiplication, $r(u - v) \in W$, and so $ru - rv \in W$; equivalently, $ru + W = rv + W$. $\square$

B.2.18 Proposition. *Let V be a vector space over F, and let W be a subspace of V. Then the quotient group $(V/W, +)$, with the scalar multiplication defined by $r(v + W) = rv + W$ for all $r \in F$ and all $v \in V$, is a vector space over F, called the quotient space V modulo W. The projection mapping $\pi_W : V \to V/W$ defined by $\pi_W(v) = v + W$ is a surjective linear transformation and $\mathrm{null}(\pi_W) = W$. Moreover, if V is finite dimensional, then $\dim_F(V/W) = \dim_F(V) - \dim_F(W)$.*

Proof. By Proposition B.2.17, the scalar multiplication operation is well defined. It remains to prove that the scalar multiplication operation has the required properties. Let $r, s \in F$ and $v + W \in V/W$. Then $(r+s)(v+W) = (r+s)v + W = rv + sv + W = (rv + W) + (sv + W) = r(v + W) + s(v + W)$. Also, $(rs)(v + W) = (rs)v + W = r(sv) + W = r(sv + W) = r(s(v + W))$, and $1_F(v + W) = 1_F v + W = v + W$. Finally, for $v, w \in V$ and $r \in F$, $r((v + W) + (w + W)) = r((v + w) + W) = r(v + w) + W = rv + rw + W = (rv + W) + (rw + W) = r(v + W) + r(w + W)$. Thus the scalar multiplication operation on V/W satisfies the requirements of the definition of a vector space. We conclude therefore that V/W is a vector space over F, with addition defined by $(v + W) + (w + W) = (v + w) + W$ and scalar multiplication defined by $r(v + W) = (rv) + W$. The projection mapping $\pi : V \to V/W$ is a surjective group homomorphism with kernel W. Let $v \in V$ and $r \in F$. Then $\pi_W(rv) = rv + W = r(v + W) = r\,\pi_W(v)$, and so π_W is a linear transformation. By Proposition B.2.14, $\dim_F(V/W) = \dim_F(\pi_W(V)) = \dim_F(V) - \dim_F(\mathrm{null}(\pi_W)) = \dim_F(V) - \dim_F(W)$. $\square$

B.2.19 Definition. *Let V and W be vector spaces over a field F. A bijective linear transformation from V to W is called an isomorphism, and if there exists an isomorphism from V to W, then V is said to be isomorphic to W.*

For example, the transpose mapping from $M_{m \times n}(F)$ to $M_{n \times m}(F)$ is an isomorphism (see Proposition B.2.6).

B.2.20 Proposition. *Let V and W be vector spaces over a field F. If $T:V \to W$ is an isomorphism, then $T^{-1}:W \to V$ is an isomorphism.*

Proof. Suppose that $T:V \to W$ is an isomorphism. Then T is an isomorphism from the group $(V,+)$ to the group $(W,+)$, and as such, the inverse mapping T^{-1} is a group homomorphism from $(W,+)$ to $(V,+)$. Let $w \in W$ and $r \in F$. We must prove that $T^{-1}(rv) = rT^{-1}(v)$. Since T is injective, this is equivalent to showing that $T(rT^{-1}(v)) = T(T^{-1}(rv))$. We have $T(T^{-1}(rv)) = rv = rT(T^{-1}(v)) = T(rT^{-1}(v))$, as required. $\square$

B.2.21 Proposition (First Isomorphism Theorem). *Let V and W be vector spaces over a field F, and let $T:V \to W$ be a linear transformation. Then $V/\operatorname{null}(T)$ is isomorphic to $T(V)$.*

Proof. Since T is a group homomorphism, the first isomorphism theorem for groups proves that the mapping $T' : W/\operatorname{null}(T) \to W$ given by $T'(v + W) = T(v)$ is a well-defined injective homomorphism from the group $(V/\operatorname{null}(T),+)$ to the group $(W,+)$ with image $T'(V)$. For linearity, let $v + W \in V/\operatorname{null}(T)$ and $r \in F$. Then $T'(r(v + W)) = T'(rv + W) = T(rv) = rT(v) = rT'(v + W)$, as required. $\square$

It is conventional to refer to an injective linear transformation as a nonsingular linear transformation, and also to refer to an isomorphism as an invertible linear transformation.

Our next result is one of the cornerstones of the theory of finite-dimensional vector spaces.

B.2.22 Proposition. *Let V be a vector space of finite dimension n over a field F, and let $B = \{v_1, v_2, \ldots, v_n\}$ be a basis for V. Then for any choice of $w_i \in W$, $i \in J_n$, there exists a unique linear transformation $T:V \to W$ for which $T(v_i) = w_i$, $i \in J_n$.*

Proof. By Proposition B.1.26, for every $v \in V$, there exist unique $r_1, r_2, \ldots, r_n \in F$ such that $v = \sum_{i=1}^{n} r_i v_i$. For $v \in V$ define $T(v) = \sum_{i=1}^{n} r_i w_i$, where $v = \sum_{i=1}^{n} r_i v_i$. By the uniqueness asserted in Proposition B.1.26. T is well-defined, and for each $i = 1, 2, \ldots, n$, $T(v_i) = w_i$. It remains to prove that T is linear. Let $v, w \in V$, and suppose that $r_i, s_i \in f$, $i \in J_n$, are such that $v = \sum_{i=1}^{n} r_i v_i$ and $w = \sum_{i=1}^{n} s_i v_i$. Then $v + w = \sum_{i=1}^{n} (r_i + s_i) v_i$, and so by definition of T, $T(v + w) = \sum_{i=1}^{n} (r_i + s_i) w_i = \sum_{i=1}^{n} (r_i w_i + s_i w_i) = \sum_{i=1}^{n} r_i w_i + \sum_{i=1}^{n} s_i w_i = T(v) + T(w)$. Similarly, for $r \in F$, $T(rv) = T(r\sum_{i=1}^{n} r_i v_i) = T(\sum_{i=1}^{n} (rr_i)v) = \sum_{i=1}^{n} (rr_i)w_i =$

$r \sum_{i=1}^{n} r_i w_i = rT(V)$, and so T is linear. Finally, if $T' : V \to W$ is any linear transformation for which $T'(v_i) = w_i$ for all $i \in J_n$, then for $v \in V$, $v = \sum_{i=1}^{n} r_i v_i$ for some $r_1, r_2, \ldots, r_n \in F$, and so $T(v) = \sum_{i=1}^{n} r_i w_i = \sum_{i=1}^{n} r_i T'(v_i) = T'(\sum_{i=1}^{n} r_i v_i) = T'(v)$. Thus $T' = T$. $\qquad\square$

B.2.23 Corollary. *If V is a vector space of finite dimension n over the field F, then V is isomorphic to F^n.*

Proof. Let $B = \{ v_1, v_2, \ldots, v_n \}$ be a basis for V. By Proposition B.2.22, there exists a linear transformation $T : V \to F^n$ such that $T(v_i) = e_i$ for each $i \in J_n$. Then $T(V) = sp(\{ e_1, e_2, \ldots, e_n \}) = F^n$, and so T is surjective. By Corollary B.2.16, T is injective and thus T is an isomorphism. $\qquad\square$

B.2.24 Proposition. *Let V and W be vector spaces over the field F. Then $\mathscr{L}(V, W)$, the set of all linear transformations from V to W, is a vector space over the field F, with addition defined by $(T + S)(v) = T(v) + S(v)$ for all $T, S \in \mathscr{L}(V, W)$, and each $v \in V$, and scalar multiplication defined by $(rT)(v) = r\,T(v)$ (using the scalar multiplication of W) for each $v \in V$, where $r \in F$ and $T \in \mathscr{L}(V, W)$.*

Proof. It was shown in Proposition 6.5.2 that $\mathscr{F}(V, W)$ is a group under the operation of addition, and it follows from Corollary 6.5.5 that $\mathscr{F}(V, W)$ is abelian. We first prove that $\mathscr{L}(V, W)$ is a subgroup of $\mathscr{F}(V, W)$. The mapping $\mathbb{0} : V \to W$ defined by $\mathbb{0}(v) = 0_W$ for all $v \in V$ was proven in Proposition 6.5.2 to be the identity element for addition in $\mathscr{F}(V, W)$, so we just need to prove that $\mathbb{0}$ is linear. We have $\mathbb{0}(rv + sw) = 0_W = r0_W + s0_W = r\mathbb{0}(v) + s\mathbb{0}(w)$ for all $r, s \in F$ and $v, w \in V$, and so $\mathbb{0} \in \mathscr{L}(V, W)$. Next, we prove that $\mathscr{L}(V, W)$ is closed under addition. Let $T, S \in \mathscr{L}(V, W)$. We must prove that $T + S \in \mathscr{L}(V, W)$; that is, we must prove that $T + S$ is a linear transformation. Let $v, w \in V$ and $r, s \in F$. Then $(T+S)(rv+sw) = T(rv+sw)+S(rv+sw) = rT(v)+sT(w)+rS(v)+sS(w) = r(T(v) + S(v)) + s(T(w) + S(w)) = r(T + S)(v)) + s(T + S)(w) = r(T + S)(v) + s(T + S)(w)$ and thus $T + S$ is linear. This proves that $\mathscr{L}(V, W)$ is closed with respect to addition. To see that it is closed with respect to inverses, we prove that the mapping $(-T) : V \to W$ defined by $(-T)(v) = -(T(v))$ for all $v \in V$ is linear, and that $T + (-T) = \mathbb{0}$. Let $r, s \in F$ and $v, w \in V$. Then $(-T)(rv + sw) = -(T(rv + sw)) = -(rT(v)+sT(w)) = r(-T(v)) + s(-T(w)) = r(-T)(v) + s(-T)(w)$, and so $-T \in \mathscr{L}(V, W)$. It was proven in Proposition 6.5.2 that $-T$ is the additive inverse of T in $\mathscr{F}(V,W)$, and so $\mathscr{L}(V,W)$ is closed under additive inverse. Thus $\mathscr{L}(V,W)$ is a subgroup of $\mathscr{F}(V,W)$, and so $\mathscr{L}(V,W)$ is an abelian group under addition. It remains to prove that scalar multiplication has all of the required properties. Let $r,s \in F$ and $T \in \mathscr{L}(V,W)$. Then $((r+s)T)(v) = (r+s)T(v) = rT(v) + sT(v) = (rT + sT)(v)$ for all $v \in V$, and so $(r+s)T = rT + sT$. As well, $((rs)T)(v) = (rs)T(v) = r(s(T(v)) = r((sT)(v)) = (r(sT))(v)$ for all $v \in V$, and thus $(rs)T = r(sT)$. Now, for $r \in F$, and $T,S \in \mathscr{L}(V,W)$, we have $(r(T+S))(v) = r(T+S)(v) = r(T(v)+S(v)) = r(T(v))+r(S(v)) = (rT)(v)+(rS)(v) = (rT+rS)(v)$ for all $v \in V$, and thus $r(T+S) = rT+rS$. Finally, $(1_F T)(v) = 1_F(T(v)) = T(v)$ for all $v \in V$ and thus $1_F T = T$. This completes the proof that $\mathscr{L}(V,W)$ is a vector space over F. $\qquad\square$

The time has come for us to establish the fundamental connection between linear transformations and matrices.

B.2.25 Definition. *Let V be a vector space of finite dimension n over the field F, and let $\alpha = (v_1, v_2, \ldots, v_n)$ be an ordered basis for V (by ordered basis, we mean a basis whose elements have been ordered so that a first basis element has been declared, a second basis element, and so on). Define a mapping $c_\alpha : V \to M_{n\times 1}(F)$, called the coordinate transformation with respect to the ordered basis α, by $c_\alpha(v) = \begin{bmatrix} r_1 \\ r_2 \\ \vdots \\ r_n \end{bmatrix}$, where $r_1, r_2, \ldots, r_n \in F$ are uniquely determined by the requirement that $v = \sum_{i=1}^n r_i v_i$.*

B.2.26 Proposition. *Let V be a vector space of finite dimension n over the field F, and let $\alpha = (v_1, v_2, \ldots, v_n)$ be an ordered basis for V. Then the mapping $c_\alpha : V \to M_{n\times 1}(F)$ is the unique linear transformation that maps v_i to e_i, $i \in J_n$.*

Proof. By Proposition B.2.22, there exists a unique linear transformation $T : V \to M_{n\times 1}(F)$ such that $T(v_i) = e_i$, $i \in J_n$. For $v \in V$, there exist unique $r_i \in F$, $i \in J_n$, such that $v = \sum_{i=1}^n r_i v_i$ and thus $T(v) = \sum_{i=1}^n r_i e_i = c_\alpha(v)$, and so $T = c_\alpha$. $\qquad\square$

B.2.27 Definition. *Let m and n be positive integers, and let F be a field. For any $A \in M_{m\times n}(F)$, and for any $i \in J_m$ and any $j \in J_n$, define $row_i(A) \in M_{1\times n}(F)$ by $(row_i(A))_{1\,k} = A_{i\,k}$ for $k \in J_n$, and define $col_j(A) \in M_{m\times 1}(F)$ by $(col_j(A))_{k\,1} = A_{k\,j}$ for each $k \in J_m$.*

We are now ready to define the matrix representatives of a linear transformation.

B.2.28 Definition. *Let V and W be finite dimensional vector spaces over a field F, and let $T\colon V \to W$ be a linear transformation. Let $n = \dim_F(V)$ and $m = \dim_F(W)$, and let $\alpha = (v_1, v_2, \ldots, v_n)$ and $\beta = (w_1, w_2, \ldots, w_m)$ be ordered bases for V and for W, respectively. Let $[T]_\alpha^\beta \in M_{m\times n}(F)$ be the matrix for which $col_j([T]_\alpha^\beta) = c_\beta(T(v_j))$, $j = 1, 2, \ldots, n$. $[T]_\alpha^\beta$ is called the matrix representative of T relative to the ordered bases α for the domain and β for the codomain of T. Finally, define the mapping $\mathscr{T}_\alpha^\beta\colon \mathscr{L}(V, W) \to M_{m\times n}(F)$ by $\mathscr{T}_\alpha^\beta(T) = [T]_\alpha^\beta$ for all $T \in \mathscr{L}(V, W)$.*

B.2.29 Proposition. *Let V and W be vector spaces of finite dimensions m and n, respectively, over the field F, and let α and β be ordered bases for V and W respectively. Then $\mathscr{T}_\alpha^\beta\colon \mathscr{L}(V, W) \to M_{m\times n}(F)$ is an isomorphism.*

Proof. First, we prove that $\mathscr{T}_\alpha^\beta$ is a linear map. Let $T, S \in \mathscr{L}(V, W)$ and $r, s \in F$. By Proposition B.2.26, c_β is linear and so for each $j \in J_n$, $c_\beta \circ (rT + sS)(v_j) = c_\beta \circ (rT(v_j) + sS(v_j)) = r\, c_\beta(T(v_j)) + s\, c_\beta(S(v_j))$. Thus for each $j \in J_n$, $col_j(\mathscr{T}_\alpha^\beta(rT + sS)) = r\, col_j(\mathscr{T}_\alpha^\beta(T)) + s\, col_j(\mathscr{T}_\alpha^\beta(S))$, and so $\mathscr{T}_\alpha^\beta(rT + sS) = r\, \mathscr{T}_\alpha^\beta(T) + s\, \mathscr{T}_\alpha^\beta(S)$; that is, $\mathscr{T}_\alpha^\beta$ is linear.

Suppose now that $T, S \in \mathscr{L}(V, W)$ are such that $\mathscr{T}_\alpha^\beta(T) = \mathscr{T}_\alpha^\beta(S) = A$, say. Then for each $j \in J_n$, $T(v_j) = \sum_{i=1}^m A_{ij} w_i = S(v_j)$, so by Proposition B.2.22, $T = S$. Thus $\mathscr{T}_\alpha^\beta$ is injective. Finally, let $A \in M_{m\times n}(F)$, and for each $j \in J_n$, let $u_j = \sum_{i=1}^m A_{ij} w_i \in W$, and let $T\colon V \to W$ be the unique linear transformation for which $T(v_j) = u_j$ for each $j \in J_n$ (the existence of T is a consequence of Proposition B.2.22). Then for each $j \in J_n$, we have $col_j(\mathscr{T}_\alpha^\beta(T)) = col_j(A)$ and so $\mathscr{T}_\alpha^\beta(T) = A$. Thus $\mathscr{T}_\alpha^\beta$ is surjective. $\qquad\square$

Let V and W be vector spaces of finite dimension n and m, respectively, over the field F, and let $T\colon V \to W$ be a linear transformation. Let α and β be ordered bases for V and W, respectively, and let $A = \mathscr{T}_\alpha^\beta(T)$. Consider the mapping diagram:

$$
\begin{array}{ccc}
V & \xrightarrow{\;\;T\;\;} & W \\
{\scriptstyle c_\alpha}\big\downarrow & & \big\downarrow{\scriptstyle c_\beta} \\
M_{n\times 1}(F) & \xrightarrow[\;\;T_A\;\;]{} & M_{m\times 1}(F)
\end{array}
$$

B.2.30 Proposition. *Let V and W be vector spaces of finite dimensions n and m, respectively, over the field F, and let $T : V \to W$ be a linear transformation. Let α and β be ordered bases for V and W, respectively, and let $A = \mathscr{T}_\alpha^\beta(T)$. Then for any $v \in V$, $c_\beta \circ T(v) = T_A \circ c_\alpha(v)$; that is, $c_\beta \circ T = T_A \circ c_\alpha$. Equivalently, for each $v \in V$, $[T]_\alpha^\beta c_\alpha(v) = c_\beta(T(v))$.*

Proof. Let $j \in J_n$. Then by definition, $c_\beta \circ T(v_j) = col_j(A)$, while we have $T_A \circ c_\alpha(v_j) = Ae_j^t = col_j(A)$, and thus the linear transformation $c_\beta \circ T$ and the linear transformation $c_\alpha \circ T_A$ each map v_j to the same element for each $j \in J_n$. By Proposition B.2.22, they are equal transformations. $\square$

It remains to see how the notion of matrix multiplication is interpretated in the context of linear transformations. In fact, it is the study of composition of linear transformations that sheds some light on the rather mysterious definition of matrix multiplication.

B.2.31 Proposition. *Let V, W, Y be finite dimensional vector spaces over the field F, and let α, β, and γ be ordered bases for V, W, and Y, respectively. Let $T : V \to W$ and $S : W \to Y$ be linear transformations. Then $\mathscr{T}_\beta^\gamma(S)\mathscr{T}_\alpha^\beta(T) = \mathscr{T}_\alpha^\gamma(S \circ T)$.*

Proof. By Proposition B.2.10 (iv), $S \circ T$ is linear, and so $\mathscr{T}_\alpha^\gamma(S \circ T)$ is defined. Suppose that $\alpha = (v_1, v_2, \ldots, v_n)$. Then for each $j \in J_n$,

$$col_j(\mathscr{T}_\alpha^\gamma(S \circ T)) = c_\gamma(S \circ T(v_j)) = c_\gamma(S(T(v_j))$$
$$= [S]_\beta^\gamma(c_\beta(T(v_j)) = [S]_\beta^\gamma([T]_\alpha^\beta(c_\alpha(v_j))$$
$$= [S]_\beta^\gamma([T]_\alpha^\beta e_j = ([S]_\beta^\gamma([T]_\alpha^\beta)e_j = col_j([S]_\beta^\gamma([T]_\alpha^\beta),$$

and so $[S \circ T]_\alpha^\gamma = [S]_\beta^\gamma[T]_\alpha^\beta$; that is, $\mathscr{T}_\alpha^\gamma(S \circ T) = \mathscr{T}_\beta^\gamma(S)\mathscr{T}_\alpha^\beta(T)$. $\square$

B.2.32 Corollary. *Let V and W be finite dimensional vector spaces over the field F, and let $T : V \to W$ be a linear transformation. Then T is invertible if, and only if, $\dim_F(V) = \dim_F(W)$ and for any ordered bases α for V and β for W, $\mathscr{T}_\alpha^\beta(T)$ is an invertible matrix, in which case, $(\mathscr{T}_\alpha^\beta(T))^{-1} = \mathscr{T}_\beta^\alpha(T^{-1})$.*

Proof. Suppose first of all that T is invertible. Then by the dimension theorem, $\dim_F(V) = \dim_F(T(V)) + \dim_F((T))$, and T is bijective, so $\dim_F((T)) = 0$ and $\dim_F(T(V)) = \dim_F(W)$; that is, $\dim_F(V) = \dim_F(W)$. Let $n = \dim_F(V)$. Then $T^{-1} : W \to V$ exists, and $T^{-1} \circ T = 1_V$. Let α and β be ordered bases for V and W, respectively. Then by Proposition B.2.31, $I_n = [1_V]_\alpha^\alpha = [T^{-1} \circ T]_\alpha^\alpha = [T^{-1}]_\beta^\alpha[T]_\alpha^\beta$, and

$[T^{-1}]^\alpha_\beta, [T]^\beta_\alpha \in M_{n\times n}(F)$, so $[T]^\beta_\alpha$ is invertible, with inverse $[T^{-1}]^\alpha_\beta$. Conversely, suppose that $\dim_F(V) = \dim_F(W) = n$, say, and that α and β are ordered bases for V and W, respectively, such that $A = [T]^\beta_\alpha$ is an invertible $n \times n$ matrix. Then A^{-1} exists, and by Proposition B.2.29, there exists $S \in \mathscr{L}(V,W)$ such that $\mathscr{T}^\beta_\alpha(S) = A^{-1}$, so by Proposition B.2.31, $\mathscr{T}^\alpha_\alpha(S \circ T) = A^{-1}A = I_n = \mathscr{T}^\alpha_\alpha(\mathbb{1}_V)$, and since $\mathscr{T}^\alpha_\alpha$ is an isomorphism, $S \circ T = \mathbb{1}_V$. Similarly, $\mathscr{T}^\beta_\beta(T \circ S) = AA^{-1} = I_n = \mathscr{T}^\beta_\beta(\mathbb{1}_W)$, and so $T \circ S = \mathbb{1}_W$. Thus T is invertible, with $T^{-1} = S$. $\qquad\square$

We remark that in the proof that $\mathscr{L}(V,W)$ is isomorphic to $M_{m\times n}(F)$, where V and W are vector spaces of finite dimensions n and m, respectively, over F, we chose an ordered basis α for V and an ordered basis β for W to construct the isomorphism. It is helpful to understand the role of the choice of bases, and so we turn our attention to this topic next.

B.2.33 Definition. *Let V be a vector space of finite dimension n over the field F, and let α, β be ordered bases for V. Then $A = \mathscr{T}^\beta_\alpha(\mathbb{1}_V) \in M_{n\times n}(F)$ is called the change of basis matrix, since for any $v \in V$, $Ac_\alpha(v) = c_\beta(v)$; that is, multiplication by A changes the coordinates of v with respect to α into the coordinates of v with respect to β.*

B.2.34 Proposition. *Let V be a vector space of finite dimension n over the field F, and let α, β be ordered bases for V. Then the change of basis matrix $A = \mathscr{T}^\beta_\alpha(\mathbb{1}_V) \in M_{n\times n}(F)$ is invertible, and A^{-1} is the change of basis matrix that changes coordinates with respect to β into coordinates with respect to α.*

Proof. This follows from Proposition B.2.32, since $\mathbb{1}_V$ is invertible with inverse $\mathbb{1}_V$. $\qquad\square$

B.2.35 Proposition. *Let V and W be vector spaces of finite dimension over the field F, and let $T:V \to W$ be a linear transformation. Let α, α' be ordered bases for V, and let β, β' be ordered bases for W. Let $P = \mathscr{T}^\beta_{\beta'}(\mathbb{1}_W)$ and $Q = \mathscr{T}^\alpha_{\alpha'}(\mathbb{1}_V)$. Then $P[T]^{\beta'}_{\alpha'} = [T]^\beta_\alpha Q$.*

Proof. This is an immediate concequence of Proposition B.2.31 since $\mathbb{1}_W \circ T = T \circ \mathbb{1}_V$. $\qquad\square$

B.2.36 Corollary. *Let V be a vector space of finite dimension over the field F, and let T be a linear operator on V. Let α, β be ordered bases for V, and let $P = \mathscr{T}^\alpha_\beta(\mathbb{1}_W)$. Then P is invertible, and $P^{-1}[T]^\alpha_\alpha P = [T]^\beta_\beta$.*

Proof. This follows immediately from Proposition B.2.35 and Proposition B.2.32. □

3 Determinants

To facilitate our discussion of determinants, we introduce the following notation.

B.3.1 Notation. *Let m and n be positive integers, and let F be a field. For any $A \in M_{m \times n}(F)$, define for each $i \in J_m$ the mapping $A^i : M_{1 \times n}(F) \to M_{m \times n}(F)$ by setting, for each $X \in M_{1 \times n}(F)$, $row_k(A^i(X)) = row_k(A)$ for all $k \in J_m$, $k \neq i$, while $row_i(A^i(X)) = X$. Similarly, for each $j \in J_n$, define the mapping $A_j : M_{m \times 1}(F) \to M_{m \times n}(F)$ by setting, for each $X \in M_{m \times 1}(F)$, $col_k(A_j(X)) = col_k(A)$ for all $k \in J_n$, $k \neq j$, while $col_j(A_j(X)) = X$.*

B.3.2 Definition. *Let n be a positive integer and let F a field. A mapping $D : M_{n \times n}(F) \to F$ is called a determinant if it satisfies the following three properties:*

(i) *for each $A \in M_{n \times n}(F)$, and for each $i \in J_n$, $D \circ A^i : M_{1 \times n}(F) \to F$ and $D \circ A_i : M_{n \times 1}(F) \to F$ are linear transformations;*

(ii) *for any $A \in M_{n \times n}(F)$, if there exist $i, j \in J_n$ with $i \neq j$ and $row_i(A) = row_j(A)$, then $D(A) = 0_F$;*

(iii) *$D(I_n) = 1_F$.*

It is not immediately clear that for a given positive integer n, there exists at least one determinant mapping. We shall ultimately show that for each positive integer n there is one, and only one, such mapping. Our first objective is to prove the existence of such a mapping.

B.3.3 Proposition. *Let n be a positive integer and F be a field. Define the mapping $\det : M_{n \times n}(F) \to F$ by $\det(A) = \sum_{\sigma \in S_n} sgn(\sigma) \prod_{i \in J_n} A_{\sigma(i) \, i}$ for each $A \in M_{n \times n}(F)$. Then $\det$ is a determinant mapping.*

Proof. Let $j \in J_n$, $r, s \in F$, and $X, Y \in M_{m \times 1}(F)$. Then

$$\det(A^j(rX + sY)) = \sum_{\sigma \in S_n} \text{sgn}(\sigma) \prod_{i \in J_n} (A^j(rX + sY))_{\sigma(i)\,i}$$

$$= \sum_{\sigma \in S_n} \text{sgn}(\sigma)(rX + sY)_{\sigma^{-1}(j)} \prod_{\substack{i \in J_n \\ \sigma(i) \neq j}} A_{\sigma(i)\,i}$$

$$= \sum_{\sigma \in S_n} \text{sgn}(\sigma) r X_{\sigma^{-1}(j)} \prod_{\substack{i \in J_n \\ \sigma(i) \neq j}} A_{\sigma(i)\,i} + \sum_{\sigma \in S_n} \text{sgn}(\sigma) s Y_{\sigma^{-1}(j)} \prod_{\substack{i \in J_n \\ \sigma(i) \neq j}} A_{\sigma(i)\,i}$$

$$= r \sum_{\sigma \in S_n} \text{sgn}(\sigma) X_{\sigma^{-1}(j)} \prod_{\substack{i \in J_n \\ \sigma(i) \neq j}} A_{\sigma(i)\,i} + s \sum_{\sigma \in S_n} \text{sgn}(\sigma) Y_{\sigma^{-1}(j)} \prod_{\substack{i \in J_n \\ \sigma(i) \neq j}} A_{\sigma(i)\,i}$$

$$= r \det(A^j(X)) + s \det(A^j(Y))$$

and so $\det \circ A^j$ is a linear transformation for each $j \in J_n$.

Next, suppose that $A \in M_{n \times n}(F)$ and there exist i, j with $1 \leq i < j \leq n$ such that $row_i(A) = row_j(A)$. Let $\tau = (i\,j) \in S_n$; that is, τ is the transposition that exchanges i and j. Then $\langle \tau \rangle = \{\tau, \tau^2 = \mathbb{1}_{J_n}\}$, and so the set of left cosets of $\langle \tau \rangle$ in S_n is a partition of S_n into sets of size 2; namely $\{\sigma, \tau \circ \sigma\}$ for $\sigma \in S_n$. From each $C \in S_n / \langle \tau \rangle$, we designate one of the two permutations in C to serve as the representative $\hat{C}$ of C, so $C = \{\hat{C}, \hat{C} \circ \tau\}$. Thus in order to sum over all permutations in S_n, we may instead sum over all cells in $S_n / \langle \tau \rangle$, and for each cell C, sum over the two elements $\hat{C}$ and $\hat{C} \circ \tau$ of the cell. We then obtain

$$\det(A) = \sum_{\sigma \in S_n} \text{sgn}(\sigma) \prod_{k \in J_n} A_{\sigma(k)\,k}$$

$$= \sum_{C \in S_n/\sigma} \left(\text{sgn}(\hat{C}) \prod_{k \in J_n} A_{\hat{C}(k)\,k} + \text{sgn}(\hat{C} \circ \tau) \prod_{k \in J_n} A_{\hat{C} \circ \tau(k)\,k} \right)$$

$$= \sum_{C \in S_n/\sigma} \text{sgn}(\hat{C})(A_{\hat{C}(i)\,i} A_{\hat{C}(j)\,j} - A_{\hat{C} \circ \tau(i)\,i} A_{\hat{C} \circ \tau(j)\,j}) \prod_{\substack{k \in J_n \\ k \neq i,j}} A_{\hat{C}(k)\,k}$$

$$= \sum_{C \in S_n/\sigma} \text{sgn}(\hat{C})(A_{\hat{C}(i)\,i} A_{\hat{C}(j)\,j} - A_{\hat{C}(j)\,i} A_{\hat{C}(i)\,j}) \prod_{\substack{k \in J_n \\ k \neq i,j}} A_{\hat{C}(k)\,k}$$

$$= 0_F$$

since $A_{i\,k} = A_{j\,k}$ for each $k \in J_n$ and thus $A_{\hat{C}(i)\,i} = A_{\hat{C}(j)\,i}$ and $A_{\hat{C}(i)\,j} = A_{\hat{C}(j)\,j}$, so $A_{\hat{C}(i)\,i} A_{\hat{C}(j)\,j} - A_{\hat{C}(j)\,i} A_{\hat{C}(i)} = 0_F$ for each $C \in S_n / \langle \tau \rangle$.

Finally, $\det(I_n) = \sum_{\sigma \in S_n} \text{sgn}(\sigma) \prod_{i \in J_n} (I_n)_{\sigma(i)\,i}$, and for any $\sigma \neq \mathbb{1}_{J_n}$, there exists $i \in J_n$ with $i \neq \sigma(i)$ and thus $\prod_{i \in J_n} (I_n)_{\sigma(i)\,i} = 0_F$ for any

$\sigma \neq \mathbb{1}_{J_n}$. On the other hand, when $\sigma = \mathbb{1}_{J_n}$, then $(I_n)_{\sigma(i)\,i} = (I_n)_{ii} = 1$ for each $i \in J_n$, and thus $\prod_{i \in J_n} (I_n)_{\sigma(i)\,i} = 1_F$ when $\sigma = \mathbb{1}_{J_n}$, which proves that $\det(I_n) = 1_F$. $\qquad\square$

For example, the calculation of the determinant of a 2×2 matrix $\left[\begin{smallmatrix} a & b \\ c & d \end{smallmatrix}\right] \in M_{2\times 2}(F)$ is $\operatorname{sgn}(\mathbb{1}_{J_2})ad + \operatorname{sgn}((1\,2))cb = ad - bc$.

Our next objective is to prove that for any positive integer n, det is the unique determinant mapping on $M_{n\times n}(F)$. We shall need to establish a few properties of a determinant mapping.

B.3.4 Proposition. *Let n be a positive integer, F a field, and let $D : M_{n\times n}(F) \to F$ be a determinant mapping. For any $A \in M_{n\times n}(F)$, and any $i, j \in J_n$ with $i < j$, if B is the matrix formed by exchanging rows i and j in A, then $D(B) = -D(A)$.*

Proof. Let C denote the matrix that is equal to A except for rows i and j, and let $row_i(C) = row_j(C) = row_i(A) + row_j(A)$. Then by the linearity in row i, we have $D(C) = D(C^i(row_i(A))) + D(C^i(row_j(A)))$. Now $C^i(row_i(A)) = A^j(row_i(A) + row_j(A))$ and so by linearity in row j, we have $D(C^i(row_i(A))) = D(A^j(row_i(A) + row_j(A))) = D(A^j(row_i(A))) + D(A^j(row_j(A))) = 0 + D(A)$, where the 0 results from the fact that $A^j(row_i(A))$ has rows i and j equal. Moreover, also by linearity in row j, $D(C^i(row_j(A))) = D(B) + D(A^i(row_j(A))) = D(B) + 0$ since $A^i(row_j(A))$ has rows i and j equal. Thus $D(C) = D(A) + D(B)$. However, C also has rows i and j equal (to $row_i(A) + row_j(A)$), and so $D(C) = 0$. Thus $D(B) + D(A) = 0$, or $D(B) = -D(A)$. $\qquad\square$

Proposition B.3.4 can be formulated in the following way. Let $A \in M_{n\times n}(F)$, and let $\tau \in S_n$ be a transposition, say $\tau = (i\,j)$. Then for any determinant mapping $D : M_{n\times n}(F) \to F$,

$$D\left(\begin{bmatrix} row_{\tau(1)}(A) \\ \hline row_{\tau(2)}(A) \\ \hline \vdots \\ \hline row_{\tau(n)}(A) \end{bmatrix}\right) = \operatorname{sgn}(\tau)D(A)$$

since $\operatorname{sgn}(\tau) = -1$.

B.3.5 Corollary. *Let n be a positive integer, F a field, and let $D : M_{n\times n}(F) \to F$ be a determinant mapping. Then for any $A \in M_{n\times n}(F)$*

and any $\sigma \in S_n$,

$$D\left(\begin{bmatrix} row_{\sigma(1)}(A) \\ \overline{row_{\sigma(2)}(A)} \\ \vdots \\ \overline{row_{\sigma(n)}(A)} \end{bmatrix}\right) = sgn(\sigma)D(A).$$

Proof. By Corollary 7.4.3, if $\sigma \in S_n$, then there are transpositions $\tau_1, \tau_2, \ldots, \tau_i$ such that $\sigma = \tau_1 \tau_2 \cdots \tau_k$. Furthermore, by Proposition 7.4.5, $sgn: S_n \to \{1, -1\}$ is a homomorphism. A simple inductive argument will then establish that

$$D\left(\begin{bmatrix} row_{\sigma(1)}(A) \\ \overline{row_{\sigma(2)}(A)} \\ \vdots \\ \overline{row_{\sigma(n)}(A)} \end{bmatrix}\right) = \mathrm{sgn}(\tau_1)\mathrm{sgn}(\tau_2) \cdots \mathrm{sgn}(\tau_k)D(A)$$

$$= \mathrm{sgn}(\tau_1 \tau_2 \cdots \tau_k)D(A) = \mathrm{sgn}(\sigma)D(A).$$

$\square$

B.3.6 Proposition. *Let n be a positive integer, F a field, and let $D : M_{n \times n}(F) \to F$ be a determinant mapping. Let $A, B \in M_{n \times n}(F)$. Then $D(AB) = \det(A)D(B)$.*

Proof. We have

$$AB = \begin{bmatrix} row_1(A)B \\ \overline{row_2(A)B} \\ \vdots \\ row_n(A)B \end{bmatrix}$$

and so using the linearity of D in the first row, then in the second row, we obtain

$$D(AB) = \sum_{j_1=1}^{n} A_{1j_1} D\left(\begin{bmatrix} row_{j_1}(B) \\ \overline{row_2(A)B} \\ \vdots \\ row_n(A)B \end{bmatrix}\right) = \sum_{j_1=1}^{n} A_{1j_1} \sum_{j_2=1}^{n} A_{2j_2} D\left(\begin{bmatrix} row_{j_1}(B) \\ \overline{row_{j_2}(B)} \\ \vdots \\ row_n(A)B \end{bmatrix}\right).$$

Continue in this fashion to obtain

$$D(AB) = \sum_{(j_1, j_2, \ldots, j_n) \in J_n^n} A_{1j_1} A_{2j_2} \cdots A_{nj_n} D\left(\begin{bmatrix} row_{j_1}(B) \\ row_{j_2}(B) \\ \vdots \\ row_{j_n}(B) \end{bmatrix}\right).$$

For any $(j_1, j_2, \ldots, j_n) \in J_n^n$ for which $j_i = j_k$ for some i, k with $i \neq k$, the matrix in the corresponding term has two identical rows and thus has determinant equal to 0_F. On the other hand, for $(j_1, j_2, \ldots, j_n) \in J_n^n$ for which $j_i \neq j_k$ for all i, k with $i \neq k$, there is a unique $\sigma \in S_n$ such that $\sigma(i) = j_i$ for $i \in J_n$. But then

$$D\left(\begin{bmatrix} row_{j_1}(B) \\ row_{j_2}(B) \\ \vdots \\ row_{j_n}(B) \end{bmatrix}\right) = D\left(\begin{bmatrix} row_{\sigma(1)}(B) \\ row_{\sigma(2)}(B) \\ \vdots \\ row_{\sigma(n)}(B) \end{bmatrix}\right)$$

$$= \text{sgn}(\sigma)D\left(\begin{bmatrix} row_1(B) \\ row_2(B) \\ \vdots \\ row_n(B) \end{bmatrix}\right) = \text{sgn}(\sigma)D(B)$$

and so

$$D(AB) = D(B) \sum_{\sigma \in S_n} \text{sgn}(\sigma)A_{1\sigma(1)}A_{2\sigma(2)} \cdots A_{n\sigma(n)}.$$

Since $\text{sgn}(\sigma) = \text{sgn}(\sigma^{-1})$ for any $\sigma \in S_n$, and the mapping $\sigma \mapsto \sigma^{-1}$ from S_n to S_n is bijectve, and for each $i \in J_n$, $A_{i\,\sigma(i)} = A_{\sigma^{-1}(\sigma(i))\,\sigma(i)}$, and $\{\sigma(i) \mid i \in J_n\} = J_n$, and multiplication in F is commutative, we have

$$D(AB) = D(B) \sum_{\sigma^{-1} \in S_n} \text{sgn}(\sigma^{-1})A_{\sigma^{-1}(1)\,1}A_{\sigma^{-1}(2)\,2} \cdots A_{\sigma^{-1}(n)\,n}$$

$$= D(B)\det(A) = \det(A)D(B),$$

as claimed. $\qquad\square$

B.3.7 Corollary. *Let n be a positive integer and F be a field. Then* det *is the unique determinant mapping from $M_{n\times n}(F)$ to F.*

Proof. If D is any determinant mapping, then by Proposition B.3.6, for any $A \in M_{n\times n}(F)$, $D(A) = D(AI_n) = \det(A)D(I_n) = \det(A)$. $\qquad\square$

B.3.8 Corollary. *Let n be a positive integer and let F be a field. Then for any $A, B \in M_{n\times n}(F)$, $\det(AB) = \det(A)\det(B)$.*

Proof. This follows immediately from Proposition B.3.6 and Corollary B.3.7. $\qquad\square$

B.3.9 Proposition. *Let n be a positive integer and let F be a field. For any $A \in M_{n\times n}(F)$, $\det(A) = \det(A^t)$.*

Proof. We have

$$\det(A^t) = \sum_{\sigma \in S_n} \mathrm{sgn}(\sigma) \prod_{i=1}^{n} (A^t)_{\sigma(i)\, i}$$

$$= \sum_{\sigma \in S_n} \mathrm{sgn}(\sigma) \prod_{i=1}^{n} A_{i\,\sigma(i)}$$

$$= \sum_{\sigma \in S_n} \mathrm{sgn}(\sigma) \prod_{i=1}^{n} A_{\sigma^{-1}(\sigma(i))\,\sigma(i)}$$

$$= \sum_{\sigma \in S_n} \mathrm{sgn}(\sigma) \prod_{i=1}^{n} A_{\sigma^{-1}(i)\, i}.$$

Since $\mathrm{sgn}(\sigma) = \mathrm{sgn}(\sigma^{-1})$ for each $\sigma \in S_n$, and as the mapping from S_n to itself that sends σ to σ^{-1} is bijective, we obtain

$$\det(A^t) = \sum_{\sigma \in S_n} \mathrm{sgn}(\sigma) \prod_{i=1}^{n} A_{\sigma^{-1}(i)\, i}$$

$$= \sum_{\sigma^{-1} \in S_n} \mathrm{sgn}(\sigma^{-1}) \prod_{i=1}^{n} A_{\sigma^{-1}(i)\, i} = \det(A).$$

$\square$

B.3.10 Corollary. *Let n be a positive integer and let F be a field. Then for any $A \in M_{n \times n}(F)$, the following hold:*

(i) *If A has two identical columns, then $\det(A) = 0_F$.*

(ii) *If A is upper or lower triangular, then $\det(A) = \prod_{i=1}^{n} A_{i\,i}$.*

(iii) *For any $i \in J_n$, $\det \circ A_i : M_{n \times 1}(F) \to F$ is a linear map; that is, for any $r, s \in F$ and any $X, Y \in M_{n \times 1}(F)$, $\det(A_i(rX + sY)) = r\det(A_i(X)) + s\det(A_i(Y))$.*

(iv) *If B is formed from A by exchanging two columns of A, then $\det(B) = -\det(A)$.*

(v) *For $r \in F$ and $i, j \in J_n$ with $i \neq j$, $\det(A^i(row_i(A) + r\,row_j(A))) = \det(A)$, and $\det(A_i(col_i(A) + r\,col_j(A))) = \det(A)$.*

Proof. (i) is immediate, since if A has two identical columns, then A^t has two identical rows and thus by Proposition B.3.9. $\det(A) = \det(A^t) = 0_F$.

For (ii), suppose that A is upper triangular. Then $A_{i\,j} = 0_F$ for all $i, j \in J_n$ with $i > j$. We have $\det(A) = \sum_{\sigma \in S_n} \mathrm{sgn}(\sigma) \prod_{i=1}^{n} A_{\sigma(i)\, i}$. For

$\sigma \in S_n$, if there exists $j \in J_n$ such that $\sigma(j) > j$, then $\prod_{i=1}^{n} A_{\sigma(i)\,k} = 0_F$, and so

$$\det(A) = \sum_{\substack{\sigma \in S_n \\ \sigma(j) \geq j,\ j \in J_n}} \operatorname{sgn}(\sigma) \prod_{i=1}^{n} A_{\sigma(i)\,i}.$$

But if $\sigma \in S_n$ satisfies $\sigma(j) \geq j$ for all $j \in J_n$, then $\sigma = \mathbb{1}_{J_n}$. To see this, suppose that $\sigma(j) \geq j$ for all $j \in J_n$, and for some $i \in J_n$, $\sigma(i) > i$. Assume i is least with this property, so that $\sigma(j) = j$ for all $j < i$. Then $i \notin \{\, \sigma(j) \mid j \in J_n \,\}$, which is not possible as σ is surjective. Thus $\det(A) = \operatorname{sgn}(\mathbb{1}_{J_n}) \prod_{i=1}^{m} A_{ii} = \prod_{i=1}^{n} A_{ii}$, as required. If A is lower triangular, then A^t is upper triangular, and $(A^t)_{ii} = A_{ii}$ for all $i \in J_n$, so $\det(A) = \prod_{i=1}^{n} A_{ii}$.

(iii) follows from Proposition B.3.9 and the fact that for any $i \in J_n$, $\det \circ (A^t)^i$ is a linear transformation from $M_{1 \times n}(F)$ to F. For if $r, s \in F$ and $X, Y \in M_{n \times 1}(F)$, we have

$$\begin{aligned}
\det(A_i(rX + sY)) &= \det((A_i(rX + sY))^t) = \det((A^t)^i(rX^t + sY^t)) \\
&= r\,\det((A^t)^i(X^t)) + s\,\det((A^t)^i(Y^t)) \\
&= r\,\det((A_i(X))^t) + s\,\det((A_i(Y))^t) \\
&= r\,\det(A_i(X)) + s\,\det(A_i(Y)).
\end{aligned}$$

(iv) is immediate from Proposition B.3.9, since if B is formed from A by exchanging two columns, then B^t is formed from A^t by exchanging two rows and thus $\det(B) = \det(B^t) = -\det(A^t = -\det(A)$.

Finally, for (v), let $r \in F$ and $i, j \in J_n$ with $i \neq j$. Then by the linearity of the determinant in row i, we have $\det(A^i(row_i(A) + r\,row_j(A))) = \det(A^i(row_i(A)) + r\,\det(A^i(row_j(A)) = \det(A) + r\,\det(A^i(row_j(A))$. Since rows i and j of $A^i(row_j(A))$ are equal, $\det(A^i(row_j(A)) = 0_F$, and thus $\det(A^i(row_i(A) + r\,row_j(A))) = \det(A)$. The other assertion follows from this result by means of Proposition B.3.9. $\qquad\square$

As a consequence of the preceding corollary, we obtain an effective method for evaluating the determinant of any matrix. By means of elementary row operations of two types (add a multiple of one row to another row, exchange two rows), we may transform any $n \times n$ matrix into an upper triangular matrix. Row operations of the first type (add a multiple of one row to another row) do not change the determinant, while exchanging two rows changes the sign of the determinant. The final determinant evaluation is then one of computing the determinant of an upper triangular matrix, and this is accomplished by multiplying the main diagonal elements together.

Another important use of the determinant is to determine whether or not an $n \times n$ matrix is invertible. We shall establish that a matrix A is invertible if, and only if, $\det(A) \neq 0_F$, but this will need the following result.

B.3.11 Proposition. *Let n be a positive integer and F be a field. For $A \in M_{n \times n}(F)$, $\{\, col_1(A), col_2(A), \ldots, col_n(A) \,\}$ is linearly dependent if, and only if, $\det(A) = 0_F$.*

Proof. Let $A \in M_{n \times n}(F)$. Suppose first that $\{\, col_i(A) \mid i \in J_n \,\}$ is linearly dependent. Then there exists $j \in J_n$ for which $col_j(A)$ is a linear combination of $\{\, col_i(A) \mid i \in J_n,\ i \neq j \,\}$, say

$$col_j(A) = \sum_{\substack{i \in J_n \\ i \neq j}} a_i col_i(A).$$

By Corollary B.3.10 (v), adding a multiple of one column to another does not change the value of the determinant, so let B be formed from A by subtracting $a_i col_i(A)$ from column j for each $i \in J_n$, $i \neq j$. Then $\det(A) = \det(B)$ and B has a zero column, so $\det(B) = 0_F$. Thus $\det(A) = 0_F$.

Conversely, suppose that $\{\, col_i(A) \mid i \in J_n \,\}$ is linearly independent. Then $M_{n \times 1}(F) = span(\{\, col_i(A) \mid i \in J_n \,\})$, and so for each $j \in J_n$, there exist $b_{ij} \in F$ for $i \in J_n$ such that $e_j^t = \sum_{i=1}^n b_{ij} col_i(A)$. Let $B \in M_{n \times n}(F)$ be the matrix for which $B_{ij} = b_{ij}$ for each $i, j \in J_n$. Then $AB = I_n$, and so by Corollary B.3.8, $\det(A)\det(B) = \det(I_n) = 1_F$, which means that $\det(A) \neq 0_F$. $\square$

B.3.12 Corollary. *Let n be a positive integer and F be a field. Then $A \in M_{n \times n}(F)$ is invertible if, and only if, $\det(A) \neq 0_F$.*

Proof. Suppose first that A is invertible. Then A^{-1} exists and $AA^{-1} = I_n$, so $\det(A)\det(A^{-1}) = \det(AA^{-1}) = det(I_n) = 1_F$, and thus $\det(A) \neq 0_F$. Conversely, suppose that $\det(A) \neq 0_F$. Then by Proposition B.3.11, the columns of A form a linearly independent subset of $M_{n \times 1}(F)$ and thus the columns of A form a basis for $M_{n \times 1}(F)$. But then $T_A(M_{n \times 1}(F)) = M_{n \times 1}(F)$, so T_A is surjective, and thus by the dimension theorem, T_A is injective and so T_A is invertible. It follows now from Corollary B.2.32 that A is invertible. $\square$

There is a useful method for the evaluation of a determinant, called the Laplace expansion. For its derivation, we require the following notation.

B.3.13 Notation. *Let F be a field, and let n be a positive integer. Let $A \in M_{n \times n}(F)$. For any $i, j \in J_n$, let $\overline{A}_{i,j}$ denote the $(n-1) \times (n-1)$ matrix obtained from A by deleting the entries of row i and of column j.*

B.3.14 Proposition. *Let F be a field, let n be a positive integer, and let $A \in M_{n \times n}(F)$. Let $i, j \in J_n$. Then $\det(A_i(e_j^t)) = \det(A^j(e_i)) = (-1_F)^{i+j} \det(\overline{A}_{i,j})$.*

Proof. We prove that $\det(A_i(e_j^t)) = (-1_F)^{i+j} \det(\overline{A}_{i,j})$, as the proof that $\det(A^j(e_i)) = (-1_F)^{i+j} \det(\overline{A}_{i,j})$ is similar (or obtained from the former by application of Proposiiton B.3.9). $A_i(e_j^t)$ is the matrix obtained from A by replacing column i by the transpose of $e_j \in M_{1 \times n}(F)$. There are $n - i - 1$ columns to the right of column i, and by performing that number of column exchanges, each of which changes the determinant by a factor of -1_F, we can move the e_j^t column to become the rightmost column. There are $n - j - 1$ rows below row j, so by performing that number of row exchanges, we can move row i to become the bottom row. The result is the matrix

$$B = \begin{bmatrix} \overline{A}_{i,j} & 0 \\ C & 1_F \end{bmatrix},$$

where 0 is the zero element of $M_{(n-1) \times 1}(F)$, and $C \in M_{1 \times (n-1)}(F)$ is equal to row i of A with the jth entry removed. We have $\det(A_i(e_j^t)) = (-1_F)^{n-i-1+n-1-j} \det(B) = (-1_F)^{i+j} \det(B)$. Now, by definition of det, we have

$$\det(B) = \sum_{\sigma \in S_n} \text{sgn}(\sigma) \prod_{k=1}^{n} B_{\sigma(k)\,k}$$

and for each σ with $\sigma(n) \neq n$, $B_{\sigma(n)\,n} = 0_F$, while if $\sigma(n) = n$, then $B_{\sigma(n)\,n} = B_{n\,n} = 1_F$, so

$$\det(B) = \sum_{\substack{\sigma \in S_n \\ \sigma(n) = n}} \text{sgn}(\sigma) \prod_{k=1}^{n-1} B_{\sigma(k)\,k}$$

$$= \sum_{\substack{\sigma \in S_n \\ \sigma(n) = n}} \text{sgn}(\sigma) \prod_{k=1}^{n-1} (\overline{A}_{i,j})_{\sigma(k)\,k}.$$

The set $\{\,\sigma \in S_n \mid \sigma(n) = n\,\}$ is a subgroup of S_n and the mapping from $\{\,\sigma \in S_n \mid \sigma(n) = n\,\}$ to S_{n-1} for which $\sigma \in S_n$ that satisfies $\sigma(n) = n$ is mapped to the restriction of σ to J_{n-1} is an isomorphism from $\{\,\sigma \in S_n \mid \sigma(n) = n\,\}$ to S_{n-1} that preserves sign. Thus we obtain

$$\det(B) = \sum_{\sigma \in S_{n-1}} \mathrm{sgn}(\sigma) \prod_{k=1}^{n-1} (\overline{A}_{i,j})_{\sigma(k)\,k} = \det(\overline{A}_{i\,j})$$

and so $\det(A_i(e_j^t)) = (-1_F)^{i+j} \det(\overline{A}_{i\,j})$.   □

B.3.15 Corollary (Laplace Expansion). *Let F be a field, n a positive integer, and $A \in M_{n \times n}(F)$. Then for any $i \in J_n$,*

$$\det(A) = \sum_{j=1}^{n} A_{i\,j}(-1_F)^{i+j} \det(\overline{A}_{i,j})$$

$$= \sum_{j=1}^{n} A_{j\,i}(-1_F)^{j+i} \det(\overline{A}_{j,i}).$$

The first expression is called the expansion of $\det(A)$ along row i, while the second expression is called the expansion of $\det(A)$ along column i.

Proof. Let $i \in J_n$. Then since the determinant mapping is linear in row i, we have $\det(A) = \sum_{j=1}^{n} A_{i\,j} \det(A^i(e_j))$, and by Proposition B.3.14, $\det(A^i(e_j)) = (-1_F)^{i+j} \det(\overline{A}_{i,j})$, so $\det(A) = \sum_{j=1}^{n} A_{i\,j}(-1_F)^{i+j} \det(\overline{A}_{i,j})$. The second expression is obtained in an entirely analogous manner.   □

B.3.16 Definition. *Let F be a field, n a positive integer, and $A \in M_{n \times n}(F)$. For each $i, j \in J_n$, $C_{i,j}(A) = (-1)^{i+j} \det(\overline{A}_{i,j})$ is called the (i,j)-cofactor of A. The $n \times n$ matrix $C(A)$ for which $C(A)_{i,j} = C_{i,j}(A)$ for all $i, j \in J_n$ is called the cofactor matrix of A, and the transpose of the cofactor matrix of A is called the classical adjoint of A, denoted by $\mathrm{adj}(A)$.*

Note that by Corollary B.3.15, for $A \in M_{n \times n}(F)$, for any $i \in J_n$, $\det(A) = \sum_{j=1}^{n} A_{i,j} C_{i,j}$.

B.3.17 Corollary. *Let F be a field, n a positive integer, and $A \in M_{n \times n}(F)$. Then $A \, \mathrm{adj}(A) = \det(A) I_n$. If $\det(A) \neq 0_F$, then $A^{-1} = \frac{1}{\det(A)} \mathrm{adj}(A)$.*

Proof. For any $i, j \in J_n$, $(A \, adj(A))_{i,j} = \sum_{k=1}^n A_{i,k} adj(A)_{k,j} = \sum_{k=1}^n A_{i,k} C(A)_{j,k} = \sum_{k=1}^n A_{i,k}(-1_F)^{j+k} \det(\overline{A}_{j,k})$. If $j = i$, this sum is $\det(A)$, while if $j \neq i$, this sum can be regarded as the expansion along row j of the determinant of the matrix obtained from A by replacing row j of A by row i, and since this is a matrix with two identical rows, its determinant is 0_F. Thus $A \, adj(A)$ is the $n \times n$ diagonal matrix with diagonal entries equal to $\det(A)$; that is, $A \, adj(A) = \det(A) I_n$. In particular, if $\det(A) \neq 0_F$, then $A(\frac{1}{\det(A)} adj(A)) = I_n$ and so $A^{-1} = \frac{1}{\det(A)} adj(A)$. □

B.3.18 Corollary. *Let n be a positive integer. Then $A \in M_{n \times n}(\mathbb{Z})$ is invertible in $M_{n \times n}(\mathbb{Z})$ if, and only if, $|\det(A)| = 1$ (that is, $A \in M_{n \times n}(\mathbb{Z}) \subseteq M_{n \times n}(\mathbb{Q})$ is invertible in $M_{n \times n}(\mathbb{Z})$ if, and only if, A is invertible in $M_{n \times n}(\mathbb{Q})$ and $\det(A) = 1$ or -1).*

Proof. Suppose that $A \in M_{n \times n}(\mathbb{Z})$ is invertible, so there exists $B \in M_{n \times n}(\mathbb{Z})$ with $AB = I_n$. But $A, B \in M_{n \times n}(\mathbb{Q})$ and so $1 = \det(I_n) = \det(A)\det(B)$. As $A, B \in M_{n \times n}(\mathbb{Z})$, both $\det(A)$ and $\det(B)$ are integers, and so $\det(A) = \det(B) = \pm 1$. Conversely, suppose that $A \in M_{n \times n}(\mathbb{Z})$ and $\det(A) = \pm 1$. Then by Corollary B.3.17, A^{-1} (in $M_{n \times n}(\mathbb{Q})$) is given by $(\pm 1) adj(A)$, and since the entries of $adj(A)$ are elements of $\mathbb{Z}$, it follows that $B = (\pm 1) adj(A) \in M_{n \times n}(\mathbb{Z})$ and $AB = I_n = BA$; that is, A has inverse B in $M_{n \times n}(\mathbb{Z})$. □

We shall finish this section with the computation of a very important determinant, called the Vandermonde determinant.

B.3.19 Proposition. *Let F be a field. For any $n \geq 1$, let $a_0, a_1, \ldots, a_n \in F$. Then the determinant of the Vandermonde matrix*

$$V(a_0, a_1, \ldots, a_n) = \begin{bmatrix} 1_F & a_0 & \cdots & a_0^n \\ 1_F & a_1 & \cdots & a_1^n \\ \vdots & \vdots & & \vdots \\ 1_F & a_n & \cdots & a_n^n \end{bmatrix}$$

is $\prod_{0 \leq j < i \leq n} (a_i - a_j)$.

Proof. Proceed by induction on n. For the base case of $n = 1$, let $a_0, a_1 \in F$ and compute $\det \begin{bmatrix} 1_F & a_0 \\ 1_F & a_1 \end{bmatrix} = 1_F a_1 - 1_F a_0 = a_1 - a_0 = \prod_{0 \leq j < i \leq 1} (a_i - a_j)$. Now suppose that $n > 1$ is an integer such that the result holds for $n - 1$, and let $a_0, a_1, \ldots, a_n \in F$. With $V(a_0, a_1, \ldots, a_n)$, subtract row 1 from each of rows 2 through $n + 1$ and expand the determinant along the first column (see Corollary B.3.15) to obtain that the determinant of the Vandermonde matrix $V(a_0, a_1, \ldots, a_n)$ is that of the matrix

$$\begin{bmatrix} a_1 - a_0 & a_1^2 - a_0^2 & \cdots & a_1^n - a_0^n \\ a_2 - a_0 & a_2^2 - a_0^2 & \cdots & a_2^n - a_0^n \\ \vdots & \vdots & & \vdots \\ a_n - a_0 & a_n^2 - a_0^2 & \cdots & a_n^n - a_0^n \end{bmatrix}.$$

Observe that for any i and k, $(a_k^{i+1} - a_0^{i+1}) - a_0(a_k^i - a_0^i) = a_k^i(a_k - a_0)$. So for $i = n-1, n-2, \ldots, n-(n-1) = 1$, subtract a_0 times column i from column $i+1$ to obtain that the determinant of $V(a_0, a_1, \ldots, a_n)$ is the same as that of the matrix

$$\begin{bmatrix} a_1 - a_0 & a_1(a_1 - a_0) & \cdots & a_1^{n-1}(a_1 - a_0) \\ a_2 - a_0 & a_2(a_2 - a_0) & \cdots & a_2^{n-1}(a_2 - a_0) \\ \vdots & \vdots & & \vdots \\ a_n - a_0 & a_n(a_n - a_0) & \cdots & a_n^{n-1}(a_n - a_0) \end{bmatrix}.$$

But the determinant of this matrix is

$$\det(V(a_1, a_2, \ldots, a_n)) \prod_{0 < j \le n} (a_j - a_0).$$

By hypothesis, we have

$$\det(V(a_0, a_1, \ldots, a_n)) = \det(V(a_1, a_2, \ldots, a_n)) \prod_{0 < j \le n} (a_j - a_0)$$

$$= \prod_{1 \le j < i \le n} (a_i - a_j) \prod_{0 < i \le n} (a_i - a_0)$$

$$= \prod_{0 \le j < i \le n} (a_i - a_j),$$

as required. The result follows now by induction. $\square$

Note that by Corollary B.3.12, $V(a_0, a_1, \ldots, a_n)$ is invertible if, and only if, $\det(V(a_0, a_1, \ldots, a_n)) = \prod_{0 \le j < i \le n} (a_i - a_j) \ne 0_F$, and since F is a field and therefore without zero divisors, this holds if, and only if, $a_i \ne a_j$ for all $i, j \in J_n$ with $i \ne j$.

B.4 Eigenvalues and Eigenvectors

In the study of linear operators on a finite dimensional vector space over a given field F, the identification of the one-dimensionsal subspaces of V that are fixed setwise under the linear operator plays an extremely important role. For example, in Chapter 15, we studied rotations in $\mathbb{R}^3$, and the problem of finding the axis of rotation amounts to finding the unique one-dimensional subspace of $\mathbb{R}^3$ that is fixed setwise (and in this case, even pointwise) under the rotation.

The key to this problem is the fact that a subspace U is one-dimensional if, and only if, any nonzero element of U provides a basis for U. Thus if $T : V \to V$ is linear, then $T(U) \subseteq U$ if, and only if for any nonzero $v \in U$, $T(v) \in U$; that is, for any $v \in U$ with $v \neq 0_V$, there exists $r \in F$ such that $T(v) = rv$.

B.4.1 Definition. *Let V be a vector space over a field F, and let T be a linear operator on V. A vector $v \in V$ is said to be an eigenvector of T if $v \neq 0_V$ and there exists $r \in F$ with $T(v) = rv$, in which case r is said to be an eigenvalue of T, and v is an eigenvector of T with associated eigenvalue r.*

For example, if T is the linear operator on $\mathbb{R}^3$ that is reflection in a plane P through the origin, then every $v \in P$ other than 0_V is an eigenvector of T with associated eigenvalue 1, and every vector w on the line through the origin that is perpendicular to P, other than 0_V, is an eigenvector of T with associated eigenvalue -1.

Note that the linear operator on $\mathbb{R}^2$ that rotates the plane by $\frac{\pi}{2}$ does not fix any one-dimensional subspace. In fact, $R_{\frac{\pi}{2}}(v) = rv$ if, and only if, $v = 0_V$, and so $R_{\frac{\pi}{2}}$ is an example of a linear operator that has no eigenvalues (as it turns out, this reflects a certain deficiency in the field, as we can think of $\mathbb{R}^2$ as a one-dimensional vector space over $\mathbb{C}$, and $R_{\frac{\pi}{2}}$ is then the linear map obtained by multiplication by i, and then $R_{\frac{\pi}{2}}(i) = -1 = i(i)$; that is, i is an eigenvalue of $R_{\frac{\pi}{2}}$.

B.4.2 Proposition. *A linear operator T on a vector space V over a field F has 0_F as an eigenvalue if, and only if, $\text{null}(T) \neq \{0_V\}$; that is, if, and only if, T is not injective.*

Proof. First, suppose that T is not injective, so $\text{null}(T) \neq \{0_V\}$. Then there exists $v \in \text{null}(T)$ with $v \neq 0_V$, so $T(v) = 0_V = 0_F v$ and thus 0_F is an eigenvalue of T. Conversely, suppose that 0_F is an eigenvalue of T. Then there exists $v \in V$ with $v \neq 0_V$ and $T(v) = 0_F v = 0_V$. Thus $v \in \text{null}(T)$ and so $\text{null}(T) \neq \{0_V\}$. $\square$

For a linear operator in general, the question is how to find the eigenvalues, if there are any. As it turns out, the determinant is an invaluable tool for this task.

B.4.3 Proposition. *Let V be a vector space over the field F, and let T be a linear operator on V. Then $v \in V$ is an eigenvector for T with associated eigenvalue $r \in F$ if, and only if, $v \neq 0_V$ and $v \in null(T - r1_V)$.*

Proof. Suppose first of all that $v \in V$ and $r \in F$ satify $v \neq 0_V$ and $T(v) = rv = (r1_V)(v)$. Then $(T - r1_V)(v) = 0_V$ so $v \in null(T - r1_V)$ and $v \neq 0_V$.

Conversely, suppose that $v \neq 0_V$ and $v \in null(T - r1_V)$. Then $0_V = (T - r1_V)(v) = T(v) - (r1_V)(v) = T(v) - rv$, so $T(v) = rv$ and $v \neq 0_V$. Thus v is an eigenvector for T with associated eigenvalue r. $\qquad\square$

B.4.4 Definition. *Let V be a finite dimensional vector space over the field F, and let T be a linear operator on V. If $r \in F$ is an eigenvalue for T, then $null(T - r1_V)$ is called the eigenspace for T with associated eigenvalue r, and is denoted by E_r.*

Note that each eigenspace of V is a subspace of V.

If V is a vector space of finite dimension n over a field F, and α and β are any ordered bases of V, then by Corollary B.2.36, for $P = \mathscr{T}^\alpha_\beta(1_V)$, P is invertible, and $P^{-1}[T]^\alpha_\alpha P = [T]^\beta_\beta$, so $\det([T]^\beta_\beta) = \det(P^{-1}[T]^\alpha_\alpha P) = \det(P^{-1})\det([T]^\alpha_\alpha)\det(P) = \det(P^{-1})\det(P)\det([T]^\alpha_\alpha) = \det(I_n)\det([T]^\alpha_\alpha) = \det([T]^\alpha_\alpha)$. We may therefore introduce the notion of the determinant of the linear operator T as follows.

B.4.5 Definition. *Let V be a finite dimensional vector space over a field F, and let T be a linear operator on V. Then $\det(T) \in F$ is defined to be $\det([T]^\alpha_\alpha)$ for any ordered basis α of V.*

B.4.6 Proposition. *Let V be a finite dimensional vector space over a field F, and let T be a linear operator on V. Then $r \in F$ is an eigenvalue of T if, and only if, $T - r1_V$ has nontrivial null space; that is, if, and only if, $\det(T - r1_V) = 0_F$.*

Proof. By Proposition B.4.3, $r \in F$ is an eigenvalue for T if, and only if, $null(T - r1_V) \neq \{0_V\}$. This holds if, and only if, $T - r1_V$ is not injective. Since V is finite dimensional, this is equivalent to $T - r1_V$ being not invertible. But by Corollary B.2.32, this is equivalent to the assertion that for any ordered basis α for V, $[T - r1_V]^\alpha_\alpha = [T]^\alpha_\alpha - rI_n$ is not invertible, and by Corollary B.3.12, this final assertion is equivalent to $0_F = \det([T - r\mathbb{1}_V]_\alpha^\alpha) = \det(T - r\mathbb{1}_V)$. □

Thus the problem of finding eigenvalues of a linear operator T on a finite dimensional vector space V can be considered as that of finding solutions $x \in F$ for the equation $\det(T - x\mathbb{1}_V) = 0_F$. It turns out that $\det(T - x\mathbb{1}_V)$ is a polynomial of degree n in the unknown x.

B.4.7 Proposition. *Let V be a finite dimensional vector space over a field F, and let T be a linear operator on V. Then $\det(T - x\mathbb{1}_V)$ is a polynomial of degree n with coefficients in F, called the characteristic polynomial of T and denoted by $c_T(x)$. Its leading coefficient is $(-1_F)^n$, while its constant term is $\det(T)$. Finally, the coefficient of x^{n-1} is $-tr(A) = -\sum_{i=1}^n A_{ii}$.*

Proof. Let $n = \dim_F(V)$, α be an ordered basis for V, and let $A = [T]_\alpha^\alpha$. Then we have

$$c_T(x) = \det(A - xI_n) = \sum_{\sigma \in S_n} \operatorname{sgn}(\sigma) \prod_{i=1}^n (A - xI_n)_{\sigma(i)\, i}.$$

Each non-diagonal element of $A - xI_n$ is an element of F, while each diagonal element of $A - xI_n$ is an element of $F[x]$ of degree 1. Thus for $\sigma \in S_n$, the term $\prod_{i=1}^n (A - xI_n)_{\sigma(i)\, i}$ is a polynomial of degree equal to the number of fixed points of σ; that is, the size of the set $\{\, i \in J_n \mid \sigma(i) = i \,\}$. If $\sigma \neq \mathbb{1}_{J_n}$, then the number of fixed points of σ is at most $n - 2$. For suppose that $\sigma \in S_n$ fixes all but one element, say i, of J_n. Then $\sigma(i) = j = \sigma(j)$ for some $j \neq i$, which contradicts the fact that σ is injective. Thus for $\sigma \neq \mathbb{1}_{J_n}$, the degree of $\prod_{i=1}^n (A - xI_n)_{\sigma(i)\, i}$ is at most $n - 2$. Let

$$f = \sum_{\substack{\sigma \in S_n \\ \sigma \neq \mathbb{1}_{J_n}}} \prod_{i=1}^n (A - xI_n)_{\sigma(i)\, i}.$$

Then $f \in F[x]$ has degree at most $n - 2$, and

$$c_T(x) = \det(A - xI_n) = f + \prod_{i=1}^n (A_{ii} - x).$$

It follows now that $c_T(x) \in F[x]$ has degree n and the coefficient of x^n is $(-1_F)^n$. Moreover, the coefficient of x^{n-1} is its coefficient in $\prod_{i=1}^n (A_{ii} - x)$, and that is $-\sum_{i=1}^n A_{ii} = -tr(A)$. Finally, the constant term is obtained by evaluating $\hat{c}_T(0_F)$, so the constant term is $\det(A)$. □

Thus the problem of finding the eigenvalues of a linear operator on a vector space of finite dimension n is that of determining the roots of the characteristic polynomial $c_T(x)$, and since $c_T(x)$ has degree n, it has at most n roots. Thus there are at most n eigenvalues for a linear operator on a vector space of finite dimension n.

B.4.8 Example. *Let $\theta \in \mathbb{R}$ be such that $0 \le \theta < 2\pi$, and consider the linear operator T on $\mathbb{R}^2$ given by $T(x,y) = (x\cos(\theta) + y\sin(\theta), x\sin(\theta) - y\cos(\theta))$ for any $(x,y) \in \mathbb{R}^2$. The matrix that represents T relative to the standard basis for $\mathbb{R}^2$ is $A = \begin{bmatrix} \cos(\theta) & \sin(\theta) \\ \sin(\theta) & -\cos(\theta) \end{bmatrix}$, and so the eigenvalues of T (if any) are the roots of the characteristic polynomial (we use the variable λ to avoid confusion with the coordinates of an element of $\mathbb{R}^2$)*

$$
\begin{aligned}
c_T(\lambda) = \det(A - \lambda I_2) &= \det \begin{bmatrix} \cos(\theta)-\lambda & \sin(\theta) \\ \sin(\theta) & -\cos(\theta)-\lambda \end{bmatrix} \\
&= -\cos^2(\theta) + \lambda^2 - \sin^2(\theta) = \lambda^2 - (\cos^2(\theta) + \sin^2(\theta)) \\
&= \lambda^2 - 1.
\end{aligned}
$$

It follows that the eigenvalues of T are 1 and -1. The eigenvectors with associated eigenvalue 1 are the nonzero elements of $\mathrm{null}(T - \mathbb{1}_{\mathbb{R}^2})$, so we must find $\{ (x,y) \in \mathbb{R}^2 \mid (T - \mathbb{1}_{\mathbb{R}^2})(x,y) = (0,0) \}$. This requires that $(x\cos(\theta) + y\sin(\theta), x\sin(\theta) - y\cos(\theta)) - (x,y) = (0,0)$, equivalently that $x(\cos(\theta) - 1) + y\sin(\theta) = 0$ and $x\sin(\theta) - y(\cos(\theta) + 1) = 0$. Suppose first that $\cos(\theta) - 1 \ne 0$. Then the first equation gives $x = -y\sin(\theta)/(\cos(\theta)-1)$, and this is a solution to the second equation, so

$$
\begin{aligned}
\{ (x,y) \in \mathbb{R}^2) \mid T(x,y) = (x,y) \} &= \{ (-y\sin(\theta)/(\cos(\theta) - 1), y) \mid y \in \mathbb{R} \} \\
&= \mathrm{span}(\{ (\sin(\theta)/(\cos(\theta) - 1), 1) \}) \\
&= \mathrm{span}(\{ (\sin(\theta), \cos(\theta) - 1) \}).
\end{aligned}
$$

On the other hand, if $\cos(\theta) - 1 = 0$, then $\cos(\theta) = 1$ and $0 \le \theta < 2\pi$ means that $\theta = 0$ and thus $\sin(\theta) = 0$, so the equations become $0 = 0$ and $-2y = 0$, and the set of solutions is then $\{ (x,0) \mid x \in \mathbb{R} \} = \mathrm{span}(\{ (1,0) \})$. Thus $E_1 = \mathrm{span}(\{ (\sin(\theta), \cos(\theta) - 1) \})$ if $\theta > 0$, while if $\theta = 0$, then $E_1 = \mathrm{span}(\{ (1,0) \})$.

The reader may recall the double angle formulas for the sine and cosine functions: $\cos(2\alpha) = \cos^2(\alpha) - \sin^2(\alpha)$ and $\sin(2\alpha) = 2\sin(\alpha)\cos(\alpha)$ for any $\alpha \in \mathbb{R}$. If we take $\alpha = \theta/2$, we obtain $\cos(\theta) = \cos^2(\theta/2) - \sin^2(\theta/2)$ and $\sin(\theta) = 2\sin(\theta/2)\cos(\theta/2)$, so $(-\sin(\theta), \cos(\theta) - 1) = (-2\sin(\theta/2)\cos(\theta/2), \cos^2(\theta/2) - \sin^2(\theta/2) - 1) = (-2\sin(\theta/2)\cos(\theta/2), -2\sin^2(\theta/2)) = -2\sin(\theta/2)(\cos(\theta/2), \sin(\theta/2))$, so if

$\cos(\theta) - 1 \neq 0$, then $E_1 = span(\{ (\cos(\theta/2), \sin(\theta/2)) \})$, *the line through the origin that makes an angle of $\theta/2$ radians relative to the positive x-axis. When $\cos(\theta) - 1 = 0$, as we observed earlier, $\theta = 0$ and $E_1 = span(\{ (1,0) \}) = span(\{ (\cos(0/2), \sin(0/2)) \})$, so in both cases, E_1 is the line through the origin making angle $\theta/2$ with the positive $x - axis$.*

Now let us determine $E_{-1} = null(T + 1_{\mathbb{R}^2})$. We see that $(x,y) \in E_{-1}$ if, and only if, $(x \cos(\theta) + y \sin(\theta), x \sin(\theta) - y \cos(\theta)) + (x,y) = (0,0)$, equivalently, if, and only if, $x(\cos(\theta) + 1) + y \sin(\theta) = 0$ and $x \sin(\theta) - y(\cos(\theta) - 1) = 0$. Suppose first that $\cos(\theta) - 1 \neq 0$. Then the second equation gives $y = x \sin(\theta)/(\cos(\theta) - 1)$, and this is a solution to the first equation, so

$$\{ (x,y) \in \mathbb{R}^2) \mid T(x,y) = -(x,y) \} = \{ (x, x \sin(\theta)/(\cos(\theta) - 1)) \mid x \in \mathbb{R} \}$$
$$= span(\{ (1, \sin(\theta)/(\cos(\theta) - 1)) \})$$
$$= span(\{ (\cos(\theta) - 1, \sin(\theta)) \}).$$

On the other hand, if $\cos(\theta) - 1 = 0$, then $\theta = 0$ and the equations become $x = 0$ and $0 = 0$, so the set of solutions is then $\{ (0,y) \mid y \in \mathbb{R} \} = span(\{ (0,1) \})$. As we argued in the determination of E_1, these two cases can be combined to give $E_{-1} = span(\{ (-\sin(\theta/2), \cos(\theta/2) \})$. Thus E_{-1} is the line through the origin that is orthogonal to E_1, and T is therefore reflection in the line E_1.

5 Exercises

1. Let $A = \begin{bmatrix} 1 & 2 \\ 3 & 1 \\ 2 & 1 \end{bmatrix} \in M_{3\times 2}(\mathbb{R})$. Prove that there are infinitely many $B \in M_{2\times 3}(\mathbb{R})$ such that $BA = I_2$, and prove that there is no $C \in M_{2\times 3}(\mathbb{R})$ such that $AC = I_3$.

2. Let F be a field, and let n be a positive integer.

   a) Prove that for any $A, B \in M_{n\times n}(F)$, $tr(AB) = tr(BA)$.

   b) Prove that for any $P, A \in M_{n\times n}(F)$ with P invertible, $tr(PAP^{-1}) = tr(A)$.

3. Let F be a field of characteristic 0, and let n be a positive integer. Prove that for all $A, B \in M_{n\times n}(F)$, $AB - BA \neq I_n$.

4. Let F be a field and let V be a vector space of finite dimension over F. Let α, β be ordered bases for V, and let $P = [1_V]_\alpha^\beta$. Prove that $c_\beta(v) = T_P \circ c_\alpha(v)$ for all $v \in V$.

5. Let F be a field, and let V be a vector space over F.

a) Prove that if U, W are subspaces of V, then $U + W$ and $U \cap W$ are each subspaces of V.

b) Let U and W be finite dimensional subspaces of V.

   (i) Prove that $U + W$ and $U \cap W$ are finite dimensional.

   (ii) Now prove that $\dim_F(U + W) = \dim_F(U) + \dim_F(W) - \dim_F(U \cap W)$.

6. a) Let $u = (1,1,1) \in \mathbb{R}^3$, $v = (0,1,1) \in \mathbb{R}^3$, and $w = (0,0,1) \in \mathbb{R}^3$. Prove that $\{u, v, w\}$ is a basis for $\mathbb{R}^3$.

   b) Let $\alpha = (u, v, w)$, so α is an ordered basis for $\mathbb{R}^3$. Find $c_\alpha(x)$ if $x = (1, 2, -3)$.

7. Let F be a field, and let U, V, and W be finite dimensional vector spaces over F. Let $T : U \to V$ and $R : V \to W$ be linear transformations. Prove that $\mathrm{rank}(R \circ T) \leq \min\{\,\mathrm{rank}(R), \mathrm{rank}(T)\,\}$. Under what conditions is $\mathrm{rank}(R \circ T) = \mathrm{rank}(T)$?

8. Let F be a field and let m, n be positive integers with $n \geq m$. Prove that for any $A \in M_{m \times n}(F)$, $\mathrm{nullity}(T_A) \geq n - m$.

9. Find all linear operators on $\mathbb{R}^3$ that map the plane with equation $x + y + z = 0$ onto the plane with equation $2x - y - 2z = 0$.

10. Let F be a field and m, n be positive integers. For $A \in M_{m \times n}(F)$ and $B \in M_{n \times m}(F)$, prove that $AB - I_m$ is invertible if, and only if, $BA - I_n$ is invertible. Hint: consider $\det(AB - I_m)$ and $\det(BA - I_n)$.

11. Let F be a field and let V be a vector space over F. A linear operator T on V is called a projection of V if $T \circ T = T$.

    a) Prove that if T is a projection on V, then $V = T(V) \oplus \mathrm{null}(T)$.

    b) Prove that if V is finite dimensional and T is a projection on V, then there exists an ordered basis α for V such that $[T]_\alpha^\alpha = \left[\begin{smallmatrix} I_k & 0 \\ 0 & 0 \end{smallmatrix}\right]$, where $k = \mathrm{rank}(T)$ and the 0's are zero matrices of appropriate sizes.

12. A square matrix A is said to be symmetric if $A = A^t$. Prove that if $A \in M_{2 \times 2}(\mathbb{R})$ is symmetric, then c_A factors as the product of two linear polynomials in $\mathbb{R}[x]$.

13. Let n be a positive integer. For $(a_1, a_2, \ldots, a_n), (b_1, b_2, \ldots, b_n) \in \mathbb{R}^n$, define the dot product $(a_1, a_2, \ldots, a_n) \cdot (b_1, b_2, \ldots, b_n) = \sum_{i=1}^{n} a_i b_i$. Prove that if $A \in M_{n \times}(\mathbb{R})$ is symmetric, and $r, s \in \mathbb{R}$ are eigenvalues of A with $r \neq s$, then for any $u \in E_r$, $v \in E_s$, $u \cdot v = 0$.

14. For each positive integer n, define $T_n \in M_{n \times n}(\mathbb{R})$ by $(T_n)_{i\,i+1} = 1$ for all $i \in J_n$ with $i < n$, $(T_n)_{i\,i-1} = 1$ for all $i \in J_n$ with $i > 1$, and all other entries equal to 0. For example, $T_1 = [0]$, $T_2 = \left[\begin{smallmatrix} 0 & 1 \\ 1 & 0 \end{smallmatrix}\right]$, and $T_3 = \left[\begin{smallmatrix} 0 & 1 & 0 \\ 1 & 0 & 1 \\ 0 & 1 & 0 \end{smallmatrix}\right]$.

    a) Prove that $c_{T_1}(x) = -x$, $c_{T_2}(x) = x^2 - 1$, and for every positive integer $n \geq 2$, $c_{T_n}(x) = -x c_{T_{n-1}}(x) - c_{T_{n-2}}(x)$.

    b) Prove that for every positive integer n,
    $$c_T(x) = \sum_{i=0}^{\lfloor \frac{n}{2} \rfloor} (-1)^{n-i} \binom{n-i}{i} x^{n-2i}.$$

15. Let F be a field, and let V be a vector space of finite dimension n over F. Let $\alpha = (v_1, v_2, \ldots, v_n)$ be an ordered basis for V and $a_1, a_2, \ldots, a_{n-1} \in F$. Then let T be the unique linear operator on V for which $T(v_i) = v_{i+1}$ for $i = 1, 2, \ldots, n-1$, and $T(v_n) = \sum_{i=1}^{n-1} a_i v_i$. Calculate the characteristic polynomial of T.

16. Let F be a field and V be a finite dimensional vector space over F.

    a) Prove that there exists a positive integer n such that $T^n(V) \cap \text{null}(T^n) = \{ \mathbb{0}_V \}$.

    b) Prove that for the integer n found in the first part of this exercise, $V = T^n(V) \oplus \text{null}(T^n)$.

17. Let F be a field and n be a positive integer.

    a) Prove that for any $A \in M_{n \times n}(F)$, $c_A(x) = c_{A^t}(x)$.

    b) By the first part of this exercise, for any $A \in M_{n \times n}(F)$, A and A^t have the same eigenvalues. Find an example of a field F, a positive integer n, and an $A \in M_{n \times n}(F)$ such that for at least one eigenvalue r of A, the eigenspace associated with r for A is different from the eigenspace associated with r for A^t.

18. Let F be a field and let V be a vector space of finite dimension n over F. Suppose that T is a linear operator on V such that the characteristic polynomial c_T factors completely as a product of linear factors. Then T has n eigenvalues (not necessarily distinct, c_T may have repeated roots). Let $r_1, r_2, \ldots, r_n$ denote the n eigenvalues of T (with repetitions if necessary). Let α be an ordered basis for V.

    a) Prove that $tr([T]_\alpha^\alpha) = \sum_{i=1}^n r_i$.

    b) Prove that $\det([T]_\alpha^\alpha) = \prod_{i=1}^n r_i$.

19. Let n be a positive integer, and suppose that $P \in M_{n \times n}(\mathbb{R})$ is such that $P^t = P^2$. Prove that if $r \in \mathbb{R}$ is an eigenvalue of P, then $r = 0$ or $r = 1$.

20. Let F be a field and n a positive integer. Prove that if $A \in M_{n \times n}(F)$ has a basis for which each element is an eigenvector of V, then there exists an invertible $P \in M_{n \times n}(F)$ such that $P^{-1}AP$ is a diagonal matrix.

Index

Printed in the United States
By Bookmasters